Classical Continuum Mechanics

APPLIED AND COMPUTATIONAL MECHANICS

A Series of Textbooks and Reference Books

Founding Editor

J.N. Reddy

Continuum Mechanics for Engineers, Forth Edition
G. Thomas Mase, Ronald E. Smelser & Jenn Stroud Rossmann

Dynamics in Engineering Practice, Eleventh Edition
Dara W. Childs, Andrew P. Conkey

Advanced Mechanics of Continua
Karan S. Surana

Physical Components of Tensors
Wolf Altman, Antonio Marmo De Oliveira

Continuum Mechanics for Engineers, Third Edition
G. Thomas Mase, Ronald E. Smelser & Jenn Stroud Rossmann

Classical Continuum Mechanics, Second Edition
Karan S. Surana

For more information about this series, please visit: https://www.crcpress.com/Applied-and-Computational-Mechanics/book-series/CRCAPPCOMMEC

Classical Continuum Mechanics

Karan S. Surana
Department of Mechanical Engineering
The University of Kansas
Lawrence, Kansas

CRC Press
Taylor & Francis Group
Boca Raton London New York

CRC Press is an imprint of the
Taylor & Francis Group, an **informa** business

Second edition published 2022
by CRC Press
6000 Broken Sound Parkway NW, Suite 300, Boca Raton, FL 33487-2742

and by CRC Press
2 Park Square, Milton Park, Abingdon, Oxon, OX14 4RN

First edition published by CRC Press 2014

CRC Press is an imprint of Taylor & Francis Group, LLC

ISBN: 978-0-367-61296-2 (hbk)
ISBN: 978-0-367-61521-5 (pbk)
ISBN: 978-1-003-10533-6 (ebk)

DOI: 10.1201/9781003105336

Typeset in CMR10
by KnowledgeWorks Global Ltd.

To

My beloved family

Abha, Deepak, Rishi, Yogini, and Riya

Contents

PREFACE

Phenomenal progress and advances in continuum theories over the last fifty years necessitate that we precisely define the specific theories of interest and study within the umbrella of continuum mechanics. Broadly speaking, all continuum theories at present fall into distinct categories: classical continuum mechanics (CCM) and non-classical continuum mechanics (NCCM). This book is devoted to the study of CCM based on three fundamental assumptions: (1) The behavior of the matter at fine-scale is neglected; instead, we employ the concept of infinitesimal volume referred to as a material point. If we assume the matter to be isotropic and homogeneous, then a continuum theory derived for a material point is valid for the entire volume of matter. (2) A material point has only three translational degrees of freedom (displacements). The displacement and forces are work conjugate in a deforming volume of matter and always coexist. (3) We assume the deforming matter to be in thermodynamic equilibrium. This allows us to use laws and principles of thermodynamics in the development of the CCM theory. CCM theory presented in this book is the development of the mathematical models of deforming continua with the limitations of these assumptions.

Non-classical continuum mechanics on the other hand considers continuum theories that are outside the scope of CCM. Couple stress theories, theories based on internal rotations and Cosserat rotations and non-local theories constitute the subjects of study in NCCM. Non-classical continuum theories are aimed at correcting some deficiencies of CCM theories and are designed to provide more enhanced continuum theories that may be better suited for advanced materials and applications.

This *classical continuum mechanics* book is the second of the *Advanced mechanics of continua*. This book, like its predecessor, is designed to be an advanced graduate-level study of the basic concepts, principles and the resulting theories within the umbrella of classical continuum mechanics. The edition retains some of the basic material from the first edition but in more compact form. Significant amount of new and more advanced material has been introduced in this edition to improve and bring completeness to the concepts, principles and theories contained in the book. Author's teaching of the subject at the University of Kansas and his extensive research and publications in NCCM theories has had profound impact in preparing this highly focused manuscript.

More advanced concepts and theories are introduced as a natural progression to the simple and commonly used concepts and theories. Curvilinear coordinates and transformation follow orthogonal frames and orthogonal transformation before tensors are introduced using curvilinear frames. Throughout the book, covariant and contravariant bases and measures in Lagrangian and Eulerian descriptions are clearly distinguished in the kinematics of deformation and stress measures. The orthogonal transformations are now introduced as Galilean as well as Euclidean or non-Galilean and their significance and applications in the convected time derivatives of covariant and contravariant tensors are presented and illustrated. Independent derivations of convected time derivatives of co- and contra-variant tensors of rank two are presented using non-Galilean orthogonal transformation. Objective tensors of various ranks are defined and the objectivity of various kinematic and stress measures are established using Galilean and non-Galilean transformations. Objectivity of rates is derived using Galilean and non-Galilean transformations.

Integral as well as differential or local form of the conservation and balance laws (CBL) are derived in Eulerian description. Assumptions are necessary to derive the differential form of the CBL are clearly stated and discussed. It is shown that the integral form of the CBL in Lagrangian description is only possible for conservation of mass; thus, the differential forms of the CBL in Lagrangian description are only derived within the restriction of the localization theorem. The total deformation of a deforming volume of matter is decomposed into volumetric and distortional deformations. It is shown that volumetric deformation physics is independent of the matter and its constitution and can be described by the constitutive theory for equilibrium stress tensor (Chapter 8). This has enabled significant reduction and streamlining the constitutive theory derivations in Chapters 9–14. Role of energy methods, the principle of virtual work and the fundamental lemma of the calculus of variations for continuous matter are considered in Chapter 16. It is shown that CBL of CCM, mathematical classification of differential operators in the CBL and the elements of the calculus of variations is a complete mathematical framework for all BVPs and IVPs and that the energy methods and the principle of virtual work are very limited and extremely restricted subset of this framework. It is shown that Hamilton's principle can only be realized within the assumptions of neglecting integrals on the open boundary of the space-time domain of the IVPs, hence it is approximate. Append A provides a list of generators and invariants for various combinations of argument tensors. Appendix B is helpful in transforming information and mathematical models from Cartesian to cylindrical or spherical coordinate systems. Reorganization of the material in the first edition and the new material introduced in this edition provides more comprehensive and more complete treatment of CCM concepts, principles and theories.

The author is grateful to many of his past and present graduate students: Dr. Daniel Nunez, Dr. Tristan Moody, Yusshy Mendoza, Dr. Aaron Joy, Dr. Michael Powell, Sai Mathi, Celso Carranza, Stephen Long, Jacob Kendall, Elie Abboud, Michael Kitchen, Thomas Ezell whose Ph.D. and M.S. thesis research in various areas of CCM and NCCM have helped me immensely in bringing the subject matter of this book to its present level of maturity. I am thankful to my graduate students Sai Mathi, Celso Carranza, Michael Kitchen, Thomas Ezell, and Elie Abboud for their efforts in proofreading the manuscript of the book. My very special thanks to my current Ph.D. student Celso H. Carranza for his interest in the subject, for his own research on NCCM, for his commitment and interest in this book project, hard work, for many discussions and suggestions and above all, for typing and retyping many times the book manuscript, single-handedly to bring it to completion. My special thanks are also to my current Ph.D student Sai Mathi whose intellectual curiosity and enormous appetite for new knowledge have always encouraged me to seek new frontiers in my research work and writings. I am very thankful for his contribution in various areas of CCM and NCCM that have resulted in significant growth in the research activities of my research group at the University of Kansas.

This book contains many involved equations, derivations and mathematical details and it is hardly possible to avoid some typographical and other errors. The Author would be grateful to those readers who are willing to draw attention to the errors using the email: kssurana@ku.edu

Karan S. Surana, *Lawrence, KS*

ABOUT THE AUTHOR

Karan S. Surana, born in India, went to undergraduate school at Birla Institute of Technology and Science (BITS), Pilani, India, and received a B.E. degree in Mechanical Engineering in 1965. He then attended the University of Wisconsin, Madison, where he obtained M.S. and Ph.D. degrees in Mechanical Engineering in 1967 and 1970, respectively. He worked in industry, in research and development in various areas of computational mechanics and software development, for fifteen years: SDRC, Cincinnati (1970–1973), EMRC, Detroit (1973–1978); and McDonnell-Douglas, St. Louis (1978–1984). In 1984, he joined the Department of Mechanical Engineering faculty at the University of Kansas, where he is currently the Deane E. Ackers University Distinguished Professor of Mechanical Engineering.

His areas of interest and expertise are computational mathematics, computational mechanics, and continuum mechanics. He is the author of over 350 research reports, conference papers, and journal articles. He has served as advisor and chairman of 50 M.S. students and 25 Ph.D. students in various areas of Computational Mathematics and Continuum Mechanics. He has delivered many plenary and keynote lectures in various national and international conferences and congresses on computational mathematics, computational mechanics, and continuum mechanics. He has served on international advisory committees of many conferences and has co-organized mini-symposia on k-version of the finite element method, computational methods, and constitutive theories at U.S. National Congresses of Computational Mechanics organized by the U.S. Association of Computational Mechanics (US-ACM). He has organized mini-symposium on classical and non-classical continuum mechanics at SES (Society of Engineering Science). He is a member of the International Association of Computational Mechanics (IACM) USACM, SES, and a fellow and life member of ASME.

Dr. Surana's most notable contributions include: large deformation finite element formulations of shells, the k-version of the finite element method, operator classification and variationally consistent integral forms in methods of approximations for BVPs and IVPs, and ordered rate constitutive theories for solid and fluent continua. His most recent and present research work is in non-classical continuum theories for solid and fluent continua and associated constitutive theories. He is the author of recently published textbooks: *Advanced Mechanics of Continua*, CRC/Taylor & France, *The Finite Element Method for Boundary Value Problems: Mathematics and Computations*, CRC/Taylor & Francis, *The Finite Element Method for Initial Value Problems: Mathematics and Computations*, CRC/Taylor & Francis, and *Numerical Methods and Methods of Approximation in Science and Engineering*, CRC/Taylor & Francis.

LIST OF ABBREVIATIONS

BAM	:	Balance of Angular Momenta
BLM	:	Balance of Linear Momenta
CCM	:	Classical Continuum Mechanics
CBL	:	Conservation and Balance Laws
CM	:	Conservation of Mass
FLT	:	First Law of Thermodynamics
GM	:	Galerkin Method
GM/WF	:	Galerkin Method with Weak form
LSM	:	Least Squares Method
LSP	:	Least Squares Process
NCCM	:	Non-classical Continuum Mechanics
PGM	:	Petrov Galerkin Method
SLT	:	Second Law of Thermodynamics
STGM	:	Space-time Galerkin Method
STGM/WF	:	Space-time Galerkin Method with Weak Form
STLSM	:	Space-time Least Squares Method
STLSP	:	Space-time Least Squares Process
STVC	:	Space-time Variationally Consistent
STVIC	:	Space-time Variationally Inconsistent
STWRM	:	Space-time Weighted Residuals Method
TVF	:	Thermoviscous Fluid
TVEF	:	Thermoviscoelastic Fluid
TES	:	Thermoelastic Solid
TVES	:	Thermoviscoelastic Solid
TVESM	:	Thermoviscoelastic Solid with Memory
VC	:	Variationally Consistent
VIC	:	Variationally Inconsistent
WRM	:	Weighted Residuals Method

1 INTRODUCTION

Continuum mechanics is the study of the physics of deformation of continuous matter constituting physical processes. All deforming physical processes are evolutions in which the states of the processes continuously change as time elapses. All processes consist of matter; hence the evolutions of the processes are, in fact, the evolutions of the states of the matter constituting the processes. The mathematical descriptions or mathematical models describing the evolving state of the deforming matter as time elapses are called initial value problems. Continuum mechanics is the study of the development of mathematical models and the evolutions described by them for the continuously evolving state of the deforming continuous matter. In continuum mechanics, we consider methodologies, concepts, theories and principles that are applicable to a large class of matter with minimum possible assumptions and approximations that may be application or problem-specific. This is obviously the most general approach to learn about the fundamentals of the entire subject. In the following, we give basic definitions that will help us in understanding the scope of study in this book.

1 Def: Mechanics

Mechanics is a much broader study in which we consider motion of nondeforming matter as well as the matter with motion and deformation. When a process or object is disturbed its state changes. The measures of the changes in the state of the process or the matter or the object may be in terms of the motion, deformation, velocity, acceleration, density, temperature, etc. Mechanics is the study of such measures of the evolving state of the processes. This field is obviously too vast and the study of all such aspects is beyond the scope of this book.

2 Def: Classical mechanics

Classical mechanics is the study of deforming and nondeforming objects or processes in which the material particles or points only have three translational degrees of freedom. Hence, in the studies of the deformation of continuous matter in classical mechanics, the influence of the rotation and rotation rates and their gradients between neighboring material particles or points is neglected in the mathematical description of the physics of deformation.

3 Def: Continuum mechanics

Stable matter at the smallest scale consists of molecules that further consist of atomic and subatomic particles. When the matter is disturbed the changes in the matter occur at all scales including molecular, atomic and subatomic scales.

DOI: 10.1201/9781003105336-1

However, there are many aspects of everyday experience regarding the behavior of matter that can be described and represented accurately with theories that pay no attention to the molecular structure of the matter or the structure of the matter at atomic and subatomic scales. The theory that describes the gross phenomenon of matter, neglecting the structure of the matter at a smaller scale, is known as continuum theory. The constitution of the matter considered at the molecular level is obviously not continuous as it consists of discrete molecules separated by the mean free path. The continuum mechanics consisting of continuum theories considers matter to be indefinitely divisible. Thus, in this theory, one accepts the concept of infinitesimal volume of matter referred to as a particle of the continuum. In every neighborhood of a particle, there are infinitely many particles present.

Whether the continuum theory is valid or not depends upon a given situation at hand. However, simple guidelines may be considered based on the molecular dimensions and the mean free path between the molecules. As a general rule, if the dimensions of the matter are between two to three orders of magnitude compared to the molecular dimensions or mean free path or larger, we can consider the continuum theory to be valid in describing the behavior of the matter.

Thus, continuum mechanics as a mathematical idealization of the matter is applicable to situations in which the fine-scale structure of the matter and its behavior is ignored. When the behavior of the fine-scale structure of the matter is important, we must use the principles of particle physics, statistical mechanics or theories that specifically address smaller scale physics.

All theories are founded on some assumptions that are fundamental in the development of the theories. Continuum mechanics theories are no exception to this. Recent developments in the continuum mechanics theories warrant that we make some distinction between different continuum theories that consider varied deformation physics in the studies of continuous media. Broadly speaking, we can classify all continuum mechanics theories into two categories: classical continuum mechanics (CCM) and non-classical continuum mechanics (NCCM).

4 Def: Classical Continuum mechanics

The classical continuum mechanics theories are founded on the basic assumption that a material point of continua has only three translation degrees of freedom, displacements and the corresponding work conjugate quantities are forces. Thus, these theories do not consider the influence of rotations and rotation rates and their gradients between the material points in describing the physics of deformation. Rotations and moments in classical continuum mechanics are consequences of displacements and forces acting on material points separated by a finite distance, hence these are not independent of displacements and forces. *In CCM theories, displacements and forces at material points must coexist in a deforming continua.*

In CCM, a material point has mass, but it is assumed to have no rotational inertial properties. Therefore, a material point only experiences linear momenta and acceleration forces due to velocities and accelerations. However, if the deformation physics results in rotations, angular velocities and angular accelerations at a material point, in CCM, there is no mechanism for incorporating these in the mathematical description of the deforming continua. Likewise, the assumption of additional unknown rotational degrees of freedom at a material point is not possible in CCM. The basic foundation of CCM is based on coexistence of displacements and forces only at each material point. Any other physics that is not supported by these conjugate quantities is beyond the scope of CCM.

5 Def: Non-classical Continuum mechanics

Broadly speaking, all continuum theories that fall outside the scope of classical continuum mechanics can be referred to as non-classical continuum theories. These constitute the subject of non-classical continuum mechanics. Couple stress theories, theories based on rotations due to deformation gradient tensor (internal rotations), and theories based on unknown Cosserat rotations as additional degrees of freedom at material points and non-local theories [116, 117] are some examples of non-classical continuum theories. The couple stress theories or the theories based on internal rotations begin with classical continuum theories but additionally incorporate influence of internal rotations in the development of the theories arising due to deformation gradient tensor (neglected in CCM). In these theories, a material point only has translational degrees of freedom as in CCM due to the fact that the internal rotations are completely defined by the components of the deformation gradient tensor, hence cannot be degrees of freedom at a material point. In the non-classical theories based on Cosserat rotations, a material point has three unknown rotational degrees of freedom at a material point about the axes of a triad located at the material point in addition to three translational degrees of freedom. Many variations of these theories exist that consider internal and/or Cosserat rotations. In the non-local theories, stress at a material point is assumed to depend upon the deformation physics at the material point as well as the stresses at the material points in its neighborhood. The non-classical continuum theories are believed to provide a more enhanced mathematical framework that may be meritorious for modern engineered materials compared to classical continuum theories.

Most textbooks and writings currently available under the title 'continuum mechanics' only address classical continuum mechanics. This book is also devoted to the study of classical continuum mechanics, but the distinction between CCM and NCCM is highlighted throughout this book. In addition, limitations of CCM are also discussed in various chapters of the book wherever appropriate.

6 Def: Homogeneous matter or Homogeneity

If the physical properties of the material in volume V do not depend upon the position $\boldsymbol{x}$ of the material points, then the material is called homogeneous. In this case, for material property $\underset{\sim}{P}$, the following holds

$$\underset{\sim}{P}(\boldsymbol{x}) = P_0 \ \forall \ \boldsymbol{x} \in V, \ P_0 \text{ is constant}$$

7 Def: Heterogeneous or Inhomogeneous or non-homogeneous matter

When $\underset{\sim}{P}$ at each material point $\boldsymbol{x} \in V$ is not the same, the material is called inhomogeneous or non-homogeneous.

8 Def: Isotropic matter

In isotropic matter, the material properties are the same in all directions at a material point. Thus, isotropic matter can be homogeneous or inhomogeneous.

9 Def: Anisotropic or non-isotropic matter

Anisotropic matter has different material properties in different directions at each material point. Thus, anisotropic matter can be homogeneous or inhomogeneous.

10 Def: Isentropic thermodynamic process

A thermodynamic process in which the entropy density η remains constant is called isentropic, thermodynamic process.

11 Def: Isothermal thermodynamic process

A thermodynamic process in which temperature θ remains constant is called isothermal thermodynamic process.

12 Def: Non-isothermal thermodynamic process

A thermodynamic process in which temperature θ does not remain constant is called non-isothermal, thermodynamic process.

13 Def: Rigid deformation

The deformation of a volume of matter V is said to be rigid if the distance between any pair of material points remains unchanged during deformation.

14 Def: Potential deformation

When the displacement field can be derived from a potential (the mean rotation vanishes), the deformation is called potential deformation. For potential deformation, the necessary and sufficient condition is

$$\boldsymbol{S}_r = \text{grad}(\boldsymbol{J})$$

$\boldsymbol{S}_r$ is right stretch and $\boldsymbol{J}$ is deformation gradient.

15 Def: Isochoric deformation, incompressible matter

The deformation under which volume V is preserved after deformation is called isochoric deformation. The necessary and sufficient conditions for isochoric deformation is given by

$$|\boldsymbol{J}| = 1 \quad ; \quad III_{C_{[0]}} = 1$$

The deformation in incompressible matter is isochoric, thus, no volumetric deformation.

16 Def: Compressible matter

Compressible matter has non-zero volumetric deformation, hence results in change of density for a deforming volume. In the Lagrangian description, conservation of mass describes density as a function of volumetric deformation.

$$\rho_0(\boldsymbol{x}) = |\boldsymbol{J}|\rho(\boldsymbol{x}, t)$$

17 Assumptions in CCM theories

In the development of the classical continuum mechanics theory, we employ three fundamental assumptions: (1) The behavior of the matter at fine scale is neglected. Instead, we employ the concept of infinitesimal volume referred to as a material particle or a material point. If we assume the deforming matter to be homogeneous and isotropic, then the continuum mechanics theory developed for a material point is valid for the entire deforming volume of the matter. This assumption permits differential form of the conservation and balance laws. This restriction of homogeneity and isotropy can be viewed in a different way by using localization theorem (Chapter 6), however, the issue of when the continuity of integrand can be ensured everywhere over the volume of deforming matter remains unresolved if we do not assume the matter to be homogeneous and isotropic. (2) A material point has only translational degrees of freedom (displacements). The displacements and forces are work conjugate and always coexist in deforming continua. (3) In classical continuum mechanics, we assume that the deforming matter is always in thermodynamic equilibrium. This assumption allows us to employ conservation and balance laws and thermodynamic principles in the development of the classical continuum mechanics theory. The development of the classical continuum mechanics theory presented in this book is founded on these three fundamental assumptions. Thus, within the limitations of these three assumptions, the classical continuum mechanics theory is the development of the mathematical descriptions of the deforming continua.

First, concise, clear, compact and unambiguous notations are necessary for giving a mathematical form to the physics of the deforming matter. Using these notations we construct basic definitions of quantities related to the deforming matter and consider various mathematical operations with these quantities. This material provides the most fundamental and basic elements to define the kinematics of motion of the deforming matter. Basic notations, concepts, orthogonal coordinate systems, orthogonal coordinate transformation, curvilinear coordinate system: co- and contra-variant bases, induced transformations, the definition of co- and contra-variant tensors of various ranks, tensors in Cartesian x-frame, orthogonal transformation of tensors in x-frame, invariants of tensors of various ranks and some other useful relations are presented in Chapter 2. Galilean and Euclidean (non-Galilean) transformations are also introduced in Chapter 2. Kinematics of motion, deformation and their measures for finite deformation are considered in Chapter 3. The change in length of a material line, change in lengths of an orthogonal group of material lines and the change in angle between them are the most fundamental aspects of the physics of deformation. Various definitions and measures are introduced that eventually lead to measures of finite strain. The definitions of strains are presented and their measures are derived using co- and contra-variant bases resulting in co- and contra-variant measures of strains that are presented in Lagrangian as well as Eulerian descriptions. These definitions and measures are somewhat a departure from the conventional approach used presently, in which strain measures are classified as either Lagrangian or Eulerian without regard to the basis. Establishing the tensorial nature of strain measures, influence of the change of frame on the strain measures, physical meaning of the components of the strain tensors, polar decomposition of strain tensors into stretch and rotation tensors, deformed and undeformed areas and volumes that are related to the kinematics of motion and deformation are all considered in Chapter 3. Definitions and measures of stress for finite deformation and finite strain, as well as for small deformation, small strain are considered in Chapter 4. Rate of deformation, area, volume, strain rate tensors, spin tensors and convected time derivatives of various stress and strain measures in covariant and contravariant bases are derived in Chapter 5. Definition of objective tensors is introduced in Chapter 5. Objective tensors and objective rates are presented using Galilean and non-Galilean transformations. Convected time derivatives of the contravariant and covariant tensors of rank two and Jaumann rates are derived using non-Galilean transformation in Chapter 5. The mathematical model for a deforming volume of matter consisting of conservation and balance laws (CBL) is derived in Chapter 6 in the Eulerian description. Development of mathematical model in Lagrangian description consisting of conservation and balance laws is derived in Chapter 7.

In this book, we only consider the thermodynamic equilibrium processes. Hence, the conservation laws and thermodynamic principles must be satisfied by the evolution of a deforming volume of matter. Since conservation of mass, balance of linear momenta, balance of angular moment and the first law of thermodynamics are independent of the specific constitution of the matter, hence are applicable to solids, liquids as well as gases, i.e., they remain valid for any continuous matter.

Chapter 8 contains axioms of constitutive theory, various approaches of deriving constitutive theories, use of representation theorem in deriving constitutive theories, including sample examples of the derivation of constitutive theories for constitutive tensors of rank two and one. Constitutive theories for finite deformation, finite strain as well as for small strain, small deformation physics for thermoelastic, thermoviscoelastic (no memory) and thermoviscoelastic solids with memory are presented in Chapters 9-11. Constitutive theories for thermoviscous and thermoviscoelastic fluent continua for compressible and incompressible physics are presented in Chapters 12 and 13. Maxwell, Oldroyd-B and Giesekus constitutive models for polymeric liquids are derived in Chapter 13. Constitutive theories for thermohypo-elastic solids are considered in Chapter 14. The equations of state, specific form of deformation dependent material coefficients, complete mathematical models in Lagrangian and Eulerian descriptions including constitutive theories are given in Chapter 15. Chapter 16 presents a short overview and discussion of energy methods, the principle of virtual work and calculus variations with the objective of establishing relationships between them. A comprehensive and general framework consisting of CBL, calculus of variations and differential operator classification provides critical assessment of the usefulness of energy methods and the principle of virtual work.

In the development of the mathematical models as well as in obtaining their solutions in structural mechanics such as for rods, beams, plates, shells, membranes, the energy methods and the principle of virtual work have played a significant role and continue to be used dominantly at present also. In Chapter 16, investigation of the consistency and validity of the mathematical models and the solution methods for boundary value problems (BVPs) and initial value problems (IVPs) based on energy methods and the principle of virtual work for homogeneous and isotropic and non-homogeneous and non-isotropic solid continua is presented. Specific model problems are also considered for illustrative purposes.

The entire book uses curvilinear coordinate systems: covariant and contravariant and fixed Cartesian frames. Use of cylindrical or spherical coordinate systems is intentionally avoided throughout the book: first to minimize unnecessary duplication and clutter, and second, the details of various measures, conservation and balance laws, etc. can be easily converted from Cartesian frame to cylindrical coordinate system or spherical frame using the material provided in Appendix B. Appendix A provides details of combined generators and invariants for symmetric and skew-symmetric tensors of rank two as well as tensors of rank one. This material is useful in deriving constitutive theories using representation theorem.

The material presented in this book provides a unified treatment of the development of mathematical descriptions of deforming solids, fluids, polymeric solids and fluids, both compressible and incompressible with clear distinction between Lagrangian and Eulerian descriptions as well as co- and contra-variant bases. Based on this short introduction of continuum mechanics, we note that this subject is mathematically demanding with many involved derivations but is very precise and rigorous within the assumptions considered. A thorough understanding and working knowledge of this subject is of paramount importance and significance in preparing graduate students for fundamental and basic research work in engineering and sciences.

2 CONCEPTS AND MATHEMATICAL PRELIMINARIES

2.1 Introduction

The material contained in this chapter is elementary but its thorough grasp and good working knowledge are essential in understanding and learning the material contained in the remaining chapters. First, we consider notations. The main purpose of notations is to present information related to the deforming matter in mathematical form. The notations must be compact, concise and consistent so that abstraction and yet accuracy of the actual information may be maintained in the mathematical description of the information. The commonly used notations in continuum mechanics are Einstein notations, index notations, matrix and vector notations. Generally, it is a matter of preference as to which notation one uses. However, there are some advantages in matrix and vector notations, especially when the solutions of mathematical models resulting from continuum mechanics principles describing evolutions are obtained numerically using computational mechanics and computational mathematics. In this chapter, we consider and use all three notations. In this book, we do not strictly adhere to one specific form of notations but rather consider a mix of these for maintaining clarity of presentation and ease of understanding, keeping in mind that this material must be easily reproducible in the classroom environment as well.

2.2 Notations

Details of Einstein notations, index notations and vector and matrix notations are presented in the following sections.

2.2.1 Einstein notations

The basic elements of the Einstein notation are: summation convention, dummy index (or indices) and free index (or indices).

2.2.1.1 Summation convention

Based on Einstein's summation convention, whenever an index is repeated once (and no more) in the same term or one or more terms in an equation, it implies summation over the specified range of the index. Thus, in such cases, there is no need to write the expanded form of the sum or use the summation sign. If the index is not specified, a value of 3 is assumed.

$$\begin{aligned}\alpha &= N_1\phi_1 + N_2\phi_2 + \dots + N_n\phi_n = \sum_{i=1}^{n} N_i\phi_i = \sum_{j=1}^{n} N_j\phi_j \\ &= N_k\phi_k = N_l\phi_l \quad ; \quad k,l = 1,2,\dots,n\end{aligned} \tag{2.1}$$

DOI: 10.1201/9781003105336-2

$$\beta = \sum_{i=1}^{3}\sum_{j=1}^{3} N_{ij}\phi_i\phi_j = \sum_{k=1}^{3}\sum_{l=1}^{3} N_{kl}\phi_k\phi_l = N_{ij}\phi_i\phi_j = N_{kl}\phi_k\phi_l \tag{2.2}$$

$$\gamma = \sum_{i=1}^{3}\sum_{j=1}^{3}\sum_{k=1}^{3} N_{ijk}\phi_i\phi_j\phi_k = N_{ijk}\phi_i\phi_j\phi_k = N_{lmn}\phi_l\phi_m\phi_n \tag{2.3}$$

$$\eta = \sum_{i=1}^{3}\sum_{j=1}^{3}\sum_{k=1}^{3} p_i q_j r_k \phi_i\phi_j\phi_k = p_i q_j r_k \phi_i\phi_j\phi_k = p_l q_m r_n \phi_l\phi_m\phi_n \tag{2.4}$$

$$\kappa = \sum_{i=1}^{3} N_i\phi_i + \sum_{j=1}^{3} P_j\phi_j = N_i\phi_i + P_j\phi_j = N_i\phi_i + P_i\phi_i = N_k\phi_k + P_l\phi_l \tag{2.5}$$

$$N_1\phi_1 + N_2\phi_2 + N_3\phi_3 = N_i\phi_i = N_j\phi_j \tag{2.6}$$

$$N_{ii} = N_{11} + N_{22} + N_{33} = N_{jj} \tag{2.7}$$

Equations (2.2)–(2.7) are examples of the use of Einstein's summation convention for compact representation of the terms in the sum in each case. Equations (2.1) and (2.3) contain a single summation term, Equations (2.2) and (2.4) are single summation terms and (2.5) is an equation containing two summation terms. In some instances, there may be a need to have an index repeated more than once in the same term. In such cases, Einstein's summation convention cannot be used and hence we must retain the summation sign for it to be meaningful. Three dots (no more, no less) in (2.1) represent $(n-3)$ remaining terms. This is a standard notation in continuum mechanics. Based on Einstein notation, the occurrence of three (or more) identical indices in each term of an expression necessitates the use of summation signs or symbols.

$$p_1q_1r_1 + p_2q_2r_2 + p_3q_3r_3 = \sum_{i=1}^{3} p_iq_ir_i \neq p_iq_ir_i \tag{2.8}$$

2.2.1.2 Dummy index and dummy variables

From (2.1), we note that the sum is independent of i and j. These are called dummy indices. In (2.2), i, j and k, l are dummy indices. Likewise, i, j, k and l, m and n are dummy indices in (2.3) and (2.4). Equation (2.1) is an example of a single dummy index, (2.2) contains two dummy indices as it is a double sum, whereas (2.3) and (2.4) are examples of the use of three dummy indices due to the fact that they are a triple sum. Thus, dummy indices are like dummy variables in the integrals.

$$G = \int_p^q g(x)dx = \int_p^q g(y)dy = \int_p^q g(z)dz = \int_p^q g(t)dt \tag{2.9}$$

In (2.9), the integrals are independent of the variables x, y, z and t. These are called dummy variables. Just like the choice of specific dummy variables does not influence the integral G, the choice of a specific variable for a dummy index also does not influence the meaning of term or terms in an equation containing the dummy index or indices.

2.2.1.3 Free indices

Free indices appear in equations. A free index appears only once in each term of the equation or expression. However, by using more than one free index, we can represent a set of equations as each free index will describe a single equation. Consider

$$p_i + q_i = r_i \tag{2.10}$$

$$p_i + q_{ij} b_j = r_i \tag{2.11}$$

$$\begin{aligned}
\phi_1 &= N_{11}\alpha_1 + N_{12}\alpha_2 + N_{13}\alpha_3 = N_{1i}\alpha_i = N_{1j}\alpha_j \\
\phi_2 &= N_{21}\alpha_1 + N_{22}\alpha_2 + N_{23}\alpha_3 = N_{2i}\alpha_i = N_{2j}\alpha_j \\
\phi_3 &= N_{31}\alpha_1 + N_{32}\alpha_2 + N_{33}\alpha_3 = N_{3i}\alpha_i = N_{3j}\alpha_j
\end{aligned} \tag{2.12}$$

or

$$\begin{aligned}
\phi_i &= N_{ij}\alpha_j \\
\phi_i &= N_{ik}\alpha_k \\
\phi_l &= N_{lm}\alpha_m
\end{aligned} \tag{2.13}$$

$$\sigma_{ij} = R_{ik}R_{jk} = \sum_{k=1}^{3} R_{ik}R_{jk} = R_{il}R_{jl} = \sum_{l=1}^{3} R_{il}R_{jl} \tag{2.14}$$

Index i in (2.10) and (2.11) is a free index as it appears once in each term of the equation. Index j in (2.11) is a dummy index. Equation (2.12) is a system of three equations in which j is a dummy index. These can be represented in compact form using (2.13) in which i, l are free indices but j, k and m are dummy indices. Equation (2.14) represents a system of nine equations. For each value of i and j, the sum over the dummy index k or l yields an equation.

2.2.2 Index notations and Kronecker delta

In the index notation, we consider a suitable coordinate frame and represent quantities of interest in this frame using the basis of the frame. We generally define a fixed orthogonal Cartesian coordinate system by ox_1, ox_2, ox_3 orthogonal directions, $ox_1x_2x_3$ for short, referred to as x-frame, o being the location or origin of the coordinate system. The use of x_1, x_2, x_3 as opposed to x,y,z is intentional so that Einstein's notation can be employed for compact representation of information. We assume that $\boldsymbol{e}_i$; $i = 1, 2, 3$ are unit vectors along ox_1, ox_2 and ox_3 axes, i.e., unit vectors $\boldsymbol{e}_i$ form an orthonormal basis in the x-frame. If $\boldsymbol{\phi}$ is a vector in x-frame with components ϕ_1, ϕ_2, ϕ_3 along ox_1, ox_2 and ox_3, then $\boldsymbol{\phi}$ has the following representation in the x-frame:

$$\boldsymbol{\phi} = \boldsymbol{e}_1\phi_1 + \boldsymbol{e}_2\phi_2 + \boldsymbol{e}_3\phi_3 = \boldsymbol{e}_i\phi_i = \vec{e}_i\phi_i \tag{2.15}$$

We note that $\boldsymbol{e}_i$ form an orthonormal basis. The following hold

$$\begin{aligned}
\boldsymbol{e}_1 \cdot \boldsymbol{e}_1 &= \boldsymbol{e}_2 \cdot \boldsymbol{e}_2 = \boldsymbol{e}_3 \cdot \boldsymbol{e}_3 = 1 \\
\boldsymbol{e}_1 \cdot \boldsymbol{e}_2 &= \boldsymbol{e}_2 \cdot \boldsymbol{e}_1 = \boldsymbol{e}_3 \cdot \boldsymbol{e}_1 = 0
\end{aligned} \tag{2.16}$$

and (2.16) can be written in more compact form using a new notation

$$\boldsymbol{e}_i \cdot \boldsymbol{e}_j = \delta_{ij} = \begin{cases} 1 & \text{if} \quad i = j \\ 0 & \text{if} \quad i \neq j \end{cases} \tag{2.17}$$

where δ_{ij} is called the Kronecker delta.

If $\boldsymbol{\psi}$ is another vector in x-frame with projections ψ_1, ψ_2, ψ_3 along ox_1, ox_2 and ox_3 axes, then

$$\boldsymbol{\psi} = \boldsymbol{e}_j \psi_j \tag{2.18}$$

Dot product of vectors $\boldsymbol{\phi}$ and $\boldsymbol{\psi}$ is defined by

$$\boldsymbol{\phi} \cdot \boldsymbol{\psi} = (\boldsymbol{e}_i \phi_i) \cdot (\boldsymbol{e}_j \psi_j) \tag{2.19}$$

and by expanding the right side of (2.19) using (2.17), we can obtain

$$\boldsymbol{\phi} \cdot \boldsymbol{\psi} = \boldsymbol{e}_i \phi_i \cdot \boldsymbol{e}_j \psi_j = \boldsymbol{e}_i \cdot \boldsymbol{e}_j \phi_i \psi_j = \delta_{ij} \phi_i \psi_j = \phi_i \psi_i \tag{2.20}$$

If we recognize the following properties of the Kronecker delta, δ_{ij}, then we can make use of δ_{ij} in many different compact representations. From (2.16) and (2.17), we note that

$$\begin{aligned} &\delta_{11} = \delta_{22} = \delta_{33} = 1 \\ &\delta_{12} = \delta_{21} = \delta_{13} = \delta_{31} = \delta_{23} = \delta_{32} = 0 \end{aligned} \tag{2.21}$$

We also note that

$$\delta_{ii} = \delta_{11} + \delta_{22} + \delta_{33} = 3 \tag{2.22}$$

$$\delta_{ij} \phi_j = \phi_i \tag{2.23}$$

$$\delta_{im} \delta_{mj} = \delta_{ij} \tag{2.24}$$

$$\delta_{im} \delta_{mk} \delta_{kj} = \delta_{ij} \tag{2.25}$$

$$\delta_{im} \sigma_{mj} = \sigma_{ij} \tag{2.26}$$

2.2.3 Vector and matrix notations

When the scalars are arranged in the form of a column we may refer to this column as a vector. From Equation (2.12), ϕ_1, ϕ_2 and ϕ_3 can be arranged as

$$\{\phi\} = \begin{Bmatrix} \phi_1 \\ \phi_2 \\ \phi_3 \end{Bmatrix} \quad \text{or} \quad \vec{\phi} \quad \text{or} \quad \boldsymbol{\phi} \tag{2.27}$$

where curly brackets, boldface character or a character with an over arrow are standard continuum mechanics notation for vectors. At this stage a vector is an ordered arrangement of quantities and has no other meaning. Just like ϕ_i, α_i appearing in (2.12) can also be arranged in the form of a column or a vector.

$$\{\alpha\} = \begin{Bmatrix} \alpha_1 \\ \alpha_2 \\ \alpha_3 \end{Bmatrix} \quad \text{or} \quad \vec{\alpha} \quad \text{or} \quad \boldsymbol{\alpha} \tag{2.28}$$

N_{ij} appearing in (2.12) or (2.13) can be arranged in rows and columns of a two-dimensional ordered arrangement, a matrix $[N]$

$$[N] = \begin{bmatrix} N_{11} & N_{12} & N_{13} \\ N_{21} & N_{22} & N_{23} \\ N_{31} & N_{32} & N_{33} \end{bmatrix} \tag{2.29}$$

where N_{ij} refers to an element of $[N]$ located at row i and column j. Using the notation for vectors in (2.27) and (2.28) and for matrix in (2.29), we can also write (2.13) as

$$\{\phi\} = [N]\{\alpha\} \quad \text{or} \quad \vec{\phi} = [N]\vec{\alpha} \tag{2.30}$$

Equations (2.13) and (2.30) are exactly equivalent but written using two different notations. In this book, we employ both notations.

Transposition of a vector, a matrix and vector, matrix operations, etc. are elementary and hence are not discussed here.

2.2.4 Remarks

Using Einstein's notations, index notations and matrix and vector notations, we can represent various quantities in many different desired forms. Some examples are given in the following.

(a) The components of the Kronecker delta δ_{ij} in (2.16) can be arranged in matrix form (i being row and j being column), then we have identity matrix $[I]$

$$\begin{bmatrix} \delta_{11} & \delta_{12} & \delta_{13} \\ \delta_{21} & \delta_{22} & \delta_{23} \\ \delta_{31} & \delta_{32} & \delta_{33} \end{bmatrix} = \begin{bmatrix} 1 & 0 & 0 \\ 0 & 1 & 0 \\ 0 & 0 & 1 \end{bmatrix} = [I] \tag{2.31}$$

(b) Using vector and matrix notations (2.24)–(2.26) can be written as

$$[I][I] = [I] \tag{2.32}$$
$$[I][I][I] = [I] \tag{2.33}$$
$$[I][\sigma] = [\sigma] \tag{2.34}$$

(c) We can also write (2.14) as follows using vector and matrix notations

$$[\sigma] = [R][R]^T \tag{2.35}$$

in which $[R]^T$ is the transpose of $[R]$.

(d) If $\boldsymbol{ds}$ is a line segment with length ds and components dx_1, dx_2 and dx_3 in x-frame, then

$$\{ds\} = \vec{ds} = \begin{Bmatrix} dx_1 \\ dx_2 \\ dx_3 \end{Bmatrix} \quad \text{or} \quad \boldsymbol{ds} = \boldsymbol{e}_i \, dx_i \tag{2.36}$$

Let $(ds)^2$ be the square of the length of $\boldsymbol{ds}$, then

$$(ds)^2 = d\boldsymbol{s} \cdot d\boldsymbol{s} = \boldsymbol{e}_i dx_i \cdot \boldsymbol{e}_j dx_j = \boldsymbol{e}_i \cdot \boldsymbol{e}_j dx_i dx_j = dx_i dx_i = \{ds\}^T \{ds\} \tag{2.37}$$

Alternatively, since $dx_i = \delta_{ij} dx_j$

$$(ds)^2 = dx_i(\delta_{ij} dx_j) = \delta_{ij} dx_i dx_j = dx_i dx_i \tag{2.38}$$

(e) If $h = h(x_1, x_2, x_3)$, then in continuum mechanics it is standard notation to use

$$\frac{\partial h}{\partial x_i} = h_{,i} \tag{2.39}$$

thus, using (2.39), total differential dh of $h(x_1, x_2, x_3)$ can be written as

$$dh = \frac{\partial h}{\partial x_i} dx_i = h_{,i}\, dx_i \tag{2.40}$$

(f) In continuum mechanics, it is standard practice to use $1, 2, 3$ to represent x_1, x_2, x_3. This leads to compact representation when $1, 2, 3$ are used instead of x_1, x_2, x_3 as subscripts or superscripts. For example, instead of $\sigma_{x_1x_1}$, $\sigma_{x_1x_2}$, $\sigma_{x_2x_3}$ ..., we could simply use σ_{11}, σ_{12}, σ_{23} ...

2.3 Permutation tensor

The permutation tensor or the permutation symbol (also called alternating tensor or Levi-Civita tensor) is denoted by ϵ_{ijk} in which if we consider counterclockwise order of ijk to be even, then it is defined as

$$\epsilon_{ijk} = \begin{Bmatrix} 1 \\ -1 \\ 0 \end{Bmatrix} \quad \text{if} \quad i,j,k \quad \text{are} \begin{Bmatrix} \text{an even} \\ \text{an odd} \\ \text{not a} \end{Bmatrix} \text{permutation of 1, 2, 3} \tag{2.41}$$

If we arrange $1, 2, 3$ in counter-clockwise fashion, then the counter clockwise order of 1, 2, 3 permutations: 1 2 3, 2 3 1, 3 1 2 are even permutations of $1, 2, 3$, hence ϵ_{123}, ϵ_{231}, ϵ_{312} each has a value of 1. On the other hand, the permutations: 1 3 2, 3 2 1, 2 1 3 of $1, 2, 3$ are clockwise permutations; thus, odd permutations, hence ϵ_{132}, ϵ_{321}, ϵ_{213} each has a value of -1. When two or more indices are the same in ϵ_{ijk}, then it is not a permutation, hence its value is 0. This gives rise to the following:

$$\begin{aligned} &\epsilon_{123} = \epsilon_{231} = \epsilon_{312} = 1 \qquad \text{(cyclic order : counter clockwise)} \\ &\epsilon_{132} = \epsilon_{213} = \epsilon_{321} = -1 \qquad \text{(cyclic order : clockwise)} \\ &\epsilon_{111} = \epsilon_{211} = \epsilon_{133} = \cdots = 0 \end{aligned} \tag{2.42}$$

or in general we can write

$$\epsilon_{ijk} = \epsilon_{jki} = \epsilon_{kij} = -\epsilon_{jik} = -\epsilon_{kji} = -\epsilon_{ikj}$$

The permutation tensor is helpful in compact representation in the cross product

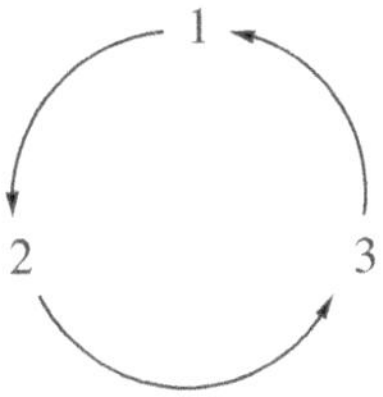

of vectors, determinant of a matrix, etc. If we consider orthonormal basis $\boldsymbol{e}_i$, then

$$\begin{aligned} \boldsymbol{e}_1 \times \boldsymbol{e}_2 &= \boldsymbol{e}_3 \quad ; \quad \boldsymbol{e}_2 \times \boldsymbol{e}_1 = -\boldsymbol{e}_3 \\ \boldsymbol{e}_2 \times \boldsymbol{e}_3 &= \boldsymbol{e}_1 \quad ; \quad \boldsymbol{e}_3 \times \boldsymbol{e}_2 = -\boldsymbol{e}_1 \\ \boldsymbol{e}_3 \times \boldsymbol{e}_1 &= \boldsymbol{e}_2 \quad ; \quad \boldsymbol{e}_1 \times \boldsymbol{e}_3 = -\boldsymbol{e}_2 \end{aligned} \tag{2.43}$$

which can be represented in compact form using the permutation symbol

$$\boldsymbol{e}_i \times \boldsymbol{e}_j = \epsilon_{ijk}\, \boldsymbol{e}_k \tag{2.44}$$

This property in (2.44) can be applied to cross products of vectors.

$$\begin{aligned} \boldsymbol{\phi} \times \boldsymbol{\psi} &= \boldsymbol{e}_i \phi_i \times \boldsymbol{e}_j \psi_j \\ &= \boldsymbol{e}_i \times \boldsymbol{e}_j \phi_i \psi_j \\ &= \epsilon_{ijk} \boldsymbol{e}_k \phi_i \psi_j = \boldsymbol{e}_k \epsilon_{ijk} \phi_i \psi_j \end{aligned} \tag{2.45}$$

Consider a 3×3 matrix $[J]$. Let $|J|$ or $\det[J]$ be the determinant of $[J]$. Then, $\det[J]$ can be obtained by using Laplace expansion. If we use the first column of $[J]$, then we have

$$\begin{aligned} |J| &= J_{11}(J_{22}J_{33} - J_{32}J_{23}) - J_{21}(J_{12}J_{33} - J_{32}J_{13}) + J_{31}(J_{12}J_{23} - J_{22}J_{13}) \\ &= J_{11}(\epsilon_{1jk}J_{j2}J_{k3}) - J_{21}(-\epsilon_{2jk}J_{j2}J_{k3}) + J_{31}(\epsilon_{3jk}J_{j2}J_{k3}) \\ &= J_{i1}(\epsilon_{ijk}J_{j2}J_{k3}) = \epsilon_{ijk}\, J_{i1}J_{j2}J_{k3} \end{aligned} \tag{2.46}$$

By considering Laplace expansion using the first row of $[J]$ we can show that

$$\det[J] = \epsilon_{ijk} J_{1i} J_{2j} J_{3k} \tag{2.47}$$

This is a compact and convenient way to define $\det[J]$.

2.4 ϵ-δ identity

The product of permutation tensors may be expressed in terms of the Kronecker delta. This is known as ϵ-δ identity.

$$\epsilon_{ijm}\epsilon_{klm} = \delta_{ik}\delta_{jl} - \delta_{il}\delta_{jk} \tag{2.48}$$

Due to the sign change property of ϵ for odd permutation, the following holds

$$\begin{aligned} \epsilon_{ijm}\epsilon_{klm} &= \epsilon_{mij}\epsilon_{klm} = \epsilon_{jmi}\epsilon_{mkl} \\ &= \epsilon_{ijm}\epsilon_{mkl} = -\epsilon_{mji}\epsilon_{lmk} \quad \text{etc.} \end{aligned} \tag{2.49}$$

Using (2.48) we can show that

$$\epsilon_{ilm}\epsilon_{klm} = 2\delta_{ik} \tag{2.50}$$

and using $k = i$ in (2.50)

$$\epsilon_{ilm}\epsilon_{ilm} = 6 \tag{2.51}$$

$\boldsymbol{\epsilon}$-$\boldsymbol{\delta}$ is an important identity that is helpful in the subsequent chapters.

2.5 Basic operations using vector, matrix and Einstein notations

There are four basic operations that are often encountered in the developments of continuum mechanics theory: multiplication, substitution, factoring and contraction when manipulating mathematical expressions. We consider these in the following.

2.5.1 Multiplication

Suppose we define scalars ϕ and ψ by

$$\phi = p_i q_i = \{p\}^T\{q\} = \boldsymbol{p}\cdot\boldsymbol{q} \tag{2.52}$$

$$\text{and} \quad \psi = r_i s_i = \{r\}^T\{s\} = \boldsymbol{r}\cdot\boldsymbol{s} \tag{2.53}$$

and wish to consider the product (multiplication) of ϕ and ψ, then we must make sure that the dummy index i in (2.52) and (2.53) is not the same. That is, the dummy index in ϕ or ψ must be changed before taking their product.

$$\phi\,\psi = p_i q_i r_j s_j = p_j q_j r_i s_i = r_i s_i p_j q_j = r_i s_i q_j p_j \tag{2.54}$$

Using vector and matrix notation we have

$$\phi\,\psi = (\{p\}^T\{q\})(\{r\}^T\{s\}) = (\boldsymbol{p}\cdot\boldsymbol{q})(\boldsymbol{r}\cdot\boldsymbol{s})$$

We note that in Einstein notations the position of quantities in the product in (2.54) does not influence the product.

2.5.2 Substitution

Suppose we define ϕ_i and q_i by

$$\phi_i = p_{ik} q_k \quad \text{or} \quad \{\phi\} = [p]\{q\} \tag{2.55}$$

$$\text{and} \quad q_i = r_{ik} s_k \quad \text{or} \quad \{q\} = [r]\{s\} \tag{2.56}$$

In (2.55) and (2.56), i is the free index and k is the dummy index. We wish to substitute q_i from (2.56) into (2.55). Before doing so, we must ensure that the free index (for q) in (2.56) is the same as the dummy index (for q) in (2.55). Furthermore, changing i into k in (2.56) will cause conflict with the dummy index k in (2.56). Thus, dummy index k in (2.56) must be changed to some other index (say l) first, followed by changing i to k. Let us change (2.56) to

$$q_k = r_{kl} s_l \tag{2.57}$$

Now we can substitute for q_k from (2.57) into (2.55)

$$\phi_i = p_{ik} r_{kl} s_l = r_{kl} p_{ik} s_l = s_l p_{ik} r_{kl} \tag{2.58}$$

and in matrix and vector notation we have

$$\{\phi\} = [p][r]\{s\}$$

Once again, we note that matrix and vector notations result in exactly the same expressions as the use of Einstein notations does, but in Einstein notations, the position of quantities in the product in (2.58) does not influence the product.

2.5.3 Factoring

Consider the following vector equation:

$$[\sigma]\{\phi\} - \lambda\{\phi\} = 0 \tag{2.59}$$

Suppose we wish to factor out $\{\phi\}$ from both terms in (2.59). This requires that the form of the second term in (2.59) must correspond to that of the first term. We can rewrite (2.59) as follows

$$[\sigma]\{\phi\} - \lambda[I]\{\phi\} = 0 \tag{2.60}$$

and now we can factor out $\{\phi\}$ in (2.60)

$$([\sigma] - \lambda[I])\{\phi\} = 0 \tag{2.61}$$

The same operation of factoring can also be done using Einstein notations. First, we rewrite (2.59).

$$\sigma_{ij}\phi_j - \lambda\phi_i = 0 \tag{2.62}$$

Since ϕ_i can also be written as

$$\phi_i = \delta_{ij}\phi_j \tag{2.63}$$

substitution of ϕ_i from (2.63) into (2.62) gives

$$\begin{aligned} \sigma_{ij}\phi_j - \lambda\delta_{ij}\phi_j &= 0 \\ \text{or} \quad (\sigma_{ij} - \lambda\delta_{ij})\phi_j &= 0 \end{aligned} \tag{2.64}$$

We note that (2.64) could have been written directly using (2.61). Once again, we note the simplicity of operation in matrix and vector notation.

2.5.4 Contraction (or trace)

The operation of contraction involves setting the two indices the same and therefore summing over the index. If we consider σ_{ij}, then the contraction of σ_{ij} would be σ_{ii}.

$$\sigma_{ii} = \sigma_{11} + \sigma_{22} + \sigma_{33} = \text{tr}[\sigma] \tag{2.65}$$

The sum of the diagonal elements of $[\sigma]$, $\text{tr}[\sigma]$ is pronounced trace of $[\sigma]$. That is, contraction results in trace. As another example consider

$$\sigma_{ij} = 2\mu D_{ij} + \kappa \delta_{ij} D_{kk} \tag{2.66}$$

then

$$\begin{aligned}\sigma_{ii} = \operatorname{tr} \boldsymbol{\sigma} &= \operatorname{tr}[\sigma] \\ &= 2\mu D_{ii} + \kappa \delta_{ii} D_{kk} \\ &= 2\mu D_{jj} + 3\kappa D_{kk} \\ &= 2\mu D_{kk} + 3\kappa D_{kk} \\ &= (2\mu + 3\kappa) D_{kk} \\ &= (2\mu + 3\kappa) D_{jj}\end{aligned} \tag{2.67}$$

Alternatively, in the matrix and vector notation, (2.66) can be written as

$$[\sigma] = 2\mu [D] + \kappa [I] \operatorname{tr}[D] \tag{2.68}$$

then

$$\begin{aligned}\operatorname{tr}[\sigma] &= 2\mu \operatorname{tr}[D] + \kappa \operatorname{tr}[I] \operatorname{tr}[D] \\ &= 2\mu \operatorname{tr}[D] + 3\kappa \operatorname{tr}[D] \\ &= (2\mu + 3\kappa) \operatorname{tr}[D]\end{aligned} \tag{2.69}$$

which is the same as (2.67).

2.5.4.1 Basic properties of trace

In the following, we list some basic properties of the trace of matrices and the trace of the products of matrices: Consider $n \times n$ square matrices $\boldsymbol{A}$ and $\boldsymbol{B}$.

(a) Trace is a linear mapping. That is

$$\begin{aligned}\operatorname{tr}(\boldsymbol{A} + \boldsymbol{B}) &= \operatorname{tr}(\boldsymbol{A}) + \operatorname{tr}(\boldsymbol{B}) \\ \operatorname{tr}(c\boldsymbol{A}) &= c \operatorname{tr}(\boldsymbol{A}) \ ; \ \forall c \in \mathbb{R} \\ \operatorname{tr}(\boldsymbol{A}) &= \operatorname{tr}(\boldsymbol{A}^T)\end{aligned}$$

(b) Trace of a product

$$\begin{aligned}\operatorname{tr}(\boldsymbol{A}^T\boldsymbol{B}) = \operatorname{tr}(\boldsymbol{A}\boldsymbol{B}^T) = \operatorname{tr}(\boldsymbol{B}^T\boldsymbol{A}) &= \operatorname{tr}(\boldsymbol{B}\boldsymbol{A}^T) \\ \operatorname{tr}(\boldsymbol{A}\boldsymbol{B}) &= \operatorname{tr}(\boldsymbol{B}\boldsymbol{A}) \\ \operatorname{tr}(\boldsymbol{A}\boldsymbol{B}) &\neq \operatorname{tr}(\boldsymbol{A}) \operatorname{tr}(\boldsymbol{B}) \\ \operatorname{tr}(\boldsymbol{A}\boldsymbol{B}) &= \operatorname{tr}\left(\boldsymbol{A}^T\boldsymbol{B}^T\right)\end{aligned}$$

(c) Cyclic property

$$\operatorname{tr}(\boldsymbol{ABCD}) = \operatorname{tr}(\boldsymbol{BCDA}) = \operatorname{tr}(\boldsymbol{CDAB}) = \operatorname{tr}(\boldsymbol{DABC})$$

That is trace is invariant of cyclic permutations.

$$\operatorname{tr}(\boldsymbol{ABC}) \neq \operatorname{tr}(\boldsymbol{ACB}).$$

Note that $(\boldsymbol{ACB})$ is not a cyclic permutation of $(\boldsymbol{ABC})$.

(d) Trace of product of symmetric matrices

$$\text{tr}(\boldsymbol{ABC}) = \text{tr}((\boldsymbol{ABC})^T) = \text{tr}(\boldsymbol{CBA}) = \text{tr}(\boldsymbol{ACB})$$

Thus, when the matrices in a product are symmetric, the trace of the product is invariant of permutation.

Proofs of these can be simply constructed by expanding or writing the expression using Einstein notations and then contraction.

2.6 Reference frame and reference frame transformation

Observing or measuring events related to the evolution of a deforming volume of matter in a fixed frame called *reference frame* is perhaps most convenient and common approach. The fixed frame is generally chosen to be orthogonal and may be Cartesian, cylindrical or spherical. The fixed frame may be viewed as *fixed observer.* However, a general case would be observations made by an observer standing on a platform that is translating and rotating in time. This notion is not far fetched. An observer standing on the surface of the earth observing an object falling towards the surface of the earth is an example. Due to translation and rotation of the earth, the falling object's path is not a straight line path. In order to determine the correct evolution of the object's motion, the observer's translation and rotation must be taken into account. Observing the equivalence between the fixed frame and the fixed observer, here also we could view the translating and rotating observer as being equivalent to a frame that is translating and rotating.

Referring to Figure 2.1 x-frame is translated to O' by $\boldsymbol{c}'(t)$ and then rotated by an orthogonal rotation tensor $\boldsymbol{Q}$ to obtain y-frame. The coordinates $\boldsymbol{x}$ and $\boldsymbol{y}$ of point P are related by

$$\{x\} = \{c'(t)\} + [Q(t)]^T\{y\} \tag{2.70}$$

$$\text{or} \quad \{y\} = \{c(t)\} + [Q(t)]\{x\} \quad ; \quad \{c(t)\} = -[Q(t)]\{c'(t)\} \tag{2.71}$$

$[Q(t)]$ is proper orthogonal rotation tensor, i.e., $\det[Q(t)] = +1$.

Newtonian mechanics is founded on the assumption that distance, time and mass are absolute, i.e., the motion is in inertial frame, which neglects the translation and rotation of the observer in time (i.e., motion of the earth on which the observer is standing). Thus, in Newtonian mechanics, i.e., inertial frame, an object falling towards the earth's surface will follow a straight line path if only gravitational forces are acting on it. Classical continuum mechanics is founded on Newton's laws in inertial frame for which (2.71) reduces to

$$\{y\} = \{c\} + [Q]\{x\} \tag{2.72}$$

in which $[Q]$ is a proper orthogonal rotation tensor (and is not a function of time). Equation (2.72) describes a rotated frame y that differs from x-frame by a rigid orthogonal rotation and a rigid translation $\{c\}$ (also not a function of time).

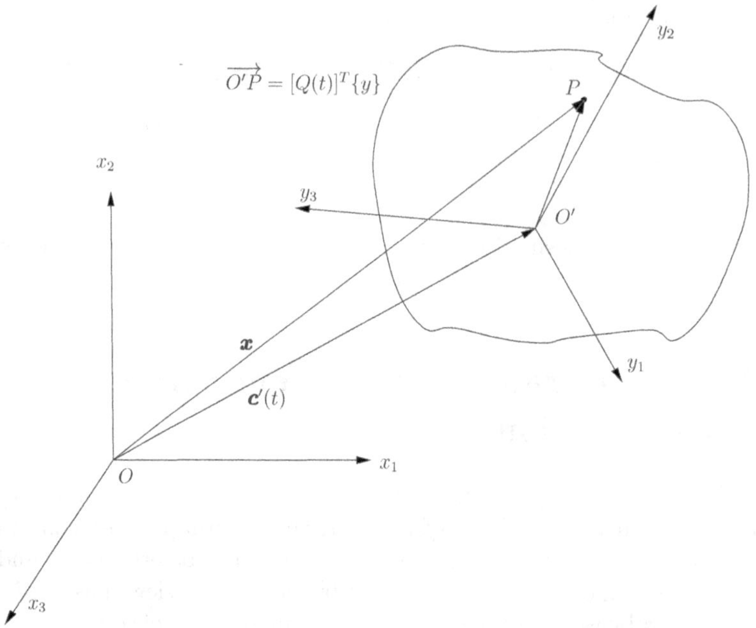

Figure 2.1: Translation and rotation of reference frame

Remarks

(1) The kinematic quantities like displacements, velocities, and strain measured in two different frames (i.e., observed by two different observers) will, in general, differ since the perspectives differ.

(2) In Chapter 5, we consider the change of frames described by (2.71) as well as (2.72) to determine which kinematic tensors are independent of the frame of measurement (i.e., the observer), *frame invariant* or *objective.*

(3) Since the constitution of the deforming matter must be independent of the observer (or frame of measurement), only the tensors that exhibit this property must be used in describing the constitution of the deforming matter (Chapters 8–14). Thus, the constitutive tensors as well as their argument tensors must be objective.

(4) For now, we consider transformation T described by (2.72) (valid only in inertial frame).

(5) The coordinate transformation (2.71) or (2.70) in which the translation of the origin of the coordinate system (or observer position) and the rigid rotation of the coordinate system (rotation of the observer) are functions of time is called *Euclidean transformation* [63] or *non-Galilean transformation.* When the translation and the rigid rotation of the frame (or observer) are not functions of time as in (2.72), the transformation is called Galilean transformation. Newtonian mechanics is based on Galilean transformation.

2.7 Coordinate frames and transformations

Orthogonal frames such as Cartesian, cylindrical and spherical are commonly used. In these frames, the axes are orthogonal and the basis (unit vectors along the orthogonal axes) is used to describe quantities of interest in the frame. Cartesian frame or x-frame is simplest of the three. The x-frame can be transformed into y-frame through a coordinate transformation T. In doing so, we also transform the quantities defined in x-frame to y-frame, referred to as induced transformation. In continuum mechanics, the nature of transformation T and the nature of induced transformation on various quantities defined in x-frame are important classes of studies. We also consider non-orthogonal curvilinear frames in which the coordinate axes are curved and are not orthogonal. An important case is a Cartesian x-frame that transforms into a non-orthogonal curvilinear frame due to coordinate transformation T. Just like orthogonal frames, a curvilinear frame has a basis. We consider details in the following sections.

2.7.1 Cartesian frame and orthogonal coordinate transformations

For now, we consider x-frame to be Cartesian frame and transformation T (2.72) to be such that it causes translation and rigid rotation of x-frame into y-frame. For simplicity, consider a two-dimensional case.

Consider two sets of Cartesian frames in a plane: the x-frame o-x_1x_2 and the y-frame o'-y_1y_2. If the y-frame is obtained from the x-frame by a shift of the origin and without a rotation of the axes, then the transformation from the x-frame to the y-frame is a translation. If the point P has coordinates (x_1, x_2) and (y_1, y_2) in x-frame and y-frame and if (c_1, c_2) are the coordinates of the origin of y-frame with respect to x-frame, then

$$\begin{aligned} x_1 &= y_1 + c_1 \\ x_2 &= y_2 + c_2 \end{aligned} \tag{2.73}$$

or in compact form

$$x_i = y_i + c_i \quad ; \quad i = 1, 2 \tag{2.74}$$

If the origin remains fixed and the y-frame is obtained by rotating the x-frame through an angle θ in the counter clockwise direction about a line perpendicular to the plane of the x-frame but at its origin (see Figure 2.2), then the transformation relating the frames can be derived as follows. Consider a fixed point P with coordinates (x_1, x_2) and (y_1, y_2) in x-frame and y-frame. From Figure 2.2, it is rather straightforward to see that

$$\begin{aligned} x_1 &= y_1 cos\theta - y_2 sin\theta \\ x_2 &= y_1 sin\theta + y_2 cos\theta \end{aligned} \tag{2.75}$$

Let us define

$$[Q] = \begin{bmatrix} Q_{11} & Q_{12} \\ Q_{21} & Q_{22} \end{bmatrix} = \begin{bmatrix} \cos\theta & \sin\theta \\ -\sin\theta & \cos\theta \end{bmatrix} \tag{2.76}$$

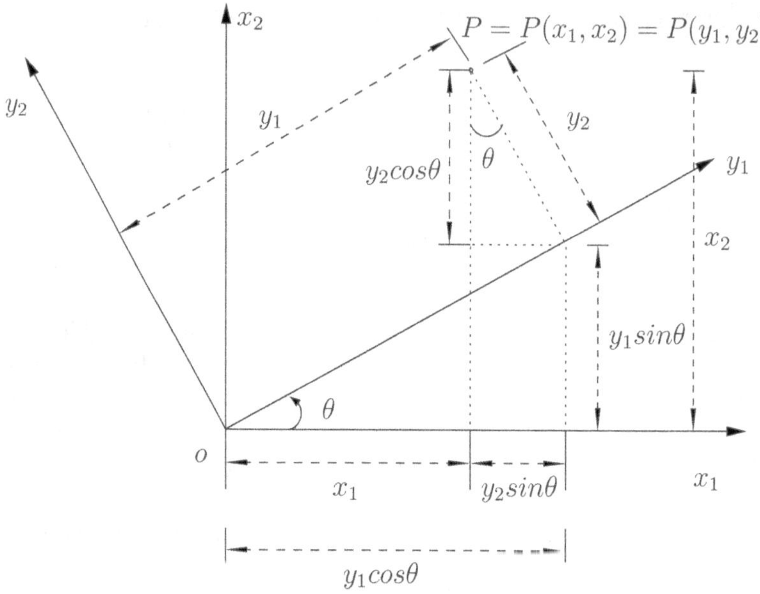

Figure 2.2: Rigid rotation of x-frame to y-frame by an angle θ

Then (2.75) can be written as

$$\begin{Bmatrix} x_1 \\ x_2 \end{Bmatrix} = [Q]^T \begin{Bmatrix} y_1 \\ y_2 \end{Bmatrix} \quad ; \quad x_i = x_i(y_j) \quad ; \quad i,j = 1,2 \tag{2.77}$$

We note that $[Q]$ defines pure rotation of x-frame to y-frame and is orthogonal, i.e.,

$$[Q]^T = [Q]^{-1} \quad \text{hence} \quad [Q]^T[Q] = [Q][Q]^T = [I] \tag{2.78}$$

Using (2.78), we obtain the following from either (2.75) or (2.77).

$$\begin{Bmatrix} y_1 \\ y_2 \end{Bmatrix} = [Q] \begin{Bmatrix} x_1 \\ x_2 \end{Bmatrix} \quad ; \quad y_i = y_i(x_j) \quad ; \quad i,j = 1,2 \tag{2.79}$$

That is, the coordinate transformation T defined by $[Q]$ in (2.79) is such that x-frame is transformed into y-frame. We note that both x-frame and y-frame are Cartesian frames and the transformation T defined by $[Q]$ in (2.79) is orthogonal.

Remarks

(1) In the rotation of the x-frame to y-frame, the point P remains fixed. Thus, if point P is a material point in a volume of matter, then every such point in the volume of matter has coordinates in x- and y-frames that are related through (2.79). It is important to note that in this process, the x-frame is rotated to y-frame, but the volume of matter containing point(s) P remains undisturbed.

(2) In contrast to (1), if we rotate the volume of matter through a rigid rotation, then its coordinate changes from x_i to y_i. Both are measured in the x-frame. The relationship between the new coordinates y_i and the original coordinates

x_i is not (2.79), but instead is given by (analogous to (2.79))

$$\begin{Bmatrix} y_1 \\ y_2 \end{Bmatrix} = [Q]^T \begin{Bmatrix} x_1 \\ x_2 \end{Bmatrix} \tag{2.80}$$

(3) Thus, the rotation of coordinate system or frame is quite different than the rigid rotation of the volume of matter. In this book, we always consider rotation of the frame and not the rigid rotation of the volume of matter regardless of the configuration.

The relationship (2.79) can also be derived in an alternate way using unit vectors (base vectors) in x-frame and y-frame. Let $\mathbf{e}_1^x$, $\mathbf{e}_2^x$ and $\mathbf{e}_1^y$, $\mathbf{e}_2^y$ be bases in x-frame and y-frame. Then the vector $\boldsymbol{OP}$ (Figure 2.2) can be written as (using bases in x- and y-frames, i.e., index notation)

$$\boldsymbol{OP} = \mathbf{e}_1^x x_1 + \mathbf{e}_2^x x_2 = \mathbf{e}_1^y y_1 + \mathbf{e}_2^y y_2 \tag{2.81}$$

Using (2.81) it is possible to derive the transformation T relating x-frame to y-frame and y-frame to x-frame. Consider the dot products $\mathbf{e}_1^x \cdot \boldsymbol{OP}$ and $\mathbf{e}_2^x \cdot \boldsymbol{OP}$.

$$\begin{aligned} \mathbf{e}_1^x \cdot \mathbf{e}_1^x x_1 + \mathbf{e}_1^x \cdot \mathbf{e}_2^x x_2 &= \mathbf{e}_1^x \cdot \mathbf{e}_1^y y_1 + \mathbf{e}_1^x \cdot \mathbf{e}_2^y y_2 \\ \mathbf{e}_2^x \cdot \mathbf{e}_1^x x_2 + \mathbf{e}_2^x \cdot \mathbf{e}_2^x x_2 &= \mathbf{e}_2^x \cdot \mathbf{e}_1^y y_1 + \mathbf{e}_2^x \cdot \mathbf{e}_2^y y_2 \end{aligned} \tag{2.82}$$

Using $\mathbf{e}_1^x \cdot \mathbf{e}_1^x = 1$, $\mathbf{e}_2^x \cdot \mathbf{e}_2^x = 1$ and $\mathbf{e}_1^x \cdot \mathbf{e}_2^x = \mathbf{e}_2^x \cdot \mathbf{e}_1^x = 0$ in (2.82) we obtain

$$\begin{Bmatrix} x_1 \\ x_2 \end{Bmatrix} = \begin{bmatrix} \mathbf{e}_1^x \cdot \mathbf{e}_1^y & \mathbf{e}_1^x \cdot \mathbf{e}_2^y \\ \mathbf{e}_2^x \cdot \mathbf{e}_1^y & \mathbf{e}_2^x \cdot \mathbf{e}_2^y \end{bmatrix} \begin{Bmatrix} y_1 \\ y_2 \end{Bmatrix} \tag{2.83}$$

In a similar fashion, if we take the dot products $\mathbf{e}_1^y \cdot \boldsymbol{OP}$ and $\mathbf{e}_2^y \cdot \boldsymbol{OP}$ and use the properties $\mathbf{e}_1^y \cdot \mathbf{e}_1^y = 1$, $\mathbf{e}_2^y \cdot \mathbf{e}_2^y = 1$ and $\mathbf{e}_1^y \cdot \mathbf{e}_2^y = \mathbf{e}_2^y \cdot \mathbf{e}_1^y = 0$, then we obtain

$$\begin{Bmatrix} y_1 \\ y_2 \end{Bmatrix} = \begin{bmatrix} \mathbf{e}_1^y \cdot \mathbf{e}_1^x & \mathbf{e}_1^y \cdot \mathbf{e}_2^x \\ \mathbf{e}_2^y \cdot \mathbf{e}_1^x & \mathbf{e}_2^y \cdot \mathbf{e}_2^x \end{bmatrix} \begin{Bmatrix} x_1 \\ x_2 \end{Bmatrix} \tag{2.84}$$

Let us define

$$[Q] = \begin{bmatrix} \mathbf{e}_1^y \cdot \mathbf{e}_1^x & \mathbf{e}_1^y \cdot \mathbf{e}_2^x \\ \mathbf{e}_2^y \cdot \mathbf{e}_1^x & \mathbf{e}_2^y \cdot \mathbf{e}_2^x \end{bmatrix} \tag{2.85}$$

Then (2.84) and (2.83) can be written as

$$\begin{Bmatrix} y_1 \\ y_2 \end{Bmatrix} = [Q] \begin{Bmatrix} x_1 \\ x_2 \end{Bmatrix} \tag{2.86}$$

$$\begin{Bmatrix} x_1 \\ x_2 \end{Bmatrix} = [Q]^T \begin{Bmatrix} y_1 \\ y_2 \end{Bmatrix} \tag{2.87}$$

That is, $[Q]$ in (2.85) is orthogonal. This form of $[Q]$ can be easily extended to the three-dimensional case as done subsequently. First, we derive a special form of $[Q]$ in (2.76) using (2.85) that corresponds to a rigid notation of the x-frame to the y-frame by an angle θ, i.e., we derive $[Q]$ of (2.76) from (2.85).

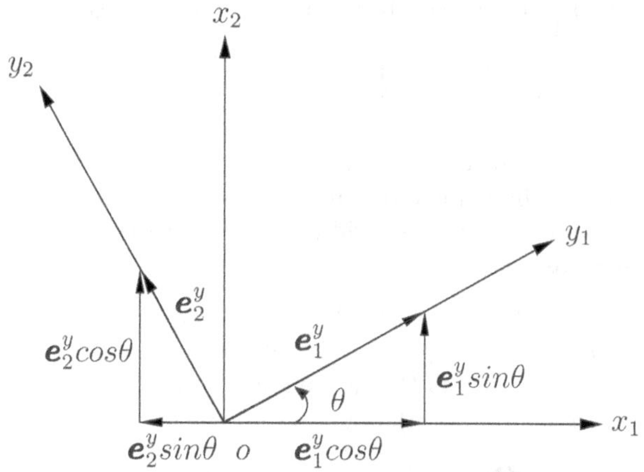

Figure 2.3: Base vectors $\mathbf{e}_1^y$, $\mathbf{e}_2^y$ of y-frame and their projections in x-frame

Consider Figure 2.3 showing unit vectors $\mathbf{e}_1^y$ and $\mathbf{e}_2^y$ of the y-frame and their projection in the x-frame. The unit vectors in the x-frame, i.e., $\mathbf{e}_1^x$ and $\mathbf{e}_2^x$, are the algebraic sum of these projections. Hence we have

$$\begin{aligned}\mathbf{e}_1^x &= \mathbf{e}_1^y \cos\theta - \mathbf{e}_2^y \sin\theta \\ \mathbf{e}_2^x &= \mathbf{e}_1^y \sin\theta + \mathbf{e}_2^y \cos\theta\end{aligned} \tag{2.88}$$

and substituting from (2.88) into (2.85) and using the properties of the dot products of $\mathbf{e}_1^y$ and $\mathbf{e}_2^y$ we obtain

$$[Q] = \begin{bmatrix} \cos\theta & \sin\theta \\ -\sin\theta & \cos\theta \end{bmatrix} \tag{2.89}$$

which is the same as (2.76). This second approach of relating x-frame and y-frame through an orthogonal transformation in (2.85) can be easily extended to three dimensions. Consider a point $P(x_1, x_2, x_3)$ in the x-frame. Let $\mathbf{e}_i^x$ be the unit base vectors in the x-frame. Let T be an orthogonal transformation such that

$$T \;:\; y_i = y_i(x_j) \tag{2.90}$$

That is, T transforms x-frame into y-frame. Let $\mathbf{e}_i^y$ be the unit base vectors in the y-frame. We consider T to be an orthogonal transformation. We want to find how x- and y-frames are related through T and the specific form of T. Let y_1, y_2, y_3 be the coordinates of the point P in the y-frame. The vector $\boldsymbol{OP}$ can be represented in x- and y-frames.

$$\boldsymbol{OP} = \mathbf{e}_i^x x_i = \mathbf{e}_i^y y_i \tag{2.91}$$

As in the two-dimensional case, here also, if we use the relations

$$\begin{aligned}\mathbf{e}_i^x \cdot \mathbf{e}_j^x &= \delta_{ij} \\ \mathbf{e}_i^y \cdot \mathbf{e}_j^y &= \delta_{ij}\end{aligned} \tag{2.92}$$

and if we take the dot product of (2.91) with $\boldsymbol{e}_1^y$, $\boldsymbol{e}_2^y$ and $\boldsymbol{e}_3^y$, we can derive the following:

$$\begin{Bmatrix} y_1 \\ y_2 \\ y_3 \end{Bmatrix} = \begin{bmatrix} \boldsymbol{e}_1^y \cdot \boldsymbol{e}_1^x & \boldsymbol{e}_1^y \cdot \boldsymbol{e}_2^x & \boldsymbol{e}_1^y \cdot \boldsymbol{e}_3^x \\ \boldsymbol{e}_2^y \cdot \boldsymbol{e}_1^x & \boldsymbol{e}_2^y \cdot \boldsymbol{e}_2^x & \boldsymbol{e}_2^y \cdot \boldsymbol{e}_3^x \\ \boldsymbol{e}_3^y \cdot \boldsymbol{e}_1^x & \boldsymbol{e}_3^y \cdot \boldsymbol{e}_2^x & \boldsymbol{e}_3^y \cdot \boldsymbol{e}_3^x \end{bmatrix} \begin{Bmatrix} x_1 \\ x_2 \\ x_3 \end{Bmatrix} \tag{2.93}$$

Let

$$[Q] = \begin{bmatrix} \boldsymbol{e}_1^y \cdot \boldsymbol{e}_1^x & \boldsymbol{e}_1^y \cdot \boldsymbol{e}_2^x & \boldsymbol{e}_1^y \cdot \boldsymbol{e}_3^x \\ \boldsymbol{e}_2^y \cdot \boldsymbol{e}_1^x & \boldsymbol{e}_2^y \cdot \boldsymbol{e}_2^x & \boldsymbol{e}_2^y \cdot \boldsymbol{e}_3^x \\ \boldsymbol{e}_3^y \cdot \boldsymbol{e}_1^x & \boldsymbol{e}_3^y \cdot \boldsymbol{e}_2^x & \boldsymbol{e}_3^y \cdot \boldsymbol{e}_3^x \end{bmatrix} \tag{2.94}$$

then

$$\begin{Bmatrix} y_1 \\ y_2 \\ y_3 \end{Bmatrix} = [Q] \begin{Bmatrix} x_1 \\ x_2 \\ x_3 \end{Bmatrix} \tag{2.95}$$

in which $[Q]$ is orthogonal (check as an exercise), i.e.,

$$[Q]^T = [Q]^{-1} \quad \text{and} \quad [Q]^T[Q] = [Q][Q]^T = [I] \tag{2.96}$$

Hence

$$\det([Q]^T[Q]) = \det[I] \tag{2.97}$$

$$\det[Q]^T \det[Q] = 1$$

$$\det[Q] \det[Q] = 1$$

$$(\det[Q])^2 = 1$$

$$\det[Q] = \pm 1 \tag{2.98}$$

If the orthogonal matrix $[Q]$ is such that $\det[Q] = +1$, then we refer to $[Q]$ as a proper orthogonal matrix and the transformation of x- to y-frame is a proper orthogonal transformation. In such transformations, a right-handed orthogonal system (say x-frame) is transformed into another right-handed orthogonal frame (say y-frame). An orthogonal matrix whose determinant is -1 is called an improper orthogonal matrix. An improper orthogonal matrix transforms a right-handed orthogonal frame into a left-handed orthogonal frame.

2.7.2 Curvilinear coordinates (or frame)

Let x_1, x_2, x_3 or x_i ($i = 1, 2, 3$) be the coordinates of a material point P (in fixed Cartesian x-frame) and y_1, y_2, y_3 or y_j ($j = 1, 2, 3$) be three variables. If we can establish a correspondence between x_i and y_j, then we say that there is a transformation T or a coordinate transformation between x_i and y_j, and we can write

$$T \ : \ y_j = y_j(x_1, x_2, x_3) \quad ; \quad j = 1, 2, 3 \tag{2.99}$$

If this correspondence is one-to-one, then there exists a unique inverse of (2.99), in the following form

$$T^{-1} \ : \ x_i = x_i(y_1, y_2, y_3) \quad ; \quad i = 1, 2, 3 \tag{2.100}$$

We can show this inverse (i.e., T^{-1}) exists, and is unique, if the determinant of the Jacobian of transformation is not equal to zero.

$$\det[J] = \det\begin{bmatrix} \dfrac{y_1, y_2, y_3}{x_1, x_2, x_3} \end{bmatrix} = \begin{vmatrix} \dfrac{\partial y_1}{\partial x_1} & \dfrac{\partial y_1}{\partial x_2} & \dfrac{\partial y_1}{\partial x_3} \\ \dfrac{\partial y_2}{\partial x_1} & \dfrac{\partial y_2}{\partial x_2} & \dfrac{\partial y_2}{\partial x_3} \\ \dfrac{\partial y_3}{\partial x_1} & \dfrac{\partial y_3}{\partial x_2} & \dfrac{\partial y_3}{\partial x_3} \end{vmatrix} \neq 0 \tag{2.101}$$

For a fixed set of values x_1, x_2, x_3, the transformation (2.100) gives three non-coincident surfaces called curvilinear surfaces. These interact at a single point P with fixed y_1, y_2, y_3. Thus, to point P, we can assign values y_1, y_2, y_3 called curvilinear coordinates of P (Figure 2.4). From Figure 2.4, we note that the intersection of any two surfaces gives a curve through P called a curvilinear coordinate line.

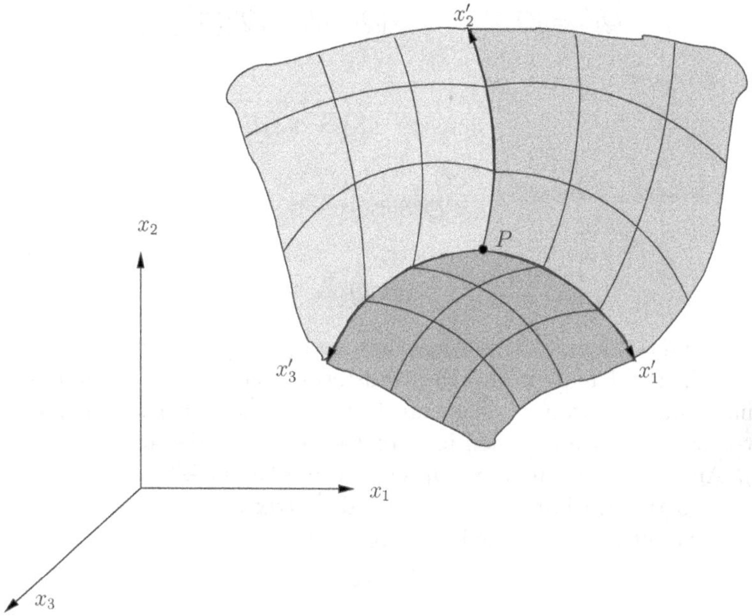

Figure 2.4: Curvilinear coordinates

2.8 Curvilinear frames, covariant and contravariant bases

To quantify the motion and deformation of matter, we need a fundamental understanding of the motion of the material particles. Consider undeformed volume

V with boundary ∂V at time t_0 as constituting the reference configuration. The reference configuration could be different than the configuration of the matter at time t_0, but assumed to be the same for the sake of simplicity. Let o-$x_1x_2x_3$ be the x-frame. Let each material particle in the reference configuration be assigned unique coordinates (x_1, x_2, x_3) in the x-frame. We assume that the matter is *homogeneous and isotropic*. Consider a material point $P(x_1, x_2, x_3)$ in this configuration. We draw lines parallel to ox_1, ox_2, ox_3 axes in the reference configuration representing material lines PA_1, PA_2, PA_3 (Figure 2.5(a)). The deformed or current configuration, i.e., volume $\bar{V}$ with boundary $\partial\bar{V}$ at time t is shown in Figure 2.5(b). Let $\bar{x}_i$ be the deformed coordinates of the material point $P(x_1, x_2, x_3)$ in the fixed coordinate x-frame. Naturally $\bar{x}_i = x_i + u_i$; u_i being displacements of the point $P(x_1, x_2, x_3)$ in the fixed x-frame. The deformed material lines $\bar{P}\tilde{A}_1$, $\bar{P}\tilde{A}_2$, $\bar{P}\tilde{A}_3$, coordinate axes $\tilde{o}\tilde{x}_1$, $\tilde{o}\tilde{x}_2$, $\tilde{o}\tilde{x}_3$ and the location of point P (i.e., $\bar{P}$) are also shown in Figure 2.5(b). The coordinate system $\tilde{o}$-$\tilde{x}_1\tilde{x}_2\tilde{x}_3$ ($\tilde{x}$-frame) is called *convected coordinate system*. This coordinate system and various measures of the state of deforming matter in this coordinate system are crucial in developing constitutive theories (considered in later chapters) because, in this coordinate system, we have deformed material lines.

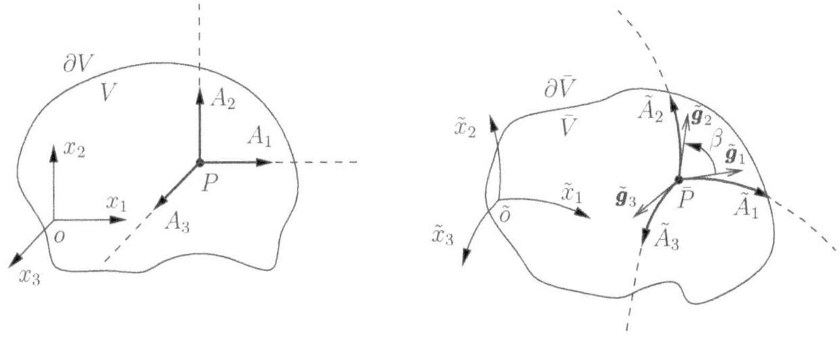

(a) Undeformed configuration

(b) Deformed configuration: covariant basis

Figure 2.5: Material lines in reference and current configurations: covariant basis

We note that at time t_0, the convected coordinate system coincides with the Cartesian coordinate system o-$x_1x_2x_3$. The convected coordinate system continuously changes in time in such a way that coordinates of a material particle in this coordinate system remain the same for all values of time. If $\tilde{x}_i$ are the coordinates of the material particle in $\tilde{o}$-$\tilde{x}_1\tilde{x}_2\tilde{x}_3$ coordinate system, then $\tilde{x}_i$ remain fixed for all values of time. An important conclusion is that coordinates x_i of a material particle $P(x_i)$ in the configuration at time $t_0 = 0$ in x-frame have exactly the same numerical values as the material particle $\bar{P}$ coordinates $\tilde{x}_i$ in $\tilde{x}$-frame. We can write the following:

$$\bar{x}_i = \bar{x}_i(\tilde{x}_1, \tilde{x}_2, \tilde{x}_3, t) \quad ; \quad i = 1, 2, 3 \tag{2.102}$$

But the material point with coordinates $\tilde{x}_i$ are the same as x_i, hence

$$\bar{x}_i = \bar{x}_i(x_1, x_2, x_3, t) \quad \text{or} \quad \bar{\boldsymbol{r}} = \bar{\boldsymbol{r}}(\boldsymbol{r}, t) \tag{2.103}$$

where $\bar{\boldsymbol{r}}$ and $\boldsymbol{r}$ are vectors with components $\bar{x}_i$ and x_i ; $i = 1, 2, 3$ respectively. Inverse of (2.103) gives

$$x_i = x_i(\bar{x}_1, \bar{x}_2, \bar{x}_3, t) \quad ; \quad \boldsymbol{r} = \boldsymbol{r}(\bar{\boldsymbol{r}}, t) \tag{2.104}$$

2.8.1 Covariant basis

Let the vectors $\tilde{\boldsymbol{g}}_1$, $\tilde{\boldsymbol{g}}_2$ and $\tilde{\boldsymbol{g}}_3$ be tangent to the material lines $\bar{P}\tilde{A}_1$, $\bar{P}\tilde{A}_2$ and $\bar{P}\tilde{A}_3$ at material point $\bar{P}$. The vectors $\tilde{\boldsymbol{g}}_1$, $\tilde{\boldsymbol{g}}_2$ and $\tilde{\boldsymbol{g}}_3$ are covariant vectors and form a non-orthogonal basis generally called *covariant basis*. Using this basis, we can define the coordinates of point $\bar{P}$ in $\tilde{x}$-frame.

In the following, we determine the explicit form of the covariant base vectors. Since $\tilde{\boldsymbol{g}}_i$ are tangent vectors to $\tilde{x}_1$, $\tilde{x}_2$ and $\tilde{x}_3$ axes, these can be obtained by differentiating the position vector $\bar{\boldsymbol{r}}$ with respect to $\tilde{x}_i$.

$$\tilde{\boldsymbol{g}}_i = \frac{\partial \bar{\boldsymbol{r}}}{\partial \tilde{x}_i} \quad ; \quad (\text{Def. of } \ \tilde{\boldsymbol{g}}_i) \tag{2.105}$$

But $\tilde{x}_i = x_i$, therefore

$$\tilde{\boldsymbol{g}}_i = \frac{\partial(\bar{\boldsymbol{r}})}{\partial x_i} \tag{2.106}$$

$$\tilde{\boldsymbol{g}}_i = \frac{\partial(\boldsymbol{e}_j \bar{x}_j)}{\partial x_i} \tag{2.107}$$

$$\text{or} \quad \tilde{\boldsymbol{g}}_i = \boldsymbol{e}_j \frac{\partial \bar{x}_j}{\partial x_i} \tag{2.108}$$

Expanding for $i = 1, 2, 3$

$$\begin{aligned}
\tilde{\boldsymbol{g}}_1 &= \frac{\partial \bar{x}_1}{\partial x_1}\boldsymbol{e}_1 + \frac{\partial \bar{x}_2}{\partial x_1}\boldsymbol{e}_2 + \frac{\partial \bar{x}_3}{\partial x_1}\boldsymbol{e}_3 \quad ; \quad \{\tilde{g}_1\}^T = \left[\frac{\partial \bar{x}_1}{\partial x_1}, \frac{\partial \bar{x}_2}{\partial x_1}, \frac{\partial \bar{x}_3}{\partial x_1}\right] \\
\tilde{\boldsymbol{g}}_2 &= \frac{\partial \bar{x}_1}{\partial x_2}\boldsymbol{e}_1 + \frac{\partial \bar{x}_2}{\partial x_2}\boldsymbol{e}_2 + \frac{\partial \bar{x}_3}{\partial x_2}\boldsymbol{e}_3 \quad ; \quad \{\tilde{g}_2\}^T = \left[\frac{\partial \bar{x}_1}{\partial x_2}, \frac{\partial \bar{x}_2}{\partial x_2}, \frac{\partial \bar{x}_3}{\partial x_2}\right] \\
\tilde{\boldsymbol{g}}_3 &= \frac{\partial \bar{x}_1}{\partial x_3}\boldsymbol{e}_1 + \frac{\partial \bar{x}_2}{\partial x_3}\boldsymbol{e}_2 + \frac{\partial \bar{x}_3}{\partial x_3}\boldsymbol{e}_3 \quad ; \quad \{\tilde{g}_3\}^T = \left[\frac{\partial \bar{x}_1}{\partial x_3}, \frac{\partial \bar{x}_2}{\partial x_3}, \frac{\partial \bar{x}_3}{\partial x_3}\right]
\end{aligned} \tag{2.109}$$

Let us introduce a new matrix notation $[J]$ containing $\tilde{\boldsymbol{g}}_i$.

$$[J] = \left[\frac{\partial\{\bar{x}\}}{\partial\{x\}}\right] = \left[\frac{\bar{\boldsymbol{x}}}{\boldsymbol{x}}\right] = \left[\frac{\bar{x}_1, \bar{x}_2, \bar{x}_3}{x_1, x_2, x_3}\right] = \begin{bmatrix} \frac{\partial \bar{x}_1}{\partial x_1} & \frac{\partial \bar{x}_1}{\partial x_2} & \frac{\partial \bar{x}_1}{\partial x_3} \\[2ex] \frac{\partial \bar{x}_2}{\partial x_1} & \frac{\partial \bar{x}_2}{\partial x_2} & \frac{\partial \bar{x}_2}{\partial x_3} \\[2ex] \frac{\partial \bar{x}_3}{\partial x_1} & \frac{\partial \bar{x}_3}{\partial x_2} & \frac{\partial \bar{x}_3}{\partial x_3} \end{bmatrix} \tag{2.110}$$

$$\text{or} \quad \boldsymbol{J} = \boldsymbol{e}_i \boldsymbol{e}_j J_{ij} = \boldsymbol{e}_i \boldsymbol{e}_j \bar{x}_{i,j}$$

This notation is due to Murnaghan [64]. In this notation, the position of the letter above the bar (–) indicates a row and those below the bar signify a column. Thus, (2.108) or (2.109) can be written as

$$\tilde{\boldsymbol{g}}_i = \boldsymbol{e}_j J_{ji} \quad \text{or} \quad \begin{bmatrix} \{\tilde{g}_1\}^T \\ \{\tilde{g}_2\}^T \\ \{\tilde{g}_3\}^T \end{bmatrix} = [J]^T \begin{Bmatrix} \boldsymbol{e}_1 \\ \boldsymbol{e}_2 \\ \boldsymbol{e}_3 \end{Bmatrix} \tag{2.111}$$

We note that columns of $[J]$ in (2.110) are the covariant base vectors.

$$[J] = \big[\{\tilde{g}_1\}, \{\tilde{g}_2\}, \{\tilde{g}_3\}\big] \tag{2.112}$$

The vectors $\tilde{\boldsymbol{g}}_i$ constitute a basis but the basis is non-orthogonal. We also note that $\tilde{\boldsymbol{g}}_i$ are not normalized. We will see in Chapter 3 that $[J]$ is, in fact, deformation gradient matrix (at a point) in Lagrangian description.

2.8.2 Contravariant basis

In terms of the basic geometric interpretation of contravariant basis, we consider orthogonal material lines $\bar{P}\bar{A}_1$, $\bar{P}\bar{A}_2$, $\bar{P}\bar{A}_3$ parallel to the axes of the x-frame in the current configuration in Figure 2.6(b). Their map in the reference configuration is naturally curvilinear ($P\tilde{A}_1$, $P\tilde{A}_2$, $P\tilde{A}_3$) as shown in Figure 2.6(a). Consider tangent vectors $\tilde{\boldsymbol{g}}^1$, $\tilde{\boldsymbol{g}}^2$ and $\tilde{\boldsymbol{g}}^3$ to the material lines $P\tilde{A}_1$, $P\tilde{A}_2$ and $P\tilde{A}_3$ at material point P. These are referred to as contravariant base vectors. There is an alternate way to visualize these basis vectors. Just like covariant base vectors, we may also define a set of convected reciprocal base vectors $\tilde{\boldsymbol{g}}^i$; $i = 1, 2, 3$ as *contravariant basis*. Based on the definition of covariant base vector (2.105), the reciprocal base vectors $\tilde{\boldsymbol{g}}^i$ can be defined using

$$\tilde{\boldsymbol{g}}^i = \frac{\partial \tilde{x}_i}{\partial \bar{\boldsymbol{r}}} \quad ; \quad (\text{Def. of } \ \tilde{\boldsymbol{g}}^i) \tag{2.113}$$

But $\tilde{x}_i = x_i$, therefore

$$\tilde{\boldsymbol{g}}^i = \frac{\partial x_i}{\partial (\bar{\boldsymbol{r}})} \tag{2.114}$$

$$\tilde{\boldsymbol{g}}^i = \left(\boldsymbol{e}_j \frac{\partial}{\partial \bar{x}_j}\right) x_i \tag{2.115}$$

$$\text{or} \quad \tilde{\boldsymbol{g}}^i = \boldsymbol{e}_j \frac{\partial x_i}{\partial \bar{x}_j} \tag{2.116}$$

Here also, we note that $\tilde{\boldsymbol{g}}^i$ are not normalized. Expanding for $i = 1, 2, 3$

$$\begin{aligned}
\tilde{\boldsymbol{g}}^1 &= \frac{\partial x_1}{\partial \bar{x}_1}\boldsymbol{e}_1 + \frac{\partial x_1}{\partial \bar{x}_2}\boldsymbol{e}_2 + \frac{\partial x_1}{\partial \bar{x}_3}\boldsymbol{e}_3 \quad ; \quad \{\tilde{g}^1\}^T = \left[\frac{\partial x_1}{\partial \bar{x}_1}, \frac{\partial x_1}{\partial \bar{x}_2}, \frac{\partial x_1}{\partial \bar{x}_3}\right] \\
\tilde{\boldsymbol{g}}^2 &= \frac{\partial x_2}{\partial \bar{x}_1}\boldsymbol{e}_1 + \frac{\partial x_2}{\partial \bar{x}_2}\boldsymbol{e}_2 + \frac{\partial x_2}{\partial \bar{x}_3}\boldsymbol{e}_3 \quad ; \quad \{\tilde{g}^2\}^T = \left[\frac{\partial x_2}{\partial \bar{x}_1}, \frac{\partial x_2}{\partial \bar{x}_2}, \frac{\partial x_2}{\partial \bar{x}_3}\right] \\
\tilde{\boldsymbol{g}}^3 &= \frac{\partial x_3}{\partial \bar{x}_1}\boldsymbol{e}_1 + \frac{\partial x_3}{\partial \bar{x}_2}\boldsymbol{e}_2 + \frac{\partial x_3}{\partial \bar{x}_3}\boldsymbol{e}_3 \quad ; \quad \{\tilde{g}^3\}^T = \left[\frac{\partial x_3}{\partial \bar{x}_1}, \frac{\partial x_3}{\partial \bar{x}_2}, \frac{\partial x_3}{\partial \bar{x}_3}\right]
\end{aligned} \tag{2.117}$$

Following the Murnaghan's notation, we can write the following using matrix notation, $[\bar{J}]$ containing $\tilde{\boldsymbol{g}}^i$.

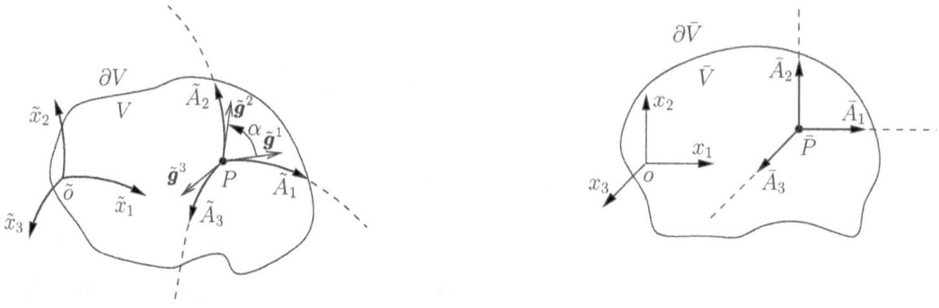

(a) Undeformed configuration: contravariant basis

(b) Deformed configuration

Figure 2.6: Map of orthogonal material lines from current to reference configuration: contravariant basis

$$[\bar{J}] = \left[\frac{\partial\{x\}}{\partial\{\bar{x}\}}\right] = \left[\frac{\boldsymbol{x}}{\bar{\boldsymbol{x}}}\right] = \left[\frac{x_1, x_2, x_3}{\bar{x}_1, \bar{x}_2, \bar{x}_3}\right] = \begin{bmatrix} \frac{\partial x_1}{\partial \bar{x}_1} & \frac{\partial x_1}{\partial \bar{x}_2} & \frac{\partial x_1}{\partial \bar{x}_3} \\ \frac{\partial x_2}{\partial \bar{x}_1} & \frac{\partial x_2}{\partial \bar{x}_2} & \frac{\partial x_2}{\partial \bar{x}_3} \\ \frac{\partial x_3}{\partial \bar{x}_1} & \frac{\partial x_3}{\partial \bar{x}_2} & \frac{\partial x_3}{\partial \bar{x}_3} \end{bmatrix} \tag{2.118}$$

$$\text{or} \quad \bar{\boldsymbol{J}} = \boldsymbol{e}_i \boldsymbol{e}_j \bar{J}_{ij} = \boldsymbol{e}_i \boldsymbol{e}_j x_{i,j}$$

Thus, (2.116) or (2.117) can be written as

$$\tilde{\boldsymbol{g}}^i = \bar{J}_{ij} \boldsymbol{e}_j \quad \text{or} \quad \begin{bmatrix} \{\tilde{g}^1\}^T \\ \{\tilde{g}^2\}^T \\ \{\tilde{g}^3\}^T \end{bmatrix} = [\bar{J}] \begin{Bmatrix} \boldsymbol{e}_1 \\ \boldsymbol{e}_2 \\ \boldsymbol{e}_3 \end{Bmatrix} \tag{2.119}$$

We note that rows of $[\bar{J}]$ in (2.117) are the contravariant base vectors.

$$[\bar{J}] = \begin{bmatrix} \{\tilde{g}^1\}^T \\ \{\tilde{g}^2\}^T \\ \{\tilde{g}^3\}^T \end{bmatrix} \tag{2.120}$$

Vectors $\tilde{\boldsymbol{g}}^i$ constitute contravariant basis but the basis is non-orthogonal. We note that $\tilde{\boldsymbol{g}}^i$ are not normalized. We note that $[\bar{J}]$ is also deformation gradient matrix but in Eulerian description.

2.8.3 Alternate way to visualize covariant and contravariant bases $\tilde{\boldsymbol{g}}_i$, $\tilde{\boldsymbol{g}}^i$ and the relationships between them

Consider an elementary tetrahedron $PA_1A_2A_3$ in the undeformed configuration whose edges PA_1, PA_2, PA_3 represent material lines (Figure 2.7(a)). The oblique plane $A_1A_2A_3$ represents part of the bounding surface ∂V (external boundary or due to cut principle of Cauchy) of volume V. $\bar{P}\bar{A}_1\bar{A}_2\bar{A}_3$ is the deformed tetrahedron in the current configuration at time t. Its edges $\bar{P}\bar{A}_1$, $\bar{P}\bar{A}_2$, $\bar{P}\bar{A}_3$ are tangent vectors to the curvilinear axes $\bar{P}\tilde{x}_1$, $\bar{P}\tilde{x}_2$, $\bar{P}\tilde{x}_3$ at point $\bar{P}$ and thus, are covariant base vectors $\tilde{\boldsymbol{g}}_i$ defined by the columns of $[J] = \left[\frac{\partial\{\bar{x}\}}{\partial\{x\}}\right]$ (see Figure 2.7(b)).

We construct vectors $\tilde{\boldsymbol{g}}^1, \tilde{\boldsymbol{g}}^2, \tilde{\boldsymbol{g}}^3$ by taking components of $\tilde{\boldsymbol{g}}_2, \tilde{\boldsymbol{g}}_3$; $\tilde{\boldsymbol{g}}_3, \tilde{\boldsymbol{g}}_1$ and $\tilde{\boldsymbol{g}}_1$, $\tilde{\boldsymbol{g}}_2$ and the volume $\bar{V}$ of the parallelepiped formed by $\tilde{\boldsymbol{g}}_i$, i.e.,

$$\tilde{\boldsymbol{g}}^1 = \frac{\tilde{\boldsymbol{g}}_2 \times \tilde{\boldsymbol{g}}_3}{\tilde{\boldsymbol{g}}_1 \cdot (\tilde{\boldsymbol{g}}_2 \times \tilde{\boldsymbol{g}}_3)} \quad ; \quad \tilde{\boldsymbol{g}}^2 = \frac{\tilde{\boldsymbol{g}}_3 \times \tilde{\boldsymbol{g}}_1}{\tilde{\boldsymbol{g}}_2 \cdot (\tilde{\boldsymbol{g}}_3 \times \tilde{\boldsymbol{g}}_1)} \quad ; \quad \tilde{\boldsymbol{g}}^3 = \frac{\tilde{\boldsymbol{g}}_1 \times \tilde{\boldsymbol{g}}_2}{\tilde{\boldsymbol{g}}_3 \cdot (\tilde{\boldsymbol{g}}_1 \times \tilde{\boldsymbol{g}}_2)} \tag{2.121}$$

We note that $\tilde{\boldsymbol{g}}^1$, $\tilde{\boldsymbol{g}}^2$ and $\tilde{\boldsymbol{g}}^3$ are normal to the faces of the deformed tetrahedron formed by vectors $\tilde{\boldsymbol{g}}_2, \tilde{\boldsymbol{g}}_3$; $\tilde{\boldsymbol{g}}_3, \tilde{\boldsymbol{g}}_1$ and $\tilde{\boldsymbol{g}}_1, \tilde{\boldsymbol{g}}_2$ (see Figure 2.7(c)), and that the volume $\bar{V}$ of the parallelepiped formed by the vectors $\tilde{\boldsymbol{g}}_i$ in the deformed configuration is given by (used in (2.121))

$$\bar{V} = \tilde{\boldsymbol{g}}_1 \cdot (\tilde{\boldsymbol{g}}_2 \times \tilde{\boldsymbol{g}}_3) = \tilde{\boldsymbol{g}}_2 \cdot (\tilde{\boldsymbol{g}}_3 \times \tilde{\boldsymbol{g}}_1) = \tilde{\boldsymbol{g}}_3 \cdot (\tilde{\boldsymbol{g}}_1 \times \tilde{\boldsymbol{g}}_2) \tag{2.122}$$

From the definitions of $\tilde{\boldsymbol{g}}_i$ and $\tilde{\boldsymbol{g}}^i$ we note that

$$\tilde{\boldsymbol{g}}^i \cdot \tilde{\boldsymbol{g}}_j = \tilde{\boldsymbol{g}}_j \cdot \tilde{\boldsymbol{g}}^i = \delta_{ij} \tag{2.123}$$

We recall that when $\bar{x}_i = \bar{x}_i(x_j, t)$, then $\tilde{\boldsymbol{g}}_i$ are defined by

$$\tilde{\boldsymbol{g}}_i = \left(\frac{\partial \bar{x}_j}{\partial x_i}\right)\boldsymbol{e}_j \quad ; \quad \text{covariant basis} \tag{2.124}$$

On the other hand if we consider $x_i = x_i(\bar{x}_j, t)$, inverse of $\bar{x}_i = \bar{x}_i(x, t)$, then

$$\tilde{\boldsymbol{g}}^i = \left(\frac{\partial x_i}{\partial \bar{x}_j}\right)\boldsymbol{e}_j \quad ; \quad \text{contravariant basis} \tag{2.125}$$

From (2.124) and (2.125), we note that

$$\begin{aligned} \tilde{\boldsymbol{g}}_i \cdot \tilde{\boldsymbol{g}}^j &= \left(\frac{\partial \bar{x}_k}{\partial x_i}\right)\boldsymbol{e}_k \cdot \left(\frac{\partial x_j}{\partial \bar{x}_l}\right)\boldsymbol{e}_l = \frac{\partial \bar{x}_k}{\partial x_i} \cdot \frac{\partial x_j}{\partial \bar{x}_l}\boldsymbol{e}_k \cdot \boldsymbol{e}_l = \frac{\partial \bar{x}_k}{\partial x_i}\frac{\partial x_j}{\partial \bar{x}_l}\delta_{kl} \\ &= \frac{\partial \bar{x}_k}{\partial x_i}\frac{\partial x_j}{\partial \bar{x}_k} = \frac{\partial x_j}{\partial \bar{x}_k}\frac{\partial \bar{x}_k}{\partial x_i} = \delta_{ji} = \delta_{ij} \end{aligned} \tag{2.126}$$

Thus, *definition of* $\tilde{\boldsymbol{g}}^i$ *in* (2.125) *is consistent with* (2.121).

2.9 Scalars, vectors and tensors

Almost all mathematical quantities in continuum mechanics related to the deformation physics of the continuous media can be uniformly defined by using what are called tensors. We consider some illustrative examples. Density, temperature, thermal conductivity, etc. (say $\underset{\sim}{Q}$) in a volume of matter can be a function of the coordinates in x-frame i.e., $\underset{\sim}{Q} = \underset{\sim}{Q}(x_1, x_2, x_3)$. Upon coordinate transformation T if x_i changes to y_i then $\underset{\sim}{Q} = \underset{\sim}{Q}(x_1(y_i), x_2(y_i), x_3(y_i))$ must have the same value as $\underset{\sim}{Q}(x_1, x_2, x_3)$ as $\underset{\sim}{Q}$ at a material point has a fixed value. Such quantities are called absolute scalars. Their values at a material point are not influenced by coordinate transformation. Description of force, velocity requires magnitude as well as direction. A force may be applied vertically, horizontally or along an arbitrary direction in the x-frame. Thus, the definition of such quantities requires magnitude as well

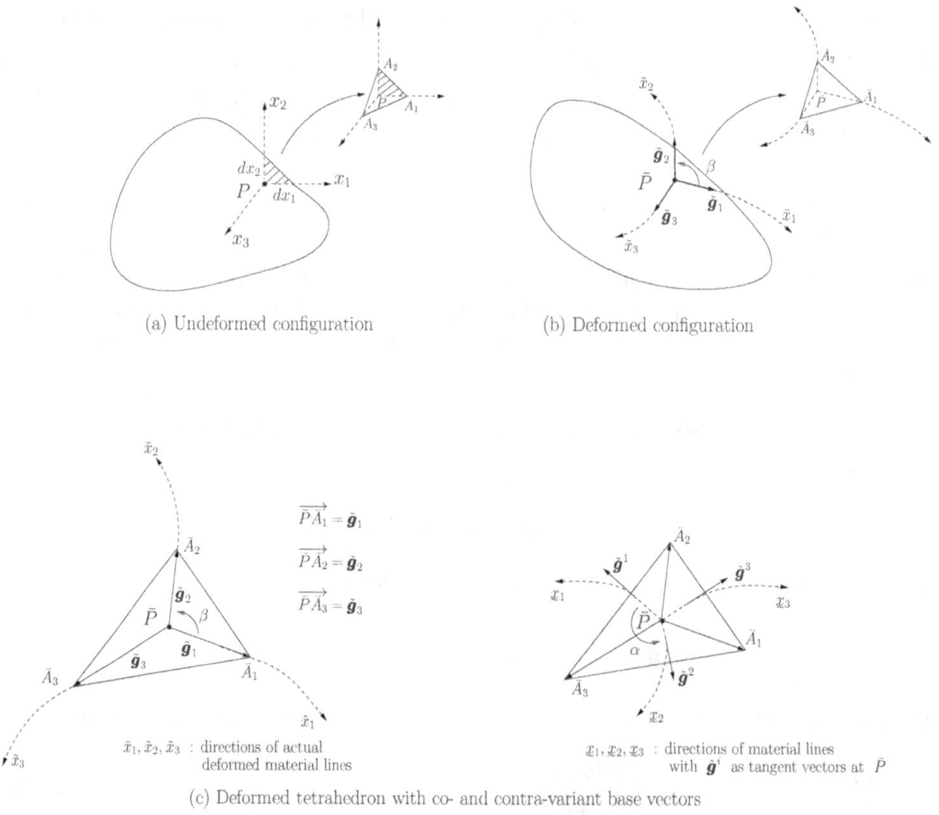

Figure 2.7: Elementary tetrahedra in undeformed and deformed configurations

as direction. Definition of average stress on a plane requires the direction of the plane (a unit vector normal to the plane) as well as the direction in which the stress is acting. There are many more such examples related to the physical quantities used in continuum mechanics. At the onset, it may appear that density, velocity, stress are totally unrelated mathematical entities. Introduction of tensors unifies the definition of these quantities as well as others used in continuum mechanics.

Consider x-frame and define quantities $\underset{\sim}{Q}_i(\boldsymbol{x})$; $i = 1, 2, \dots, n$ in it. If we transform x-frame to y-frame through a coordinate transformation T, then $\underset{\sim}{Q}_i(\boldsymbol{x})$ defined in x-frame will also transform and will have their new description in y-frame. Transformation of coordinates and, as a consequence, transformation of the quantities defined in them is fundamental in the definitions and study of tensors. We consider details in the following section.

2.10 Coordinate transformation and definitions, operations, transformations and calculus of tensors

In continuum mechanics we choose a frame of reference, say x-frame (remains fixed), and define coordinates of the material points in this frame. Description of the quantities of interest at the material points is then established using the coordinates of material point and time [95].

Suppose we apply a transformation T to the x-coordinates (x-frame) such that the transformation T changes them to y-coordinates (y-frame), i.e., the transformation transforms material point x-coordinates to y-coordinates, then we want to know how the quantities defined in the x-frame transform into the y-frame or perhaps a better way to state this is to say we would like to know their correspondence in the y-frame. Transformation T is a general coordinate transformation, hence y-frame can be curvilinear.

If we consider a frame in which the axes are curved and the tangent vectors to the axes at the origin are not orthogonal, then we have a frame with curvilinear axes or simply a "curvilinear frame". If the axes are straight lines but are not orthogonal to each other, then we have a "non-orthogonal Cartesian frame". If the axes are straight lines and are orthogonal to each other, then we have an "orthogonal Cartesian frame". At present in most writings, the Cartesian frame or rectangular frame implies "orthogonal Cartesian frame". Thus, in this book, we shall use rectangular frame or Cartesian frame to mean orthogonal or rectangular Cartesian frame. In the material presented in this book, both "curvilinear frames" and "rectangular or Cartesian frames" are used. Cartesian frames are, of course, the simplest kind and are generally preferred for this reason. However, non-orthogonal frames are also quite common and, in many instances, are necessitated by the physics of the deforming matter. For example, an orthogonal Cartesian frame etched in reference configuration defining the material lines becomes a curvilinear non-orthogonal frame in the current configuration if the deformation is finite. This curvilinear frame is of significance due to the fact that it describes the deformation of the material lines. The frame formed by the tangent vectors to the deformed material lines at the origin is a non-orthogonal Cartesian frame. In the material presented in the following section, the coordinate systems are not necessarily orthogonal. Orthogonal frames, their translations and rigid rotations of the orthogonal frames have been considered (in Section 2.7). Material in Section 2.7 is a special case of the material presented in this section.

2.10.1 Coordinate transformation T

Let $x_1, x_2, \ldots, x_n$ be the coordinates of a point in x-frame (fixed) and let T transform this point into $y_1, y_2, \ldots, y_n$ in y-frame. Thus

$$T \; : \; y_i = y_i(x_1, x_2, \ldots, x_n) \quad ; \quad i = 1, 2, \ldots, n \tag{2.127}$$

where transformation (2.127) expresses y_i coordinates as a function of x_i coordinates.

Theorem

If a transformation of coordinates T possesses an inverse T^{-1} and if J and K are the Jacobians, i.e., determinants of T and T^{-1} respectively, then $JK = 1$.

Proof

Since T^{-1} exists, then $x_i = x_i(y_1, y_2, \dots, y_n)$

$$\therefore \quad y_i = y_i(x_1(y_1, y_2, \dots, y_n), \dots, x_n(y_1, y_2, \dots, y_n)) \quad ; \quad i = 1, 2, \dots, n \tag{2.128}$$

Hence, differentiation with respect to y_j gives

$$\frac{\partial y_i}{\partial y_j} = \delta_{ij} = \frac{\partial y_i}{\partial x_\alpha}\frac{\partial x_\alpha}{\partial y_j} \quad ; \quad i, j, \alpha = 1, 2, \dots, n \tag{2.129}$$

and

$$\left|\frac{\partial y_i}{\partial y_j}\right| = \left|\frac{\partial y_i}{\partial x_\alpha}\frac{\partial x_\alpha}{\partial y_j}\right| = \left|\frac{\partial y_i}{\partial x_\alpha}\right|\left|\frac{\partial x_\alpha}{\partial y_j}\right| = JK \tag{2.130}$$

but $\left|\dfrac{\partial y_i}{\partial y_j}\right| = |\delta_{ij}| = 1$

$$\therefore \quad JK = 1$$

■

Remarks

(1) Consider the following:

$$\left.\begin{array}{l} T_1 \, : \, y_i = y_i(x_1, x_2, \dots, x_n) \\ T_2 \, : \, z_i = z_i(y_1, y_2, \dots, y_n) \\ T_3 \, : \, z_i = z_i(y_1(x_1, x_2, \dots, x_n), \dots, y_n(x_1, x_2, \dots, x_n)) \end{array}\right\} \; ; \; i = 1, 2, \dots, n$$

T_3 is called the product of transformation T_2 and T_1 and we denote $T_3 = T_2T_1$. If the Jacobian of T_3 is denoted by J_3, it follows that

$$J_3 = \left|\frac{\partial z_i}{\partial x_j}\right| = \left|\frac{\partial z_i}{\partial y_\alpha}\frac{\partial y_\alpha}{\partial x_j}\right| = \left|\frac{\partial z_i}{\partial y_\alpha}\right|\left|\frac{\partial y_\alpha}{\partial x_j}\right| = J_2 \, J_1 \quad ; \quad i, j, \alpha = 1, 2, \dots, n$$

where J_2 and J_1 are Jacobians of transformations T_2 and T_1.

(2) Transformation T at this stage is a general coordinate transformation without any specific properties.

2.10.2 Induced transformations

Consider x-coordinate with points x_i ; $i = 1, 2, \dots, n$ in x-frame and y-coordinate with points y_i ; $i = 1, 2, \dots, n$ in y-frame. Let x-frame and y-frame be related through a transformation T such that

$$T \, : \, x_i = x_i(y_1, y_2, \dots, y_n) \quad ; \quad i = 1, 2, \dots, n \tag{2.131}$$

Consider a set of functions $f_i(x_1, x_2, \dots, x_n)$; $i = 1, 2, \dots, m$ in the x-frame. Due to transformation T, the functions $f_i(x_1, x_2, \dots, x_n)$ defined in the x-frame transform into functions $g_i(y_1, y_2, \dots, y_n)$ in the y-frame, and we have

$$f_i(x_1(y_1, y_2, \dots, y_n), \dots, x_n(y_1, y_2, \dots, y_n)) = g_i(y_1, y_2, \dots, y_n) \tag{2.132}$$

in which $i = 1, 2, \dots, m$. Thus, as a consequence of coordinate transformation T, $f_i(\cdot)$ in the x-frame transform into $g_i(\cdot)$ in the y-frame according to (2.132). We formally write

$$G^0 \ : \ f_i(\boldsymbol{x}(\boldsymbol{y})) = g_i(\boldsymbol{y}) \quad ; \quad i = 1, 2, \dots, m \tag{2.133}$$

G^0 is called *induced transformation* due to T. The coordinate transformation T induces a transformation on $f_i(x_j)$ such that they transform into $g_i(y_j)$.

Generalization

We can generalize the concept of induced transformation. Consider a coordinate transformation T such that

$$T \ : \ x_i = x_i(y_1, y_2, \dots, y_n) \quad ; \quad i = 1, 2, \dots, n \tag{2.134}$$

Let $\underset{\sim}{Q}{}_i^x(x_1, x_2, \dots, x_n)$; $i = 1, 2, \dots, m$ be the quantities defined in the x-frame. Let $\underset{\sim}{Q}{}_i^y(y_1, y_2, \dots, y_n)$; $i = 1, 2, \dots, m$ be their maps in the y-frame due to coordinate transformation T. Then we can write

$$G_i^0 \ : \ \underset{\sim}{Q}{}_i^x(x_1, x_2, \dots, x_n) = \underset{\sim}{Q}{}_i^y(y_1, y_2, \dots, y_n) \quad ; \quad i = 1, 2, \dots, m \tag{2.135}$$

G_i^0 ; $i = 1, 2, \dots, m$ are induced transformations for $\underset{\sim}{Q}{}_i^x(x_1, x_2, \dots, x_n)$; $i = 1, 2, \dots, m$ defined in the x-frame.

2.10.3 Isomorphism between coordinate transformations and induced transformations

Let T_1, T_2 and T_3 be coordinate transformations such that

$$\begin{aligned} T_1 \ &: \ x\text{-coordinates} \rightarrow y\text{-coordinates} \quad \text{(i.e., } x\text{-frame to } y\text{-frame)} \\ T_2 \ &: \ y\text{-coordinates} \rightarrow z\text{-coordinates} \quad \text{(i.e., } y\text{-frame to } z\text{-frame)} \\ T_3 \ &: \ x\text{-coordinates} \rightarrow z\text{-coordinates} \quad \text{(i.e., } x\text{-frame to } z\text{-frame)} \end{aligned} \tag{2.136}$$

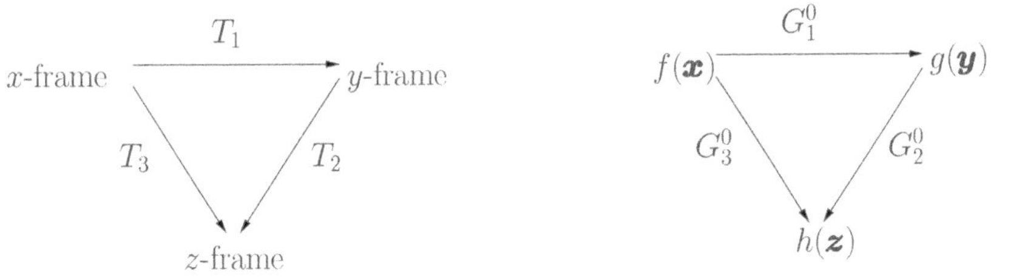

(a) Coordinate transformations (b) Induced transformations

Figure 2.8: Coordinate transformations and induced transformations

Let G_1^0, G_2^0 and G_3^0 be the induced transformations due to T_1, T_2 and T_3 such that

$$\begin{aligned} G_1^0 \ &: \ f(\boldsymbol{x}) \rightarrow g(\boldsymbol{y}) = f(\boldsymbol{x}(\boldsymbol{y})) \\ G_2^0 \ &: \ g(\boldsymbol{y}) \rightarrow h(\boldsymbol{z}) = g(\boldsymbol{y}(\boldsymbol{z})) \\ G_3^0 \ &: \ f(\boldsymbol{x}) \rightarrow h(\boldsymbol{z}) = f(\boldsymbol{x}(\boldsymbol{y}(\boldsymbol{z}))) \end{aligned} \tag{2.137}$$

which are graphically illustrated in Figure 2.8.

If we define the determinants of the transformations T_1, T_2 and T_3 by T_1, T_2 and T_3 also (henceforth assumed so), then from (2.136) we note that $T_3 = T_2T_1$ and from (2.137) it is clear that $G_3^0 = G_2^0G_1^0$. We note that transformations T map coordinates whereas the transformations G_i^0 map functions.

When the relationship between the successive transformations T_i and G_i^0 is such that if $T_3 = T_2T_1$ holds then $G_3^0 = G_2^0G_1^0$ also holds, then the two groups of transformations T_i and G_i^0 are said to be *isomorphic.* We note the transformation of functions by G_i^0 is induced due to the transformation T_i of coordinates.

The isomorphism between the transformation of coordinates and the transformations of functions induced by the transformation of coordinates is an important characteristic of a class of quantities called *tensors.*

2.10.4 Transformations by covariance and contravariance

2.10.4.1 Def: Covariant law or transformation by covariance

Covariant law of transformation transforms covariant quantities from one frame to another. In other words, this law of transformation relates covariant quantities in two frames. Consider a continuously differentiable function $f(x_1, x_2, \dots, x_n)$ and a transformation of coordinates

$$T \, : \, x_i = x_i(y_1, y_2, \dots, y_n) \quad ; \quad i = 1, 2, \dots, n \tag{2.138}$$

and form a set of partial derivatives of $f(x_1, x_2, \dots, x_n)$ with respect to $x_1, x_2, \dots, x_n$, i.e.,

$$\frac{\partial f}{\partial x_1}, \frac{\partial f}{\partial x_2}, \dots, \frac{\partial f}{\partial x_n} = \{f_{,x_i}\} \quad ; \quad i = 1, 2, \dots, n \tag{2.139}$$

What does the set $\{f_{,x_i}\}$ become when x_i are subjected to transformation T, i.e., when x_i are transformed to y_i? The answer requires establishing a law of transformation of derivatives under the transformation T. For example, we might substitute in each function $f_{,x_i}(x_1, x_2, \dots, x_n)$ the values x_i from the transformation T. This will yield a set of n functions

$$g_1(y_1, y_2, \dots, y_n) \, , \; g_2(y_1, y_2, \dots, y_n) \, , \; \dots \, , \; g_n(y_1, y_2, \dots, y_n) \tag{2.140}$$

On the other hand, if we wish to examine the gradient of $f(\cdot)$, it is necessary to say that the set of functions corresponding to $\{f_{,x_i}\}$ in x-frame are not $g_i(y)$ in y-frame, but the set of n partial derivatives

$$\frac{\partial f}{\partial y_1}, \frac{\partial f}{\partial y_2}, \dots, \frac{\partial f}{\partial y_n} \tag{2.141}$$

obtained by using the chain rule of differentiation

$$\frac{\partial f}{\partial y_i} = \frac{\partial f}{\partial x_\alpha}\frac{\partial x_\alpha}{\partial y_i} \quad ; \quad i, \alpha = 1, 2, \dots, n \tag{2.142}$$

$$\text{or} \quad G^1 \, : \, \frac{\partial f}{\partial y_i} = \left(\frac{\partial x_\alpha}{\partial y_i}\right)\frac{\partial f}{\partial x_\alpha} \quad ; \quad i, \alpha = 1, 2, \dots, n \tag{2.143}$$

The term G^1 is known as *covariant law* or *transformation by covariance*. Let $\frac{\partial f}{\partial x_i}$ and $\frac{\partial f}{\partial y_i}$ be arranged in the form of vectors $\left\{\frac{\partial f}{\partial x}\right\}$ and $\left\{\frac{\partial f}{\partial y}\right\}$, then (2.143) can be written as

$$G^1 \; : \; \left\{\frac{\partial f}{\partial y}\right\} = [\bar{J}]^T \left\{\frac{\partial f}{\partial x}\right\} \quad \text{or} \quad f_{,\boldsymbol{y}} = (\bar{\boldsymbol{J}})^T \cdot f_{,\boldsymbol{x}} \tag{2.144}$$

where

$$[\bar{J}] = \begin{bmatrix} \frac{\partial x_1}{\partial y_1} & \frac{\partial x_1}{\partial y_2} & \cdots & \frac{\partial x_1}{\partial y_n} \\ \frac{\partial x_2}{\partial y_1} & \frac{\partial x_2}{\partial y_2} & \cdots & \frac{\partial x_2}{\partial y_n} \\ \vdots & & & \\ \frac{\partial x_n}{\partial y_1} & \frac{\partial x_n}{\partial y_2} & \cdots & \frac{\partial x_n}{\partial y_n} \end{bmatrix} = \begin{bmatrix} \{\tilde{g}^1\}^T \\ \{\tilde{g}^2\}^T \\ \vdots \\ \{\tilde{g}^n\}^T \end{bmatrix} \quad ; \quad \{\tilde{g}^i\} = \left\{\frac{\partial x_i}{\partial y}\right\} = \begin{Bmatrix} \frac{\partial x_i}{\partial y_1} \\ \frac{\partial x_i}{\partial y_2} \\ \vdots \\ \frac{\partial x_i}{\partial y_n} \end{Bmatrix} \tag{2.145}$$

and $i = 1, 2, \dots, n$. That is, *the rows of* $[\bar{J}]$ *are the components of the* $\{\tilde{g}^i\}$ *vectors.* The $\{\tilde{g}^i\}$ are called *contravariant base vectors* (see definition of contravariant base vectors in Section 2.8.2). The components of $\left\{\frac{\partial f}{\partial y}\right\}$ and $\left\{\frac{\partial f}{\partial x}\right\}$ are rate of change of the function f at $\boldsymbol{y}$ and $\boldsymbol{x}$ locations in the directions tangent to y_i and x_i axes, hence their components are *covariant components.* Thus, based on the definition (2.144), the covariant components in x- and y-frame are related through components of contravariant base vectors. An important point to note is that based on this definition, *transformation of covariant components requires contravariant base vectors.*

2.10.4.2 Generalization

We can generalize (2.143). Let $A_1(\boldsymbol{x}), A_2(\boldsymbol{x}), \dots, A_n(\boldsymbol{x})$ be n quantities in the x-coordinate system and if we agree to calculate the corresponding quantities $B_1(\boldsymbol{y})$, $B_2(\boldsymbol{y})$, $\dots, B_n(\boldsymbol{y})$ in the y-coordinate system by G^1 law

$$G^1 \; : \; B_i(\boldsymbol{y}) = \frac{\partial x_\alpha}{\partial y_i} A_\alpha(\boldsymbol{x}) \quad ; \quad i, \alpha = 1, 2, \dots, n \tag{2.146}$$

then we can say that set $A_\alpha(\boldsymbol{x})$; $\alpha = 1, 2, \dots, n$ represents the *covariant components of a vector* in the x-frame, and the set $B_i(\boldsymbol{y})$; $i = 1, 2, \dots, n$ represents the *covariant components of the same vector* in the y-frame. For simplicity, we drop arguments $\boldsymbol{y}$ and $\boldsymbol{x}$ in $B_i(\cdot)$ and $A_\alpha(\cdot)$.

$$G^1 \; : \; B_i = \frac{\partial x_\alpha}{\partial y_i} A_\alpha \quad ; \quad i, \alpha = 1, 2, \dots, n \tag{2.147}$$

Using definition of contravariant basis, we can write (2.147) we

$$\boldsymbol{B} = \tilde{\boldsymbol{g}}^\alpha A_\alpha \quad ; \quad \alpha = 1, 2, \dots, n \tag{2.148}$$

$$\text{or} \quad \boldsymbol{B} = \boldsymbol{e}_i \frac{\partial x_\alpha}{\partial y_i} A_\alpha = \boldsymbol{e}_i B_i \quad ; \quad i = 1, 2, \dots, n \tag{2.149}$$

In matrix and vector notation, (2.147) can be written as

$$G^1 \; : \; \{B\} = [\bar{J}]^T\{A\} \tag{2.150}$$

Thus, if v_α are covariant components of a vector $\boldsymbol{v}$, then

$$\boldsymbol{v} = \tilde{\boldsymbol{g}}^\alpha \; v_\alpha = \boldsymbol{e}_i \frac{\partial x_\alpha}{\partial y_i} v_\alpha = \boldsymbol{e}_i v_i \tag{2.151}$$

Clearly, components v_α are different than the physical components v_i of $\boldsymbol{v}$.

2.10.4.3 Contravariant law or transformation by contravariance

Contravariant law of transformation transforms contravariant quantities from one frame to another. In other words, this law of transformation relates contravariant quantities in two frames. Consider a coordinate transformation T

$$T \; : \; x_i = x_i(y_1, y_2, \dots, y_n) \quad ; \quad i = 1, 2, \dots, n \tag{2.152}$$

expressing x_i coordinates as a function of y_i coordinates. Consider two points $P_1^x(x_1, x_2, \dots, x_n)$ and $P_2^x(x_1 + dx_1, x_2 + dx_2, \dots, x_n + dx_n)$ in the x-frame (Figure 2.9). Let $P_1^y(y_1, y_2, \dots, y_n)$ and $P_2^y(y_1 + dy_1, y_2 + dy_2, \dots, y_n + dy_n)$ be the maps of $P_1^x(\cdot)$ and $P_2^x(\cdot)$ in the y-frame (Figure 2.9). The differentials $dx_1, dx_2, \dots, dx_n$ represent displacement of point $P_1^x(\cdot)$ to $P_2^x(\cdot)$ in the x-frame and the differentials $dy_1, dy_2, \dots, dy_n$ are the displacement of point $P_1^y(\cdot)$ to $P_2^y(\cdot)$ in the y-frame. We can write

$$dy_i = \frac{\partial y_i}{\partial x_\alpha} dx_\alpha \quad ; \quad i, \alpha = 1, 2, \dots, n \tag{2.153}$$

$$\text{or} \quad G^2 \; : \; dy_i = \frac{\partial y_i}{\partial x_\alpha} dx_\alpha \quad ; \quad i, \alpha = 1, 2, \dots, n \tag{2.154}$$

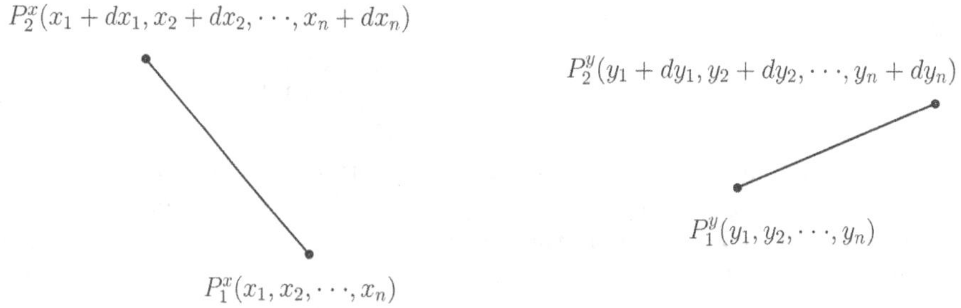

(a) x-coordinate system (b) y-coordinate frame

Figure 2.9: Differential lengths in x- and y-coordinate systems

The term G^2 is known as *contravariant law* or *transformation by contravariance.* Let dx_i and dy_i be arranged in the form of vectors $\{dx\}$ and $\{dy\}$, then (2.154) can be written as

$$G^2 \; : \; \{dy\} = [J]\{dx\} \quad \text{or} \quad \boldsymbol{dy} = \boldsymbol{J} \cdot \boldsymbol{dx} \tag{2.155}$$

where

$$[J] = \begin{bmatrix} \frac{\partial y_1}{\partial x_1} & \frac{\partial y_1}{\partial x_2} & \cdots & \frac{\partial y_1}{\partial x_n} \\ \frac{\partial y_2}{\partial x_1} & \frac{\partial y_2}{\partial x_2} & \cdots & \frac{\partial y_2}{\partial x_n} \\ \vdots \\ \frac{\partial y_n}{\partial x_1} & \frac{\partial y_n}{\partial x_2} & \cdots & \frac{\partial y_n}{\partial x_n} \end{bmatrix} = \left[\{\tilde{g}_1\},\ \{\tilde{g}_2\},\ \dots \{\tilde{g}_n\} \right]$$

$$\{\tilde{g}_i\} = \left\{ \frac{\partial y}{\partial x_i} \right\} = \begin{bmatrix} \frac{\partial y_1}{\partial x_i} \\ \\ \frac{\partial y_2}{\partial x_i} \\ \vdots \\ \frac{\partial y_n}{\partial x_i} \end{bmatrix} \tag{2.156}$$

and $i = 1, 2, \dots, n$. That is, *the columns of* $[J]$ *are the components of the* $\{\tilde{g}_i\}$ *vectors*. Vectors $\{\tilde{g}_i\}$ are called *covariant base vectors* (also see definition of covariant base vectors in Section 2.8.3). The components of $\{dy\}$ and $\{dx\}$ are defined as *contravariant components*. Thus, based on the definition (2.155), the contravariant components in x- and y-frame are related through components of covariant base vectors, i.e., *transformation of contravariant components require covariant base vectors.*

2.10.4.4 Generalization

We can generalize (2.154). Let $A^1(\boldsymbol{x}), A^2(\boldsymbol{x}), \dots, A^n(\boldsymbol{x})$ be n quantities in the x-coordinate system and if we agree to calculate the corresponding quantities $B^1(\boldsymbol{y})$, $B^2(\boldsymbol{y})$, $\dots, B^n(\boldsymbol{y})$ in the y-coordinate system by G^2 law

$$G^2 \ :\ B^i(\boldsymbol{y}) = \frac{\partial y_i}{\partial x_\alpha} A^\alpha(\boldsymbol{x}) \quad ; \quad i, \alpha = 1, 2, \dots, n \tag{2.157}$$

then we can say that the set $A^\alpha(\boldsymbol{x})$; $\alpha = 1, 2, \dots, n$ represents the *contravariant components of a vector* in the x-coordinate system and the set $B^i(\boldsymbol{y})$; $i = 1, 2, \dots, n$ represents the *contravariant components of the same vector* in the y-coordinate system. For simplicity we drop the arguments $\boldsymbol{y}$ and $\boldsymbol{x}$ in $B^i(\cdot)$ and $A^\alpha(\cdot)$.

$$G^2 \ :\ B^i = \frac{\partial y_i}{\partial x_\alpha} A^\alpha \quad ; \quad i, \alpha = 1, 2, \dots, n \tag{2.158}$$

Using definition of covariant basis we can write (2.158) as

$$\boldsymbol{B} = \tilde{\boldsymbol{g}}_\alpha A^\alpha \quad ; \quad \alpha = 1, 2, \dots, n \tag{2.159}$$

$$\text{or} \quad \boldsymbol{B} = \boldsymbol{e}_i \frac{\partial y_i}{\partial x_\alpha} A^\alpha = \boldsymbol{e}_i B^i \tag{2.160}$$

Using matrix and vector notation, we can write (2.158)

$$G^2 \,:\, \{B\} = [J]\{A\} \tag{2.161}$$

Thus, if v^α are contravariant components of a vector $\boldsymbol{v}$, then

$$\boldsymbol{v} = \tilde{\boldsymbol{g}}_\alpha \, v^\alpha = \boldsymbol{e}_i \frac{\partial y_i}{\partial x_\alpha} v^\alpha = \boldsymbol{e}_i v_i \tag{2.162}$$

Here, also components v^α are different than the physical components v_i of $\boldsymbol{v}$.

Remarks

(1) Note that in the case of covariant components we use subscripts for B and A while in the case of contravariant components we use superscripts. This is done intentionally for clear distinction between co- and contra-variant components.

(2) Obviously, the contravariant law is quite different than covariant law. We reiterate that transformation of covariant quantities requires contravariant base vectors, while the transformation of contravariant quantities requires covariant basis.

2.10.5 The tensor concept: Covariant and contravariant tensors

Consider an admissible coordinate transformation

$$T \,:\, y_i = y_i(x_1, x_2, \dots, x_n) \quad ; \quad i = 1, 2, \dots, n \tag{2.163}$$

Let $\underset{\sim}{Q}{}_i^x(\boldsymbol{x})$ be m continuous quantities defined in x-frame. Let G be the induced transformation due to coordinate transformation T. Then G transforms $\underset{\sim}{Q}{}_i^x(\boldsymbol{x})$; $i = 1, 2, \dots, m$ in the x-frame to $\underset{\sim}{Q}{}_i^y(\boldsymbol{y})$; $i = 1, 2, \dots, m$ in the y-frame.

$$G \,:\, \underset{\sim}{Q}{}_i^x(\boldsymbol{x}) \rightarrow \underset{\sim}{Q}{}_i^y(\boldsymbol{y}) \quad ; \quad i = 1, 2, \dots, m \tag{2.164}$$

2.10.5.1 Def: Tensors

Let us assume that the induced transformation G is a function of coordinate transformation T. Let us also assume that the transformation T and G have the following properties.

(a) When T is an identity transformation, then G is an identity transformation. This means that if

$$\begin{aligned} T \,&:\, y_i = x_i \quad ; \quad i = 1, 2, \dots, n \\ \text{then} \quad G \,&:\, \underset{\sim}{Q}{}_i^x(\cdot) \rightarrow \underset{\sim}{Q}{}_i^y(\cdot) = \underset{\sim}{Q}{}_i^x(\cdot) \quad ; \quad i = 1, 2, \dots, m \end{aligned}$$

(b) If T_1, T_2, T_3 are three transformations of the type T and G_1, G_2, G_3 are the corresponding induced transformations of type G, and if $T_3 = T_2 T_1$, then $G_3 = G_2 G_1$. That is, transformations T_i and G_i are isomorphic.

If $\underset{\sim}{Q}{}_i^x(\cdot)$ and $\underset{\sim}{Q}{}_i^y(\cdot)$; $i = 1, 2, \dots, m$ satisfy conditions (a) and (b), then the sets $\{\underset{\sim}{Q}{}_i^x(\cdot)\}$ and $\{\underset{\sim}{Q}{}_i^y(\cdot)\}$ contain the components $\underset{\sim}{Q}{}_i^x(\cdot)$ and $\underset{\sim}{Q}{}_i^y(\cdot)$ of the tensors $\{\underset{\sim}{Q}{}_i^x(\cdot)\}$ and $\{\underset{\sim}{Q}{}_i^y(\cdot)\}$ in x- and y-frames.

Remarks

(1) The term tensor was used by A. Einstein [29] in connection with quantities transforming in accordance with covariant and contravariant laws.
(2) Covariant and contravariant laws as well as algebra and calculus of tensors are due to G. Ricci [86].
(3) Characterization of tensors by an isomorphism of transformations of coordinates and induced transformations is due to H. Weyl [138] and O. Veblen [131].
(4) The term tensor is generally used in the same sense as contemplated by Einstein.

2.10.5.2 Def: Tensors of rank zero

Consider a coordinate transformation T.

$$T \; : \; x_i = x_i(y_1, y_2, \dots, y_n) \quad ; \quad i = 1, 2, \dots, n \tag{2.165}$$

Let $\{f_i(x)\}$ be a set of m continuous functions in the x-frame, i.e.,

$$f_i(x_1, x_2, \dots, x_n) \quad ; \quad i = 1, 2, \dots, m \tag{2.166}$$

Consider an induced transformation of type G^0, i.e.,

$$G^0 \; : \; f_i(x_j) \rightarrow g_i(y_j) \quad ; \quad i = 1, 2, \dots, m \tag{2.167}$$

If the transformations T and G^0 are isomorphic and if G^0 is an identity transformation when T is identity transformation, then each f_i in the set $\{f_i\}$ is a tensor of rank zero in the x-frame, and likewise, each g_i in the set $\{g_i\}$ is a corresponding tensor of rank zero in the y-frame. That is, functions are tensors of rank zero. Obviously, there is no contravariant and covariant concept for tensors of rank zero. Tensors of rank zero are scalars and are sometimes called absolute scalars.

2.10.5.3 Def: Covariant tensors of rank one

Let T be an admissible coordinate transformation

$$T \; : \; x_i = x_i(y_1, y_2, \dots, y_n) \quad ; \quad i = 1, 2, \dots, n \tag{2.168}$$

Let the set of quantities A_α ; $\alpha = 1, 2, \dots, n$ in the x-frame transform into the quantities B_i ; $i = 1, 2, \dots, n$ in the y-frame due to transformation T by the following law

$$G^1 \; : B_i = \frac{\partial x_\alpha}{\partial y_i} A_\alpha \quad ; \quad i, \alpha = 1, 2, \dots, n \tag{2.169}$$

Using definition of contravariant basis we can write (2.169) as

$$\boldsymbol{B} = \tilde{\boldsymbol{g}}^\alpha A_\alpha \quad ; \quad \alpha = 1, 2, \dots, n \tag{2.170}$$

$$\text{or} \quad \boldsymbol{B} = \boldsymbol{e}_i \frac{\partial x_\alpha}{\partial y_i} A_\alpha = \boldsymbol{e}_i B_i \tag{2.171}$$

In matrix and vector notation we can write (2.169) as

$$\{B\} = [\bar{J}]^T \{A\} \tag{2.172}$$

The components of set $\{A_\alpha\}$ is the representation of a covariant tensor of rank one in the x-frame and the components of set $\{B_i\}$ is its representation in the y-frame, i.e., the set $\{B_i\}$ is the covariant tensor of rank one in the y-frame.

Remarks

(1) Vectors are tensors of rank one.

(2) Thus, we could view the sets $\{A_\alpha\}$ and $\{B_i\}$ as covariant vectors in x- and y-frames.

(3) The rule of transformation for the covariant vectors requires contravariant base vectors that are rows of $[\bar{J}]$.

(4) We can write (2.171) as

$$B_i = C^{i\alpha} A_\alpha \quad ; \quad C^{i\alpha} = \frac{\partial x_\alpha}{\partial y_i} \quad ; \quad \alpha, i = 1, 2, \dots, n \tag{2.173}$$

This form of (2.169) is sometimes convenient.

2.10.5.4 Def: Contravariant tensors of rank one

Let T be an admissible coordinate transformation

$$T \,:\, x_i = x_i(y_1, y_2, \dots, y_n) \quad ; \quad i = 1, 2, \dots, n \tag{2.174}$$

Let the set of quantities A^α ; $\alpha = 1, 2, \dots, n$ in the x-frame transform into the quantities B^i ; $i = 1, 2, \dots, n$ in the y-frame due to transformation T by the following law

$$G^2 \,:\, B^i = \frac{\partial y_i}{\partial x_\alpha} A^\alpha \quad ; \quad i, \alpha = 1, 2, \dots, n \tag{2.175}$$

Using definition of covariant basis we can write (2.175) as

$$\boldsymbol{B} = \tilde{\boldsymbol{g}}_\alpha A^\alpha \tag{2.176}$$

$$\text{or} \quad \boldsymbol{B} = \boldsymbol{e}_i \frac{\partial y_i}{\partial x_\alpha} A^\alpha = \boldsymbol{e}_i B^i \tag{2.177}$$

In matrix and vector form (2.175) can be written as

$$\{B\} = [J]\{A\} \tag{2.178}$$

The components of set $\{A^\alpha\}$ is the representation of a contravariant tensor of rank one in the x-frame and the components of set $\{B^i\}$ is its representation in the y-frame, i.e., the set $\{B^i\}$ is the contravariant tensor of rank one in the y-frame.

Remarks

(1) We can also write (2.175) as

$$B^i = C_{i\alpha} A^\alpha \quad ; \quad C_{i\alpha} = \frac{\partial y_i}{\partial x_\alpha} \quad ; \quad i, \alpha = 1, 2, \dots, n \tag{2.179}$$

(2) The concept of the tensor of rank one can be extended to the tensors of ranks higher than one [95].

(3) In the following we present the definitions of co- and contra-variant tensors of rank two, three and four in three dimensional space (used commonly) and tensors of rank r in n-dimensional space.

2.10.5.5 Def: Co- and contra-variant tensors of rank two in three dimensional space

Let T be an admissible coordinate transformation

$$T \ : \ x_i = x_i(y_1, y_2, y_3) \tag{2.180}$$

Let $A_{\alpha\beta}$ and $A^{\alpha\beta}$; $\alpha, \beta = 1,2,3$ be the quantities in the x-frame and let B_{ij} and B^{ij} ; $i,j = 1,2,3$ be their corresponding maps in the y-frame related by the following laws. First, we consider covariant tensors of rank two in x- and y-frames

$$B_{ij} = \frac{\partial x_\alpha}{\partial y_i}\frac{\partial x_\beta}{\partial y_j} A_{\alpha\beta} \quad ; \quad \alpha, \beta = 1,2,3 \qquad i,j = 1,2,3 \tag{2.181}$$

Using definition of contravariant basis we can write (2.181) as

$$\boldsymbol{B} = \tilde{\boldsymbol{g}}^\alpha \tilde{\boldsymbol{g}}^\beta A_{\alpha\beta} = \tilde{\boldsymbol{g}}^\alpha \otimes \tilde{\boldsymbol{g}}^\beta A_{\alpha\beta} \tag{2.182}$$

$$\boldsymbol{B} = \boldsymbol{e}_i \frac{\partial x_\alpha}{\partial y_i} \boldsymbol{e}_j \frac{\partial x_\beta}{\partial y_j} A_{\alpha\beta} = \boldsymbol{e}_i \frac{\partial x_\alpha}{\partial y_i} \otimes \boldsymbol{e}_j \frac{\partial x_\beta}{\partial y_j} A_{\alpha\beta} \tag{2.183}$$

$$\text{or} \quad \boldsymbol{B} = \boldsymbol{e}_i \boldsymbol{e}_j \left(\frac{\partial x_\alpha}{\partial y_i}\frac{\partial x_\beta}{\partial y_j} A_{\alpha\beta} \right) = \boldsymbol{e}_i \boldsymbol{e}_j B_{ij} = \boldsymbol{e}_i \otimes \boldsymbol{e}_j B_{ij} \tag{2.184}$$

We can also write B_{ij} in (2.181) as

$$B_{ij} = C^{ij\alpha\beta} A_{\alpha\beta} \tag{2.185}$$

In matrix and vector notation (2.181) can be written as

$$[B] = [\bar{J}]^T [A] [\bar{J}] \tag{2.186}$$

Now we consider contravariant tensors of rank two in x- and y-frames

$$B^{ij} = \frac{\partial y_i}{\partial x_\alpha}\frac{\partial y_j}{\partial x_\beta} A^{\alpha\beta} \tag{2.187}$$

Using definition of covariant basis we can write (2.187) as

$$\boldsymbol{B} = \tilde{\boldsymbol{g}}_\alpha \tilde{\boldsymbol{g}}_\beta A^{\alpha\beta} = \tilde{\boldsymbol{g}}_\alpha \otimes \tilde{\boldsymbol{g}}_\beta A^{\alpha\beta} \tag{2.188}$$

$$\boldsymbol{B} = \boldsymbol{e}_i \frac{\partial y_i}{\partial x_\alpha} \boldsymbol{e}_j \frac{\partial y_i}{\partial x_\alpha} A^{\alpha\beta} = \boldsymbol{e}_i \frac{\partial y_i}{\partial x_\alpha} \otimes \boldsymbol{e}_j \frac{\partial y_j}{\partial x_\alpha} A^{\alpha\beta} \tag{2.189}$$

$$\text{or} \quad \boldsymbol{B} = \boldsymbol{e}_i \boldsymbol{e}_j \left(\frac{\partial y_i}{\partial x_\alpha}\frac{\partial y_j}{\partial x_\beta} A^{\alpha\beta} \right) = \boldsymbol{e}_i \boldsymbol{e}_j B^{ij} = \boldsymbol{e}_i \otimes \boldsymbol{e}_j B^{ij} \tag{2.190}$$

We can also write (2.187) as

$$B^{ij} = C_{ij\alpha\beta} A^{\alpha\beta} \tag{2.191}$$

In matrix and vector notations we can write

$$[B] = [J][A][J]^T \tag{2.192}$$

$A_{\alpha\beta}$ and B_{ij} are the components of a covariant tensor of rank two in the x- and the y-frames. Likewise, $A^{\alpha\beta}$ and B^{ij} are the components of a contravariant tensor of rank two in the x- and the y-frames.

2.10.5.6 Def: Co- and contra-variant tensors of rank three in three-dimensional space

Let T be an admissible coordinate transformation

$$T \ : \ x_i = x_i(y_1, y_2, y_3) \tag{2.193}$$

Let $A_{\alpha\beta\gamma}$ and $A^{\alpha\beta\gamma}$; $\alpha, \beta, \gamma = 1, 2, 3$ be quantities in the x-frame and let B_{ijk} and B^{ijk} ; $i, j, k = 1, 2, 3$ be their maps in the y-frame related by the following laws. First, we consider covariant tensors of rank three in x- and y-frames.

$$B_{ijk} = \frac{\partial x_\alpha}{\partial y_i}\frac{\partial x_\beta}{\partial y_j}\frac{\partial x_\gamma}{\partial y_k} A_{\alpha\beta\gamma} \tag{2.194}$$

Using definition of contravariant basis we can write (2.194) as

$$\boldsymbol{B} = \tilde{\boldsymbol{g}}^\alpha \tilde{\boldsymbol{g}}^\beta \tilde{\boldsymbol{g}}^\gamma A_{\alpha\beta\gamma} = \tilde{\boldsymbol{g}}^\alpha \otimes \tilde{\boldsymbol{g}}^\beta \otimes \tilde{\boldsymbol{g}}^\gamma A_{\alpha\beta\gamma} \tag{2.195}$$

$$\boldsymbol{B} = \boldsymbol{e}_i \frac{\partial x_\alpha}{\partial y_i} \boldsymbol{e}_j \frac{\partial x_\beta}{\partial y_j} \boldsymbol{e}_k \frac{\partial x_\gamma}{\partial y_k} A_{\alpha\beta\gamma} = \boldsymbol{e}_i \frac{\partial x_\alpha}{\partial y_i} \otimes \boldsymbol{e}_j \frac{\partial x_\beta}{\partial y_j} \otimes \boldsymbol{e}_k \frac{\partial x_\gamma}{\partial y_k} A_{\alpha\beta\gamma} \tag{2.196}$$

$$\boldsymbol{B} = \boldsymbol{e}_i \boldsymbol{e}_j \boldsymbol{e}_k \left(\frac{\partial x_\alpha}{\partial x_i}\frac{\partial x_\beta}{\partial y_j}\frac{\partial x_\gamma}{\partial y_k} A_{\alpha\beta\gamma} \right) = \boldsymbol{e}_i \boldsymbol{e}_j \boldsymbol{e}_k B_{ijk} = \boldsymbol{e}_i \otimes \boldsymbol{e}_j \otimes \boldsymbol{e}_k B_{ijk} \tag{2.197}$$

We can also write B_{ijk} in (2.194) as

$$B_{ijk} = C^{ijk\alpha\beta\gamma} A_{\alpha\beta\gamma} \tag{2.198}$$

Now we consider contravariant tensors of rank three in x- and y-frames.

$$B^{ijk} = \frac{\partial y_i}{\partial x_\alpha}\frac{\partial y_j}{\partial x_\beta}\frac{\partial y_k}{\partial x_\gamma} A^{\alpha\beta\gamma} \tag{2.199}$$

Using definition of covariant basis (2.199) can be written as

$$\boldsymbol{B} = \tilde{\boldsymbol{g}}_\alpha \tilde{\boldsymbol{g}}_\beta \tilde{\boldsymbol{g}}_\gamma A^{\alpha\beta\gamma} = \tilde{\boldsymbol{g}}_\alpha \otimes \tilde{\boldsymbol{g}}_\beta \otimes \tilde{\boldsymbol{g}}_\gamma A^{\alpha\beta\gamma} \tag{2.200}$$

$$\boldsymbol{B} = \boldsymbol{e}_i \frac{\partial y_i}{\partial x_\alpha} \boldsymbol{e}_j \frac{\partial y_j}{\partial x_\beta} \boldsymbol{e}_k \frac{\partial y_k}{\partial x_\gamma} A^{\alpha\beta\gamma} = \boldsymbol{e}_i \frac{\partial y_i}{\partial x_\alpha} \otimes \boldsymbol{e}_j \frac{\partial y_j}{\partial x_\beta} \otimes \boldsymbol{e}_k \frac{\partial y_k}{\partial x_\gamma} \tag{2.201}$$

or $$\boldsymbol{B} = \boldsymbol{e}_i \frac{\partial y_i}{\partial x_\alpha} \boldsymbol{e}_j \frac{\partial y_j}{\partial x_\beta} \boldsymbol{e}_k \frac{\partial y_k}{\partial x_\gamma} A^{\alpha\beta\gamma} = \boldsymbol{e}_i \frac{\partial y_i}{\partial x_\alpha} \otimes \boldsymbol{e}_j \frac{\partial y_j}{\partial x_\beta} \otimes \boldsymbol{e}_k \frac{\partial y_k}{\partial x_\gamma} A^{\alpha\beta\gamma} \tag{2.202}$$

$$\boldsymbol{B} = \boldsymbol{e}_i \boldsymbol{e}_j \boldsymbol{e}_k \left(\frac{\partial y_i}{\partial x_\alpha}\frac{\partial y_j}{\partial x_\beta}\frac{\partial y_k}{\partial x_\gamma} A^{\alpha\beta\gamma} \right) = \boldsymbol{e}_i \boldsymbol{e}_j \boldsymbol{e}_k B^{ijk} = \boldsymbol{e}_i \otimes \boldsymbol{e}_j \otimes \boldsymbol{e}_k B^{ijk} \tag{2.203}$$

$$B^{ijk} = C_{ijk\alpha\beta\gamma} A^{\alpha\beta\gamma} \tag{2.204}$$

$A_{\alpha\beta\gamma}$ and B_{ijk} are components of a covariant tensor of rank three in x- and y-frames. Likewise, $A^{\alpha\beta\gamma}$ and B^{ijk} are components of the contravariant tensors of rank three in x- and y-frames.

2.10.5.7 Def: Co- and contra-variant tensors of rank four in three dimensional space

Let T be an admissible coordinate transformation

$$T \; : \; x_i = x_i(y_1, y_2, y_3) \tag{2.205}$$

Let $A_{\alpha\beta\gamma\zeta}$ and $A^{\alpha\beta\gamma\zeta}$; $\alpha, \beta, \gamma, \zeta = 1, 2, 3$ be quantities in the x-frame and let B_{ijkl} and B^{ijkl} ; $i, j, k, l = 1, 2, 3$ be their maps in the y-frame related by the following laws. First, we consider covariant tensors of rank four in x- and y-frames

$$B_{ijkl} = \frac{\partial x_\alpha}{\partial y_i} \frac{\partial x_\beta}{\partial y_j} \frac{\partial x_\gamma}{\partial y_k} \frac{\partial x_\zeta}{\partial y_l} A_{\alpha\beta\gamma\zeta} \tag{2.206}$$

Using definition of contravariant basis we can write (2.206) as

$$\boldsymbol{B} = \tilde{\boldsymbol{g}}^\alpha \tilde{\boldsymbol{g}}^\beta \tilde{\boldsymbol{g}}^\gamma \tilde{\boldsymbol{g}}^\zeta A_{\alpha\beta\gamma\zeta} = \tilde{\boldsymbol{g}}^\alpha \otimes \tilde{\boldsymbol{g}}^\beta \otimes \tilde{\boldsymbol{g}}^\gamma \otimes \tilde{\boldsymbol{g}}^\zeta A_{\alpha\beta\gamma\zeta} \tag{2.207}$$

$$\begin{aligned} \boldsymbol{B} &= \boldsymbol{e}_i \frac{\partial x_\alpha}{\partial y_i} \boldsymbol{e}_j \frac{\partial x_\beta}{\partial y_j} \boldsymbol{e}_k \frac{\partial x_\gamma}{\partial y_k} \boldsymbol{e}_l \frac{\partial x_\zeta}{\partial y_l} A_{\alpha\beta\gamma\zeta} \\ &= \boldsymbol{e}_i \frac{\partial x_\alpha}{\partial y_i} \otimes \boldsymbol{e}_j \frac{\partial x_\beta}{\partial y_j} \otimes \boldsymbol{e}_k \frac{\partial x_\gamma}{\partial y_k} \otimes \boldsymbol{e}_l \frac{\partial x_\zeta}{\partial y_l} A_{\alpha\beta\gamma\zeta} \end{aligned} \tag{2.208}$$

$$\begin{aligned} \boldsymbol{B} &= \boldsymbol{e}_i \boldsymbol{e}_j \boldsymbol{e}_k \boldsymbol{e}_l \left(\frac{\partial x_\alpha}{\partial x_i} \frac{\partial x_\beta}{\partial y_j} \frac{\partial x_\gamma}{\partial y_k} \frac{\partial x_\zeta}{\partial y_l} A_{\alpha\beta\gamma\zeta} \right) \\ &= \boldsymbol{e}_i \boldsymbol{e}_j \boldsymbol{e}_k \boldsymbol{e}_l B_{ijkl} = \boldsymbol{e}_i \otimes \boldsymbol{e}_j \otimes \boldsymbol{e}_k \otimes \boldsymbol{e}_l B_{ijkl} \end{aligned} \tag{2.209}$$

We can also write (2.206) as

$$B_{ijkl} = C^{ijkl\alpha\beta\gamma\zeta} A_{\alpha\beta\gamma\zeta} \tag{2.210}$$

Now consider contravariant tensors of rank four in x- and y-frames

$$B^{ijkl} = \frac{\partial y_i}{\partial x_\alpha} \frac{\partial y_j}{\partial x_\beta} \frac{\partial y_k}{\partial x_\gamma} \frac{\partial y_l}{\partial x_\zeta} A^{\alpha\beta\gamma\zeta} \tag{2.211}$$

Using definition of covariant basis (2.211) can be written as

$$\boldsymbol{B} = \tilde{\boldsymbol{g}}_\alpha \tilde{\boldsymbol{g}}_\beta \tilde{\boldsymbol{g}}_\gamma \tilde{\boldsymbol{g}}_\zeta A^{\alpha\beta\gamma\zeta} = \tilde{\boldsymbol{g}}_\alpha \otimes \tilde{\boldsymbol{g}}_\beta \otimes \tilde{\boldsymbol{g}}_\gamma \otimes \tilde{\boldsymbol{g}}_\zeta A^{\alpha\beta\gamma\zeta} \tag{2.212}$$

$$\begin{aligned} \boldsymbol{B} &= \boldsymbol{e}_i \frac{\partial y_i}{\partial x_\alpha} \boldsymbol{e}_j \frac{\partial y_j}{\partial x_\beta} \boldsymbol{e}_k \frac{\partial y_k}{\partial x_\gamma} \boldsymbol{e}_l \frac{\partial y_l}{\partial x_\zeta} A^{\alpha\beta\gamma\zeta} \\ &= \boldsymbol{e}_i \frac{\partial y_i}{\partial x_\alpha} \otimes \boldsymbol{e}_j \frac{\partial y_j}{\partial x_\beta} \otimes \boldsymbol{e}_k \frac{\partial y_k}{\partial x_\gamma} \otimes \boldsymbol{e}_l \frac{\partial y_l}{\partial x_\zeta} \end{aligned} \tag{2.213}$$

or
$$\begin{aligned} \boldsymbol{B} &= \boldsymbol{e}_i \frac{\partial y_i}{\partial x_\alpha} \boldsymbol{e}_j \frac{\partial y_j}{\partial x_\beta} \boldsymbol{e}_k \frac{\partial y_k}{\partial x_\gamma} \boldsymbol{e}_l \frac{\partial y_k}{\partial x_\zeta} A^{\alpha\beta\gamma\zeta} \\ &= \boldsymbol{e}_i \frac{\partial y_i}{\partial x_\alpha} \otimes \boldsymbol{e}_j \frac{\partial y_j}{\partial x_\beta} \otimes \boldsymbol{e}_k \frac{\partial y_k}{\partial x_\gamma} \otimes \boldsymbol{e}_l \frac{\partial y_l}{\partial x_\zeta} A^{\alpha\beta\gamma\zeta} \end{aligned} \tag{2.214}$$

$$\begin{aligned} \boldsymbol{B} &= \boldsymbol{e}_i \boldsymbol{e}_j \boldsymbol{e}_k \boldsymbol{e}_l \left(\frac{\partial y_i}{\partial x_\alpha} \frac{\partial y_j}{\partial x_\beta} \frac{\partial y_k}{\partial x_\gamma} \frac{\partial y_l}{\partial x_\zeta} A^{\alpha\beta\gamma\zeta} \right) = \boldsymbol{e}_i \boldsymbol{e}_j \boldsymbol{e}_k \boldsymbol{e}_l B^{ijkl} \\ &= \boldsymbol{e}_i \otimes \boldsymbol{e}_j \otimes \boldsymbol{e}_k \otimes \boldsymbol{e}_l B^{ijkl} \end{aligned} \tag{2.215}$$

We can also write (2.211) as

$$B^{ijkl} = C_{ijkl\alpha\beta\gamma\zeta} A^{\alpha\beta\gamma\zeta} \tag{2.216}$$

2.10.5.8 Def: Co- and contra-variant tensors of rank r in n-dimensional space

Let T be an admissible coordinate transformation

$$T \;:\; y_i = y_i(x_1, x_2, \dots, x_n) \quad ; \quad i = 1, 2, \dots, n \tag{2.217}$$

Consider a set of n^r quantities $A_{\alpha_1\alpha_2\dots\alpha_r}$ and $A^{\alpha_1\alpha_2\dots\alpha_r}$; $\alpha_1, \alpha_2, \dots, \alpha_r = 1, 2, \dots, n$ associated with x-frame and $B_{i_1 i_2 \dots i_r}$ and $B^{i_1 i_2 \dots i_r}$; $i_1, i_2, \dots, i_r = 1, 2, \dots, n$. Their corresponding maps in the y-frame are related by the following laws. First, consider covariant tensors of rank r in x- and y-frames.

$$B_{i_1 i_2 \dots i_r} = \frac{\partial x_{\alpha_1}}{\partial y_{i_1}} \frac{\partial x_{\alpha_2}}{\partial y_{i_2}} \cdots \frac{\partial x_{\alpha_r}}{\partial y_{i_r}} A_{\alpha_1\alpha_2\dots\alpha_r} \tag{2.218}$$

Using contravariant basis we can write (2.218) as

$$\boldsymbol{B} = \tilde{\boldsymbol{g}}^{\alpha_1}\tilde{\boldsymbol{g}}^{\alpha_2} \dots \tilde{\boldsymbol{g}}^{\alpha_r} A_{\alpha_1\alpha_2\dots\alpha_r} = \tilde{\boldsymbol{g}}^{\alpha_1} \otimes \tilde{\boldsymbol{g}}^{\alpha_2} \otimes \cdots \otimes \tilde{\boldsymbol{g}}^{\alpha_r} A_{\alpha_1\alpha_2\dots\alpha_r} \tag{2.219}$$

$$\boldsymbol{B} = \boldsymbol{e}_{i_1} \frac{\partial x_{\alpha_1}}{\partial y_{i_1}} \boldsymbol{e}_{i_2} \frac{\partial x_{\alpha_1}}{\partial y_{i_2}} \dots \boldsymbol{e}_{i_r} \frac{\partial x_{\alpha_r}}{\partial y_{i_r}} A_{\alpha_1\alpha_2\dots\alpha_r} \tag{2.220}$$

$$\boldsymbol{B} = \boldsymbol{e}_{i_1}\boldsymbol{e}_{i_2} \dots \boldsymbol{e}_{i_r} \left(\frac{\partial x_{\alpha_1}}{\partial y_{i_1}} \frac{\partial x_{\alpha_2}}{\partial y_{i_2}} \cdots \frac{\partial x_{\alpha_r}}{\partial y_{i_r}} A_{\alpha_1\alpha_2\dots\alpha_r} \right) \tag{2.221}$$

$$\boldsymbol{B} = \boldsymbol{e}_{i_1}\boldsymbol{e}_{i_2} \dots \boldsymbol{e}_{i_r} B_{i_1 i_2 \dots i_r} = \boldsymbol{e}_{i_1} \otimes \boldsymbol{e}_{i_2} \otimes \cdots \otimes \boldsymbol{e}_{i_r} B_{i_1 i_2 \dots i_r} \tag{2.222}$$

We can also write (2.222) as

$$B_{i_1 i_2 \dots i_r} = C^{i_1 i_2 \dots i_r \alpha_1\alpha_2\dots\alpha_r} A_{\alpha_1\alpha_2\dots\alpha_r} \tag{2.223}$$

Next we consider contravariant tensors of rank r in x- and y-frames.

$$B^{i_1 i_2 \dots i_r} = \frac{\partial y_{i_1}}{\partial x_{\alpha_1}} \frac{\partial y_{i_2}}{\partial x_{\alpha_2}} \cdots \frac{\partial y_{i_r}}{\partial x_{\alpha_r}} A^{\alpha_1\alpha_2\dots\alpha_r} \tag{2.224}$$

Using covariant basis we can write (2.224) as

$$\boldsymbol{B} = \tilde{\boldsymbol{g}}_{\alpha_1}\tilde{\boldsymbol{g}}_{\alpha_2} \cdots \tilde{\boldsymbol{g}}_{\alpha_r} A^{\alpha_1\alpha_2\dots\alpha_r} = \tilde{\boldsymbol{g}}_{\alpha_1} \otimes \tilde{\boldsymbol{g}}_{\alpha_2} \cdots \otimes \tilde{\boldsymbol{g}}_{\alpha_r} A^{\alpha_1\alpha_2\dots\alpha_r} \tag{2.225}$$

$$\boldsymbol{B} = \boldsymbol{e}_{i_1} \frac{\partial y_{i_1}}{\partial x_{\alpha_1}} \boldsymbol{e}_{i_2} \frac{\partial y_{i_2}}{\partial x_{\alpha_2}} \dots \boldsymbol{e}_{i_r} \frac{\partial y_{i_r}}{\partial x_{\alpha_r}} A^{\alpha_1\alpha_2\dots\alpha_r} \tag{2.226}$$

$$\boldsymbol{B} = \boldsymbol{e}_{i_1}\boldsymbol{e}_{i_2} \dots \boldsymbol{e}_{i_r} \left(\frac{\partial y_{i_1}}{\partial x_{\alpha_1}} \frac{\partial y_{i_2}}{\partial x_{\alpha_2}} \cdots \frac{\partial y_{i_r}}{\partial x_{\alpha_r}} A^{\alpha_1\alpha_2\dots\alpha_r} \right) \tag{2.227}$$

$$\boldsymbol{B} = \boldsymbol{e}_{i_1}\boldsymbol{e}_{i_2} \dots \boldsymbol{e}_{i_r} B^{i_1 i_2 \dots i_r} = \boldsymbol{e}_{i_1} \otimes \boldsymbol{e}_{i_2} \cdots \otimes \boldsymbol{e}_{i_r} B^{i_1 i_2 \dots i_r} \tag{2.228}$$

We can also write (2.224) as

$$B^{i_1 i_2 \dots i_r} = C_{i_1 i_2 \dots i_r \alpha_1\alpha_2\dots\alpha_r} A^{\alpha_1\alpha_2\dots\alpha_r} \tag{2.229}$$

α_j is $1, 2, \dots, n$ and each $\boldsymbol{e}_k$ is also $1, 2, \dots, n$

Equation (2.218) is the covariant law of rank r and (2.224) is the contravariant law of rank r. Quantities $A_{\alpha_1\alpha_2\ldots\alpha_r}$ and $B_{i_1 i_2\ldots i_r}$ are the components of the covariant tensors of rank r in the x- and the y-frames. Likewise $A^{\alpha_1\alpha_2\ldots\alpha_r}$ and $B^{i_1 i_2\ldots i_r}$ are the components of the contravariant tensors of rank r in the x- and the y-frames.

2.11 Tensors in three-dimensional x-frame, tensor operations, orthogonal coordinate transformations and invariance

2.11.1 Tensors in Cartesian x-frame

From the definition of tensors of various ranks in Section 2.10, we have seen that tensors can be covariant or contravariant. Covariant tensors transform according to the contravariant law of transformation that requires contravariant basis and vice versa. We have also seen that upon substituting the definitions of covariant or contravariant basis, both co- and contra-variant tensors ultimately can be represented in terms of unit dyads in the x-frame. This permits us to define tensors directly in orthogonal x-frame using unit vectors or basis $\boldsymbol{e}_i$ (or unit dyads) and the components of the tensor in the x-frame.

Def: Tensor or dyadic product

Consider x-frame with basis vectors (unit dyads) $\boldsymbol{e}_i$. Then, the tensor or dyadic product of $\boldsymbol{e}_i$, $\boldsymbol{e}_j$ is indicated by symbol $\otimes$ and is defined as $\boldsymbol{e}_i \otimes \boldsymbol{e}_j$ or simply $\boldsymbol{e}_i\boldsymbol{e}_j$. The ordered pair of coordinate directions $\boldsymbol{e}_i\boldsymbol{e}_j$ (or $\boldsymbol{e}_i \otimes \boldsymbol{e}_j$) are called unit dyads. These are graphically illustrated in Figure 2.10. The solid arrow represent first direction and the dotted arrows represent the second direction. The dyadic product of $\boldsymbol{e}_i$, $\boldsymbol{e}_j$, $\boldsymbol{e}_k$ is $\boldsymbol{e}_i \otimes \boldsymbol{e}_j \otimes \boldsymbol{e}_k$ or simply $\boldsymbol{e}_i\boldsymbol{e}_j\boldsymbol{e}_k$, etc.

2.11.1.1 Tensors of rank zero

Tensors of rank zero have no dyads. Thus, functions are tensors of rank zero. Given $f(\boldsymbol{x})$ in x-frame and $T : \boldsymbol{y} = \boldsymbol{y}(\boldsymbol{x})$ or $T^{-1} : \boldsymbol{x} = \boldsymbol{x}(\boldsymbol{y})$

$$g(\boldsymbol{y}) = f(\boldsymbol{x}(\boldsymbol{y})) \tag{2.230}$$

$g(\boldsymbol{y})$ is the map of $f(\boldsymbol{x})$ in the y-frame

2.11.1.2 Tensors of rank one

Definition of tensors of rank one in x-frame requires basis $\boldsymbol{e}_i$ in x-frame. Thus, if $\boldsymbol{q}$ is a tensor of rank one with components q_i along o-x_i axes, then

$$\boldsymbol{q} = \boldsymbol{e}_i q_i \tag{2.231}$$

2.11.1.3 Tensors of rank two

Definition of tensors of rank two in x-frame with components τ_{ij} require dyadic product of $\boldsymbol{e}_i$, $\boldsymbol{e}_j$, i.e., $\boldsymbol{e}_i \otimes \boldsymbol{e}_j$ or $\boldsymbol{e}_i\boldsymbol{e}_j$ and we can write

$$\boldsymbol{\tau} = \boldsymbol{e}_i \otimes \boldsymbol{e}_j \tau_{ij} = \boldsymbol{e}_i\boldsymbol{e}_j \tau_{ij} \tag{2.232}$$

Expansion of (2.232) gives

$$\begin{aligned}\boldsymbol{\tau} = \boldsymbol{e}_1\boldsymbol{e}_1\tau_{11} + \boldsymbol{e}_1\boldsymbol{e}_2\tau_{12} + \boldsymbol{e}_1\boldsymbol{e}_3\tau_{13} \\ + \boldsymbol{e}_2\boldsymbol{e}_1\tau_{21} + \boldsymbol{e}_2\boldsymbol{e}_2\tau_{22} + \boldsymbol{e}_2\boldsymbol{e}_3\tau_{23} \\ + \boldsymbol{e}_3\boldsymbol{e}_1\tau_{13} + \boldsymbol{e}_3\boldsymbol{e}_2\tau_{32} + \boldsymbol{e}_3\boldsymbol{e}_3\tau_{33}\end{aligned} \tag{2.233}$$

unit dyads in (2.233) as shown in Figure 2.10.

Special tensors of rank two

(1) If $\tau_{ij} = \tau_{ji}$ then the tensor $\boldsymbol{\tau}$ is said to be a symmetric tensor.
(2) If $\tau_{ij} = -\tau_{ji}$ or $\boldsymbol{\tau} = -\boldsymbol{\tau}^T$ or $[\tau] = -[\tau]^T$ then the tensor $\boldsymbol{\tau}$ is called an antisymmetric or skew-symmetric tensor.
(3) If the components of a tensor $\boldsymbol{\tau}$ are taken to be the components of $\boldsymbol{\tau}$ but with indices interchanged, the resulting tensor is called transpose of $\boldsymbol{\tau}$.

$$\text{If} \quad \boldsymbol{\tau} = \boldsymbol{e}_i\boldsymbol{e}_j\tau_{ij} \tag{2.234}$$

$$\text{then} \quad \boldsymbol{\tau}^T = \boldsymbol{e}_i\boldsymbol{e}_j\tau_{ji} \tag{2.235}$$

Note that in $\boldsymbol{\tau}^T$ only the indices i and j switch for τ_{ij} and not for $\boldsymbol{e}_i$, $\boldsymbol{e}_j$.

2.11.1.4 Tensors of rank r

Let $\tau_{i_1 i_2 \dots i_r}$ by components of the tensor of rank r in x-frame. Then

$$\boldsymbol{\tau} = \boldsymbol{e}_{i_1} \otimes \boldsymbol{e}_{i_2} \cdots \otimes \boldsymbol{e}_{i_r}\tau_{i_1 i_2 \dots i_r} \quad ; \quad i_j = 1,2,3 \quad ; \quad j = 1,2,\dots,r \tag{2.236}$$

2.11.2 Tensor operations

In the following, we present various commonly encountered operations with tensors of ranks zero, one and two.

2.11.2.1 Def: Tensors of rank two from dyadic products of tensors of rank one

If the components of a tensor of rank two are formed by the ordered pairs of the components of two tensors of rank one, say $\boldsymbol{\phi}$ and $\boldsymbol{\psi}$, the resulting tensor of rank two is called the dyadic product of $\boldsymbol{\phi}$ and $\boldsymbol{\psi}$ and is represented by

$$\boldsymbol{\phi} \otimes \boldsymbol{\psi} \quad \text{or} \quad \boldsymbol{\phi}\boldsymbol{\psi}$$

$$\text{where} \quad \boldsymbol{\phi} \otimes \boldsymbol{\psi} = \boldsymbol{e}_i\phi_i \otimes \boldsymbol{e}_j\psi_j = \boldsymbol{e}_i \otimes \boldsymbol{e}_j\phi_i\psi_j \tag{2.237}$$

We note that $\boldsymbol{\phi} \otimes \boldsymbol{\psi} \neq \boldsymbol{\psi} \otimes \boldsymbol{\phi}$ but $(\boldsymbol{\phi} \otimes \boldsymbol{\psi})^T = \boldsymbol{\psi} \otimes \boldsymbol{\phi}$ as $\boldsymbol{\phi} \otimes \boldsymbol{\psi} = \boldsymbol{e}_i \otimes \boldsymbol{e}_j\phi_i\psi_j$; $\boldsymbol{\psi} \otimes \boldsymbol{\phi} = \boldsymbol{e}_i \otimes \boldsymbol{e}_j\psi_i\phi_j$, hence $(\boldsymbol{\phi} \otimes \boldsymbol{\psi})^T = \boldsymbol{e}_i \otimes \boldsymbol{e}_j\psi_i\phi_j = \boldsymbol{\psi} \otimes \boldsymbol{\phi}$. If the components of a second rank tensor are given by Kronecker delta δ_{ij}, the resulting tensor is called unit tensor of rank two.

$$\boldsymbol{I} = \boldsymbol{e}_i\boldsymbol{e}_j\delta_{ij} = \boldsymbol{e} \tag{2.238}$$

Remark

In the definitions of tensor operation in the following sections, the tensors of rank two can be unsymmetric, symmetric or skew-symmetric.

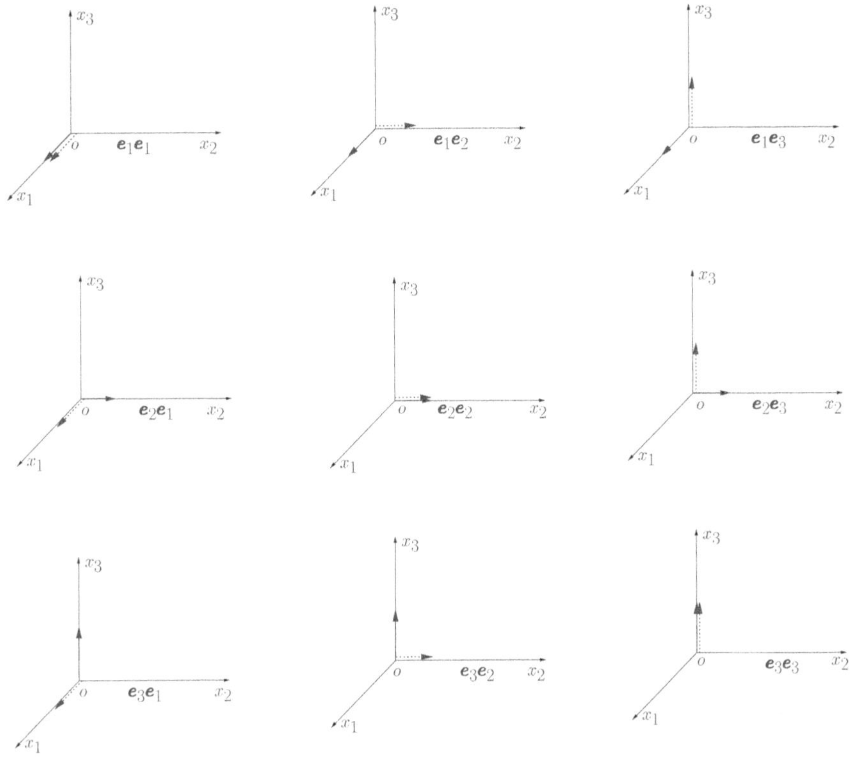

Figure 2.10: The unit dyad $\boldsymbol{e}_i\boldsymbol{e}_j$

2.11.2.2 Def: Scalar product or double dot product of tensors of rank two

If $\boldsymbol{\sigma}$ and $\boldsymbol{\tau}$ are tensors of rank two, then we indicate their scalar product or double dot product by $\boldsymbol{\sigma} : \boldsymbol{\tau}$ and it is defined as

$$
\begin{aligned}
\boldsymbol{\sigma} : \boldsymbol{\tau} &= \boldsymbol{e}_i\boldsymbol{e}_j\sigma_{ij} : \boldsymbol{e}_k\boldsymbol{e}_l\tau_{kl} \\
&= (\boldsymbol{e}_i\boldsymbol{e}_j : \boldsymbol{e}_k\boldsymbol{e}_l)\sigma_{ij}\tau_{kl} \\
&= \underbrace{\delta_{jk}}_{k=j}\underbrace{\delta_{il}}_{l=i}\sigma_{ij}\tau_{kl} \\
&= \sigma_{ij}\tau_{ji}
\end{aligned}
$$

$$
\therefore \quad \boldsymbol{\sigma} : \boldsymbol{\tau} = \sigma_{ij}\tau_{ji} \quad ; \quad \text{trace of } [\sigma][\tau] \text{ (tensor of rank zero)} \tag{2.239}
$$

Thus, scalar product or double dot product of two tensors of rank two is a tensor of rank zero. Similarly

$$
\begin{aligned}
\boldsymbol{\tau} : (\boldsymbol{\phi} \otimes \boldsymbol{\psi}) &= \boldsymbol{e}_i\boldsymbol{e}_j\tau_{ij} : (\boldsymbol{e}_k\boldsymbol{e}_l\phi_k\psi_l) \\
&= \boldsymbol{e}_i\boldsymbol{e}_j : \boldsymbol{e}_k\boldsymbol{e}_l\tau_{ij}\phi_k\psi_l \\
&= \underbrace{\delta_{jk}}_{k=j}\underbrace{\delta_{il}}_{l=i}\tau_{ij}\phi_k\psi_l \\
&= \tau_{ij}\phi_j\psi_i
\end{aligned}
$$

$$\therefore \quad \boldsymbol{\tau} : (\boldsymbol{\phi} \otimes \boldsymbol{\psi}) = \tau_{ij}\phi_j\psi_i \qquad \text{is a tensor of rank zero} \tag{2.240}$$

and

$$\begin{aligned}
(\boldsymbol{\phi} \otimes \boldsymbol{\psi}) : (\boldsymbol{\alpha} \otimes \boldsymbol{\beta}) &= (\boldsymbol{e}_i\boldsymbol{e}_j\phi_i\psi_j) : (\boldsymbol{e}_k\boldsymbol{e}_l\alpha_k\beta_l) \\
&= (\boldsymbol{e}_i\boldsymbol{e}_j : \boldsymbol{e}_k\boldsymbol{e}_l)\phi_i\psi_j\alpha_k\beta_l \\
&= \underbrace{\delta_{jk}}_{k=j}\ \underbrace{\delta_{il}}_{l=i}\ \phi_i\psi_j\alpha_k\beta_l \\
&= \phi_i\psi_j\alpha_j\beta_i
\end{aligned}$$

$$\therefore \quad (\boldsymbol{\phi} \otimes \boldsymbol{\psi}) : (\boldsymbol{\alpha} \otimes \boldsymbol{\beta}) = \phi_i\psi_j\alpha_j\beta_i \qquad \text{is a tensor of rank zero} \tag{2.241}$$

We note that

$$(\boldsymbol{e}_i\boldsymbol{e}_j) : (\boldsymbol{e}_k\boldsymbol{e}_l) = \delta_{jk}\delta_{il} \tag{2.242}$$

2.11.2.3 Def: Tensor product or single dot product of tensors of rank two

If $\boldsymbol{\sigma}$ and $\boldsymbol{\tau}$ are two tensors of rank two, then the tensor product or single dot product of $\boldsymbol{\sigma}$ and $\boldsymbol{\tau}$ is indicated by $\boldsymbol{\sigma} \cdot \boldsymbol{\tau}$ and is defined as

$$\begin{aligned}
\boldsymbol{\sigma} \cdot \boldsymbol{\tau} &= \boldsymbol{e}_i\boldsymbol{e}_j\sigma_{ij} \cdot \boldsymbol{e}_k\boldsymbol{e}_l\tau_{kl} \\
&= (\boldsymbol{e}_i\boldsymbol{e}_j \cdot \boldsymbol{e}_k\boldsymbol{e}_l)\sigma_{ij}\tau_{kl} \\
&= \boldsymbol{e}_i(\boldsymbol{e}_j \cdot \boldsymbol{e}_k)\boldsymbol{e}_l\sigma_{ij}\tau_{kl} \\
&= \boldsymbol{e}_i\ \underbrace{\delta_{jk}}_{k=j}\ \boldsymbol{e}_l\sigma_{ij}\tau_{kl} \\
&= \boldsymbol{e}_i\boldsymbol{e}_l\sigma_{ij}\tau_{jl}
\end{aligned}$$

$$\therefore \quad \boldsymbol{\sigma} \cdot \boldsymbol{\tau} = \boldsymbol{e}_i\boldsymbol{e}_l\sigma_{ij}\tau_{jl} \tag{2.243}$$

Thus, $\boldsymbol{\sigma} \cdot \boldsymbol{\tau}$ is a tensor of rank two.

2.11.2.4 Def: Dot product of a tensor of rank two with a tensor of rank one

If $\boldsymbol{\tau}$ is a tensor of rank two and $\boldsymbol{\phi}$ is a tensor of rank one then we define

$$\begin{aligned}
\boldsymbol{\tau} \cdot \boldsymbol{\phi} &= (\boldsymbol{e}_i\boldsymbol{e}_j\tau_{ij}) \cdot \boldsymbol{e}_k\phi_k \\
&= \boldsymbol{e}_i(\boldsymbol{e}_j \cdot \boldsymbol{e}_k)\tau_{ij}\phi_k \\
&= \boldsymbol{e}_i\ \underbrace{\delta_{jk}}_{k=j}\ \tau_{ij}\phi_k
\end{aligned}$$

$$\therefore \quad \boldsymbol{\tau} \cdot \boldsymbol{\phi} = \boldsymbol{e}_i(\tau_{ij}\phi_j) \tag{2.244}$$

Thus, $\boldsymbol{\tau} \cdot \boldsymbol{\phi}$ is a tensor of rank one. We note that $\boldsymbol{\tau} \cdot \boldsymbol{\phi} \neq \boldsymbol{\phi} \cdot \boldsymbol{\tau}$ unless $\boldsymbol{\tau}$ is symmetric.

2.11.2.5 Def: Cross product of tensor of rank two with tensor of rank one

If $\boldsymbol{\tau}$ is a tensor of rank two and $\boldsymbol{\phi}$ is a tensor of rank one, then the cross product of $\boldsymbol{\tau}$ and $\boldsymbol{\phi}$ is indicated by $\boldsymbol{\tau} \times \boldsymbol{\phi}$ and is defined as

$$\begin{aligned} \boldsymbol{\tau} \times \boldsymbol{\phi} &= (\boldsymbol{e}_i \boldsymbol{e}_j \tau_{ij}) \times \boldsymbol{e}_k \phi_k \\ &= \boldsymbol{e}_i (\boldsymbol{e}_j \times \boldsymbol{e}_k) \tau_{ij} \phi_k \\ &= \boldsymbol{e}_i (\epsilon_{jkl} \boldsymbol{e}_l) \tau_{ij} \phi_k \\ \therefore \quad \boldsymbol{\tau} \times \boldsymbol{\phi} &= \epsilon_{jkl} \boldsymbol{e}_i \boldsymbol{e}_l \tau_{ij} \phi_k \end{aligned} \tag{2.245}$$

Thus, $\boldsymbol{\tau} \times \boldsymbol{\phi}$ is a tensor of rank two and the il component of $\boldsymbol{\tau} \times \boldsymbol{\phi}$ is $\epsilon_{jkl} \tau_{ij} \phi_k$. Consider the cross product $\boldsymbol{\phi} \times \boldsymbol{\tau}$

$$\begin{aligned} \boldsymbol{\phi} \times \boldsymbol{\tau} &= \boldsymbol{e}_i \phi_i \times (\boldsymbol{e}_j \boldsymbol{e}_k \tau_{jk}) \\ &= \boldsymbol{e}_i \times \boldsymbol{e}_j \boldsymbol{e}_k \phi_i \tau_{jk} \\ &= \epsilon_{ijl} \boldsymbol{e}_l \boldsymbol{e}_k \phi_i \tau_{jk} \\ \therefore \quad \boldsymbol{\phi} \times \boldsymbol{\tau} &= \epsilon_{ijl} \boldsymbol{e}_l \boldsymbol{e}_k \phi_i \tau_{jk} \end{aligned} \tag{2.246}$$

Thus, the lk component of $\boldsymbol{\phi} \times \boldsymbol{\tau}$ tensor is $\epsilon_{ijl} \phi_i \tau_{jk}$.

2.11.2.6 Def: Addition of tensors of rank two

If $\boldsymbol{\sigma}$ and $\boldsymbol{\tau}$ are both tensors of rank two, then

$$\boldsymbol{\sigma} + \boldsymbol{\tau} = \boldsymbol{e}_i \boldsymbol{e}_j \sigma_{ij} + \boldsymbol{e}_i \boldsymbol{e}_j \tau_{ij} = \boldsymbol{e}_i \boldsymbol{e}_j (\sigma_{ij} + \tau_{ij}) \tag{2.247}$$

Thus, the sum of two tensors of rank two is also a tensor of rank two whose components are the sums of the corresponding components of the two tensors.

2.11.2.7 Def: Multiplication of a tensor of rank two by a scalar

$$p\,\boldsymbol{\tau} = p(\boldsymbol{e}_i \boldsymbol{e}_j \tau_{ij}) = \boldsymbol{e}_i \boldsymbol{e}_j (p \tau_{ij}) \tag{2.248}$$

That is, multiplication of a tensor of rank two $\boldsymbol{\tau}$ by a scalar p corresponds to multiplying each component of the tensor of rank two by the scalar.

2.11.2.8 Def: Magnitude of a tensor of rank two

If $\boldsymbol{\tau}$ is a tensor of rank two, then its magnitude is a tensor of rank zero indicated by $|\boldsymbol{\tau}|$ and defined as

$$|\boldsymbol{\tau}| = \sqrt{\frac{1}{2}(\boldsymbol{\tau} : \boldsymbol{\tau}^T)} = \sqrt{\frac{1}{2}\,\mathrm{tr}([\tau][\tau]^T)} \tag{2.249}$$

2.11.3 Transformations of tensors defined in orthogonal frames due to orthogonal coordinate transformation

Consider a Cartesian frame, say x-frame, with basis $\boldsymbol{e}_i^x$. Let x-frame be transformed into y-frame through an orthogonal coordinate transformation $[Q]$. Let $\boldsymbol{e}_i^y$ be the basis in the y-frame. Then we have

$$\{\boldsymbol{e}^y\} = [Q]\{\boldsymbol{e}^x\} \quad ; \quad \boldsymbol{e}_i^y = Q_{ij} \boldsymbol{e}_j^x \tag{2.250}$$

where $\{\mathbf{e}^x\}^T = [\mathbf{e}_1^x, \mathbf{e}_2^x, \mathbf{e}_3^x]$ and $\{\mathbf{e}^y\}^T = [\mathbf{e}_1^y, \mathbf{e}_2^y, \mathbf{e}_3^y]$. $[Q]$ is orthogonal, hence $[Q]^T = [Q]^{-1}$ and we can write

$$\{\mathbf{e}^x\} = [Q]^T\{\mathbf{e}^y\} \quad ; \quad \mathbf{e}_i^x = Q_{ji}\mathbf{e}_j^y \tag{2.251}$$

We refer to this transformation defined by $[Q]$ as transformation T i.e.,

$$T \,:\, y_i = y_i(x_j) \quad \text{through} \quad [Q] \tag{2.252}$$

We consider induced transformation of type G^0, G^1 ..., etc. due to T. That is, we consider tensors of rank zero, one and two in x-frame and their correspondence in y-frame due to transformations induced by the coordinate transformation T defined by $[Q]$.

2.11.3.1 Tensors of rank zero

Let $f(x)$ be a tensor of rank zero defined in the x-frame and let $g(y)$ represent the corresponding tensor of rank zero in the y-frame. Let transformations T and T^{-1} be given by

$$T \,:\, y_i = x_i(x_j) \quad ; \quad T^{-1} \,:\, x_i = x_i(y_j) \tag{2.253}$$
$$f(\boldsymbol{x}) = f(\boldsymbol{x}(\boldsymbol{y})) = g(\boldsymbol{y}) \tag{2.254}$$

We notes that the expressions of $f(\boldsymbol{x})$ and $g(\boldsymbol{y})$ are different, but their values are same. Tensors of rank zero are scalars or absolute scalar. Density, temperature, specific heat are examples.

2.11.3.2 Tensors of rank one

Let $\boldsymbol{\phi}^x$ be a tensor of rank one in the x-frame and $\boldsymbol{\phi}^y$ be the corresponding tensor of rank one in y-frame.

$$\boldsymbol{\phi}^x = \mathbf{e}_i^x\phi_i^x = \{\mathbf{e}^x\}^T\{\phi^x\} \tag{2.255}$$
$$\boldsymbol{\phi}^y = \mathbf{e}_i^y\phi_i^y = \{\mathbf{e}^y\}^T\{\phi^y\} \tag{2.256}$$

Consider $\boldsymbol{\phi}^y$. Substituting for $\mathbf{e}^y$ from (2.250)

$$\boldsymbol{\phi}^y = \{\mathbf{e}^y\}^T\{\phi^y\} = \{\mathbf{e}^x\}^T[Q]^T\{\phi^y\} = \{\mathbf{e}^x\}^T([Q]^T\{\phi^y\}) = \{\mathbf{e}^x\}^T\{\phi^x\} \tag{2.257}$$
$$\therefore \quad \{\phi^x\} = [Q]^T\{\phi^y\} \tag{2.258}$$

Consider $\boldsymbol{\phi}^x$. Substituting for $\mathbf{e}^x$ from (2.251)

$$\boldsymbol{\phi}^x = \{\mathbf{e}^x\}^T\{\phi^x\} = \{\mathbf{e}^y\}^T[Q]\{\phi^x\} = \{\mathbf{e}^y\}^T([Q]\{\phi^x\}) = \{\mathbf{e}^y\}^T\{\phi^y\} \tag{2.259}$$
$$\therefore \quad \{\phi^y\} = [Q]\{\phi^x\} \tag{2.260}$$

We can derive the same relations using Einstein's notations. Consider $\boldsymbol{\phi}^y$. Substituting for $\mathbf{e}_i^y$ from (2.250)

$$\boldsymbol{\phi}^y = \mathbf{e}_i^y\phi_i^y = Q_{ij}\mathbf{e}_j^x\phi_i^y = \mathbf{e}_j^x(Q_{ij}\phi_i^y) \tag{2.261}$$
$$\therefore \quad \phi_j^x = Q_{ij}\phi_i^y \quad \text{or} \quad \{\phi^x\} = [Q]^T\{\phi^y\} \tag{2.262}$$

Consider $\boldsymbol{\phi}^x$. Substituting for $\boldsymbol{e}_i^x$ from (2.251)

$$\boldsymbol{\phi}^x = \boldsymbol{e}_i^x \phi_i^x = (Q_{ji}\boldsymbol{e}_j^y)\phi_i^x = \boldsymbol{e}_j^y(Q_{ji}\phi_i^x) \tag{2.263}$$

$$\therefore \quad \phi_j^y = Q_{ji}\phi_i^x \quad \text{or} \quad \{\phi^y\} = [Q]\{\phi^x\} \tag{2.264}$$

2.11.3.3 Tensors of rank two

Let $\boldsymbol{\tau}^x$ be a tensor of rank two in the x-frame and let $\boldsymbol{\tau}^y$ be a corresponding tensor of rank two in the y-frame due to transformation induced on $\boldsymbol{\tau}^x$ by the coordinate transformation $T\ :\ y_i = y_i(x_j)$ through $[Q]$. Then

$$\boldsymbol{\tau}^x = \boldsymbol{e}_i^x\boldsymbol{e}_j^x\tau_{ij}^x = \boldsymbol{e}_i^x \otimes \boldsymbol{e}_j^x\tau_{ij}^x \tag{2.265}$$

and

$$\boldsymbol{\tau}^y = \boldsymbol{e}_i^y\boldsymbol{e}_j^y\tau_{ij}^y = \boldsymbol{e}_i^y \otimes \boldsymbol{e}_j^y\tau_{ij}^y \tag{2.266}$$

Tensors $\boldsymbol{\tau}^x$, $\boldsymbol{\tau}^y$ can be unsymmetric, symmetric or skew-symmetric. Terms τ_{ij}^x and τ_{ij}^y are the components of tensors $\boldsymbol{\tau}^x$ and $\boldsymbol{\tau}^y$ in x- and y-frames. Consider $\boldsymbol{\tau}^y$ in (2.266). Substituting for $\boldsymbol{e}_i^y$ and $\boldsymbol{e}_j^y$ from (2.250)

$$\boldsymbol{\tau}^y = (Q_{ik}\boldsymbol{e}_k^x) \otimes (Q_{jl}\boldsymbol{e}_l^x)\tau_{ij}^y = (\boldsymbol{e}_k^x \otimes \boldsymbol{e}_l^x)Q_{ik}Q_{jl}\tau_{ij}^y \tag{2.267}$$

but

$$\boldsymbol{\tau}^x = (\boldsymbol{e}_k^x \otimes \boldsymbol{e}_l^x)\tau_{kl}^x \tag{2.268}$$

$$\therefore \quad \tau_{kl}^x = Q_{ik}Q_{jl}\tau_{ij}^y \tag{2.269}$$

Alternatively, we can consider $\boldsymbol{\tau}^x$ in (2.265). Substituting for $\boldsymbol{e}_i^x$ and $\boldsymbol{e}_j^x$ from (2.251)

$$\begin{aligned}\boldsymbol{\tau}^x = \boldsymbol{e}_i^x \otimes \boldsymbol{e}_j^x\tau_{ij}^x &= (Q_{ki}\boldsymbol{e}_k^y) \otimes (Q_{lj}\boldsymbol{e}_l^y)\tau_{ij}^x \\ &= (\boldsymbol{e}_k^y \otimes \boldsymbol{e}_l^y)Q_{ki}Q_{lj}\tau_{ij}^x\end{aligned} \tag{2.270}$$

$$\text{but} \quad \boldsymbol{\tau}^y = (\boldsymbol{e}_k^y \otimes \boldsymbol{e}_l^y)\tau_{kl}^y \tag{2.271}$$

$$\therefore \quad \tau_{kl}^y = Q_{ki}Q_{lj}\tau_{ij}^x \tag{2.272}$$

We can also derive (2.269) and (2.272) using matrix and vector notations. First, rewrite $\boldsymbol{\tau}^x$ and $\boldsymbol{\tau}^y$ in matrix and vector notations.

$$\boldsymbol{\tau}^x = \{\boldsymbol{e}^x\}^T[\tau^x]\{\boldsymbol{e}^x\} \tag{2.273}$$

$$\boldsymbol{\tau}^y = \{\boldsymbol{e}^y\}^T[\tau^y]\{\boldsymbol{e}^y\} \tag{2.274}$$

Consider $\boldsymbol{\tau}^y$ in (2.274) and substitute for $\boldsymbol{e}^y$ from (2.250).

$$\boldsymbol{\tau}^y = \{\boldsymbol{e}^x\}^T([Q]^T[\tau^y][Q])\{\boldsymbol{e}^x\} \tag{2.275}$$

$$\therefore \quad [\tau^x] = [Q]^T[\tau^y][Q] \tag{2.276}$$

Alternatively, consider $\boldsymbol{\tau}^x$ in (2.273) and substitute for $\boldsymbol{e}^x$ from (2.251)

$$\boldsymbol{\tau}^x = \{\boldsymbol{e}^y\}^T([Q][\tau^x][Q]^T)\{\boldsymbol{e}^y\} \tag{2.277}$$

$$\therefore \quad [\tau^y] = [Q][\tau^x][Q]^T \tag{2.278}$$

2.11.3.4 Tensors of rank three

Let $\boldsymbol{\tau}^x$ be tensor of rank three in the x-frame and let $\boldsymbol{\tau}^y$ be the corresponding tensor of rank three in y-frame. Let the coordinate transformations T and T^{-1} be defined by

$$T \; : \; y_i = y_i(x_j) \quad ; \quad T^{-1} \; : \; x_i = x_i(y_j) \tag{2.279}$$

Then

$$\boldsymbol{\tau}^x = \boldsymbol{e}_i^x \boldsymbol{e}_j^x \boldsymbol{e}_k^x \tau_{ijk}^x = \boldsymbol{e}_i^x \otimes \boldsymbol{e}_j^x \otimes \boldsymbol{e}_k^x \tau_{ijk}^x \tag{2.280}$$

$$\text{and} \quad \boldsymbol{\tau}^y = \boldsymbol{e}_i^y \boldsymbol{e}_j^y \boldsymbol{e}_k^y \tau_{ijk}^y = \boldsymbol{e}_i^y \otimes \boldsymbol{e}_j^y \otimes \boldsymbol{e}_k^y \tau_{ijk}^y \tag{2.281}$$

Consider $\boldsymbol{\tau}^y$ in (2.280) and substitute $\boldsymbol{e}_i^y$, $\boldsymbol{e}_j^y$ and $\boldsymbol{e}_k^y$ from (2.250).

$$\boldsymbol{\tau}^y = (Q_{il}\boldsymbol{e}_l^x)(Q_{jm}\boldsymbol{e}_m^x)(Q_{kn}\boldsymbol{e}_n^x)\tau_{ijk}^y \tag{2.282}$$

$$= \boldsymbol{e}_l^x \boldsymbol{e}_m^x \boldsymbol{e}_n^x (Q_{il}Q_{jm}Q_{kn}\tau_{ijk}^y) \tag{2.283}$$

$$\begin{aligned} \boldsymbol{\tau}^x &= \boldsymbol{e}_l^x \boldsymbol{e}_m^x \boldsymbol{e}_n^x \tau_{lmn}^x \\ \therefore \quad \tau_{lmn}^x &= Q_{il}Q_{jm}Q_{kn}\tau_{ijk}^y \end{aligned} \tag{2.284}$$

Alternatively

$$\boldsymbol{\tau}^x = \boldsymbol{e}_i^x \boldsymbol{e}_j^x \boldsymbol{e}_k^x \tau_{ijk}^x = \boldsymbol{e}_i^x \otimes \boldsymbol{e}_j^x \otimes \boldsymbol{e}_k^x \tau_{ijk}^x \tag{2.285}$$

Substituting for $\boldsymbol{e}_i^x$, $\boldsymbol{e}_j^x$ and $\boldsymbol{e}_k^x$ from (2.251)

$$\boldsymbol{\tau}^x = (Q_{li}\boldsymbol{e}_l^y) \otimes (Q_{mj}\boldsymbol{e}_m^y) \otimes (Q_{nk}\boldsymbol{e}_n^y)\tau_{ijk}^x \tag{2.286}$$

$$= \boldsymbol{e}_l^y \otimes \boldsymbol{e}_m^y \otimes \boldsymbol{e}_n^y (Q_{li}Q_{mj}Q_{nk}\tau_{ijk}^x) \tag{2.287}$$

$$\text{but} \quad \boldsymbol{\tau}^y = \boldsymbol{e}_l^y \otimes \boldsymbol{e}_m^y \otimes \boldsymbol{e}_n^y \tau_{lmn}^y \tag{2.288}$$

$$\therefore \quad \tau_{lmn}^y = Q_{li}Q_{mj}Q_{nk}\tau_{ijk}^x \tag{2.289}$$

Similarly tensors of higher ranks in the two frames can be related due to orthogonal transformation of the frames.

2.11.4 Invariants of tensors

2.11.4.1 Def: Invariants of a tensor

Invariants of a tensor are the quantities related to the tensor that does not change due to induced transformations as a result of transformation of coordinates.

If the material point coordinates in the x-frame are transformed into coordinates in y-frame due to an orthogonal transformation T due to a rotation matrix $[Q]$, then a transformation is induced on the tensors defined in the x-frame to define their map in the y-frame.

In this section, we determine quantities related to tensors of rank zero, one and two that do not change due to orthogonal transformation of x-frame into y-frame. Such quantities are called invariants of a tensor. Thus, invariants of a tensor are independent of the induced transformation due to rotation of the frame. Let T be an orthogonal transformation through a rotation matrix $[Q]$ that transforms material point coordinates in the x-frame to y-frame.

$$\begin{aligned} T \ : \ y_i &= y_i(x_j) \\ \{y\} &= [Q]\{x\} \end{aligned} \tag{2.290}$$

Thus, y-frame is obtained from x-frame through a rigid rotation of the x-frame.

2.11.4.2 Tensors of rank zero

Consider tensor $f(\boldsymbol{x})$ of rank zero in the x-frame and $g(\boldsymbol{y})$ be the corresponding tensor of rank zero in the y-frame, then

$$g(\boldsymbol{y}) = g(\boldsymbol{y}(\boldsymbol{x})) = f(\boldsymbol{x}) \tag{2.291}$$

Thus, numerical value of $f(\boldsymbol{x})$ and $g(\boldsymbol{y})$ is same, but their forms are different. Hence, tensors of rank zero have no invariants.

2.11.4.3 Tensors of rank one

Consider a tensor $\boldsymbol{\phi}^x$ of rank one in the x-frame and let $\boldsymbol{\phi}^y$ be a corresponding tensor of rank one in the y-frame. Then

$$\{\phi^y\} = [Q]\{\phi^x\} \quad \text{or} \quad \phi_i^y = Q_{ij}\phi_j^x \tag{2.292}$$

Consider the dot product $\boldsymbol{\phi}^y \cdot \boldsymbol{\phi}^y = \{\phi^y\}^T\{\phi^y\} = \phi_i^y\phi_i^y$. Using matrix and vector notations

$$\{\phi^y\}^T\{\phi^y\}=([Q]\{\phi^x\})^T([Q]\{\phi^x\})=\{\phi^x\}^T[Q]^T[Q]\{\phi^x\}=\{\phi^x\}^T\{\phi^x\} \tag{2.293}$$

Thus, the dot product of a tensor of rank one with itself representing the square of the magnitude of the tensor of rank one is an invariant of the orthogonal transformation T through the rotation matrix $[Q]$. That is, the magnitude of a tensor of rank one is invariant of the orthogonal transformation.

2.11.4.4 Tensors of rank two: invariants based on trace

The material presented in this section applies to square matrices that can be unsymmetric or symmetric or skew-symmetric. However, in this course, we generally are concerned with symmetric tensors.

Let $\boldsymbol{\tau}^x$ be a tensor of rank two in the x-frame and $\boldsymbol{\tau}^y$ be the corresponding tensor in the y-frame obtained due to the induced transformation. Both $\boldsymbol{\tau}^x$ and $\boldsymbol{\tau}^y$ can be either unsymmetric or symmetric. Then

$$[\tau^y] = [Q][\tau^x][Q]^T \quad \text{or} \quad \tau_{ij}^y = Q_{ik}\tau_{kl}^x Q_{jl} = Q_{ik}Q_{jl}\tau_{kl}^x \tag{2.294}$$

Contracting indices i and j

$$\tau_{ii}^y = Q_{ik}Q_{il}\tau_{kl}^x = \underbrace{\delta_{kl}}_{l=k}\tau_{kl}^x = \tau_{kk}^x \tag{2.295}$$

$$\therefore \quad \operatorname{tr}\boldsymbol{\tau}^y = \operatorname{tr}\boldsymbol{\tau}^x \tag{2.296}$$

Thus, the trace of the second order tensor $\boldsymbol{\tau}^x$ in the x-frame is the same as the trace of the second order tensor $\boldsymbol{\tau}^y$ in the y-frame. Hence, the trace of a second order tensor is the invariant of the orthogonal coordinate transformation T.

Consider $\boldsymbol{\tau}^y \cdot \boldsymbol{\tau}^y$ or $[\tau^y][\tau^y]$

$$[\tau^y][\tau^y] = ([Q][\tau^x][Q]^T)([Q][\tau^x][Q]^T) \tag{2.297}$$

$$\text{or} \quad [\tau^y]^2 = [Q][\tau^x]^2[Q]^T \quad ; \quad [\tau^x]^2 = [\tau^x][\tau^x] \quad ; \quad [\tau^y]^2 = [\tau^y][\tau^y] \tag{2.298}$$

$$\text{or} \quad ([\tau^y]^2)_{ij} = Q_{ik}([\tau^x]^2)_{kl}Q_{jl} = Q_{ik}Q_{jl}([\tau^x]^2)_{kl} \tag{2.299}$$

Contracting indices i and j

$$([\tau^y]^2)_{ii} = Q_{ik}Q_{il}([\tau^x]^2)_{kl} = \underbrace{\delta_{kl}}_{l=k}([\tau^x]^2)_{kl} \tag{2.300}$$

$$\text{or} \quad ([\tau^y]^2)_{ii} = ([\tau^x]^2)_{kk} \tag{2.301}$$

$$\text{tr}(\boldsymbol{\tau}^y)^2 = \text{tr}(\boldsymbol{\tau}^x)^2 \tag{2.302}$$

Thus, the trace of $(\boldsymbol{\tau}^x)^2$ in x-frame and the trace of $(\boldsymbol{\tau}^y)^2$ in y-frame are the same, hence the trace of the square of a second order tensor is the invariant of the orthogonal coordinate transformation T. Consider $\boldsymbol{\tau}^y \cdot \boldsymbol{\tau}^y \cdot \boldsymbol{\tau}^y$ or $[\tau^y][\tau^y][\tau^y]$

$$[\tau^y][\tau^y][\tau^y] = ([Q][\tau^x][Q]^T)([Q][\tau^x][Q]^T)([Q][\tau^x][Q]^T) \tag{2.303}$$

$$\text{or} \quad [\tau^y]^3 = [Q][\tau^x]^3[Q]^T \; ; \; [\tau^x]^3 = [\tau^x][\tau^x][\tau^x] \, , \, [\tau^y]^3 = [\tau^y][\tau^y][\tau^y] \tag{2.304}$$

$$\text{or} \quad ([\tau^y]^3)_{ij} = Q_{ik}([\tau^x]^3)_{kl}Q_{jl} = Q_{ik}Q_{jl}([\tau^x]^3)_{kl} \tag{2.305}$$

Contracting indices i and j

$$([\tau^y]^3)_{ii} = Q_{ik}Q_{il}([\tau^x]^3)_{kl} = \underbrace{\delta_{kl}}_{l=k}([\tau^x]^3)_{kl} \tag{2.306}$$

$$\text{or} \quad ([\tau^y]^3)_{ii} = ([\tau^x]^3)_{kk} \tag{2.307}$$

$$\text{tr}(\boldsymbol{\tau}^y)^3 = \text{tr}(\boldsymbol{\tau}^x)^3 \tag{2.308}$$

Hence the trace of the cube of a second order tensor is invariant of the orthogonal coordinate transformation T.

Using similar procedure we can show that

$$\text{tr}\,(\boldsymbol{\tau}^y)^i = \text{tr}(\boldsymbol{\tau}^x)^i \quad ; \quad i = 4, 5, \dots, n \tag{2.309}$$

Thus, if $\boldsymbol{\tau}$ is a second order tensor defined in an orthogonal frame of reference, then $\text{tr}(\boldsymbol{\tau}), \text{tr}(\boldsymbol{\tau})^2, \dots, \text{tr}(\boldsymbol{\tau})^n$ are invariant of the induced transformation on $\boldsymbol{\tau}$ due to the orthogonal transformation of coordinates.

First three of invariants of $\boldsymbol{\tau}$, i.e., $\text{tr}(\boldsymbol{\tau})$, $\text{tr}(\boldsymbol{\tau})^2$, $\text{tr}(\boldsymbol{\tau})^3$ are denoted by

$$i_\tau = \text{tr}\,\boldsymbol{\tau} \quad ; \quad ii_\tau = \text{tr}\,\boldsymbol{\tau}^2 \quad ; \quad iii_\tau = \text{tr}\,\boldsymbol{\tau}^3 \tag{2.310}$$

Expansions for the three invariants in terms of the components of $\boldsymbol{\tau}$ can be easily obtained by expanding right-hand side of (2.310) and are given

$$i_\tau = \tau_{ii} \tag{2.311}$$

$$ii_\tau = \tau_{ij}\tau_{ji} \tag{2.312}$$

$$iii_\tau = \tau_{ij}\tau_{jk}\tau_{ki} \tag{2.313}$$

Remark

We note that since

$$[\tau^y] = [Q][\tau^x][Q]^T \tag{2.314}$$

taking the determinant of both sides in (2.314), we have

$$\det[\tau^y] = \det[Q]\det[\tau^x]\det[Q]^T \tag{2.315}$$

but

$$\det[Q] = \det[Q]^T = \pm 1 \tag{2.316}$$

hence

$$\det[\tau^y] = \det[\tau^x] \tag{2.317}$$

Thus, if $\boldsymbol{\tau}$ is a tensor of rank two, then its determinant is invariant of the transformation induced by the orthogonal coordinate transformation.

2.11.4.5 Tensors of rank two: Principal invariants

The invariants of a tensor of rank two (assumed unsymmetric for generality) can also be determined using an alternate approach. If $\boldsymbol{\tau}$ is a tensor of rank two and $\boldsymbol{\phi}$ is an eigenvector of $\boldsymbol{\tau}$, then there exists a scalar λ called eigenvalue of $\boldsymbol{\tau}$ such that the following holds.

$$[\tau]\{\phi\} - \lambda\{\phi\} = 0 \quad \text{or} \quad [\tau]\{\phi\} - \lambda[I]\{\phi\} = 0 \tag{2.318}$$

Equation (2.318) represents a system of homogeneous algebraic equations in $\{\phi\}$. Hence the necessary and sufficient conditions for λ to be unique eigenvalue of $\boldsymbol{\tau}$ is that

$$\det\left([\tau - \lambda I]\right) = 0 \tag{2.319}$$

Using Laplace expansion we obtain the following from (2.319):

$$p(\lambda) = \lambda^3 - I_\tau\lambda^2 + II_\tau\lambda - III_\tau = 0 \tag{2.320}$$

Equation (2.320) is known as the characteristic equation of the second order tensor $\boldsymbol{\tau}$. The expressions for I_τ, II_τ and III_τ in terms of the components of the tensor $\boldsymbol{\tau}$ can be easily obtained. The eigenvalues of a tensor of rank two are invariant of the induced transformation due to orthogonal coordinate transformation, that is the eigenvalues λ of tensor $\boldsymbol{\tau}$ do not change if its frame of reference is subjected to orthogonal coordinate transformation. Proof is straightforward (given below). Let

$$\{\Phi\} = [Q]\{\psi\} \tag{2.321}$$

represent the change of basis (orthogonal coordinate transformation). Substituting (2.321) in (2.318) and premultiply (2.321) by $[Q]^T$.

$$[Q]^T\left[[\tau] - \lambda[I]\right][Q]\{\psi\} = 0 \tag{2.322}$$

Equation (2.322) represents new eigenvalue, hence for unique eigenvalues λ

$$\det\left([Q]^T\,[[\tau]-\lambda[I]]\,[Q]\right)=0 \tag{2.323}$$

Since $[Q]$ is orthogonal, (2.323) reduces to

$$\det\left([\tau]-\lambda[I]\right)=0 \tag{2.324}$$

Equation (2.324) yields the same characteristic equation as (2.320) confirming that eigenvalues λ are invariant of orthogonal transformation. This is only possible if I_τ, $I\!I_\tau$ and $I\!I\!I_\tau$ are invariant of the orthogonal transformation. Thus, I_τ, $I\!I_\tau$ and $I\!I\!I_\tau$ are invariants of the second order tensor $\boldsymbol{\tau}$ and are known as the principal invariants of $\boldsymbol{\tau}$. Explicit expressions of the principal invariants are given by

$$I_\tau=\tau_{11}+\tau_{22}+\tau_{33}=\operatorname{tr}\boldsymbol{\tau} \tag{2.325}$$

$$\begin{aligned} I\!I_\tau &= \begin{vmatrix}\tau_{11} & \tau_{12}\\ \tau_{21} & \tau_{22}\end{vmatrix}+\begin{vmatrix}\tau_{11} & \tau_{13}\\ \tau_{31} & \tau_{33}\end{vmatrix}+\begin{vmatrix}\tau_{11} & \tau_{23}\\ \tau_{32} & \tau_{33}\end{vmatrix}\\ &=\tau_{11}\tau_{22}+\tau_{22}\tau_{33}+\tau_{33}\tau_{11}-\tau_{12}\tau_{21}-\tau_{13}\tau_{31}-\tau_{23}\tau_{32}\end{aligned} \tag{2.326}$$

$$\begin{aligned} I\!I\!I_\tau &= \det\boldsymbol{\tau}=\tau_{11}\tau_{22}\tau_{33}-\tau_{11}\tau_{23}\tau_{32}-\tau_{22}\tau_{13}\tau_{31}-\tau_{33}\tau_{12}\tau_{21}+\tau_{12}\tau_{23}\tau_{31}+\tau_{13}\tau_{32}\tau_{21}\\ &=\epsilon_{ijk}\tau_{1i}\tau_{2j}\tau_{3k}=\epsilon_{ijk}\tau_{i1}\tau_{j2}\tau_{k3}\end{aligned} \tag{2.327}$$

Remarks

(1) We note that $p(\lambda)=0$, the characteristic polynomial, is a polynomial of degree three in λ. Thus, it has three roots λ_i ; $i=1,2,3$ that are eigenvalues of $\boldsymbol{\tau}$. Corresponding to each λ_i we have $\{\phi\}_i$ called eigenvector, thus, have three eigenpairs $(\lambda_i,\{\phi_i\})$.

(2) When $[\tau]$ is not symmetric, the eigenvalues can repeat and can be complex, even if components τ_{ij} of $[\tau]$ are real. However, if $[\tau]$ is real, any complex eigenvalues must occur in complex-conjugate pairs implying one real eigenvalue and two complex eigenvalues.

(3) The eigenvalues of $[\tau]$ are called the spectrum of $[\tau]$. We note that $\operatorname{tr}[\tau]=\operatorname{tr}[\Lambda]$ where $[\Lambda]$ is a diagonal matrix of λ_i and $\det[\tau]=\lambda_1\lambda_2\lambda_3$.

(4) Eigenvectors of $[\tau]$ form an orthogonal basis. When $[\tau]$ is not symmetric, two of the three $\{\phi\}_i$ can be complex, hence in this case, the eigenvectors do form an orthogonal basis, but the basis is not real.

(5) Regardless of whether $[\tau]$ is unsymmetric or symmetric, the invariants of i_τ, ii_τ, iii_τ and I_τ, $I\!I_\tau$, $I\!I\!I_\tau$ are always real.

(6) Since the invariants i_τ, ii_τ, iii_τ and I_τ, $I\!I_\tau$, $I\!I\!I_\tau$ ((2.325)–(2.327)) are the invariants of same tensor $\boldsymbol{\tau}$, and they are obviously related. By using the expression for the invariants in (2.310) and the principal invariants I_τ, $I\!I_\tau$ and $I\!I\!I_\tau$ from the characteristic equation (2.320) of tensor $\boldsymbol{\tau}$, we can show that the following relations hold.

$$
\begin{aligned}
I_\tau &= \operatorname{tr}\boldsymbol{\tau} = i_\tau \\
II_\tau &= \frac{1}{2}\left((\operatorname{tr}\boldsymbol{\tau})^2 - \operatorname{tr}\boldsymbol{\tau}^2\right) = \frac{1}{2}\left((i_\tau)^2 - ii_\tau\right) \\
III_\tau &= \det\boldsymbol{\tau} = \epsilon_{ijk}\tau_{1i}\tau_{2j}\tau_{3k} = \epsilon_{ijk}\tau_{i1}\tau_{j2}\tau_{k3} \\
&= \frac{1}{3}\left(\operatorname{tr}\boldsymbol{\tau}^3 + \frac{1}{2}(\operatorname{tr}\boldsymbol{\tau})^3 - \frac{3}{2}\operatorname{tr}\boldsymbol{\tau}\operatorname{tr}\boldsymbol{\tau}^2\right) \\
&= \frac{1}{3}\left(iii_\tau + \frac{1}{2}(i_\tau)^3 - \frac{3}{2}i_\tau ii_\tau\right)
\end{aligned}
\tag{2.328}
$$

Alternatively

$$
\begin{aligned}
i_\tau &= I_\tau \\
ii_\tau &= I_\tau^2 - 2II_\tau \\
iii_\tau &= 3III_\tau + I_\tau^3 - 3I_\tau II_\tau
\end{aligned}
\tag{2.329}
$$

(7) From the characteristic equation we note that a tensor of rank two can only have three invariants and no more. Hence, i_τ, ii_τ and iii_τ are the invariants of $\boldsymbol{\tau}$ even though traces $\boldsymbol{\tau}^4$, $\boldsymbol{\tau}^5, \dots$ also do not change under orthogonal coordinate transformation.

(8) The relations (2.328) and (2.329) are important in the development of constitutive theory for symmetric and skew-symmetric constitutive tensors. For example, if a scalar $\phi = \phi(i_\tau, ii_\tau, iii_\tau)$, then using (2.329) we can also express $\phi = \phi(I_\tau, II_\tau, III_\tau)$ or vice-versa. Suppose ϕ is a scalar such that $\phi = \phi(I_\tau, II_\tau, III_\tau)$, then $\frac{\partial\phi}{\partial I_\tau}, \frac{\partial\phi}{\partial II_\tau}, \frac{\partial\phi}{\partial III_\tau}$ are known as functions of I_τ, II_τ and III_τ. Now using (2.328) and (2.329), $\frac{\partial\phi}{\partial i_\tau}, \frac{\partial\phi}{\partial ii_\tau}, \frac{\partial\phi}{\partial iii_\tau}$ can be determined as functions of i_τ, ii_τ and iii_τ.

2.11.5 Hamilton-Cayley theorem

The Hamilton-Cayley theorem states that a matrix $[\tau]$ satisfies its own characteristic equation. That is in the characteristic equation (2.320) of $[\tau]$ we can replace λ with $[\tau]$.

$$
[\tau]^3 - I_\tau[\tau]^2 + II_\tau[\tau] - III_\tau[I] = 0 \tag{2.330}
$$

Proof is rather trivial. Since the characteristic equation (2.320) is derived using eigenvalue problem (2.318), we can substitute of $\lambda = [\tau]$ in (2.318), then check if this choice is admissible

$$
\left[[\tau] - [\tau][I]\right]\{\Phi\} = 0 \tag{2.331}
$$

$$
\text{or} \quad \left[[\tau] - [\tau]\right]\{\Phi\} = 0 \tag{2.332}
$$

Thus, $\lambda = [\tau]$ is an admissible choice in (2.320). The Hamilton-Cayley theorem is useful in the derivation of the constitutive theories for symmetric and skew-symmetric constitutive tensors. Using (2.330), we can obtain an expression for $[\tau]^{-1}$ provided $III_\tau = \det[\tau] \neq 0$. We can premultiply (2.330) by $[\tau]^{-1}$ and then solve for $[\tau]^{-1}$.

$$[\tau]^{-1} = \frac{[\tau]^2 - I_\tau[\tau] + II_\tau[I]}{III_\tau} \tag{2.333}$$

If $[\tau]$ is nonsingular, then $III_\tau = \det[\tau] \neq 0$ and hence (2.333) is valid. If we premultiply (2.330) by $[\tau]^{-2}$ then we can solve for $[\tau]$ in terms of $[\tau]^{-1}$ and $[\tau]^{-2}$.

$$[\tau] = I_\tau[I] - II_\tau[\tau]^{-1} + III_\tau[\tau]^{-2} \tag{2.334}$$

If we take the trace of (2.333) and use II_τ in (2.328), then we obtain

$$II_\tau = III_\tau \operatorname{tr}([\tau]^{-1}) \tag{2.335}$$

Equations (2.330)–(2.335) are useful in the development of the constitutive theories considered in the later chapters.

2.11.6 Differential calculus of tensors

In this section, we present some helpful details of the differential calculus of tensors in orthogonal Cartesian frame (say x-frame).

2.11.6.1 Def: The differential operator or the gradient operator $\vec{\nabla}$ or $\boldsymbol{\nabla}$

In the orthogonal Cartesian coordinate system (x-frame) the differential operator $\boldsymbol{\nabla}$ is defined as

$$\boldsymbol{\nabla} = \boldsymbol{e}_i \frac{\partial}{\partial x_i} \tag{2.336}$$

2.11.6.2 Def: Gradient of a scalar field, i.e., a tensor of rank zero

If $\alpha = \alpha(x_i)$ is a scalar field, then the operator $\boldsymbol{\nabla}$ acting on α produces

$$\boldsymbol{\nabla}\alpha = \boldsymbol{e}_i \frac{\partial \alpha}{\partial x_i} \tag{2.337}$$

We note the following properties:

$$\begin{aligned} &\boldsymbol{\nabla}\alpha \neq \alpha\boldsymbol{\nabla} && ; \quad \text{not commutative} \\ &(\boldsymbol{\nabla}\alpha)\beta \neq \boldsymbol{\nabla}(\alpha\beta) && ; \quad \text{not associative} \\ &\boldsymbol{\nabla}(\alpha+\beta) = \boldsymbol{\nabla}\alpha + \boldsymbol{\nabla}\beta && ; \quad \text{distributive} \end{aligned} \tag{2.338}$$

2.11.6.3 Def: Divergence of tensor of rank one

If $\boldsymbol{\phi}(x_i)$ is a tensor of rank one, then the scalar product of $\boldsymbol{\nabla}$ and $\boldsymbol{\phi}$ is called divergence of the tensor of rank one $\boldsymbol{\phi}$ and is defined by

$$\begin{aligned} \boldsymbol{\nabla}\cdot\boldsymbol{\phi} &= (\boldsymbol{e}_i \frac{\partial}{\partial x_i}) \cdot (\boldsymbol{e}_j \phi_j) \\ &= \boldsymbol{e}_i \cdot \boldsymbol{e}_j \frac{\partial \phi_j}{\partial x_i} \\ &= \underbrace{\delta_{ij}}_{j=i} \frac{\partial \phi_j}{\partial x_i} = \frac{\partial \phi_i}{\partial x_i} \end{aligned}$$

$$\therefore \quad \boldsymbol{\nabla} \cdot \boldsymbol{\phi} = \frac{\partial \phi_i}{\partial x_i} = \text{div}\boldsymbol{\phi} \tag{2.339}$$

We note the following properties:

$$\begin{aligned}
&\boldsymbol{\nabla} \cdot \boldsymbol{\phi} \neq \boldsymbol{\phi} \cdot \boldsymbol{\nabla} &&; \quad \text{not commutative} \\
&\boldsymbol{\nabla} \cdot (\alpha \boldsymbol{\phi}) \neq \boldsymbol{\nabla}\alpha \cdot \boldsymbol{\phi} &&; \quad \text{not associative; } \alpha \text{ is a scalar} \\
&\boldsymbol{\nabla} \cdot (\boldsymbol{\phi} + \boldsymbol{\psi}) = \boldsymbol{\nabla} \cdot \boldsymbol{\phi} + \boldsymbol{\nabla} \cdot \boldsymbol{\psi} &&; \quad \text{distributive}
\end{aligned} \tag{2.340}$$

2.11.6.4 Def: Cross product of the operator $\boldsymbol{\nabla}$ with tensor of rank one (curl of a tensor of rank one)

If $\boldsymbol{\phi}$ is a tensor of rank one, then the curl of $\boldsymbol{\phi}$ is defined as

$$\begin{aligned}
\boldsymbol{\nabla} \times \boldsymbol{\phi} &= (\boldsymbol{e}_i \frac{\partial}{\partial x_i}) \times (\boldsymbol{e}_j \phi_j) \\
&= \boldsymbol{e}_i \times \boldsymbol{e}_j \frac{\partial \phi_j}{\partial x_i} \\
&= \epsilon_{ijk} \, \boldsymbol{e}_k \frac{\partial \phi_j}{\partial x_i}
\end{aligned}$$

$$\therefore \quad \boldsymbol{\nabla} \times \boldsymbol{\phi} = \text{curl}\boldsymbol{\phi} = \epsilon_{ijk} \boldsymbol{e}_k \frac{\partial \phi_j}{\partial x_i} \tag{2.341}$$

We also note the following:

$$\boldsymbol{\nabla} \times \boldsymbol{\phi} = \begin{vmatrix} \boldsymbol{e}_1 & \boldsymbol{e}_2 & \boldsymbol{e}_3 \\ \dfrac{\partial}{\partial x_1} & \dfrac{\partial}{\partial x_2} & \dfrac{\partial}{\partial x_3} \\ \phi_1 & \phi_2 & \phi_3 \end{vmatrix} \tag{2.342}$$

$$\therefore \quad \boldsymbol{\nabla} \times \boldsymbol{\phi} = \boldsymbol{e}_1 \Big(\frac{\partial \phi_3}{\partial x_2} - \frac{\partial \phi_2}{\partial x_3} \Big) + \boldsymbol{e}_2 \Big(\frac{\partial \phi_1}{\partial x_3} - \frac{\partial \phi_3}{\partial x_1} \Big) + \boldsymbol{e}_3 \Big(\frac{\partial \phi_2}{\partial x_1} - \frac{\partial \phi_1}{\partial x_2} \Big) \tag{2.343}$$

2.11.6.5 Def: Gradient of a tensor of rank one

Gradient of $\boldsymbol{\phi}$, a tensor of rank one, is the dyadic product of the operator $\boldsymbol{\nabla}$ and $\boldsymbol{\phi}$ and is denoted by $\boldsymbol{\nabla}\boldsymbol{\phi}$ or $\boldsymbol{\nabla} \otimes \boldsymbol{\phi}$.

$$\begin{aligned}
\boldsymbol{\nabla}\boldsymbol{\phi} = \boldsymbol{\nabla} \otimes \boldsymbol{\phi} &= \boldsymbol{e}_i \frac{\partial}{\partial x_i} \otimes \boldsymbol{e}_j \phi_j \\
&= \boldsymbol{e}_i \otimes \boldsymbol{e}_j \frac{\partial \phi_j}{\partial x_i}
\end{aligned}$$

$$\therefore \quad \boldsymbol{\nabla}\boldsymbol{\phi} = \boldsymbol{\nabla} \otimes \boldsymbol{\phi} = \boldsymbol{e}_i \otimes \boldsymbol{e}_j \frac{\partial \phi_j}{\partial x_i} = \boldsymbol{e}_i \boldsymbol{e}_j \frac{\partial \phi_j}{\partial x_i} \tag{2.344}$$

$\boldsymbol{\nabla}\boldsymbol{\phi}$ is a tensor of rank two. We note that $\boldsymbol{\nabla}\boldsymbol{\phi} \neq \boldsymbol{\phi}\boldsymbol{\nabla}$ and $(\boldsymbol{\nabla}\boldsymbol{\phi})^T \neq \boldsymbol{\phi}\boldsymbol{\nabla}$.

2.11.6.6 Def: Divergence of a tensor of rank two

If $\boldsymbol{\tau}(x_i)$ is a tensor of rank two, then its divergence is denoted by $\boldsymbol{\nabla} \cdot \boldsymbol{\tau}$ and is defined as

$$\begin{aligned}
\boldsymbol{\nabla} \cdot \boldsymbol{\tau} &= \left(\boldsymbol{e}_i \frac{\partial}{\partial x_i}\right) \cdot (\boldsymbol{e}_j \boldsymbol{e}_k \tau_{jk}) \\
&= \boldsymbol{e}_i \cdot \boldsymbol{e}_j \boldsymbol{e}_k \frac{\partial \tau_{jk}}{\partial x_i} \\
&= \underbrace{\delta_{ij}}_{j=i} \boldsymbol{e}_k \frac{\partial \tau_{jk}}{\partial x_i} = \boldsymbol{e}_k \frac{\partial \tau_{ik}}{\partial x_i}
\end{aligned}$$

$$\therefore \quad \boldsymbol{\nabla} \cdot \boldsymbol{\tau} = \boldsymbol{e}_k \frac{\partial \tau_{ik}}{\partial x_i} = \text{div}\boldsymbol{\tau} \tag{2.345}$$

Thus, divergence of a tensor of rank two is a tensor of rank one.

2.11.6.7 Def: Laplacian of a scalar field, i.e., a tensor of rank zero

Laplacian of a scalar field is defined as the divergence of the gradient of the scalar field. If $\alpha(x_i)$ is a scalar field then

$$\begin{aligned}
\boldsymbol{\nabla} \cdot (\boldsymbol{\nabla} \alpha) &= \left(\boldsymbol{e}_i \frac{\partial}{\partial x_i}\right) \cdot \left(\boldsymbol{e}_j \frac{\partial \alpha}{\partial x_j}\right) \\
&= \boldsymbol{e}_i \cdot \boldsymbol{e}_j \frac{\partial}{\partial x_i}\left(\frac{\partial \alpha}{\partial x_j}\right) \\
&= \underbrace{\delta_{ij}}_{j=i} \frac{\partial}{\partial x_i}\left(\frac{\partial \alpha}{\partial x_j}\right) \\
&= \frac{\partial}{\partial x_i}\left(\frac{\partial \alpha}{\partial x_i}\right) \quad ; \quad \left(\sum_{i=1}^{3} \text{not needed}\right)
\end{aligned}$$

$$\therefore \quad \boldsymbol{\nabla} \cdot (\boldsymbol{\nabla} \alpha) = \sum_{i=1}^{3} \frac{\partial^2 \alpha}{\partial x_i^2} \quad ; \quad \left(\sum_{i=1}^{3} \text{needed}\right) \tag{2.346}$$

If we define the Laplacian operator $\boldsymbol{\Delta}$ by

$$\boldsymbol{\nabla} \cdot \boldsymbol{\nabla} = \boldsymbol{\nabla}^2 = \boldsymbol{\Delta} = \frac{\partial^2}{\partial {x_1}^2} + \frac{\partial^2}{\partial {x_2}^2} + \frac{\partial^2}{\partial {x_3}^2} \tag{2.347}$$

then

$$\boldsymbol{\nabla} \cdot \boldsymbol{\nabla} \alpha = \boldsymbol{\nabla}^2 \alpha = \boldsymbol{\Delta} \alpha = \sum_{i=1}^{3} \frac{\partial^2 \alpha}{\partial {x_i}^2} \tag{2.348}$$

Thus, Laplacian of a scalar field is a scalar or tensor of rank zero.

2.11.6.8 Def: Laplacian of a tensor of rank one

Laplacian of a tensor of rank one is defined as the divergence of the gradient of a tensor of rank one. If $\boldsymbol{\phi}(x_i)$ is a tensor of rank one then

$$\boldsymbol{\nabla}\cdot(\boldsymbol{\nabla}\boldsymbol{\phi}) = \left(\boldsymbol{e}_i\frac{\partial}{\partial x_i}\right)\cdot\left(\boldsymbol{e}_j\frac{\partial}{\partial x_j}(\boldsymbol{e}_k\phi_k)\right) = \left(\boldsymbol{e}_i\frac{\partial}{\partial x_i}\right)\cdot\left(\boldsymbol{e}_j\boldsymbol{e}_k\frac{\partial\phi_k}{\partial x_j}\right)$$

$$= (\boldsymbol{e}_i\cdot\boldsymbol{e}_j)\boldsymbol{e}_k\frac{\partial}{\partial x_i}\left(\frac{\partial\phi_k}{\partial x_j}\right)$$

$$= \underbrace{\delta_{ij}}_{j=i}\ \boldsymbol{e}_k\frac{\partial}{\partial x_i}\left(\frac{\partial\phi_k}{\partial x_j}\right)$$

$$= \boldsymbol{e}_k\frac{\partial}{\partial x_i}\left(\frac{\partial\phi_k}{\partial x_i}\right) \quad ; \quad \left(\sum_{i=1}^{3}\text{ not needed}\right)$$

$$\therefore\quad \boldsymbol{\nabla}\cdot(\boldsymbol{\nabla}\boldsymbol{\phi}) = \sum_{i=1}^{3}\boldsymbol{e}_k\frac{\partial^2\phi_k}{\partial x_i^2} \quad ; \quad \left(\sum_{i=1}^{3}\text{ needed}\right) \tag{2.349}$$

Using the Laplacian notation defined in (2.347), we can write

$$\boldsymbol{\nabla}\cdot(\boldsymbol{\nabla}\boldsymbol{\phi}) = \boldsymbol{\nabla}^2\boldsymbol{\phi} = \boldsymbol{\Delta}\boldsymbol{\phi} = \sum_{i=1}^{3}\boldsymbol{e}_k\frac{\partial^2\phi_k}{\partial {x_i}^2} \tag{2.350}$$

Thus, Laplacian of a tensor of rank one is a tensor of rank one.

2.11.6.9 Def: Gradient theorem

Consider a volume V bounded by a closed surface A and let $\boldsymbol{n}$ be the exterior unit normal to the surface A. Then for a scalar function $\phi(\boldsymbol{x},t)$ the following holds:

$$\int_V \boldsymbol{\nabla}\phi\, dV = \oint_A \boldsymbol{n}\,\phi\, dA \tag{2.351}$$

We can use integration by parts in the integral over V to transfer one order of differentiation with respect to $\boldsymbol{x}$ from ϕ to $\boldsymbol{e}_i$ and since the derivative of $\boldsymbol{e}_i$ with respect to $\boldsymbol{x}$ is zero we obtain the following.

$$\int_V \left(\boldsymbol{e}_1\frac{\partial\phi}{\partial x_1} + \boldsymbol{e}_2\frac{\partial\phi}{\partial x_2} + \boldsymbol{e}_3\frac{\partial\phi}{\partial x_3}\right)dV = \oint_A (n_{x_1}\boldsymbol{e}_1 + n_{x_2}\boldsymbol{e}_2 + n_{x_3}\boldsymbol{e}_3)\phi\, dA \tag{2.352}$$

In (2.352), the equality of the vectors is possible only if the following hold:

$$\int_V \frac{\partial\phi}{\partial x_1}\, dV = \oint_A n_{x_1}\phi\, dA\ ;\ \int_V \frac{\partial\phi}{\partial x_2}\, dV = \oint_A n_{x_2}\phi\, dA\ ;\ \int_V \frac{\partial\phi}{\partial x_3}\, dV = \oint_A n_{x_3}\phi\, dA \tag{2.353}$$

2.11.6.10 Def: Divergence of a tensor of rank one: Gauss's divergence theorem

If $\boldsymbol{\psi}$ is a tensor of rank one, then the following holds:

$$\int_V \boldsymbol{\nabla}\cdot\boldsymbol{\psi}\, dV = \oint_A \boldsymbol{n}\cdot\boldsymbol{\psi}\, dA = \oint_A \boldsymbol{\psi}\cdot\boldsymbol{n}\, dA = \oint_A \boldsymbol{\psi}\cdot d\boldsymbol{A} \tag{2.354}$$

Def: Divergence theorem for non-symmetric tensors of rank two: Generalized Gauss's divergence theorem

Consider a mixed tensor $\boldsymbol{\tau}$ (including tensors of rank two or higher). Then the following holds (Generalized Gauss's divergence theorem):

$$\int_V \boldsymbol{\nabla} \cdot \boldsymbol{\tau}^T \, dV = \oint_A \boldsymbol{n} \cdot \boldsymbol{\tau}^T dA \tag{2.355}$$

Now, consider inner products of a tensor of rank one $\boldsymbol{n}$ with a non-symmetric tensor of rank two $\boldsymbol{\tau}$.

$$\boldsymbol{n} \cdot \boldsymbol{\tau} = \boldsymbol{\tau}^T \cdot \boldsymbol{n} \tag{2.356}$$

$$\therefore \quad \boldsymbol{n} \cdot \boldsymbol{\tau}^T = (\boldsymbol{\tau}^T)^T \cdot \boldsymbol{n} = \boldsymbol{\tau} \cdot \boldsymbol{n} \tag{2.357}$$

Hence, when $\boldsymbol{\tau}$ is a non-symmetric tensor of rank two, then we obtain the following two expressions for the Gauss's divergence theorem for non-symmetric tensors of rank two:

$$\oint_A \boldsymbol{\tau} \cdot \boldsymbol{n} \, dA = \oint_A \boldsymbol{\tau} \cdot d\boldsymbol{A} = \int_V \boldsymbol{\nabla} \cdot \boldsymbol{\tau}^T \, dV \tag{2.358}$$

$$\therefore \quad \oint_A \boldsymbol{\tau}^T \cdot \boldsymbol{n} \, dA = \oint_A \boldsymbol{\tau}^T \cdot d\boldsymbol{A} = \int_V \boldsymbol{\nabla} \cdot \boldsymbol{\tau} \, dV \tag{2.359}$$

2.11.6.11 Def: Divergence theorem for symmetric tensors of rank two

In this case $\boldsymbol{\tau} = \boldsymbol{\tau}^T$, hence $\boldsymbol{\tau}^T$ can be replaced by $\boldsymbol{\tau}$ in (2.358).

$$\oint_A \boldsymbol{\tau} \cdot \boldsymbol{n} \, dA = \oint_A \boldsymbol{\tau} \cdot d\boldsymbol{A} = \int_V \boldsymbol{\nabla} \cdot \boldsymbol{\tau} \, dV \tag{2.360}$$

2.11.6.12 Def: Integration by parts: Transfer of differentiation in the integrand

Let $\bar{\Omega}$ be a closed domain with boundary Γ such that $\bar{\Omega} = \Omega \cup \Gamma$ in which Ω is an open domain in $\mathbb{R}^1$ or $\mathbb{R}^2$ or $\mathbb{R}^3$. Let ϕ and ψ be functions of $\boldsymbol{x}$ and let $\boldsymbol{n}$ be the unit exterior normal to the boundary Γ. Then

(a) ***Line integrals in*** $\mathbb{R}^1$

$$\int_{\bar{\Omega}} \psi \frac{d\phi}{dx_1} \, dx_1 = \int_a^b \psi \frac{d\phi}{dx_1} \, dx_1 = -\int_a^b \frac{d\psi}{dx_1} \phi \, dx_1 + (\psi \, \phi)\Big|_a^b \tag{2.361}$$

$$\int_{\bar{\Omega}} \psi \frac{d^2\phi}{dx_1^2} \, dx_1 = \int_a^b \psi \frac{d^2\phi}{dx_1^2} \, dx_1 = -\int_a^b \frac{d\psi}{dx_1} \frac{d\phi}{dx_1} \, dx_1 + \left(\psi \, \frac{d\phi}{dx_1}\right)\Big|_a^b \tag{2.362}$$

(b) *Area integrals in* $\mathbb{R}^2$

$$\int_{\bar{\Omega}} \left(\psi \frac{\partial \phi}{\partial x_1} + \psi \frac{\partial \phi}{\partial x_2} \right) dx_1\, dx_2 = - \int_{\Omega} \left(\frac{\partial \psi}{\partial x_1} \phi + \frac{\partial \psi}{\partial x_2} \phi \right) dx_1\, dx_2 + \oint_{\Gamma} \left(\psi \phi\, n_{x_1} + \psi \phi\, n_{x_2} \right) d\Gamma \tag{2.363}$$

$$\int_{\bar{\Omega}} \left(\psi \frac{\partial^2 \phi}{\partial x_1^2} + \psi \frac{\partial^2 \phi}{\partial x_2^2} \right) dx_1\, dx_2 = - \int_{\Omega} \left(\frac{\partial \psi}{\partial x_1} \frac{\partial \phi}{\partial x_1} + \frac{\partial \psi}{\partial x_2} \frac{\partial \phi}{\partial x_2} \right) dx_1\, dx_2 + \oint_{\Gamma} \left(\psi \frac{\partial \phi}{\partial x_1} n_{x_1} + \psi \frac{\partial \phi}{\partial x_2} n_{x_2} \right) d\Gamma \tag{2.364}$$

(c) *Volume integrals in* $\mathbb{R}^3$

$$\int_{\bar{\Omega}} \left(\psi \frac{\partial \phi}{\partial x_1} + \psi \frac{\partial \phi}{\partial x_2} + \psi \frac{\partial \phi}{\partial x_3} \right) dx_1 dx_2 dx_3 = - \int_{\Omega} \left(\frac{\partial \psi}{\partial x_1} \phi + \frac{\partial \psi}{\partial x_2} \phi + \frac{\partial \psi}{\partial x_3} \phi \right) dx_1 dx_2 dx_3 + \oint_{\Gamma} \left(\psi \phi\, n_{x_1} + \psi \phi\, n_{x_2} + \psi \phi\, n_{x_3} \right) d\Gamma \tag{2.365}$$

$$\int_{\bar{\Omega}} \left(\psi \frac{\partial^2 \phi}{\partial x_1^2} + \psi \frac{\partial^2 \phi}{\partial x_2^2} + \psi \frac{\partial^2 \phi}{\partial x_3^2} \right) dx_1 dx_2 dx_3 = - \int_{\Omega} \left(\frac{\partial \psi}{\partial x_1} \frac{\partial \phi}{\partial x_1} + \frac{\partial \psi}{\partial x_2} \frac{\partial \phi}{\partial x_2} + \frac{\partial \psi}{\partial x_3} \frac{\partial \psi}{\partial x_3} \right) dx_1 dx_2 dx_3 + \oint_{\Gamma} \left(\psi \frac{\partial \phi}{\partial x_1} n_{x_1} + \psi \frac{\partial \phi}{\partial x_2} n_{x_2} + \psi \frac{\partial \phi}{\partial x_3} n_{x_3} \right) d\Gamma \tag{2.366}$$

2.12 Some useful relations

If $\boldsymbol{\sigma}$ is a non-symmetric tensor of rank two and $\boldsymbol{n}$ is a tensor of rank one then

(a) $\boldsymbol{\sigma} \cdot \boldsymbol{n}$ can be written as $[\sigma]\{n\}$.

$$\begin{aligned} \boldsymbol{\sigma} \cdot \boldsymbol{n} &= \boldsymbol{e}_i \boldsymbol{e}_j \sigma_{ij} \cdot \boldsymbol{e}_k n_k \\ &= \boldsymbol{e}_i (\boldsymbol{e}_j \cdot \boldsymbol{e}_k) \sigma_{ij}\, n_k \\ &= \boldsymbol{e}_i\, \delta_{jk}\, \sigma_{ij}\, n_k \\ &= \boldsymbol{e}_i\, \sigma_{ij}\, n_j \end{aligned} \tag{2.367}$$

By expanding $[\sigma]\{n\}$ we can show that the components of $[\sigma]\{n\}$ are the same as those in (2.367).

(b) $\boldsymbol{\sigma}^T \cdot \boldsymbol{n}$ can be written as $[\sigma]^T\{n\}$.

$$\begin{aligned}\boldsymbol{\sigma}^T \cdot \boldsymbol{n} &= \boldsymbol{e}_i\boldsymbol{e}_j\sigma_{ji} \cdot \boldsymbol{e}_k n_k \\ &= \boldsymbol{e}_i(\boldsymbol{e}_j \cdot \boldsymbol{e}_k)\sigma_{ji}\, n_k \\ &= \boldsymbol{e}_i\, \delta_{jk}\, \sigma_{ji}\, n_k \\ &= \boldsymbol{e}_i\, \sigma_{ji}\, n_j \end{aligned} \tag{2.368}$$

Components of the vector in (2.368) are the same as those in $[\sigma]^T\{n\}$.

(c) $\boldsymbol{n} \cdot \boldsymbol{\sigma}^T = \boldsymbol{\sigma} \cdot \boldsymbol{n}$ can be written as $[\sigma]\{n\}$.

$$\begin{aligned}\boldsymbol{n} \cdot \boldsymbol{\sigma}^T &= \boldsymbol{e}_k n_k \cdot \boldsymbol{e}_j\boldsymbol{e}_i\sigma_{ij} \\ &= (\boldsymbol{e}_k \cdot \boldsymbol{e}_j)\boldsymbol{e}_i\sigma_{ij}\, n_k \\ &= \boldsymbol{e}_i\, \delta_{kj}\, \sigma_{ij}\, n_k \\ &= \boldsymbol{e}_i\, \sigma_{ij}\, n_j \end{aligned} \tag{2.369}$$

We note that (2.369) is the same as (2.367).

(d) $\boldsymbol{n} \cdot \boldsymbol{\sigma} = \boldsymbol{\sigma}^T \cdot \boldsymbol{n}$ can be written as $[\sigma]^T\{n\}$.

$$\begin{aligned}\boldsymbol{n} \cdot \boldsymbol{\sigma} &= \boldsymbol{e}_k n_k \cdot \boldsymbol{e}_j\boldsymbol{e}_i\sigma_{ji} \\ &= (\boldsymbol{e}_k \cdot \boldsymbol{e}_j)\boldsymbol{e}_i\sigma_{ji}\, n_k \\ &= \boldsymbol{e}_i\, \delta_{kj}\, \sigma_{ji}\, n_k \\ &= \boldsymbol{e}_i\, \sigma_{ji}\, n_j \end{aligned} \tag{2.370}$$

We note that (2.370) is the same as (2.368). *The relations* (2.367)–(2.370) *are helpful in deriving balance laws in Chapters 6 and 7.*

2.13 Summary

In this chapter, preliminary material related to Einstein's notations, matrix and vector notations and indexed notations as well as various basic operations using these notations have been presented. Orthogonal and curvilinear coordinate frames are introduced. Reference coordinate frame, reference frame transformation using Galilean and non-Galilean transformations, concept of induced transformation, definition of a tensor, definition of covariant and contravariant tensors of rank zero, one, two and higher, definitions of tensors of various ranks in Cartesian frame, transformation of tensors of various ranks are defined in x-frame due to rigid rotation of x-frame, invariants of tensors of various ranks based on trace as well as principal invariants, Hamlilton-Cayley theorem and its proof and differential calculus of tensors are presented in this chapter. The material in this chapter is self-contained and is also well suited for self-study.

Problems

In Problems 2.1 to 2.9, derive the expanded forms. Indices in each case range from 1 to 3. Simplify the resulting expressions using properties of Kronecker delta δ_{ij}.

2.1

$$\delta_{ij}a^j$$

2.2

$$\delta_{ij}x^i y^j$$

2.3

$$p_{kl}q_{lm} = \delta_{km}$$

2.4

$$Q_{ijm}y^m$$

2.5

$$\frac{\partial g_k}{\partial y_m}dy_m$$

2.6

$$\delta_{kk}$$

2.7

$$F_i = \rho\frac{\partial u_i}{\partial t} + u_k\frac{\partial u_i}{\partial x_k} + \frac{\partial p}{\partial x_i} - \frac{\partial \tau_{im}}{\partial x_m}$$

2.8

$$\frac{\partial \rho}{\partial t} + \rho\frac{\partial u_i}{\partial x_i} + u_k\frac{\partial \rho}{\partial x_k} = 0$$

2.9 What does the product $\delta_{pq}\delta_{qr}\delta_{rs}$ mean?

2.10 Let $\hat{e}_1$, $\hat{e}_2$ and $\hat{e}_3$ be the components of a vector in a new coordinate system denoted by $(\hat{\ })$, which are related to the components e_1, e_2 and e_3 of the same vector in an existing coordinate system by

$$\hat{e}_1 = \frac{1}{3}(2e_1 + 2e_2 + e_3)$$
$$\hat{e}_2 = \frac{1}{\sqrt{2}}(e_1 - e_2)$$
$$\hat{e}_3 = \frac{1}{3\sqrt{2}}(e_1 + e_2 - 4e_3)$$

Determine the coordinate transform matrix $[R]$, i.e., R_{ij} (a 3×3 matrix) relating the original coordinate system to the new coordinate system (with $\hat{\ }$).

2.11 Consider a vector $\boldsymbol{v}$ with components $v_1 = 2$, $v_2 = 1$ and $v_3 = -1$. A vector $\boldsymbol{\omega}$ has components $\omega_1 = 2$, $\omega_2 = -4$ and $\omega_3 = 2$. Evaluate

(a)

$$\boldsymbol{v} \cdot \boldsymbol{\omega}$$

(b)

$$\text{the length of} \quad \boldsymbol{v} \quad \text{and} \quad \boldsymbol{\omega}$$

(c)

$$\boldsymbol{e}_1 \times \boldsymbol{\omega}$$

(d)

$$\boldsymbol{v} \times \boldsymbol{\omega}$$

(e)

$$\boldsymbol{e}_1 \cdot \boldsymbol{v}$$

(f)

$$\boldsymbol{r} \cdot \boldsymbol{v}$$

where $\boldsymbol{r}$ is the positive vector with components r_1, r_2 and r_3.

2.12 If the components of a symmetric tensor $\boldsymbol{\sigma}$ are

$$\begin{array}{lll} \sigma_{11} = 4 & \sigma_{12} = 1 & \sigma_{13} = -1 \\ \sigma_{21} = 1 & \sigma_{22} = 3 & \sigma_{23} = 1 \\ \sigma_{31} = -1 & \sigma_{32} = 1 & \sigma_{33} = 2 \end{array}$$

and the components of a vector $\boldsymbol{v}$ are given by $v_1 = 2$, $v_2 = 1$ and $v_3 = -1$, evaluate

(a)

$$\boldsymbol{\sigma} \cdot \boldsymbol{v}$$

(b)

$$\boldsymbol{v} \cdot \boldsymbol{\sigma}$$

(c)

$$\boldsymbol{\sigma} : \boldsymbol{\sigma}$$

(d)

$$\boldsymbol{v} \cdot (\boldsymbol{\sigma} \cdot \boldsymbol{v})$$

(e)

$$\boldsymbol{v}\,\boldsymbol{v} = \boldsymbol{v} \otimes \boldsymbol{v}$$

(f)

$$\boldsymbol{\sigma} \cdot \boldsymbol{e}_1$$

(g)

$$\boldsymbol{\sigma} \cdot \boldsymbol{e}$$

where $\boldsymbol{e}$ is a unit tensor.

2.13 Evaluate the following in which $\boldsymbol{e}$ is a unit tensor:

(a)

$$(\boldsymbol{e}_1\, \boldsymbol{e}_2 \cdot \boldsymbol{e}_2) \times \boldsymbol{e}_1$$

(b)

$$\boldsymbol{e} : \boldsymbol{e}_1\, \boldsymbol{e}_2$$

(c)

$$\boldsymbol{e} : \boldsymbol{e}$$

(d)

$$\boldsymbol{e} \cdot \boldsymbol{e}$$

2.14 Consider $[\tau]$, a symmetric tensor of rank two. Derive explicit expressions for the invariants i_τ, ii_τ and iii_τ.

2.15 Consider $[\tau]$, a symmetric tensor of rank two. Derive explicit expressions for the principal invariants I_τ, II_τ and III_τ using the characteristic equation of $[\tau]$.

2.16 Prove that the following relations hold between the invariants i_τ, ii_τ, iii_τ and I_τ, II_τ, III_τ.

$$\text{(a)}\quad \begin{aligned} I_\tau &= i_\tau \\ II_\tau &= \frac{1}{2}((i_\tau)^2 - ii_\tau) \\ III_\tau &= \frac{1}{3}(iii_\tau + \frac{1}{2}(i_\tau)^3 - \frac{3}{2}i_\tau ii_\tau) \end{aligned}$$

$$\text{(b)}\quad \begin{aligned} i_\tau &= I_\tau \\ ii_\tau &= I_\tau^2 - 2II_\tau \\ iii_\tau &= 3III_\tau + I_\tau^3 - 3I_\tau II_\tau \end{aligned}$$

2.17 Show that if the coordinate transformation $T \;:\; y_i = y_i(x_j)$ is orthogonal, then the distinction between covariant and contravariant laws disappears.

2.18 Given

$$\begin{aligned} \sigma_{ij} &= \lambda \varepsilon_{kk}\delta_{ij} + 2\mu\varepsilon_{ij} \\ \beta &= \frac{1}{2}\sigma_{kl}\varepsilon_{kl} \\ \Phi &= \sigma_{mn}\sigma_{mn} \end{aligned}$$

(a) determine β as a function of λ, μ and ε_{qp};

(b) determine Φ as a function of λ, μ and ε_{qp}.

Give the most simplified forms.

2.19 Show that the tensor

$$W_{ij} = \epsilon_{ijk}\, u_k$$

is skew-symmetric. Determine the specific form of W_{ij} for this case.

2.20 Let

$$\sigma_{ij} = \alpha\, \varepsilon_{kk}\, \delta_{ij} + 2\, \beta\, \varepsilon_{ij}$$

where α and β are positive constants. Solve for ε_{ij} in terms of σ_{ij}, i.e., determine ε_{ij} explicitly in terms of σ_{ij}: (a) using Einstein notation and (b) using matrix and vector notation (Voigt's notation).

2.21 Consider an orthogonal coordinate transformation given by

$$\bar{\boldsymbol{e}}_i = R_{ij}\boldsymbol{e}_j$$

A tensor of rank two, σ_{ij}, is transformed as follows due to this transformation:

$$\bar{\sigma}_{ij} = R_{ik}R_{jl}\, \sigma_{kl}$$

Show that

(a) $\det[\bar{\sigma}] = \det[\sigma]$
(b) $\bar{\sigma}_{ij}\bar{\sigma}_{ji} = \sigma_{kl}\sigma_{lk}$
(c) $\bar{\sigma}_{ii} = \sigma_{kk}$

3 KINEMATICS OF MOTION, DEFORMATION AND THEIR MEASURES

If a volume of matter undergoes finite deformation and finite strain during evolution, then kinematics of motion, deformation, definition of strains and their measures differ significantly from infinitesimal theories, such as linear theory of elasticity for solid matter. The type of description (Lagrangian or Eulerian) and the types of measures (covariant and contravariant) play a significant role in deriving the mathematical descriptions of the kinematics of motion and deformation as well as possible definitions and measures of strains and stresses. In this chapter, we consider mathematical details, the kinematics of motion, deformation and their measures in the deforming matter experiencing finite motion and finite deformation as well as specialize these for small deformation, small strain.

3.1 Description of motion

Let a volume of matter occupy a region V of physical space with closed boundary ∂V at time t_0, the reference time at which we begin to monitor its motion and deformation. This configuration of matter can be defined with respect to another convenient reference configuration if so desired. For convenience and simplicity, we consider the configuration of the matter at time t_0 to be the reference configuration. To initiate measures of motion of the matter, we need to identify the infinitely many material points of the matter. The most convenient way to do this is to consider a *fixed Cartesian coordinate system*. The origin of this coordinate system may be located within the matter or outside of it. Let o-x_i or o-$x_1x_2x_3$ or simply *x-frame* be this coordinate system (Figure 3.1(a)). In this coordinate system, we assign each material point (an infinitesimal) of the matter unique coordinates. Let the material particle P in the reference configuration at time t_0 have coordinates x_i or (x_1, x_2, x_3). Since each material particle has unique coordinates, we now have a scheme in which all material particles of the matter are uniquely identified. If we wish, we can think of these coordinates of the material particles as labels or names that are unique for each material particle in the reference configuration at time t_0.

When the volume of matter in the reference configuration at time t_0 is disturbed, its material particles experience motion and as a consequence at current time $t > t_0$ the volume of matter (in equilibrium) occupies a different configuration $\bar{V}$ with closed boundary $\partial\bar{V}$ referred to as present or current configuration (Figure 3.1(b)). A typical *material particle* P in the reference configuration with coordinates x_i or (x_1, x_2, x_3) or $\boldsymbol{x}$ occupies position $\bar{P}$ with coordinates $\bar{x}_i$ or $(\bar{x}_1, \bar{x}_2, \bar{x}_3)$ or $\bar{\boldsymbol{x}}$ in the current configuration measured with respect to the same fixed orthogonal coordinate system o-x_i or x-frame that was used to define the material particle position in the

DOI: 10.1201/9781003105336-3

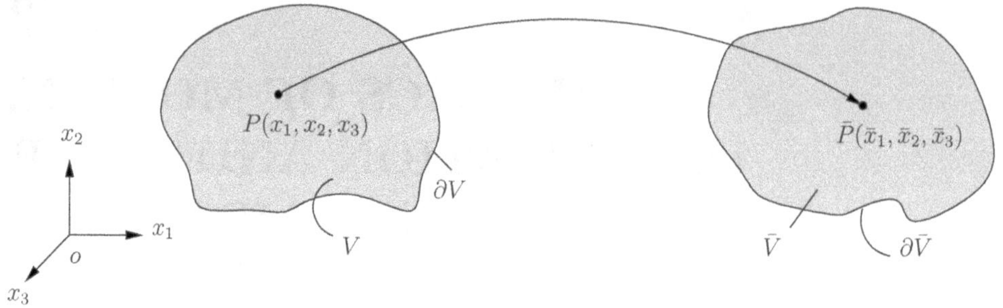

(a) Reference configuration at time t_0 (b) Current configuration at time t

Figure 3.1: Configurations of a volume of matter at reference time t_0 and current time t

reference configuration at time t_0. Then

$$\bar{x}_i = \bar{x}_i(x_1, x_2, x_3, t) = \bar{x}_i(x_j, t) \tag{3.1}$$

describes the path or the trajectory of the particle $P(x_i)$ in the reference configuration to $\bar{P}(\bar{x}_i)$ in the current configuration. Clearly, (3.1) is a description of the motion of the particle P in time. From (3.1), we note that if we choose t to be t_0, then we recover the coordinates of the material particle $\bar{P}(\bar{x}_i)$ in the reference configuration at time t_0, i.e.,

$$x_i = \bar{x}_i(x_1, x_2, x_3, t_0) = \bar{x}_i(x_j, t_0) \tag{3.2}$$

We observe that the right side of Equation (3.1) defines $\bar{x}_i(\cdot)$ as a function of x_1, x_2, x_3 and time t whereas the left side is the value of this function. This is a standard notation in continuum mechanics, i.e., we use the same symbol ($\bar{x}_i$ in this case) for the function as well as its value. From (3.1), we note that we have three equations

$$\bar{x}_1 = \bar{x}_1(x_1, x_2, x_3, t) \quad ; \quad \bar{x}_2 = \bar{x}_2(x_1, x_2, x_3, t) \quad ; \quad \bar{x}_3 = \bar{x}_3(x_1, x_2, x_3, t) \tag{3.3}$$

where $\bar{x}_1, \bar{x}_2$ and $\bar{x}_3$ are expressed as a function of x_1, x_2, x_3 and t. From (3.3), we could solve for x_1, x_2, x_3 in terms of $\bar{x}_1, \bar{x}_2$, $\bar{x}_3$ and t and write

$$x_i = x_i(\bar{x}_1, \bar{x}_2, \bar{x}_3, t) = x_i(\bar{x}_j, t) \tag{3.4}$$

Clearly (3.1) and (3.4) are inverse of each other. In (3.1), the coordinates of point $\bar{P}$ in the current configuration are expressed in terms of its coordinates in the reference configuration and time t, whereas (3.4) expresses coordinates of point P in the reference configuration in terms of its coordinates in the current configuration and time t. Clearly, (3.1) and (3.4) are the inverse of each other. If we know one, the other can be explicitly determined by taking the inverse of the one we know.

3.2 Material particle displacements

When the matter is disturbed, its particles experience motion regardless of whether the matter is solid, liquid or gas. The concept of the material particles being displaced when the matter is disturbed from its original configuration (at time t_0) is central in continuum mechanics.

By definition, the displacement u_i of a material particle $P(x_i)$ is the difference between its coordinates $\bar{x}_i$ at time t (current configuration) and those in the reference configuration (x_i) at time t_0.

$$u_i = \bar{x}_i - x_i \tag{3.5}$$

In Lagrangian description of motion, the displacements u_i of a particle $P(x_i)$ are defined or specified as a function of x_i and time t.

$$u_i = u_i(x_1, x_2, x_3, t) = u_i(x_j, t) \tag{3.6}$$

Equation (3.6) is a Lagrangian description of the displacements. Using (3.4), we can also express (3.6) as

$$\bar{u}_i = \bar{u}_i(\bar{x}_1, \bar{x}_2, \bar{x}_3, t) = \bar{u}_i(\bar{x}_j, t) \tag{3.7}$$

Equation (3.7) is an Eulerian description of the displacements. Thus, using (3.5)–(3.7)

$$\bar{x}_i = x_i + u_i(x_j, t) \tag{3.8}$$
$$x_i = \bar{x}_i - \bar{u}_i(\bar{x}_j, t) \tag{3.9}$$

That is, knowing the undeformed coordinates and displacements, we can obtain the deformed coordinates by using (3.8). Likewise, knowing the deformed position coordinates and displacements, we can obtain the undeformed coordinates by using (3.9).

3.3 Lagrangian, Eulerian descriptions and descriptions in fluid mechanics

Descriptions (3.1) and (3.4) suggest two possible ways to define the motion of the deforming matter. The choice of one over the other is important in describing the motion and deformation of matter while keeping in mind that the two choices are equivalent. We introduce the concept of reference frame. A *reference frame* is a choice of coordinate system in which the motion and deformation of a continuum is described. When matter is disturbed, its state changes. The measures of the changes in the state of the matter over time can be in terms of velocities, density, temperature, etc. associated with the material particles. The choice of a reference frame or coordinate system to measure these is the topic of discussion here.

Consider a material point P with coordinates x_i in the x-frame in the reference configuration (volume V with closed boundary ∂V). Upon deformation, the material particle P occupies position $\bar{P}$ in the current configuration (volume $\bar{V}$ with closed boundary $\partial\bar{V}$) with position coordinates $\bar{x}_i$ in the x-frame.

3.3.1 Lagrangian or referential description of motion

Let $\underset{\sim}{Q}$ be the quantity of interest at a material particle P located at x_i in the reference configuration. Let $\underset{\sim}{Q}_0$ be its value at x_i at time $t = t_0 = 0$. In the current configuration at time t, the material point P occupies location $\bar{P}(\bar{x}_i)$ and $\underset{\sim}{Q}_0$ changes

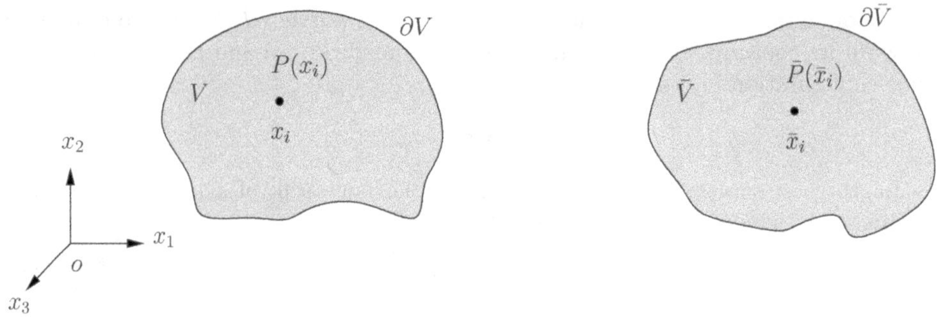

(a) Reference configuration at time $t = t_0 = 0$

(b) Current configuration at time $t > t_0$

Figure 3.2: Reference and current configurations

to a new value $\underset{\sim}{Q}$. If we define $\underset{\sim}{Q}$ at $\bar{P}(\bar{\boldsymbol{x}})$ using

$$\underset{\sim}{Q} = \underset{\sim}{Q}(x_1, x_2, x_3, t) = \underset{\sim}{Q}(x_i, t) \quad ; \quad \underset{\sim}{Q}(x_i, 0) = \underset{\sim}{Q}_0(x_i) \tag{3.10}$$

then (3.10) is known as the *Lagrangian or referential description* of $\underset{\sim}{Q}$. In Lagrangian description, the description of the state of matter in the current configuration requires initial position coordinates of the material points in a fixed frame of reference and the time t corresponding to the current configuration. In other words, in this description, all quantities of interest at point $\bar{P}(\bar{x}_i)$ are expressed as functions of x_i and time t.

3.3.2 Eulerian or spatial description of motion

If we define the quantity of interest $\underset{\sim}{Q}$ at $\bar{P}(\bar{x}_i)$ using deformed coordinates $\bar{x}_i$ and time t, then

$$\bar{\underset{\sim}{Q}} = \bar{\underset{\sim}{Q}}(\bar{x}_1, \bar{x}_2, \bar{x}_3, t) = \bar{\underset{\sim}{Q}}(\bar{x}_i, t) \tag{3.11}$$

Description (3.11) is known as the *Eulerian or spatial description* of $\underset{\sim}{Q}$. We have intentionally used $\bar{\underset{\sim}{Q}}$ to distinguish it from the Lagrangian description. Thus, when quantities are described in the Eulerian or spatial description, we use an over bar on them (as in (3.11)) to distinguish them from their Lagrangian description.

Remarks

(1) Lagrangian and Eulerian descriptions of a quantity of interest $\underset{\sim}{Q}$ are two convenient ways of representing the same information at a unique material point $\bar{P}(\bar{x}_i)$ in the current configuration at time t. Thus, *regardless of whether we use Lagrangian description* (3.10) *or Eulerian description* (3.11), *the numerical value of the quantity of interest $\underset{\sim}{Q}$ at $\bar{P}(\bar{x}_i)$ at time t remains unchanged.*

(2) Motion of material points described by displacements u_i is fundamental in continuum mechanics regardless of the composition of matter and the choice of description. Without u_i, it is not possible to obtain $\bar{x}_i$ from x_i and likewise, x_i

from $\bar{x}_i$, i.e., *without displacements, the transparency and equivalency between Lagrangian and Eulerian descriptions is lost.*

3.3.3 Descriptions in fluid mechanics

In fluid mechanics, the motion of the material particles can be quite complex. The description of their displacements may be difficult and may not be of any real value either. In such situations, we generally take the approach of monitoring changes in the quantities of interest describing the state of the matter over time at a fixed position in space. Thus, in this approach, the observer is stationed at a fixed location in space and monitors the changes in the state of the matter over time at this fixed location. As the state of the matter changes over time, the fixed location of the observer is occupied by different material particles over time. In this approach, the observer never knows which particles are occupying this position but simply monitors the state of the particle that happens to be at this location at time t. Thus, in this method, the position of the observer is the position of some material particle (not known) in the current configuration. If we denote the fixed position of the observer by ${}^f\bar{x}_i$, then the evolution of a quantity of interest $\underset{\sim}{Q}$ (say temperature, velocity, etc.) at location ${}^f\bar{x}_i$ can be described by

$$ {}^f\underset{\sim}{\bar{Q}} = {}^f\underset{\sim}{\bar{Q}}({}^f\bar{x}_1, {}^f\bar{x}_2, {}^f\bar{x}_3, t) = {}^f\underset{\sim}{\bar{Q}}({}^f\bar{x}_i, t) \tag{3.12} $$

The back subscript f on ${}^f\bar{\boldsymbol{x}}$ refers to a fixed location in space and the overbar on ${}^f\bar{\boldsymbol{x}}$ refers to the fact that this position ${}^f\bar{\boldsymbol{x}}$ is, in fact, the position of some material particle in the current configuration at time t. In this approach, position ${}^f\bar{\boldsymbol{x}}$ is occupied by different material particles at different values of time and the material particle displacements are not measured. This is by choice, as the motion of material particles may be too complex to describe or may not be of any real value in the description of deformation of the fluid. Thus, in such descriptions, we lose the ability to follow material particles over time.

Remarks

(1) In Sections 3.3.1 and 3.3.2, we have seen that material particle displacements allow us to transform quantities such as $\underset{\sim}{Q}$ from Lagrangian to Eulerian description and vice-versa. The $\bar{x}_i$ are true deformed coordinates of material particle $\bar{P}(\bar{x}_i)$.

(2) In view of (3.10), (3.11) and Remark (1), we conclude that the description (3.12) of the state of $\underset{\sim}{Q}$ at time t at a fixed location ${}^f\bar{\boldsymbol{x}}$ is neither Lagrangian nor Eulerian, but is a useful and practical way to monitor the changing state of matter without the knowledge of material point displacements.

(3) The physics and kinematics of fluids reveal that for such matter, the material particles may experience large motion relative to their immediate neighbors. Secondly, for the concept of strain to have meaning, a group of material particles must experience deformation relative to each other without rigid body motion as in the case of solid matter. This aspect of kinematics of deformation does not exist in fluids. The influence of strain on the stress field is very minimal in fluids. Hence, the measures of strain are not important and can be neglected in fluids. This further substantiates the importance of the lack

of material point displacements in fluid mechanics. Since the strain measures require displacement gradients, and the strains are not needed in fluids, displacement gradients are not needed. Hence, no need to monitor displacements in fluids. Thus, *a description like* (3.12) *that ignores material point displacements is considered suitable for fluids.*

3.3.4 Notations

(1) Since the position coordinates of a material particle P in the reference configuration are x_i, we define dx_1, dx_2 and dx_3 as elemental lengths along o-$x_1x_2x_3$ axes in the reference configuration. Length ds represents the length of an infinitesimal line segment with components dx_1, dx_2 and dx_3 in the x-frame. We use $\boldsymbol{A}$, $d\boldsymbol{A}$ to define vector areas and V, dV to define volumes in the reference configuration. All these quantities are related to the geometry of the reference configuration in some manner. *Lack of overbar on these (geometric) quantities is not to be confused with Lagrangian description*, i.e., geometric quantities without overbar imply that they are defined in the reference configuration.

(2) Analogous to (1), since $\bar{x}_i$ are the position coordinates of the material particle $\bar{P}$ in the current configuration, we use $d\bar{x}_1$, $d\bar{x}_2$ and $d\bar{x}_3$ to define line segments in the x-frame in the current configuration. Length $d\bar{s}$ represents the length of a line segment with component $d\bar{x}_1$, $d\bar{x}_2$ and $d\bar{x}_3$ (in the x-frame) in the current configuration. We use $\bar{\boldsymbol{A}}$, $d\bar{\boldsymbol{A}}$ to define vector areas and $\bar{V}$, $d\bar{V}$ to define volumes in the current configuration. Here also, all these quantities are related to the geometry of the current configuration at time t in some manner. *Presence of overbar on these (geometric) quantities is not to be confused with Eulerian description*, i.e., the geometric quantities with overbar imply that they are defined in the current configuration.

(3) In (3.10) and (3.11), we note that both $\underset{\sim}{Q}$ and $\underset{\sim}{\bar{Q}}$ are in the current configuration. The distinction is that $\underset{\sim}{Q}$ is a Lagrangian description whereas $\underset{\sim}{\bar{Q}}$ is an Eulerian description. We follow this notation in this chapter as well as the subsequent chapters of this book. In rare instances, we may simply use $\underset{\sim}{Q}$ and let its arguments or the context in which it is used dictate whether it is a Lagrangian or an Eulerian description (this is obviously not preferable).

(4) When a dependent variable is a measure in the reference configuration ($t = t_0 = 0$) we indicate this by using subscript zero. Thus, $\underset{\sim}{Q}(x_1, x_2, x_3, t = 0) = \underset{\sim}{Q}(x_1, x_2, x_3) = \underset{\sim}{Q}_0$ refers to the state $\underset{\sim}{Q}$ of the matter in the reference configuration. Density $\rho(x_1, x_2, x_3, t = 0) = \rho(x_1, x_2, x_3) = \rho_0$ is the density at a material point $P(x_i)$ in the reference configuration. Likewise $\rho = \rho(x_i, t)$ and $\bar{\rho} = \bar{\rho}(\bar{x}_i, t)$ both define density at a material point $\bar{x}_i$ in the current configuration at time t, the distinction being that ρ is a Lagrangian description whereas $\bar{\rho}$ is an Eulerian description.

(5) Often we encounter integrals (as in conservation laws) over the volume V or $\bar{V}(t)$ when considering Lagrangian or Eulerian descriptions. In such integrals, the quantities in the integrals are functions of x_i and t in case of V, and functions of $\bar{x}_i$ and t in case of $\bar{V}(t)$. When converting the integrals from V to $\bar{V}(t)$ or vice-versa, care must be taken to ensure the quantities in the integrand are appropriately converted to the desired description by using the

transformation laws. For example, density $\rho(x_i,t)$, velocity $\boldsymbol{v}(x_i,t)$, etc. can simply be changed to $\bar{\rho}(x_i,t)$, velocity $\bar{\boldsymbol{v}}(x_i,t)$, etc. when switching from V to $\bar{V}(t)$. However, this is not the case for temperature gradient $\boldsymbol{g}$, stress tensor $\boldsymbol{\sigma}$, etc. Changing these to $\bar{\boldsymbol{g}}$ and $\bar{\boldsymbol{\sigma}}$ requires the use of proper transformation laws (as used in Chapter 7).

(6) At this stage, we only know that $\boldsymbol{x} = \boldsymbol{x}(\bar{\boldsymbol{x}})$ and $\bar{\boldsymbol{x}} = \bar{\boldsymbol{x}}(\boldsymbol{x})$ permit us to convert quantities of interest (tensors of rank zero) from Lagrangian to Eulerian description and vice-versa (G^0 induced transformation). Tensors of rank higher than zero require transformations based on covariant law (G^1 transformation) or contravariant law (G^2 transformation) (see Section 2.10.4).

3.4 Material derivative

The time rate of change of a quantity, say $\underset{\sim}{Q}$ (such as density, velocity, temperature, etc.), at a material particle is called material derivative of that quantity. The material derivative is denoted by $\frac{D}{Dt}$ or $\dot{D}$. Thus, material derivative of $\underset{\sim}{Q}$ is $\frac{D\underset{\sim}{Q}}{Dt}$ or $\dot{D}\underset{\sim}{Q}$ or simply $\dot{\underset{\sim}{Q}}$. The material derivative is sometimes also called the *total derivative.* In fluid mechanics, it is common to refer to the material derivative as *substantial derivative.*

The explicit form of the material derivative of $\underset{\sim}{Q}$ depends upon the choice of the type of description used in the definition of the quantity $\underset{\sim}{Q}$, i.e., Lagrangian (or referential) or Eulerian (or spatial) or in general we say, the choice of the description of the motion of the matter.

3.4.1 Material derivative in Lagrangian or referential description

In Lagrangian description of motion, we employ material particle coordinates in the reference configuration, i.e., x_i and time t. The coordinates x_i are described with respect to a fixed x-frame. Thus, if $\underset{\sim}{Q}$ is the quantity of interest in the current configuration at time t, then

$$\underset{\sim}{Q} = \underset{\sim}{Q}(x_1, x_2, x_3, t) \tag{3.13}$$

keeping in mind that x_i coordinates of a particle $P(x_i)$ remain fixed during the evolution of the state of matter. Thus, we can write

$$\frac{D\underset{\sim}{Q}}{Dt} = \dot{D}\underset{\sim}{Q} = \frac{d\underset{\sim}{Q}}{dt} = \left.\frac{\partial \underset{\sim}{Q}}{\partial t}\right|_{x_i;\ \text{fixed}} = \dot{\underset{\sim}{Q}} \tag{3.14}$$

Hence, when we employ Lagrangian description, the material derivative of a quantity $\underset{\sim}{Q}$ is the same as its partial derivative with respect to time (it is understood that x_i remain fixed).

3.4.2 Material derivative in Eulerian or spatial description

In Eulerian description of the evolution of the state of matter, we employ deformed coordinates $\bar{x}_i$ of the material particles and time t. Thus, if $\underset{\sim}{Q}$ is the quantity of interest in the current configuration at time t, then

$$\bar{\underset{\sim}{Q}} = \bar{\underset{\sim}{Q}}(\bar{x}_1, \bar{x}_2, \bar{x}_3, t) \tag{3.15}$$

keeping in mind that

$$\bar{x}_i = \bar{x}_i(x_1, x_2, x_3, t) \tag{3.16}$$

We note that x_i are fixed in time while $\bar{x}_i$ are changing in time as evident from (3.16). Then, using the chain rule of differentiation, i.e., treating $\bar{\underset{\sim}{Q}}$ as a function of $\bar{x}_1, \bar{x}_2, \bar{x}_3$ and t, and in turn, treating $\bar{x}_i$ as a function of x_1, x_2, x_3 and t, we can write

$$\frac{D\bar{\underset{\sim}{Q}}}{Dt} = \dot{D}\bar{\underset{\sim}{Q}} = \dot{\bar{\underset{\sim}{Q}}} = \frac{d\bar{\underset{\sim}{Q}}}{dt} = \left.\frac{\partial \bar{\underset{\sim}{Q}}}{\partial t}\right|_{\bar{x}_i;\ \text{fixed}} + \left.\frac{\partial \bar{\underset{\sim}{Q}}}{\partial \bar{x}_j}\frac{\partial \bar{x}_j}{\partial t}\right|_{x_j;\ \text{fixed}} \tag{3.17}$$

But, in (orthogonal) x-frame

$$v_j = \left.\frac{\partial \bar{x}_j}{\partial t}\right|_{x_i;\ \text{fixed}} = v_j(x_1, x_2, x_3, t) = \bar{v}_j(\bar{x}_1, \bar{x}_2, \bar{x}_3, t) \tag{3.18}$$

In (3.18), v_j are the components of the velocity of a material particle in the current configuration at position $\bar{x}_i$. We note that $v_j(x_1, x_2, x_3, t)$ is the Lagrangian description and $\bar{v}_j(\bar{x}_1, \bar{x}_2, \bar{x}_3, t)$ is the Eulerian description of the velocity components in the coordinate system o-$x_1x_2x_3$ (fixed orthogonal Cartesian coordinate system, x-frame). Substituting from (3.18) in (3.17)

$$\frac{D\bar{\underset{\sim}{Q}}}{Dt} = \dot{D}\bar{\underset{\sim}{Q}} = \dot{\bar{\underset{\sim}{Q}}} = \frac{d\bar{\underset{\sim}{Q}}}{dt} = \left.\frac{\partial \bar{\underset{\sim}{Q}}}{\partial t}\right|_{\bar{x}_i;\ \text{fixed}} + \left.\frac{\partial \bar{\underset{\sim}{Q}}}{\partial \bar{x}_j}\bar{v}_j\right|_{x_j;\ \text{fixed}} \tag{3.19}$$

or

$$\frac{D\bar{\underset{\sim}{Q}}}{Dt} = \frac{d\bar{\underset{\sim}{Q}}}{dt} = \left.\frac{\partial \bar{\underset{\sim}{Q}}}{\partial t}\right|_{\bar{x}_i;\ \text{fixed}} + \bar{v}_j \left.\frac{\partial \bar{\underset{\sim}{Q}}}{\partial \bar{x}_j}\right|_{x_j;\ \text{fixed}} \tag{3.20}$$

Henceforth, we always assume $\bar{x}_i$; fixed and x_j; fixed in (3.20) to be understood. Thus, (3.20) can be written as

$$\frac{D\bar{\underset{\sim}{Q}}}{Dt} = \frac{d\bar{\underset{\sim}{Q}}}{dt} = \frac{\partial \bar{\underset{\sim}{Q}}}{\partial t} + \bar{v}_j \frac{\partial \bar{\underset{\sim}{Q}}}{\partial \bar{x}_j} \tag{3.21}$$

From (3.21), we can extract the material derivative operator $\frac{D}{Dt}$ or $\frac{d}{dt}$

$$\frac{D}{Dt} = \frac{d}{dt} = \frac{\partial}{\partial t} + \bar{v}_j \frac{\partial}{\partial \bar{x}_j} \tag{3.22}$$

or

$$\frac{D}{Dt} = \frac{d}{dt} = \frac{\partial}{\partial t} + (\bar{\boldsymbol{v}} \cdot \bar{\boldsymbol{\nabla}})$$

and we can write

$$\frac{D\underset{\sim}{\bar{Q}}}{Dt} = \frac{d\underset{\sim}{\bar{Q}}}{dt} = \frac{\partial \underset{\sim}{\bar{Q}}}{\partial t} + (\bar{\boldsymbol{v}} \cdot \bar{\boldsymbol{\nabla}})\underset{\sim}{\bar{Q}} \tag{3.23}$$

If $[\bar{\tau}]$ is a tensor of rank two, then

$$\frac{D[\bar{\tau}]}{Dt} = \frac{d[\bar{\tau}]}{dt} = \frac{\partial [\bar{\tau}]}{\partial t} + (\bar{\boldsymbol{v}} \cdot \bar{\boldsymbol{\nabla}})[\bar{\tau}] \tag{3.24}$$

We note that $\bar{\boldsymbol{v}} \cdot \bar{\boldsymbol{\nabla}}$ is a scalar as it is the dot product of two vectors.

3.5 Acceleration of a material particle

Acceleration of a material particle is the rate of change of the velocity field. We consider Lagrangian and Eulerian descriptions of motion to describe the acceleration of a material particle.

3.5.1 Lagrangian or referential description

Let the motion of the deforming matter or continuum be given by

$$\bar{x}_i = \bar{x}_i(x_1, x_2, x_3, t)$$

This is Lagrangian description of the material particle positions in the current configuration at time t. Then the velocities v_i at time t of a material particle $\underset{\sim}{Q}(x_i)$ are given by

$$v_i = \left.\frac{\partial \bar{x}_i}{\partial t}\right|_{x_j;\ \text{fixed}} = v_i(x_1, x_2, x_3, t) \tag{3.25}$$

and the accelerations a_i in x_i directions at time t of material particle $\underset{\sim}{Q}(x_i)$ are given by

$$a_i = \frac{Dv_i}{Dt} = \dot{D}v_i = \dot{v}_i = \frac{dv_i}{dt} = \left.\frac{\partial v_i}{\partial t}\right|_{x_j;\ \text{fixed}} = \frac{\partial v_i}{\partial t} \tag{3.26}$$

Thus, when the velocities v_i of a material particle $\underset{\sim}{Q}(x_i)$ are known as a function of x_i and time t, then the acceleration a_i of the material particle $\underset{\sim}{Q}(x_i)$ can be found by taking the partial derivative of the velocities $v_i(x_1, x_2, x_3, t)$ with respect to time t.

3.5.2 Eulerian or spatial description

When the description of the velocity is spatial or Eulerian, we have

$$\bar{v}_i = \bar{v}_i(\bar{x}_1, \bar{x}_2, \bar{x}_3, t) \tag{3.27}$$

In this case, we need to take the material derivative of $\bar{v}_i$ due to the fact that acceleration of a material particle is the time rate of change of the velocity of the material particle. Thus

$$\bar{a}_i = \frac{D\bar{v}_i}{Dt} = \dot{D}\bar{v}_i = \frac{d\bar{v}_i}{dt} = \dot{\bar{v}}_i = \left.\frac{\partial \bar{v}_i}{\partial t}\right|_{\bar{x}_j;\ \text{fixed}} + \left.\frac{\partial \bar{v}_i}{\partial \bar{x}_j}\frac{\partial \bar{x}_j}{\partial t}\right|_{x_i;\ \text{fixed}} \tag{3.28}$$

Substituting $\frac{\partial \bar{x}_j}{\partial t} = \bar{v}_j$ in (3.28),

$$\bar{a}_i = \frac{D\bar{v}_i}{Dt} = \frac{d\bar{v}_i}{dt} = \left.\frac{\partial \bar{v}_i}{\partial t}\right|_{\bar{x}_j;\text{ fixed}} + \bar{v}_j \left.\frac{\partial \bar{v}_i}{\partial \bar{x}_j}\right|_{x_i;\text{ fixed}} \tag{3.29}$$

which can be written as

$$\begin{aligned} \bar{a}_i &= \frac{D\bar{v}_i}{Dt} = \frac{d\bar{v}_i}{dt} = \frac{\partial \bar{v}_i}{\partial t} + \bar{v}_j \frac{\partial \bar{v}_i}{\partial \bar{x}_j} \\ \text{or} \quad \bar{a}_i &= \frac{D\bar{v}_i}{Dt} = \frac{d\bar{v}_i}{dt} = \frac{\partial \bar{v}_i}{\partial t} + (\bar{\boldsymbol{v}} \cdot \bar{\boldsymbol{\nabla}})\bar{v}_i \end{aligned} \tag{3.30}$$

The term $\bar{v}_j \frac{\partial \bar{v}_i}{\partial \bar{x}_j}$ is called *convective acceleration* and $\frac{\partial \bar{v}_i}{\partial t}$ is called *local acceleration.* As a special case, if the motion is purely along x_1-axis, then $\bar{v}_2 = \bar{v}_3 = 0$ and $\bar{v}_1 = \bar{v}_1(\bar{x}_1, t)$ and we have $\bar{a}_1 = \frac{\partial \bar{v}_1}{\partial t} + \bar{v}_1 \frac{\partial \bar{v}_1}{\partial \bar{x}_1}$.

3.6 Deformation Gradient Tensor

Consider material particles $P(x_i)$ and $Q(x_i + dx_i)$ in the reference configuration at time t_0. Upon deformation, these particles move to positions $\bar{P}(\bar{x}_i, t)$ and $\bar{Q}(\bar{x}_i + d\bar{x}_i, t)$ in the current configuration at time t. The lengths and directions of vectors $\bar{\boldsymbol{P}}\bar{\boldsymbol{Q}}$ and $\boldsymbol{PQ}$ are not the same. Let the motion or deformation of the matter be given by

$$\bar{x}_i = \bar{x}_i(x_1, x_2, x_3, t) \tag{3.31}$$

We assume that $\bar{x}_i(x_1, x_2, x_3, t)$ are continuous and differentiable functions in their arguments, then

$$\bar{\boldsymbol{P}}\bar{\boldsymbol{Q}} = \big(\bar{x}_i(x_1 + dx_1, x_2 + dx_2, x_3 + dx_3, t) - \bar{x}_i(x_1, x_2, x_3, t)\big)\boldsymbol{e}_i \tag{3.32}$$

Using Taylor series expansion of the first term in (3.32) at $P(x_i)$ and retaining up to second order term in dx_i

$$\begin{aligned} \bar{\boldsymbol{P}}\bar{\boldsymbol{Q}} =& \Big(\bar{x}_i(x_1, x_2, x_3, t) + \frac{\partial \bar{x}_i}{\partial x_j} dx_j + \frac{1}{2}\frac{\partial^2 \bar{x}_i}{\partial x_j \partial x_k} dx_j dx_k \\ &+ O(||d\boldsymbol{x}||)^3 - \bar{x}_i(x_1, x_2, x_3, t)\Big)\boldsymbol{e}_i \end{aligned} \tag{3.33}$$

where $||d\boldsymbol{x}|| = \sqrt{d\boldsymbol{x} \cdot d\boldsymbol{x}}$ is norm of $d\boldsymbol{x}$. Neglecting $O(||d\boldsymbol{x}||)^3$ in (3.33)

$$\bar{\boldsymbol{P}}\bar{\boldsymbol{Q}} = \frac{\partial \bar{x}_i}{\partial x_j} dx_j \boldsymbol{e}_i + \frac{1}{2}\frac{\partial^2 \bar{x}_i}{\partial x_j \partial x_k} dx_j dx_k \boldsymbol{e}_i = \boldsymbol{e}_i d\bar{x}_i \tag{3.34}$$

in which $d\bar{x}_1$, $d\bar{x}_2$, $d\bar{x}_3$ are components of $\bar{\boldsymbol{P}}\bar{\boldsymbol{Q}}$ in x-frame. Equating the coefficients of $\boldsymbol{e}_i$ in (3.34)

$$d\bar{x}_i = \frac{\partial \bar{x}_i}{\partial x_j} dx_j + \frac{1}{2}\frac{\partial^2 \bar{x}_i}{\partial x_j \partial x_k} dx_j dx_k \tag{3.35}$$

Equation (3.35) is a fundamental relationship that tells us how deformed lengths $d\bar{x}_i$ are related to undeformed lengths dx_i with the assumption that infinitesimals of order three or higher in lengths are negligible. We note that $\bar{\boldsymbol{P}}\bar{\boldsymbol{Q}}$ is a nonlinear function of dx_1, dx_2, dx_3. Similarly if we use $x_i = x_i(\bar{x}_1, \bar{x}_2, \bar{x}_3, t)$, inverse of (3.31) and

$$\boldsymbol{PQ} = (x_i(\bar{x}_1 + d\bar{x}_1, \bar{x}_2 + d\bar{x}_2, \bar{x}_3 + d\bar{x}_3, t) - x_i(\bar{x}_1, \bar{x}_2, \bar{x}_3, t))\boldsymbol{e}_i \tag{3.36}$$

and follow the same procedure as used for $\bar{\boldsymbol{P}}\bar{\boldsymbol{Q}}$, we can obtain

$$\boldsymbol{PQ} = \frac{\partial x_i}{\partial \bar{x}_j} d\bar{x}_j \boldsymbol{e}_i + \frac{1}{2} \frac{\partial^2 x_i}{\partial \bar{x}_j \partial \bar{x}_k} d\bar{x}_j d\bar{x}_k \boldsymbol{e}_i + O(||d\bar{\boldsymbol{x}}||)^3 \tag{3.37}$$

If we neglect $O(||d\bar{\boldsymbol{x}}||)^3$ in (3.37), we obtain

$$\boldsymbol{PQ} = \frac{\partial x_i}{\partial \bar{x}_j} d\bar{x}_j \boldsymbol{e}_i + \frac{1}{2} \frac{\partial^2 x_i}{\partial \bar{x}_j \partial \bar{x}_k} d\bar{x}_j d\bar{x}_k \boldsymbol{e}_i = \boldsymbol{e}_i dx_i \tag{3.38}$$

in which dx_1, dx_2, dx_3 are the components of $\boldsymbol{PQ}$ in x-frame. Equating coefficients of $\boldsymbol{e}_i$ in (3.38)

$$dx_i = \frac{\partial x_i}{\partial \bar{x}_j} d\bar{x}_j + \frac{1}{2} \frac{\partial^2 x_i}{\partial \bar{x}_j \partial \bar{x}_k} d\bar{x}_j d\bar{x}_k \tag{3.39}$$

This is another fundamental relationship similar to (3.35). In this case, $\boldsymbol{PQ}$ is a nonlinear function of $d\bar{x}_1$, $d\bar{x}_2$, $d\bar{x}_3$. Due to nonlinear relationships in (3.35) and (3.39), the mapping between $d\boldsymbol{x}$ and $d\bar{\boldsymbol{x}}$ expressed by (3.35) and (3.39) is not assured to be one-to-one and onto. Thus, these relationships cannot be used in the further development of the kinematics of motion and deformation. However, if we assume that the deformation is such that infinitesimals of order two and higher, i.e., $O(||d\boldsymbol{x}||)^2$ and $O(||d\bar{\boldsymbol{x}}||)^2$ can be neglected, then we obtain the following from (3.35) and (3.39):

$$\begin{aligned} d\bar{x}_i &= \frac{\partial \bar{x}_i}{\partial x_j} dx_j \quad ; \quad dx_i = \frac{\partial x_i}{\partial \bar{x}_j} d\bar{x}_j \\ \text{or} \quad d\bar{\boldsymbol{x}} &= \boldsymbol{J} \cdot d\boldsymbol{x} \quad ; \quad d\boldsymbol{x} = \bar{\boldsymbol{J}} \cdot d\bar{\boldsymbol{x}} \end{aligned} \tag{3.40}$$

or in matrix notation

$$\{d\bar{x}\} = \begin{Bmatrix} d\bar{x}_1 \\ d\bar{x}_2 \\ d\bar{x}_3 \end{Bmatrix} = [J] \begin{Bmatrix} dx_1 \\ dx_2 \\ dx_3 \end{Bmatrix} = [J]\{dx\} \tag{3.41}$$

and

$$\{dx\} = \begin{Bmatrix} dx_1 \\ dx_2 \\ dx_3 \end{Bmatrix} = [\bar{J}] \begin{Bmatrix} d\bar{x}_1 \\ d\bar{x}_2 \\ d\bar{x}_3 \end{Bmatrix} = [\bar{J}]\{d\bar{x}\} \tag{3.42}$$

As shown earlier, columns of $[J]$ are covariant base vectors $\tilde{\boldsymbol{g}}_i$ and the rows of $[\bar{J}]$ are the contravariant base vectors. The matrices $[J]$ and $[\bar{J}]$ are called deformation gradient tensors. $[J]$ is a covariant measure in Lagrangian description and $[\bar{J}]$ is a contravariant measure in Eulerian description. We note

$$
\begin{aligned}
\frac{\partial \bar{x}_i}{\partial \bar{x}_j} &= \delta_{ij} = \frac{\partial \bar{x}_i}{\partial x_k}\frac{\partial x_k}{\partial \bar{x}_j} = J_{ik}\bar{J}_{kj} \\
\text{or} \quad [I] &= [J][\bar{J}] \implies [J] = [\bar{J}]^{-1} \text{ and } [\bar{J}] = [J]^{-1}
\end{aligned}
\tag{3.43}
$$

Alternatively

$$
\begin{aligned}
\frac{\partial x_i}{\partial x_j} &= \delta_{ij} = \frac{\partial x_i}{\partial \bar{x}_k}\frac{\partial \bar{x}_k}{\partial x_j} = \bar{J}_{ik}J_{kj} \\
\text{or} \quad [I] &= [\bar{J}][J] \implies [\bar{J}] = [J]^{-1} \text{ and } [J] = [\bar{J}]^{-1}
\end{aligned}
\tag{3.44}
$$

Based on (3.43) and (3.44), mapping of dx to $d\bar{x}_i$ and those of $d\bar{x}_i$ to dx_i is one to one and onto. These relations are fundamental in the further development of the measures for finite deformation.

We remark that (3.41) and (3.42) are based on the assumption that infinitesimals of the lengths of order two and higher can be neglected, but (3.43) and (3.44) hold regardless of the magnitude of the deformation as these contain information about the gradients at a material point which is independent of the magnitude of deformation.

Remarks

(1) We note that for (3.43) and (3.44) to hold, the inverse of $[J]$ and $[\bar{J}]$ must exist and must be unique.

(2) Uniqueness of mapping for x_i to $\bar{x}_i$ and for $\bar{x}_i$ to x_i must ensure that $[J]^{-1}$ and $[\bar{J}]^{-1}$ are unique.

(3) Thus, at this stage (2) is satisfied, if $\det[J] \neq 0$ and $\det[\bar{J}] \neq 0$.

(4) We note that components of $[J]$ are functions of x_i and time and its columns are covariant base vectors $\tilde{\boldsymbol{g}}_i$. Thus, $[J]$ is covariant measure in Lagrangian description

(5) On the other hand, rows of $[\bar{J}]$ are contravariant base vectors $\tilde{\boldsymbol{g}}^i$ and its components are functions of $\bar{x}_i$ and time t. Thus, $[\bar{J}]$ is a contravariant measure in Eulerian description.

3.7 Continuous deformation of matter, restrictions on the description of motion

In continuum mechanics, we assume that different material particles always occupy distinct locations for all values of time. For example, in the reference configuration at time t_0 each material particle has unique coordinates x_i (or labels). Upon deformation these unique labels or coordinates occupy positions $\bar{x}_i$ in the current configuration at time t. The point $P(x_i)$ at time t_0 maps into $\bar{P}(\bar{x}_i)$ at time t. Thus, in a manner of speaking, material particles are immortal. We can represent $[J]$ and $[\bar{J}]$ in (3.41) and (3.42) using Murnaghan's notation

$$[J] = \left[\frac{\bar{x}_1, \bar{x}_2, \bar{x}_3}{x_1, x_2, x_3}\right] = [J] = \begin{bmatrix} \frac{\partial \bar{x}_1}{\partial x_1} & \frac{\partial \bar{x}_1}{\partial x_2} & \frac{\partial \bar{x}_1}{\partial x_3} \\ \frac{\partial \bar{x}_2}{\partial x_1} & \frac{\partial \bar{x}_2}{\partial x_2} & \frac{\partial \bar{x}_2}{\partial x_3} \\ \frac{\partial \bar{x}_3}{\partial x_1} & \frac{\partial \bar{x}_3}{\partial x_2} & \frac{\partial \bar{x}_3}{\partial x_3} \end{bmatrix} \tag{3.45}$$

$$[J] = \left[\frac{x_1, x_2, x_3}{\bar{x}_1, \bar{x}_2, \bar{x}_3}\right] = [\bar{J}] = \begin{bmatrix} \frac{\partial x_1}{\partial \bar{x}_1} & \frac{\partial x_1}{\partial \bar{x}_2} & \frac{\partial x_1}{\partial \bar{x}_3} \\ \frac{\partial x_2}{\partial \bar{x}_1} & \frac{\partial x_2}{\partial \bar{x}_2} & \frac{\partial x_3}{\partial \bar{x}_3} \\ \frac{\partial x_3}{\partial \bar{x}_1} & \frac{\partial x_3}{\partial \bar{x}_2} & \frac{\partial x_3}{\partial \bar{x}_3} \end{bmatrix} \tag{3.46}$$

Remarks

(1) At time t_0, $\bar{x}_i = x_i$, hence $\dfrac{\partial \bar{x}_i}{\partial x_j} = \dfrac{\partial x_i}{\partial x_j} = \delta_{ij}$. Therefore $\det[J]_{t_0} = \det[I] = 1$.

(2) $\det[J]$ is a continuous function of time and the mapping between $\bar{x}$ and x is one-to-one and onto.

(3) $\det[J]$ must be positive for all values of time t and strictly greater than zero. This is a necessary condition for a continuous deformation to be physically admissible. However, it does not mean that all deformations satisfying this requirement can be produced in a deformable volume of matter. $\det[J]$ relates volumes in two configurations (shown in a later section) hence must be greater than zero.

(4) A deformation field satisfying the condition $\det[J] > 0$ is called proper and admissible deformation or simply *admissible deformation.*

(5) We can also relate $[J]$ to gradients of displacements. Since

$$\bar{x}_i = x_i + u_i \tag{3.47}$$

$$\frac{\partial \bar{x}_i}{\partial x_j} = \frac{\partial x_i}{\partial x_j} + \frac{\partial u_i}{\partial x_j} \tag{3.48}$$

or

$$\frac{\partial \bar{x}_i}{\partial x_j} = \delta_{ij} + \frac{\partial u_i}{\partial x_j} \tag{3.49}$$

Hence

$$\det[J] = \det\left[\delta_{ij} + \frac{\partial u_i}{\partial x_j}\right] \tag{3.50}$$

From (3.50), we can conclude that for an admissible deformation of matter in a continuum, the displacement field u_i must satisfy the condition $\det[J] > 0$.

3.8 Change of description, co- and contra-variant measures

When we want to change a description from Lagrangian to Eulerian and vice-versa, we use the following:

$$\boldsymbol{x} = \boldsymbol{x}(\bar{\boldsymbol{x}}, t) \quad ; \quad \bar{\boldsymbol{x}} = \bar{\boldsymbol{x}}(\boldsymbol{x}, t) \quad ; \quad \{d\bar{x}\} = [J]\{dx\} \quad ; \quad \{dx\} = [\bar{J}]\{d\bar{x}\} \tag{3.51}$$

Consider Q, a tensor of rank zero. If $Q = Q(\boldsymbol{x}, t)$ is a Lagrangian description of quantity Q, then its Eulerian description $\bar{Q} = \bar{Q}(\bar{\boldsymbol{x}}, t)$ can be obtained by using (3.51) and replacing $\boldsymbol{x}$ by $\boldsymbol{x}(\bar{\boldsymbol{x}}, t)$, i.e., $\bar{Q} = \bar{Q}(\boldsymbol{x}(\bar{\boldsymbol{x}}, t), t) = \bar{Q}(\bar{\boldsymbol{x}}, t)$. Likewise $\bar{Q}(\bar{\boldsymbol{x}}, t)$, Eulerian description of Q can be converted to Lagrangian description $Q = Q(\boldsymbol{x}, t)$ by using $\bar{\boldsymbol{x}} = \bar{\boldsymbol{x}}(\boldsymbol{x}, t)$. In converting tensors of rank one and higher from one description to another, we must make use of $[J]$ and $[\bar{J}]$ and the fact that $[\bar{J}] = [J]^{-1}$ and $[J] = [\bar{J}]^{-1}$. We note that $[J]$ is a Lagrangian description whereas $[\bar{J}]$ is an Eulerian description. Thus, if $\boldsymbol{\sigma}([J], t)$ is a Lagrangian description of $\boldsymbol{\sigma}$, then $\bar{\boldsymbol{\sigma}} = \bar{\boldsymbol{\sigma}}([\bar{J}]^{-1}, t)$ is its corresponding Eulerian description. Likewise, if $\bar{\boldsymbol{\sigma}} = \bar{\boldsymbol{\sigma}}([\bar{J}], t)$, then this Eulerian description can be converted to the corresponding Lagrangian description using $[\bar{J}] = [J]^{-1}$ and we have $\boldsymbol{\sigma}([J]^{-1}, t)$. In this approach, the Eulerian and Lagrangian descriptions are completely equivalent. That is, the numerical values of the quantities at a material point in the current configuration remain the same (as expected) irrespective of the choice of description. In the remainder of the book, we use these transformations for changing descriptions.

We also note that columns of $[J]$ are covariant base vectors $\tilde{\boldsymbol{g}}_i$ in Lagrangian description and the rows of $[\bar{J}]$ are contravariant base vectors $\tilde{\boldsymbol{g}}^i$ in Eulerian description. These aspects are important in deriving measures of deformation and strains. At this stage, we consider some guidelines that we follow in developing measures of deformation, length, change in length and strains.

(1) Measures utilizing $[J]$ are covariant measures in Lagrangian description. These can be converted to Eulerian description by replacing $[J]$ with $[\bar{J}]^{-1}$, keeping in mind that this substitution does not change the measure from covariant to contravariant.

(2) Measures utilizing $[\bar{J}]$ are contravariant measures in Eulerian description. These can be converted to Lagrangian descriptions by replacing $[\bar{J}]$ with $[J]^{-1}$, keeping in mind that this substitution does not change the measure from contravariant to covariant.

(3) Thus, based on (1) and (2), the measures of deformation, length, change in length and strains naturally result as covariant or contravariant measures if these are derived using $[J]$ or $[\bar{J}]$. Both of these can be expressed either as Lagrangian or Eulerian description using $[J]$ and $[\bar{J}]$, but their covariant and contravariant nature remains unaffected.

3.9 Notations for covariant and contravariant measures

In Chapter 2, covariant measures are expressed using subscripts and the contravariant measures are denoted using superscripts. Thus, if $\boldsymbol{\sigma}$ is a tensor of rank two, then σ^{ij} and σ_{ij} are its contravariant and covariant components. Instead of using this notation, we introduce a new notation in the following. If $\boldsymbol{\sigma}$ a contravariant tensor of rank two, then we define its material derivative of order n by

$$\boldsymbol{\sigma}^{[n]} = \frac{D^n(\boldsymbol{\sigma})}{Dt^n} \quad \text{or} \quad [\sigma^{[n]}] = \frac{D^n[\sigma]}{Dt^n} \tag{3.52}$$

When $n = 0$, we have material derivative of order zero, i.e., the tensor itself, thus, a contravariant tensor $\boldsymbol{\sigma}$ of rank two will be denoted by $\boldsymbol{\sigma}^{[0]}$ or $[\sigma^{[0]}]$ and its components by $(\sigma^{[0]})_{ij}$. Superscript signifies that it is a contravariant measure and zero in square brackets implies material derivative of order zero. Likewise, for a covariant tensor $\boldsymbol{\sigma}$ of rank two, we define its material derivative of order n by

$$\boldsymbol{\sigma}_{[n]} = \frac{D^n(\boldsymbol{\sigma})}{Dt^n} \quad \text{or} \quad [\sigma_{[n]}] = \frac{D^n[\sigma]}{Dt^n} \tag{3.53}$$

Here also, when $n = 0$, we have material derivative of order zero, i.e., the tensor itself, thus, a covariant tensor $\boldsymbol{\sigma}$ of rank two will be denoted by $\boldsymbol{\sigma}_{[0]}$ or $[\sigma_{[0]}]$ and its components by $(\sigma_{[0]})_{ij}$. Subscript signifies that it is a covariant measure and zero in square brackets implies material derivative of order zero.

We note that this notation regarding co- and contra-variant measures can be used in general. For example, $[J]$ is the same as $[J_{[0]}]$ and likewise $[\bar{J}]$ is the same as $[\bar{J}^{[0]}]$. The usual convention of Lagrangian descriptions without overbar and Eulerian descriptions with overbar continues to hold. Thus, $\boldsymbol{\sigma}^{[0]}$ and $\bar{\boldsymbol{\sigma}}^{[0]}$ are Lagrangian and Eulerian descriptions of contravariant tensor $\boldsymbol{\sigma}$. Likewise, $\boldsymbol{\sigma}_{[0]}$ and $\bar{\boldsymbol{\sigma}}_{[0]}$ are Lagrangian and Eulerian descriptions of covariant tensor $\boldsymbol{\sigma}$.

3.10 Deformation, measures of length and change in length

Consider Figure 3.3, in which $\boldsymbol{PQ}$ deforms into $\bar{\boldsymbol{P}}\bar{\boldsymbol{Q}}$. Let the lengths of $\boldsymbol{PQ}$ and $\bar{\boldsymbol{P}}\bar{\boldsymbol{Q}}$ be ds and $d\bar{s}$. In the following development we use the matrix and vector notations. Recall

$$\bar{x}_i = \bar{x}_i(x_j, t) \quad ; \quad \{\bar{x}\} = \{\bar{x}(\{x\}, t)\} = \{x\} + \{u(\{x\}, t)\} \tag{3.54}$$

where $\{u\} = [u_1, u_2, u_3]^T$ defines displacements of a material point. The inverse of (3.54) is given by

$$\{x\} = \{x(\{\bar{x}\}, t)\} = \{\bar{x}\} - \{\bar{u}(\{\bar{x}\}, t)\} \tag{3.55}$$

Definitions of $[J]$ and $[\bar{J}]$ in (3.41) and (3.42) hold using (3.54) and (3.55).

3.10.1 Covariant measures of length and change in length

The covariant measures must use $[J]$ as the columns of $[J]$ are covariant base vectors.

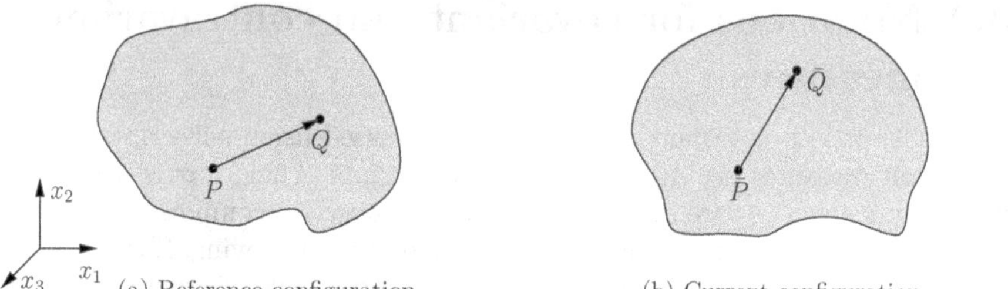

(a) Reference configuration (b) Current configuration

Figure 3.3: Undeformed and deformed line segments in reference and current configurations

3.10.1.1 Covariant measure, Lagrangian description

In covariant Lagrangian description, measures of deformations are derived using dx_i (components of $d\boldsymbol{s}$ in x-frame) as reference lengths and $[J]$. A line segment $d\boldsymbol{s}$ in the reference configuration with component dx_1, dx_2, dx_3 in x-frame deforms into the line segment $d\bar{\boldsymbol{s}}$ in the current configuration with its components $d\bar{x}_1, d\bar{x}_2, d\bar{x}_3$ in x-frame.

$$\{dx\} = [dx_1, dx_2, dx_3]^T \quad ; \quad \{d\bar{x}\} = [d\bar{x}_1, d\bar{x}_2, d\bar{x}_3]^T \tag{3.56}$$

and

$$\{d\bar{x}\} = [J]\{dx\} \tag{3.57}$$

The squares of the lengths ds and $d\bar{s}$ can be defined by

$$(ds)^2 = \{dx\}^T\{dx\} \quad ; \quad (d\bar{s})^2 = \{d\bar{x}\}^T\{d\bar{x}\} \tag{3.58}$$

A measure of the change in length can be represented by $\big((d\bar{s})^2 - (ds)^2\big)$.

$$(d\bar{s})^2 - (ds)^2 = \{d\bar{x}\}^T\{d\bar{x}\} - \{dx\}^T\{dx\} \tag{3.59}$$

Since dx_i are reference lengths we substitute $\{d\bar{x}\}$ from (3.56) in (3.59).

$$\begin{aligned}(d\bar{s})^2 - (ds)^2 &= \{dx\}^T[J]^T[J]\{dx\} - \{dx\}^T[I]\{dx\} \\ &= \{dx\}^T([J]^T[J] - [I])\{dx\}\end{aligned} \tag{3.60}$$

Let us define

$$[M_{[0]}] = [J]^T[J] - [I] \qquad \text{(Def.)} \tag{3.61}$$

Then

$$(d\bar{s})^2 - (ds)^2 = \{dx\}^T[M_{[0]}]\{dx\} \tag{3.62}$$

where $[M_{[0]}]$ contains details regarding the deformation related to $\big((d\bar{s})^2 - (ds)^2\big)$. We can also express $[M_{[0]}]$ in terms of displacement gradients. Since $\{\bar{x}\} = \{x\} + \{u\}$

$$[J] = \left[\frac{\partial\{\bar{x}\}}{\partial\{x\}}\right] = \left[\frac{\partial\{x\}}{\partial\{x\}}\right] + \left[\frac{\partial\{u\}}{\partial\{x\}}\right] = [I] + [{}^dJ] \tag{3.63}$$

where

$$[^dJ] = \left[\frac{\partial\{u\}}{\partial\{x\}}\right] = \left[\frac{u_1, u_2, u_3}{x_1, x_2, x_3}\right] \tag{3.64}$$

Therefore

$$[M_{[0]}] = [J]^T[J] - [I] = [^dJ] + [^dJ]^T + [^dJ]^T[^dJ] \tag{3.65}$$

We note that $[M_{[0]}]$ in (3.61) and hence in (3.65) is a *covariant measure in Lagrangian description.* Components of $[M_{[0]}]$ are derived using $[J]$, whose columns consist of covariant base vectors in Lagrangian description. From (3.62) we note that in the deformation measure $[M_{[0]}]$, the reference lengths are dx_i in the undeformed configuration, which are the components of $d\boldsymbol{s}$ in x-frame.

3.10.1.2 Covariant measure, Eulerian description

We can obtain Eulerian description of $[M_{[0]}]$ in (3.61), denoted as $[\bar{M}_{[0]}]$ by using (3.61) and replacing $[J]$ by $[\bar{J}]^{-1}$.

$$[\bar{M}_{[0]}] = \left[[\bar{J}]^{-1}\right]^T \left[[\bar{J}]^{-1}\right] - [I] \tag{3.66}$$

$[\bar{M}_{[0]}]$ in (3.66) can be expressed in terms of gradients of displacements using $\{x\} = \{\bar{x}\} - \{\bar{u}\}$

$$[\bar{J}] = \left[\frac{\partial\{x\}}{\partial\{\bar{x}\}}\right] = \left[\frac{\partial\{\bar{x}\}}{\partial\{\bar{x}\}}\right] - \left[\frac{\partial\{\bar{u}\}}{\partial\{\bar{x}\}}\right] = [I] - [^d\bar{J}] \tag{3.67}$$

where

$$[^d\bar{J}] = \frac{\partial\{\bar{u}\}}{\partial\{\bar{x}\}} = \left[\frac{\bar{u}_1, \bar{u}_2, \bar{u}_3}{\bar{x}_1, \bar{x}_2, \bar{x}_3}\right] \tag{3.68}$$

Thus, $[\bar{M}_{[0]}]$ in (3.66) can be written as

$$[\bar{M}_{[0]}] = \left[[\,[I] - [^d\bar{J}]\,]^{-1}\right]^T \left[[I] - [^d\bar{J}]\right]^{-1} - [I] \tag{3.69}$$

where $[\bar{M}_{[0]}]$ is the *covariant measure in Eulerian description.* We remark that numerical values of $[M_{[0]}]$ and $[\bar{M}_{[0]}]$ remain the same at a material point and the substitution $[J] = [\bar{J}]^{-1}$ does not change the measures from co- to contra-variant. We note that the reference lengths are dx_i even though the description is Eulerian.

3.10.2 Contravariant measures of length and change in length

The contravariant measures must use $[\bar{J}]$ as its rows are contravariant base vectors.

3.10.2.1 Contravariant measure, Eulerian description

In the contravariant Eulerian description, measures of deformation are derived using $d\bar{x}_i$ (components of $d\bar{\boldsymbol{s}}$ in x-frame) as reference length and $[\bar{J}]$. Components of $[\bar{J}]$ are functions of $\bar{x}$ and t.

Consider the measure of change in length defined by

$$(d\bar{s})^2 - (ds)^2 = \{d\bar{x}\}^T\{d\bar{x}\} - \{dx\}^T\{dx\} \tag{3.70}$$

Since $d\bar{x}_i$ are reference lengths, we substitute for $\{dx\} = [\bar{J}]\{d\bar{x}\}$ in (3.70)

$$\begin{aligned}(d\bar{s})^2 - (ds)^2 &= \{d\bar{x}\}^T[I]\{d\bar{x}\} - \{d\bar{x}\}^T[\bar{J}]^T[\bar{J}]\{d\bar{x}\} \\ &= \{d\bar{x}\}^T([I] - [\bar{J}]^T[\bar{J}])\{d\bar{x}\}\end{aligned} \tag{3.71}$$

Let us define

$$[\bar{M}^{[0]}] = [I] - [\bar{J}]^T[\bar{J}] \qquad \text{(Def.)} \tag{3.72}$$

then

$$(d\bar{s})^2 - (ds)^2 = \{d\bar{x}\}^T[\bar{M}^{[0]}]\{d\bar{x}\} \tag{3.73}$$

where $[\bar{M}^{[0]}]$ contains details regarding the deformation related to $((d\bar{s})^2 - (ds)^2)$. We can also express $[\bar{M}^{[0]}]$ in terms of displacement gradients using (3.67).

$$\begin{aligned}[\bar{M}^{[0]}] &= [I] - \big[[I] - [{}^d\bar{J}]\big]^T\big[[I] - [{}^d\bar{J}]\big] \\ \text{or} \quad [\bar{M}^{[0]}] &= [{}^d\bar{J}] + [{}^d\bar{J}]^T - [{}^d\bar{J}]^T[{}^d\bar{J}]\end{aligned} \tag{3.74}$$

We note that $[\bar{M}^{[0]}]$ is a *contravariant measure in Eulerian description.*

3.10.2.2 Contravariant measure, Lagrangian description

We can also derive Lagrangian description for $[\bar{M}^{[0]}]$ in (3.72) denoted as $[M^{[0]}]$ by using $[\bar{J}] = [J]^{-1}$ in (3.72)

$$[M^{[0]}] = [I] - \big[[J]^{-1}\big]^T[J]^{-1} \tag{3.75}$$

where $[M^{[0]}]$ is the *contravariant measure in Lagrangian description.* $[M^{[0]}]$ can also be expressed in terms of gradients of displacements using $[J]$ from (3.63) in (3.75).

$$[M^{[0]}] = [I] - \big[([I] + [{}^dJ])^{-1}\big]^T \big[[I] + [{}^dJ]\big]^{-1} \tag{3.76}$$

As in the case of covariant measures, in this case also, numerical values of $[\bar{M}^{[0]}]$ and $[M^{[0]}]$ are the same at a material point. Definitions of $[M_{[0]}]$, $[\bar{M}_{[0]}]$ and $[M^{[0]}]$, $[\bar{M}^{[0]}]$ play an important role in defining the covariant and contravariant measures of strains in Lagrangian and Eulerian descriptions.

3.11 Covariant and contravariant measures of finite strain in Lagrangian and Eulerian descriptions

In this section, we derive various measures of finite strain and discuss their physical meanings and limitations. In order to derive measures of strain, we must consider material lines in the reference configuration and their map in the current configuration. Consider orthogonal material lines PA_1, PA_2 and PA_3 in the reference configuration at a material point $P(\boldsymbol{x})$ parallel to ox_i axes of a fixed x-frame (Figure 2.5(a)). At time t, these material lines in the current configuration will become non-orthogonal curvilinear (Figure 2.5(b)) at material point $\bar{P}(\bar{\boldsymbol{x}}, t)$, deformed position of material point $P(\boldsymbol{x})$. The measures of strain at the material point $\bar{P}(\bar{\boldsymbol{x}})$

are measures of elongation of the material lines PA_1, PA_2 and PA_3 and change in the 90 degree angle between them. This information is extracted from the deformed curvilinear material lines and the angles between them. It is difficult to deal with the curvilinear material lines at point $\bar{P}(\bar{\boldsymbol{x}})$, thus, we construct tangent vectors to these material lines at point $\bar{P}(\bar{\boldsymbol{x}})$. These are covariant base vectors $\tilde{\boldsymbol{g}}_i$. We assume that in the neighborhood of the material point $\bar{P}(\bar{\boldsymbol{x}})$, the strain measures can be extracted using $\tilde{\boldsymbol{g}}_i$, whose components are columns of $[J]$, and the angles between them. The strain measures so derived have two important aspects: (1) such measures are in agreement with the physics of deformation as these consider true tangent vectors at a material point $\bar{P}(\bar{\boldsymbol{x}})$ in the deformed configuration to the actual deformed material lines and (2) these measures of strain are undoubtedly *covariant measures of strain as the covariant basis forms the foundation of their derivations.* These covariant measures of strain are naturally in Lagrangian description but can be easily changed to Eulerian description. Change in description obviously does not affect their numerical values, it only alters the functional relationships.

Since the contravariant basis is reciprocal to the covariant basis, it appears that strain measures can possibly be derived using contravariant basis $\tilde{\boldsymbol{g}}^i$ or $[\bar{J}]$. We discuss this possibility and details of this approach. Based on what has been described for covariant measures of strain, if contravariant strain measures are to be extracted using $\tilde{\boldsymbol{g}}^i$ (i.e., rows of $[\bar{J}]$) and the angles between them, then undoubtedly the contravariant base vectors $\tilde{\boldsymbol{g}}^i$ must form tangent vectors to some deformed material lines at point $\bar{P}(\bar{\boldsymbol{x}})$. Figure 2.7(c) shows actual deformed material lines with tangent vectors $\tilde{\boldsymbol{g}}_i$ at point $\bar{P}(\bar{\boldsymbol{x}})$ as well as contravariant base vectors $\tilde{\boldsymbol{g}}^i$ that are tangent to material lines coincident with $\bar{P}\underset{\sim}{x}_1$, $\bar{P}\underset{\sim}{x}_2$, $\bar{P}\underset{\sim}{x}_3$ curvilinear axes. Thus, if we use $\tilde{\boldsymbol{g}}^i$ base vectors and the angles between them to extract measures of strain, then we assume the deformed material lines to be coincident with $\bar{P}\underset{\sim}{x}_1$, $\bar{P}\underset{\sim}{x}_2$, $\bar{P}\underset{\sim}{x}_3$ curvilinear lines, which are obviously different than the actual deformed material lines that are coincident with curvilinear lines $\bar{P}\tilde{x}_1$, $\bar{P}\tilde{x}_2$, $\bar{P}\tilde{x}_3$ when the deformation is finite. We remark that *when the deformation is infinitesimal, the distinction between the covariant and contravariant bases disappears* as the deformed tetrahedron faces are orthogonal to each other. Thus, it is quite clear that in the case of finite deformation, the contravariant measures of strain are quite different than the covariant measures, and are non-physical as these measures use $\tilde{\boldsymbol{g}}^i$ base vectors that are tangent to non-physical deformed material lines. It is obvious that as the deformation approaches the infinitesimal case, contravariant measures of strain will approach covariant measures. We consider contravariant measures of strain primarily (1) due to completeness reasons, (2) because these measures are commonly used in currently published works and writings and hence need to be included here and (3) to demonstrate how these measures can become totally non-physical with progressively increasing deformation. The contravariant strain measures can also be in Eulerian or Lagrangian description depending on your choice. As in the case of covariant measures, here also, change in description only changes the functional relationships but the numerical values of the strains remain unaffected.

Strain measures must be dimensionless and must be extracted using ds and $d\bar{s}$, undeformed and deformed lengths whose components are dx_1, dx_2, dx_3 and $d\bar{x}_1$, $d\bar{x}_2$, $d\bar{x}_3$ in a fixed x-frame. In defining and deriving measures of strain, it is preferable to use $(ds)^2$ and $(d\bar{s})^2$ as opposed to ds and $d\bar{s}$ in order to avoid square roots. Second, strain measures must be derived from dimensionless quantities containing $(ds)^2$ and $(d\bar{s})^2$. Except these, there are no other restrictions in deriving strain measures (at least at this stage). This flexibility gives rise to more than one possibility to define and derive strain measures. We consider covariant as well as contravariant measures of strain in Lagrangian as well as Eulerian descriptions. Table 3.1 shows the type of strain measures: Cauchy, Finger, Green's, Almansi, and the basis for their derivations and notations used to denote them in Lagrangian and Eulerian descriptions. Details of the derivations are presented in the following.

3.11.1 Covariant measures of finite strains

In deriving the covariant measures of finite strain that are naturally Lagrangian description, we use $[J]$ containing covariant base vectors $\tilde{\boldsymbol{g}}_i$ as its columns. In these derivations, reference lengths are dx_i that are components of $d\boldsymbol{s}$. These can also be converted to Eulerian description, but they remain covariant measures. We present details of Cauchy strain, Finger strain and Green's strain based on covariant measures in Lagrangian and Eulerian descriptions (columns 1-4 in Table 3.1).

3.11.1.1 Covariant Cauchy strain or right Cauchy strain: $[C_{[0]}]$, $[\bar{C}_{[0]}]$

(a) Covariant Lagrangian description: $[C_{[0]}]$

The covariant measure of strain $[C_{[0]}]$ can be derived purely using deformation gradient tensor $[J]$. We consider

$$\frac{(d\bar{s})^2}{(ds)^2} = \frac{\{d\bar{x}\}^T\{d\bar{x}\}}{\{dx\}^T\{dx\}} \tag{3.77}$$

Substituting for $\{d\bar{x}\}$ from (3.41) we can write

$$\therefore \quad \frac{(d\bar{s})^2}{(ds)^2} = \frac{\{dx\}^T[J]^T[J]\{dx\}}{\{dx\}^T\{dx\}} \tag{3.78}$$

Since dx_i are fixed reference lengths, the measure of strain must be contained in those terms in (3.78) that do not contain $\{dx\}$ and $\{dx\}^T$. Let us define

$$[C_{[0]}] = [J]^T[J] \quad ; \quad \text{(Def.)} \tag{3.79}$$

We can express $[C_{[0]}]$ in terms of $[M_{[0]}]$ defined by (3.61)

$$[C_{[0]}] = [I] + [M_{[0]}] \tag{3.80}$$

where

$$[M_{[0]}] = [J]^T[J] - [I] \tag{3.81}$$

If we wish to express $[C_{[0]}]$ in terms of displacement gradients, then we can use the definition of $[M_{[0]}]$ from (3.65) in (3.80), i.e., $[M_{[0]}] = [{}^dJ] + [{}^dJ]^T + [{}^dJ]^T[{}^dJ]$. $[C_{[0]}]$ is called Cauchy strain, Cauchy deformation or *right Cauchy strain* (see polar decomposition in Section 3.17.2). $[C_{[0]}]$ is symmetric and is a covariant measure in Lagrangian description.

Table 3.1: Types of strains, basis for their derivations and the notations used to represent them

Covariant measures: Based on $[J]$				**Contravariant measures: Based on $[\bar{J}]$**			
		Notation				Notation	
Type of strain measure	Basis for derivation	Lagrangian description	Eulerian description	Type of strain measure	Basis for derivation	Lagrangian description	Eulerian description
Cauchy	$\frac{(d\bar{s})^2}{(ds)^2}$	$[C_{[0]}]$	$[\bar{C}_{[0]}]$	Cauchy	$\frac{(ds)^2}{(d\bar{s})^2}$	$[C^{[0]}]$	$[\bar{C}^{[0]}]$
Finger	inverse of Cauchy	$[F_{[0]}] = [C_{[0]}]^{-1}$	$[\bar{F}_{[0]}]$	Finger	inverse of Cauchy	$[F^{[0]}] = [B^{[0]}]$	$[\bar{F}^{[0]}] = [\bar{B}^{[0]}] = [\bar{C}^{[0]}]^{-1}$
Green	$\frac{(d\bar{s})^2 - (ds)^2}{2(ds)^2}$	$[\varepsilon_{[0]}]$	$[\bar{\varepsilon}_{[0]}]$	Almansi	$\frac{(d\bar{s})^2 - (ds)^2}{2(d\bar{s})^2}$	$[\varepsilon^{[0]}]$	$[\bar{\varepsilon}^{[0]}]$

(b) Covariant Eulerian description: $[\bar{C}_{[0]}]$

We can also derive the Cauchy strain in Eulerian description using (3.80) and by replacing $[J]$ by $[\bar{J}]^{-1}$.

$$[\bar{C}_{[0]}] = \left[[\bar{J}]^{-1}\right]^T [\bar{J}]^{-1} \tag{3.82}$$

We can also express $[\bar{C}_{[0]}]$ in terms of $[\bar{M}_{[0]}]$ given by (3.66)

$$[\bar{C}_{[0]}] = [I] + [\bar{M}_{[0]}] \tag{3.83}$$

and in terms of displacement gradients we have

$$[\bar{C}_{[0]}] = \left[\ \left[[I] - [{}^d\bar{J}]\right]^{-1}\ \right]^T \left[[I] - [{}^d\bar{J}]\right]^{-1} \tag{3.84}$$

This is a symmetric covariant Cauchy strain measure in Eulerian description. Numerical values of $[C_{[0]}]$ and $[\bar{C}_{[0]}]$ remain the same at a material point, but their expression may be different.

3.11.1.2 Covariant Finger strain: $[F_{[0]}]$, $[\bar{F}_{[0]}]$

(a) Covariant Lagrangian description: $[F_{[0]}]$

The Finger strain $[F_{[0]}]$ is the inverse of Cauchy deformation or strain $[C_{[0]}]$

$$[F_{[0]}] = [C_{[0]}]^{-1} \quad ; \quad \text{(Def.)} \tag{3.85}$$

or

$$[F_{[0]}] = \left[[J]^T[J]\right]^{-1} = [J]^{-1}\left[[J]^T\right]^{-1} = [J]^{-1}\left[[J]^{-1}\right]^T \tag{3.86}$$

which can also be written as

$$[F_{[0]}] = \left[[I] + [M_{[0]}]\right]^{-1} \tag{3.87}$$

$$\text{or} \quad [F_{[0]}] = \left[[{}^dJ] + [{}^dJ]^T + [{}^dJ]^T[{}^dJ] + [I]\right]^{-1} \tag{3.88}$$

The strain measure $[F_{[0]}]$ is symmetric covariant measure in Lagrangian description.

(b) Covariant Eulerian description: $[\bar{F}_{[0]}]$

The strain measure $[\bar{F}_{[0]}]$ can be obtained by taking Lagrangian description $[F_{[0]}]$ and replacing $[J]$ by $[\bar{J}]^{-1}$ or $[J]^{-1}$ by $[\bar{J}]$. Thus, from (3.86), we obtain

$$[\bar{F}_{[0]}] = [\bar{J}][\bar{J}]^T \tag{3.89}$$

In terms of displacement gradients we can write

$$[\bar{F}_{[0]}] = \left[[I] - [{}^d\bar{J}]\right]\left[[I] - [{}^d\bar{J}]\right]^T \tag{3.90}$$

$$\text{or} \quad [\bar{F}_{[0]}] = [I] - [{}^d\bar{J}] - [{}^d\bar{J}]^T + [{}^d\bar{J}][{}^d\bar{J}]^T \tag{3.91}$$

The strain measure $[\bar{F}_{[0]}]$ is symmetric covariant measure in Eulerian description.

3.11.1.3 Covariant Green's strain: $[\varepsilon_{[0]}]$, $[\bar{\varepsilon}_{[0]}]$

(a) Covariant Lagrangian description: $[\varepsilon_{[0]}]$

To derive the strain measure $[\varepsilon_{[0]}]$ we consider

$$\frac{(d\bar{s})^2 - (ds)^2}{2(ds)^2} = \frac{\{d\bar{x}\}^T\{d\bar{x}\} - \{dx\}^T\{dx\}}{2\{dx\}^T\{dx\}} \tag{3.92}$$

Substituting $\{d\bar{x}\}$ from (3.41) and factoring $\{dx\}^T$ and $\{dx\}$.

$$\therefore \quad \frac{(d\bar{s})^2 - (ds)^2}{2(ds)^2} = \frac{\{dx\}^T[\frac{1}{2}([J]^T[J] - [I])]\{dx\}}{\{dx\}^T\{dx\}} \tag{3.93}$$

Since $\{dx\}$ are fixed lengths, the strain measure $[\varepsilon_{[0]}$ is given by

$$[\varepsilon_{[0]}] = \frac{1}{2}([J]^T[J] - [I]) \quad ; \quad \text{(Def.)} \tag{3.94}$$

The strain measure $[\varepsilon_{[0]}]$ is called Green's strain. This is a covariant measure in Lagrangian description. The Green's strain can also be expressed in terms of the gradients of the displacements. Substituting $[J]$ from (3.63) in (3.94) we obtain

$$[\varepsilon_{[0]}] = \frac{1}{2}([^dJ] + [^dJ]^T + [^dJ]^T[^dJ]) \tag{3.95}$$

We can also express Green's strain $[\varepsilon_{[0]}]$ in terms of $[M_{[0]}]$

$$[\varepsilon_{[0]}] = \frac{1}{2}[M_{[0]}] \tag{3.96}$$

The Green's strain $[\varepsilon_{[0]}]$ is symmetric covariant measure in Lagrangian description.

(b) Covariant Eulerian description: $[\bar{\varepsilon}_{[0]}]$

The strain measure $[\bar{\varepsilon}_{[0]}]$ can be obtained by replacing $[J]$ with $[\bar{J}]^{-1}$ in Lagrangian description, thus

$$[\bar{\varepsilon}_{[0]}] = \frac{1}{2}[\bar{M}_{[0]}] = \frac{1}{2}(\ [[\bar{J}]^{-1}]^T[\bar{J}]^{-1} - [I]\ ) \tag{3.97}$$

which can also be written as

$$[\bar{\varepsilon}_{[0]}] = \frac{1}{2}(\ [\ [[I] - [^d\bar{J}]\ ]^{-1}]^T[\ [I] - [^d\bar{J}]\ ]^{-1} - [I]\ ) \tag{3.98}$$

The strain $[\bar{\varepsilon}_{[0]}]$ is symmetric covariant measure in Eulerian description.

The covariant measures of strain are derived using covariant base vectors that are tangent to the actual deformed material lines at a point $\bar{P}(\bar{\boldsymbol{x}})$ in the current configuration, hence *covariant measures of strain are in agreement with the physics of finite deformation.* Physical meaning of the components of the strain measures will be considered in a later section.

3.11.2 Contravariant measures of finite strains

In deriving contravariant measures of strains that are naturally Eulerian description, we must use $[\bar{J}]$ containing contravariant base vectors $\tilde{\boldsymbol{g}}^i$ as its rows. In these derivations, reference lengths are $d\bar{x}_i$ that are components of $d\bar{\boldsymbol{s}}$. These can also be converted to Lagrangian description. We present details of Cauchy strain, Finger strain and Almansi strain based on contravariant measures in Eulerian and Lagrangian descriptions (columns 5 to 8 in Table 3.1).

3.11.2.1 Contravariant Cauchy strain or Cauchy deformation: $[\bar{C}^{[0]}]$, $[C^{[0]}]$

(a) Contravariant Eulerian description: $[\bar{C}^{[0]}]$

We consider

$$\frac{(ds)^2}{(d\bar{s})^2} = \frac{\{dx\}^T\{dx\}}{\{d\bar{x}\}^T\{d\bar{x}\}} \tag{3.99}$$

Since $d\bar{x}_i$ are fixed reference lengths, the measure of strain must be contained in those terms in (3.101) that do not contain $\{d\bar{x}\}$ and $\{d\bar{x}\}^T$. Substituting $\{dx\}$ from (3.42) we can write

$$\{dx\} = [\bar{J}]\{d\bar{x}\} \tag{3.100}$$

$$\therefore \quad \frac{(ds)^2}{(d\bar{s})^2} = \frac{\{d\bar{x}\}^T[\bar{J}]^T[\bar{J}]\{d\bar{x}\}}{\{d\bar{x}\}^T\{d\bar{x}\}} \tag{3.101}$$

Let us define

$$[\bar{C}^{[0]}] = [\bar{J}]^T[\bar{J}] \quad ; \quad \text{(Def.)} \tag{3.102}$$

which can also be written as

$$[\bar{C}^{[0]}] = [I] - [\bar{M}^{[0]}] \tag{3.103}$$

The strain measure $[\bar{C}^{[0]}]$ in (3.102) can also be expressed in terms of displacement gradients. Using $[\bar{J}]$ in (3.67).

$$[\bar{C}^{[0]}] = \big[[I] - [{}^d\bar{J}]\big]^T \big[[I] - [{}^d\bar{J}]\big] \tag{3.104}$$

$$\text{or} \quad [\bar{C}^{[0]}] = [I] - [{}^d\bar{J}] - [{}^d\bar{J}]^T + [{}^d\bar{J}]^T[{}^d\bar{J}] \tag{3.105}$$

The strain measure $[\bar{C}^{[0]}]$ is symmetric and is the contravariant Cauchy strain in Eulerian description because $[\bar{J}]$ and $[{}^d\bar{J}]$ are functions of $\bar{\boldsymbol{x}}$ and t, hence $[\bar{C}^{[0]}]$ is a function of $\bar{\boldsymbol{x}}$ and t. The strain measure $[\bar{C}^{[0]}]$ is symmetric.

(b) Contravariant Lagrangian description: $[C^{[0]}]$

We can derive the Lagrangian description $[C^{[0]}]$ by replacing $[\bar{J}]$ with $[J]^{-1}$ in (3.102), thus

$$[C^{[0]}] = \big[[J]^{-1}\big]^T[J]^{-1} = \Big[\,\big[[I] + [{}^dJ]\big]^{-1}\,\Big]^T\big[[I] + [{}^dJ]\big]^{-1} \tag{3.106}$$

Using $[M^{[0]}]$ in (3.75) we can write

$$[C^{[0]}] = [I] - [M^{[0]}] \tag{3.107}$$

The strain $[C^{[0]}]$ is symmetric contravariant measure in Lagrangian description.

3.11.2.2 Contravariant Finger strain or left Cauchy strain: $[\bar{F}^{[0]}]$, $[F^{[0]}]$

(a) Contravariant Eulerian description: $[\bar{F}^{[0]}]$

The Finger strain $[\bar{F}^{[0]}]$ is the inverse of Cauchy deformation or Cauchy strain.

$$[\bar{F}^{[0]}] = [\bar{C}^{[0]}]^{-1} \quad ; \quad \text{(Def.)} \tag{3.108}$$

or

$$[\bar{F}^{[0]}] = \left[[\bar{J}]^T[\bar{J}]\right]^{-1} = [\bar{J}]^{-1}\left[[\bar{J}]^T\right]^{-1} = [\bar{J}]^{-1}\left[[\bar{J}]^{-1}\right]^T$$

which can also be written as

$$[\bar{F}^{[0]}] = \left[[I] - [\bar{M}^{[0]}]\right]^{-1} \tag{3.109}$$

and in terms of displacement gradients we can write

$$[\bar{F}^{[0]}] = \left[\ [I] - [{}^d\bar{J}] - [{}^d\bar{J}]^T + [{}^d\bar{J}]^T[{}^d\bar{J}]\ \right]^{-1} \tag{3.110}$$

The strain measure $[\bar{F}^{[0]}]$ is obviously symmetric and is contravariant measure in Eulerian description.

(b) Contravariant Lagrangian description: $[F^{[0]}]$ *or* $[B^{[0]}]$

If we replace $[\bar{J}]$ by $[J]^{-1}$ or $[\bar{J}]^{-1}$ by $[J]$, we obtain the corresponding Lagrangian description.

$$[F^{[0]}] = [J][J]^T = [B^{[0]}] \tag{3.111}$$

Using $[J]$ from (3.63) in (3.111) we can write

$$\therefore \quad [F^{[0]}] = \left[[I] + [{}^dJ]\right]\left[[I] + [{}^dJ]^T\right] \tag{3.112}$$

$$\text{or} \quad [F^{[0]}] = [I] + [{}^dJ] + [{}^dJ]^T + [{}^dJ][{}^dJ]^T \tag{3.113}$$

The strain measure $[F^{[0]}]$ (or $[B^{[0]}]$) is symmetric and is a contravariant Lagrangian description. This measure of strain is also called *left Cauchy strain tensor* (see polar decomposition section).

3.11.2.3 Contravariant Almansi strain: $[\bar{\varepsilon}^{[0]}]$, $[\varepsilon^{[0]}]$

(a) Contravariant Eulerian description: $[\bar{\varepsilon}^{[0]}]$

To derive the strain measure $[\bar{\varepsilon}^{[0]}]$, we consider

$$\frac{(d\bar{s})^2 - (ds)^2}{2(d\bar{s})^2} = \frac{\{d\bar{x}\}^T\{d\bar{x}\} - \{dx\}^T\{dx\}}{2\{d\bar{x}\}^T\{d\bar{x}\}} \tag{3.114}$$

Substituting $\{dx\}$ from (3.42) and factoring $\{d\bar{x}\}^T$ and $\{d\bar{x}\}$

$$\therefore \quad \frac{(d\bar{s})^2 - (ds)^2}{2(d\bar{s})^2} = \frac{\{d\bar{x}\}^T[\frac{1}{2}([I] - [\bar{J}]^T[\bar{J}])]\{d\bar{x}\}}{\{d\bar{x}\}^T\{d\bar{x}\}} \tag{3.115}$$

Since $\{d\bar{x}\}$ are fixed length, the strain measure $[\varepsilon^{[0]}]$ is given by

$$[\bar{\varepsilon}^{[0]}] = \frac{1}{2}([I] - [\bar{J}]^T[\bar{J}]) \quad ; \quad \text{(Def.)} \tag{3.116}$$

which can also be written as

$$[\bar{\varepsilon}^{[0]}] = \frac{1}{2}[\bar{M}^{[0]}] \tag{3.117}$$

The strain measure $[\bar{\varepsilon}^{[0]}]$ can also be expressed in terms of the derivatives of the displacement using (3.67) for $[\bar{J}]$.

$$[\bar{\varepsilon}^{[0]}] = \frac{1}{2}([I] - [[I] - [{}^d\bar{J}]]^T[[I] - [{}^d\bar{J}]]) \tag{3.118}$$

$$\text{or} \quad [\bar{\varepsilon}^{[0]}] = \frac{1}{2}([{}^d\bar{J}] + [{}^d\bar{J}]^T - [{}^d\bar{J}]^T[{}^d\bar{J}]) \tag{3.119}$$

The strain measure $[\bar{\varepsilon}^{[0]}]$ is symmetric and is a contravariant measure in Eulerian description.

(b) Contravariant Lagrangian description: $[\varepsilon^{[0]}]$

We can also derive Lagrangian description of Almansi strain by replacing $[\bar{J}]$ with $[J]^{-1}$ or $[\bar{J}]^{-1}$ by $[J]$ in Eulerian description (3.116).

$$[\varepsilon^{[0]}] = \frac{1}{2}[[I] - [[J]^{-1}]^T[J]^{-1}] = \frac{1}{2}[M^{[0]}] \tag{3.120}$$

In terms of gradients of displacements we can write using $[J]$ defined in (3.63).

$$[\varepsilon^{[0]}] = \frac{1}{2}\left[[I] - \left([[I] + [{}^dJ]]^{-1}\right)^T\left([[I] + [{}^dJ]]^{-1}\right)\right] \tag{3.121}$$

This strain measure is symmetric and is contravariant measure in Lagrangian description.

Remarks on strain measures

(1) The covariant and contravariant measures of finite strain in Lagrangian and Eulerian descriptions are derived. It is shown that covariant measures of strain are in agreement with the physics of deformation of the material lines in the current configuration. The contravariant measures of strain utilize contravariant base vectors that are tangent to the new set of curvilinear material lines in the deformed configuration that are different than the actual deformed material lines utilized in the derivations of the covariant measures of strain. The curvilinear material lines used in the contravariant measures do not correspond to the undeformed tetrahedron with its edges parallel to the axes of the x-frame when the deformation is finite. *With progressively increasing*

Table 3.2: Summary of strain measures used currently

Lagrangian description		Eulerian description	
Strain	Measure	Strain	Measure
Cauchy or right Cauchy	$[C] = [J]^T[J]$	Cauchy	$[\bar{C}] = [\bar{J}]^T[\bar{J}]$
Finger $[F]$	$[F] = [C]^{-1}$ $= [J]^{-1}[[J]^{-1}]^T$ or $[F] = [\bar{J}][\bar{J}]^T$	Finger or left Cauchy $[\bar{F}]$ or $[B]$	$[\bar{F}] = [\bar{C}]^{-1}$ $= [\bar{J}]^{-1}[[\bar{J}]^{-1}]^T$ or $[\bar{F}] = [B] = [J][J]^T$
Green	$[\varepsilon] = \frac{1}{2}([J]^T[J] - [I])$	Almansi	$[\bar{\varepsilon}] = \frac{1}{2}([I] - [\bar{J}]^T[\bar{J}])$

deformation, the contravariant measures of strain become progressively more spurious while the covariant measures remain valid.

(2) For a given strain measure in co- or contra-variant basis, their Lagrangian and Eulerian descriptions are precisely equivalent and hence yield the same numerical values at a material point in the current configuration. For example, if we consider $[\varepsilon_{[0]}]$ and $[\bar{\varepsilon}_{[0]}]$ that are covariant measures of strain in Lagrangian and Eulerian descriptions (Green's strain), their numerical values at a material point will be identical using the measures derived here. This holds for all strain measures derived using covariant and contravariant bases.

(3) It is rather important to discuss the strain measures used currently in continuum mechanics [30, 31, 42].

(a) The strain measures are strictly classified as Lagrangian and Eulerian without regard to covariant and contravariant bases that are essential in relating the strain measures to the physics of deformation. For example $[C] = [J]^T[J]$ and $[\bar{C}] = [\bar{J}]^T[\bar{J}]$ are defined as Lagrangian and Eulerian Cauchy strain measures, yet it is obvious that numerical values of $[C]$ and $[\bar{C}]$ at a material point are going to be different. This is obviously in contradiction to the fact that Lagrangian and Eulerian descriptions must be equivalent. If we use the co- and contra-variant measure concepts in deriving strain measures, then it becomes clear that $[C]$ and $[\bar{C}]$ are related to different material lines, hence ought to produce different results when the deformation is finite.

(b) Table 3.2 provides a summary of strain measures used currently. These are strictly classified as Lagrangian and Eulerian without regard to the basis used in deriving them. We note that there is no equivalence between

$[C]$ and $[\bar{C}]$, $[F]$ and $[\bar{F}]$, $[\varepsilon]$ and $[\bar{\varepsilon}]$ at a material point. Based on the measures derived here, $[C]$, $[F]$ and $[\varepsilon]$ are covariant measures in Lagrangian description and $[\bar{C}]$, $[\bar{F}]$ and $[\bar{\varepsilon}]$ are contravariant measures in Eulerian description.

(4) If the deformation is such that $[J][J]^T = [J]^T[J]$, then this deformation represents *rigid body rotation* of the matter of volume as $[J][J]^T = [J]^T[J]$ can only hold if $[J]$ is orthogonal.

(5) If the deformation is such that $[J] = [J]^T$, then this type of deformation represents *pure stretch* (see Section 3.17.2, remarks).

(6) At this stage, we only have a matrix representation of strain measures. Their tensorial nature and rank need to be established. This is done in Section 3.12.

(7) Table 3.3 provides a summary of covariant and contravariant strain measures in Lagrangian and Eulerian descriptions derived here.

3.12 Changes in strain measures due to rigid rotation of frames

In the development of the measures of strain, we had considered covariant and contravariant measures in Lagrangian and in Eulerian descriptions. In Lagrangian description, all quantities are a function of material point coordinates x_i in the reference configuration and time t. In Eulerian description, we use deformed coordinates $\bar{x}_i$ of the material points and time t to describe behaviors of the quantities of interest.

In this section, we consider two cases of the change of material coordinates through rigid rotation of the appropriate frames to determine how the Lagrangian and the Eulerian measures of strain are affected due to the change of coordinates through rigid rotations of the frames.

(1) In the first case we consider the reference configuration with coordinates of the material points x_i in x-frame. We change x_i coordinates to x_i' say in x'-frame by a rigid rotation of x_i coordinates (x-frame). Thus, the coordinates x_i and x_i' are related through a rigid rotation matrix obtained by a rigid rotation of x-frame to x'-frame. In this process, $\bar{x}_i$-coordinates of the material points in the current configuration defined in x-frame remain unchanged. Due to this change of frame, we study which Lagrangian and Eulerian descriptions of strain measures are affected and establish their tensorial nature.

(2) In the second case, we consider the current configuration with material point coordinates $\bar{x}_i$ in the x-frame. We change coordinates $\bar{x}_i$ to $\bar{x}_i'$ say in $\bar{x}'$-frame obtained by a rigid rotation of $\bar{x}_i$ coordinates ($\bar{x}$-frame). Thus, the coordinates $\bar{x}_i'$ and $\bar{x}_i$ are related through a rotation matrix obtained from a rigid rotation of $\bar{x}$-frame to $\bar{x}'$-frame. The x_i-coordinates of the material points remain unaffected. Due to this change of coordinates of the current configuration, we study which Lagrangian and Eulerian descriptions of the strains are affected and establish their tensorial nature.

Table 3.3: Summary of co- and contra-variant strain measures in Lagrangian and Eulerian descriptions

Covariant measures: Based on $[J]$			Contravariant measures: Based on $[\bar{J}]$		
Type of strain	Lagrangian description	Eulerian description	Type of strain	Lagrangian description	Eulerian description
Cauchy	$[C_{[0]}]=[J]^T[J]$	$[\bar{C}_{[0]}]=\left[[\bar{J}]^{-1}\right]^T[\bar{J}]^{-1}$	Cauchy	$[C^{[0]}]=\left[[J]^{-1}\right]^T[J]^{-1}$	$[\bar{C}^{[0]}]=[\bar{J}]^T[\bar{J}]$
Finger	$[F_{[0]}]=[C_{[0]}]^{-1}$ $=[J]^{-1}\left[[J]^{-1}\right]^T$	$[\bar{F}_{[0]}]=[\bar{J}][\bar{J}]^T$	Finger	$[F^{[0]}]=[B^{[0]}]=[J][J]^T$	$[\bar{F}^{[0]}]=[\bar{B}^{[0]}]=[\bar{C}^{[0]}]^{-1}$ $=[\bar{J}]^{-1}\left[[\bar{J}]^{-1}\right]^T$
Green	$[\varepsilon_{[0]}]=\frac{1}{2}([J]^T[J]-[I])$	$[\bar{\varepsilon}_{[0]}]=\frac{1}{2}(\left[[\bar{J}]^{-1}\right]^T[\bar{J}]^{-1}-[I])$	Almansi	$[\varepsilon^{[0]}]=\frac{1}{2}([I]-\left[[J]^{-1}\right]^T[J]^{-1})$	$[\bar{\varepsilon}^{[0]}]=\frac{1}{2}([I]-[\bar{J}]^T[\bar{J}])$

(3) The covariant measures of strain in Lagrangian description use $d\boldsymbol{s}$ and dx_i as reference lengths in their derivations. Thus, if x_i coordinates are changed to x'_i coordinates through rigid rotation, we expect these measures to be affected. But, the contravariant measures of strain that use $d\bar{\boldsymbol{s}}$ and $d\bar{x}_i$ as reference lengths will remain unaffected.

(4) The contravariant measures of strain in Eulerian description use $d\bar{\boldsymbol{s}}$ and $d\bar{x}_i$ as reference lengths in their derivations. Thus, if $\bar{x}_i$ coordinates are changed to $\bar{x}'_i$ through rigid rotation, we expected these measures to be affected. But, the covariant measures of strain that use $d\boldsymbol{s}$ and dx_i reference lengths will remain unaffected.

3.12.1 Change in covariant Lagrangian descriptions of strain when x_i are changed to x'_i due to rigid rotation

Since x'_i are obtained from x_i by rigid rotation, we have the following in which $[Q]$ is orthogonal (Galilean transformation, i.e., $[Q]$ is not a function of time).

$$\begin{aligned} \{x'\} &= [Q]\{x\} \quad ; \quad \{dx'\} = [Q]\{dx\} \\ \{x\} &= [Q]^T\{x'\} \quad ; \quad \{dx\} = [Q]^T\{dx'\} \end{aligned} \tag{3.122}$$

$$\{d\bar{x}\} = [J]\{dx\} = [J][Q]^T\{dx'\} \tag{3.123}$$

$$\{d\bar{x}\}^T = \{dx'\}^T[Q][J]^T \tag{3.124}$$

In the following we consider how the strain measures in Lagrangian description are affected by rigid rotation of the x-frame to x'-frame for the reference configuration. We only consider some typical strain measures to illustrate how these are affected. Findings derived from these apply to similar other measures of strain as well.

3.12.1.1 Covariant Cauchy strain in Lagrangian description: $[C_{[0]}]$

We note that lengths $ds = ds'$ as x'_i are due to rigid rotation of x_i. First, using ds

$$\begin{aligned} \frac{d\bar{s}^2}{ds^2} &= \frac{\{d\bar{x}\}^T\{d\bar{x}\}}{\{dx\}^T\{dx\}} = \frac{\{dx\}^T[J]^T[J]\{dx\}}{\{dx\}^T\{dx\}} = \frac{\{dx'\}^T\left([Q][J]^T[J][Q]^T\right)\{dx'\}}{\{dx'\}^T[Q][Q]^T\{dx'\}} \\ \therefore \quad \frac{d\bar{s}^2}{ds^2} &= \frac{\{dx'\}^T[C'_{[0]}]\{dx'\}}{\{dx'\}^T\{dx'\}} \quad ; \quad [C'_{[0]}] = [Q][J]^T[J][Q]^T \end{aligned} \tag{3.125}$$

Alternatively using ds' and (3.123), (3.124)

$$\frac{d\bar{s}^2}{d(s')^2} = \frac{\{d\bar{x}\}^T\{d\bar{x}\}}{\{dx'\}^T\{dx'\}} = \frac{\{dx'\}^T\left([Q][J]^T[J][Q]^T\right)\{dx'\}}{\{dx'\}^T[Q][Q]^T\{dx'\}} = \frac{\{dx'\}^T[C'_{[0]}]\{dx'\}}{\{dx'\}^T\{dx'\}} \tag{3.126}$$

use of ds or ds' yield the same results for $[C'_{[0]}]$, the covariant Cauchy strain measure in x'-frame. From (3.125), we have

$$[C'_{[0]}] = [Q][J]^T[J][Q]^T = [Q][C_{[0]}][Q]^T \tag{3.127}$$

$$\text{or} \quad [C_{[0]}] = [Q]^T[C'_{[0]}][Q] \tag{3.128}$$

From (3.126) and (3.127), we conclude that $[C_{[0]}]$ and $[C'_{[0]}]$ are symmetric Cauchy strain tensors of rank two in x- and x'-frames. Thus, covariant Cauchy strain tensor in Lagrangian description is affected by changing x_i to x'_i through rigid rotation.

3.12.1.2 Covariant Finger strain in Lagrangian description: $[F_{[0]}]$

Since in x-frame

$$[F_{[0]}] = [C_{[0]}]^{-1} \tag{3.129}$$

in x'-frame, we can write (using (3.127))

$$[F'_{[0]}] = [C'_{[0]}]^{-1} = [Q][C_{[0]}]^{-1}[Q]^T = [Q][F_{[0]}][Q]^T \tag{3.130}$$

and

$$[F_{[0]}] = [Q]^T[F'_{[0]}][Q] \tag{3.131}$$

From (3.131), we note $[F_{[0]}]$ and $[F'_{[0]}]$ are second order tensors in the x-frame and the x'-frame respectively. Thus, covariant Finger strain in Lagrangian description is affected due to the rigid rotation of x_i to x'_i.

3.12.1.3 Covariant Green's strain in Lagrangian description: $[\varepsilon_{[0]}]$

We recall that lengths $ds = ds'$, hence, we can use either in the derivation. First, using ds

$$\frac{(d\bar{s})^2 - (ds)^2}{2(ds)^2} = \frac{\{dx\}^T\left(\frac{1}{2}([J]^T[J] - [I])\right)\{dx\}}{\{dx\}^T\{dx\}} = \frac{\{dx\}^T[\varepsilon_{[0]}]\{dx\}}{\{dx\}^T\{dx\}} \tag{3.132}$$

Substituting for $\{dx\}$ from (3.122)

$$\frac{(d\bar{s})^2 - (ds)^2}{2(ds)^2} = \frac{\{dx'\}^T[Q][\varepsilon_{[0]}][Q]^T\{dx'\}}{\{dx'\}^T[Q][Q]^T\{dx'\}} = \frac{\{dx'\}^T[\varepsilon'_{[0]}]\{dx'\}}{\{dx'\}^T\{dx'\}} \tag{3.133}$$

in which

$$[\varepsilon'_{[0]}] = [Q][\varepsilon_{[0]}][Q]^T \quad ; \quad [\varepsilon_{[0]}] = [Q]^T[\varepsilon'_{[0]}][Q] \tag{3.134}$$

Alternatively, using ds'

$$\frac{(d\bar{s})^2 - (ds')^2}{2(ds')^2} = \frac{\{d\bar{x}\}^T\{d\bar{x}\} - \{dx'\}^T\{dx'\}}{2\{dx'\}^T\{dx'\}} \tag{3.135}$$

using (3.123) and (3.124) and grouping terms

$$\frac{(d\bar{s})^2 - (ds')^2}{2(ds')^2} = \frac{\{dx'\}^T\left([Q][\varepsilon_{[0]}][Q]^T\right)\{dx'\}}{\{dx'\}^T\{dx'\}} = \frac{\{dx'\}^T[\varepsilon'_{[0]}]\{dx'\}}{\{dx'\}^T\{dx'\}} \tag{3.136}$$

which is same as (3.133) obtained using ds. Thus, $[\varepsilon_{[0]}]$ and $[\varepsilon'_{[0]}]$ are symmetric tensors of rank two in x- and x'-frame respectively.

Remarks

(1) Based on the orthogonal transformations presented here, we conclude $[C_{[0]}]$, $[F_{[0]}]$, $[\varepsilon_{[0]}]$ are covariant symmetric tensors of rank two in Lagrangian description.

(2) Lagrangian descriptions are sensitive when x_i are changed to x_i' by rigid rotation as these descriptions utilize reference lengths dx_i, thus, when dx_i change, the measures utilizing them change too.

3.12.2 Changes in contravariant Eulerian measures of strains when x_i are changed to x_i' due to rigid rotation

In addition to (3.122), we also have

$$\begin{aligned} \{d\bar{x}\} &= [J]\{dx\} = [J][Q]^T\{dx'\} \\ \{dx'\} &= [Q][J]^{-1}\{d\bar{x}\} \\ \{dx'\}^T &= \{d\bar{x}\}^T[[J]^{-1}]^T[Q]^T = \{d\bar{x}\}^T[\bar{J}]^T[Q]^T \end{aligned} \tag{3.137}$$

In the following we consider $[\bar{C}^{[0]}]$, $[\bar{F}^{[0]}]$ and $[\bar{\varepsilon}^{[0]}]$ as typical examples of strain measures to illustrate the influence of the change of frame. In all derivations that follow, we use $ds = ds'$.

3.12.2.1 Contravariant Cauchy strain in Eulerian description: $[\bar{C}^{[0]}]$

First, using ds'

$$\frac{(ds')^2}{(d\bar{s})^2} = \frac{\{dx'\}^T\{dx'\}}{\{d\bar{x}\}^T\{d\bar{x}\}} \tag{3.138}$$

using $\{dx'\}$ from (3.137)

$$\begin{aligned} \frac{(ds')^2}{(d\bar{s})^2} &= \frac{\left(\{d\bar{x}\}^T[\bar{J}]^T[Q]^T\right)\left([Q][\bar{J}]\{d\bar{x}\}\right)}{\{d\bar{x}\}^T\{d\bar{x}\}} \\ &= \frac{\{d\bar{x}\}^T[\bar{J}]^T[\bar{J}]\{d\bar{x}\}}{\{d\bar{x}\}^T\{d\bar{x}\}} = \frac{\{d\bar{x}\}^T[(\bar{C}_{[0]})']\{d\bar{x}\}}{\{d\bar{x}\}^T\{d\bar{x}\}} \end{aligned} \tag{3.139}$$

$$[(\bar{C}^{[0]})'] = [\bar{J}]^T[\bar{J}] = [\bar{C}^{[0]}] \tag{3.140}$$

Alternatively, using ds

$$\frac{(ds)^2}{(d\bar{s})^2} = \frac{\{dx\}^T\{dx\}}{\{d\bar{x}\}^T\{d\bar{x}\}} = \frac{\{d\bar{x}\}\left([\bar{J}]^T[\bar{J}]\right)\{d\bar{x}\}}{\{d\bar{x}\}^T\{d\bar{x}\}} = \frac{\{d\bar{x}\}^T[(\bar{C}_{[0]})]\{d\bar{x}\}}{\{d\bar{x}\}^T\{d\bar{x}\}} \tag{3.141}$$

which gives us (3.139) obtained using ds'. Thus, contravariant Cauchy strain in Eulerian description remains unaffected due to change of x_i to x_i' by a rigid rotation.

3.12.2.2 Contravariant Finger strain in Eulerian description or left Cauchy strain in Eulerian description: $[\bar{F}^{[0]}]$ or $[\bar{B}^{[0]}]$

Since $[\bar{F}^{[0]}] = [\bar{C}^{[0]}]^{-1}$ and $[\bar{C}^{[0]}] = [(\bar{C}^{[0]})']$, it is rather obvious that

$$[\bar{F}^{[0]}] = [(\bar{F}^{[0]})'] \tag{3.142}$$

That is, contravariant Finger strain in Eulerian description is unaffected by a rigid rotation of the x_i to x'_i.

3.12.2.3 Contravariant Almansi strain in Eulerian description: $[\bar{\varepsilon}^{[0]}]$

First, using ds'

$$\frac{(d\bar{s})^2 - (ds')^2}{2(d\bar{s})^2} = \frac{\{d\bar{x}\}^T\{d\bar{x}\} - \{dx'\}^T\{dx'\}}{2\{d\bar{x}\}^T\{d\bar{x}\}} \tag{3.143}$$

using $\{dx'\}$ from (3.137)

$$\begin{aligned}\frac{(d\bar{s})^2 - (ds')^2}{2(d\bar{s})^2} &= \frac{\{d\bar{x}\}^T[I]\{d\bar{x}\} - \left(\{d\bar{x}\}^T[\bar{J}]^T[Q]^T\right)\left([Q][\bar{J}]\{d\bar{x}\}\right)}{2\{d\bar{x}\}^T\{d\bar{x}\}}\\ &= \frac{\{d\bar{x}\}^T\left([\frac{1}{2}([I] - [\bar{J}]^T[\bar{J}])]\right)\{d\bar{x}\}}{\{d\bar{x}\}^T\{d\bar{x}\}} = \frac{\{d\bar{x}\}^T[(\bar{\varepsilon}^{[0]})']\{d\bar{x}\}}{\{d\bar{x}\}^T\{d\bar{x}\}}\end{aligned} \tag{3.144}$$

$$[(\bar{\varepsilon}^{[0]})'] = \frac{1}{2}([I] - [\bar{J}]^T[\bar{J}]) = [\bar{\varepsilon}^{[0]}] \tag{3.145}$$

Alternatively using ds (and (3.42) for $\{dx\}$

$$\begin{aligned}\frac{(d\bar{s})^2 - (ds)^2}{2(d\bar{s})^2} &= \frac{\{d\bar{x}\}^T\{d\bar{x}\} - \{dx\}^T\{dx\}}{\{d\bar{x}\}^T\{d\bar{x}\}}\\ &= \frac{\{d\bar{x}\}^T[I]\{d\bar{x}\} - \left(\{d\bar{x}\}^T[\bar{J}]^T\right)\left([\bar{J}]\{d\bar{x}\}\right)}{2\{d\bar{x}\}^T\{d\bar{x}\}}\\ &= \frac{\{d\bar{x}\}^T\left([\frac{1}{2}([I] - [\bar{J}]^T[\bar{J}])]\right)\{d\bar{x}\}}{\{d\bar{x}\}^T\{d\bar{x}\}} = \frac{\{d\bar{x}\}^T[(\bar{\varepsilon}^{[0]})']\{d\bar{x}\}}{\{d\bar{x}\}^T\{d\bar{x}\}}\end{aligned} \tag{3.146}$$

which is same as (3.145) obtained using ds'.

3.12.3 Change in covariant Lagrangian measures of strain when $\bar{x}_i$ are changed to $\bar{x}'_i$ by rigid rotation $[\bar{Q}]$

$$\begin{aligned}\{\bar{x}'\} = [\bar{Q}]\{\bar{x}\} \quad &; \quad \{\bar{x}\} = [\bar{Q}]^T\{\bar{x}'\}\\ \{d\bar{x}'\} = [\bar{Q}]\{d\bar{x}\} \quad &; \quad \{d\bar{x}\} = [\bar{Q}]^T\{d\bar{x}'\}\end{aligned} \tag{3.147}$$

We consider strain measures $[C_{[0]}]$, $[F_{[0]}]$ and $[\varepsilon_{[0]}]$, covariant measures in Lagrangian description. We note in this case $d\bar{s} = d\bar{s}'$ (as lengths are invariant under rigid rotation).

3.12.3.1 Covariant Cauchy strain in Lagrangian description: $[C_{[0]}]$

First, consider $d\bar{s}'$

$$\frac{(d\bar{s}')^2}{(ds)^2} = \frac{\{d\bar{x}'\}^T\{d\bar{x}'\}}{\{dx\}^T\{dx\}} = \frac{\{d\bar{x}\}^T[\bar{Q}]^T[\bar{Q}]\{d\bar{x}\}}{\{dx\}\{dx\}} = \frac{\{d\bar{x}\}^T\{d\bar{x}\}}{\{dx\}^T\{dx\}} \tag{3.148}$$

using (3.41) for $\{d\bar{x}\}$

$$\frac{(d\bar{s})^2}{(ds)^T} = \frac{\{dx\}^T\left([J]^T[J]\right)\{dx\}}{\{dx\}^T\{dx\}} = \frac{\{dx\}^T[\underline{C}'_{[0]}]\{dx\}}{\{dx\}^T\{dx\}} \tag{3.149}$$

$$\therefore \quad [\underline{C}'_{[0]}] = [J]^T[J] = [C_{[0]}] \tag{3.150}$$

$[\underline{C}'_{[0]}]$ is the covariant Cauchy strain in Lagrangian description due to change of $\bar{x}_i$ to $\bar{x}'_i$ by a rigid rotation.

Alternatively, using $\bar{ds}$

$$\frac{(d\bar{s})^2}{(ds)^2} = \frac{\{d\bar{x}\}^T\{d\bar{x}\}}{\{dx\}\{dx\}} \tag{3.151}$$

Since (3.151) is same as (3.148), the rest of the derivation follows and we conclude (3.150). Thus, the covariant Cauchy strain $[C_{[0]}]$ in Lagrangian description remains unaffected when $\bar{x}_i$ are change to $\bar{x}_i$ by rigid rotation.

3.12.3.2 Covariant Finger strain in Lagrangian description: $[F_{[0]}]$

Since

$$[\underline{F}'_{[0]}] = [\underline{C}'_{[0]}]^{-1} = [C_{[0]}]^{-1} = [F_{[0]}] \tag{3.152}$$

where $[\underline{F}'_{[0]}]$ is the covariant Finger strain in Lagrangian description in $\bar{x}'$-frame. Clearly $[F_{[0]}]$ remains unaffected due to this change of frame.

3.12.3.3 Covariant Green's strain in Lagrangian description: $[\varepsilon_{[0]}]$

First, consider $d\bar{s}'$

$$\frac{(d\bar{s}')^2 - (ds)^2}{2(ds)^2} = \frac{\{d\bar{x}'\}^T\{d\bar{x}'\} - \{dx\}^T\{dx\}}{2\{dx\}^T\{dx\}} \tag{3.153}$$

using (3.147) and (3.58)

$$\frac{(ds')^2 - (ds)^2}{2(ds)^2} = \frac{\{d\bar{x}\}^T[\bar{Q}]^T[\bar{Q}]\{d\bar{x}\} - \{dx\}^T\{dx\}}{2\{dx\}^T\{dx\}} = \frac{\{d\bar{x}\}^T\{d\bar{x}\} - \{dx\}^T\{dx\}}{2\{dx\}\{dx\}} \tag{3.154}$$

$$\therefore \quad \frac{(d\bar{s}')^2 - (ds)^2}{2(ds)^2} = \frac{\{dx\}^T\left[\frac{1}{2}\left([J]^T[J] - [I]\right)\right]\{dx\}}{\{dx\}^T\{dx\}} = \frac{\{dx\}^T[\underline{\varepsilon}'_{[0]}]\{dx\}}{\{dx\}^T\{dx\}} \tag{3.155}$$

$$[\underline{\varepsilon}'_{[0]}] = \frac{1}{2}\left([J]^T[J] - [I]\right) = [\varepsilon_{[0]}] \tag{3.156}$$

Alternatively, using $d\bar{s}$

$$\frac{(d\bar{s})^2 - (ds)^2}{2(ds)^2} = \frac{\{d\bar{x}\}^T\{d\bar{x}\} - \{dx\}^T\{dx\}}{2\{dx\}^T\{dx\}} \tag{3.157}$$

Since (3.157) is same as (3.154), rest of the derivations follows and we can conclude (3.157). Thus, covariant Green's strain in Lagrangian description is unaffected due to $\bar{x}_i$ changing to $\bar{x}'_i$ by rigid rotation.

3.12.4 Changes in Eulerian measures of strain when $\bar{x}_i$ are changed to $\bar{x}'_i$ by rigid rotation $[Q]$

In this section, we consider how the Eulerian descriptions of strains are affected when $\bar{x}_i$ are changed to $\bar{x}'_i$ by rigid rotation. We consider strain measures $[\bar{C}^{[0]}]$, $[\bar{F}^{[0]}]$ and $[\bar{\varepsilon}^{[0]}]$. In the derivations that follow, in addition to (3.147) we also use the following

$$\{dx\} = [\bar{J}]\{d\bar{x}\} = [\bar{J}][\bar{Q}]^T\{d\bar{x}'\} \quad ; \quad \{dx\}^T = \{d\bar{x}'\}^T[\bar{Q}][\bar{J}]^T \tag{3.158}$$

3.12.4.1 Contravariant Cauchy strain in Eulerian description: $[\bar{C}^{[0]}]$

First, consider $d\bar{s}'$ and we use (3.158)

$$\frac{(ds)^2}{(d\bar{s}')^2} = \frac{\{dx\}^T\{dx\}}{\{d\bar{x}'\}^T\{d\bar{x}'\}} = \frac{\left(\{d\bar{x}'\}^T[\bar{Q}][\bar{J}]^T\right)\left([\bar{J}][\bar{Q}]^T\{d\bar{x}'\}\right)}{\{d\bar{x}'\}^T\{d\bar{x}'\}} \tag{3.159}$$

$$= \frac{\{d\bar{x}'\}^T[(\bar{C}^{[0]})']\{d\bar{x}\}}{\{d\bar{x}'\}^T\{d\bar{x}'\}} \tag{3.160}$$

$$\begin{aligned} [(\bar{C}^{[0]})'] &= [\bar{Q}]\left([\bar{J}]^T[\bar{J}]\right)[\bar{Q}]^T = [\bar{Q}][\bar{C}^{[0]}][\bar{Q}]^T \\ \text{and} \quad [\bar{C}^{[0]}] &= [\bar{Q}]^T[(\bar{C}^{[0]})'][\bar{Q}] \end{aligned} \tag{3.161}$$

Alternatively, using $d\bar{s}$ and (3.147) for $\{d\bar{x}\}$

$$\frac{(ds)^2}{(d\bar{s})^2} = \frac{\{dx\}^T\{dx\}}{\{d\bar{x}\}^T\{d\bar{x}\}} = \frac{\{dx\}^T\{dx\}}{\{d\bar{x}'\}^T[\bar{Q}][\bar{Q}]^T\{d\bar{x}'\}} = \frac{\{dx\}^T\{dx\}}{\{d\bar{x}'\}^T\{d\bar{x}'\}} \tag{3.162}$$

Since (3.162) is same as (3.159), rest of the derivations follows and we conclude (3.161). Equations (3.161) confirm that $[\bar{C}^{[0]}]$ and $[(\bar{C}^{[0]})']$ both are symmetric tensors of rank two in $\bar{x}$- and $\bar{x}'$-frames and that the contravariant Eulerian Cauchy strain tensor is effected when $\bar{x}_i$ are changed to $\bar{x}'_i$ by rigid rotation $[\bar{Q}]$.

3.12.4.2 Contravariant Finger strain in Eulerian description: $[\bar{F}^{[0]}]$

Consider

$$[(\bar{F}^{[0]})'] = [(\bar{C}^{[0]})']^{-1} = [[\bar{Q}][\bar{C}^{[0]}][\bar{Q}]^T]^{-1} \tag{3.163}$$

Hence

$$[(\bar{F}^{[0]})'] = [\bar{Q}][\bar{C}^{[0]}]^{-1}[\bar{Q}]^T = [\bar{Q}][\bar{J}]^{-1}[[\bar{J}]^T]^{-1}[\bar{Q}]^T \tag{3.164}$$

or

$$[(\bar{F}^{[0]})'] = [\bar{Q}][\bar{F}^{[0]}][\bar{Q}]^T \tag{3.165}$$

From equation (3.160) we conclude that $[\bar{F}^{[0]}]$ and $[(\bar{F}^{[0]})']$ are symmetric tensors of rank two in $\bar{x}$- and $\bar{x}'$-frames and that contravariant Eulerian Finger strain is effected when $\bar{x}_i$ are changed to $\bar{x}'_i$ by rigid rotation $[\bar{Q}]$.

3.12.4.3 Contravariant Almansi strain in Eulerian description: $[\bar{\varepsilon}^{[0]}]$

First, consider $d\bar{s}'$

$$
\begin{aligned}
\frac{(d\bar{s}')^2 - (ds)^2}{2(d\bar{s}')^2} &= \frac{\{d\bar{x}'\}^T\{d\bar{x}'\} - \{dx\}^T\{dx\}}{2\{d\bar{x}'\}^T\{d\bar{x}'\}} \\
&= \frac{\{d\bar{x}'\}^T\left([\bar{Q}][I][\bar{Q}]^T\right)\{d\bar{x}'\} - \left(\{d\bar{x}'\}^T[\bar{Q}][\bar{J}]^T\right)\left([\bar{J}][\bar{Q}]^T\{d\bar{x}'\}\right)}{2\{d\bar{x}'\}^T\{d\bar{x}'\}}
\end{aligned}
\tag{3.166}
$$

$$
\begin{aligned}
&= \frac{\{d\bar{x}'\}^T\left([\bar{Q}]\left(\frac{1}{2}\left([I] - [\bar{J}]^T[\bar{J}]\right)\right)[\bar{Q}]^T\right)\{d\bar{x}'\}}{\{d\bar{x}'\}^T\{d\bar{x}'\}} \\
&= \frac{\{d\bar{x}'\}^T[(\bar{\varepsilon}^{[0]})']\{d\bar{x}'\}}{\{d\bar{x}'\}^T\{d\bar{x}'\}}
\end{aligned}
\tag{3.167}
$$

$$
\begin{aligned}
[(\bar{\varepsilon}^{[0]})'] &= [\bar{Q}]\left(\frac{1}{2}\left([I] - [\bar{J}]^T[\bar{J}]\right)\right)[\bar{Q}]^T = [\bar{Q}][\bar{\varepsilon}^{[0]}][\bar{Q}]^T \\
\text{and} \quad [\bar{\varepsilon}^{[0]}] &= [\bar{Q}]^T[(\bar{\varepsilon}^{[0]})'][\bar{Q}]
\end{aligned}
\tag{3.168}
$$

Alternatively, using $d\bar{s}$ and (3.147) for $\{d\bar{x}\}$

$$
\begin{aligned}
\frac{(d\bar{s})^2 - (ds)^2}{2(d\bar{s})^2} &= \frac{\{d\bar{x}\}^T\{d\bar{x}\} - \{dx\}^T\{dx\}}{2\{d\bar{x}\}^T\{d\bar{x}\}} \\
&= \frac{\{d\bar{x}'\}^T[\bar{Q}]^T[\bar{Q}]\{d\bar{x}'\} - \{dx\}^T\{dx\}}{2\{d\bar{x}'\}^T[\bar{Q}]^T[\bar{Q}]\{d\bar{x}'\}} \\
&= \frac{\{d\bar{x}'\}^T\{d\bar{x}'\} - \{dx\}^T\{dx\}}{2\{d\bar{x}'\}^T\{d\bar{x}'\}}
\end{aligned}
\tag{3.169}
$$

Since (3.169) is same as (3.166), rest of the derivations follow and we conclude (3.168). Equations (3.168) confirm that $[\bar{\varepsilon}^{[0]}]$ and $[(\bar{\varepsilon}^{[0]})']$ both are symmetric tensors of rank two in $\bar{x}$- and $\bar{x}$-frames and that contravariant Almansi strain tensor is effected when $\bar{x}_i$ are change to $\bar{x}'_i$ by rigid rotation $[\bar{Q}]$.

Remarks

(1) $[C_{[0]}], [F_{[0]}]$ and $[\varepsilon_{[0]}]$ strain tensors are covariant measures in Lagrangian description. These are affected by rigid rotation of the reference configuration x-frame to x'-frame. $[C_{[0]}], [F_{[0]}]$ and $[\varepsilon_{[0]}]$ in x-frame and $[C'_{[0]}], [F'_{[0]}]$ and $[\varepsilon'_{[0]}]$ in x'-frame are all symmetric tensors of rank two.

(2) $[\bar{C}^{[0]}], [\bar{F}^{[0]}]$ and $[\bar{\varepsilon}^{[0]}]$ strain tensors are contravariant measures in Eulerian description. These are not affected by rigid rotation of the x-frame to x'-frame.

(3) The strain tensors $[C_{[0]}], [F_{[0]}]$ and $[\varepsilon_{[0]}]$ in Lagrangian description are not affected by rigid rotation of the $\bar{x}$-frame to $\bar{x}'$-frame.

(4) The strain tensors $[\bar{C}^{[0]}], [\bar{F}^{[0]}]$ and $[\bar{\varepsilon}^{[0]}]$ in Eulerian description are affected by rigid rotation of the $\bar{x}$-frame to $\bar{x}'$-frame. $[\bar{C}^{[0]}], [\bar{F}^{[0]}]$ and $[\bar{\varepsilon}^{[0]}]$ in $\bar{x}$-frame and $[(\bar{C}^{[0]})'], [(\bar{F}^{[0]})']$ and $[(\bar{\varepsilon}^{[0]})']$ in $\bar{x}'$-frame are all symmetric tensors of rank two.

(5) The Green's strain tensor $[\varepsilon_{[0]}]$ in Lagrangian description and Almansi strain tensor $[\bar{\varepsilon}^{[0]}]$ in Eulerian description are used predominately for finite deformation in the current literature.

(6) In the material presented in Section 3.12, we have only considered strain measures $[C_{[0]}]$, $[F_{[0]}]$ and $[\varepsilon_{[0]}]$ that are covariant measures in Lagrangian description and the strain measures $[\bar{C}^{[0]}]$, $[\bar{F}^{[0]}]$ and $[\bar{\varepsilon}^{[0]}]$ that are contravariant measures in Eulerian description to establish their tensorial nature and the influence of change of frame. The principles used in Section 3.12 for these strain measures also hold for other strain measures in Table 3.1, i.e., all strain measures in Table 3.1 are symmetric tensors of rank two and obey transformation rules due to change of frame similar to those discussed in Section 3.12.

3.13 Invariants of strain tensors

Since strain tensors are tensors of rank two, the material presented in Chapter 2 on invariants of a tensor of rank two is applicable to each one of these strain tensors as well. Thus, if $[\tilde{\varepsilon}]$ represents any one of the strain tensors, then we have the following:

(1) *Invariants using trace:*

$\text{tr}([\tilde{\varepsilon}])$, $\text{tr}([\tilde{\varepsilon}][\tilde{\varepsilon}]) = \text{tr}([\tilde{\varepsilon}]^2)$, $\text{tr}([\tilde{\varepsilon}][\tilde{\varepsilon}][\tilde{\varepsilon}]) = \text{tr}([\tilde{\varepsilon}]^3)$ are invariants of $[\tilde{\varepsilon}]$. These have the same values in any two orthogonal coordinate systems that are related through a rigid rotation. For example, if the x-frame is rotated into x'-frame, the invariants of $[C]$ and $[C']$ remain the same. Likewise if the $\bar{x}$-frame is rotated into $\bar{x}'$-frame, the invariants of $[\bar{C}]$ and $[\bar{C}']$ remain unaffected. We denote these invariants of $[\tilde{\varepsilon}]$ by $i_{\tilde{\varepsilon}}$, $ii_{\tilde{\varepsilon}}$ and $iii_{\tilde{\varepsilon}}$, i.e., $i_{\tilde{\varepsilon}} = \text{tr}([\tilde{\varepsilon}])$, $ii_{\tilde{\varepsilon}} = \text{tr}([\tilde{\varepsilon}]^2)$ and $iii_{\tilde{\varepsilon}} = \text{tr}([\tilde{\varepsilon}]^3)$.

(2) *Principal invariants of $[\tilde{\varepsilon}]$ using the characteristic equation:*

If a vector $\{\phi\}$ is an eigenvector of $[\tilde{\varepsilon}]$, then there exists a scalar λ such that

$$[\tilde{\varepsilon}]\{\phi\} = \lambda\{\phi\} \quad \text{or} \quad \tilde{\varepsilon}_{ij}\phi_j = \lambda\phi_j \quad ; \quad \text{(Def.)} \tag{3.170}$$

The scalar λ is called the eigenvalue of $[\tilde{\varepsilon}]$ corresponding to the eigenvector $\{\phi\}$. Equations (3.170) imply that

$$\det([\tilde{\varepsilon}] - \lambda[I]) = 0 \tag{3.171}$$

This can be written as

$$\lambda^3 - I_{\tilde{\varepsilon}}\lambda^2 + II_{\tilde{\varepsilon}}\lambda - III_{\tilde{\varepsilon}} = 0 \tag{3.172}$$

Equation (3.172) is called the characteristic equation of the second order tensor $[\tilde{\varepsilon}]$ and has three roots λ_1, λ_2 and λ_3. Corresponding to each root there is an eigenvector $\{\phi\}_i$; $i = 1, 2, 3$. The pairs $(\lambda_i, \{\phi\}_i)$ are called eigenpairs of (3.172). The λ_i are principal strains and $\{\phi\}_i$; $i = 1, 2, 3$ are the directions of the principal strains. In (3.172), we have

$$\begin{aligned} I_{\tilde{\varepsilon}} &= \operatorname{tr}([\tilde{\varepsilon}]) \\ II_{\tilde{\varepsilon}} &= \frac{1}{2}(\operatorname{tr}([\tilde{\varepsilon}])^2 - \operatorname{tr}([\tilde{\varepsilon}]^2)) \\ III_{\tilde{\varepsilon}} &= \det([\tilde{\varepsilon}]) \end{aligned} \tag{3.173}$$

Since the eigenvalues λ_i are invariant of orthogonal transformation, $I_{\tilde{\varepsilon}}$, $II_{\tilde{\varepsilon}}$ and $III_{\tilde{\varepsilon}}$ are invariants of the orthogonal transformation as well.

(3) The invariants $i_{\tilde{\varepsilon}}$, $ii_{\tilde{\varepsilon}}$ and $iii_{\tilde{\varepsilon}}$ and the principal invariants $I_{\tilde{\varepsilon}}$, $II_{\tilde{\varepsilon}}$ and $III_{\tilde{\varepsilon}}$ are related.

$$\begin{aligned} i_{\tilde{\varepsilon}} &= I_{\tilde{\varepsilon}} \\ ii_{\tilde{\varepsilon}} &= I_{\tilde{\varepsilon}}^2 - 2II_{\tilde{\varepsilon}} \\ iii_{\tilde{\varepsilon}} &= 3III_{\tilde{\varepsilon}} + I_{\tilde{\varepsilon}}^3 - 3I_{\tilde{\varepsilon}}II_{\tilde{\varepsilon}} \end{aligned} \tag{3.174}$$

(4) The Hamilton-Cayley theorem also holds for $[\tilde{\varepsilon}]$. That is, the strain tensor $[\tilde{\varepsilon}]$ satisfies its own characteristic equation. See Chapter 2 for more details.

(5) Expanded forms of the two sets of invariants can be obtained using the expanded form of the invariants given in Chapter 2 for a tensor of rank two.

3.14 Expanded form of strain tensors

The expanded forms of the covariant and contravariant strain tensors in Lagrangian as well as Eulerian descriptions can be obtained.

By using definitions of deformation gradient tensors $[J]$ and $[\bar{J}]$ expressed in terms of displacement gradients, i.e., using

$$[J] = \left[\frac{\bar{x}_1, \bar{x}_2, \bar{x}_3}{x_1, x_2, x_3}\right] = [I] + \left[\frac{u_1, u_2, u_3}{x_1, x_2, x_3}\right] = [I] + [\,{}^dJ] \tag{3.175}$$

$$[\bar{J}] = \left[\frac{x_1, x_2, x_3}{\bar{x}_1, \bar{x}_2, \bar{x}_3}\right] = [I] - \left[\frac{\bar{u}_1, \bar{u}_2, \bar{u}_3}{\bar{x}_1, \bar{x}_2, \bar{x}_3}\right] = [I] - [{}^d\bar{J}] \tag{3.176}$$

In the following, we only consider expanded forms of Green's strain $[\varepsilon_{[0]}]$ and Almansi strain $[\bar{\varepsilon}^{[0]}]$. Other strain measures can be expanded using the same approach.

3.14.1 Green's strain measure: $[\varepsilon_{[0]}]$

In matrix and vector notations we can express components of the Green's strain measure as

$$[\varepsilon_{[0]}] = \begin{bmatrix} (\varepsilon_{[0]})_{11} & (\varepsilon_{[0]})_{12} & (\varepsilon_{[0]})_{13} \\ (\varepsilon_{[0]})_{21} & (\varepsilon_{[0]})_{22} & (\varepsilon_{[0]})_{23} \\ (\varepsilon_{[0]})_{31} & (\varepsilon_{[0]})_{32} & (\varepsilon_{[0]})_{33} \end{bmatrix} \tag{3.177}$$

Using (3.175), Green's strain can be expressed

$$[\varepsilon_{[0]}] = \frac{1}{2}([{}^dJ] + [{}^dJ]^T + [{}^dJ]^T[{}^dJ]) \tag{3.178}$$

Substituting for $[\,{}^dJ]$ in terms of gradients of u_i (3.175) in (3.178) and separating linear and nonlinear terms in the gradients in u_i for each component of $[\varepsilon_{[0]}]$, we obtain the following.

$$
\begin{aligned}
(\varepsilon_{[0]})_{11} &= \frac{\partial u_1}{\partial x_1} + \frac{1}{2}\left(\left(\frac{\partial u_1}{\partial x_1}\right)^2 + \left(\frac{\partial u_2}{\partial x_1}\right)^2 + \left(\frac{\partial u_3}{\partial x_1}\right)^2\right) \\
(\varepsilon_{[0]})_{22} &= \frac{\partial u_2}{\partial x_2} + \frac{1}{2}\left(\left(\frac{\partial u_1}{\partial x_2}\right)^2 + \left(\frac{\partial u_2}{\partial x_2}\right)^2 + \left(\frac{\partial u_3}{\partial x_2}\right)^2\right) \\
(\varepsilon_{[0]})_{33} &= \frac{\partial u_3}{\partial x_3} + \frac{1}{2}\left(\left(\frac{\partial u_1}{\partial x_3}\right)^2 + \left(\frac{\partial u_2}{\partial x_3}\right)^2 + \left(\frac{\partial u_3}{\partial x_3}\right)^2\right) \\
(\varepsilon_{[0]})_{12} &= \frac{1}{2}\left(\frac{\partial u_2}{\partial x_1} + \frac{\partial u_1}{\partial x_2}\right) + \frac{1}{2}\left(\frac{\partial u_1}{\partial x_1}\frac{\partial u_1}{\partial x_2} + \frac{\partial u_2}{\partial x_1}\frac{\partial u_2}{\partial x_2} + \frac{\partial u_3}{\partial x_1}\frac{\partial u_3}{\partial x_2}\right) \\
(\varepsilon_{[0]})_{23} &= \frac{1}{2}\left(\frac{\partial u_2}{\partial x_3} + \frac{\partial u_3}{\partial x_2}\right) + \frac{1}{2}\left(\frac{\partial u_1}{\partial x_2}\frac{\partial u_1}{\partial x_3} + \frac{\partial u_2}{\partial x_2}\frac{\partial u_2}{\partial x_3} + \frac{\partial u_3}{\partial x_2}\frac{\partial u_3}{\partial x_3}\right) \\
(\varepsilon_{[0]})_{31} &= \frac{1}{2}\left(\frac{\partial u_1}{\partial x_3} + \frac{\partial u_3}{\partial x_1}\right) + \frac{1}{2}\left(\frac{\partial u_1}{\partial x_1}\frac{\partial u_1}{\partial x_3} + \frac{\partial u_2}{\partial x_1}\frac{\partial u_2}{\partial x_3} + \frac{\partial u_3}{\partial x_1}\frac{\partial u_3}{\partial x_3}\right)
\end{aligned}
\tag{3.179}
$$

Remarks

(1) Strains are nonlinear functions of the components of the displacement gradient tensors $[\,^d\!J]$.

(2) Decomposing $[\varepsilon_{[0]}]$ into linear and nonlinear components reveals the precise nature of linear and nonlinear terms.

(3) Simplification of (3.179) for small deformation (infinitesimal), small strain physics will be considered in a later section.

3.14.2 Almansi strain measure : $[\bar{\varepsilon}^{[0]}]$

In the matrix form we can write $[\bar{\varepsilon}^{[0]}]$ as

$$
[\bar{\varepsilon}^{[0]}] = \begin{bmatrix} (\bar{\varepsilon}^{[0]})_{11} & (\bar{\varepsilon}^{[0]})_{12} & (\bar{\varepsilon}^{[0]})_{13} \\ (\bar{\varepsilon}^{[0]})_{21} & (\bar{\varepsilon}^{[0]})_{22} & (\bar{\varepsilon}^{[0]})_{23} \\ (\bar{\varepsilon}^{[0]})_{31} & (\bar{\varepsilon}^{[0]})_{32} & (\bar{\varepsilon}^{[0]})_{33} \end{bmatrix} \tag{3.180}
$$

using (3.176) Almansi strain $[\bar{\varepsilon}^{[0]}]$ can be expressed as

$$
[\bar{\varepsilon}^{[0]}] = \frac{1}{2}\left([I] - [\bar{J}]^T[\bar{J}]\right) = \frac{1}{2}\left([\,^d\bar{J}] + [\,^d\bar{J}]^T - [\,^d\bar{J}]^T[\,^d\bar{J}]\right) \tag{3.181}
$$

substituting $[\,^d\bar{J}]$ in terms of the gradients $\bar{u}_i$ with respect to $\bar{x}_j$ (equation (3.176)) and separating linear and nonlinear terms in the gradients of $\bar{u}_i$ for each element of $[\bar{\varepsilon}^{[0]}]$ we can obtain expression for $[\bar{\varepsilon}^{[0]}]$ (by replacing u_i and x_j with $\bar{u}_i$ and $\bar{x}_j$ and changing the signs of the non-linear terms on the right side to negative).

3.15 Physical meaning of strains

In the following we consider Green's strain $\boldsymbol{\varepsilon}_{[0]}$, in Lagrangian description, a covariant measure of strain, and Almansi strain $\bar{\boldsymbol{\varepsilon}}^{[0]}$, in Eulerian description, a contravariant measure of strain. Before we present mathematical details, we describe the physics of deformation. For simplicity, we consider 2D case (i.e., $\mathbb{R}^2$). Consider a pair of orthogonal material lines ϕ and θ (i.e., $\boldsymbol{PQ}$ and $\boldsymbol{PR}$) at a material point $P(x_1, x_2)$ in the reference configuration (Figure 3.4(a)). In the current configuration at time $t > 0$ (t_0) these deform into curvilinear material lines $\bar{\phi}$ and $\bar{\theta}$ (i.e., $\bar{\boldsymbol{P}}\bar{\boldsymbol{Q}}$

and $\bar{\boldsymbol{P}}\bar{\boldsymbol{R}}$) at $\bar{P}(\bar{x}_1, \bar{x}_2)$ (Figure 3.4(b)). The tangent vectors to deformed material lines $\bar{\phi}$ and $\bar{\theta}$ in Figure 3.4(b) are covariant base vectors. Covariant measures of strain are measures of the extension of the material lines ϕ and θ in terms of $\tilde{\boldsymbol{g}}_1$, $\tilde{\boldsymbol{g}}_2$ (and $\tilde{\boldsymbol{g}}_3$ in $\mathbb{R}^3$) and a measure of the change in 90 degree angle between them through angle β (in $\mathbb{R}^2$). In $\mathbb{R}^3$, these amount to a measure of extensions of three orthogonal material lines in the reference configuration and the change in 90 degree angle between them in terms of $\tilde{\boldsymbol{g}}_i$ and the three angles between pairs of $\tilde{\boldsymbol{g}}_i$.

The physics of contravariant measures of strain can be viewed in two ways. We could consider a pair of orthogonal material lines $\bar{\boldsymbol{P}}\bar{\boldsymbol{Q}}$ and $\bar{\boldsymbol{P}}\bar{\boldsymbol{R}}$, i.e., $\bar{\phi}$ and $\bar{\theta}$ in the current configuration (Figure 3.4(d)). Their map in the reference configuration will result in curvilinear material lines $\boldsymbol{PQ}$ and $\boldsymbol{PR}$, i.e., ϕ and θ (Figure 3.4(c)). The tangent vectors to ϕ and θ at point P are indeed contravariant base vectors. Contravariant measures of strain are measures of extensions of $\bar{\phi}$ and $\bar{\theta}$ (Figure 3.4(d)) and the change in 90 degree angle between them using $\tilde{\boldsymbol{g}}^i$ and angle α. In $\mathbb{R}^3$, we naturally have three extensions and three angles between pairs of $\tilde{\boldsymbol{g}}^i$.

Another way to visualize the contravariant measure of strain is to refer to Figures 2.7(a)–(c). The contravariant base vectors are perpendicular to the faces of deformed tetrahedron whose straight edges are formed by covariant base vectors. The angles β and α shown in Figure 2.7 are exactly the same as those in Figure 3.4. For illustrating the details of what the strain components represent, we consider $\boldsymbol{\varepsilon}_{[0]}$, Green's strain in Lagrangian description, (covariant measure) and $\bar{\boldsymbol{\varepsilon}}^{[0]}$, Almansi strain in Eulerian description (contravariant measure).

Consider an arbitrary line segment $\boldsymbol{AB}$ of length ds in the reference configuration at time t_0 with its components dx_1, dx_2, dx_3 in the x-frame. Let this line segment $\boldsymbol{AB}$ be deformed into $\bar{\boldsymbol{A}}\bar{\boldsymbol{B}}$ in the current configuration at time t with its length $d\bar{s}$. Let $d\bar{x}_1, d\bar{x}_2, d\bar{x}_3$ be the components of $d\bar{s}$ in the x-frame.

Then $\{l\}$, the direction cosines of $\boldsymbol{AB}$, and $\{\bar{l}\}$, the direction cosines of $\bar{\boldsymbol{A}}\bar{\boldsymbol{B}}$, are given by

$$\{l\} = [l_1, l_2, l_3]^T = \left\{ \frac{dx}{ds} \right\} = \left[\frac{dx_1}{ds}, \frac{dx_2}{ds}, \frac{dx_3}{ds} \right]^T \tag{3.182}$$

$$\text{and} \quad \{\bar{l}\} = [\bar{l}_1, \bar{l}_2, \bar{l}_3]^T = \left\{ \frac{d\bar{x}}{d\bar{s}} \right\} = \left[\frac{d\bar{x}_1}{d\bar{s}}, \frac{d\bar{x}_2}{d\bar{s}}, \frac{d\bar{x}_3}{d\bar{s}} \right]^T \tag{3.183}$$

such that

$$\{l\}^T\{l\} = 1 \quad \text{and} \quad \{\bar{l}\}^T\{\bar{l}\} = 1 \tag{3.184}$$

Hence

$$\{dx\} = \{l\}ds \quad ; \quad \{d\bar{x}\} = \{\bar{l}\}d\bar{s} \tag{3.185}$$

Consider Green's strain tensor $[\varepsilon_{[0]}]$ in Lagrangian description, which is derived using

$$\frac{(d\bar{s})^2 - (ds)^2}{2(ds)^2} = \frac{\{dx\}^T[\frac{1}{2}([J]^T[J] - [I])]\{dx\}}{\{dx\}^T\{dx\}} = \frac{\{dx\}^T[\varepsilon_{[0]}]\{dx\}}{\{dx\}^T\{dx\}} \tag{3.186}$$

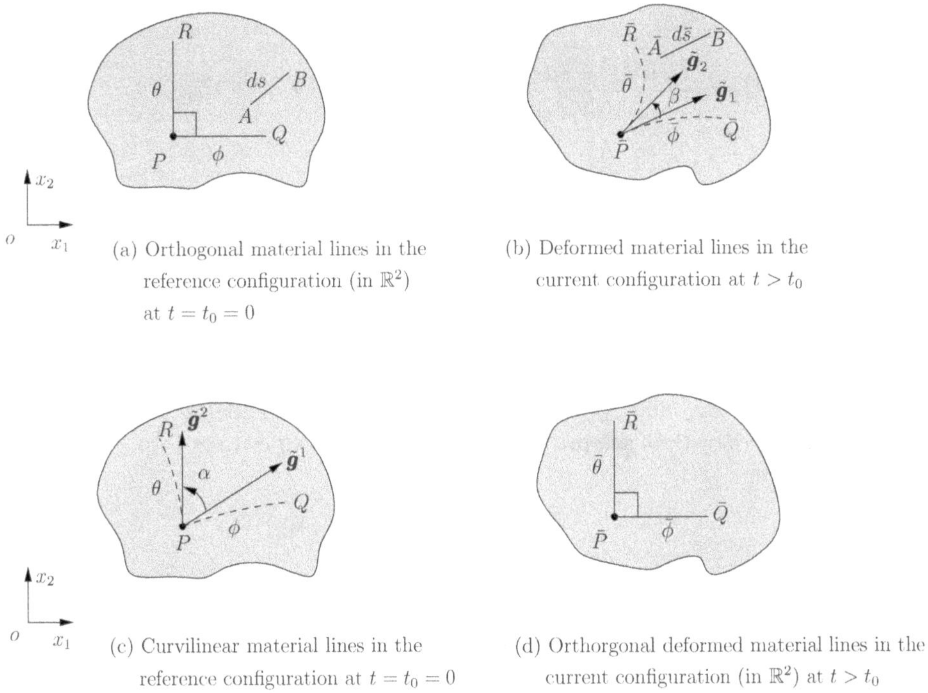

(a) Orthogonal material lines in the reference configuration (in $\mathbb{R}^2$) at $t = t_0 = 0$

(b) Deformed material lines in the current configuration at $t > t_0$

(c) Curvilinear material lines in the reference configuration at $t = t_0 = 0$

(d) Orthorgonal deformed material lines in the current configuration (in $\mathbb{R}^2$) at $t > t_0$

Figure 3.4: Co- and contra-variant bases used in co- and contra-variant measures of strain

Substituting for $\{dx\}$ in (3.186) from (3.185)

$$\frac{(d\bar{s})^2-(ds)^2}{2(ds)^2} = \frac{ds\{l\}^T[\varepsilon_{[0]}]\{l\}ds}{ds\{l\}^T\{l\}ds} = \frac{\{l\}^T[\varepsilon_{[0]}]\{l\}}{\{l\}^T\{l\}} = \{l\}^T[\varepsilon_{[0]}]\{l\} \tag{3.187}$$

$$\therefore \quad \frac{1}{2}\Big(\frac{(d\bar{s})^2-(ds)^2}{(ds)^2}\Big) = \{l\}^T[\varepsilon_{[0]}]\{l\} \tag{3.188}$$

Let us define elongation of $\boldsymbol{AB}$, E_{AB} per unit length of $\boldsymbol{AB}$ (also called extension of $\boldsymbol{AB}$) and λ_{AB}, the stretch of $\boldsymbol{AB}$ as

$$E_{AB} = \frac{d\bar{s}-ds}{ds} \tag{3.189}$$

$$\lambda_{AB} = \frac{d\bar{s}}{ds} \tag{3.190}$$

From (3.189) and (3.190), we can write

$$E_{AB} = \frac{d\bar{s}}{ds} - 1 = \lambda_{AB} - 1 \tag{3.191}$$

$$\lambda_{AB} = E_{AB} + 1 \tag{3.192}$$

$$\frac{d\bar{s}}{ds} = E_{AB} + 1 \tag{3.193}$$

$$\therefore \quad d\bar{s} = (E_{AB}+1)ds = \lambda_{AB}\, ds \tag{3.194}$$

Consider (3.188)

$$\{l\}^T[\varepsilon_{[0]}]\{l\} = \frac{1}{2}\Big(\frac{(d\bar{s})^2-(ds)^2}{(ds)^2}\Big) = \frac{1}{2}\Big(\Big(\frac{d\bar{s}}{ds}\Big)^2-1\Big) = \frac{1}{2}(\lambda_{AB}^2-1) \tag{3.195}$$

Using (3.192) for λ_{AB}, (3.195) can be written as

$$\{l\}^T[\varepsilon_{[0]}]\{l\} = \frac{1}{2}(\lambda_{AB}^2-1) = \frac{1}{2}((E_{AB}+1)^2-1) = E_{AB}\Big(1+\frac{1}{2}E_{AB}\Big) \tag{3.196}$$

3.15.1 Extensions and stretches parallel to the axes of x-frame using $[\varepsilon_{[0]}]$

Let $E_{x_1}, E_{x_2}, E_{x_3}$ and $\lambda_{x_1}, \lambda_{x_2}, \lambda_{x_3}$ be the extension and the stretches of the fibers dx_1, dx_2, dx_3 that are projections of ds parallel to ox_1, ox_2, ox_3 axes of the x-frame, and if

$$[\varepsilon_{[0]}] = \begin{bmatrix} (\varepsilon_{[0]})_{11} & (\varepsilon_{[0]})_{12} & (\varepsilon_{[0]})_{13} \\ (\varepsilon_{[0]})_{21} & (\varepsilon_{[0]})_{22} & (\varepsilon_{[0]})_{23} \\ (\varepsilon_{[0]})_{31} & (\varepsilon_{[0]})_{32} & (\varepsilon_{[0]})_{33} \end{bmatrix} \tag{3.197}$$

then using (3.196) we can write the following:

$$\Big(1+\frac{1}{2}E_{x_1}\Big)E_{x_1} = \begin{Bmatrix}1\\0\\0\end{Bmatrix}^T [\varepsilon_{[0]}] \begin{Bmatrix}1\\0\\0\end{Bmatrix} = (\varepsilon_{[0]})_{11} \tag{3.198}$$

or

$$E_{x_1}^2 + 2E_{x_1} - 2(\varepsilon_{[0]})_{11} = 0 \tag{3.199}$$

Solving the quadratic equation in E_{x_1} given by (3.199)

$$E_{x_1} = (1+2(\varepsilon_{[0]})_{11})^{1/2} - 1 \tag{3.200}$$

Similarly

$$\Big(1+\frac{1}{2}E_{x_2}\Big)E_{x_2} = \begin{Bmatrix}0\\1\\0\end{Bmatrix}^T [\varepsilon_{[0]}] \begin{Bmatrix}0\\1\\0\end{Bmatrix} = (\varepsilon_{[0]})_{22} \tag{3.201}$$

and

$$\Big(1+\frac{1}{2}E_{x_3}\Big)E_{x_3} = \begin{Bmatrix}0\\0\\1\end{Bmatrix}^T [\varepsilon_{[0]}] \begin{Bmatrix}0\\0\\1\end{Bmatrix} = (\varepsilon_{[0]})_{33} \tag{3.202}$$

Solving for E_{x_2} and E_{x_3} from (3.201) and (3.202) we obtain

$$\begin{aligned} E_{x_2} &= (1+2(\varepsilon_{[0]})_{22})^{1/2} - 1 \\ E_{x_3} &= (1+2(\varepsilon_{[0]})_{33})^{1/2} - 1 \end{aligned} \tag{3.203}$$

Since

$$E_{x_1} = \lambda_{x_1} - 1 \quad ; \quad E_{x_2} = \lambda_{x_2} - 1 \quad ; \quad E_{x_3} = \lambda_{x_3} - 1 \tag{3.204}$$

we can write (3.200) and (3.203) as follows:

$$\begin{aligned} \lambda_{x_1} &= (1 + 2(\varepsilon_{[0]})_{11})^{1/2} \quad ; \quad \lambda_{x_2} = (1 + 2(\varepsilon_{[0]})_{22})^{1/2} \\ \lambda_{x_3} &= (1 + 2(\varepsilon_{[0]})_{33})^{1/2} \end{aligned} \tag{3.205}$$

We note the following relationship:

$$\{\bar{l}\} = \left\{\frac{d\bar{x}}{d\bar{s}}\right\} = \left[\frac{d\bar{x}}{dx}\right]\left\{\frac{dx}{ds}\right\}\frac{ds}{d\bar{s}} = [J]\,\{l\}\,\frac{1}{\lambda_{AB}} \tag{3.206}$$

when

$$\begin{aligned} \{l\} &= \begin{Bmatrix} 1 \\ 0 \\ 0 \end{Bmatrix} \quad ; \quad \lambda_{AB} = \lambda_{x_1} \quad \text{and} \quad \{\bar{l}\} = \{\bar{l}_{x_1}\} \\ \{l\} &= \begin{Bmatrix} 0 \\ 1 \\ 0 \end{Bmatrix} \quad ; \quad \lambda_{AB} = \lambda_{x_2} \quad \text{and} \quad \{\bar{l}\} = \{\bar{l}_{x_2}\} \\ \{l\} &= \begin{Bmatrix} 0 \\ 0 \\ 1 \end{Bmatrix} \quad ; \quad \lambda_{AB} = \lambda_{x_3} \quad \text{and} \quad \{\bar{l}\} = \{\bar{l}_{x_3}\} \end{aligned} \tag{3.207}$$

Using (3.206)) and (3.207)), we can write

$$[\{\bar{l}_{x_1}\}, \{\bar{l}_{x_2}\}, \{\bar{l}_{x_3}\}] = [J]\begin{bmatrix} 1/\lambda_{x_1} & 0 & 0 \\ 0 & 1/\lambda_{x_2} & 0 \\ 0 & 0 & 1/\lambda_{x_3} \end{bmatrix} = [J][\lambda]^{-1} \tag{3.208}$$

or

$$[\{\bar{l}_{x_1}\}, \{\bar{l}_{x_2}\}, \{\bar{l}_{x_3}\}] = [J][\lambda]^{-1} \tag{3.209}$$

Remarks

(1) The terms $\{\bar{l}_{x_1}\}, \{\bar{l}_{x_2}\}, \{\bar{l}_{x_3}\}$ are the direction cosines of the tangent vectors to the deformed curved fibers corresponding to dx_1, dx_2, dx_3 in the reference or undeformed configuration, i.e., the covariant base vector. These vectors are normalized due to appearance of $[\lambda]^{-1}$.

(2) From (3.200) and (3.203), we note that extensions parallel to coordinate axes in the x-frame are nonlinear functions (due to finite deformation) of the diagonal components of the Green's strains. In other words, the diagonals of the Green's strain tensor in Lagrangian description are a measure of the extensions of line segments in the reference configuration that are parallel to the coordinate axes in the x-frame.

3.15.2 Extensions and stretches parallel to $\bar{x}$-frame axes using $[\bar{\varepsilon}^{[0]}]$

Consider

$$\frac{(d\bar{s})^2-(ds)^2}{2(d\bar{s})^2}=\frac{\{d\bar{x}\}^T\{d\bar{x}\}-\{dx\}^T\{dx\}}{2\{d\bar{x}\}^T\{d\bar{x}\}} \tag{3.210}$$

or

$$\frac{(d\bar{s})^2-(ds)^2}{2(d\bar{s})^2}=\frac{\{d\bar{x}\}^T[\frac{1}{2}([I]-[\bar{J}]^T[\bar{J}])]\{d\bar{x}\}}{\{d\bar{x}\}^T\{d\bar{x}\}}=\frac{\{d\bar{x}\}^T[\bar{\varepsilon}^{[0]}]\{d\bar{x}\}}{\{d\bar{x}\}^T\{d\bar{x}\}} \tag{3.211}$$

But

$$\{d\bar{x}\}=\{\bar{l}\}d\bar{s} \quad ; \quad \{d\bar{x}\}^T=\{\bar{l}\}^T d\bar{s} \tag{3.212}$$

Substituting from (3.212) into (3.211), we obtain

$$\begin{aligned}\frac{(d\bar{s})^2-(ds)^2}{2(d\bar{s})^2}&=\frac{d\bar{s}\{\bar{l}\}^T[\bar{\varepsilon}^{[0]}]\{\bar{l}\}d\bar{s}}{d\bar{s}\{\bar{l}\}^T\{\bar{l}\}d\bar{s}}=\frac{\{\bar{l}\}^T[\bar{\varepsilon}^{[0]}]\{\bar{l}\}}{\{\bar{l}\}^T\{\bar{l}\}}\\&=\frac{\{\bar{l}\}^T[\bar{\varepsilon}^{[0]}]\{\bar{l}\}}{\{\bar{l}\}^T\{\bar{l}\}}=\{\bar{l}\}^T[\bar{\varepsilon}^{[0]}]\{\bar{l}\}\end{aligned} \tag{3.213}$$

Let $\lambda_{\bar{A}\bar{B}}=\lambda_{AB}=\dfrac{d\bar{s}}{ds}$, then

$$\frac{(d\bar{s})^2-(ds)^2}{2(d\bar{s})^2}=\frac{1}{2}\Big(1-\Big(\frac{ds}{d\bar{s}}\Big)^2\Big)=\frac{1}{2}\Big(1-\frac{1}{(\lambda_{\bar{A}\bar{B}})^2}\Big) \tag{3.214}$$

Equating right hand side of (3.213) and (3.214)

$$\frac{1}{2}\Big(1-\frac{1}{(\lambda_{\bar{A}\bar{B}})^2}\Big)=\{\bar{l}\}^T[\bar{\varepsilon}^{[0]}]\{\bar{l}\} \tag{3.215}$$

The extensions and the stretches of the fibers $d\bar{x}_1, d\bar{x}_2, d\bar{x}_3$ that are projections of $d\bar{s}$ in x-frame can be obtained using

$$\{\bar{l}\}^T=[1,0,0] \quad ; \quad [0,1,0] \quad ; \quad [0,0,1] \tag{3.216}$$

and

$$[\bar{\varepsilon}^{[0]}]=\begin{bmatrix}\bar{\varepsilon}_{11}^{[0]} & \bar{\varepsilon}_{12}^{[0]} & \bar{\varepsilon}_{13}^{[0]}\\ \bar{\varepsilon}_{21}^{[0]} & \bar{\varepsilon}_{22}^{[0]} & \bar{\varepsilon}_{23}^{[0]}\\ \bar{\varepsilon}_{31}^{[0]} & \bar{\varepsilon}_{32}^{[0]} & \bar{\varepsilon}_{33}^{[0]}\end{bmatrix} \tag{3.217}$$

Using (3.215) we can write

$$\frac{1}{2}\Big(1-\frac{1}{(\lambda_{\bar{x}_1})^2}\Big)=\begin{Bmatrix}1\\0\\0\end{Bmatrix}^T[\bar{\varepsilon}^{[0]}]\begin{Bmatrix}1\\0\\0\end{Bmatrix}=\bar{\varepsilon}_{11}^{[0]}$$

$$\frac{1}{2}\left(1-\frac{1}{(\lambda_{\bar{x}_2})^2}\right)=\begin{Bmatrix}0\\1\\0\end{Bmatrix}^T[\bar{\varepsilon}^{[0]}]\begin{Bmatrix}0\\1\\0\end{Bmatrix}=\bar{\varepsilon}_{22}^{[0]}$$
$$\frac{1}{2}\left(1-\frac{1}{(\lambda_{\bar{x}_3})^2}\right)=\begin{Bmatrix}0\\0\\1\end{Bmatrix}^T[\bar{\varepsilon}^{[0]}]\begin{Bmatrix}0\\0\\1\end{Bmatrix}=\bar{\varepsilon}_{33}^{[0]} \tag{3.218}$$

or

$$\lambda_{\bar{x}_1}=(1-2\bar{\varepsilon}_{11}^{[0]})^{-1/2}\quad;\quad\lambda_{\bar{x}_2}=(1-2\bar{\varepsilon}_{22}^{[0]})^{-1/2}$$
$$\lambda_{\bar{x}_3}=(1-2\bar{\varepsilon}_{33}^{[0]})^{-1/2} \tag{3.219}$$

If we define

$$E_{\bar{A}\bar{B}}=E_{AB}=\frac{d\bar{s}-ds}{ds}=\lambda_{\bar{A}\bar{B}}-1 \tag{3.220}$$

then (3.219) and (3.220) give

$$E_{\bar{x}_1}=(1-2\bar{\varepsilon}_{11}^{[0]})^{-1/2}-1\quad;\quad E_{\bar{x}_2}=(1-2\bar{\varepsilon}_{22}^{[0]})^{-1/2}-1$$
$$E_{\bar{x}_3}=(1-2\bar{\varepsilon}_{33}^{[0]})^{-1/2}-1 \tag{3.221}$$

Recall (3.206), i.e.

$$\{\bar{l}\}=[J]\{l\}\frac{1}{\lambda_{AB}}\quad;\quad\{l\}=\lambda_{\bar{A}\bar{B}}[J]^{-1}\{\bar{l}\}\quad;\quad\text{since}\quad\lambda_{AB}=\lambda_{\bar{A}\bar{B}} \tag{3.222}$$

when

$$\{\bar{l}\}^T=\begin{bmatrix}1&0&0\end{bmatrix}\quad;\quad\lambda_{\bar{A}\bar{B}}=\lambda_{\bar{x}_1}\quad\text{and}\quad\{l\}=\{l_{x_1}\}$$
$$\{\bar{l}\}^T=\begin{bmatrix}0&1&0\end{bmatrix}\quad;\quad\lambda_{\bar{A}\bar{B}}=\lambda_{\bar{x}_2}\quad\text{and}\quad\{l\}=\{l_{x_2}\} \tag{3.223}$$
$$\{\bar{l}\}^T=\begin{bmatrix}0&0&1\end{bmatrix}\quad;\quad\lambda_{\bar{A}\bar{B}}=\lambda_{\bar{x}_3}\quad\text{and}\quad\{l\}=\{l_{x_3}\}$$

Hence, using (3.222) and (3.223), we can write

$$[\{l_{x_1}\},\{l_{x_2}\},\{l_{x_3}\}]=[J]^{-1}\begin{bmatrix}\lambda_{x_1}&0&0\\0&\lambda_{x_2}&0\\0&0&\lambda_{x_3}\end{bmatrix}=[\bar{J}][\lambda] \tag{3.224}$$

Thus, if $d\bar{x}_1, d\bar{x}_2, d\bar{x}_3$ are orthogonal fibers parallel to the axes of the x-frame, then $\{l_{x_1}\}$, $\{l_{x_2}\}$, $\{l_{x_3}\}$ are the direction cosines of the tangent vectors to the corresponding curvilinear configurations in the reference frame. Components of these vectors are indeed components of $\tilde{\boldsymbol{g}}^i$, the rows of $[\bar{J}]$, noting that the columns of the transpose of $[\bar{J}]$ are the contravariant vectors $\tilde{\boldsymbol{g}}^i$.

As in the case of Green's strain tensor in Lagrangian description, here also we note that the extensions in the x-frame along coordinate axes are nonlinear functions (due to finite deformation) of the diagonals of the Almansi strain tensor. In other words the diagonal elements of the Almansi strain tensor are a contravariant measure of the extensions in Eulerian description in the x-frame.

3.15.3 Angles between the fibers or material lines

Consider fibers or material lines parallel to coordinate axes in the x-frame in the reference configuration. After deformation the fibers are no longer straight (Figure 3.4(a) and (b)). Let the tangent vectors to the deformed fibers make an angle β. We denote these deformed fibers by $\bar{\phi}$ and $\bar{\theta}$. Let $\{\bar{l}_\phi\}$ and $\{\bar{l}_\theta\}$ be the direction cosines of the tangent vectors to the deformed fibers. Similarly, we consider two fibers parallel to ox_1 and ox_2 in the current configuration(Figure 3.4(c) and (d)). The map of these fibers in the reference configuration will result in curvilinear axes. Let ϕ and θ be these fibers and α be the angle formed by the tangent vectors to these fibers. Let $\{l_\phi\}$ and $\{l_\theta\}$ be the direction cosines of the tangent vectors to these fibers.

$$\cos\alpha = \{l_\phi\}^T\{l_\theta\} \tag{3.225}$$

$$\cos\beta = \{\bar{l}_\phi\}^T\{\bar{l}_\theta\} \tag{3.226}$$

But

$$\begin{aligned} \{\bar{l}_\theta\} &= \frac{1}{\lambda_\theta}[J]\{l_\theta\} \\ \{\bar{l}_\phi\} &= \frac{1}{\lambda_\phi}[J]\{l_\phi\} \end{aligned} \tag{3.227}$$

and

$$\begin{aligned} \{l_\theta\} &= \lambda_\theta[J]^{-1}\{\bar{l}_\theta\} = \lambda_{\bar{\theta}}[J]^{-1}\{\bar{l}_\theta\} \\ \{l_\phi\} &= \lambda_\phi[J]^{-1}\{\bar{l}_\phi\} = \lambda_{\bar{\phi}}[J]^{-1}\{\bar{l}_\phi\} \end{aligned} \tag{3.228}$$

$$\begin{aligned} \therefore \quad \cos\alpha &= (\lambda_{\bar{\phi}}[J]^{-1}\{\bar{l}_\phi\})^T(\lambda_{\bar{\theta}}[J]^{-1}\{\bar{l}_\theta\}) \\ &= \lambda_{\bar{\theta}}\lambda_{\bar{\phi}}\{\bar{l}_\phi\}^T[J^{-1}]^T[J]^{-1}\{\bar{l}_\theta\} \\ &= \lambda_{\bar{\theta}}\lambda_{\bar{\phi}}\{\bar{l}_\phi\}^T([\bar{J}]^T[\bar{J}])\{\bar{l}_\theta\} \end{aligned} \tag{3.229}$$

But

$$[\bar{\varepsilon}^{[0]}] = \frac{1}{2}([I] - [\bar{J}]^T[\bar{J}]) \tag{3.230}$$

$$\therefore \quad [\bar{J}]^T[\bar{J}] = [I] - 2[\bar{\varepsilon}^{[0]}] \tag{3.231}$$

Substituting from (3.231) into (3.229)

$$\cos\alpha = \lambda_{\bar{\theta}}\lambda_{\bar{\phi}}\{\bar{l}_\phi\}^T([I] - 2[\bar{\varepsilon}^{[0]}])\{\bar{l}_\theta\} \tag{3.232}$$

Similarly, using (3.226) and (3.227), we can obtain

$$\cos\beta = \Big(\frac{1}{\lambda_\theta}[J]\{l_\phi\}\Big)^T\Big(\frac{1}{\lambda_\phi}[J]\{l_\theta\}\Big) \tag{3.233}$$

$$= \frac{1}{\lambda_\theta}\frac{1}{\lambda_\phi}\{l_\phi\}^T([J]^T[J])\{l_\theta\}$$

Since

$$[\varepsilon_{[0]}] = \frac{1}{2}([J]^T[J] - [I]) \tag{3.234}$$

$$\therefore \quad [J]^T[J] = [I] + 2[\varepsilon_{[0]}] \tag{3.235}$$

Substituting from (3.235) into (3.233)

$$\cos\beta = \frac{1}{\lambda_\theta}\frac{1}{\lambda_\phi}\{l_\phi\}^T([I] + 2[\varepsilon_{[0]}])\{l_\theta\} \tag{3.236}$$

Using (3.232) and (3.236), we can derive the expression for shear angles between the material lines.

Shear angles

In this section, we derive the expression for the shear angles. Consider a pair of perpendicular fibers or material lines in the undeformed or reference configuration. Let (dx_1, dx_2), (dx_2, dx_3), (dx_3, dx_1) be such pairs at material point P. In the current configuration, these deform into curved fibers or material lines $\bar{\phi}, \bar{\theta}, \bar{\zeta}$ forming corresponding pairs $(\bar{\phi}, \bar{\theta}), (\bar{\theta}, \bar{\zeta})$, $(\bar{\zeta}, \bar{\phi})$ at material point $\bar{P}$ corresponding to (dx_1, dx_2), (dx_2, dx_3), (dx_3, dx_1) at material point P in the reference configuration.

Let β_{12} be the angle between the tangent vectors to $\bar{\phi}$ and $\bar{\theta}$ material lines, β_{23} be the angle between the tangent vectors to θ and $\bar{\zeta}$ material lines, and β_{31} be the angle between the tangent vectors to $\bar{\zeta}$ and $\bar{\phi}$ material lines. Let us define

$$\beta_{12} = 90 - \phi_{12} \quad ; \quad \beta_{23} = 90 - \phi_{23} \quad ; \quad \beta_{31} = 90 - \phi_{31} \tag{3.237}$$

in which ϕ_{12}, ϕ_{23} and ϕ_{31} are the corresponding changes in the 90 degree angles between the material lines in the undeformed configuration. We use (3.236), in which $\{l_\theta\}$, $\{l_\phi\}$ are replaced by $\{l_{x_1}\}$, $\{l_{x_2}\}$; $\{l_{x_2}\}$, $\{l_{x_3}\}$ and $\{l_{x_3}\}$, $\{l_{x_1}\}$ where

$$\{l_{x_1}\}^T = [1,0,0] \quad ; \quad \{l_{x_2}\}^T = [0,1,0] \quad ; \quad \{l_{x_3}\}^T = [0,0,1] \tag{3.238}$$

and correspondingly β is replaced by β_{12}, β_{23} and β_{31} respectively. Using $\{l_{x_1}\}$, $\{l_{x_2}\}$ and β_{12} we can write

$$\cos\beta_{12} = \cos(90-\phi_{12}) = \sin\phi_{12} = \frac{1}{\lambda_{x_1}}\frac{1}{\lambda_{x_2}}\begin{Bmatrix}1\\0\\0\end{Bmatrix}^T([I]+2[\varepsilon_{[0]}])\begin{Bmatrix}0\\1\\0\end{Bmatrix} \tag{3.239}$$

or

$$\sin\phi_{12} = \frac{2(\varepsilon_{[0]})_{12}}{\lambda_{x_1}\lambda_{x_2}} \tag{3.240}$$

Using (3.205) for λ_{x_1} and λ_{x_2} in terms of ε_{12} and ε_{23} we can write (3.240) as follows.

$$\sin\phi_{12} = \frac{2(\varepsilon_{[0]})_{12}}{\sqrt{(1 + 2(\varepsilon_{[0]})_{11})(1 + 2(\varepsilon_{[0]})_{22})}} \tag{3.241}$$

Similarly using $\{l_{x_2}\}$, $\{l_{x_3}\}$, β_{23} and $\{l_{x_3}\}$, $\{l_{x_1}\}$, β_{31} we can write

$$\cos\beta_{23} = \cos(90 - \phi_{23}) = \sin\phi_{23} = \frac{1}{\lambda_{x_2}}\frac{1}{\lambda_{x_3}} \begin{Bmatrix} 0 \\ 1 \\ 0 \end{Bmatrix}^T ([I] + 2[\varepsilon_{[0]}]) \begin{Bmatrix} 0 \\ 0 \\ 1 \end{Bmatrix}$$

$$\cos\beta_{31} = \cos(90 - \phi_{31}) = \sin\phi_{31} = \frac{1}{\lambda_{x_3}}\frac{1}{\lambda_{x_1}} \begin{Bmatrix} 0 \\ 0 \\ 1 \end{Bmatrix}^T ([I] + 2[\varepsilon_{[0]}]) \begin{Bmatrix} 1 \\ 0 \\ 0 \end{Bmatrix}$$

which can be written as

$$\sin\phi_{23} = \frac{2(\varepsilon_{[0]})_{23}}{\lambda_{x_2}\lambda_{x_3}} = \frac{2(\varepsilon_{[0]})_{23}}{\sqrt{(1 + 2(\varepsilon_{[0]})_{22})(1 + 2(\varepsilon_{[0]})_{33})}} \tag{3.242}$$

$$\sin\phi_{31} = \frac{2(\varepsilon_{[0]})_{31}}{\lambda_{x_3}\lambda_{x_1}} = \frac{2(\varepsilon_{[0]})_{31}}{\sqrt{(1 + 2(\varepsilon_{[0]})_{33})(1 + 2(\varepsilon_{[0]})_{11})}} \tag{3.243}$$

The terms ϕ_{12}, ϕ_{23} and ϕ_{31} are called shear angles. The shear angles are nonlinear functions of the corresponding off-diagonal elements of the Green's strain tensor as well as the corresponding diagonal elements. In other words the off-diagonal elements of the Green's strain tensor are a measure of the shear angles. We note that the sine of the change in 90 degree angles is a linear function of the off-diagonal elements of the Green's strain tensor but a nonlinear function of the corresponding diagonal elements.

Similarly, the perpendicular fibers $d\bar{x}_1$, $d\bar{x}_2$, $d\bar{x}_3$ in the current configuration have angles α_{12}, α_{23}, α_{31} in the reference configuration that are $90 - \bar{\phi}_{12}$, $90 - \bar{\phi}_{23}$, $90 - \bar{\phi}_{31}$ in which $\bar{\phi}_{12}$, $\bar{\phi}_{23}$, $\bar{\phi}_{31}$ are shear angles. We use (3.232) in which $\{\bar{l}_\theta\}$, $\{\bar{l}_\phi\}$ are replaced by $\{\bar{l}_{x_1}\}$, $\{\bar{l}_{x_2}\}$; $\{\bar{l}_{x_2}\}$, $\{\bar{l}_{x_3}\}$ and $\{\bar{l}_{x_3}\}$, $\{\bar{l}_{x_1}\}$ where

$$\{\bar{l}_{x_1}\}^T = [1,0,0] \quad ; \quad \{\bar{l}_{x_2}\}^T = [0,1,0] \quad ; \quad \{\bar{l}_{x_3}\}^T = [0,0,1]$$

and correspondingly α is replaced by α_{12}, α_{23} and α_{31}, respectively. Using $\{\bar{l}_{x_1}\}$, $\{\bar{l}_{x_2}\}$ and α_{12} we can write

$$\cos\alpha_{12} = \cos(90 - \bar{\phi}_{12}) = \sin\bar{\phi}_{12} = \lambda_{\bar{x}_1}\lambda_{\bar{x}_2} \begin{Bmatrix} 1 \\ 0 \\ 0 \end{Bmatrix}^T ([I] - 2[\bar{\varepsilon}^{[0]}]) \begin{Bmatrix} 0 \\ 1 \\ 0 \end{Bmatrix}$$

or

$$\sin\bar{\phi}_{12} = -2\lambda_{\bar{x}_1}\lambda_{\bar{x}_2}\bar{\varepsilon}_{12}^{[0]} \tag{3.244}$$

Using the definition of $\lambda_{\bar{x}_1}$ and $\lambda_{\bar{x}_2}$ in terms of $\bar{\varepsilon}_{12}^{[0]}$, $\bar{\varepsilon}_{11}^{[0]}$ and $\bar{\varepsilon}_{22}^{[0]}$ we obtain

$$\sin\bar{\phi}_{12} = \frac{-2\bar{\varepsilon}_{12}^{[0]}}{\sqrt{(1 - 2\bar{\varepsilon}_{11}^{[0]})(1 - 2\bar{\varepsilon}_{22}^{[0]})}} \tag{3.245}$$

Similarly, using $\{\bar{l}_{x_2}\}$, $\{\bar{l}_{x_3}\}$ and $\alpha_{x_2x_3}$ and $\{\bar{l}_{x_3}\}$, $\{\bar{l}_{x_1}\}$ and $\alpha_{x_3x_1}$ we can write

$$\cos\alpha_{23} = \cos(90 - \bar{\phi}_{23}) = \sin\bar{\phi}_{23} = \lambda_{\bar{x}_2}\lambda_{\bar{x}_3} \begin{Bmatrix} 0 \\ 1 \\ 0 \end{Bmatrix}^T ([I] - 2[\bar{\varepsilon}^{[0]}]) \begin{Bmatrix} 0 \\ 0 \\ 1 \end{Bmatrix}$$

$$\cos\alpha_{31} = \cos(90 - \bar{\phi}_{31}) = \sin\bar{\phi}_{31} = \lambda_{\bar{x}_3}\lambda_{\bar{x}_1} \begin{Bmatrix} 0 \\ 0 \\ 1 \end{Bmatrix}^T ([I] - 2[\bar{\varepsilon}^{[0]}]) \begin{Bmatrix} 1 \\ 0 \\ 0 \end{Bmatrix}$$

which can be written as

$$\sin\bar{\phi}_{23} = \frac{-2\bar{\varepsilon}_{23}^{[0]}}{\sqrt{(1 - 2\bar{\varepsilon}_{22}^{[0]})(1 - 2\bar{\varepsilon}_{33}^{[0]})}} \tag{3.246}$$

$$\sin\bar{\phi}_{31} = \frac{-2\bar{\varepsilon}_{31}^{[0]}}{\sqrt{(1 - 2\bar{\varepsilon}_{33}^{[0]})(1 - 2\bar{\varepsilon}_{11}^{[0]})}} \tag{3.247}$$

The terms $\bar{\phi}_{12}$, $\bar{\phi}_{23}$, $\bar{\phi}_{31}$ are shear angles in Eulerian description.

These shear angles are the deviations of the material lines from 90 degree angle in the reference configuration so that in the current configuration they are orthogonal and parallel to the coordinate axes in the x-frame. As in the case of Lagrangian description, here also we note that the shear angles are functions of the corresponding off-diagonal elements of the Almansi strain tensor as well as the corresponding diagonal elements of the Almansi strain tensor. That is, the off-diagonal elements of the Almansi strain tensor are a measure of the shear angles in the Eulerian description. As in the case of Lagrangian description, here also we note that the sine of the change in the 90 degree angles is a linear function of the off-diagonal elements of the Almansi strain tensor but is a nonlinear function of the corresponding diagonal elements.

3.16 Small deformation, small strain deformation physics

In this deformation physics, the deformed coordinates $\bar{x}_i$ are not appreciably different than the undeformed coordinates x_i for every material point $P(x_i)$ i.e., $\bar{x}_i \simeq x_i$. We consider Green's strain and Almansi strain as these measures reduce to well known definition of strain in linear elasticity.

3.16.1 Green's strain: Lagrangian description

3.16.1.1 Strain measure: $[\varepsilon]$

When $\bar{x}_i \simeq x_i$, hence $d\bar{s} \simeq ds$ can be used to simplify the expression used in deriving Green's strain measure.

$$\frac{d\bar{s}^2 - ds^2}{2ds^2} = \frac{(d\bar{s} - ds)(d\bar{s} + ds)}{2ds^2} = \frac{(d\bar{s} - ds)2ds}{2ds^2} = \frac{d\bar{s} - ds}{ds} \tag{3.248}$$

This is of course change in length per unit length, a definition of strain for infinitesimal deformation in $\mathbb{R}^1$. The components of the Green's strain tensor $\boldsymbol{\varepsilon}_{[0]}$ for small deformation, small strain are defined by $\boldsymbol{\varepsilon}$ and are derived in the following. Recall

$$[\varepsilon_{[0]}] = \frac{1}{2}\left([\,{}^d\!J]^T + [\,{}^d\!J]^T + [\,{}^d\!J]^T[\,{}^d\!J]\right) \tag{3.249}$$

when $\bar{x}_i \simeq x_i$, quadratic terms in the components of $[\,{}^d\!J]$ in (3.249) can be neglected compared to the components of $[\,{}^d\!J]$.

$$[\varepsilon_{[0]}] = \frac{1}{2}\left([\,{}^d\!J] + [\,{}^d\!J]^T\right) = [\varepsilon] \quad \text{(Def.)} \tag{3.250}$$

Components of $[\varepsilon]$ in (3.250) are given by

$$\begin{aligned}
\varepsilon_{11} &= \frac{\partial u_1}{\partial x_1} \quad ; \quad \varepsilon_{22} = \frac{\partial u_2}{\partial x_2} \quad ; \quad \varepsilon_{33} = \frac{\partial u_3}{\partial x_3} \\
\varepsilon_{12} &= \frac{1}{2}\left(\frac{\partial u_1}{\partial x_2} + \frac{\partial u_2}{\partial x_1}\right) \quad ; \quad \varepsilon_{23} = \frac{1}{2}\left(\frac{\partial u_2}{\partial x_3} + \frac{\partial u_3}{\partial x_2}\right) \\
\varepsilon_{31} &= \frac{1}{2}\left(\frac{\partial u_3}{\partial x_1} + \frac{\partial u_1}{\partial x_3}\right)
\end{aligned} \tag{3.251}$$

We note that (3.251) can also be obtained by neglecting the quadratic terms in the expression for $[\varepsilon_{[0]}]$.

3.16.1.2 Elongations and stretches

When $\bar{x}_i \simeq x_i$, then $d\bar{s} + ds \simeq 2ds$ and we have

$$\frac{(d\bar{s})^2 - (ds)^2}{2ds^2} = \frac{(d\bar{s} - ds)(d\bar{s} + ds)}{2ds^2} = \frac{(d\bar{s} - ds)(2ds)}{2ds^2} = \frac{d\bar{s} - ds}{ds} = \frac{d\bar{s}}{ds} - 1 \tag{3.252}$$

Thus

$$E_{AB} = \lambda_{AB} - 1 \tag{3.253}$$

and in x-frame

$$E_{x_1} = \lambda_{x_1} - 1 \quad ; \quad E_{x_2} = \lambda_{x_2} - 1 \quad ; \quad E_{x_3} = \lambda_{x_3} - 1 \tag{3.254}$$

Which are same as used earlier for finite deformation finite strain. It is also obvious from the definitions of stretch and elongation that these are not dependent on magnitude of deformation. We note that

$$E_{x_1} = \frac{d\bar{x}_1 - dx_1}{dx_1} \quad ; \quad E_{x_2} = \frac{d\bar{x}_2 - dx_2}{dx_2} \quad ; \quad E_{x_3} = \frac{d\bar{x}_3 - dx_3}{dx_3} \tag{3.255}$$

Thus, for infinitesimal deformation E_{x_1}, E_{x_2}, E_{x_3} are change in length per unit length along axes of x-frame, hence, the following holds.

$$E_{x_1} \simeq \varepsilon_{11} = \frac{\partial u_1}{\partial x_1} \quad ; \quad E_{x_2} \simeq \varepsilon_{22} = \frac{\partial u_2}{\partial x_2} \quad ; \quad E_{x_3} \simeq \varepsilon_{33} = \frac{\partial u_3}{\partial x_3} \tag{3.256}$$

That is, elongations along axes of x-frame are strains in ox_i directions (direct strains or normal strains).

3.16.1.3 Shear angles: shear strains

Recall that the shear angles ϕ_{12}, ϕ_{23} and ϕ_{31} representing the change in the 90° angles after deformation between the material lines parallel to x_1, x_2 ; x_2, x_3 ; x_3, x_1 axes in the undeformed configurations are given by (equations (3.242)–(3.243)).

$$\sin\phi_{12} = \frac{2(\varepsilon_{[0]})_{12}}{\sqrt{\left(1+2(\varepsilon_{[0]})_{11}\right)\left(1+2(\varepsilon_{[0]})_{22}\right)}} \tag{3.257}$$

$$\sin\phi_{23} = \frac{2(\varepsilon_{[0]})_{23}}{\sqrt{\left(1+2(\varepsilon_{[0]})_{22}\right)\left(1+2(\varepsilon_{[0]})_{33}\right)}} \tag{3.258}$$

$$\sin\phi_{31} = \frac{2(\varepsilon_{[0]})_{31}}{\sqrt{\left(1+2(\varepsilon_{[0]})_{33}\right)\left(1+2(\varepsilon_{[0]})_{11}\right)}} \tag{3.259}$$

For small deformation, small strain (infinitesimal deformation) $2(\varepsilon_{[0]})_{11}$, $2(\varepsilon_{[0]})_{22}$, $2(\varepsilon_{[0]})_{33}$ can be neglected compared to one in the denominators of (3.257)–(3.259) and the sine of the angle can be approximated by the angle itself. Thus, (3.257)–(3.259) can be approximated as (also replacing $(\varepsilon_{[0]})_{11}$, $(\varepsilon_{[0]})_{22}$, $(\varepsilon_{[0]})_{33}$ by ε_{11}, ε_{12}, ε_{33})

$$\phi_{12} = 2\varepsilon_{12} \quad ; \quad \phi_{23} = 2\varepsilon_{23} \quad ; \quad \phi_{31} = 2\varepsilon_{31} \tag{3.260}$$

or

$$\varepsilon_{12} = \varepsilon_{21} = \frac{\phi_{12}}{2} \quad ; \quad \varepsilon_{23} = \varepsilon_{32} = \frac{\phi_{23}}{2} \quad ; \quad \varepsilon_{31} = \varepsilon_{13} = \frac{\phi_{31}}{2} \tag{3.261}$$

Thus, the off diagonal elements ε_{12}, ε_{23} and ε_{31} of the linear strain tensor $[\varepsilon]$ represent half the change in the 90° between x_1, x_2 ; x_2, x_3 ; x_3, x_1 axes.

3.16.2 Almansi strain tensor: Eulerian description

3.16.2.1 Strain measure: $[\bar{\varepsilon}]$

In case of Almansi strain tensor $[\bar{\varepsilon}^{[0]}]$ we begin with

$$\frac{(d\bar{s})^2-(ds)^2}{2(d\bar{s})^2} = \frac{(d\bar{s}-ds)(d\bar{s}+ds)}{2(d\bar{s})^2} \simeq \frac{(d\bar{s}-ds)(2d\bar{s})}{2(d\bar{s})^2} = \frac{d\bar{s}-ds}{d\bar{s}} \tag{3.262}$$

Expression (3.262) implies that a reference length $d\bar{s}$ in the current configuration becomes ds in the reference configuration upon removing the external disturbance acting on the volume of matter. Thus, (3.262) represents change in length per unit length, but for small strain, small deformation in $\mathbb{R}^1$. The components of the Almansi strain tensor $[\bar{\varepsilon}^{[0]}]$ for small deformation, small strain $[\bar{\varepsilon}]$ are derived in the following. Recall

$$[\bar{\varepsilon}^{[0]}] = \frac{1}{2}\left([\,{}^{d}\bar{J}] + [\,{}^{d}\bar{J}]^T - [\,{}^{d}\bar{J}]^T[\,{}^{d}\bar{J}]\right) \tag{3.263}$$

when $\bar{x}_i \simeq x_i$, the quadratic term in the components of $[\,{}^d\!\bar{J}]$ in (3.263) can be neglected compared to $[\,{}^d\!\bar{J}]$.

$$[\bar{\varepsilon}^{[0]}] \simeq \frac{1}{2}\left([\,{}^d\!\bar{J}] + [\,{}^d\!\bar{J}]^T\right) = [\bar{\varepsilon}] \quad \text{(Def.)} \tag{3.264}$$

Components of $[\bar{\varepsilon}]$ are given by

$$\begin{aligned}
\bar{\varepsilon}_{11} &= \frac{\partial \bar{u}_1}{\partial \bar{x}_1} \quad ; \quad \bar{\varepsilon}_{22} = \frac{\partial \bar{u}_2}{\partial \bar{x}_2} \quad ; \quad \bar{\varepsilon}_{33} = \frac{\partial \bar{u}_3}{\partial \bar{x}_3} \\
\bar{\varepsilon}_{12} &= \frac{1}{2}\left(\frac{\partial \bar{u}_1}{\partial \bar{x}_2} + \frac{\partial \bar{u}_2}{\partial \bar{x}_1}\right) \quad ; \quad \bar{\varepsilon}_{23} = \frac{1}{2}\left(\frac{\partial \bar{u}_2}{\partial \bar{x}_3} + \frac{\partial \bar{u}_3}{\partial \bar{x}_2}\right) \\
\bar{\varepsilon}_{31} &= \frac{1}{2}\left(\frac{\partial \bar{u}_3}{\partial \bar{x}_1} + \frac{\partial \bar{u}_1}{\partial \bar{x}_3}\right)
\end{aligned} \tag{3.265}$$

3.16.2.2 Elongations and stretches

When $\bar{x}_i \simeq x_i$ and $d\bar{s} + ds \simeq 2d\bar{s}$ we have

$$\frac{(d\bar{s})^2 - (ds)^2}{2(d\bar{s})^2} = \frac{d\bar{s} - ds}{d\bar{s}} = 1 - \frac{ds}{d\bar{s}} \tag{3.266}$$

Let $\lambda_{\bar{A}\bar{B}} = \dfrac{ds}{d\bar{s}} = \dfrac{1}{\lambda_{AB}}$ also

$$E_{\bar{x}_1} = 1 - \frac{1}{\lambda_{x_1}} \quad ; \quad E_{\bar{x}_2} = 1 - \frac{1}{\lambda_{x_2}} \quad ; \quad E_{\bar{x}_3} = 1 - \frac{1}{\lambda_{x_3}} \tag{3.267}$$

$E_{\bar{x}_1}$, $E_{\bar{x}_2}$, $E_{\bar{x}_3}$ are elongations per unit length along the axes of the x-frame.

3.16.2.3 Shear angles, shear strains

Recall from Section 3.15.3 that the shear angles $\bar{\phi}_{12}$, $\bar{\phi}_{23}$, $\bar{\phi}_{31}$ representing the change in the 90° angles between the orthogonal material lines $\bar{x}_1, \bar{x}_2$; $\bar{x}_2, \bar{x}_3$; $\bar{x}_3, \bar{x}_1$ in the current configuration when the external disturbance is removed are given by (3.245)–(3.247).

$$\sin\bar{\phi}_{12} = \frac{-2\bar{\varepsilon}_{12}^{[0]}}{\sqrt{\left(1 - 2\bar{\varepsilon}_{11}^{[0]}\right)\left(1 - 2\bar{\varepsilon}_{22}^{[0]}\right)}} \tag{3.268}$$

$$\sin\bar{\phi}_{23} = \frac{-2\bar{\varepsilon}_{23}^{[0]}}{\sqrt{\left(1 - 2\bar{\varepsilon}_{22}^{[0]}\right)\left(1 - 2\bar{\varepsilon}_{33}^{[0]}\right)}} \tag{3.269}$$

$$\sin\bar{\phi}_{31} = \frac{-2\bar{\varepsilon}_{31}^{[0]}}{\sqrt{\left(1 - 2\bar{\varepsilon}_{33}^{[0]}\right)\left(1 - 2\bar{\varepsilon}_{11}^{[0]}\right)}} \tag{3.270}$$

For small deformation, small strain, the strains in the radical sign can be neglected compared to one and the sine of the angles can be approximated by the angle itself, (3.268)–(3.270) reduce to (also replacing $\bar{\varepsilon}_{12}^{[0]}$, $\bar{\varepsilon}_{23}^{[0]}$, $\bar{\varepsilon}_{31}^{[0]}$ by $\bar{\varepsilon}_{12}$, $\bar{\varepsilon}_{23}$, $\bar{\varepsilon}_{31}$).

$$
\begin{aligned}
\bar{\phi}_{12} = -2\bar{\varepsilon}_{12} \quad &; \quad \bar{\phi}_{23} = -2\bar{\varepsilon}_{23} \quad ; \quad \bar{\phi}_{31} = -2\bar{\varepsilon}_{31} \\
\bar{\varepsilon}_{12} = -\frac{\bar{\phi}_{12}}{2} \quad &; \quad \bar{\varepsilon}_{23} = -\frac{\bar{\phi}_{23}}{2} \quad ; \quad \bar{\varepsilon}_{31} = -\frac{\bar{\phi}_{31}}{2}
\end{aligned}
\tag{3.271}
$$

Thus, the off diagonal elements $\bar{\varepsilon}_{12}$, $\bar{\varepsilon}_{23}$, $\bar{\varepsilon}_{31}$ of $[\bar{\varepsilon}]$ represent half of the change (reduction, hence negative) in the 90° angle between the orthogonal material lines in the deformed configuration.

3.17 Additive and multiplicative decompositions of deformation gradient tensor $[J]$

At each material point the deformation gradient tensors contains details of the deformation of material lines and rigid rotation of the material lines. This information can be extracted from $[J]$ either through additive decomposition or through multiplicative decomposition.

3.17.1 Additive decomposition of $[J]$

We decompose $[J]$, a tensor of rank two, into symmetric and skew-symmetric tensors

$$[J] = [{}_sJ] + [{}_aJ] \tag{3.272}$$

$$[{}_sJ] = \frac{1}{2}\left([J] + [J]^T\right) \quad ; \quad [{}_aJ] = \frac{1}{2}\left([J] - [J]^T\right) \tag{3.273}$$

$$
[{}_sJ] = [\varepsilon] \quad ; \quad [{}_aJ] = \begin{bmatrix} 0 & \frac{1}{2}(u_{1,2} - u_{2,1}) & \frac{1}{2}(u_{1,3} - u_{3,1}) \\ \frac{1}{2}(u_{2,1} - u_{1,2}) & 0 & \frac{1}{2}(u_{2,3} - u_{3,2}) \\ \frac{1}{2}(u_{3,1} - u_{1,3}) & \frac{1}{2}(u_{3,2} - u_{2,3}) & 0 \end{bmatrix} \tag{3.274}
$$

Consider

$$\boldsymbol{\nabla} \times \boldsymbol{u} = \boldsymbol{e}_1\Theta_1 + \boldsymbol{e}_2\Theta_2 + \boldsymbol{e}_3\Theta_3 \tag{3.275}$$

$$\Theta_1 = u_{3,2} - u_{2,3} \quad ; \quad \Theta_2 = u_{1,3} - u_{3,1} \quad ; \quad \Theta_3 = u_{2,1} - u_{1,2} \tag{3.276}$$

Θ_1, Θ_2, Θ_3 are rotations about x_1, x_2, x_3 axes of fixed x-frame, positive when counterclockwise. Using (3.276), we can write $[{}_aJ]$ in (3.274) as

$$
[{}_aJ] = \begin{bmatrix} 0 & -\frac{\Theta_3}{2} & \frac{\Theta_2}{2} \\ \frac{\Theta_3}{2} & 0 & -\frac{\Theta_1}{2} \\ -\frac{\Theta_2}{2} & \frac{\Theta_1}{2} & 0 \end{bmatrix} \tag{3.277}
$$

Thus, $[{}_sJ]$ contains information about the strain components (small deformation, small strain) and $[{}_aJ]$ contains rotations angles at the material point about the axes of a triad located at the material whose axes are parallel to x-frame. Additive decomposition (3.272) helps us separate deformation due to stretch and due to pure rotation at a material point.

We shall see in Chapter 7 from the entropy inequality derived from the second law of thermodynamics that if $[\sigma]$ is Cauchy stress tensor then $\operatorname{tr}\left([\sigma][{}_s\dot{J}]\right) = \operatorname{tr}\left([\sigma][\dot{\varepsilon}]\right)$ represents rate of mechanical work for small deformation, small strain theory in CCM. Thus, $[\sigma]$ and $[\dot{\varepsilon}]$ are rate of work conjugate pair, thus, we can derive constitutive theory for $[\sigma]$ using $[\varepsilon]$ as its argument tensor. In CCM $[{}_aJ]$ containing information regarding rotation of the material lines due to deformation is completely ignored. Even though complete deformation details are contained in $[J]$ at a material point, CCM only considers $[{}_sJ]$, hence $[\varepsilon]$ is only due to symmetric part of $[J]$ and ignores the skew-symmetric part of $[J]$ completely. This is obviously a limitation of the current CCM theories.

Since $[J]$ relates element lengths in the reference and current configuration, we can write

$$\{d\bar{x}\} = [J]\{dx\} = [[{}_sJ] + [{}_aJ]]\,\{dx\} = [{}_sJ]\{dx\} + [{}_aJ]\{dx\} \tag{3.278}$$

Equation (3.278) shows an additive operation that $\{dx\}$ goes through to transform to $\{d\bar{x}\}$.

3.17.2 Multiplicative decomposition of $[J]$: Polar decomposition into stretch and rotation tensor

The deformation gradient tensors $[J]$ and $[\bar{J}]$ are invertible, i.e., $[J] = [\bar{J}]^{-1}$; $[\bar{J}] = [J]^{-1}$ and their inverse are unique. The polar decomposition allows any invertible matrix to be decomposed into product of two matrices. We state a theorem.

Theorem 3.1

The polar decomposition holds for every invertible matrix. Thus, the deformation gradient $[J]$ can be decomposed into the product of two matrices

$$[J] = [R][S_r] = [S_l][R] \tag{3.279}$$

in which $[S_r]$ and $[S_l]$ are symmetric and positive definite, that is

$$[S_r] = [S_r]^T \tag{3.280}$$

$$[S_l] = [S_l]^T \tag{3.281}$$

and

$$\{w\}^T[S_r]\{w\} > 0 \quad ; \quad \{w\}^T[S_l]\{w\} > 0 \quad \forall \{w\} \neq \{0\} \tag{3.282}$$

The matrix $[R]$ is orthogonal, i.e., $[R]^T[R] = [R][R]^T = [I]$. $[S_r]$ and $[S_l]$ are called right and left stretch matrices and $[R]$ is called the rotation tensor. $[S_r]$ and $[S_l]$ as well as $[R]$ are uniquely defined. Thus, polar decomposition allows us to decompose the deformation of the matter in $[J]$ into pure rotation and stretch. Using (3.279) we can write

$$\{d\bar{x}\} = [J]\{dx\} = [R][S_r]\{dx\} = ([R][S_r])\{dx\} \tag{3.283}$$

$$\text{and} \quad \{d\bar{x}\} = [J]\{dx\} = [S_l][R]\{dx\} = ([S_l][R])\{dx\} \tag{3.284}$$

In equation (3.283), we note that rotation is followed by stretch and in the case of (3.284), stretch is followed by rotation. The following theorem forms the basis for the proof of theorem 3.1.

Theorem 3.2

If $[H]$ is a symmetric positive definite matrix, then $[H] = [S]^2$ exists and is unique and $[S]$ is symmetric and positive definite. We also say that $[S] = \sqrt{[H]}$ exists and is unique.

Proof

Since $[H]$ is symmetric and positive definite, $[H]\{\phi\} = \lambda\{\phi\}$, the eigenvalue problem yields unique $(\lambda_i, \{\phi\}_i)$; $i = 1, 2, \dots$ eigenpairs in which $\lambda_i > 0$; $i = 1, 2, \dots$ and $\{\phi\}_i$ are orthonormal, i.e., $\{\phi\}_i^T\{\phi\}_j = \delta_{ij}$. Then, $[H]$ can be represented by

$$[H] = [\Phi][\lambda][\Phi]^T; \text{ columns of } [\Phi] \text{ are orthonormal eigenvectors } \{\phi\}_i \tag{3.285}$$

If we choose

$$[S] = [\Phi][\sqrt{\lambda}][\Phi]^T \tag{3.286}$$

then

$$[S][S] = [\Phi][\sqrt{\lambda}][\Phi]^T[\Phi][\sqrt{\lambda}][\Phi]^T \tag{3.287}$$

or

$$[S]^2 = [\Phi][\sqrt{\lambda}][\sqrt{\lambda}][\Phi]^T = [\Phi][\lambda][\Phi]^T = [H] \tag{3.288}$$

$$\therefore \quad [S] = \sqrt{[H]} \tag{3.289}$$

Uniqueness of $[S]$ is ensured due to uniqueness of $(\lambda_i, \{\phi\}_i)$, the eigenpairs of $[H]$. ■

Remarks

(1) Consider covariant Cauchy stress strain tensor $[C_{[0]}] = [J]^T[J]$

(a) $[C_{[0]}]$ is symmetric.

(b) $[C_{[0]}]$ is positive definite as

$$\{w\}^T[C_{[0]}]\{w\} = \{w\}^T[J]^T[J]\{w\} = ([J]\{w\})^T([J]\{w\}) > 0 \ \forall \{w\} \neq \{0\}$$

(c)

$$
\begin{aligned}
[C_{[0]}] &= [J]^T[J] = ([R][S_r])^T([R][S_r]) \\
&= [S_r]^T[R]^T[R][S_r] = [S_r]^T[S_r] = [S_r][S_r] = [S_r]^2 \\
\text{or} \quad [S_r] &= \sqrt{[C_{[0]}]}
\end{aligned}
$$

Thus, $[C_{[0]}]$ can take the place of $[H]$ and $[S_r]$ can take the place of $[S]$ in theorem 3.2.

(2) Consider contravariant strain tensor $[B^{[0]}] = [J][J]^T$

(a) $[B^{[0]}]$ is symmetric.
(b) $[B^{[0]}]$ is positive definite as

$$
\begin{aligned}
\{w\}^T[B^{[0]}]\{w\} &= \{w\}^T[J][J]^T\{w\} \\
&= ([J]^T\{w\})^T([J]^T\{w\}) > 0 \ \forall\{w\} \neq \{0\}
\end{aligned}
$$

This confirms that $[B^{[0]}]$ is positive definite.
(c) We note that

$$
\begin{aligned}
[B^{[0]}] &= [J][J]^T = ([S_l][R])([S_l][R])^T \\
&= [S_l][R][R]^T[S_l]^T = [S_l][S_l]^T = [S_l][S_l] = [S_l]^2 \\
\text{or} \quad [S_l] &= \sqrt{[B^{[0]}]}
\end{aligned}
$$

Thus, $[B^{[0]}]$ can take the place of $[H]$ and $[S_l]$ can take the place of $[S]$ in theorem 3.2.

Proof of Theorem 3.1

(I) *Using* $[J] = [R][S_r]$ *decomposition*

Using Remark (1) we can construct a proof of Theorem 3.1. Consider (based on (3.279))

$$
[J] = [R][S_r] \tag{3.290}
$$

We need to show that: (i) $[S_r]$ is symmetric ; (ii) $[S_r]$ is positive definite ; (iii) $[S_r]$ is unique ; (iv) $[R]$ is orthogonal ; (v) $[R]$ is unique. Consider

$$
[C_{[0]}] = [J]^T[J] = [S_r]^2 \quad ; \quad [S_r] = \sqrt{[C_{[0]}]} \tag{3.291}
$$

From (3.290)

$$
[R]=[J][S_r]^{-1}, \text{ i.e., knowing } [J] \text{ and } [S_r],\ [R] \text{ can be determined.} \tag{3.292}
$$

Based on Theorem 3.2, we can treat $[C_{[0]}]$ as $[H]$ (as it is symmetric and positive definite, the requirements based on Theorem 3.2). Since $[C_{[0]}]$ is symmetric and positive definite, $[C_{[0]}]\{\phi\} = \lambda\{\phi\}$, the eigenvalue problem yields unique $(\lambda_i, \{\phi\}_i)$; $i = 1, 2, 3$ eigenpairs in which $\lambda_i > 0$; $i = 1, 2, 3$ and $\{\phi\}_i$ are orthonormal, i.e.,

$\{\phi\}_i^T\{\phi\}_j = \delta_{ij}$. Then $[C_{[0]}]$ can be represented by

$$[C_{[0]}] = [\Phi][\lambda][\Phi]^T \tag{3.293}$$

Columns of $[\Phi]$ are orthonormal eigenvectors $\{\phi\}_i$. In (3.293), $[\lambda]$ is a diagonal matrix containing λ_i ; $i = 1, 2, 3$. If we choose

$$[S_r] = [\Phi][\sqrt{\lambda}][\Phi]^T, \quad \text{then} \tag{3.294}$$

$$[S_r][S_r] = ([\Phi][\sqrt{\lambda}][\Phi]^T)([\Phi][\sqrt{\lambda}][\Phi]^T) = [\Phi][\sqrt{\lambda}][\sqrt{\lambda}][\Phi]^T \tag{3.295}$$

$$\text{or} \quad [S_r]^2 = [\Phi][\lambda][\Phi]^T = [C_{[0]}] \quad ; \quad [S_r] = \sqrt{[C_{[0]}]} \tag{3.296}$$

(a) Symmetry, positive definiteness and uniqueness of $[S_r]$

$[S_r]$ *is symmetric* due to (3.294). Positive definiteness of $[S_r]$ can be easily established. Consider

$$\begin{aligned}\{w\}^T[S_r]\{w\} &= \{w\}^T[\Phi][\sqrt{\lambda}][\Phi]^T\{w\} \\ &= ([\lambda^{\frac{1}{4}}][\Phi]^T\{w\})^T([\lambda^{\frac{1}{4}}][\Phi]^T\{w\}) > 0\,;\ \forall\{w\} \neq \{0\}\end{aligned} \tag{3.297}$$

Hence $[S_r]$ *is positive definite.* Furthermore, *uniqueness of $[S_r]$ is ensured due to uniqueness of the eigenpairs of $[C_{[0]}]$.*

(b) Orthogonality and uniqueness of $[R]$

First, we consider orthogonality of $[R]$. Using (3.292)

$$[R][R]^T = ([J][S_r]^{-1})([J][S_r]^{-1})^T = [J][S_r]^{-1}([S_r]^{-1})^T[J]^T \tag{3.298}$$

Since $[S_r]$ is symmetric

$$[R][R]^T = [J][S_r]^{-1}[S_r]^{-1}[J]^T = [J][S_r]^{-2}[J]^T \tag{3.299}$$

Recall that

$$[J]^T[J] = [S_r]^2 \quad ; \quad [S_r]^{-2} = ([J]^T[J])^{-1} = [J]^{-1}([J]^T)^{-1} \tag{3.300}$$

Substituting from (3.300) into (3.299)

$$[R][R]^T = [J][J]^{-1}([J]^T)^{-1}[J]^T = [I][I] = [I] \tag{3.301}$$

Hence $[R]$ *is orthogonal.*

Next we consider uniqueness of $[R]$. We construct proof of the uniqueness of $[R]$ by contradiction. Let us assume that $[R]$ is not unique, then based on Theorem 3.1, we must have

$$[J] = [R][S_r] = [S_l][\underset{\sim}{R}] \quad ; \quad [\underset{\sim}{R}] = [S_l]^{-1}[J] \tag{3.302}$$

$$\begin{aligned}[\underset{\sim}{R}][\underset{\sim}{R}]^T &= [S_l]^{-1}([J][J]^T)([S_l]^{-1})^T \\ &= [S_l]^{-1}([J][J]^T)[S_l]^{-1} \quad \text{due to symmety of } [S_l]\end{aligned} \tag{3.303}$$

But $[J][J]^T = [S_l]^2 = [S_l][S_l]$

$$\therefore \quad [\underset{\sim}{R}][\underset{\sim}{R}]^T = [S_l]^{-1}[S_l][S_l][S_l]^{-1} = [I] \tag{3.304}$$

Thus, $[\underset{\sim}{R}]$ is orthogonal and therefore

$$[J] = [S_l][\underset{\sim}{R}] \tag{3.305}$$

holds at this stage. If $[R] \neq [\underset{\sim}{R}]$, then we have (since $[\underset{\sim}{R}]$ is orthogonal, hence $[\underset{\sim}{R}][\underset{\sim}{R}]^T = [I]$)

$$[J] = [R][S_r] = [S_l][\underset{\sim}{R}] = [I][S_l][\underset{\sim}{R}] = ([\underset{\sim}{R}][\underset{\sim}{R}]^T)([S_l][\underset{\sim}{R}]) \tag{3.306}$$

$$= [\underset{\sim}{R}]([\underset{\sim}{R}]^T[S_l][\underset{\sim}{R}]) = [\underset{\sim}{R}][\underset{\sim}{S_r}] \tag{3.307}$$

$$\therefore \quad [J]^T[J] = [\underset{\sim}{S_r}][\underset{\sim}{R}]^T[\underset{\sim}{R}][\underset{\sim}{S_r}] = [\underset{\sim}{S_r}]^2 = [S_r]^2 \tag{3.308}$$

Equation (3.308) implies that $[J]^T[J]$ is not unique, a contradiction, hence

$$[\underset{\sim}{R}] = [R] \quad \text{must hold.} \tag{3.309}$$

That is, $[R]$ *is unique*. This completes the proof of the theorem. ■

Since the properties of $[R]$ and $[S_r]$ are established, the required properties of $[S_l]$ in $[J] = [R][S_r] = [S_l][R]$ are automatically established.

(II) *Using* $[J] = [S_l][R]$ *decomposition*

In the proof of Theorem 3.1 provided in the preceding section we had considered $[J] = [R][S_r]$ and $[J]^T[J] = [C_{[0]}] = [S_r]^2$. We can also construct the proof of Theorem 3.1 using

$$[J] = [S_l][R] \quad \text{and} \quad [J][J]^T = [B^{[0]}] = [S_l]^2 \tag{3.310}$$

We consider details in the following. We need to show that: (i) $[S_l]$ is symmetric ; (ii) $[S_l]$ is positive definite ; (iii) $[S_l]$ is unique ; (iv) $[R]$ is orthogonal ; (v)$[R]$ is unique. From (3.310)

$$[R] = [S_l]^{-1}[J] \tag{3.311}$$

Based on Theorem 3.2, we can treat $[B^{[0]}]$ as $[H]$ (as it is symmetric and positive definite, the requirements based on Theorem 3.2).

Since $[B^{[0]}]$ is symmetric and positive definite, $[B^{[0]}]\{\phi\} = \lambda\{\phi\}$, the eigenvalue problem yields unique $(\lambda_i, \{\phi\}_i)$; $i = 1, 2, 3$ eigenpairs in which $\lambda_i > 0$; $i = 1, 2, 3$ and $\{\phi\}_i$ are orthonormal, i.e., $\{\phi\}_i^T\{\phi\}_j = \delta_{ij}$. Then $[B^{[0]}]$ can be represented by

$$[B^{[0]}] = [\Phi][\lambda][\Phi]^T \tag{3.312}$$

Columns of $[\Phi]$ are orthonormal eigenvectors $\{\phi\}_i$. Matrix $[\lambda]$ is diagonal containing

λ_i ; $i = 1, 2, 3$. If we choose

$$[S_l] = [\Phi][\sqrt{\lambda}][\Phi]^T \tag{3.313}$$

$$[S_l][S_l] = ([\Phi][\sqrt{\lambda}][\Phi]^T)([\Phi][\sqrt{\lambda}][\Phi]^T) = [\Phi][\sqrt{\lambda}][\sqrt{\lambda}][\Phi]^T \tag{3.314}$$

$$[S_l][S_l] = [S_l]^2 = [\Phi][\lambda][\Phi]^T = [B^{[0]}] \quad ; \quad [S_l] = \sqrt{[B^{[0]}]} \tag{3.315}$$

(a) Symmetry, positive definiteness and uniqueness of $[S_l]$

$[S_l]$ *is obviously symmetric due to* (3.313). Positive definiteness of $[S_l]$ can be easily established. Consider

$$\begin{aligned}\{w\}^T[S_l]\{w\} &= \{w\}^T[\Phi][\sqrt{\lambda}][\Phi]^T\{w\} \\ &= ([\lambda^{\frac{1}{4}}][\Phi]^T\{w\})^T([\lambda^{\frac{1}{4}}][\Phi]^T\{w\}) > 0\,;\ \forall\{w\} \neq \{0\}\end{aligned} \tag{3.316}$$

Hence $[S_l]$ *is positive definite.* Furthermore, *uniqueness of $[S_l]$ is ensured due to uniqueness of the eigenpairs of $[B^{[0]}]$.*

(b) Orthogonality and uniqueness of $[R]$

First, we consider the orthogonality of $[R]$.

$$[R][R]^T = ([S_l]^{-1}[J])([S_l]^{-1}[J])^T = [S_l]^{-1}[J][J]^T([S_l]^{-1})^T \tag{3.317}$$

Since $[S_l]$ is symmetric

$$[R][R]^T = [S_l]^{-1}[J][J]^T[S_l]^{-1} \tag{3.318}$$

Recall that

$$[J][J]^T = [S_l]^2 = [S_l][S_l] \tag{3.319}$$

Substituting from (3.319) into (3.317)

$$[R][R]^T = [S_l]^{-1}[S_l][S_l][S_l]^{-1} = [I] \tag{3.320}$$

Hence $[R]$ *is orthogonal.*

Next we consider uniqueness of $[R]$. We construct proof of the uniqueness of $[R]$ by contradiction. Let us assume that $[R]$ is not unique, then based on Theorem 3.1, we must have

$$[J] = [\underset{\sim}{R}][S_r] = [S_l][R] \quad ; \quad [\underset{\sim}{R}] = [J][S_r]^{-1} \tag{3.321}$$

$$\begin{aligned}[\underset{\sim}{R}][\underset{\sim}{R}]^T &= ([J][S_r]^{-1})([J][S_r]^{-1})^T = [J][S_r]^{-1}[S_r]^{-1}[J]^T \\ &= [J][S_r]^{-2}[J]^T \quad \text{due to symmety of } [S_r]\end{aligned} \tag{3.322}$$

But $[J]^T[J] = [S_r]^2$

$$\therefore \quad [S_r]^{-2} = [J]^{-1}([J]^T)^{-1} \tag{3.323}$$

$$[\underset{\sim}{R}][\underset{\sim}{R}]^T = [J][J]^{-1}([J]^T)^{-1}[J]^T = [I] \tag{3.324}$$

Thus, $[\underset{\sim}{R}]$ is orthogonal and therefore

$$[J] = [S_l][\underset{\sim}{R}] \tag{3.325}$$

holds at this stage. If $[R] \neq [\underset{\sim}{R}]$, then we have (since $[\underset{\sim}{R}]$ is orthogonal, $[\underset{\sim}{R}][\underset{\sim}{R}]^T = [I]$)

$$[J] = [\underset{\sim}{R}][S_r] = ([S_l][R]) \tag{3.326}$$

$$= [\underset{\sim}{R}][S_r]([\underset{\sim}{R}]^T[\underset{\sim}{R}]) = ([\underset{\sim}{R}][S_r][\underset{\sim}{R}]^T)[\underset{\sim}{R}] = [\underset{\sim}{S_l}][\underset{\sim}{R}] \tag{3.327}$$

$$\therefore \quad [J][J]^T = [\underset{\sim}{S_l}][\underset{\sim}{R}][\underset{\sim}{R}]^T[\underset{\sim}{S_l}]^T = [\underset{\sim}{S_l}][\underset{\sim}{S_l}]^T = [\underset{\sim}{S_l}]^2 = [S_l]^2 \tag{3.328}$$

Equation (3.328) implies that the positive square root of $[J][J]^T$ is not unique, a contradiction, hence

$$[\underset{\sim}{R}] = [R] \quad \text{must hold.} \tag{3.329}$$

That is, $[R]$ *is unique.* This completes the proof of the theorem. ■

Remarks

(1) Determination of $[S_r]$ and $[R]$ in $[J] = [R][S_r]$.

(i) Construct $[C_{[0]}] = [J]^T[J]$ using $[J]$.

(ii) Find eigenpairs of $[C_{[0]}]$; $(\lambda_i, \{\phi\}_i)$; $i = 1, 2, 3$ in which $\{\phi\}_i^T\{\phi\}_j = \delta_{ij}$. Construct matrix $[\Phi]$ using $\{\phi\}_i$ as columns of $[\Phi]$.

(iii) Then, $[S_r] = [\Phi][\sqrt{\lambda}][\Phi]^T$ in which $\sqrt{\lambda}$ is a diagonal matrix containing $\sqrt{\lambda_i}$.

(iv) Determine $[R]$ using $[R] = [J][S_r]^{-1}$.

(v) $[S_l]$ can now be obtained using $[S_l] = [J][R]^T$.

(2) Determination of $[S_l]$ and $[R]$ in $[J] = [S_l][R]$.

(i) Construct $[B^{[0]}] = [J]^T[J]$ using $[J]$.

(ii) Find eigenpairs of $[B^{[0]}]$; $(\lambda_i, \{\phi\}_i)$; $i = 1, 2, 3$ in which $\{\phi\}_i^T\{\phi\}_j = \delta_{ij}$. Construct matrix $[\Phi]$ using $\{\phi\}_i$ as columns of $[\Phi]$.

(iii) Then, $[S_l] = [\Phi][\sqrt{\lambda}][\Phi]^T$ in which $\sqrt{\lambda}$ is a diagonal matrix containing $\sqrt{\lambda_i}$.

(iv) Determine $[R]$ using $[R] = [S_l]^{-1}[J]$.

(v) $[S_r]$ can now be obtained using $[S_r] = [R]^T[J]$

(3) In continuum mechanics we define

$$
\begin{aligned}
[S_r] &= \sqrt{[C_{[0]}]} = \sqrt{[J]^T[J]} &&; \quad \text{right stretch matrix} \\
[S_l] &= \sqrt{[B^{[0]}]} = \sqrt{[J][J]^T} &&; \quad \text{left stretch matrix} \\
[C_{[0]}] &= [J]^T[J] = [S_r]^2 &&; \quad \text{right Cauchy - Green tensor} \\
[B^{[0]}] &= [J][J]^T = [S_l]^2 &&; \quad \text{left Cauchy - Green tensor}
\end{aligned}
\tag{3.330}
$$

We note that *right Cauchy strain tensor* is completely defined by right stretch $[S_r]$, hence the name. Likewise, *left Cauchy strain tensor* is completely defined by left stretch $[S_l]$, hence the name.

(4) From (3.279), we note that

$$
\begin{aligned}
[S_r] &= [R]^T[S_l][R] \\
[S_l] &= [R][S_r][R]^T
\end{aligned}
\tag{3.331}
$$

(5) When $[J] = [J]^T$, $[R][S_r] = [S_r][R]^T$, which can only hold if $[R] = [I]$, also $[S_l][R] = [R]^T[S_l]$ which can only hold if $[R] = [I]$. Both of these imply that $[J] = [J]^T = [S_r] = [S_l]$, i.e., when $[J] = [J]^T$, $[J]$ represents pure stretch.

3.17.3 Strain measures in terms of $[S_r]$, $[S_l]$ and $[R]$

In this section, we consider all strain measures derived earlier and express these in terms of stretches and rotation tensor.

3.17.3.1 Covariant Cauchy strain tensor in Lagrangian description or right Cauchy strain: $[C_{[0]}]$

Consider

$$
[C_{[0]}] = [J]^T[J] = [S_r]^T[R]^T[R][S_r] = [S_r]^T[S_r] = [S_r]^2 \tag{3.332}
$$

3.17.3.2 Covariant Finger strain tensor in Lagrangian description: $[F_{[0]}]$

From definition of $[F_{[0]}]$

$$
[F_{[0]}] = [C_{[0]}]^{-1} = [S_r]^{-2} \tag{3.333}
$$

$$
[S_r] = [R]^T[S_l][R] \quad ; \quad [S_r]^{-1} = [R]^T[S_l]^{-1}[R] \tag{3.334}
$$

$$
\therefore \quad [F_{[0]}] = [R]^T[S_l]^{-2}[R] \tag{3.335}
$$

3.17.3.3 Covariant Green's strain tensor in Lagrangian description $[\varepsilon_{[0]}]$

$$
\begin{aligned}
[\varepsilon_{[0]}] &= \frac{1}{2}\left([J]^T[J] - [I]\right) = \frac{1}{2}\left(([R][S_r])^T([R][S_r]) - [I]\right) \\
\text{or} \quad [\varepsilon_{[0]}] &= \frac{1}{2}\left([S_r]^2 - [I]\right)
\end{aligned}
\tag{3.336}
$$

3.17.3.4 Contravariant Cauchy strain tensor in Eulerian description: $[\bar{C}^{[0]}]$

$$
[\bar{J}] = [J]^{-1} = ([R][S_r])^{-1} = [S_r]^{-1}[R]^{-1} = [S_r]^{-1}[R]^T
$$

$$
\therefore \quad [\bar{C}^{[0]}] = [\bar{J}]^T[\bar{J}] = [R][S_r]^{-2}[R]^T \tag{3.337}
$$

Alternatively,

$$[\bar{J}] = [J]^{-1} = ([S_l][R])^{-1} = [R]^T[S_l]^{-1} \tag{3.338}$$

$$\therefore \quad [\bar{C}^{[0]}] = [\bar{J}]^T[\bar{J}] = ([R]^T[S_l]^{-1})^T([R]^T[S_l]^{-1}) = [S_l]^{-2} \tag{3.339}$$

3.17.3.5 Contravariant Finger strain tensor in Eulerian description: $[\bar{F}^{[0]}]$ or left Cauchy strain: $[B^{[0]}]$

Since $[\bar{F}^{[0]}] = [\bar{C}^{[0]}]^{-1}$, we can obtain the following using (3.338) and (3.339):

$$[B^{[0]}] = [\bar{F}^{[0]}] = [[S_l]^{-2}]^{-1} = [S_l]^2$$

$$\text{also} \quad [\bar{F}^{[0]}] = ([R][S_r]^{-2}[R^T])^{-1} = [R][S_r]^2[R]^T$$

Since the strain tensor $[B^{[0]}]$ can be expressed purely in terms of left stretch tensor $[S_l]$, it is also referred to as *left Cauchy strain.*

3.17.3.6 Contravariant Almansi strain tensor in Eulerian description: $[\bar{\varepsilon}^{[0]}]$

Consider

$$[\bar{\varepsilon}^{[0]}] = \frac{1}{2}([I] - [\bar{J}]^T[\bar{J}]) = \frac{1}{2}([I] - [S_l]^{-2}) = \frac{1}{2}([I] - [R][S_r]^{-2}[R^T]) \tag{3.340}$$

Remarks

(1) In the following we present some additional helpful details.

$$\det[J] = \det([S_l][R]) = \det[S_l]\det[R] = \det[S_l] \quad ; \quad \det[R] = 1$$

$$\text{also} \quad \det[J] = \det[R]\det[S_r] = \det[R]\det[S_r] = \det[S_r]$$

(2) Based on Remark (1) and $\det[J] = \det[S_r] = \det[S_l]$, $[J]$, $[S_l]$ and $[S_r]$ have the same eigenvalues.

3.18 Invariants of $[C_{[0]}]$, $[B^{[0]}]$, $[S_r]$ and $[S_l]$ in terms of principal stretches of $[S_r]$ and $[S_l]$

We note that since $[C_{[0]}]$, $[B^{[0]}]$, $[S_r]$ and $[S_l]$ are symmetric and positive definite. Their invariants using traces or their principal invariant can be easily obtained. Expressing invariants of these in terms of principal stretches of $[S_r]$ and $[S_l]$ leads to much simplified expressions that are especially helpful in designing experiments.

3.18.1 Principal stretches of $[S_r]$ and $[S_l]$

Since $[S_r]$ and $[S_l]$ are symmetric and positive definite, we can construct the following eigenvalue problems.

$$[S_r]\{\Phi^r\} - \lambda^r[I]\{\Phi^r\} = 0 \tag{3.341}$$

$$[S_l]\{\Phi^l\} - \lambda^l[I]\{\Phi^l\} = 0 \tag{3.342}$$

From (3.341), we can obtain eigenpairs $(\lambda_i^r, \{\Phi^r\}_i)$; $i = 1, 2, 3$ in which λ_i^r are the right principal stretches and $\{\Phi^r\}_i$ are the directions of the right principal stretches.

These eigenpairs $(\lambda_i^r, \{\Phi^r\}_i$ obviously satisfy (3.341) (no sum over i in (3.343)).

$$[S_r]\{\Phi^r\}_i - \lambda_i^r[I]\{\Phi^r\}_i = 0 \quad ; \quad i = 1, 2, 3 \tag{3.343}$$

Likewise, from eigenvalue problem (3.342) we can obtain eigenpairs $(\lambda_i^l, \{\Phi^l\}_i)$; $i = 1, 2, 3$ in which λ_i^l are left principal stretches and $\{\Phi^l\}_i$ are the directions of the left principal stretches. These eigenpairs satisfy (3.342) (no sum over i in (3.344)).

$$[S_l]\{\Phi^l\}_i - \lambda_i^l[I]\{\Phi^l\}_i = 0 \tag{3.344}$$

3.18.2 Principal invariants of $[C_{[0]}]$ in terms of λ_i^r

Recall

$$\begin{aligned}
&[C_{[0]}] = [S_r]^2 = [S_r][S_r] \\
\therefore\ &[C_{[0]}]\{\Phi^r\}_i = [S_r][S_r]\{\Phi^r\}_i = [S_r]\left((\lambda_i^r)\{\Phi^r\}_i\right) = (\lambda_i^r)[S_r]\{\Phi^r\}_i \\
&[C_{[0]}]\{\Phi^r\}_i = (\lambda_i^r)^2\{\Phi^r\}_i
\end{aligned} \tag{3.345}$$

which implies that $(\lambda_i^r)^2$ is an eigenvalue of $[C_{[0]}]$ and $\{\Phi^r\}_i$ is the corresponding eigenvector. The eigenvalues of $[C_{[0]}]$ are squares of the eigenvalues of $[S_r]$ but eigenvectors of $[C_{[0]}]$ and $[S_r]$ are the same. Let

$$[\Phi^r] = [\{\Phi^r\}_1, \{\Phi^r\}_2, \{\Phi^r\}_3] \tag{3.346}$$

Since $\{\Phi^r\}_i^T\{\Phi^r\}_j = \delta_{ij}$;

$$[\Phi^r]^T[\Phi^r] = [\Phi^r][\Phi^r]^T = [I] \tag{3.347}$$

we can represent $[C_{[0]}]$ using $(\lambda_i^r)^2$ and $[\Phi^r]$

$$[C_{[0]}] = [\Phi^r][(\lambda^r)^2][\Phi^r]^T \tag{3.348}$$

in which $[(\lambda^r)^2]$ is a diagonal matrix of $(\lambda_i^r)^2$.

$$[S_r] = [\Phi^r][\lambda^r][\Phi^r]^T \tag{3.349}$$

$[\lambda^r]$ being diagonal matrix of λ_i^r.

Consider the principal invariants of $[C_{[0]}]$: $I_{C_{[0]}}$, $II_{C_{[0]}}$ and $III_{C_{[0]}}$. Using (3.348)

$$(C_{[0]})_{im} = \Phi_{ij}^r\left((\lambda^r)^2\right)_{jk}\Phi_{mk}^r = \left((\lambda^r)^2\right)_{jk}\Phi_{ij}^r\Phi_{mk}^r \tag{3.350}$$

$$\therefore\ I_{C_{[0]}} = (C_{[0]})_{ii} = \left((\lambda^r)^2\right)_{jk}\Phi_{ij}^r\Phi_{ik}^r = \left((\lambda^r)^2\right)_{jk}\delta_{jk} = \left((\lambda^r)^2\right)_{kk} \tag{3.351}$$

$$I_{C_{[0]}} = (\lambda_1^r)^2 + (\lambda_2^r)^2 + (\lambda_3^r)^2 \quad ; \quad (\lambda^r)_{11} = \lambda_1^r \text{ etc. is implied} \tag{3.352}$$

Consider $II_{C_{[0]}}$

$$II_{C_{[0]}} = \frac{1}{2}\left(-(C_{[0]})_{il})(C_{[0]})_{li} + (C_{[0]})_{kk})(C_{[0]})_{ll}\right) \tag{3.353}$$

$$(C_{[0]})_{il} = \Phi_{ij}^r\left((\lambda^r)^2\right)_{jk}\Phi_{lk}^r \quad ; \quad (C_{[0]})_{li} = \Phi_{lm}^r\left((\lambda^r)^2\right)_{mn}\Phi_{in}^r \tag{3.354}$$

$$\therefore\ (C_{[0]})_{il}(C_{[0]})_{li} = \left(\Phi_{ij}^r\left((\lambda^r)^2\right)_{jk}\Phi_{lk}^r\right)\left(\Phi_{lm}^r\left((\lambda^r)^2\right)_{mn}\Phi_{in}^r\right) \tag{3.355}$$

$$\begin{aligned} \therefore \quad (C_{[0]})_{il}(C_{[0]})_{li} &= \Phi^r_{ij}(\lambda^r)^2_{jk}\Phi^r_{lk}\Phi^r_{lm}(\lambda^r)^2_{mn}\Phi^r_{in} \\ &= \Phi^r_{ij}\Phi^r_{in}\Phi^r_{lk}\Phi^r_{lm}(\lambda^r)^2_{jk}(\lambda^r)^2_{mn} \\ &= \delta_{jn}\delta_{km}(\lambda^r)^2_{jk}(\lambda^r)^2_{mn} \end{aligned} \tag{3.356}$$

Since $(\lambda^r)^2_{jk} = 0$ when $j \neq k$ we can write

$$\begin{aligned} (C_{[0]})_{il}(C_{[0]})_{li} &= (\lambda^r)^2_{nk}(\lambda^r)^2_{kn} \\ &= (\lambda^r_1)^2(\lambda^r_1)^2 + (\lambda^r_2)^2(\lambda^r_2)^2 + (\lambda^r_3)^2(\lambda^r_3)^2 \\ &= (\lambda^r_1)^4 + (\lambda^r_2)^4 + (\lambda^r_3)^4 \end{aligned} \tag{3.357}$$

and we obtain the following expressions:

$$(C_{[0]})_{il}(C_{[0]})_{li} = (\lambda^r_1)^4 + (\lambda^r_2)^4 + (\lambda^r_3)^4 \tag{3.358}$$

$$(C_{[0]})_{kk}(C_{[0]})_{ll} = ((\lambda^r_1)^2 + (\lambda^r_2)^2 + (\lambda^r_3)^2)((\lambda^r_1)^2 + (\lambda^r_2)^2 + (\lambda^r_3)^2) \tag{3.359}$$

Therefore

$$\begin{aligned} I\!I_{C_{[0]}} &= \frac{1}{2}\big(- ((\lambda^r_1)^4 + (\lambda^r_2)^4 + (\lambda^r_3)^4) + ((\lambda^r_1)^2 \\ &\quad + (\lambda^r_2)^2 + (\lambda^r_3)^2)((\lambda^r_1)^2 + (\lambda^r_2)^2 + (\lambda^r_3)^2)\big) \end{aligned} \tag{3.360}$$

$$\therefore \quad I\!I_{C_{[0]}} = (\lambda^r_1)^2(\lambda^r_2)^2 + (\lambda^r_2)^2(\lambda^r_3)^2 + (\lambda^r_3)^2(\lambda^r_1)^2 \tag{3.361}$$

Consider $I\!I\!I_{C_{[0]}}$

$$I\!I\!I_{C_{[0]}} = \det[C_{[0]}] = \det[\Phi^r]\det[(\lambda^r)^2]\det[\Phi^r]^T = \det[(\lambda^r)^2] \tag{3.362}$$

$$\therefore \quad I\!I\!I_{C_{[0]}} = (\lambda^r_1)^2(\lambda^r_2)^2(\lambda^r_3)^2 \tag{3.363}$$

Thus, we have

$$\begin{aligned} I_{C_{[0]}} &= (\lambda^r_1)^2 + (\lambda^r_2)^2 + (\lambda^r_3)^2 \\ I\!I_{C_{[0]}} &= (\lambda^r_1)^2(\lambda^r_2)^2 + (\lambda^r_2)^2(\lambda^r_3)^2 + (\lambda^r_3)^2(\lambda^r_1)^2 \\ I\!I\!I_{C_{[0]}} &= (\lambda^r_1)^2(\lambda^r_2)^2(\lambda^r_3)^2 \end{aligned} \tag{3.364}$$

3.18.3 Principal Invariants of $[S_r]$

Using

$$[S_r] = [\Phi^r][\lambda^r][\Phi^r]^T \tag{3.365}$$

Since the right side of $[S_r]$ is exactly the same as the right side of $[C_{[0]}]$ in (3.348) except that $[S_r]$ has $[\lambda^r]$ and $[C_{[0]}]$ has $[(\lambda^r)^2]$, hence following exactly same procedure as used for $[C_{[0]}]$ we can obtain (or by inspection we can immediately write).

$$\begin{aligned} I_{S_r} &= \lambda^r_1 + \lambda^r_2 + \lambda^r_3 \\ I\!I_{S_r} &= \lambda^r_1\lambda^r_2 + \lambda^r_2\lambda^r_3 + \lambda^r_3\lambda^r_1 \end{aligned} \tag{3.366}$$

$$III_{Sr} = \lambda_1^r \lambda_2^r \lambda_3^r = \sqrt{III_{C_{[0]}}}$$
$$\text{or} \quad III_{C_{[0]}} = (III_{Sr})^2$$

3.18.4 Principal invariants of $[B^{[0]}]$ in terms of λ_i^l

Recall

$$[B^{[0]}] = [J][J]^T = [S_l]^2 = [S_l][S_l] \tag{3.367}$$

$$\begin{aligned} &[B^{[0]}]\{\Phi^l\}_i = [S_l][S_l]\{\Phi^l\}_i = [S_l](\lambda_i^l\{\Phi^l\}_i) = (\lambda_i^l)[S_l]\{\Phi^l\}_i \\ \therefore \quad &[B^{[0]}]\{\Phi^l\}_i = (\lambda^l)^2\{\Phi^l\}_i \end{aligned} \tag{3.368}$$

which implies that $(\lambda_i^l)^2$ are the eigenvalues of $[B^{[0]}]$ and $\{\Phi^l\}_i$ are the corresponding eigenvector. The eigenvalues of $[B^{[0]}]$ are squares of the eigenvalues of $[S_l]$ but the eigenvectors of $[B^{[0]}]$ and $[S_l]$ are the same.

$$\text{Let} \quad [\Phi^l] = \big[\{\Phi^l\}_1, \{\Phi^l\}_2, \{\Phi^l\}_3\big] \tag{3.369}$$

$$\text{Since} \quad \{\Phi^l\}_i^T\{\Phi^l\}_j = \delta_{ij} \tag{3.370}$$

$$[\Phi^l][\Phi^l]^T = [\Phi^l]^T[\Phi^l] = [I] \tag{3.371}$$

We can express $[B^{[0]}]$ using $[\Phi^l]$ and $(\lambda_i^l)^2$

$$[B^{[0]}] = [\Phi^l][(\lambda^l)^2][\Phi^l]^T \tag{3.372}$$

in which $[(\lambda^l)^2]$ is a diagonal matrix of $(\lambda_i^l)^2$ and

$$[S_l] = [\Phi^l][\lambda^l][\Phi^l]^T \tag{3.373}$$

Consider principal invariants of $[B^{[0]}]$, $I_{B^{[0]}}$, $II_{B^{[0]}}$, $III_{B^{[0]}}$. Using (3.372)

$$(B^{[0]})_{im} = \Phi_{ij}^l(\lambda^l)_{jk}^2\Phi_{mk}^l = (\lambda^l)_{jk}^2\Phi_{ij}^l\Phi_{mk}^l \tag{3.374}$$

$$(B^{[0]})_{ii} = (\lambda^l)_{jk}^2\Phi_{ij}^l\Phi_{ik}^l = (\lambda^l)_{jk}^2\delta_{jk} = (\lambda^l)_{kk}^2 \tag{3.375}$$

$$\therefore \quad I_{B^{[0]}} = (B^{[0]})_{ii} = (\lambda^l)_{kk}^2 = (\lambda_1^l)^2 + (\lambda_2^l)^2 + (\lambda_3^l)^2 \tag{3.376}$$

Consider $II_{B^{[0]}}$

$$II_{B^{[0]}} = \frac{1}{2}\Big(-((B^{[0]})_{il})(B^{[0]})_{li} + ((B^{[0]})_{kk})(B^{[0]})_{ll}\Big) \tag{3.377}$$

$$(B^{[0]})_{il} = \Phi_{ij}^l\left((\lambda^l)^2\right)_{jk}\Phi_{lk}^l \quad ; \quad (B^{[0]})_{li} = \Phi_{lm}^l\left((\lambda^l)^2\right)_{mn}\Phi_{in}^l \tag{3.378}$$

$$\therefore \quad (B^{[0]})_{il}(B^{[0]})_{li} = \left(\Phi_{ij}^l\left((\lambda^l)^2\right)_{jk}\Phi_{lk}^l\right)\left(\Phi_{lm}^l(\lambda^l)_{mn}^2\Phi_{in}^l\right) \tag{3.379}$$

$$\begin{aligned} &= \Phi_{ij}^l(\lambda^l)_{jk}^2\Phi_{lk}^l\Phi_{lm}^l(\lambda^l)_{mn}^2\Phi_{in}^l \\ &= \Phi_{ij}^l\Phi_{in}^l\Phi_{lk}^l\Phi_{lm}^l(\lambda^l)_{jk}^2(\lambda^l)_{mn}^2 \\ &= \delta_{jn}\delta_{km}(\lambda^l)_{jk}^2(\lambda^l)_{mn}^2 \end{aligned} \tag{3.380}$$

Since $(\lambda^l)_{jk}^2 = 0$ when $j \neq k$ we can write

$$
\begin{aligned}
(B^{[0]})_{il}(B^{[0]})_{li} &= (\lambda^l)^2_{nk}(\lambda^l)^2_{kn} \\
&= (\lambda^l_1)^2(\lambda^l_1)^2 + (\lambda^l_2)^2(\lambda^l_2)^2 + (\lambda^l_3)^2(\lambda^l_3)^2 \\
&= (\lambda^l_1)^4 + (\lambda^l_2)^4 + (\lambda^l_3)^4
\end{aligned} \tag{3.381}
$$

and we obtain the following expressions:

$$
(B^{[0]})_{il}(B^{[0]})_{li} = (\lambda^l_1)^4 + (\lambda^l_2)^4 + (\lambda^l_3)^4 \tag{3.382}
$$

$$
(B^{[0]})_{kk}(B^{[0]})_{ll} = ((\lambda^l_1)^2 + (\lambda^l_2)^2 + (\lambda^l_3)^2)((\lambda^l_1)^2 + (\lambda^l_2)^2 + (\lambda^l_3)^2) \tag{3.383}
$$

$$
\begin{aligned}
II_{B^{[0]}} = \frac{1}{2}\big(&-((\lambda^l_1)^4 + (\lambda^l_2)^4 + (\lambda^l_3)^4) \\
&+ ((\lambda^l_1)^2 + (\lambda^l_2)^2 + (\lambda^l_3)^2)((\lambda^l_1)^2 + (\lambda^l_2)^2 + (\lambda^l_3)^2)\big)
\end{aligned} \tag{3.384}
$$

$$
\therefore \quad II_{B^{[0]}} = (\lambda^l_1)^2(\lambda^l_2)^2 + (\lambda^l_2)^2(\lambda^l_3)^2 + (\lambda^l_3)^2(\lambda^l_1)^2 \tag{3.385}
$$

Consider $III_{B^{[0]}}$

$$
III_{B^{[0]}} = \det[B^{[0]}] = \det[\Phi^l]\det[(\lambda^l)^2]\det[\Phi^l]^T = \det[(\lambda^l)^2] \tag{3.386}
$$

$$
\therefore \quad III_{B^{[0]}} = (\lambda^l_1)^2(\lambda^l_2)^2(\lambda^l_3)^2 \tag{3.387}
$$

Thus, we have

$$
\begin{aligned}
I_{B^{[0]}} &= (\lambda^l_1)^2 + (\lambda^l_2)^2 + (\lambda^l_3)^2 \\
II_{B^{[0]}} &= (\lambda^l_1)^2(\lambda^l_2)^2 + (\lambda^l_2)^2(\lambda^l_3)^2 + (\lambda^l_3)^2(\lambda^l_1)^2 \\
III_{B^{[0]}} &= (\lambda^l_1)^2(\lambda^l_2)^2(\lambda^l_3)^2
\end{aligned} \tag{3.388}
$$

3.18.5 Principal Invariants of $[S_l]$

Using

$$
[S_l] = [\Phi][\lambda^l][\Phi]^T \tag{3.389}
$$

Since the right side of $[S_l]$ is exactly the same as the right side of $[B^{[0]}]$ except that $[S_l]$ has $[\lambda^l]$ and $[B^{[0]}]$ has $[(\lambda^l)^2]$, hence following exactly same procedure as used for $[B^{[0]}]$, we can obtain (or by inspection we can immediately write).

$$
\begin{aligned}
I_{S_l} &= \lambda^l_1 + \lambda^l_2 + \lambda^l_3 \\
II_{S_l} &= \lambda^l_1\lambda^l_2 + \lambda^l_2\lambda^l_3 + \lambda^l_3\lambda^l_1 \\
III_{S_l} &= \lambda^l_1\lambda^l_2\lambda^l_3 = \sqrt{III_{B^{[0]}}} \\
\text{or} \quad III_{B^{[0]}} &= (III_{S_l})^2
\end{aligned} \tag{3.390}
$$

3.19 Deformation of areas and volumes

3.19.1 Areas

Consider material lines $\boldsymbol{PQ}$, $\boldsymbol{PR}$ and $\boldsymbol{PS}$ originating from the material point P in the reference configuration (Figure 3.5).

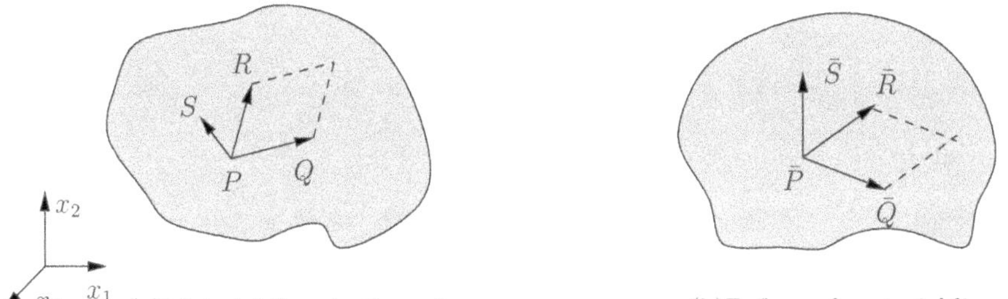

(a)Material lines in the reference configuration at a material point P

(b)Deformed material lines at the deformed material point $\bar{P}$

Figure 3.5: Material lines in the reference and current configurations

During the deformation, lines $\boldsymbol{PQ}$, $\boldsymbol{PR}$ and $\boldsymbol{PS}$ are deformed into $\boldsymbol{\bar{P}\bar{Q}}$, $\boldsymbol{\bar{P}\bar{R}}$ and $\boldsymbol{\bar{P}\bar{S}}$ respectively in the current configuration. Hence, the parallelogram whose adjacent sides are $\boldsymbol{PQ}$ and $\boldsymbol{PR}$ in the reference configuration deforms into the one with adjacent sides $\boldsymbol{\bar{P}\bar{Q}}$ and $\boldsymbol{\bar{P}\bar{R}}$. Let $\boldsymbol{dA}$ and $\boldsymbol{d\bar{A}}$ be the areas of these parallelograms. Then

$$\boldsymbol{dA} = \boldsymbol{PQ} \times \boldsymbol{PR}$$

which can be written as

$$\begin{aligned} &\boldsymbol{dA} = \boldsymbol{e}_i PQ_i \times \boldsymbol{e}_j PR_j = \boldsymbol{e}_i \times \boldsymbol{e}_j PQ_i PR_j = \epsilon_{ijk} \boldsymbol{e}_k PQ_i PR_j \\ \therefore\ & dA_k = \epsilon_{ijk} PQ_i PR_j \end{aligned}$$

Consider the deformed area

$$\boldsymbol{d\bar{A}} = \boldsymbol{\bar{P}\bar{Q}} \times \boldsymbol{\bar{P}\bar{R}}$$

which can be written as

$$\begin{aligned} &\boldsymbol{d\bar{A}} = \boldsymbol{e}_l \bar{P}\bar{Q}_l \times \boldsymbol{e}_m \bar{P}\bar{R}_m = \boldsymbol{e}_l \times \boldsymbol{e}_m \bar{P}\bar{Q}_l \bar{P}\bar{R}_m = \epsilon_{lmn} \boldsymbol{e}_n \bar{P}\bar{Q}_l \bar{P}\bar{R}_m \\ \therefore\ & d\bar{A}_n = \epsilon_{lmn} \bar{P}\bar{Q}_l \bar{P}\bar{R}_m \end{aligned} \tag{3.391}$$

Also we note that

$$\begin{aligned} \bar{P}\bar{Q}_l &= J_{lp} PQ_p \\ \bar{P}\bar{R}_m &= J_{mq} PR_q \end{aligned} \tag{3.392}$$

Substituting from (3.392) into (3.391)

$$d\bar{A}_n = \epsilon_{lmn} J_{lp} J_{mq} PQ_p PR_q \tag{3.393}$$

but

$$\delta_{sn} = \begin{cases} 1 & \text{when} \quad s = n \\ 0 & \text{when} \quad s \neq n \end{cases} \tag{3.394}$$

Introduce δ_{sn} on the right side of (3.393)

$$d\bar{A}_n = \epsilon_{lms} J_{lp} J_{mq} \delta_{sn} PQ_p PR_q \tag{3.395}$$

but

$$[J][J]^{-1} = [I] = \delta_{sn} = J_{st}(J^{-1})_{tn} \tag{3.396}$$

Substituting from (3.396) for δ_{sn} into (3.395)

$$d\bar{A}_n = \epsilon_{lms} J_{lp} J_{mq} J_{st} (J^{-1})_{tn} PQ_p PR_q \tag{3.397}$$

We note that

$$\det[J] = \epsilon_{ijk} J_{i1} J_{j2} J_{k3}$$

$$\epsilon_{lms} J_{lp} J_{mq} J_{st} = \det[J] \epsilon_{pqt} \tag{3.398}$$

Substituting from (3.398) into (3.397)

$$d\bar{A}_n = \det[J] \epsilon_{pqt} [J]_{tn}^{-1} PQ_p PR_q \tag{3.399}$$

Also

$$\epsilon_{pqt} PQ_p PR_q = dA_t \tag{3.400}$$

Substituting from (3.400) into (3.399)

$$d\bar{A}_n = \det[J][J]_{tn}{}^{-1} dA_t = \det[J][[J]_{nt}{}^{-1}]^T dA_t \tag{3.401}$$

$$\therefore \quad \{d\bar{A}\} = \det[J][[J]^{-1}]^T \{dA\} \tag{3.402}$$

Similarly, areas formed by ($\boldsymbol{PR}, \boldsymbol{PS}$) and ($\boldsymbol{PS}, \boldsymbol{PQ}$) also relate to those formed by ($\bar{\boldsymbol{P}}\bar{\boldsymbol{R}}, \bar{\boldsymbol{P}}\bar{\boldsymbol{S}}$) and ($\bar{\boldsymbol{P}}\bar{\boldsymbol{S}}, \bar{\boldsymbol{P}}\bar{\boldsymbol{Q}}$) according to (3.402).

3.19.2 Volumes

Consider the parallelepiped formed by the line segments $\boldsymbol{PQ}, \boldsymbol{PR}, \boldsymbol{PS}$ in the reference configuration. The material line segments $\boldsymbol{PQ}, \boldsymbol{PR}, \boldsymbol{PS}$ are all originating at the material point P. This parallelepiped is deformed into the one formed by the deformed material line segments $\bar{\boldsymbol{P}}\bar{\boldsymbol{Q}}, \bar{\boldsymbol{P}}\bar{\boldsymbol{R}}, \bar{\boldsymbol{P}}\bar{\boldsymbol{S}}$, all originating at the deformed material particle $\bar{P}$ in the current configuration. Let dV and $d\bar{V}$ be the volume of these parallelepipeds in the reference and current configurations. Then

$$dV = (\boldsymbol{PQ} \times \boldsymbol{PR}) \cdot \boldsymbol{PS}$$

which can be written as

$$\begin{aligned} dV &= (\boldsymbol{e}_i PQ_i \times \boldsymbol{e}_j PR_j) \cdot \boldsymbol{e}_k PS_k = (\boldsymbol{e}_i \times \boldsymbol{e}_j) \cdot \boldsymbol{e}_k PQ_i PR_j PS_k \\ &= \epsilon_{ijl} \boldsymbol{e}_l \cdot \boldsymbol{e}_k PQ_i PR_j PS_k \\ \text{or} \quad dV &= \epsilon_{ijk} PQ_i PR_j PS_k \end{aligned} \tag{3.403}$$

Consider the deformed volume

$$d\bar{V} = (\bar{\boldsymbol{P}}\bar{\boldsymbol{Q}} \times \bar{\boldsymbol{P}}\bar{\boldsymbol{R}}) \cdot \bar{\boldsymbol{P}}\bar{\boldsymbol{S}} \tag{3.404}$$

Following the same procedure as for dV we obtain

$$d\bar{V} = \epsilon_{lmn}\bar{P}\bar{Q}_l\bar{P}\bar{R}_m\bar{P}\bar{S}_n \tag{3.405}$$

but

$$\begin{aligned} \bar{P}\bar{Q}_l &= J_{lp}PQ_p \\ \bar{P}\bar{R}_m &= J_{mq}PR_q \\ \bar{P}\bar{S}_n &= J_{nr}PS_r \end{aligned} \tag{3.406}$$

Substituting from (3.406) into (3.405)

$$d\bar{V} = \epsilon_{lmn}J_{lp}J_{mq}J_{nr}PQ_pPR_qPS_r \tag{3.407}$$

but

$$\epsilon_{lmn}J_{lp}J_{mq}J_{nr} = \det[J]\epsilon_{pqr} \tag{3.408}$$

therefore

$$d\bar{V} = \det[J]\epsilon_{pqr}PQ_pPR_qPS_r \tag{3.409}$$

Substituting from (3.403) into (3.409), we obtain

$$d\bar{V} = \det[J]dV \tag{3.410}$$

From (3.410), we note that $\det[J] > 0$ *must hold for the deformation to be admissible.*

3.19.3 Integral form of $\{d\bar{A}\}$ over $\partial\bar{V}$

Consider a deformed volume $\bar{V}$ with closed boundary $\partial\bar{V}$ (surface), then

$$\int_{\partial\bar{V}} \{d\bar{A}\} = \int_{\partial\bar{V}} \{\bar{n}\}d\bar{A} = \int_{\partial\bar{V}} \{\bar{n}\}\,1d\bar{A} \tag{3.411}$$

Using divergence theorem

$$\int_{\partial\bar{V}} \{d\bar{A}\} = \int_{\bar{V}} \bar{\boldsymbol{\nabla}}\,1d\bar{V} = \int_{\bar{V}} 0d\bar{V} = 0 \tag{3.412}$$

Thus,

$$\int_{\partial\bar{V}} \{d\bar{A}\} = 0 \tag{3.413}$$

That is integral of $\{d\bar{A}\}$ over closed boundary $\partial\bar{V}$ of a deformed volume $\bar{V}$ is zero. This is helpful in some derivations and simplifications.

3.20 Summary

In this chapter, two fundamental descriptions of motion, Lagrangian and Eulerian are discussed. Using these two descriptions, deformation gradient is derived and used to present derivations of measures of length. Contravariant and covariant bases are introduced as fundamental means of defining measures of finite strains in Lagrangian and Eulerian descriptions. The tensorial nature of strain measures and the influence of change of frame are considered. Physical meaning of the components of the strain tensors for finite deformation is established. Details of polar decomposition as a means of separating rotation and stretch in the deformation gradient including theorems and their proofs are presented. Invariants of strain tensors and representations of strain measures in terms of left and right stretches are also considered. Deformation of areas and volumes are presented for finite deformation.

Problems

3.1 Let the deformation of a unit area (1×1) in the Lagrangian description be given by

$$\left.\begin{aligned} \bar{x}_1 &= x_1 + 0.15tx_2 \\ \bar{x}_2 &= x_2 \end{aligned}\right\} \quad \bar{x}_i = \bar{x}_i(x_1, x_2, t) \quad ; \quad i = 1, 2 \tag{1}$$

where x_1, x_2 are the position coordinates of a material point at $t = 0$, i.e., in the reference configuration.

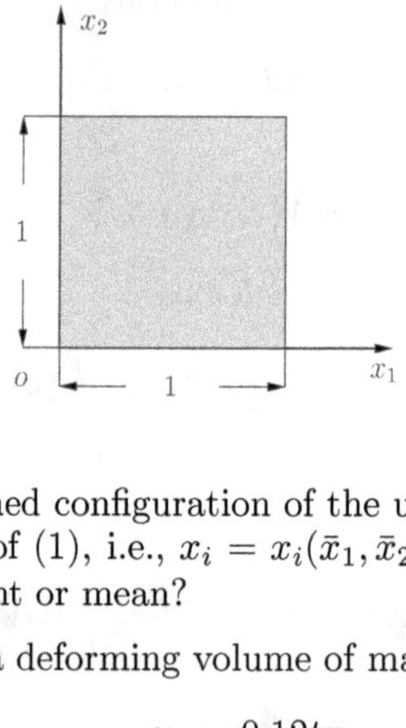

(a) Sketch the deformed configuration of the unit square at $t = 2.0$.

(b) Find the inverse of (1), i.e., $x_i = x_i(\bar{x}_1, \bar{x}_2, t)$; $i = 1, 2$. What do these equations represent or mean?

3.2 Let the kinematics of a deforming volume of matter be given by

$$\begin{aligned} u_1 &= 0.12tx_2 \\ u_2 &= 0 \\ u_3 &= 0 \end{aligned}$$

Let Q be a dependent variable (some quantity of interest) in the current configuration, described by

$$\bar{Q} = 3\bar{x}_1 + \frac{1}{2}(\bar{x}_2)^2$$

(a) Find the material or Lagrangian or referential description of Q.

(b) Find the rate of change Q at a material particle located at $(1, 2, 1)$ at $t = 0$, i.e., in the reference configuration using

$$Q = Q(x_i, t) \tag{1}$$

$$\bar{Q} = \bar{Q}(\bar{x}_i, t) \tag{2}$$

3.3 Consider the following for a deforming volume of matter

$$\begin{aligned}
\bar{x}_1 &= x_1 + x_1 t^3 + 2tx_2 \\
\bar{x}_2 &= x_2 + 2x_2 t^3 + tx_1 \\
\bar{x}_3 &= x_3 + t^2 x_3
\end{aligned}$$

Determine components of the displacement field in

(a) Lagrangian description;
(b) Eulerian description.

3.4 Consider the following for a deforming volume of matter

$$\begin{aligned}
u_1 &= 2\alpha t x_2 \\
u_2 &= -2\alpha t x_1 \\
u_3 &= 0
\end{aligned} \tag{1}$$

Let the dependent variable Q (some quantity of interest) in the current configuration be defined by

$$\bar{Q}(\bar{x}_i, t) = \frac{1}{2}\bar{x}_1 + \frac{1}{3}t^2\bar{x}_2 \tag{2}$$

(a) Determine $\bar{x}_i$ as a function of x_i and time t using (1). Also determine the inverse of the mapping.
(b) Obtain the velocity components.
(c) Determine the material derivative of Q in the two descriptions (Lagrangian and Eulerian).
(d) Sketch the reference and the current configurations of a unit cube.

3.5 Consider one-dimensional deformation given by

$$u_1 = (\alpha t + 3\beta t^2)x_1 \tag{1}$$

Let the material description of a quantity of interest Q be given by

$$Q = 2\gamma x_1 + 3\delta t^2 x_1 \tag{2}$$

(a) Determine $\bar{x}_1$ as a function of x_1 and time t using (1) and the inverse of the mapping.
(b) Express Q in spatial description.
(c) Determine the displacement of a material particle in

(i) Lagrangian description;
(ii) Eulerian description.

3.6 Consider the following deformation field:

$$\begin{aligned}
u_1 &= 4 + 1.5x_2 \\
u_2 &= 2 \\
u_3 &= 0
\end{aligned} \tag{1}$$

(a) Determine $\bar{x}_i$ as a function of x_i and time t using (1) and the inverse of the mapping.
(b) Determine the deformation Jacobian $[J]$ and $[\bar{J}]$.
(c) Show that $[J]^{-1} = [\bar{J}]$ and $[\bar{J}]^{-1} = [J]$.
(d) Determine

(i) how unit vector in x-frame deforms into $\bar{x}$-frame;
(ii) how unit vector in $\bar{x}$-frame deforms into x-frame.

3.7 Consider the dependent variable Q and its description

$$Q = 2(\bar{x}_1)^2 + 3(\bar{x}_2)^2$$

where $\bar{x}_i = x_i(1 + 0.1t)$. Find the rate of change of Q at the material point at $t = 1$, which in the reference configuration was at $(x_1, x_2, x_3) = (2, 1, 2)$, using

(a) Lagrangian description;
(b) Eulerian description.

3.8 Consider the motion of a continuum described by

$$u_1 = 0.16tx_2$$
$$u_2 = 0$$
$$u_3 = 0$$

Find the acceleration of the motion described by (1) using Lagrangian as well as Eulerian descriptions.

3.9 Consider the deformation of a volume of matter given by

$$u_1 = 4 + 1.5x_2$$
$$u_2 = 2$$
$$u_3 = 0$$

(a) Determine the Cauchy strain, Finger strain and Green's strain tensors in Lagrangian description, i.e., covariant measures in Lagrangian description.
(b) Also determine Cauchy strain, Finger strain and Almansi strain tensors in Eulerian description, i.e., contravariant measures in Eulerian description.
(c) Express your results in matrix form.

3.10 Consider the deformation of a square block of sides two units centered at $(x_1, x_2) = (0, 0)$ in the reference configuration. The deformation of the block is described by

$$\begin{aligned} u_1 &= 4.5 + 1.5x_2 \\ u_2 &= 3.5 + 0.5x_2 \\ u_3 &= 0 \end{aligned} \tag{1}$$

(a) Sketch the deformed configuration.
(b) Compute the components of the Jacobian of deformation $[J]$ and its inverse. Verify that the inverse of $[J]$ is in fact $[\bar{J}]$ using (1).

(c) Compute $[C_{[0]}]$, $[F_{[0]}]$, $[\varepsilon_{[0]}]$ and $[\bar{C}^{[0]}]$, $[\bar{F}^{[0]}]$, $[\bar{\varepsilon}^{[0]}]$. Express these in matrix form.

3.11 In Problem 3.10 if the reference configuration is rotated about x_3-axis through 45 degrees to obtain a new configuration with coordinates x' (x'-frame), then determine $[C'_{[0]}]$, $[F'_{[0]}]$ and $[\varepsilon'_{[0]}]$ in x'-frame.

3.12 The state of strain at a material point in the deformed volume of matter is given by

$$[\varepsilon^{[0]}] = 10^{-3} \begin{bmatrix} 4 & -4 & 0 \\ -4 & 0 & 0 \\ 0 & 0 & 3 \end{bmatrix}$$

Determine the principal strains and their directions.

3.13 Consider the deformation of a volume of matter defined by

$$\begin{aligned} \bar{x}_1 &= x_1 + x_2 + 2x_1^2 \\ \bar{x}_2 &= x_2 + x_3 + 1.5x_2^2 \\ \bar{x}_3 &= x_1 + x_3 + 2.5x_3^2 \end{aligned}$$

Consider a vector $\boldsymbol{PQ}$ in the reference configuration with components $\delta[1/6, 1/6, 1/6]^T$ in which δ is an infinitesimal real number passing through point $(2, 2, 2)$. Find $\bar{\boldsymbol{P}}\bar{\boldsymbol{Q}}$, the map of $\boldsymbol{PQ}$ in the current configuration.

3.14 The deformation of a volume of matter is given by

$$\left.\begin{aligned} \bar{x}_1 &= x_1 + x_1^2 + 3x_1x_2 \\ \bar{x}_2 &= x_2 + 2x_2^2 \\ \bar{x}_3 &= x_3 \end{aligned}\right\} \quad \text{for } x_1 \geq 0 \quad ; \quad x_2 \geq 0$$

(a) Find a vector in the reference configuration that is deformed into a vector parallel to the x_1-axis and passing through the material point $(1, 0, 0)$ in the current configuration.

(b) Find the stretch of a line segment that is deformed into a vector parallel to the x_1-axis through the point $(1, 0, 0)$ in the current configuration.

3.15 Let the deformation of a volume of matter be given by

$$\begin{aligned} \bar{x}_1 &= x_1 + 4x_1^2 + x_2 \\ \bar{x}_2 &= x_2 + 3x_2^2 + x_3 \\ \bar{x}_3 &= x_3 + 2x_3^2 + x_1 \end{aligned}$$

Use the covariant Cauchy measure of strain in Lagrangian description for the following.

(a) Find the principal axial strains at a material point $(1/2, 1/2, 1/2)$ in the reference configuration.

(b) Find the direction of the maximum axial strain through the material point $(1/2, 1/2, 1/2)$ in the reference configuration. Also find the direction of the maximum principal strain in the current configuration.

3.16 Consider the simple shear deformation defined by

$$u_1 = 0 \quad , \quad u_2 = \beta x_1 \quad , \quad u_3 = 0$$

(a) Find the principal stretches.
(b) Determine $[S_r]$ and $[S_l]$.

3.17 For a simple shear deformation

$$\begin{aligned} u_1 &= 0 \\ u_2 &= \beta x_3 \\ u_3 &= \beta x_2 \end{aligned}$$

where β is a constant, find the orthogonal matrix $[R]$ and the right and left stretch tensors $[S_r]$ and $[S_l]$ in the polar decomposition of the deformation gradient $[J]$. Also find the principal stretches and their directions.

3.18 Consider the one-dimensional deformation

$$u_1 = 0.5tx_1 \tag{1}$$

describing the motion of a rod of length 4 centimeters. The rod experiences a temperature distribution θ given by the material description $\theta = 3x_1t^2$.

(a) Determine $\bar{x}_i$ as a function of x_i and time t using (1) and the inverse of the mapping.
(b) Determine the temperature distribution in the spatial description.
(c) Determine the displacement of a material particle in Lagrangian and Eulerian descriptions.
(d) Plot a graph of x_1 or $\bar{x}_1$ versus t to illustrate the deformation of the rod and the corresponding temperature values of the middle of the rod and at the end of the rod (i.e., $x_1 = 2.0$ and $x_1 = 4.0$).

3.19 Consider the temperature field defined by

$$\bar{\theta} = 2\bar{x}_1 + 3\bar{x}_2^2 \quad \text{where} \quad \bar{x}_i = x_i(1 + 0.1t)$$

Find the rate of change of temperature of the material particle at $t = 2$, which in the reference configuration was at $(x_1, x_2, x_3) = (1/2, 1/2, 1/2)$, using

(a) Lagrangian description;
(b) Eulerian description.

3.20 Consider a cube formed by orthogonal vectors $\bar{\mathbf{e}}_1$, $\bar{\mathbf{e}}_2$, $\bar{\mathbf{e}}_3$ where $\bar{\mathbf{e}}_i = c_{ij}\mathbf{e}_j$. Determine the conditions on the coefficients c_{ij} if the volume of the cube in the current configuration is 1.

3.21 Consider the motion

$$\begin{aligned} u_1 &= 2x_2 \\ u_2 &= 0.5x_1 + 3x_2 \\ u_3 &= 0.95x_3 \end{aligned}$$

Is this admissible in a continuously deforming matter?

3.22 Let the displacement components for a deforming matter be given by

$$u_1 = 2x_1 + x_2$$
$$u_2 = x_3$$
$$u_3 = x_3 - x_2$$

(a) Verify if this displacement field is admissible in a continuously deforming matter.
(b) Is the deformation homogeneous?
(c) Is this deformation isochoric?

3.23 Consider the deformation given by

$$u_1 = (\lambda_1 - 1)x_1$$
$$u_2 = (\lambda_2 - 1)x_2$$
$$u_3 = (\lambda_3 - 1)x_3$$

Consider Green's strain $[\varepsilon]$. Let I_ε, II_ε and III_ε be the invariants (or coefficients) in the characteristic equation of $[\varepsilon]$. Determine I_ε, II_ε and III_ε for the given deformation.

3.24 Let

$$u_1 = 2x_1^2 + 3x_1x_2$$
$$u_2 = 4x_2^2$$
$$u_3 = 0$$

be the displacement field of a deforming matter. At material point $(1/2, 0, 0)$ in the reference configuration, consider an infinitesimal plane (i.e., area) formed by the vectors $\delta(1, 0, 0)$ and $\delta(1, 2, 0)$. Find the map of this vector area in the current configuration, i.e., find a vector area in the current configuration into which this vector area in the reference configuration maps. The term δ is a small arbitrary constant.

3.25 Consider the deformation of a volume of matter given by

$$\begin{aligned} u_1 &= (\alpha_1 - 1)x_1 \\ u_2 &= -x_2 - \alpha_3 x_3 \\ u_3 &= \alpha_2 x_2 - x_3 \end{aligned} \tag{1}$$

(a) Find the deformed volume of a unit cube located at $(0, 0, 0)$ in x-frame, i.e., $(0, 0, 1) \times (0, 1, 0) \times (0, 0, 1)$ cube.
(b) Find $[S_r]$, $[S_l]$ and $[R]$ in the polar decomposition of $[J]$ associated with (1).

4
DEFINITIONS AND MEASURES OF STRESSES

In this chapter, we consider various definitions and measures of stresses. The concept of stress is defined through the concept of force and area in the current configuration of a deforming volume of continua. When the deformation and strains are small, the deformed coordinates are approximately same as the undeformed coordinates of the material points (i.e., $\bar{x}_i \simeq x_i$). In this deformation physics undeformed length $d\boldsymbol{s}$, area $d\boldsymbol{A}$ and volume V virtually remain the same after deformation, i.e., $d\bar{\boldsymbol{s}} \simeq d\boldsymbol{s}$, $d\bar{\boldsymbol{A}} \simeq d\boldsymbol{A}$ and $\bar{V} \simeq V$. This deformation physics simplifies definitions and measure of stress as the distinction between reference and current configurations need not be considered. When the deformation and the strains are finite, the orientation and the magnitude of deformed areas and the directions of the forces acting on the deformed areas can be significantly different compared to their maps in the reference configuration. Thus, the definitions and measures of stresses in this case are more complicated compared to small deformation, small strain, hence require more careful considerations.

In the following, first we consider some basic concepts that can be illustrated more clearly if we assume small deformation, small strain deformation physics. Hence, we assume $\bar{x}_i \simeq x_i$, i.e., we do not distinguish between the reference and the deformed configurations. We establish basic definitions using this assumption which are then subsequently extended for finite deformation, finite strain physics.

4.1 Concept of stress

When a volume of continuous matter is subject to time dependent disturbance, the points of application of the disturbance experience time dependent motion resulting in rate of work. This, in turn, imparts time dependent motion to the material points of the continua, resulting in relative motion between them (measured in terms of strain tensor). The resistance offered by the continua to this relative motion results in a force field inside the volume of the deforming continua. A measure of this internal force field relative to some area is the notion of stress. Thus, inherently stresses are internal to the deforming volume of matter. On the bounding surface of the volume, the stresses are referred to as tractions, meaning externally applied disturbance. An infinitesimal distance inward into the volume from the bounding surface or boundary, we have stresses.

4.2 Cut Principle of Cauchy

Any deforming volume of matter V in equilibrium at any time t (Figure 4.1) can be cut into two volumes V_1 and V_2 by a hypothetical diaphragm or cut plane S. The forces acting on the cut areas $d\bar{A}_n$ of $\bar{V}_1$ and $\bar{V}_2$ and the forces exerted by volumes $\bar{V}_1$ and $\bar{V}_2$ on both sides of diaphragm S with area $d\bar{A}_n$ maintain equilibrium of volumes $\bar{V}_1$, $\bar{V}_2$ and the diaphragm. Clearly forces on the two sides of cut plane are

DOI: 10.1201/9781003105336-4

equal and opposite. When volumes $\bar{V}_1$ and $\bar{V}_2$ are brought together using diaphragm S, the equilibrium of the uncut volume $\bar{V}$ is restored.

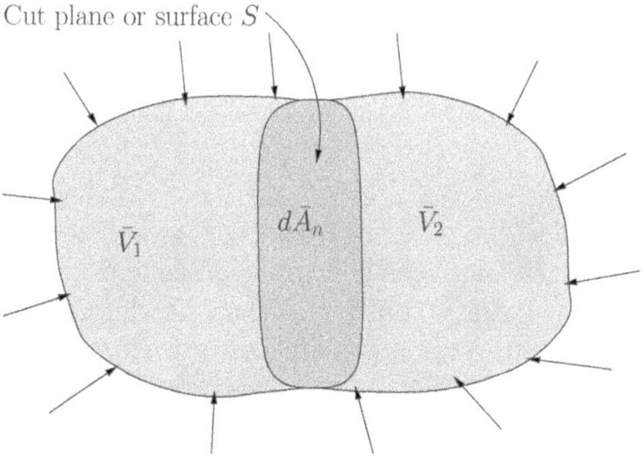

(a) Deformed volume in $\bar{V}$ with boundary $\partial\bar{V}$ in stable equilibrium at time t (current configuration) and the cut plane or surface or diaphragm

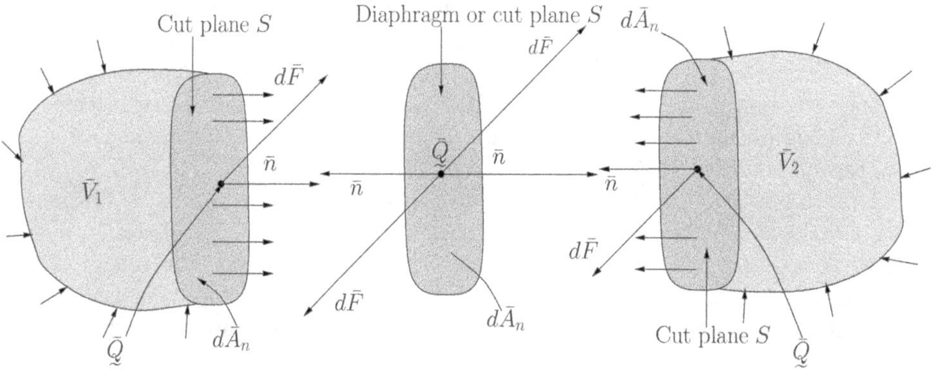

(b) Volumes $\bar{V}_1$, $\bar{V}_2$ and cut plane S with forces acting on cut areas $d\bar{A}_n$ of volumes $\bar{V}_1$ and $\bar{V}_2$ and the diaphragm

Figure 4.1: Cut principle of Cauchy

4.3 Definition of stress on area $d\bar{A}_n$

Since the cut surface area $d\bar{A}_n$ is interior of the volume and is exposed only due to cut plane S, it is possible to define stress on it. Consider volume $\bar{V}_1$ ($\bar{V}_2$ is similar). Let $\bar{n}$ be unit normal to the area (assumed cut plane S is flat). Area $d\bar{A}_n$ has distribution of forces acting on it whose resultant is force $d\bar{\boldsymbol{F}}$ acting at point $\underset{\sim}{Q}$. Directions of $\bar{\boldsymbol{n}}$ and $d\bar{\boldsymbol{F}}$ are not the same. Let $\bar{\boldsymbol{P}}$ be average force per unit area on area $d\bar{A}_n$. Let

$$d\bar{\boldsymbol{F}} = \boldsymbol{e}_1 d\bar{F}_1 + \boldsymbol{e}_2 d\bar{F}_2 + \boldsymbol{e}_3 d\bar{F}_3 \tag{4.1}$$

$$\text{or} \quad \{d\bar{F}\}^T = [d\bar{F}_1, d\bar{F}_2, d\bar{F}_3] \tag{4.2}$$

$$\text{and} \quad \bar{\boldsymbol{P}} = \boldsymbol{e}_1 \bar{P}_1 + \boldsymbol{e}_2 \bar{P}_2 + \boldsymbol{e}_3 \bar{P}_3 \tag{4.3}$$

$$\text{or} \quad \{\bar{P}\}^T = [\bar{P}_1, \bar{P}_2, \bar{P}_3] \tag{4.4}$$

A traditional way of defining stress (as done in elementary mechanics courses) is

$$\{\bar{P}\} = \lim_{d\bar{A}_n \to 0} \frac{1}{d\bar{A}_n}\{d\bar{F}\} \tag{4.5}$$

where $\{\bar{P}\}$ is the stress at point $\underset{\sim}{Q}$ (Figure 4.1(b)).

This definition is obviously nonphysical due to the fact that $d\bar{A}_n \to 0$ will not be able to support a force $\{d\bar{F}\}$. In the second definition of stress, we define $\bar{P}$ as force per unit area on area $d\bar{A}_n$ using resulting force $d\bar{\boldsymbol{F}}$ and area $d\bar{A}_n$.

$$\{\bar{P}\} = \begin{Bmatrix} \bar{P}_1 \\ \bar{P}_2 \\ \bar{P}_3 \end{Bmatrix} = \frac{1}{d\bar{A}_n}\begin{Bmatrix} d\bar{F}_1 \\ d\bar{F}_2 \\ d\bar{F}_3 \end{Bmatrix} \tag{4.6}$$

$$\text{or} \quad d\bar{A}_n\{\bar{P}\} = \{d\bar{F}\} \tag{4.7}$$

Stresses on area $d\bar{A}_n$ of volume V_1 (likewise of volume V_2) are completely defined by $\{\bar{P}\}$ along O-x_1, O-x_2, O-x_3 directions of x-frame.

Instead of choosing $\{\bar{P}\}$ as a measure of stress in the x-frame, we could also choose normal direction $\{\bar{n}\}$ and two other mutually perpendicular tangential directions $\{\bar{l}\}$ and $\{\bar{m}\}$ in the plane containing area $d\bar{A}_n$ giving rise to three components of stress $\{\bar{\sigma}_n\}$.

$$\{\bar{\sigma}_n\}^T = [\bar{\sigma}_{nn}, \bar{\sigma}_{nl}, \bar{\sigma}_{nm}] \tag{4.8}$$

$\{\bar{P}\}$ and $\{\bar{\sigma}_n\}$ representation of stress are equivalent in the sense that

$$\{\bar{P}\}^T\{\bar{P}\} = \{\bar{\sigma}_n\}^T\{\bar{\sigma}_n\} \tag{4.9}$$

That is, L_2-norm of $\{\bar{P}\}$ and $\{\bar{\sigma}\}$ are exactly the same. This obviously must hold because $\{d\bar{F}\}$ and $d\bar{A}_n$ used in both measures are same.

4.4 Cauchy stress tensor

Definition of stress in Section 4.3 serves as a basic definition. However, we need to consider a broader view and a more general definition of stress. Consider a volume of matter V bounded by ∂V in the reference configuration (Figure 4.2(a)). Consider tractions acting on ∂V (for simplicity, other loadings are admissible too), Figure 4.2(b) shows deformed current configuration with volume $\bar{V}$ bounded by $\partial\bar{V}$. The definition of the stress must be such that the deformed volume in the current configuration under the action of tractions on $\partial\bar{V}$ and the internal stress field in the volume $\bar{V}$ must remain in equilibrium, i.e., no rigid translations and no rigid rotations of the volume $\bar{V}$ are allowed to occur.We must define stress such that these two criteria are satisfied. As stated earlier, for simplicity, we assume small deformation and small strain so that $\bar{x}_i \simeq x_i$, i.e., we do not have to distinguish between reference and current configurations. An elementary volume $OA_1A_2A_3$ (tetrahedron T) in the reference configuration deforms into $O\bar{A}_1\bar{A}_2\bar{A}_3$ (tetrahedron $\bar{T}$) in the current configuration. Due to $\bar{x}_i \simeq x_i$, the two deformed volumes in reference and the current configuration can be assumed to be same in shape, size and orientation. In defining the stress concept in $\mathbb{R}^3$, we must consider the following.

(i) Consider an elementary volume (a tetrahedron). The elementary volume must contain externally applied tractions as well as internal stress field in order to establish relationship between them.

(ii) The elementary volume must be representative of all possible choices within volume $\bar{V}$ including its boundary $\partial\bar{V}$. This aspect is essential to ensure that the equilibrium of the elementary volume is assurance of the equilibrium of the entire volume $\bar{V}$.

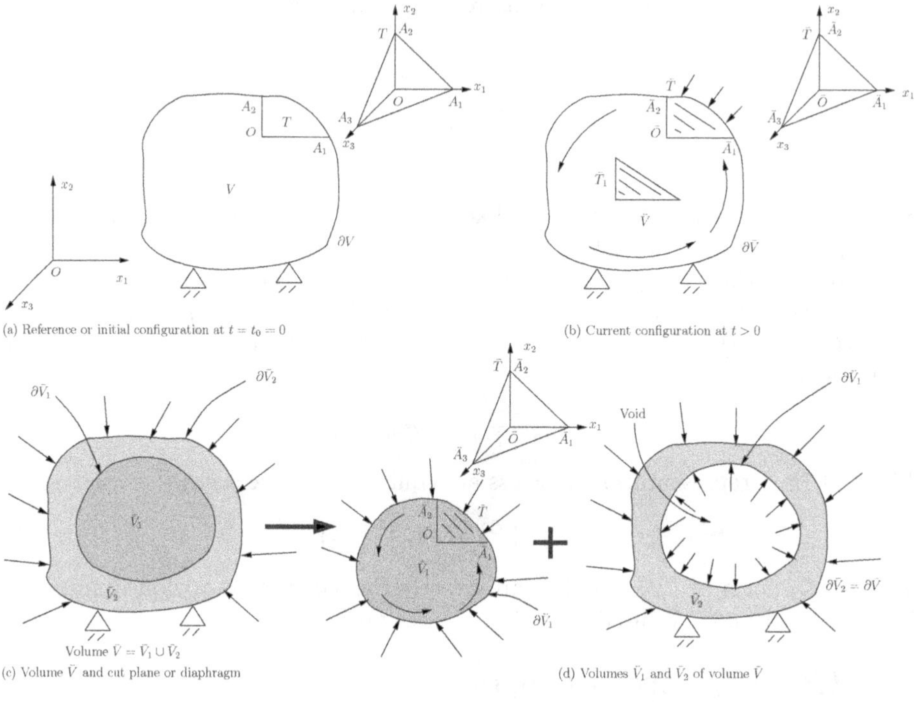

Figure 4.2: Consideration of elementary volume for defining Cauchy stress

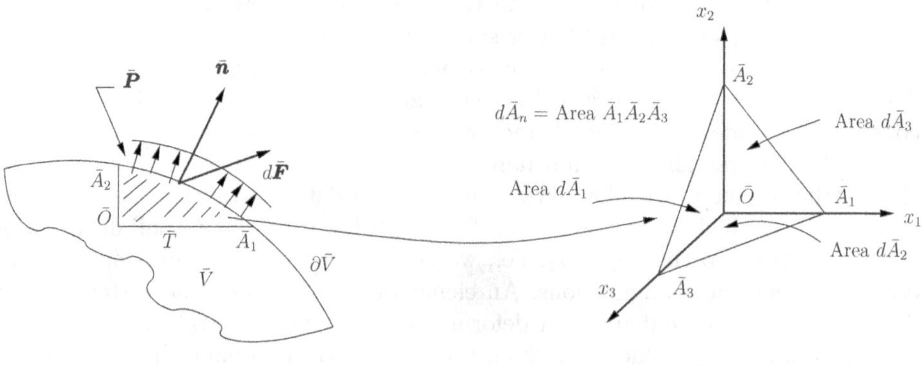

Figure 4.3: Elementary volume, tetrahedron $\bar{T}$

Referring to Figure 4.2(b), consider a tetrahedron $O\bar{A}_1\bar{A}_2\bar{A}_3$ (elementary volume) whose oblique plane $\bar{A}_1\bar{A}_2\bar{A}_3$ has applied tractions on it whereas the planes

$O\bar{A}_1\bar{A}_2$, $O\bar{A}_2\bar{A}_3$, $O\bar{A}_3\bar{A}_1$ have internal force field, i.e., stresses acting on them. This is a valid choice for elementary volume as it contains both tractions and stresses. This holds true for every tetrahedron in which oblique plan $\bar{A}_1\bar{A}_2\bar{A}_3$ is part of $\partial\bar{V}$. However, if we choose a tetrahedron $\bar{T}_1$ 4.2(b) for which $\bar{A}_1\bar{A}_2\bar{A}_3$ is not part of $\partial\bar{V}$, then we could conclude that $\bar{A}_1\bar{A}_2\bar{A}_3$ does not have applied tractions. This issue is resolved using cut principle of Cauchy. Figure 4.2(c) shows a diaphragm or cut plane separating $\bar{V}$ into $\bar{V}_1$ and $\bar{V}_2$ as shown in Figure 4.2(d). The boundary $\partial\bar{V}_1$ of $\bar{V}_1$ has tractions due to removed material constituting volume $\bar{V}_2$. Volume $\bar{V}_1$ with bounding surface $\partial\bar{V}_1$ is exactly like the original volume $\bar{V}$ with bounding surface $\partial\bar{V}$, hence the concept of elementary volume constituting of tetrahedron $\bar{O}\bar{A}_1\bar{A}_2\bar{A}_3$ with $\bar{A}_1\bar{A}_2\bar{A}_3$ being part of $\partial\bar{V}_1$ is valid. Following the cut principle of Cauchy, every tetrahedron in volume $\bar{V}$ has oblique plane $\bar{A}_1\bar{A}_2\bar{A}_3$ with tractions acting on it. Thus, to establish the definition of Cauchy stress tensor we only need to assume a tetrahedron $\bar{O}\bar{A}_1\bar{A}_2\bar{A}_3$ (Figure 4.3) that has tractions on oblique plane $\bar{A}_1\bar{A}_2\bar{A}_3$ and stress on the planes $\bar{O}\bar{A}_1\bar{A}_2$, $\bar{O}\bar{A}_2\bar{A}_3$, $\bar{O}\bar{A}_3\bar{A}_1$. Let $d\bar{A}_1$, $d\bar{A}_2$, $d\bar{A}_3$ be areas of planes $\bar{O}\bar{A}_2\bar{A}_3$ (x_1 plane), $\bar{O}\bar{A}_3\bar{A}_1$ (x_2 plane) and $\bar{O}\bar{A}_1\bar{A}_2$ (x_3 plane). Let $\bar{O}\bar{N}$ be the length of the perpendicular from point $\bar{O}$ to plane $\bar{A}_1\bar{A}_2\bar{A}_3$. We note that the oblique plane $\bar{A}_1\bar{A}_2\bar{A}_3$ can be assumed flat due to the fact that the elementary volume may be chosen as small as we wish so that this assumption holds. On the oblique plane $\bar{A}_1\bar{A}_2\bar{A}_3$ of the tetrahedron $\bar{T}$ of Figure 4.3, we have resultant force $d\bar{\boldsymbol{F}}$, unit exterior normal $\bar{\boldsymbol{n}}$ and average force per unit area $\bar{\boldsymbol{P}}$, similar to those in Section 4.3. Thus, equations (4.6)–(4.9) hold here as well. The average stress (force per unit area) acting on planes $\bar{O}\bar{A}_1\bar{A}_2$, $\bar{O}\bar{A}_2\bar{A}_3$, $\bar{O}\bar{A}_3\bar{A}_1$ must equilibrate with the average tractions (force per unit area) $\bar{\boldsymbol{P}}$ acting on oblique plane $\bar{A}_1\bar{A}_2\bar{A}_3$. This requires the following in the x-frame.

(a) Force balance in x-frame. This eliminates rigid translation of the volume $\bar{V}$.
(b) Sum of moments of resultant forces about any point must be zero. This eliminates rigid rotation of the volume $\bar{V}$.

4.4.1 Force balance

Let $d\bar{A}_n$, $d\bar{A}_1$, $d\bar{A}_2$ and $d\bar{A}_3$ be the areas of the triangles $\bar{A}_1\bar{A}_2\bar{A}_3$, $\bar{O}\bar{A}_2\bar{A}_3$, $\bar{O}\bar{A}_1\bar{A}_3$ and $\bar{O}\bar{A}_2\bar{A}_1$ in which $\bar{O}\bar{A}_1$, $\bar{O}\bar{A}_2$ and $\bar{O}\bar{A}_3$ are perpendicular to the faces of the tetrahedron $\bar{O}\bar{A}_1\bar{A}_2\bar{A}_3$. Considering the volume of the tetrahedron $\bar{O}\bar{A}_1\bar{A}_2\bar{A}_3$ we can write

$$\frac{1}{2}d\bar{A}_n\bar{O}\bar{N} = \frac{1}{2}d\bar{A}_1\bar{O}\bar{A}_1 = \frac{1}{2}d\bar{A}_2\bar{O}\bar{A}_2 = \frac{1}{2}d\bar{A}_3\bar{O}\bar{A}_3 \tag{4.10}$$

or

$$\begin{Bmatrix} \dfrac{\bar{O}\bar{N}}{\bar{O}\bar{A}_1} \\ \dfrac{\bar{O}\bar{N}}{\bar{O}\bar{A}_2} \\ \dfrac{\bar{O}\bar{N}}{\bar{O}\bar{A}_3} \end{Bmatrix} = \frac{1}{d\bar{A}_n}\begin{Bmatrix} d\bar{A}_1 \\ d\bar{A}_2 \\ d\bar{A}_3 \end{Bmatrix} = \begin{Bmatrix} \bar{n}_1 \\ \bar{n}_2 \\ \bar{n}_3 \end{Bmatrix} = \{\bar{n}\} \tag{4.11}$$

where $\{\bar{n}\}$ are the direction cosines of the normal $\bar{O}\bar{N}$ to the plane $\bar{A}_1\bar{A}_2\bar{A}_3$.

Consider the stresses acting on the three perpendicular faces of the tetrahedron consisting of areas $d\bar{A}_1$, $d\bar{A}_2$ and $d\bar{A}_3$ represented by the two forms $\{\bar{P}\}$ and $\{\bar{\sigma}_n\}$. These two forms are the same if the tangential directions in each area are along coordinate axes in the x-frame. Thus, we have (3×1) stress vector on each area consisting of a component normal to the area and the other two in the plane along two mutually perpendicular directions. Hence, we have the following

$$\begin{aligned} &\text{On area} \quad d\bar{A}_1 : \quad \bar{\sigma}_{11} \quad , \quad \bar{\sigma}_{12} \quad , \quad \bar{\sigma}_{13} \\ &\text{On area} \quad d\bar{A}_2 : \quad \bar{\sigma}_{21} \quad , \quad \bar{\sigma}_{22} \quad , \quad \bar{\sigma}_{23} \\ &\text{On area} \quad d\bar{A}_3 : \quad \bar{\sigma}_{31} \quad , \quad \bar{\sigma}_{32} \quad , \quad \bar{\sigma}_{33} \end{aligned} \tag{4.12}$$

In denoting the stress components, *the first subscript is the direction of the normal to the plane, i.e., it identifies the plane on which this component acts, and the second subscript is the direction of the stress.* Throughout the book we use this notation. These components can be arranged in the form of a (3×3) matrix $[\bar{\sigma}]$ (stress matrix). The components of the stress in each row of $[\bar{\sigma}]$ corresponds to each of the three areas.

$$[\bar{\sigma}] = \begin{bmatrix} \bar{\sigma}_{11} & \bar{\sigma}_{12} & \bar{\sigma}_{13} \\ \bar{\sigma}_{21} & \bar{\sigma}_{22} & \bar{\sigma}_{23} \\ \bar{\sigma}_{31} & \bar{\sigma}_{32} & \bar{\sigma}_{33} \end{bmatrix} \begin{matrix} \leftarrow \text{On area} \quad d\bar{A}_1 \\ \leftarrow \text{On area} \quad d\bar{A}_2 \\ \leftarrow \text{On area} \quad d\bar{A}_3 \end{matrix} \tag{4.13}$$

For the deformed tetrahedron to be in equilibrium (1) there must be equilibrium of forces to stop rigid translation and (2) there must be equilibrium of the moment of the forces to stop rigid rotation. If we consider the equilibrium of the forces in x_i directions on the tetrahedron $\bar{O}\bar{A}_1\bar{A}_2\bar{A}_3$, then we can write

$$\begin{aligned} d\bar{A}_n\bar{P}_1 &= d\bar{A}_1\bar{\sigma}_{11} + d\bar{A}_2\bar{\sigma}_{21} + d\bar{A}_3\bar{\sigma}_{31} \\ d\bar{A}_n\bar{P}_2 &= d\bar{A}_1\bar{\sigma}_{12} + d\bar{A}_2\bar{\sigma}_{22} + d\bar{A}_3\bar{\sigma}_{32} \\ d\bar{A}_n\bar{P}_3 &= d\bar{A}_1\bar{\sigma}_{13} + d\bar{A}_2\bar{\sigma}_{23} + d\bar{A}_3\bar{\sigma}_{33} \end{aligned} \tag{4.14}$$

Dividing each equation of (4.14) by $d\bar{A}_n$ and noting that

$$\frac{d\bar{A}_1}{d\bar{A}_n} = \bar{n}_1 \quad ; \quad \frac{d\bar{A}_2}{d\bar{A}_n} = \bar{n}_2 \quad ; \quad \frac{d\bar{A}_3}{d\bar{A}_n} = \bar{n}_3 \tag{4.15}$$

then we can write (4.14) as follows

$$\{\bar{P}\} = \begin{Bmatrix} \bar{P}_1 \\ \bar{P}_2 \\ \bar{P}_3 \end{Bmatrix} = [\bar{\sigma}]^T\{\bar{n}\} \quad ; \quad \{\bar{n}\} = \begin{Bmatrix} \bar{n}_1 \\ \bar{n}_2 \\ \bar{n}_3 \end{Bmatrix} \quad \text{or} \quad \bar{\boldsymbol{P}} = \bar{\boldsymbol{\sigma}}^T \cdot \bar{\boldsymbol{n}} \tag{4.16}$$

Equation (4.16) is known as the *Cauchy principle* in which $\{\bar{P}\}$ is referred to as the average stress on the oblique plane of the elementary volume, i.e., tetrahedron. In the second form, the normal and shearing stresses on the surface $d\bar{A}_n$ are given by

$$\begin{aligned} \bar{\sigma}_{nn} &= \bar{P}_1\bar{n}_1 + \bar{P}_2\bar{n}_2 + \bar{P}_3\bar{n}_3 \\ \bar{\sigma}_{nl} &= \bar{P}_1\bar{l}_1 \ + \bar{P}_2\bar{l}_2 \ + \bar{P}_3\bar{l}_3 \\ \bar{\sigma}_{nm} &= \bar{P}_1\bar{m}_1 + \bar{P}_2\bar{m}_2 + \bar{P}_3\bar{m}_3 \end{aligned} \tag{4.17}$$

in which $\{\bar{n}\}$, $\{\bar{l}\}$ and $\{\bar{m}\}$ are the direction cosines of the normal $\bar{O}\bar{N}$ and the tangent vectors in the plane $\bar{A}_1\bar{A}_2\bar{A}_3$. Equation (4.17) can also be written as

$$\{\bar{\sigma}_n\} = \begin{bmatrix} \{\bar{n}\}^T \\ \{\bar{l}\}^T \\ \{\bar{m}\}^T \end{bmatrix} \{\bar{P}\} \tag{4.18}$$

Substituting for $\{\bar{P}\}$ from (4.16) into (4.18)

$$\{\bar{\sigma}_n\} = [\{\bar{n}\}, \{\bar{l}\}, \{\bar{m}\}]^T [\bar{\sigma}]^T \{\bar{n}\} \tag{4.19}$$

We remark that Cauchy principle is due to force balance, hence it ensures that this definition of stress eliminates rigid translation of the deforming volume $\bar{V}$. Clearly Cauchy principle is not a postulate as the postulate implies "*assumed truth for reasoning*" which is not the case here. The Cauchy principle is based on the force balance.

4.4.2 Moment of Forces

Moment of the forces acting on the tetrahedron must be zero to stop rigid rotation of the tetrahedron. A more general derivation is presented in Chapter 6 based on the principle of angular momenta in which the time rate of the moment of momenta for a continuum is equal to the vector sum of the moments of the forces acting on the continuum. In the following derivation we only consider $\bar{\boldsymbol{P}}$ acting on $d\bar{A}_n$ which equilibrates with $\bar{\sigma}_{ij}$ acting on $d\bar{A}_1$, $d\bar{A}_2$ and $d\bar{A}_3$. Considering average stress $\bar{\boldsymbol{P}}$ acting on $d\bar{A}_n$, we can write

$$\int_{\bar{A}_n} \bar{\boldsymbol{x}} \times \bar{\boldsymbol{P}} \, d\bar{A}_n = 0 \tag{4.20}$$

$$\text{or} \quad \int_{\bar{A}_n} \epsilon_{ijk} \, \bar{x}_i \, \bar{P}_j \, d\bar{A}_n = 0 \tag{4.21}$$

Using the Cauchy principle we can write

$$\bar{P}_j = \bar{\sigma}_{mj} \, \bar{n}_m \tag{4.22}$$

Substituting from (4.22) into (4.21)

$$\int_{\bar{A}_n} \epsilon_{ijk} \, \bar{x}_i \, \bar{\sigma}_{mj} \, \bar{n}_m \, d\bar{A}_n = 0 \tag{4.23}$$

Using Gauss's divergence theorem to convert the integral over $\bar{A}_n$ to $\bar{V}$ (volume of the deformed tetrahedron) yields

$$\int_{\bar{V}} \left(\epsilon_{ijk} \, \bar{x}_i \, \bar{\sigma}_{mj}\right)_{,m} d\bar{V} = 0 \tag{4.24}$$

$$\text{or} \quad \int_{\bar{V}} \epsilon_{ijk} \big(\bar{x}_i \, \bar{\sigma}_{mj,m} + \bar{x}_{i,m} \, \bar{\sigma}_{mj}\big) \, d\bar{V} = 0 \tag{4.25}$$

which can be written as

$$\int_{\bar{V}} \epsilon_{ijk}\big(\bar{x}_i\,\bar{\sigma}_{mj,m} + \delta_{ip}\,\bar{\sigma}_{pj}\big)\,d\bar{V} = 0 \tag{4.26}$$

$$\text{or} \quad \int_{\bar{V}} \epsilon_{ijk}\big(\bar{x}_i\,\bar{\sigma}_{mj,m} + \bar{\sigma}_{ij}\big)\,d\bar{V} = 0 \tag{4.27}$$

In the absence of body forces and inertial forces, the balance of moment of momenta gives $\bar{\sigma}_{mj,m} = 0$. Substituting this in (4.27) we conclude the following (as $\bar{V}$ is arbitrary for isotropic, homogeneous matter).

$$\epsilon_{ijk}\bar{\sigma}_{ij} = 0 \tag{4.28}$$

From (4.28) we have

$$\begin{aligned} \bar{\sigma}_{12} - \bar{\sigma}_{21} &= 0 \\ \bar{\sigma}_{23} - \bar{\sigma}_{32} &= 0 \\ \bar{\sigma}_{31} - \bar{\sigma}_{13} &= 0 \end{aligned} \tag{4.29}$$

Equation (4.29) implies that the Cauchy stress tensor $\bar{\boldsymbol{\sigma}}$ is symmetric, i.e., $[\bar{\sigma}] = [\bar{\sigma}]^T$. *In the above derivation, we have assumed infinitesimal deformation* for which $\bar{\boldsymbol{x}} \approx \boldsymbol{x}$ and $\bar{V} \approx V$. See Chapters 6 and 7 for the general case that consider finite strain and finite deformation.

4.4.3 Cauchy principle

Def: Cauchy principle

The stress matrix $[\bar{\sigma}]$ in x-frame related to the deformed tetrahedron in the current configuration is called Cauchy stress matrix or Eulerian stress matrix. Equation (4.16) is known as the Cauchy principle. The Cauchy principle requires a tetrahedron with its faces parallel to the planes of the x-frame and the oblique plane $\bar{A}_1\bar{A}_2\bar{A}_3$ to be part of the boundary on which the resultant forces (i.e., tractions) act, producing average stress $\{\bar{P}\}$ on plane $\bar{A}_1\bar{A}_2\bar{A}_3$. Thus, it relates stresses values on the faces of this tetrahedron to the average stress on the oblique plane through unit exterior normal $\{\bar{n}\}$ to the oblique plane.

Transformation of stress matrix $[\bar{\sigma}]$

If $\bar{x}_i$ coordinates are changed to $\bar{x}'_i$ through rigid rotation, then we determine how $[\bar{\sigma}]$ in $\bar{x}$-frame transforms to $[\bar{\sigma}']$ in $\bar{x}'$-frame

Let $\bar{x}_i$ and $\bar{x}'_i$ be related through an orthogonal rotation matrix $[\bar{Q}]$ (Galilean transformation)

$$\{\bar{x}'\} = [\bar{Q}]\{\bar{x}\} \quad ; \quad [\bar{Q}]^T = [\bar{Q}]^{-1} \tag{4.30}$$

then

$$\{\bar{n}'\} = [\bar{Q}]\{\bar{n}\} \quad ; \quad \{\bar{n}\} = [\bar{Q}]^T\{\bar{n}'\} \tag{4.31}$$

Also

$$\{\bar{P}'\} = [\bar{Q}]\{\bar{P}\} \quad ; \quad \{\bar{P}\} = [\bar{Q}]^T\{\bar{P}'\} \tag{4.32}$$

Consider (4.16)

$$\{\bar{P}\} = [\bar{\sigma}]\{\bar{n}\} \quad ; \quad [\bar{\sigma}]^T = [\bar{\sigma}] \tag{4.33}$$

which, using (4.31), can be written as

$$\{\bar{P}\} = [\bar{\sigma}][\bar{Q}]^T\{\bar{n}'\} \tag{4.34}$$

Premultiply (4.34) by $[\bar{Q}]$.

$$[\bar{Q}]\{\bar{P}\} = [\bar{Q}][\bar{\sigma}][\bar{Q}]^T\{\bar{n}'\} \tag{4.35}$$

Using (4.32)

$$\{\bar{P}'\} = [\bar{Q}][\bar{\sigma}][\bar{Q}]^T\{\bar{n}'\} = [\bar{\sigma}']\{\bar{n}'\} \tag{4.36}$$

$$\therefore \quad [\bar{\sigma}'] = [\bar{Q}][\bar{\sigma}][\bar{Q}]^T \tag{4.37}$$

Thus, $[\bar{\sigma}]$ and $[\bar{\sigma}']$ both are tensors of rank two in $\bar{x}$- and $\bar{x}'$-frames. That is, the Cauchy stress matrix $[\bar{\sigma}]$ is a symmetric tensor of rank two.

Remarks

(1) We note that $\bar{\sigma}_{11}$, $\bar{\sigma}_{12}$ and $\bar{\sigma}_{13}$ act on x_1-plane (a plane with ox_1 direction normal to it) in x_1, x_2, x_3 directions, respectively. Thus, $\boldsymbol{e}_1 \otimes \boldsymbol{e}_1$, $\boldsymbol{e}_1 \otimes \boldsymbol{e}_2$ and $\boldsymbol{e}_1 \otimes \boldsymbol{e}_3$ are dyads associated with the components $\bar{\sigma}_{11}$, $\bar{\sigma}_{12}$ and $\bar{\sigma}_{13}$. Likewise $\boldsymbol{e}_2 \otimes \boldsymbol{e}_1$, $\boldsymbol{e}_2 \otimes \boldsymbol{e}_2$ and $\boldsymbol{e}_2 \otimes \boldsymbol{e}_3$ are dyads for $\bar{\sigma}_{21}$, $\bar{\sigma}_{22}$ and $\bar{\sigma}_{23}$ and $\boldsymbol{e}_3 \otimes \boldsymbol{e}_1$, $\boldsymbol{e}_3 \otimes \boldsymbol{e}_2$ and $\boldsymbol{e}_3 \otimes \boldsymbol{e}_3$ are dyads for $\bar{\sigma}_{31}$, $\bar{\sigma}_{32}$ and $\bar{\sigma}_{33}$. Therefore, using index and Einstein notations we can represent stress tensor $\bar{\boldsymbol{\sigma}}$ as

$$\bar{\boldsymbol{\sigma}} = \boldsymbol{e}_i \otimes \boldsymbol{e}_j \bar{\sigma}_{ij} \tag{4.38}$$

in which $\bar{\sigma}_{ij}$ are the components of the tensor $\bar{\boldsymbol{\sigma}}$ in x-frame and $\boldsymbol{e}_i \otimes \boldsymbol{e}_j$ are the dyads.

(2) Based on Remark (1), the dyadic product of the basis in which the components of a tensor are defined can be used to represent the tensor in the indexed and Einstein notations.

(3) In the following we extend the concept of stress for finite deformation and finite strain physics.

4.5 Stress measures: finite deformation, finite strain

Measures of stresses for finite deformation, finite strain require special consideration over and beyond small deformation, small strain considered thus far. A material point $\underset{\sim}{Q}$ at x_i in the reference configuration with orthogonal material lines $\underset{\sim}{Q}A_1$, $\underset{\sim}{Q}A_2$ and $\underset{\sim}{Q}A_3$ deforms (translates) to new location $\bar{\underset{\sim}{Q}}$ with coordinates $\bar{x}_i$. The material lines become non-orthogonal and curvilinear, i.e., $\underset{\sim}{Q}A_1$, $\underset{\sim}{Q}A_2$ and $\underset{\sim}{Q}A_3$ rotate into $\bar{\underset{\sim}{Q}}\bar{A}_1$, $\bar{\underset{\sim}{Q}}\bar{A}_2$ and $\bar{\underset{\sim}{Q}}\bar{A}_3$. The tangent vectors to $\bar{\underset{\sim}{Q}}\bar{A}_1$, $\bar{\underset{\sim}{Q}}\bar{A}_2$ and $\bar{\underset{\sim}{Q}}\bar{A}_3$ are covariant base vectors $\tilde{\boldsymbol{g}}_i$ to the deformed material lines at point $\bar{\underset{\sim}{Q}}$ in the current configuration. The location $\bar{x}_i$ of $\bar{\underset{\sim}{Q}}$ and orientation of $\tilde{\boldsymbol{g}}_i$ continuously change during the evolution. In other words, the position of the deformed tetrahedron and its orientation as well as angles between $\tilde{\boldsymbol{g}}_i$ and $\tilde{\boldsymbol{g}}_j$ are continuously changing in time. Definition of Cauchy stress using deformed tetrahedron at material point $\bar{x}_i$ in the current configuration at an instant of time for a stationary observer is in fact a measure of Cauchy stress in a continuously translating and rotating frame in which the translations of the material point and rotations of the material lines, both are functions of time, thus a Euclidean transformation [63] or non-Galilean transformation. Fortunately, we always determine components of Cauchy stress tensor in x-frame.

Figure 4.4(a) and 4.4(b) show elementary tetrahedron in the reference and the deformed configuration. Due to finite deformation the orthogonal material lines $\underset{\sim}{Q}A_1$, $\underset{\sim}{Q}A_2$, $\underset{\sim}{Q}A_3$ become curvilinear in the current configuration at $\bar{\underset{\sim}{Q}}$, location of material point $\underset{\sim}{Q}$ in the current configuration. The non-orthogonal covariant vectors $\tilde{\boldsymbol{g}}_i$ at the material point $\bar{\underset{\sim}{Q}}$ are tangent to the material lines O-$\tilde{x}_i$ in the current configuration and form the edges $\bar{\underset{\sim}{Q}}\bar{A}_1$, $\bar{\underset{\sim}{Q}}\bar{A}_2$, $\bar{\underset{\sim}{Q}}\bar{A}_3$ of the deformed tetrahedron $\bar{\underset{\sim}{Q}}\bar{A}_1\bar{A}_2\bar{A}_3$. We can also define a reciprocal basis formed by the contravariant vectors $\tilde{\boldsymbol{g}}^i$ (normal to the faces of the deformed tetrahedron) given by the cross products of the covariant base vectors $\tilde{\boldsymbol{g}}_i$, i.e., $\tilde{\boldsymbol{g}}^1 = \tilde{\boldsymbol{g}}_2 \times \tilde{\boldsymbol{g}}_3$, $\tilde{\boldsymbol{g}}^2 = \tilde{\boldsymbol{g}}_3 \times \tilde{\boldsymbol{g}}_1$ and $\tilde{\boldsymbol{g}}^3 = \tilde{\boldsymbol{g}}_1 \times \tilde{\boldsymbol{g}}_2$.

Since the edges of the deformed tetrahedron represent material lines, the most natural way to define stresses is to consider faces of the deformed tetrahedron (as in section for small deformation, small strain). We keep in mind that the *contravariant components transform according contravariant law of transformation, hence they require covariant basis or dyads.* For example, on the face of the tetrahedron formed by $\tilde{\boldsymbol{g}}_2$, $\tilde{\boldsymbol{g}}_3$ base vectors, we could define a normal stress component $\underset{\sim}{\sigma}_{11}^{(0)}$ with dyad $\tilde{\boldsymbol{g}}_1 \otimes \tilde{\boldsymbol{g}}_1$ and the two other components $\underset{\sim}{\sigma}_{12}^{(0)}$, $\underset{\sim}{\sigma}_{13}^{(0)}$ with dyads $\tilde{\boldsymbol{g}}_1 \otimes \tilde{\boldsymbol{g}}_2$, $\tilde{\boldsymbol{g}}_1 \otimes \tilde{\boldsymbol{g}}_3$. Thus, for each deformed face of the tetrahedron we have three components of stress in the contravariant directions with dyads defined by $\tilde{\boldsymbol{g}}_i$. These can be arranged in the form of a contravariant stress matrix $[\underset{\sim}{\sigma}^{(0)}]$. The dyads of $\underset{\sim}{\boldsymbol{\sigma}}^{(0)}$ are from the dyadic product of the covariant base vectors $\tilde{\boldsymbol{g}}_i$ and $\tilde{\boldsymbol{g}}_j$.

We recall that columns of $[J]$ are covariant base vectors and the rows of $[\bar{J}]$ are contravariant base vectors defined by

$$\tilde{\boldsymbol{g}}_i = \frac{\partial \bar{x}_j}{\partial x_i}\boldsymbol{e}_j = J_{ji}\boldsymbol{e}_j \quad ; \quad \text{covariant basis} \tag{4.39}$$

$$\tilde{\boldsymbol{g}}^i = \frac{\partial x_i}{\partial \bar{x}_j}\boldsymbol{e}_j = \bar{J}_{ij}\boldsymbol{e}_j \quad ; \quad \text{contravariant basis} \tag{4.40}$$

4.5.1 Contravariant Cauchy stress tensor $\underset{\sim}{\boldsymbol{\sigma}}^{(0)}$ and $\bar{\underset{\sim}{\boldsymbol{\sigma}}}^{(0)}$ in Lagrangian and Eulerian descriptions

In this section, we present definitions of contravariant Cauchy stress tensors $\underset{\sim}{\boldsymbol{\sigma}}^{(0)}$ and $\bar{\underset{\sim}{\boldsymbol{\sigma}}}^{(0)}$ in Lagrangian and Eulerian descriptions. The contravariant stress components $\underset{\sim}{\sigma}_{ij}^{(0)}$ acting on the faces of the deformed tetrahedron require the contravariant law of transformation that uses covariant basis. The Cartesian components of $[\underset{\sim}{\sigma}^{(0)}]$

in x-frame can be obtained using

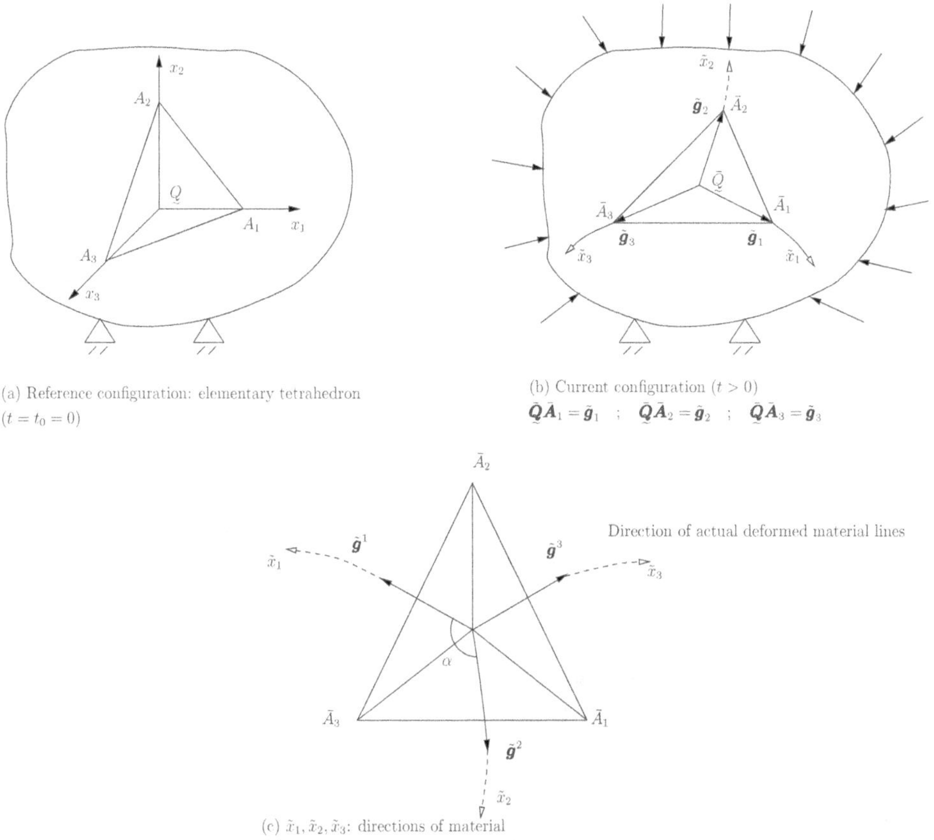

(a) Reference configuration: elementary tetrahedron ($t = t_0 = 0$)

(b) Current configuration ($t > 0$) $\bar{\underset{\sim}{\boldsymbol{Q}}}\bar{\boldsymbol{A}}_1 = \tilde{\boldsymbol{g}}_1$; $\bar{\underset{\sim}{\boldsymbol{Q}}}\bar{\boldsymbol{A}}_2 = \tilde{\boldsymbol{g}}_2$; $\bar{\underset{\sim}{\boldsymbol{Q}}}\bar{\boldsymbol{A}}_3 = \tilde{\boldsymbol{g}}_3$

(c) $\bar{x}_1, \bar{x}_2, \bar{x}_3$: directions of material lines with $\tilde{\boldsymbol{g}}^i$ as tangent vectors at $\bar{P}$

Figure 4.4: Elementary tetrahedron in reference and current configurations, co- and contra-variant bases

$$\boldsymbol{\sigma}^{(0)} = \tilde{\boldsymbol{g}}_i \otimes \tilde{\boldsymbol{g}}_j (\underset{\sim}{\sigma}^{(0)})_{ij} \tag{4.41}$$

Since $\tilde{\boldsymbol{g}}_i$ are in Lagrangian description, thus, *we define* $\boldsymbol{\sigma}^{(0)}$ *in* (4.41) *as contravariant Cauchy stress tensor in Lagrangian description.* By substituting definitions of $\tilde{\boldsymbol{g}}_i$ from (4.39) in (4.41) we can obtain components of $\boldsymbol{\sigma}^{(0)}$ with dyads in the x-frame.

$$\begin{aligned}
\boldsymbol{\sigma}^{(0)} &= \frac{\partial \bar{x}_m}{\partial x_i}\boldsymbol{e}_m \otimes \frac{\partial \bar{x}_n}{\partial x_j}\boldsymbol{e}_n (\underset{\sim}{\sigma}^{(0)})_{ij} \\
&= (\boldsymbol{e}_m \otimes \boldsymbol{e}_n)\frac{\partial \bar{x}_m}{\partial x_i}\frac{\partial \bar{x}_n}{\partial x_j}(\underset{\sim}{\sigma}^{(0)})_{ij} \\
&= (\boldsymbol{e}_m \otimes \boldsymbol{e}_n)\left(J_{mi}J_{nj}(\underset{\sim}{\sigma}^{(0)})_{ij}\right) \\
&= (\boldsymbol{e}_m \otimes \boldsymbol{e}_n)\left(J_{mi}(\underset{\sim}{\sigma}^{(0)})_{ij}J_{nj}\right)
\end{aligned} \tag{4.42}$$

which can be written as

$$\boldsymbol{\sigma}^{(0)} = \boldsymbol{e}_m \otimes \boldsymbol{e}_n \sigma_{mn}^{(0)} \tag{4.43}$$

in which

$$\sigma_{mn}^{(0)} = J_{mi} \underset{\sim}{\sigma}_{ij}^{(0)} J_{nj} \quad ; \quad \text{(Def.)} \tag{4.44}$$

In matrix notation

$$[\sigma^{(0)}] = [J][\underset{\sim}{\sigma}^{(0)}][J]^T \quad ; \quad \text{(Def.)} \tag{4.45}$$

Definition of $\boldsymbol{\sigma}^{(0)}$ in (4.45) needs to be modified when the matter is compressible as in this case; columns of $[J]$ are components of non-normalized covariant base vectors. We make some remarks that are helpful in better understanding of what $\boldsymbol{\sigma}^{(0)}$ means .

Remarks

(1) Tensor $\boldsymbol{\sigma}^{(0)}$ is in Lagrangian description. This is obvious from (4.45) as the components of $[J]$ are functions of $\boldsymbol{x}$ and t.

(2) The components $\sigma_{ij}^{(0)}$ are in x-frame. This is also clear from (4.43) in which the dyads of the components of $\boldsymbol{\sigma}^{(0)}$ are $\boldsymbol{e}_i \otimes \boldsymbol{e}_j$, in the x-frame.

(3) Based on (1) and (2) we could view $[\sigma^{(0)}]$ as a stress tensor acting on the faces of a tetrahedron in the current configuration such that the faces of this tetrahedron are orthogonal to each other and are parallel to the planes of the x-frame but with the same oblique plane $\bar{A}_1\bar{A}_2\bar{A}_3$ as in the case of the original deformed tetrahedron with non-orthogonal faces. Furthermore, the areas of the faces of this new tetrahedron are assumed to be same as those of the original deformed tetrahedron. This is essential for the correct force equilibrium of the tetrahedron with orthogonal faces. Thus, for this new tetrahedron in the current configuration with orthogonal faces on which $[\bar{\sigma}^{(0)}]$ (Eulerian description) acts, the Cauchy principle is applicable between $\boldsymbol{\sigma}^{(0)}$, $\bar{\boldsymbol{n}}$ (unit exterior normal to plane $\bar{A}_1$, $\bar{A}_2$, $\bar{A}_2$) and $\bar{\boldsymbol{P}}$, the resultant force per unit area on the plane $\bar{A}_1\bar{A}_2\bar{A}_3$, i.e., average stress.

$$\bar{\boldsymbol{P}} = (\bar{\boldsymbol{\sigma}}^{(0)})^T \cdot \bar{\boldsymbol{n}} \tag{4.46}$$

(4) Sum of moments of the forces (to be zero) on the new tetrahedron with orthogonal faces requires $[\bar{\sigma}^{(0)}]$ to be symmetric.

(5) Since $[\bar{\sigma}^{(0)}]$ is a symmetric tensor of rank two (as shown by considering rigid rotation of the frame), from (4.45) we can conclude that $[\underset{\sim}{\sigma}^{(0)}]$ is also a symmetric tensor of rank two.

(6) Contravariant Cauchy stress tensor $\bar{\boldsymbol{\sigma}}^{(0)}$ in Eulerian description can be defined using (4.45) and by replacing $[J]$ with $[\bar{J}]^{-1}$ and $[J]^T$ with $[[\bar{J}]^{-1}]^T$.

$$[\bar{\sigma}^{(0)}] = [\bar{J}]^{-1}[\underset{\sim}{\sigma}^{(0)}][[\bar{J}]^{-1}]^T \tag{4.47}$$

4.5.2 Covariant Cauchy stress tensor in $\bar{\boldsymbol{\sigma}}_{(0)}$ and $\boldsymbol{\sigma}_{(0)}$ in Eulerian and Lagrangian descriptions

Since covariant and contravariant bases are reciprocal to each other, we could consider covariant directions for defining stress tensor and the contravariant directions for measure of strain. We note that use of covariant directions to define stress requires deformed tetrahedron such that covariant vectors are normal to the faces of the new tetrahedron. The edges of this tetrahedron representing contravariant base vectors must be used for defining strain measure. The covariant stress components $(\underset{\sim}{\sigma}_{(0)})_{ij}$ require the covariant law of transformation, i.e., contravariant dyads. Thus, we can define

$$\bar{\boldsymbol{\sigma}}_{(0)} = \tilde{\boldsymbol{g}}^i \otimes \tilde{\boldsymbol{g}}^j (\underset{\sim}{\sigma}_{(0)})_{ij} \tag{4.48}$$

Since $\tilde{\boldsymbol{g}}^i$ are in Eulerian description, *we define* $\bar{\boldsymbol{\sigma}}_{(0)}$ *as covariant Cauchy stress tensor in Eulerian description.* By substituting definitions of $\tilde{\boldsymbol{g}}^i$ from (4.40) in (4.48) we can obtain a explicit form of the components of $\bar{\boldsymbol{\sigma}}_{(0)}$ with dyads in x-frame.

$$\begin{aligned} \bar{\boldsymbol{\sigma}}_{(0)} &= \frac{\partial x_i}{\partial \bar{x}_m} \boldsymbol{e}_m \otimes \frac{\partial x_j}{\partial \bar{x}_n} \boldsymbol{e}_n (\underset{\sim}{\sigma}_{(0)})_{ij} \\ &= (\boldsymbol{e}_m \otimes \boldsymbol{e}_n) \frac{\partial x_i}{\partial \bar{x}_m} \frac{\partial x_j}{\partial \bar{x}_n} (\underset{\sim}{\sigma}_{(0)})_{ij} \\ &= (\boldsymbol{e}_m \otimes \boldsymbol{e}_n) \left(\bar{J}_{im} \bar{J}_{jn} (\underset{\sim}{\sigma}_{(0)})_{ij} \right) \\ &= (\boldsymbol{e}_m \otimes \boldsymbol{e}_n) \left(\bar{J}_{im} (\underset{\sim}{\sigma}_{(0)})_{ij} \bar{J}_{jn} \right) \end{aligned} \tag{4.49}$$

which can be written as

$$\bar{\boldsymbol{\sigma}}_{(0)} = \boldsymbol{e}_m \otimes \boldsymbol{e}_n (\bar{\sigma}_{(0)})_{mn} \tag{4.50}$$

where

$$(\bar{\sigma}_{(0)})_{mn} = \bar{J}_{im} (\underset{\sim}{\sigma}_{(0)})_{ij} \bar{J}_{jn} \quad ; \quad \text{(Def.)} \tag{4.51}$$

In matrix notation

$$[\bar{\sigma}_{(0)}] = [\bar{J}]^T [\underset{\sim}{\sigma}_{(0)}] [\bar{J}] \quad ; \quad \text{(Def.)} \tag{4.52}$$

We make some remarks that are helpful in better understanding of what $[\bar{\sigma}_{(0)}]$ means.

Remarks

(1) Tensor $\bar{\boldsymbol{\sigma}}_{(0)}$ is an Eulerian description. This is obvious from (4.52) as the components of $[\bar{J}]$ are functions of $\bar{\boldsymbol{x}}$ and t.

(2) The components $(\bar{\sigma}_{(0)})_{ij}$ are in x-frame. This is also clear from (4.50) in which the dyads of $\bar{\boldsymbol{\sigma}}_{(0)}$ are $\boldsymbol{e}_i \otimes \boldsymbol{e}_j$.

(3) Definition of $[\underset{\sim}{\sigma}_{(0)}]$ tensor requires the true deformed tetrahedron in the current configuration to be further deformed and/or reoriented such that $\tilde{\boldsymbol{g}}_i$ vectors are normal to the faces of the new tetrahedron. This obviously is non-physical when the deformation is finite. In this process the area of the plane $\bar{A}_1\bar{A}_2\bar{A}_3$, the resultant average force per unit area $\bar{\boldsymbol{P}}$ acting on it and the normal $\bar{\boldsymbol{n}}$ may

be affected adversely.

(4) Following the reasoning similar to the contravariant case, in this case also, we could view $[\bar{\sigma}_{(0)}]$ as a stress tensor acting on the faces of a tetrahedron in the current configuration such that the faces of this tetrahedron are orthogonal to each other and are parallel to the planes of the x-frame but not necessarily with the same oblique plane $\bar{A}_1\bar{A}_2\bar{A}_3$ as in the case of the actual deformed tetrahedron with covariant base vectors forming its edges. This new tetrahedron with orthogonal faces and new oblique plane $\underset{\sim}{\bar{A}}_1\underset{\sim}{\bar{A}}_2\underset{\sim}{\bar{A}}_3$ with average stress $\underset{\sim}{\bar{\boldsymbol{P}}}$ and the normal $\underset{\sim}{\bar{\boldsymbol{n}}}$ to the plane $\underset{\sim}{\bar{A}}_1\underset{\sim}{\bar{A}}_2\underset{\sim}{\bar{A}}_3$ also allows us to use the Cauchy principle, i.e.

$$\underset{\sim}{\bar{\boldsymbol{P}}} = (\bar{\boldsymbol{\sigma}}_{(0)})^T \cdot \underset{\sim}{\bar{\boldsymbol{n}}} \quad \text{or} \quad \{\underset{\sim}{\bar{P}}\} = [\bar{\sigma}_{(0)}]^T \{\underset{\sim}{\bar{n}}\} \tag{4.53}$$

(5) Sum of moments of the forces (to be zero) on the final tetrahedron with orthogonal faces requires $[\bar{\sigma}_{(0)}]$ to be symmetric.

(6) Since, $[\bar{\sigma}_{(0)}]$ is a symmetric tensor of rank two (can be shown by considering rigid rotation of the frame), based on (4.52), $[\underset{\sim}{\bar{\sigma}}_{(0)}]$ is also a symmetric tensor of rank two.

(7) In view of (4.45) and (4.52), we observe that stress measures $[\bar{\sigma}^{(0)}]$ and $[\bar{\sigma}_{(0)}]$ are not the same.

(8) The covariant Cauchy stress tensor $\boldsymbol{\sigma}_{(0)}$ in Lagrangian description can be defined using (4.52) and replacing $[\bar{J}]$ with $[J]^{-1}$ and $[\bar{J}]^T$ with $[[J]^{-1}]^T$.

$$[\sigma_{(0)}] = [[J]^{-1}]^T [\underset{\sim}{\sigma}_{(0)}][J]^{-1} \tag{4.54}$$

4.5.3 Mixed stress tensors: Jaumann stress tensor

Since both contravariant and covariant Cauchy stress tensors components are in x-frame, their definitions contain dyads $\boldsymbol{e}_i \otimes \boldsymbol{e}_j$ after substituting definitions of $\tilde{\boldsymbol{g}}_i$ and $\tilde{\boldsymbol{g}}^i$. We can define mixed tensors $\sigma^{(0)}_{(0)}$ or $\bar{\sigma}^{(0)}_{(0)}$ by considering linear combinations of the corresponding contravariant and covariant tensors.

$$\sigma^{(0)}_{(0)} = \frac{1}{2} e_i \otimes e_j \left(\alpha(\sigma^{(0)})_{ij} + \beta(\sigma_{(0)})_{ij} \right) \tag{4.55}$$

$$\bar{\sigma}^{(0)}_{(0)} = \frac{1}{2} e_i \otimes e_j \left(\alpha(\bar{\sigma}^{(0)})_{ij} + \beta(\bar{\sigma}_{(0)})_{ij} \right) \tag{4.56}$$

when $\alpha = 1$ and $\beta = 1$, we have Jaumann stress tensors

$${}^{(0)}\boldsymbol{\sigma}^J = \frac{1}{2} e_i \otimes e_j \left((\sigma^{(0)})_{ij} + (\sigma_{(0)})_{ij} \right) \tag{4.57}$$

$${}^{(0)}\bar{\boldsymbol{\sigma}}^J = \frac{1}{2} e_i \otimes e_j \left((\bar{\sigma}^{(0)})_{ij} + (\bar{\sigma}_{(0)})_{ij} \right) \tag{4.58}$$

Tensor ${}^{(0)}\sigma^J$ and ${}^{(0)}\bar{\sigma}^J$ are called Jaumman stress tensors, Lagrangian and Eulerian descriptions, respectively. We recall that $\boldsymbol{\sigma}^{(0)}$, $\bar{\boldsymbol{\sigma}}^{(0)}$ and $\boldsymbol{\sigma}_{(0)}$, $\bar{\boldsymbol{\sigma}}_{(0)}$ tensors correspond to two completely different tetrahedra. The physical meaning of mixed tensors and Jaumann tensors is naturally lost. Jaumann tensors have been used often in the published works, especially for solid matter in finite deformation studies [127, 130], hence the reason for including them here. We note that ${}^{(0)}\boldsymbol{\sigma}^J$ and ${}^{(0)}\bar{\boldsymbol{\sigma}}^J$ are symmetric tensors of rank two.

4.6 Contravariant Second Piola-Kirchhoff stress tensor $\boldsymbol{\sigma}^{[0]}$ or $\bar{\boldsymbol{\sigma}}^{[0]}$

In Lagrangian description of motion and deformation of continua, all quantities are measured in the reference configuration at initial time $t = t_0 = 0$. In such description, all quantities are eventually functions of coordinates $\boldsymbol{x}$ and time t. All integrals in the derivation of conservation and balance laws (Chapter 7) in Lagrangian description must be over lengths, areas and volume in the reference configuration. If we closely examine the derivations and definitions of Cauchy stress $\boldsymbol{\sigma}^{(0)}$, $\boldsymbol{\sigma}_{(0)}$ and $\bar{\boldsymbol{\sigma}}^{(0)}$, $\bar{\boldsymbol{\sigma}}_{(0)}$ we find that their definitions use deformed tetrahedron in the current configuration. Thus, during the evolution, the Cauchy stress tensor for each instant of time is defined using the configuration for that value of time, resulting in different deformed tetrahedron for each value of time. Obviously these quantities cannot be used in the definitions of the integrands of the integrals that are over the undeformed lengths, areas and volume in the reference or undeformed configuration. This necessitates that we consider Cauchy stress tensor $\bar{\boldsymbol{\sigma}}^{(0)}$ or $\boldsymbol{\sigma}^{(0)}$ and $\bar{\boldsymbol{\sigma}}_{(0)}$ or $\boldsymbol{\sigma}_{(0)}$ in the current configuration and find its correspondence in the reference configuration using a map of the deformed tetrahedron in the reference configuration which of course remains fixed during the evolution. This process requires a correspondence rule that allows us to describe a hypothetical stress tensor corresponding to $\bar{\boldsymbol{\sigma}}^{(0)}$ or $\boldsymbol{\sigma}^{(0)}$ and $\bar{\boldsymbol{\sigma}}_{(0)}$ or $\boldsymbol{\sigma}_{(0)}$ on the undeformed tetrahedron in the reference configuration. Precise definition of the stress on the undeformed tetrahedron depends upon the correspondence rule. The correspondence rule is an assumed relationship between force $d\bar{\boldsymbol{F}}$ acting on $d\bar{A}_n$ and the corresponding hypothetical force $d\boldsymbol{F}$ acting on dA_n, map of $d\bar{A}_n$ in the reference configuration. Let $[\sigma^{[0]}]$ be the hypothetical stress tensor acting on the planes of the undeformed tetrahedron that corresponds to $\boldsymbol{\sigma}^{(0)}$ or $\bar{\boldsymbol{\sigma}}^{(0)}$ contravariant Cauchy stress tensor in the current configuration. Recall

$$\{\bar{n}\} = \frac{1}{d\bar{A}_n}\{d\bar{A}\} \quad ; \quad \{n\} = \frac{1}{dA_n}\{dA\} \quad ; \quad \{d\bar{A}\} = |J|\,[J^T]^{-1}\{dA\} \tag{4.59}$$

We assume that force $\{dF\}$ acting on the undeformed area dA_n and the force $\{d\bar{F}\}$ acting on the deformed area $d\bar{A}_n$ are related through the deformation gradient tensor $[J]$. This is known as the correspondence rule as it establishes a correspondence between $\{dF\}$ and $\{d\bar{F}\}$.

$$\{d\bar{F}\} = [J]\{dF\} \tag{4.60}$$

Let the application of $[\sigma^{[0]}]$ stress tensor on areas $\{dA\}$ of the tetrahedron in the reference configuration yield the force vector $\{dF\}$ on area dA_n. Tensor $[\sigma^{[0]}]$ is called the *second Piola-Kirchhoff stress tensor* (Lagrangian description). Using forces $\{dF\}$ and $\{d\bar{F}\}$ on the undeformed and deformed tetrahedra and the Cauchy principle we can write

$$\{dF\} = dA_n\{P\} = dA_n[\sigma^{[0]}]^T\{n\} = [\sigma^{[0]}]^T\{dA\} \tag{4.61}$$

$$\{d\bar{F}\} = d\bar{A}_n\{\bar{P}\} = d\bar{A}_n[\bar{\sigma}^{(0)}]^T\{\bar{n}\} = [\bar{\sigma}^{(0)}]^T\{d\bar{A}\} \tag{4.62}$$

Substituting $\{dF\}$ and $\{d\bar{F}\}$ from (4.61) and (4.62) into (4.60)

$$[\bar{\sigma}^{(0)}]^T\{d\bar{A}\} = [J][\sigma^{[0]}]^T\{dA\} \tag{4.63}$$

Substituting $\{d\bar{A}\}$ from (4.59) in (4.63)

$$[\sigma^{(0)}]^T\,|J|\,[J^T]^{-1}\{dA\} = [J][\sigma^{[0]}]^T\{dA\} \tag{4.64}$$

$$\therefore \quad [\sigma^{(0)}]^T|J|[J^T]^{-1} = [J][\sigma^{[0]}]^T \tag{4.65}$$

Premultiply (4.65) by $[J]^{-1}$ and

$$[\sigma^{[0]}]^T = |J|\,[J]^{-1}[\sigma^{(0)}]^T[J^T]^{-1} \tag{4.66}$$

which can also be written as (Eulerian description)

$$[\bar{\sigma}^{[0]}]^T = |[\bar{J}]^{-1}|[\bar{J}][\bar{\sigma}^{(0)}]^T[\bar{J}]^T \quad ; \quad \text{Def. (compressible)} \tag{4.67}$$

Taking the transpose of (4.67) we can write

$$[\bar{\sigma}^{[0]}] = |[\bar{J}]^{-1}|[\bar{J}][\bar{\sigma}^{(0)}][\bar{J}]^T \tag{4.68}$$

When the matter is incompressible $|J| = |[\bar{J}]^{-1}| = 1$, hence for this case, (4.67) reduces to

$$[\bar{\sigma}^{[0]}]^T = [\bar{J}][\bar{\sigma}^{(0)}]^T[\bar{J}]^T \quad ; \quad \text{Def. (incompressible)} \tag{4.69}$$

Remarks

The stress $[\sigma^{[0]}]$ is hypothetical as it acts on the undeformed tetrahedron in the reference configuration based on a correspondence rule.

4.7 Contravariant First Piola-Kirchhoff or Lagrange stress tensor $\boldsymbol{\sigma}^*$

In this case, instead of using the correspondence rule defined by (4.60), *we assume the following correspondence rule.*

$$\{dF\} = \{d\bar{F}\} \tag{4.70}$$

That is, the force $\{d\bar{F}\}$ acting on the deformed area $d\bar{A}_n$ is exactly the same as force $\{dF\}$ acting on the undeformed area dA_n in the reference configuration. Let the application of $[\sigma^*]$ stress tensor on areas $\{dA\}$ of the faces of the tetrahedron in the reference configuration yield the force vector $\{dF\}$ on area dA_n. Stress $[\sigma^*]$ is called first *Piola-Kirchhoff or Lagrange stress tensor*. Using forces $\{dF\}$ and $\{d\bar{F}\}$ on the undeformed and deformed tetrahedra and Cauchy principle we can write

$$\{dF\} = dA_n\{P\} = dA_n[\sigma^*]^T\{n\} = [\sigma^*]^T\{dA\} \tag{4.71}$$

$$\{d\bar{F}\} = d\bar{A}_n\{\bar{P}\} = d\bar{A}_n[\bar{\sigma}^{(0)}]^T\{\bar{n}\} = [\bar{\sigma}^{(0)}]^T\{d\bar{A}\} \tag{4.72}$$

We note (4.67) utilizes true deformed tetrahedron with contravariant measure of stress $\bar{\boldsymbol{\sigma}}^{(0)}$. Substituting from (4.71) and (4.72) into (4.70)

$$[\sigma^*]^T\{dA\} = [\bar{\sigma}^{(0)}]^T\{d\bar{A}\} \tag{4.73}$$

Substituting for $\{d\bar{A}\}$ from (4.59) into (4.73)

$$[\sigma^*]^T\{dA\} = [\sigma^{(0)}]^T\,|J|\,[J^T]^{-1}\{dA\} \tag{4.74}$$

$$\therefore\quad [\sigma^*]^T = |J|\,[\sigma^{(0)}]^T[J^T]^{-1} \quad ; \quad \text{Def. (compressible)} \tag{4.75}$$

Taking the transpose of (4.75)

$$[\sigma^*] = |J|\,[J]^{-1}[\sigma^{(0)}] \tag{4.76}$$

For incompressible matter $|J| = 1$, hence for this case, (4.75) reduces to

$$[\sigma^*]^T = [\sigma^{(0)}]^T[J^T]^{-1} \quad ; \quad \text{Def. (incompressible)} \tag{4.77}$$

We note that $[\sigma^*]$ is not symmetric, i.e., $[\sigma^*] \neq [\sigma^*]^T$.

Remarks

(1) Stress $[\sigma^*]^T$ in (4.75) and (4.77) is Lagrangian description.

(2) Eulerian description of $[\sigma^*]^T$, i.e., $[\bar{\sigma}^*]^T$ can be obtained by replacing $[J]^{-1}$ with $[\bar{J}]$ and $[\sigma^{(0)}]^T$ with $[\bar{\sigma}^{(0)}]^T$ in (4.75).

$$[\bar{\sigma}^*]^T = |[\bar{J}]^{-1}|[\bar{\sigma}^{(0)}]^T[\bar{J}]^T \tag{4.78}$$

(3) Stress $[\sigma^*]^T$ and $[\bar{\sigma}^*]^T$ are hypothetical for the same reasons as for $[\sigma^{[0]}]$ and $[\bar{\sigma}^{[0]}]$.

4.8 Covariant Second Piola-Kirchhoff stress tensor $\boldsymbol{\sigma}_{[0]}$ or $\bar{\boldsymbol{\sigma}}_{[0]}$

Recall the relations used in Section 4.6 (equations (4.59) and (4.60))

$$\{d\bar{A}\} = |J|\,[\bar{J}]^T\{dA\} \tag{4.79}$$

$$\{d\bar{F}\} = [J]\{dF\} \tag{4.80}$$

In the following derivation we replace (4.79) and (4.80) by

$$\{d\bar{A}\} = |J|\,[J]\{dA\} \tag{4.81}$$

$$\{d\bar{F}\} = [\bar{J}]^T\{dF\} \tag{4.82}$$

Equation (4.82) is obviously the new correspondence rule. Thus, we have

$$\{dF\} = dA_n\{P\} = dA_n[\sigma_{[0]}]^T\{n\} = [\sigma_{[0]}]^T dA_n\{n\} = [\sigma_{[0]}]^T\{dA\} \tag{4.83}$$

$$\{d\bar{F}\} = d\bar{A}_n\{\bar{P}\} = d\bar{A}_n[\bar{\sigma}_{(0)}]^T\{\bar{n}\} = [\bar{\sigma}_{(0)}]^T d\bar{A}_n\{\bar{n}\} = [\bar{\sigma}_{(0)}]^T\{d\bar{A}\} \tag{4.84}$$

Substituting from (4.83) and (4.84) into (4.82) and using (4.81), we can write

$$[\sigma_{(0)}]^T\,|J|\,[J]\{dA\} = [\bar{J}]^T[\sigma_{[0]}]^T\{dA\} \tag{4.85}$$

Premultiply (4.85) by $[\bar{J}^T]^{-1}$ or $[J]^T$

$$[J]^T [\sigma_{(0)}]^T \, |J| \, [J]\{dA\} = [\sigma_{[0]}]^T \{dA\} \tag{4.86}$$

Hence

$$[\sigma_{[0]}]^T = |J| \, [J]^T [\sigma_{(0)}]^T [J] \quad ; \quad \text{Def. (compressible)} \tag{4.87}$$

Taking the transpose of (4.87) we obtain

$$[\sigma_{[0]}] = |J| \, [J]^T [\sigma_{(0)}][J] \tag{4.88}$$

We refer to $[\sigma_{[0]}]$, the *second Piola-Kirchhoff stress in covariant basis.* When the matter is incompressible $\det[J]$ or $|J| = 1$, hence (4.87) reduces to

$$[\sigma_{[0]}]^T = [J]^T [\sigma_{(0)}]^T [J] \quad ; \quad \text{Def. (incompressible)}. \tag{4.89}$$

Remarks

(1) Stress $[\sigma_{[0]}]$ is a covariant measure in Lagrangian description.

(2) Eulerian description of $[\sigma_{[0]}]$, i.e., $[\bar{\sigma}_{[0]}]$ can be obtained by replacing $[J]$ with $[\bar{J}]^{-1}$ and $[\sigma_{(0)}]$ with $[\bar{\sigma}_{(0)}]$ in (4.87) for compressible matter.

$$[\bar{\sigma}_{[0]}]^T = |[\bar{J}]^{-1}| [[\bar{J}]^{-1}]^T [\bar{\sigma}_{(0)}]^T [\bar{J}]^{-1} \tag{4.90}$$

(3) Tensors $[\bar{\sigma}^{[0]}]$ or $[\sigma^{[0]}]$, $[\bar{\sigma}_{[0]}]$ or $[\sigma_{[0]}]$ and $[\sigma^*]$ are hypothetical as these are derived based on an assumed correspondence rule.

4.9 General Remarks

(1) We consider a tetrahedron T_1 in the reference configuration with its oblique plane being part of the boundary ∂V of the volume of matter V. The edges of the tetrahedron are parallel to the axes of the x-frame, hence its orthogonal faces are also parallel to the planes formed by the axes of the x-frame. This tetrahedron obviously remains fixed during deformation.

(2) In the current configuration, tetrahedron T_1 deforms into T_2. The edges of T_2 are covariant base vectors $\tilde{\boldsymbol{g}}_i$, its planes are flat but are neither orthogonal to each other nor are parallel to the planes formed by the axes of the fixed x-frame. Vectors $\tilde{\boldsymbol{g}}_i$ form non-orthogonal covariant basis.

(3) If the edges of the tetrahedron T_1 are material lines in the undeformed or reference configuration, then covariant base vectors $\tilde{\boldsymbol{g}}_i$ defining the edges of T_2 are tangent to deformed material lines in the current configuration.

(4) Contravariant base vectors $\tilde{\boldsymbol{g}}^i$, which are normal to the faces of the deformed tetrahedron T_2, form a non-orthogonal contravariant basis that is reciprocal to the covariant basis.

(5) *Defining strain measures using covariant basis (deformed material lines) and stresses using contravariant basis (normal to the faces of the deformed tetrahedron T_2) is the natural way to proceed as these are related to the deformed volume of the tetrahedron T_2.* But there are other possibilities as well. We could choose contravariant measures of strain and a covariant basis for stress.

This would imply consideration of a new tetrahedron T_3 in the current configuration such that its edges are defined by the contravariant base vectors $\tilde{\boldsymbol{g}}^i$ that are tangent to the deformed material lines and covariant base vectors are normal to the faces of this new tetrahedron formed by $\tilde{\boldsymbol{g}}^i$.

(6) Definitions of hypothetical stress tensors that act on the faces of the undeformed tetrahedron T are possible by establishing correspondence rules between $d\boldsymbol{F}$ and $d\bar{\boldsymbol{F}}$ acting on dA_n and $d\bar{A}_n$. First and second Piola-Kirchhoff stress tensors are examples.

(7) In simple deformation physics, the strain measures that are conjugate to the chosen stress measures can be determined from physical reasoning. In general first and second law of thermodynamics need to be considered for establishing rate of work conjugate pairs when the deformation is finite.

(8) We have seen that the first and second Piola-Kirchhoff stress tensors based on contravariant and covariant Cauchy stress tensors act on undeformed tetrahedron in the undeformed configuration and are derived using correspondence rules. These stress measures are obviously hypothetical, i.e., non-physical. The importance of the second Piola-Kirchhoff stress tensor is realized when deriving convected time derivatives of the contravariant and covariant Cauchy stress tensors in contravariant and covariant bases (Chapter 5). The first and second Piola-Kirchhoff stresses play a significant role in deriving balance laws and in establishing work conjugate stress and strain pair. This is essential in deriving constitutive theories (Chapters 8–13).

4.10 Summary of stress measures

A summary of the stress measures is given in the following.

4.10.1 Cauchy stress tensors

$$
\begin{aligned}
[\bar{\sigma}^{(0)}] &= [\bar{J}]^{-1}[\underset{\sim}{\sigma}^{(0)}][\bar{J}^T]^{-1} && ; \quad \text{contravariant, Eulerian : using } T_2 \\
[\sigma^{(0)}] &= [J][\underset{\sim}{\sigma}^{(0)}][J]^T && ; \quad \text{contravariant, Lagrangian : using } T_2 \\
[\bar{\sigma}_{(0)}] &= [\bar{J}]^T[\underset{\sim}{\sigma}_{(0)}][\bar{J}] && ; \quad \text{covariant, Eulerian : using } T_3 \\
[\sigma_{(0)}] &= [J^T]^{-1}[\underset{\sim}{\sigma}_{(0)}][J]^{-1} && ; \quad \text{covariant, Lagrangian : using } T_3
\end{aligned}
\tag{4.91}
$$

4.10.2 Jaumann stress tensors

$$
\begin{aligned}
[{}^{(0)}\bar{\sigma}^J] &= \frac{1}{2}\big([\bar{\sigma}^{(0)}] + [\bar{\sigma}_{(0)}]\big) && ; \quad \text{Eulerian} \\
[{}^{(0)}\sigma^J] &= \frac{1}{2}\big([\sigma^{(0)}] + [\sigma_{(0)}]\big) && ; \quad \text{Lagrangian}
\end{aligned}
\tag{4.92}
$$

4.10.3 Second Piola-Kirchhoff stress tensors

$$
\begin{aligned}
[\bar{\sigma}^{[0]}]^T &= |[\bar{J}]^{-1}|[\bar{J}][\bar{\sigma}^{(0)}]^T[\bar{J}]^T && ; \quad \text{contravariant, Eulerian : } \bar{\boldsymbol{\sigma}}^{(0)} \text{ on } T_2 \\
[\sigma^{[0]}]^T &= |J|[J]^{-1}[\sigma^{(0)}]^T[J^T]^{-1} && ; \quad \text{contravariant, Lagrangian : } \boldsymbol{\sigma}^{(0)} \text{ on } T_2 \\
[\bar{\sigma}_{[0]}]^T &= |[\bar{J}]^{-1}|[\bar{J}^T]^{-1}[\bar{\sigma}_{(0)}]^T[\bar{J}]^{-1} && ; \quad \text{covariant, Eulerian : } \bar{\boldsymbol{\sigma}}_{(0)} \text{ on } T_3 \\
[\sigma_{[0]}]^T &= |J|[J]^T[\sigma_{(0)}]^T[J] && ; \quad \text{covariant, Lagrangian : } \boldsymbol{\sigma}_{(0)} \text{ on } T_3
\end{aligned}
\tag{4.93}
$$

4.10.4 First Piola-Kirchhoff stress tensor

$$[\sigma^*]^T = |J|\,[\sigma^{(0)}]^T[J^T]^{-1} \quad ; \quad \text{Lagrangian using using } T_2,\ T_1 \text{ and } \boldsymbol{\sigma}^{(0)} \tag{4.94}$$

4.11 Conjugate strain measures

The strain or strain rate measures that can be used to define energy or rate of energy using a chosen stress measure are called conjugate strain or strain rate measures. Conjugate stress and strain or stress and strain rate measures are derived in Chapters 6 and 7. These play a significant role in the derivations of the constitutive theories. In general, *for contravariant stress measures, the conjugate strain or strain rate measures must be covariant. Likewise, contravariant strain or strain rate measures are conjugate to the covariant stress measures.*

4.12 Relations between different stress measures and some other useful relations

(1) *We list relations between different stress measures that allow us to convert one measure into others.* We consider the deforming matter to be compressible. For simplicity, consider Lagrangian description only. For the incompressible case, details follow the compressible case with $|J| = 1$.

$$[\sigma^{[0]}]^T = |J|\,[J]^{-1}[\sigma^{(0)}]^T[J^T]^{-1} \tag{4.95}$$

$$[\sigma_{[0]}]^T = |J|\,[J]^T[\sigma_{(0)}]^T[J] \tag{4.96}$$

$$[\sigma^*]^T = |J|\,[\sigma^{(0)}]^T[J^T]^{-1} \tag{4.97}$$

The stress measures (4.95)–(4.97) are commonly used. We recall that $[\sigma^{[0]}]$, the second Piola-Kirchhoff stress tensor, is based on the contravariant Cauchy stress tensor. Stress $[\sigma_{[0]}]$ is also a second Piola-Kirchhoff stress tensor but is based on the covariant Cauchy stress tensor. The stress $[\sigma^*]$, the first Piola-Kirchhoff stress tensor, is not symmetric but has been used in many important derivations in the development of the constitutive theory in the published works. From (4.95)

$$[\sigma^{(0)}]^T = |J|^{-1}\,[J][\sigma^{[0]}]^T[J]^T \tag{4.98}$$

Also from (4.97)

$$[\sigma^{(0)}]^T = |J|^{-1}\,[\sigma^*]^T[J]^T \tag{4.99}$$

Equating (4.98) and (4.99)

$$[\sigma^*]^T = [J][\sigma^{[0]}]^T \tag{4.100}$$

$$\text{or} \quad [\sigma^{[0]}]^T = [J]^{-1}[\sigma^*]^T \tag{4.101}$$

(2) In the following, *we derive the Lagrangian equivalent of* $\bar{\boldsymbol{P}}d\bar{A}$. Consider $\bar{\boldsymbol{P}}d\bar{A}$ and apply the Cauchy principle, i.e., $\bar{\boldsymbol{P}} = (\bar{\boldsymbol{\sigma}}^{(0)})^T \cdot \bar{\boldsymbol{n}}$.

$$\bar{\boldsymbol{P}}\,d\bar{A} = (\bar{\boldsymbol{\sigma}}^{(0)})^T \cdot \bar{\boldsymbol{n}}\,d\bar{A} = (\bar{\boldsymbol{\sigma}}^{(0)})^T \cdot d\bar{\boldsymbol{A}} \quad \text{or} \quad \{\bar{P}\}d\bar{A} = [\bar{\sigma}^{(0)}]^T\{d\bar{A}\} \tag{4.102}$$

But

$$[\sigma^{(0)}]^T = |J|^{-1}\,[\sigma^*]^T[J]^T \quad \text{and} \quad \{d\bar{A}\} = |J|\,[J^{-1}]^T\{dA\} \tag{4.103}$$

Using $\boldsymbol{\sigma}^{(0)}$ (Lagrangian description) in (4.102) in place of $\bar{\boldsymbol{\sigma}}^{(0)}$, and substituting from (4.103)

$$\begin{aligned}[\sigma^{(0)}]^T\{d\bar{A}\} &= |J|^{-1}\,[\sigma^*]^T[J]^T\,|J|\,[J^{-1}]^T\{dA\}\\ &= [\sigma^*]^T\{dA\} = [\sigma^*]^T\{n\}dA\end{aligned} \tag{4.104}$$

which can be written as

$$(\boldsymbol{\sigma}^{(0)})^T \cdot d\bar{\boldsymbol{A}} = (\boldsymbol{\sigma}^*)^T \cdot \boldsymbol{n}\,dA = (\boldsymbol{\sigma}^*)^T \cdot d\boldsymbol{A} \tag{4.105}$$

Thus, we have

$$\begin{aligned}\bar{\boldsymbol{P}}\,d\bar{A} &= (\bar{\boldsymbol{\sigma}}^{(0)})^T \cdot \bar{\boldsymbol{n}}\,d\bar{A} = (\bar{\boldsymbol{\sigma}}^{(0)})^T \cdot d\bar{\boldsymbol{A}}\\ &= (\boldsymbol{\sigma}^*)^T \cdot \boldsymbol{n}\,dA = (\boldsymbol{\sigma}^*)^T \cdot d\boldsymbol{A}\end{aligned} \tag{4.106}$$

The left side of (4.106) is in Eulerian description in the current configuration, whereas the right side is in Lagrangian description in the reference configuration. *The relation* (4.106) *is useful when deriving momentum equations in Lagrangian description.* We could verify that $(\bar{\boldsymbol{\sigma}}^{(0)})^T \cdot \bar{\boldsymbol{n}}\,d\bar{A} = (\boldsymbol{\sigma}^*)^T \cdot d\boldsymbol{A}$ is in fact due to $d\bar{\boldsymbol{F}} = d\boldsymbol{F}$ used in deriving $\boldsymbol{\sigma}^*$ from $(\bar{\boldsymbol{\sigma}}^{(0)})^T$. Proof is simple. First, consider

$$\begin{aligned}(\bar{\boldsymbol{\sigma}}^{(0)})^T \cdot d\bar{\boldsymbol{A}} &= (\bar{\boldsymbol{\sigma}}^{(0)})^T \cdot \bar{\boldsymbol{n}}\,d\bar{A}\\ (\boldsymbol{\sigma}^*)^T \cdot d\boldsymbol{A} &= (\boldsymbol{\sigma}^*)^T \cdot \boldsymbol{n}\,dA\end{aligned} \tag{4.107}$$

Hence

$$(\bar{\boldsymbol{\sigma}}^{(0)})^T \cdot \bar{\boldsymbol{n}}\,d\bar{A} = (\boldsymbol{\sigma}^*)^T \cdot \boldsymbol{n}\,dA \tag{4.108}$$

Using the Cauchy principle for undeformed and deformed tetrahedra yields

$$\begin{aligned}(\boldsymbol{\sigma}^*)^T \cdot \boldsymbol{n} &= \boldsymbol{P}\\ (\bar{\boldsymbol{\sigma}}^{(0)})^T \cdot \bar{\boldsymbol{n}} &= \bar{\boldsymbol{P}}\end{aligned} \tag{4.109}$$

Using (4.109) in (4.108)

$$\bar{\boldsymbol{P}}\,d\bar{A} = \boldsymbol{P}\,dA \quad \text{or} \quad d\bar{\boldsymbol{F}} = d\boldsymbol{F} \tag{4.110}$$

(3) In the following, *we derive Lagrangian and Eulerian equivalents of* $\bar{\boldsymbol{P}} \cdot \bar{\boldsymbol{v}}\,d\bar{A}$. Consider the Cauchy principle $\bar{\boldsymbol{P}} = (\bar{\boldsymbol{\sigma}}^{(0)})^T \cdot \bar{\boldsymbol{n}}$. Hence

$$\bar{\boldsymbol{P}} \cdot \bar{\boldsymbol{v}}\,d\bar{A} = \big((\bar{\boldsymbol{\sigma}}^{(0)})^T \cdot \bar{\boldsymbol{n}}\big) \cdot \bar{\boldsymbol{v}}\,d\bar{A} = \bar{\boldsymbol{v}} \cdot \big((\bar{\boldsymbol{\sigma}}^{(0)})^T \cdot \bar{\boldsymbol{n}}\,d\bar{A}\big) \tag{4.111}$$

Using (4.106) in the last term of (4.111) and changing $\bar{\boldsymbol{v}}$ to $\boldsymbol{v}$ (Lagrangian description)

$$\bar{\boldsymbol{P}} \cdot \bar{\boldsymbol{v}}\,d\bar{A} = \boldsymbol{v} \cdot \big((\boldsymbol{\sigma}^*)^T \cdot \boldsymbol{n}\,dA\big) \tag{4.112}$$

Thus, we have

$$\bar{\boldsymbol{P}} \cdot \bar{\boldsymbol{v}}\, d\bar{A} = \left(\boldsymbol{v} \cdot (\boldsymbol{\sigma}^*)^T\right) \cdot \boldsymbol{n}\, dA = \left(\boldsymbol{v} \cdot (\boldsymbol{\sigma}^*)^T\right) \cdot d\boldsymbol{A} \tag{4.113}$$

Equation (4.113) *is useful in deriving the energy equation in Lagrangian description.* Also from (4.111) (analogous to (4.113)), we have

$$\bar{\boldsymbol{P}} \cdot \bar{\boldsymbol{v}}\, d\bar{A} = \left(\bar{\boldsymbol{v}} \cdot (\bar{\boldsymbol{\sigma}}^{(0)})^T\right) \cdot \bar{\boldsymbol{n}}\, d\bar{A} = \left(\bar{\boldsymbol{v}} \cdot (\bar{\boldsymbol{\sigma}}^{(0)})^T\right) \cdot d\bar{\boldsymbol{A}} \tag{4.114}$$

Equation (4.114) *is useful in deriving the energy equation in Eulerian description.* From (4.113) and (4.114) we have

$$\left(\bar{\boldsymbol{v}} \cdot (\bar{\boldsymbol{\sigma}}^{(0)})^T\right) \cdot d\bar{\boldsymbol{A}} = \left(\boldsymbol{v} \cdot (\boldsymbol{\sigma}^*)^T\right) \cdot d\boldsymbol{A} \tag{4.115}$$

4.13 Summary

The concept of stress is introduced and the Cauchy principle is derived using the deformed tetrahedron in the current configuration with the assumptions of infinitesimal deformation. The tensorial nature of the Cauchy stress matrix is established by considering rigid rotation of the frame and its symmetry is established by using the balance of angular momenta (also see Chapters 6 and 7). By considering the stress components acting on the faces of the deformed tetrahedron (contravariant directions), i.e., contravariant stress components and the covariant basis, the contravariant Cauchy stress is established in Lagrangian as well as Eulerian descriptions for finite deformation. Likewise, covariant Cauchy stress is defined in Lagrangian and Eulerian descriptions using covariant stress components and the contravariant basis. Jaumann stress tensor is defined using contra- and co-variant Cauchy stress tensor. Physical meanings of these various stress measures are discussed. It is shown that in the case of finite deformation, only contravariant Cauchy stress is the correct representation of the physics. The first and second Piola-Kirchhoff stress tensors are derived using contra- and co-variant Cauchy stress tensors in Lagrangian as well as Eulerian descriptions. Since the components of these stress tenors act on the undeformed tetrahedron in the reference configuration, these are non-physical. The importance of these measures is realized when deriving convected time derivatives of Cauchy stress tensors in contra- and co-variant bases, deriving mathematical descriptions using balance laws and establishing conjugate stress and strain measures that are essential in deriving constitutive theories.

Problems

4.1 Consider the contravariant Cauchy stress tensor $[\sigma^{(0)}]$ in x-frame acting on the orthogonal faces of a deformed tetrahedron whose oblique plane is defined by $2x_1 + 4x_2 + 4x_3 = \text{constant}$.

$$[\sigma^{(0)}] = \begin{bmatrix} 2 & 4 & 3 \\ 4 & 0 & 0 \\ 3 & 0 & -1 \end{bmatrix}$$

Determine the stress vector $\boldsymbol{P}$ on the oblique plane and its components normal and tangential to the oblique plane.

4.2 Consider the following deformation:

$$\bar{\boldsymbol{x}} = \alpha x_1 \boldsymbol{e}_1 - \beta x_3 \boldsymbol{e}_2 + \gamma x_2 \boldsymbol{e}_3$$

where α, β and γ are constants. Let the contravariant Cauchy stress tensor be given by

$$[\sigma^{(0)}] = \begin{bmatrix} \sigma_0 & 0 & 0 \\ 0 & 0 & 0 \\ 0 & 0 & 0 \end{bmatrix} \quad ; \quad [\sigma^{(0)}] = \begin{bmatrix} 0 & 0 & 0 \\ 0 & \sigma_0 & 0 \\ 0 & 0 & 0 \end{bmatrix} \quad ; \quad [\sigma^{(0)}] = \begin{bmatrix} 0 & 0 & 0 \\ 0 & 0 & 0 \\ 0 & 0 & \sigma_0 \end{bmatrix}$$

where σ_0 is a constant. For each $[\sigma^{(0)}]$, determine

(a) the first Piola-Kirchhoff stress tensor $[\sigma^*]$;
(b) the second Piola-Kirchhoff stress tensor $[\sigma^{[0]}]$.

4.3 Consider the following deformation field:

$$\begin{aligned} u_1 &= (\alpha \cos\theta - 1)x_1 - \beta x_2 \sin\theta \\ u_2 &= \alpha x_1 \sin\theta + (\beta \cos\theta - 1)x_2 \\ u_3 &= 0 \end{aligned}$$

The components of the Cauchy stress tensor $[\sigma^{(0)}]$ are given by

$$[\sigma^{(0)}] = \sigma_0 \begin{bmatrix} \cos^2\theta & \cos\theta\sin\theta & 0 \\ \cos\theta\sin\theta & \cos^2\theta & 0 \\ 0 & 0 & 0 \end{bmatrix}$$

where σ_0 is a constant. Determine

(a) the Lagrange stress tensor $[\sigma^*]$;
(b) the second Piola-Kirchhoff stress tensor $[\sigma^{[0]}]$.

4.4 Let $[\sigma]$ be a symmetric tensor of rank two.

(a) How many invariants does $[\sigma]$ have? Clearly substantiate your answer.
(b) What is the meaning of invariants of a second rank tensor?
(c) How many ways can you establish these invariants? If there is more than one way to establish these, then are the invariants from the two or more approaches related? Explain the procedure to establish these invariants from each of the approaches you advocate.
(d) What are the invariants of $[\sigma]$? Derive general expressions for them.

4.5 Consider a symmetric stress tensor $[\sigma]$ given by

$$[\sigma] = \begin{bmatrix} 2.5 & 1 & 0 \\ 1 & 1 & 0 \\ 0 & 0 & 1.5 \end{bmatrix}$$

Find principal stresses and their directions. Show that the normalized principle directions constitute orthonormal basis.

4.6 Let the Cauchy stress tensor be given by

$$[\bar{\sigma}] = \begin{bmatrix} 3 & -9 & 6 \\ -9 & 0 & 12 \\ 6 & 12 & -3 \end{bmatrix} \tag{1}$$

(a) Consider an oblique plane of the tetrahedron whose normal $\{\bar{n}\}$ is given by

$$\{\bar{n}\}^T = \frac{1}{\sqrt{12.25}}\left[1\boldsymbol{e}_1 - 1.5\boldsymbol{e}_2 + 3\boldsymbol{e}_3\right]$$

(1) Determine average force/unit area $\{\bar{P}\}$ on the oblique plane.
(2) Find direction of $\{\bar{P}\}$.

(b) Consider the same Cauchy stress tensor as in (1), but let the unit normal to the oblique plane of the tetrahedron be given by

$$\{\bar{n}\}^T = \frac{1}{\sqrt{2.25}}\left[\boldsymbol{e}_1 + \boldsymbol{e}_2 + 0.5\boldsymbol{e}_3\right]$$

(1) Determine $\{\bar{P}\}$ on the oblique plane.
(2) Find direction of $\{\bar{P}\}$.

Clearly $\{\bar{P}\}$ in **(a)** and **(b)** is different for the same Cauchy stress tensor $[\bar{\sigma}]$. Discuss your results.

4.7 Consider same Cauchy stress tensor $[\bar{\sigma}]$ as in Problem 4.6. Let $\{\bar{n}\}$ be the unit vector normal to the oblique plane of the tetrahedron.

(a) Find $\{\bar{n}\}$ for which $\{\bar{P}\}$, the average stress on the oblique plane is zero.
(b) Find $\{\bar{n}\}$ for which $\bar{P}_{x_1} = \bar{P}_{x_2} = \bar{P}_{x_3} = 3$

4.8 Show that if the Cauchy stress tensor $[\bar{\sigma}]$ is singular, i.e., if $\det[\bar{\sigma}] = 0$, then one of the principle stresses is zero.

4.9 Consider bounding surface $\partial\bar{V}$ of deformed volume $\bar{V}(t)$. Consider an elementary tetrahedron with oblique planes (part of $\partial\bar{V}$) of areas $\{d\bar{A}_1\}$ and $\{d\bar{A}_2\}$ with unit exterior normals $\{\bar{n}_1\}$ and $\{\bar{n}_2\}$, respectively. If $\{\bar{P}_1\}$ and $\{\bar{P}_2\}$ are the average stresses on oblique areas $\{d\bar{A}_1\}$ and $\{d\bar{A}_2\}$, then show that the following holds.

$$\bar{\boldsymbol{n}} \cdot \bar{\boldsymbol{P}}_1 = \bar{\boldsymbol{n}}_2 \cdot \bar{\boldsymbol{P}}_2$$

5

RATES, CONVECTED TIME DERIVATIVES, OBJECTIVE TENSORS AND OBJECTIVE RATES

Since the deformation of a volume of matter is continuously evolving, i.e., the state of the matter changes as time elapses, all measures describing the deformation are naturally functions of time in addition to coordinates x_i (in Lagrangian description) or $\bar{x}_i$ (in Eulerian description). Thus, rate of deformation, i.e., of length, area and volume, strain rate tensors, spin tensors and convected time derivatives of stress and strain tensors in co- and contra-variant bases need to be considered. This material is of utmost significance and importance in the development of the constitutive theories as well as the mathematical description of the deforming volume of matter using conservation and balance laws. In this chapter, we consider rate of deformation of length, area and volume, strain rate tensors, spin tensors and convected time derivatives of various orders of stress and strain tensors. Definitions of objective tensors and objective rates are given. Derivations are presented to establish objectivity and objective rates of various tensors commonly encountered in classical continuum mechanics. Galilean as well as Euclidean [63] or non-Galilean transformations are considered.

5.1 Rate of deformation

Let $P(x_1, x_2, x_3)$ be the location of a material point in the reference or undeformed configuration and $Q(x_1+dx_1, x_2+dx_2, x_3+dx_3)$ be a neighboring material point such that $\{dx\} = [dx_1, dx_2, dx_3]^T$ are the components of the vector $\boldsymbol{PQ}$ with length ds. After deformation, material points P and Q occupy positions $\bar{P}$, $\bar{Q}$ in the current configuration at time t. The components of the vector $\bar{\boldsymbol{P}}\bar{\boldsymbol{Q}}$ with length $d\bar{s}$ are $\{d\bar{x}\} = [d\bar{x}_1, d\bar{x}_2, d\bar{x}_3]^T$ (see Figure 5.1). If $\boldsymbol{PQ}$ and $\bar{\boldsymbol{P}}\bar{\boldsymbol{Q}}$ are undeformed and deformed material lines, then we are interested in determining the time rate of change of $\bar{\boldsymbol{P}}\bar{\boldsymbol{Q}}$ (i.e., rate of deformation) in Lagrangian and Eulerian descriptions.

5.1.1 Lagrangian description

$$d\bar{x}_i = \bar{x}_i(x_k + dx_k, t) - \bar{x}_i(x_k, t) \tag{5.1}$$

Taking the material derivative,

$$\begin{aligned}
&\frac{D}{Dt}(d\bar{x}_i) = \frac{D}{Dt}\bar{x}_i(x_k + dx_k, t) - \frac{D}{Dt}\bar{x}_i(x_k, t) \\
\text{or}\quad &\frac{D}{Dt}(d\bar{x}_i) = \frac{d}{dt}\bar{x}_i(x_k + dx_k, t) - \frac{d}{dt}\bar{x}_i(x_k, t) \\
\text{or}\quad &\frac{D}{Dt}(d\bar{x}_i) = v_i(x_k + dx_k, t) - v_i(x_k, t)
\end{aligned} \tag{5.2}$$

DOI: 10.1201/9781003105336-5

Expand $v_i(x_k + dx_k, t)$ in Taylor series about x_k and retain only up to linear terms

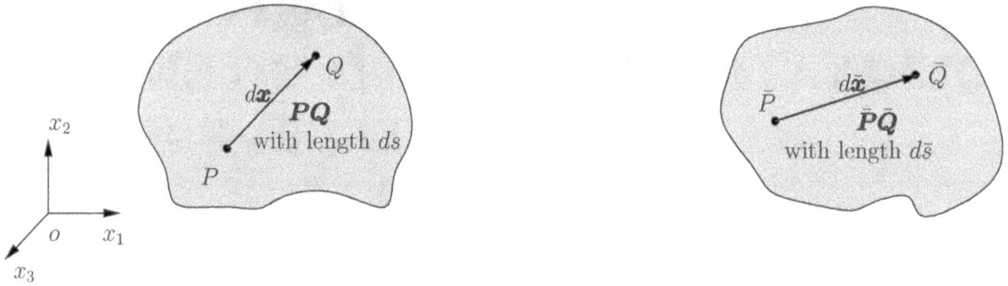

(a) Reference configuration (b) Current configuration

Figure 5.1: Deformation of a line segment

in dx_k and velocity gradients. Neglecting second and higher degree terms in dx_k as well as higher order terms in velocity gradients

$$\begin{aligned} \frac{D}{Dt}(d\bar{x}_i) &= v_i(x_k, t) + \frac{\partial v_i}{\partial x_k} dx_k - v_i(x_k, t) \\ \text{or} \quad \frac{D}{Dt}(d\bar{x}_i) &= \frac{\partial v_i}{\partial x_k} dx_k \end{aligned} \tag{5.3}$$

$$\frac{\partial v_i}{\partial x_k} = \frac{\partial}{\partial t}\frac{\partial \bar{x}_i}{\partial x_k} = \frac{\partial J_{ik}}{\partial t} = \dot{J}_{ik} \tag{5.4}$$

$$\therefore \quad \frac{D}{Dt}\{d\bar{x}\} = [\dot{J}]\{dx\} \quad \text{or} \quad \frac{D}{Dt}(d\bar{x}_i) = \dot{J}_{ij} dx_j \quad : \quad \text{(Lagrangian)} \tag{5.5}$$

5.1.2 Eulerian description

The simplest way is to begin with (5.3) in which we can define velocities in the Eulerian description (i.e., we replace v_i by $\bar{v}_i$ in (5.3)).

$$\frac{D}{Dt}(d\bar{x}_i) = \left(\frac{\partial \bar{v}_i}{\partial \bar{x}_j} \frac{\partial \bar{x}_j}{\partial x_k} \right) dx_k = \frac{\partial \bar{v}_i}{\partial \bar{x}_j} \frac{\partial \bar{x}_j}{\partial x_k} dx_k \tag{5.6}$$

Let

$$\frac{\partial \bar{v}_i}{\partial \bar{x}_j} = \bar{L}_{ij} \quad \text{and} \quad \frac{\partial \bar{x}_j}{\partial x_k} = J_{jk} \tag{5.7}$$

Substituting from (5.7) into (5.6) we obtain

$$\frac{D}{Dt}(d\bar{x}_i) = \bar{L}_{ij} J_{jk} dx_k \tag{5.8}$$

or

$$\left.\begin{aligned} \frac{D}{Dt}\{d\bar{x}\} = [\bar{L}][J]\{dx\} = [\bar{L}]\{d\bar{x}\} \quad &\text{or} \quad \frac{D}{Dt}(d\bar{x}_i) = \bar{L}_{ij} d\bar{x}_j \\ \therefore \quad \{d\bar{v}\} = [\bar{L}]\{d\bar{x}\} \quad &\text{or} \quad d\bar{v}_i = \bar{L}_{ij} d\bar{x}_j \quad \text{(Eulerian)} \\ \text{also} \quad \{\bar{v}\} = [\bar{L}]\{\bar{x}\} \quad &\text{or} \quad \bar{v}_i = \bar{L}_{ij} \bar{x}_j \end{aligned}\right\} \tag{5.9}$$

We note that both (5.5) and (5.9) are material derivatives of $\{d\bar{x}\}$, (5.5) being Lagrangian description whereas (5.9) is Eulerian description. Hence the appearance of $\{dx\}$ and $\{d\bar{x}\}$ on the right sides of (5.5) and (5.9). The coefficients of $\{dx\}$ and $\{d\bar{x}\}$ on the right sides of (5.5) and (5.9) are fundamental quantities. From (5.5) and (5.9), we conclude that coefficients of $\{dx\}$ must be the same as the material derivative has a unique value.

$$\therefore \quad [\dot{J}] = [\bar{L}][J] \quad \text{or} \quad [\bar{L}] = [\dot{J}][J]^{-1} \quad \text{(Def.)} \tag{5.10}$$

$[\bar{L}]$ is known as spatial velocity gradient matrix or tensor. $[\dot{J}]$ is called material velocity gradient or tensor. The relations (5.10) are fundamental and will be used frequently in the subsequent sections.

5.2 Additive decomposition of the spatial velocity gradient tensor $[\bar{L}]$

We note that

$$\bar{L}_{ij} = \frac{\partial \bar{v}_i}{\partial \bar{x}_j} = \bar{v}_{i,j} \quad \text{or} \quad [\bar{L}] = \left[\frac{\partial\{\bar{v}\}}{\partial\{\bar{x}\}}\right] \tag{5.11}$$

$$\begin{aligned} \bar{L}_{ij} &= \frac{1}{2}(\bar{v}_{i,j} + \bar{v}_{j,i}) + \frac{1}{2}(\bar{v}_{i,j} - \bar{v}_{j,i}) \\ \text{or} \quad [\bar{L}] &= \frac{1}{2}\left[[\bar{L}] + [\bar{L}]^T\right] + \frac{1}{2}\left[[\bar{L}] - [\bar{L}]^T\right] \end{aligned} \tag{5.12}$$

Let us define

$$\bar{D}_{ij} = \frac{1}{2}(\bar{v}_{i,j} + \bar{v}_{j,i}) \quad \text{or} \quad [\bar{D}] = \frac{1}{2}\left([\bar{L}] + [\bar{L}]^T\right) \tag{5.13}$$

$$\bar{W}_{ij} = \frac{1}{2}(\bar{v}_{i,j} - \bar{v}_{j,i}) \quad \text{or} \quad [\bar{W}] = \frac{1}{2}\left([\bar{L}] - [\bar{L}]^T\right) \tag{5.14}$$

Thus

$$[\bar{L}] = [\bar{D}] + [\bar{W}] \tag{5.15}$$

Consider

$$\bar{\boldsymbol{\nabla}} \times \bar{\boldsymbol{v}} = \boldsymbol{e}_1({}_i^r\bar{\Theta}_1) + \boldsymbol{e}_2({}_i^r\bar{\Theta}_2) + \boldsymbol{e}_3({}_i^r\bar{\Theta}_3) \tag{5.16}$$

where ${}_i^r\bar{\Theta}_1$, ${}_i^r\bar{\Theta}_2$, ${}_i^r\bar{\Theta}_3$ are rotation rates about ox_i axes of the x-frame and are given by

$${}_i^r\bar{\Theta}_1 = \left(\frac{\partial \bar{v}_3}{\partial \bar{x}_2} - \frac{\partial \bar{v}_2}{\partial \bar{x}_3}\right) \; ; \; {}_i^r\bar{\Theta}_2 = \left(\frac{\partial \bar{v}_1}{\partial \bar{x}_3} - \frac{\partial \bar{v}_3}{\partial \bar{x}_1}\right) \; ; \; {}_i^r\bar{\Theta}_3 = \left(\frac{\partial \bar{v}_2}{\partial \bar{x}_1} - \frac{\partial \bar{v}_1}{\partial \bar{x}_2}\right) \tag{5.17}$$

From (5.15), the non-zero components of $[\bar{W}]$ can be expressed in terms of rotation rates.

$$[\bar{W}] = \begin{bmatrix} 0 & \bar{W}_{12} & \bar{W}_{13} \\ \bar{W}_{21} & 0 & \bar{W}_{23} \\ \bar{W}_{31} & \bar{W}_{32} & 0 \end{bmatrix} = \frac{1}{2}\begin{bmatrix} 0 & -{}_i^r\bar{\Theta}_3 & {}_i^r\bar{\Theta}_2 \\ {}_i^r\bar{\Theta}_3 & 0 & -{}_i^r\bar{\Theta}_1 \\ -{}_i^r\bar{\Theta}_2 & {}_i^r\bar{\Theta}_1 & 0 \end{bmatrix}$$

Remarks

(1) We note that $[\bar{D}]$ is the symmetric part of the spatial velocity gradient tensor or strain rate tensor (shown later), whereas $[\bar{W}]$ is the antisymmetric part of the *spatial velocity gradient tensor* and is known as the *spin tensor*, it contains rotation rates.

(2) In deriving the above relations, we have assumed that ds and $d\bar{s}$ are such that infinitesimals of order two and higher of dx_i and $d\bar{x}_i$ can be neglected, but we have made no assumption regarding $\frac{\partial \bar{v}_i}{\partial \bar{x}_j}$ or $\bar{v}_{i,j}$ other than the fact that second and higher order terms in the velocity gradients are neglected.

(3) In (5.3), the velocities are expressed as a function of x_i and t (Lagrangian description), whereas in (5.6), the velocities are expressed as a function of $\bar{x}_i$ and time t (Eulerian description).

(4) We note that $v_i(x_k, t)$ and $\bar{v}_i(\bar{x}_k, t)$ are in general different functions but their numerical values are the same.

5.3 Interpretation of the components of $[\bar{D}]$

5.3.1 Diagonal components of $[\bar{D}]$

Let $d\bar{s}$ be the length of $\bar{\boldsymbol{P}}\bar{\boldsymbol{Q}}$ or $\{d\bar{x}\}$ and let $\bar{\boldsymbol{n}}$ be the directions cosines of $\bar{\boldsymbol{P}}\bar{\boldsymbol{Q}}$, then

$$\{d\bar{x}\} = d\bar{s}\{\bar{n}\} \tag{5.18}$$

Let $\{d\bar{x}\}$ or $d\bar{x}_i$ be the components of the length $d\bar{s}$ in x-frame, then

$$\begin{aligned} (d\bar{s})^2 &= d\bar{x}_i d\bar{x}_i \\ \therefore \quad \frac{D}{Dt}(d\bar{s})^2 &= \frac{D}{Dt}(d\bar{x}_i d\bar{x}_i) \\ 2d\bar{s}\frac{D}{Dt}(d\bar{s}) &= 2d\bar{x}_i\frac{D}{Dt}(d\bar{x}_i) \\ \text{or} \quad d\bar{s}\frac{D}{Dt}(d\bar{s}) &= d\bar{x}_i\frac{D}{Dt}(d\bar{x}_i) \end{aligned} \tag{5.19}$$

But from (5.9)

$$\frac{D}{Dt}(d\bar{x}_i) = \bar{L}_{ij}d\bar{x}_j = (\bar{D}_{ij} + \bar{W}_{ij})d\bar{x}_j \tag{5.20}$$

Substituting from (5.20) into (5.19)

$$d\bar{s}\frac{D}{Dt}(d\bar{s}) = d\bar{x}_i(\bar{D}_{ij} + \bar{W}_{ij})d\bar{x}_j \tag{5.21}$$

Since $\bar{W}_{ij} = -\bar{W}_{ji}$ (antisymmetric)

$$d\bar{x}_i(\bar{W}_{ij})d\bar{x}_j = 0$$

$$\therefore \quad d\bar{s}\frac{D}{Dt}(d\bar{s}) = d\bar{x}_i\bar{D}_{ij}d\bar{x}_j \tag{5.22}$$

We note that

$$d\bar{x}_i = d\bar{s}\bar{n}_i \quad ; \quad d\bar{x}_j = d\bar{s}\bar{n}_j \tag{5.23}$$

$$\therefore \quad d\bar{s}\frac{D}{Dt}(d\bar{s}) = (d\bar{s})^2\bar{n}_i\bar{D}_{ij}\bar{n}_j \tag{5.24}$$

$$\frac{1}{d\bar{s}}\frac{D}{Dt}(d\bar{s}) = \bar{n}_i\bar{D}_{ij}\bar{n}_j \tag{5.25}$$

We note that the left side of (5.25) is the rate of change of length $d\bar{s}$ per unit length. We can specifically examine the physical meaning of the diagonal components of $\bar{D}_{ij}$ using (5.25). If we let $\{\bar{n}\} = [1,0,0]^T$, i.e we consider the deformed material line $d\bar{s}$ along ox_1 axis, then (5.25) gives

$$\frac{1}{d\bar{s}}\frac{D}{Dt}(d\bar{s}) = \bar{D}_{11} = \frac{\partial \bar{v}_1}{\partial \bar{x}_1} \tag{5.26}$$

Equation (5.26) gives the rate of change of length per unit length known as rate of extension (or stretching) of a material line in the current configuration parallel to o-x_1 axis.

Similarly, if we choose

$$\{\bar{n}\} = [0,1,0]^T \quad \text{and} \quad \{\bar{n}\} = [0,0,1]^T \tag{5.27}$$

then we obtain the following from (5.25):

$$\begin{aligned} \frac{1}{d\bar{s}}\frac{D}{Dt}(d\bar{s}) &= \bar{D}_{22} = \frac{\partial \bar{v}_2}{\partial \bar{x}_2} \quad ; \quad \text{when} \quad \{\bar{n}\} = [0,1,0]^T \\ \frac{1}{d\bar{s}}\frac{D}{Dt}(d\bar{s}) &= \bar{D}_{33} = \frac{\partial \bar{v}_3}{\partial \bar{x}_3} \quad ; \quad \text{when} \quad \{\bar{n}\} = [0,0,1]^T \end{aligned} \tag{5.28}$$

These are rate of extensions of material lines parallel to o-x_2 and o-x_3 axes of the x-frame in the current configuration.

5.3.2 Off diagonal components of $[\bar{D}]$: physical interpretation

Let $\boldsymbol{PQ}$ or $d\boldsymbol{x}^1$ and $\boldsymbol{PR}$ or $d\boldsymbol{x}^2$ be two line segments at a material point P in the reference configuration (Figure 5.2(a)). Let $\bar{\boldsymbol{P}}\bar{\boldsymbol{Q}}$ or $d\bar{\boldsymbol{x}}^1$ and $\bar{\boldsymbol{P}}\bar{\boldsymbol{R}}$ or $d\bar{\boldsymbol{x}}^2$ be their deformed positions in the current configuration at the material point $\bar{P}$ (Figure 5.2(b)) in the deformed configuration.

Let $d\bar{s}_1$ and $d\bar{s}_2$ be the scalar lengths of $d\bar{\boldsymbol{x}}^1$ and $d\bar{\boldsymbol{x}}^2$ and $\bar{\boldsymbol{n}}$ and $\bar{\boldsymbol{m}}$ be the direction cosines of $d\bar{\boldsymbol{x}}^1$ and $d\bar{\boldsymbol{x}}^2$. Then

$$d\bar{\boldsymbol{x}}^1 = d\bar{s}_1\,\bar{\boldsymbol{n}} \quad \text{and} \quad d\bar{\boldsymbol{x}}^2 = d\bar{s}_2\,\bar{\boldsymbol{m}} \tag{5.29}$$

Consider

$$\frac{1}{d\bar{s}_1 d\bar{s}_2}\frac{D}{Dt}(d\bar{\boldsymbol{x}}^1 \cdot d\bar{\boldsymbol{x}}^2) \tag{5.30}$$

The quantity in (5.30) can be expanded in two ways. First

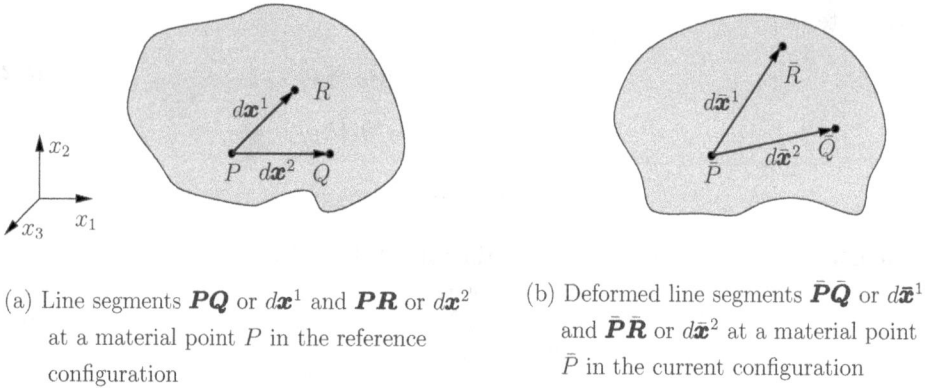

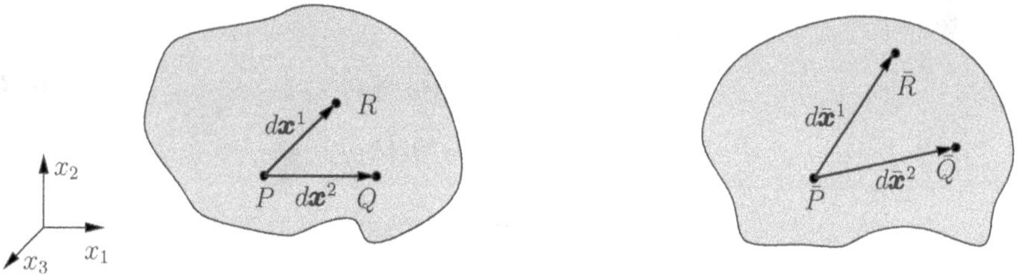

(a) Line segments $\boldsymbol{PQ}$ or $d\boldsymbol{x}^1$ and $\boldsymbol{PR}$ or $d\boldsymbol{x}^2$ at a material point P in the reference configuration

(b) Deformed line segments $\bar{\boldsymbol{P}}\bar{\boldsymbol{Q}}$ or $d\bar{\boldsymbol{x}}^1$ and $\bar{\boldsymbol{P}}\bar{\boldsymbol{R}}$ or $d\bar{\boldsymbol{x}}^2$ at a material point $\bar{P}$ in the current configuration

Figure 5.2: Deformation of material line segments at a material point

$$\begin{aligned} \frac{1}{d\bar{s}_1 d\bar{s}_2}\frac{D}{Dt}(d\bar{\boldsymbol{x}}^1 \cdot d\bar{\boldsymbol{x}}^2) &= \frac{1}{d\bar{s}_1 d\bar{s}_2}\frac{D}{Dt}(d\bar{s}_1\, \bar{\boldsymbol{n}} \cdot d\bar{s}_2\, \bar{\boldsymbol{m}}) \\ &= \frac{1}{d\bar{s}_1 d\bar{s}_2}\frac{D}{Dt}(d\bar{s}_1 d\bar{s}_2\, \bar{\boldsymbol{n}} \cdot \bar{\boldsymbol{m}}) \\ &= \frac{1}{d\bar{s}_1 d\bar{s}_2}\frac{D}{Dt}(d\bar{s}_1 d\bar{s}_2\, \cos\theta_{nm}) \end{aligned} \tag{5.31}$$

where θ_{nm} is the angle between the direction cosine vectors $\bar{\boldsymbol{n}}$ and $\bar{\boldsymbol{m}}$ or $d\bar{\boldsymbol{x}}^1$ and $d\bar{\boldsymbol{x}}^2$ in the current configuration. Expanding the right side of (5.31) using the product rule

$$\begin{aligned} \frac{1}{d\bar{s}_1 d\bar{s}_2}\frac{D}{Dt}(d\bar{\boldsymbol{x}}^1 \cdot d\bar{\boldsymbol{x}}^2) = &\left(\frac{1}{d\bar{s}_1}\frac{D}{Dt}(d\bar{s}_1) + \frac{1}{d\bar{s}_2}\frac{D}{Dt}(d\bar{s}_2)\right)\cos\theta_{nm} \\ &+ \frac{D}{Dt}(\cos\theta_{nm}) \end{aligned} \tag{5.32}$$

Using (5.25) we can write

$$\frac{1}{d\bar{s}_1}\frac{D}{Dt}(d\bar{s}_1) = \bar{n}_i \bar{D}_{ij} \bar{n}_j \quad ; \quad \frac{1}{d\bar{s}_2}\frac{D}{Dt}(d\bar{s}_2) = \bar{m}_i \bar{D}_{ij} \bar{m}_j \tag{5.33}$$

$$\text{and} \quad \frac{D}{Dt}(\cos\theta_{nm}) = -\sin\theta_{nm}\dot{\theta}_{nm} \tag{5.34}$$

Substituting from (5.33) and (5.34) into (5.32) we obtain

$$\frac{1}{d\bar{s}_1 d\bar{s}_2}\frac{D}{Dt}(d\bar{\boldsymbol{x}}^1 \cdot d\bar{\boldsymbol{x}}^2) = (\bar{n}_i \bar{D}_{ij} \bar{n}_j + \bar{m}_i \bar{D}_{ij} \bar{m}_j)\cos\theta_{nm} - \sin\theta_{nm}\dot{\theta}_{nm} \tag{5.35}$$

Secondly

$$\frac{1}{d\bar{s}_1 d\bar{s}_2}\frac{D}{Dt}(d\bar{\boldsymbol{x}}^1 \cdot d\bar{\boldsymbol{x}}^2) = \frac{1}{d\bar{s}_1 d\bar{s}_2}\frac{D}{Dt}(d\bar{x}_i^1 d\bar{x}_i^2) \tag{5.36}$$

in which dx_i^1 and dx_i^2 are the components of $d\bar{s}_1$ and $d\bar{s}_2$. Equation (5.36) can be written as

$$\frac{1}{d\bar{s}_1 d\bar{s}_2}\frac{D}{Dt}(d\bar{\boldsymbol{x}}^1 \cdot d\bar{\boldsymbol{x}}^2) = \frac{1}{d\bar{s}_1 d\bar{s}_2}\left(d\bar{x}_i^2 \frac{D}{Dt}(d\bar{x}_i^1) + d\bar{x}_i^1 \frac{D}{Dt}(d\bar{x}_i^2)\right) \tag{5.37}$$

But using (5.9) and (5.15) we can write

$$\frac{D}{Dt}(d\bar{x}_i^1) = (\bar{D}_{ij} + \bar{W}_{ij})d\bar{x}_j^1 \tag{5.38}$$

$$\frac{D}{Dt}(d\bar{x}_i^2) = (\bar{D}_{ij} + \bar{W}_{ij})d\bar{x}_j^2 \tag{5.39}$$

Substituting from (5.38) and (5.39) into (5.37) we obtain

$$\begin{aligned}\frac{1}{d\bar{s}_1 d\bar{s}_2}\frac{D}{Dt}(d\bar{\boldsymbol{x}}^1 \cdot d\bar{\boldsymbol{x}}^2) = &\frac{1}{d\bar{s}_1 d\bar{s}_2} d\bar{x}_i^2(\bar{D}_{ij} + \bar{W}_{ij})d\bar{x}_j^1 \\ &+ \frac{1}{d\bar{s}_1 d\bar{s}_2} d\bar{x}_i^1(\bar{D}_{ij} + \bar{W}_{ij})d\bar{x}_j^2\end{aligned} \tag{5.40}$$

or $$\frac{1}{d\bar{s}_1 d\bar{s}_2}\frac{D}{Dt}(d\bar{\boldsymbol{x}}^1 \cdot d\bar{\boldsymbol{x}}^2) = \frac{d\bar{x}_i^2}{d\bar{s}_2}(\bar{D}_{ij} + \bar{W}_{ij})\frac{d\bar{x}_j^1}{d\bar{s}_1} + \frac{d\bar{x}_i^1}{d\bar{s}_1}(\bar{D}_{ij} + \bar{W}_{ij})\frac{d\bar{x}_j^2}{d\bar{s}_2} \tag{5.41}$$

or $$\frac{1}{d\bar{s}_1 d\bar{s}_2}\frac{D}{Dt}(d\bar{\boldsymbol{x}}^1 \cdot d\bar{\boldsymbol{x}}^2) = \bar{m}_i(\bar{D}_{ij} + \bar{W}_{ij})\bar{n}_j + \bar{n}_i(\bar{D}_{ij} + \bar{W}_{ij})\bar{m}_j \tag{5.42}$$

Since $\bar{W}_{ij}$ is skew-symmetric symmetric, we have

$$\bar{m}_i\bar{W}_{ij}\bar{n}_j = -\bar{n}_i\bar{W}_{ij}\bar{m}_j \tag{5.43}$$

$$\therefore \quad \frac{1}{d\bar{s}_1 d\bar{s}_2}\frac{D}{Dt}(d\bar{\boldsymbol{x}}^1 \cdot d\bar{\boldsymbol{x}}^2) = \bar{m}_i(\bar{D}_{ij})\bar{n}_j + \bar{n}_i(\bar{D}_{ij})\bar{m}_j \tag{5.44}$$

Furthermore, since $\bar{D}_{ij} = \bar{D}_{ji}$

$$\therefore \quad \frac{1}{d\bar{s}_1 d\bar{s}_2}\frac{D}{Dt}(d\bar{\boldsymbol{x}}^1 \cdot d\bar{\boldsymbol{x}}^2) = 2\bar{m}_i\bar{D}_{ij}\bar{n}_j \tag{5.45}$$

Equating the right hand sides of (5.35) and (5.45)

$$(\bar{n}_i\bar{D}_{ij}\bar{n}_j + \bar{m}_i\bar{D}_{ij}\bar{m}_j)\cos\theta_{nm} - \sin\theta_{nm}\dot{\theta}_{nm} = 2\bar{m}_i\bar{D}_{ij}\bar{n}_j \tag{5.46}$$

If we choose $\{\bar{n}\} = [1,0,0]^T$ and $\{\bar{m}\} = [0,1,0]^T$, i.e., $\bar{\boldsymbol{P}}\bar{\boldsymbol{Q}}$, $\bar{\boldsymbol{P}}\bar{\boldsymbol{R}}$ parallel to o-x_1 and o-x_2 axes and hence at $\pi/2$ to each other, then using (5.46) with $\theta_{nm} = \pi/2$, we obtain

$$-\dot{\theta}_{12} = 2\bar{D}_{12} = \frac{\partial\bar{v}_1}{\partial\bar{x}_2} + \frac{\partial\bar{v}_2}{\partial\bar{x}_1} \tag{5.47}$$

That is, $2\bar{D}_{12}$ is equal to the rate of decrease of the angle from $\pi/2$ of the two material line elements parallel to o-x_1 and o-x_2 axes in the current configuration. This is referred to as the rate of shear.

Similarly, we have

$$-\dot{\theta}_{23} = 2\bar{D}_{23} = \frac{\partial\bar{v}_2}{\partial\bar{x}_3} + \frac{\partial\bar{v}_3}{\partial\bar{x}_2} \tag{5.48}$$

$$-\dot{\theta}_{31} = 2\bar{D}_{31} = \frac{\partial \bar{v}_3}{\partial \bar{x}_1} + \frac{\partial \bar{v}_1}{\partial \bar{x}_3} \tag{5.49}$$

Remarks

(1) Since $\bar{D}_{ij}$ is symmetric, we could find its eigenpairs and thereby establish the maximum and minimum rates of extensions as well as their directions. The eigenvalues of $\bar{D}_{ij}$ are principal rates of extension and the corresponding eigenvectors are their directions.

(2) We remark again that $[\bar{D}]$ or $\bar{D}_{ij}$ is the strain rate tensor. Obviously the rate of strain tensor is different than the strain rate tensor. We consider some details in the following.

5.4 Rate of change or material derivative of strain tensors $[C_{[0]}]$ and $[\varepsilon_{[0]}]$

In the following we consider material derivatives or rate of change of $[C_{[0]}]$ and $[\varepsilon_{[0]}]$ at a material point. Consider

$$[C_{[0]}] = [J]^T[J] \tag{5.50}$$

Taking material derivative of $[C_{[0]}]$ in (5.50)

$$\frac{D}{Dt}([C_{[0]}]) = [\dot{C}_{[0]}] = [\dot{J}]^T[J] + [J]^T[\dot{J}] \tag{5.51}$$

But

$$[\dot{J}] = [\bar{L}][J] \tag{5.52}$$

Hence

$$[\dot{C}_{[0]}] = [J]^T[\bar{L}]^T[J] + [J]^T[\bar{L}][J] \tag{5.53}$$

$$\text{or} \quad [\dot{C}_{[0]}] = [J]^T([\bar{L}]^T + [\bar{L}])[J]$$

$$[\dot{C}_{[0]}] = [J]^T(2[\bar{D}])[J] \tag{5.54}$$

$$\therefore \quad \frac{D}{Dt}([C_{[0]}]) = [\dot{C}_{[0]}] = 2[J]^T[\bar{D}][J]$$

Equation (5.54) defines the rate of strain tensor whereas $[\bar{D}]$ is the strain rate tensor. These two are obviously not the same. Also, see Section 5.13.5 on convected time derivatives of the strain tensors.

Material derivative of Green's strain tensor $[\varepsilon_{[0]}]$ can be easily found using (5.54)

$$[\varepsilon_{[0]}] = \frac{1}{2}[[J]^T[J] - [I]] = \frac{1}{2}[[C_{[0]}] - [I]] \tag{5.55}$$

$$\therefore \quad \frac{D}{Dt}([\varepsilon_{[0]}]) = [\dot{\varepsilon}_{[0]}] = \frac{1}{2}\frac{D[C_{[0]}]}{Dt} = \frac{1}{2}[\dot{C}_{[0]}] \tag{5.56}$$

and using (5.54) we obtain

$$\frac{D}{Dt}([\varepsilon_{[0]}]) = [\dot{\varepsilon}_{[0]}] = [J]^T[\bar{D}][J] \tag{5.57}$$

We note that $[\varepsilon_{[0]}] = \frac{1}{2}\left([C_{[0]}] - [I]\right)$, hence $[\dot{\varepsilon}_{[0]}] = \frac{1}{2}[\dot{C}_{[0]}]$.

5.5 Physical meaning of spin tensor $[\bar{W}]$

Let $\{\bar{n}\}$ be a unit eigenvector of $[\bar{D}]$ and λ be the corresponding eigenvalue. Then

$$[\bar{D}]\{\bar{n}\} = \lambda\{\bar{n}\} \tag{5.58}$$

Consider $d\bar{s}\bar{n}_i = d\bar{x}_i$, a vector along $\bar{\boldsymbol{n}}$, and take its material derivative.

$$\frac{D}{Dt}(d\bar{s}\bar{n}_i) = \frac{D}{Dt}(d\bar{s})\bar{n}_i + d\bar{s}\frac{D}{Dt}(\bar{n}_i) \tag{5.59}$$

Based on (5.20), i.e.

$$\frac{D}{Dt}(d\bar{x}_i) = (\bar{D}_{ij} + \bar{W}_{ij})d\bar{x}_j \tag{5.60}$$

We can write the following using $d\bar{x}_i = d\bar{s}\bar{n}_i$ and $d\bar{x}_j = d\bar{s}\bar{n}_j$

$$\frac{D}{Dt}(d\bar{s}\bar{n}_i) = (\bar{D}_{ij} + \bar{W}_{ij})d\bar{s}\bar{n}_j \tag{5.61}$$

From (5.59) and (5.61), we obtain

$$\frac{D}{Dt}(d\bar{s})\bar{n}_i + d\bar{s}\frac{D}{Dt}(\bar{n}_i) = (\bar{D}_{ij} + \bar{W}_{ij})d\bar{s}\bar{n}_j \tag{5.62}$$

Based on (5.25)

$$\frac{D}{Dt}(d\bar{s}) = (\bar{n}_k\bar{D}_{kl}\bar{n}_l)d\bar{s} \tag{5.63}$$

and

$$\frac{D}{Dt}(\bar{n}_i) = \dot{\bar{n}}_i \tag{5.64}$$

Substituting (5.63) and (5.64) in (5.62)

$$(\bar{n}_k\bar{D}_{kl}\bar{n}_l)d\bar{s}\bar{n}_i + d\bar{s}\dot{\bar{n}}_i = (\bar{D}_{ij} + \bar{W}_{ij})d\bar{s}\bar{n}_j \tag{5.65}$$

Divide throughout by $d\bar{s}$.

$$(\bar{n}_k\bar{D}_{kl}\bar{n}_l)\bar{n}_i + \dot{\bar{n}}_i = (\bar{D}_{ij} + \bar{W}_{ij})\bar{n}_j \tag{5.66}$$

$$\therefore\quad \dot{\bar{n}}_i = \bar{D}_{ij}\bar{n}_j + \bar{W}_{ij}\bar{n}_j - (\bar{n}_k\bar{D}_{kl}\bar{n}_l)\bar{n}_i \tag{5.67}$$

$$\dot{\bar{n}}_i = \lambda\bar{n}_i + \bar{W}_{ij}\bar{n}_j - \lambda\bar{n}_i \tag{5.68}$$

$$\dot{\bar{n}}_i = \bar{W}_{ij}\bar{n}_j \tag{5.69}$$

Remarks

(1) Equation (5.69) states that the spin tensor acting on a unit eigenvector of $[\bar{D}]$ gives the rate of change of that unit vector.

(2) Since

$$\bar{n}_i \bar{n}_i = 1 \quad \therefore \quad \dot{\bar{n}}_i \bar{n}_i = 0 \tag{5.70}$$

Equation (5.70) implies that $\dot{\bar{n}}_i$ and $\bar{n}_j$ are orthogonal to each other. Thus, based on (5.70) and (5.69), we can conclude that $\bar{W}_{ij}$, i.e the spin tensor, transforms $\bar{n}_j$ (a unit eigenvector of $[\bar{D}]$) into a vector $\dot{\bar{n}}_i$ normal to $\bar{n}_j$. Thus

$$\dot{\bar{n}}_i = \bar{W}_{ij} \bar{n}_j = \alpha \bar{n}'_i \tag{5.71}$$

where α is a constant and $\bar{n}'_i$ is a unit vector perpendicular to $\bar{n}_i$.

5.6 Vorticity vector and vorticity

Let

$$\bar{\boldsymbol{\omega}} = \bar{\boldsymbol{\nabla}} \times \bar{\boldsymbol{v}} = \text{curl}\bar{\boldsymbol{v}} \tag{5.72}$$

be the vorticity vector. Expanding the cross product

$$\bar{\boldsymbol{\omega}} = \boldsymbol{e}_j \frac{\partial}{\partial \bar{x}_j} \times \boldsymbol{e}_k \bar{v}_k = \boldsymbol{e}_j \times \boldsymbol{e}_k \frac{\partial \bar{v}_k}{\partial \bar{x}_j} = \epsilon_{ijk} \boldsymbol{e}_i \frac{\partial \bar{v}_k}{\partial \bar{x}_j} \tag{5.73}$$

$$\therefore \quad \bar{\omega}_i = \epsilon_{ijk} \bar{v}_{k,j} \tag{5.74}$$

Using the definition of permutation symbol and spin tensor, we can write (5.74) in the following form:

$$\bar{\omega}_i = -\epsilon_{ijk} \bar{W}_{jk} \tag{5.75}$$

Let $||\bar{\boldsymbol{\omega}}||$ be the magnitude of $\bar{\boldsymbol{\omega}}$. Then

$$||\bar{\boldsymbol{\omega}}|| = \sqrt{\bar{\omega}_i \bar{\omega}_i} \tag{5.76}$$

where

$$\bar{\omega}_i \bar{\omega}_i = (-\epsilon_{ijk} \bar{W}_{jk})(-\epsilon_{ilm} \bar{W}_{lm}) = \epsilon_{ijk} \epsilon_{ilm} \bar{W}_{jk} \bar{W}_{lm} \tag{5.77}$$

$$\therefore \quad \bar{\omega}_i \bar{\omega}_i = \epsilon_{jki} \epsilon_{lmi} \bar{W}_{jk} \bar{W}_{lm} \tag{5.78}$$

If we note the following identity

$$\epsilon_{jki} \epsilon_{lmi} = \delta_{jl} \delta_{km} - \delta_{jm} \delta_{kl} \tag{5.79}$$

and substitute from (5.79) into (5.78) we obtain

$$\bar{\omega}_i \bar{\omega}_i = \delta_{jl} \delta_{km} \bar{W}_{jk} \bar{W}_{lm} - \delta_{jm} \delta_{kl} \bar{W}_{jk} \bar{W}_{lm} \tag{5.80}$$

$$\bar{\omega}_i \bar{\omega}_i = \bar{W}_{jk} \bar{W}_{jk} - \bar{W}_{jk} \bar{W}_{kj} \tag{5.81}$$

But $\bar{W}_{kj} = -\bar{W}_{jk}$, hence

$$\bar{\omega}_i \bar{\omega}_i = 2\bar{W}_{jk}\bar{W}_{jk} = 2\bar{W}_{ij}\bar{W}_{ij} \tag{5.82}$$

$$\therefore \quad ||\bar{\boldsymbol{\omega}}|| = \sqrt{2\bar{W}_{ij}\bar{W}_{ij}} \tag{5.83}$$

From (5.75)

$$\bar{\omega}_k = -\epsilon_{kij}\bar{W}_{ij} = -\epsilon_{ijk}\bar{W}_{ij} \tag{5.84}$$

$$\epsilon_{ijk}\bar{\omega}_k = \epsilon_{ijk}(-\epsilon_{lmk}\bar{W}_{lm}) = -\epsilon_{kij}\epsilon_{klm}\bar{W}_{lm} \tag{5.85}$$

Using identity (5.79) and following the simplification, (5.85) reduces to

$$\epsilon_{ijk}\bar{\omega}_k = -2\bar{W}_{ij} \tag{5.86}$$

$$\therefore \quad \bar{W}_{ij} = -\frac{1}{2}\epsilon_{ijk}\bar{\omega}_k \tag{5.87}$$

Recall that

$$\dot{\bar{n}}_i = \bar{W}_{ij}\bar{n}_j \tag{5.88}$$

Substitute $\bar{W}_{ij}$ from (5.87) into (5.88)

$$\dot{\bar{n}}_i = \Big(-\frac{1}{2}\epsilon_{ijk}\bar{\omega}_k\Big)\bar{n}_j = -\frac{1}{2}\epsilon_{ijk}\bar{n}_j\bar{\omega}_k \tag{5.89}$$

We note that $\epsilon_{ijk} = -\epsilon_{ikj}$

$$\dot{\bar{n}}_i = \frac{1}{2}\epsilon_{ikj}\bar{n}_j\bar{\omega}_k = \frac{1}{2}\epsilon_{ijk}\bar{n}_k\bar{\omega}_j \tag{5.90}$$

Hence

$$2\dot{\bar{\boldsymbol{n}}} = \bar{\boldsymbol{\omega}} \times \bar{\boldsymbol{n}} \tag{5.91}$$

This is a fundamental relationship.

Remarks

(1) A motion for which $\bar{\boldsymbol{\omega}} = 0$ is called irrotational.

(2) Thus, for irrotational motion $\dot{\bar{\boldsymbol{n}}} = 0$, that is rate of change of $\bar{\boldsymbol{n}}$ is zero, hence the eigenvectors $\bar{\boldsymbol{n}}$ of the strain rate tensor are stationary at a spatial location or point. However, their orientation may change from point to point in the current configuration.

(3) In an irrotational motion $\bar{\boldsymbol{\omega}} = \bar{\boldsymbol{\nabla}} \times \bar{\boldsymbol{v}} = \text{curl}\bar{\boldsymbol{v}} = 0$.

5.7 Rate of change of $[J]$, i.e., material derivative of $[J]$

Since

$$[J] = \left[\frac{\partial\{\bar{x}\}}{\partial\{x\}}\right] \tag{5.92}$$

$$\frac{D[J]}{Dt} = \frac{D}{Dt}\left[\frac{\partial\{\bar{x}\}}{\partial\{x\}}\right] = \left[\frac{\partial\{\frac{D}{Dt}\{\bar{x}\}\}}{\partial\{x\}}\right] \tag{5.93}$$

$$= \left[\frac{\partial\{\bar{v}\}}{\partial\{x\}}\right] = \left[\frac{\partial\{\bar{v}\}}{\partial\{\bar{x}\}}\right]\left[\frac{\partial\{\bar{x}\}}{\partial\{x\}}\right] = [\bar{L}][J] \tag{5.94}$$

$$\therefore \quad \frac{D[J]}{Dt} = [\dot{J}] = [\bar{L}][J] \tag{5.95}$$

5.8 Rate of change of $[\bar{J}]$, i.e., material derivative of $[\bar{J}]$

We begin with

$$[J][\bar{J}] = [I] \tag{5.96}$$

Taking material derivative

$$[\dot{J}][\bar{J}] + [J][\dot{\bar{J}}] = 0 \tag{5.97}$$

$$\text{or} \quad [J][\dot{\bar{J}}] = -[\dot{J}][\bar{J}] \tag{5.98}$$

$$\text{or} \quad [\dot{\bar{J}}] = -[J]^{-1}[\dot{J}][\bar{J}] \tag{5.99}$$

Using $[\dot{J}] = [\bar{L}][J]$

$$[\dot{\bar{J}}] = -[\bar{J}][\bar{L}][J][\bar{J}] = -[\bar{J}][\bar{L}] \tag{5.100}$$

$$\therefore \quad [\dot{\bar{J}}] = -[\bar{J}][\bar{L}] \tag{5.101}$$

5.9 Rate of change of $\det[J]$, i.e material derivative of $|J|$

Recall that

$$[J] = \left[\frac{\partial\{\bar{x}\}}{\partial\{x\}}\right] = \left[\frac{\partial \bar{x}_i}{\partial x_j}\right] \tag{5.102}$$

and

$$|J| = \epsilon_{ijk} J_{1i} J_{2j} J_{3k} \tag{5.103}$$

$$|\dot{J}| = \frac{D|J|}{Dt} = \epsilon_{ijk}(\dot{J}_{1i} J_{2j} J_{3k} + J_{1i}\dot{J}_{2j} J_{3k} + J_{1i} J_{2j}\dot{J}_{3k}) \tag{5.104}$$

Consider $\epsilon_{ijk}\dot{J}_{1i} J_{2j} J_{3k}$

$$\epsilon_{ijk}\dot{J}_{1i} J_{2j} J_{3k} = \epsilon_{ijk}\frac{\partial \dot{\bar{x}}_1}{\partial x_i} J_{2j} J_{3k} \tag{5.105}$$

Since $\dot{\bar{x}}_1 = v_1$, then

$$\epsilon_{ijk}\dot{J}_{1i}J_{2j}J_{3k} = \epsilon_{ijk}\frac{\partial v_1}{\partial x_i}J_{2j}J_{3k} \tag{5.106}$$

Since $v_1(x_i,t) = \bar{v}_1(\bar{x}_i,t)$ we have

$$\epsilon_{ijk}\dot{J}_{1i}J_{2j}J_{3k} = \epsilon_{ijk}\left(\frac{\partial \bar{v}_1}{\partial \bar{x}_1}\frac{\partial \bar{x}_1}{\partial x_i} + \frac{\partial \bar{v}_1}{\partial \bar{x}_2}\frac{\partial \bar{x}_2}{\partial x_i} + \frac{\partial \bar{v}_1}{\partial \bar{x}_3}\frac{\partial \bar{x}_3}{\partial x_i}\right)J_{2j}J_{3k} \tag{5.107}$$

$$\text{or} \quad \epsilon_{ijk}\dot{J}_{1i}J_{2j}J_{3k} = \epsilon_{ijk}\left(\frac{\partial \bar{v}_1}{\partial \bar{x}_1}J_{1i} + \frac{\partial \bar{v}_1}{\partial \bar{x}_2}J_{2i} + \frac{\partial \bar{v}_1}{\partial \bar{x}_3}J_{3i}\right)J_{2j}J_{3k} \tag{5.108}$$

But

$$\epsilon_{ijk}J_{2i}J_{2j}J_{3k} = 0 \quad ; \quad \epsilon_{ijk}J_{3i}J_{2j}J_{3k} = 0 \tag{5.109}$$

Equation (5.109) holds because the determinant of a matrix is zero when any two rows or any two columns are the same. Therefore

$$\epsilon_{ijk}\dot{J}_{1i}J_{2j}J_{3k} = \epsilon_{ijk}\frac{\partial \bar{v}_1}{\partial \bar{x}_1}J_{1i}J_{2j}J_{3k} = \epsilon_{ijk}J_{1i}J_{2j}J_{3k}\frac{\partial \bar{v}_1}{\partial \bar{x}_1} \tag{5.110}$$

$$\epsilon_{ijk}\dot{J}_{1i}J_{2j}J_{3k} = |J|\,\frac{\partial \bar{v}_1}{\partial \bar{x}_1} \tag{5.111}$$

Similarly

$$\epsilon_{ijk}J_{1i}\dot{J}_{2j}J_{3k} = |J|\,\frac{\partial \bar{v}_2}{\partial \bar{x}_2} \quad ; \quad \epsilon_{ijk}J_{1i}J_{2j}\dot{J}_{3k} = |J|\,\frac{\partial \bar{v}_3}{\partial \bar{x}_3} \tag{5.112}$$

Substituting from (5.111)–(5.112) into (5.104)

$$|\dot{J}| = |J|\left(\frac{\partial \bar{v}_1}{\partial \bar{x}_1} + \frac{\partial \bar{v}_2}{\partial \bar{x}_2} + \frac{\partial \bar{v}_3}{\partial \bar{x}_3}\right) \tag{5.113}$$

$$\text{or} \quad |\dot{J}| = \frac{D}{Dt}(|J|) = |J|\,\frac{\partial \bar{v}_i}{\partial \bar{x}_i} = |J|\operatorname{tr}[\bar{D}] = |J|\,i_{\bar{D}} = |J|\operatorname{tr}[\bar{L}] = |J|\,\bar{\boldsymbol{\nabla}}\cdot\bar{\boldsymbol{v}} \tag{5.114}$$

5.10 Rate of change of det[$\bar{J}$], i.e., material derivative of det[$\bar{J}$]

Using

$$[J][\bar{J}] = [I] \tag{5.115}$$

$$\therefore \quad \det\left([J][\bar{J}]\right) = \det[I] \tag{5.116}$$

$$\text{or} \quad |J||\bar{J}| = 1 \tag{5.117}$$

Taking material derivative

$$|\dot{J}||\bar{J}| + |J||\dot{\bar{J}}| = 0 \tag{5.118}$$

$$\text{or} \quad |\dot{\bar{J}}| = -|J|^{-1}|\dot{J}||\bar{J}| \tag{5.119}$$

Using $|\dot{J}| = |J|\bar{\boldsymbol{\nabla}}\cdot\bar{\boldsymbol{v}}$

$$
\begin{aligned}
|\dot{\bar{J}}| &= -|J|^{-1}\left(|J|\left(\bar{\boldsymbol{\nabla}} \cdot \bar{\boldsymbol{v}}\right)\right)|\bar{J}| \\
&= -\left(|J|^{-1}|J|\right)|\bar{J}|\left(\bar{\boldsymbol{\nabla}} \cdot \bar{\boldsymbol{v}}\right)
\end{aligned}
\tag{5.120}
$$

$$
\therefore \quad |\dot{\bar{J}}| = -|\bar{J}|\bar{\boldsymbol{\nabla}} \cdot \bar{\boldsymbol{v}} = -|\bar{J}|\,\mathrm{tr}[\bar{L}] \tag{5.121}
$$

5.11 Rate of change of volume $\bar{V}$, i.e., material derivative of volume $\bar{V}$

In the following we consider material derivative, i.e., rate of change of volume $\bar{V}$ at a material point in the current configuration. Recall

$$
d\bar{V} = |J|\,dV \tag{5.122}
$$

$$
\therefore \quad \frac{D}{Dt}(d\bar{V}) = \frac{D}{Dt}(|J|\,dV) = \frac{D}{Dt}(|J|)dV = |\dot{J}|dV \tag{5.123}
$$

Substituting for $|\dot{J}|$ from (5.114)

$$
\frac{D}{Dt}(d\bar{V}) = |J|\,\frac{\partial \bar{v}_i}{\partial \bar{x}_i}dV = \frac{\partial \bar{v}_i}{\partial \bar{x}_i}(|J|\,dV) = \frac{\partial \bar{v}_i}{\partial \bar{x}_i}d\bar{V} \tag{5.124}
$$

$$
\therefore \quad \frac{D}{Dt}(d\bar{V}) = \frac{\partial \bar{v}_i}{\partial \bar{x}_i}d\bar{V} \tag{5.125}
$$

Remarks

(1) From (5.125), we note that

$$
\frac{1}{d\bar{V}}\frac{D}{Dt}(d\bar{V}) = \frac{\partial \bar{v}_i}{\partial \bar{x}_i} = \bar{D}_{ii} = i_{\bar{D}} = \mathrm{tr}[\bar{D}] = \mathrm{tr}[\bar{L}] \tag{5.126}
$$

where $i_{\bar{D}}$ is the first invariant of $[\bar{D}]$. It is also called volumetric strain rate.

(2) For isochoric deformation,

$$
|J| = 1 \tag{5.127}
$$

Hence

$$
|\dot{J}| = 0 \quad \text{and} \quad i_{\bar{D}} = 0 \tag{5.128}
$$

(3) Consider the components of the deviatoric strain rate tensor $[\bar{D}^d]$

$$
\bar{D}^d_{ij} = \bar{D}_{ij} - \frac{1}{3}\delta_{ij}\bar{D}_{kk} \tag{5.129}
$$

We note that

$$
\bar{D}^d_{ii} = \mathrm{tr}[\bar{D}^d] = i_{\bar{D}^d} \tag{5.130}
$$

The terms $\bar{D}^d_{ij}$ describe purely distortional deformation, that is deformation without change of volume.

(4) For isochoric deformation $i_{\bar{D}} = \bar{D}_{kk} = 0$, hence, from (5.129)

$$[\bar{D}^d] = [\bar{D}] \quad \text{or} \quad \bar{D}^d_{ij} = \bar{D}_{ij} \tag{5.131}$$

(5) Based on (4), for isochoric deformation ($|J| = 1$)

$$|\dot{J}| = |J|\, i_{\bar{D}} = i_{\bar{D}} = 0 \tag{5.132}$$

This is also obvious from $|J| = 1$ (as its material derivative is naturally zero).

5.12 Rate of change of area: material derivative of area

In the following we consider material derivative, i.e., rate of change of area $\bar{\boldsymbol{A}}$ at a material point in the current configuration. Recall that

$$\{d\bar{A}\} = |J|\,[J^{-1}]^T\{dA\} \tag{5.133}$$

$$\therefore \quad [J]^T\{d\bar{A}\} = |J|\,\{dA\} \tag{5.134}$$

Taking the material derivative of both sides,

$$\frac{D}{Dt}([J]^T\{d\bar{A}\}) = \frac{D}{Dt}(|J|\,\{dA\}) = |\dot{J}|\{dA\} \quad \text{as} \quad \frac{D}{Dt}\{dA\} = 0 \tag{5.135}$$

$$\therefore \quad |\dot{J}|\{dA\} = \frac{D}{Dt}([J]^T)\{d\bar{A}\} + [J]^T\frac{D}{Dt}(\{d\bar{A}\}) \tag{5.136}$$

Also

$$\frac{D}{Dt}[J] = [\bar{L}][J] \quad ; \quad |\dot{J}| = |J|\, i_{\bar{D}} \tag{5.137}$$

Substituting from (5.137) into (5.136)

$$|J|\, i_{\bar{D}}\{dA\} = [J]^T[\bar{L}]^T\{d\bar{A}\} + [J]^T\frac{D}{Dt}(\{d\bar{A}\}) \tag{5.138}$$

$$\therefore \quad [J]^T\frac{D}{Dt}(\{d\bar{A}\}) = |J|\, i_{\bar{D}}\{dA\} - [J]^T[\bar{L}]^T\{d\bar{A}\} \tag{5.139}$$

Premultiply by $[J^T]^{-1}$ and

$$\frac{D}{Dt}(\{d\bar{A}\}) = |J|\,[J^T]^{-1} i_{\bar{D}}\{dA\} - [\bar{L}]^T\{d\bar{A}\} \tag{5.140}$$

$$\text{or} \quad \frac{D}{Dt}(\{d\bar{A}\}) = (|J|\,[J^T]^{-1}\{dA\}) i_{\bar{D}} - [\bar{L}]^T\{d\bar{A}\} \tag{5.141}$$

$$\text{or} \quad \frac{D}{Dt}(\{d\bar{A}\}) = \{d\bar{A}\} i_{\bar{D}} - [\bar{L}]^T\{d\bar{A}\} \tag{5.142}$$

$$\therefore \quad \frac{D}{Dt}(\{d\bar{A}\}) = (i_{\bar{D}}[I] - [\bar{L}]^T)\{d\bar{A}\} \tag{5.143}$$

$$\begin{aligned} \text{or} \quad \frac{D}{Dt}(d\bar{A}_i) &= \left(i_{\bar{D}}\delta_{ij} - \frac{\partial \bar{v}_j}{\partial \bar{x}_i}\right) d\bar{A}_j = (i_{\bar{D}}\delta_{ij} - \bar{L}_{ji}) d\bar{A}_j \\ &= i_{\bar{D}}\delta_{ij} d\bar{A}_j - \bar{L}_{ji} d\bar{A}_j \end{aligned} \tag{5.144}$$

$$\begin{aligned} \therefore \quad & \frac{D}{Dt}(d\bar{A}_i) = \bar{D}_{kk} d\bar{A}_i - \bar{L}_{ji} d\bar{A}_j \\ \text{or} \quad & \frac{D}{Dt}\{d\bar{A}\} = (\text{tr}[\bar{D}])\{d\bar{A}\} - [\bar{L}]^T\{d\bar{A}\} \end{aligned} \tag{5.145}$$

Remarks

For isochoric deformation ($|J| = 1$)

$$\frac{D}{Dt}(d\bar{V}) = 0 \tag{5.146}$$

However

$$\frac{D}{Dt}(d\bar{A}_i) = -\bar{L}_{ji} d\bar{A}_j \quad ; \quad \text{need not vanish} \tag{5.147}$$

Equation (5.147) is obtained from (5.145) using $i_{\bar{D}} = 0$ for isochoric deformation.

5.13 Convected time derivatives of stress and strain tensors

In case of finite deformation and finite strain (as discussed in Chapter 4) a material point $\underset{\sim}{Q}$ with position x_i in the reference configuration translates (moves or deforms) to position $\underset{\sim}{\bar{Q}}$ in the current configuration with position coordinates $\bar{x}_i$. Orthogonal set of material lines at $\underset{\sim}{Q}$ become curvilinear at point $\underset{\sim}{\bar{Q}}$. This can be viewed as rotation of the material lines in the current configuration compared to those in the reference configuration. During continuous evolution of the deformation of the volume of matter, the translations (displacements) of $\underset{\sim}{\bar{Q}}$ are a function of time. The rotations of the material axes are also a function of time. Thus, to a stationary observer, the evolution of material point $\underset{\sim}{Q}$ in time is in a continuously translating and rotating frame in which the translations and rotations of the frame are functions of time at each material point. *This is a Euclidean transformation* [63] or *non-Galilean transformation.*

At material point $\underset{\sim}{\bar{Q}}$ the tangent vectors to the deformed material lines are covariant base vectors $\tilde{\boldsymbol{g}}_i$, hence are ideal to define measures of strain (e.g., Cauchy strain, Green's strain). We consider a tetrahedron located at $\underset{\sim}{\bar{Q}}$ with $\tilde{\boldsymbol{g}}_i$ as its edges and the tips of $\tilde{\boldsymbol{g}}_i$ forming its oblique plane. The vectors perpendicular to the faces of the tetrahedron (deformed) are contravariant base vectors $\tilde{\boldsymbol{g}}^i$. These form a contravariant basis. As seen in Chapter 2, covariant and contravariant bases are reciprocal to each other. The choice of $\tilde{\boldsymbol{g}}_i$ as basis for strain measures is natural as these are tangent to the deformed material lines. This necessitates that we use contravariant directions perpendicular to the faces of the deformed tetrahedron formed by $\tilde{\boldsymbol{g}}_i$ for Cauchy stress measure. Since $\tilde{\boldsymbol{g}}_i$ are reciprocal to $\tilde{\boldsymbol{g}}^i$, we could also use $\tilde{\boldsymbol{g}}^i$ for strain measure and $\tilde{\boldsymbol{g}}_i$ for stress measure. Mathematically this is plausible.

Using the deformed tetrahedron we define $\bar{\boldsymbol{\sigma}}^{(0)}$, $\boldsymbol{\sigma}^{(0)}$ as contravariant and $\bar{\boldsymbol{\sigma}}_{(0)}$, $\boldsymbol{\sigma}_{(0)}$ as covariant Cauchy stress tensors and likewise we define $\boldsymbol{\varepsilon}_{[0]}$ as covariant (Green's strain) and $\bar{\boldsymbol{\varepsilon}}^{[0]}$ as contravariant (Almansi strain) strain measures.

Obviously these stress and strain measures are crucial in describing the constitution of the matter.

Remarks

(1) At a material point, $\tilde{\boldsymbol{g}}_i$ and $\tilde{\boldsymbol{g}}^i$ directions are important as these relate to deformed material lines and definition of Cauchy stress tensor on the planes of the deformed tetrahedron.
(2) During evolution of the volume of matter, a material point is continuously translating in time and the orientation of the material lines are rotating (thus continuously changing $\tilde{\boldsymbol{g}}_i$ and $\tilde{\boldsymbol{g}}^i$). To a fixed observer this is non-Galilean transformation.
(3) If the constitutive theories require time rate of change of quantities at a material point, then these must be in covariant and contravariant directions. The rate of change of quantities in $\tilde{\boldsymbol{g}}_i$ and $\tilde{\boldsymbol{g}}^i$ directions are referred to as convected time derivatives in covariant and contravariant bases. Thus, we need to consider convected time derivatives of Cauchy stress tensor and convected strain tensors.
(4) We also note that Lagrangian description require transformation of $\bar{\boldsymbol{\sigma}}^{(0)}$ (or $\boldsymbol{\sigma}^{(0)}$) and $\bar{\boldsymbol{\sigma}}_{(0)}$ (or $\boldsymbol{\sigma}_{(0)}$) through a correspondence rule giving rise to first and second Piola-Kirchoff stress tensors.
(5) The convected time derivatives of a tensor (covariant and contravariant) are related to continuously translating and rotating frame, hence can also be derived using non-Galilean transformation (see Section 5.15.2.6–5.15.2.9).

5.13.1 Stress and strain measures for convected time derivatives

The definitions and measures of stresses and strains in contra- and co- variant bases are of interest because in these coordinate systems the deformed material lines are identified. If one chooses directions ox_1, ox_2 and ox_3 of the x-frame with base vectors $\boldsymbol{e}_1$, $\boldsymbol{e}_2$ and $\boldsymbol{e}_3$ then we can define a stress tensor $\boldsymbol{T}$ using base vectors $\boldsymbol{e}_i, \boldsymbol{e}_j$ and its components T_{ij} as (since x-frame is orthogonal)

$$\boldsymbol{T} = T_{ij}\, \boldsymbol{e}_i \boldsymbol{e}_j \tag{5.148}$$

On the other hand, a more natural way to define stresses is to use the current configuration and to consider the deformed volume, that is to consider the deformed tetrahedron. As shown in Chapter 4, there are many alternate ways to define stress measures. In the following we consider the second Piola-Kirchhoff stress tensor based on contra- and co-variant Cauchy stress measures. Thus, if $[\bar{T}^{(0)}]$ and $[\bar{T}_{(0)}]$ are contra- and co-variant Cauchy stress tensors, then the corresponding second Piola-Kirchhoff stress tensors $[\bar{T}^{[0]}]$ and $[\bar{T}_{[0]}]$ for compressible matter are given by

$$[\bar{T}^{[0]}] = [\bar{T}^{[0]}]^T = |J|\, [\bar{J}][\bar{T}^{(0)}]^T[\bar{J}]^T = |J|\, [\bar{J}][\bar{T}^{(0)}][\bar{J}]^T \quad \text{(Def.)} \tag{5.149}$$

$$[\bar{T}_{[0]}] = [\bar{T}_{[0]}]^T = |J|\, [J]^T[\bar{T}_{(0)}]^T[J] = |J|\, [J]^T[\bar{T}_{(0)}][J] \quad \text{(Def.)} \tag{5.150}$$

Both (5.149) and (5.150) also hold for incompressible matter with $|J| = 1$. We also consider convected time derivatives of the Jaumann stress tensor. The reason for choosing second Piola-Kirchhoff stress tensors to derive convective time derivatives of $[\bar{T}^{(0)}]$ and $[\bar{T}_{(0)}]$ is explained in Section 5.13.2.

For measures of strain we consider Green's strain tensor (a covariant measure) and Almansi strain tensor (a contravariant measure). These measures naturally reduce to the well known definitions of strain in linear theory of elasticity for infinitesimal deformation. A straightforward derivation of Green's strain and Almansi strain based on undeformed and deformed line segments ds and $d\bar{s}$ in the reference and current configurations using $[J]$ and $[\bar{J}]$ yields the following for Green's strain tensor $[\varepsilon_{[0]}]$ and Almansi strain tensor $[\bar{\varepsilon}^{[0]}]$ (see Chapter 3):

$$[\varepsilon_{[0]}] = \frac{1}{2}([J]^T[J] - [I]) \quad \text{(Def.)} \tag{5.151}$$

$$[\bar{\varepsilon}^{[0]}] = \frac{1}{2}([I] - [\bar{J}]^T[\bar{J}]) \tag{5.152}$$

In deriving strain measure $\boldsymbol{\varepsilon}_{[0]}$ we have used $[J]$. Columns of $[J]$ are $\tilde{\boldsymbol{g}}_i$. In case of $\bar{\boldsymbol{\varepsilon}}^{[0]}$ we use $[\bar{J}]$, the rows of $[\bar{J}]$ are $\tilde{\boldsymbol{g}}^i$, but both $\boldsymbol{J}$ and $\bar{\boldsymbol{J}}$ have same dyads $(\boldsymbol{e}_i \otimes \boldsymbol{e}_j)$. To correct this situation we must use a form of $[\bar{J}]$ in the definition of $[\bar{\varepsilon}^{[0]}]$ in which the columns are the contravariant base vectors. This of course can be done by using the transpose of $[\bar{J}]$ in place of $[\bar{J}]$ and vice versa. Thus, in (5.152), $[\bar{J}]$ must be replaced with $[\bar{J}]^T$ and $[\bar{J}]^T$ with $[\bar{J}]$, which yields the following definition of Almansi strain tensor:

$$[\bar{\varepsilon}^{[0]}] = \frac{1}{2}([I] - [\bar{J}][\bar{J}]^T) \quad \text{(Def.)} \tag{5.153}$$

Now, the columns of $[J]$ and $[\bar{J}]$ contain $\tilde{\boldsymbol{g}}_i$ and $\tilde{\boldsymbol{g}}^i$. In deriving convected rates we must ensure that definitions of $[\varepsilon_{[0]}]$ and $[\bar{\varepsilon}^{[0]}]$ in (5.151) and (5.153) are used instead of $[\bar{\varepsilon}^{[0]}]$ defined by (5.152). We note that relations $\{d\bar{x}\} = [J]\{dx\}$ and $\{dx\} = [\bar{J}]\{d\bar{x}\}$ have the assumption that infinitesimals of order two and higher in lengths $\{d\bar{x}\}$ can be neglected but $[J] = [\bar{J}]^{-1}$ and $[\bar{J}] = [J]^{-1}$ hold regardless of the magnitude of deformation. Jaumann strain rates are also derived using co- and contra-variant strain rates.

5.13.2 Convected time derivatives of the Cauchy stress tensor: compressible matter

To make the presentation general, we consider $[T]$ to be a symmetric stress tensor of rank two. Then, $[\bar{T}^{(0)}]$ and $[\bar{T}_{(0)}]$ are contravariant and covariant Cauchy stress tensors in Eulerian description. This choice of description becomes clear when we examine the derivations of the mathematical models presented in the following chapters. We are interested in convected time derivatives of up to orders n of $[\bar{T}^{(0)}]$ and $[\bar{T}_{(0)}]$ in contra- and co-variant bases. Using definitions of $[\bar{T}^{(0)}]$ and $[\bar{T}_{(0)}]$ in terms of their contravariant and covariant components and transformations by contravariance and covariance, it is not possible to determine $[\bar{T}^{(i)}]$ and $[\bar{T}_{(i)}]$; $i = 1, 2, \dots, n$. However, by taking material derivatives of $[\bar{T}^{[0]}]$, second Piola-Kirchhoff stress tensor based on contravariant Cauchy stress tensor, and of $[\bar{T}_{[0]}]$, second Piola-Kirchhoff stress tensor based on covariant Cauchy stress tensor, the convected time derivatives of $[\bar{T}^{(0)}]$ and $[\bar{T}_{(0)}]$ can be extracted. Recall the following from Sections 5.9 and 5.10

$$\frac{D}{Dt}[J] = [\bar{L}][J] \quad ; \quad \frac{D}{Dt}[J]^T = [J]^T[\bar{L}]^T \tag{5.154}$$

$$\frac{D}{Dt}[\bar{J}] = -[\bar{J}][\bar{L}] \quad ; \quad \frac{D}{Dt}[\bar{J}]^T = -[\bar{L}]^T[\bar{J}]^T \tag{5.155}$$

Equations (5.154) and (5.155) are the key expressions that are used in deriving convected time derivatives of the stress tensors $[\bar{T}^{(0)}]$ and $[\bar{T}_{(0)}]$. For incompressible matter $|J| = 1$.

We derive convected time derivatives of contravariant and covariant Cauchy stress tensors $[\bar{T}^{(0)}]$ and $[\bar{T}_{(0)}]$ for compressible matter. Recall (5.149) and (5.150),

$$[\bar{T}^{[0]}] = |J|[\bar{J}][\bar{T}^{(0)}][\bar{J}]^T \tag{5.156}$$

$$[\bar{T}_{[0]}] = |J|[J]^T[\bar{T}_{(0)}][J] \tag{5.157}$$

5.13.2.1 Contravariant measure: compressible

Consider the material derivative of $[\bar{T}^{[0]}]$ in (5.156), the second Piola-Kirchhoff stress tensor derived using contravariant Cauchy stress tensor $[\bar{T}^{(0)}]$.

$$\begin{aligned}\frac{D}{Dt}[\bar{T}^{[0]}] &= \frac{D}{Dt}\left(|J|\,[\bar{J}][\bar{T}^{(0)}][\bar{J}]^T\right)\\ &= \frac{D}{Dt}(|J|)[\bar{J}][\bar{T}^{(0)}][\bar{J}]^T + |J|\,\frac{D}{Dt}([\bar{J}])[\bar{T}^{(0)}][\bar{J}]^T\\ &+ |J|\,[\bar{J}]\frac{D}{Dt}([\bar{T}^{(0)}])[\bar{J}]^T + |J|\,[\bar{J}][\bar{T}^{(0)}]\frac{D}{Dt}([\bar{J}]^T)\end{aligned} \tag{5.158}$$

Substituting from (5.155) into (5.158) and noting that

$$\frac{D}{Dt}|J| = |J|\,\mathrm{tr}([\bar{L}]) \tag{5.159}$$

and regrouping the terms in (5.158)

$$\frac{D}{Dt}[\bar{T}^{[0]}] = |J|\,[\bar{J}]\left(\frac{D}{Dt}[\bar{T}^{(0)}] - [\bar{L}][\bar{T}^{(0)}] - [\bar{T}^{(0)}][\bar{L}]^T + [\bar{T}^{(0)}]\,\mathrm{tr}([\bar{L}])\right)[\bar{J}]^T \tag{5.160}$$

If we define

$$\frac{D}{Dt}[\bar{T}^{[0]}] = [\bar{T}^{[1]}] \quad \text{(Def.)} \tag{5.161}$$

$$[\bar{T}^{(1)}] = \frac{D}{Dt}[\bar{T}^{(0)}] - [\bar{L}][\bar{T}^{(0)}] - [\bar{T}^{(0)}][\bar{L}]^T + [\bar{T}^{(0)}]\,\mathrm{tr}([\bar{L}]) \quad \text{(Def.)} \tag{5.162}$$

then we obtain the following from (5.160):

$$[\bar{T}^{[1]}] = |J|[\bar{J}][\bar{T}^{(1)}][\bar{J}]^T \tag{5.163}$$

$[\bar{T}^{(1)}]$ is the first convected time derivative of the contravariant Cauchy stress $[\bar{T}^{(0)}]$.

To obtain the second convected time derivative of the contravariant Cauchy stress we take the material derivative of (5.163), and if we follow exactly the same steps as in the case of $[T^{[1]}]$, then we obtain the following:

$$[\bar{T}^{[2]}] = \frac{D}{Dt}[\bar{T}^{[1]}] = |J|[\bar{J}][\bar{T}^{(2)}][\bar{J}]^T \tag{5.164}$$

$$\text{where} \quad [\bar{T}^{(2)}] = \frac{D}{Dt}[\bar{T}^{(1)}] - [\bar{L}][\bar{T}^{(1)}] - [\bar{T}^{(1)}][\bar{L}]^T + [\bar{T}^{(1)}]\,\text{tr}([\bar{L}]) \tag{5.165}$$

In general, we can write the following recursive relations that can be used to obtain the convected time derivative of any desired order k of the contravariant Cauchy stress $[\bar{T}^{(0)}]$ for compressible matter:

$$\left.\begin{aligned} &\frac{D}{Dt}[\bar{T}^{[k-1]}] = [\bar{T}^{[k]}] \\ &[\bar{T}^{[k]}] = |J|[\bar{J}][\bar{T}^{(k)}][\bar{J}]^T \\ &[\bar{T}^{(k)}] = \frac{D}{Dt}[\bar{T}^{(k-1)}] - [\bar{L}][\bar{T}^{(k-1)}] - [\bar{T}^{(k-1)}][\bar{L}]^T \\ &\qquad + [\bar{T}^{(k-1)}]\,\text{tr}([\bar{L}]) \end{aligned}\right\} \quad k = 1, 2, \dots \tag{5.166}$$

The first convected time derivative $[\bar{T}^{(1)}]$ of the Cauchy stress tensor $[\bar{T}^{(0)}]$ is also referred to as *upper convected time derivative* and is denoted by $[\overset{\triangledown}{T}]$ or $\overset{\triangledown}{\frac{D}{Dt}}[T]$.

$$[\overset{\triangledown}{T}] = \overset{\triangledown}{\frac{D}{Dt}}[T] = [\bar{T}^{(1)}] \tag{5.167}$$

5.13.2.2 Covariant measure: compressible

In the following, we derive convected time derivatives of covariant Cauchy stress tensors $([\bar{T}_{(0)}])$ for compressible matter.

Consider the material derivative of $[\bar{T}_{[0]}]$, i.e., second Piola-Kirchhoff stress tensor derived using covariant Cauchy stress tensor $[\bar{T}_{(0)}]$.

$$\begin{aligned} \frac{D}{Dt}[\bar{T}_{[0]}] &= \frac{D}{Dt}\left(|J|\,[J]^T[\bar{T}_{(0)}][J]\right) \\ &= \frac{D}{Dt}(|J|)[J]^T[\bar{T}_{(0)}][J] + |J|\,\frac{D}{Dt}([J]^T)[\bar{T}_{(0)}][J] \\ &+ |J|\,[J]^T\frac{D}{Dt}([\bar{T}_{(0)}])[J] + |J|\,[J]^T[\bar{T}_{(0)}]\frac{D}{Dt}([J]) \end{aligned} \tag{5.168}$$

Substituting from (5.154) and (5.159) into (5.168) and regrouping the terms

$$\frac{D}{Dt}[\bar{T}_{[0]}] = |J|\,[J]^T\left(\frac{D}{Dt}[\bar{T}_{(0)}] + [\bar{L}]^T[\bar{T}_{(0)}] + [\bar{T}_{(0)}][\bar{L}] + [\bar{T}_{(0)}]\,\text{tr}([\bar{L}])\right)[J] \tag{5.169}$$

If we define

$$\frac{D}{Dt}[\bar{T}_{[0]}] = [\bar{T}_{[1]}] \quad \text{(Def.)} \tag{5.170}$$

$$[\bar{T}_{(1)}] = \frac{D}{Dt}[\bar{T}_{(0)}] + [\bar{L}]^T[\bar{T}_{(0)}] + [\bar{T}_{(0)}][\bar{L}] + [\bar{T}_{(0)}]\,\text{tr}([\bar{L}]) \quad \text{(Def.)} \tag{5.171}$$

then we obtain the following from (5.169):

$$[\bar{T}_{[1]}] = |J|[J]^T[\bar{T}_{(1)}][J] \tag{5.172}$$

where $[\bar{T}_{(1)}]$ is the first convected time derivative of the covariant Cauchy stress $[\bar{T}_{(0)}]$ or lower convected time derivative of the covariant Cauchy stress for compressible matter. To obtain the second convected time derivative of the covariant Cauchy stress we take the material derivative of (5.172), and if we follow exactly the same steps as in the case of $[T_{[1]}]$, we obtain the following:

$$[\bar{T}_{[2]}] = \frac{D}{Dt}[\bar{T}_{[1]}] = |J|[J]^T[\bar{T}_{(2)}][J] \tag{5.173}$$

$$\text{where} \quad [\bar{T}_{(2)}] = \frac{D}{Dt}[\bar{T}_{(1)}] + [\bar{L}]^T[\bar{T}_{(1)}] + [\bar{T}_{(1)}][\bar{L}] + [\bar{T}_{(1)}]\,\text{tr}([\bar{L}]) \tag{5.174}$$

In general, we can write the following recursive relations that can be used to obtain the convected time derivative of any desired order k of the covariant Cauchy stress $[\bar{T}_{(0)}]$ for compressible matter:

$$\left.\begin{aligned} &\frac{D}{Dt}[\bar{T}_{[k-1]}] = [\bar{T}_{[k]}] \\ &\quad [\bar{T}_{[k]}] = |J|[J]^T[\bar{T}_{(k)}][J] \\ &\quad [\bar{T}_{(k)}] = \frac{D}{Dt}[\bar{T}_{(k-1)}] + [\bar{L}]^T[\bar{T}_{(k-1)}] + [\bar{T}_{(k-1)}][\bar{L}] \\ &\qquad\quad + [\bar{T}_{(k-1)}]\,\text{tr}([\bar{L}]) \end{aligned}\right\} \quad k = 1, 2, \dots \tag{5.175}$$

The first convected time derivative $[\bar{T}_{(1)}]$ of the Cauchy stress tensor $[\bar{T}_{(0)}]$ is also referred to as *lower convected time derivative* and is denoted by $[\overset{\Delta}{T}]$ or $\frac{\overset{\Delta}{D}}{Dt}[T]$.

$$[\overset{\Delta}{T}] = \frac{\overset{\Delta}{D}}{Dt}[T] = [\bar{T}_{(1)}] \tag{5.176}$$

5.13.2.3 Jaumann stress rates: compressible

Jaumann rates are the average of contra- and co-variant convected time derivatives, when $[\bar{T}^{(k)}] = [\bar{T}_{(k)}]$ (for small strain, small deformation).

$$[{}^{(k)}\bar{T}^J] = \frac{1}{2}([\bar{T}^{(k)}] + [\bar{T}_{(k)}]) \quad ; \quad [{}^{(k)}\bar{T}^J] = [\bar{T}^{(k)}] = [\bar{T}_{(k)}] \quad k = 1, 2, \dots \tag{5.177}$$

Substituting for $[\bar{T}^{(k)}]$ and $[\bar{T}_{(k)}]$ from (5.166) and (5.175) in (5.177), assuming the velocity field to be the same in $[\bar{T}^{(k)}]$ and $[\bar{T}_{(k)}]$ definitions (i.e., same $[\bar{L}]$), and using the definition of $[\bar{W}]$, we can derive the following:

$$\begin{aligned} [{}^{(k)}\bar{T}^J] =& \frac{D}{Dt}[{}^{(k-1)}\bar{T}^J] - [\bar{W}][{}^{(k-1)}\bar{T}^J] + [{}^{(k-1)}\bar{T}^J][\bar{W}] \\ &+ [{}^{(k-1)}\bar{T}^J]\,\text{tr}([\bar{L}]) \quad ; \quad k = 1, 2, \dots \end{aligned} \tag{5.178}$$

The first convected time derivative of the Jaumann stress tensor, i.e., the Jaumann stress rate for compressible matter, can be written as

$$\frac{{}^J D}{Dt}[T] = [{}^{(1)}\bar{T}^J] = \frac{D}{Dt}[\bar{T}] - [\bar{W}][\bar{T}] + [\bar{T}][\bar{W}] + [\bar{T}]\,\text{tr}([\bar{L}]) \tag{5.179}$$

in which $\frac{{}^J D}{Dt}[T]$ is the Jaumann stress rate of the Jaumann stress tensor $[T]$ for compressible matter. The terms $[{}^{(k)}\bar{T}^J]$; $k = 1, 2, \dots$ are the convected time derivatives of orders $1, 2, \dots$ of the Jaumann stress tensor $[{}^{(0)}\bar{T}^J]$.

5.13.3 Convected time derivatives of the Cauchy stress tensor: incompressible matter

For incompressible matter

$$|J| = 1 \text{ and } \operatorname{tr}[\bar{L}] = 0 \tag{5.180}$$

Thus, (5.156) and (5.157) for $[\bar{T}^{(0)}]$ and $[\bar{T}_{(0)}]$ reduce to the following

$$[\bar{T}^{[0]}] = [\bar{J}][\bar{T}^{(0)}][\bar{J}]^T \tag{5.181}$$

$$[\bar{T}_{[0]}] = [J]^T[\bar{T}_{(0)}][J] \tag{5.182}$$

Using (5.181) and (5.182) and following the same procedure as in Section 5.13.2 we can easily derive the convected time derivative of $[\bar{T}^{(0)}]$ and $[\bar{T}_{(0)}]$ for incompressible matter as well as Jaumann stress rates.

The other simpler alternative is to use convected time derivatives of $[\bar{T}^{(0)}]$, $[\bar{T}_{(0)}]$ for compressible matter and the Jaumann rates in Section 5.13.2 and introduce $|J| = 1$ and $\operatorname{tr}[\bar{L}] = 0$ to obtain the convected time derivatives of contravariant and covariant Cauchy stress tensors and Jaumann stress rates for incompressible matter.

5.13.3.1 Contravariant measure: incompressible matter

In (5.166), we substitute $|J| = 1$ and $\operatorname{tr}[\bar{L}] = 0$ to obtain the convected time derivatives of the contravariant Cauchy stress tensor $[\bar{T}^{(0)}]$ of orders $1, 2, \dots$.

$$\left.\begin{aligned} &\frac{D}{Dt}[\bar{T}^{[k-1]}] = [\bar{T}^{[k]}] \\ &[\bar{T}^{[k]}] = [\bar{J}][\bar{T}^{(k)}][\bar{J}]^T \\ &[\bar{T}^{(k)}] = \frac{D}{Dt}[\bar{T}^{(k-1)}] - [\bar{L}][\bar{T}^{(k-1)}] - [\bar{T}^{(k-1)}][\bar{L}]^T \end{aligned}\right\} \quad k = 1, 2, \dots \tag{5.183}$$

As in compressible matter, in this case also the first convected time derivative of $[\bar{T}^{(0)}]$ is called *upper convected time derivative* and is denoted by $\overset{\triangledown}{[\bar{T}]}$ or $\frac{\overset{\triangledown}{D}}{Dt}[\bar{T}]$

$$\overset{\triangledown}{[T]} = \frac{\overset{\triangledown}{D}}{Dt}[T] = [\bar{T}^{(1)}] \tag{5.184}$$

5.13.3.2 Covariant measure: incompressible matter

In (5.175), we can substitute $|J| = 1$ and $\operatorname{tr}[\bar{L}] = 0$ to obtain the convected time derivatives of covariant Cauchy stress tensor $[\bar{T}_{(0)}]$ for incompressible matter.

$$\left.\begin{aligned} &\frac{D}{Dt}[\bar{T}_{[k-1]}] = [\bar{T}_{[k]}] \\ &[\bar{T}_{[k]}] = [J]^T[\bar{T}_{(k)}][J] \\ &[\bar{T}_{(k)}] = \frac{D}{Dt}[\bar{T}_{(k-1)}] + [\bar{L}]^T[\bar{T}_{(k-1)}] + [\bar{T}_{(k-1)}][\bar{L}] \end{aligned}\right\} \quad k = 1, 2, \dots \tag{5.185}$$

The first convected time derivative of $[\bar{T}_{(0)}]$ is called *lower convected time derivative* and is denoted by $[\overset{\Delta}{\bar{T}}]$ or $\frac{\overset{\Delta}{D}}{Dt}[\bar{T}]$

$$[\overset{\Delta}{\bar{T}}] = \frac{\overset{\Delta}{D}}{Dt}[T] = [\bar{T}_{(1)}] \tag{5.186}$$

5.13.3.3 Jaumann stress rates: incompressible matter

In (5.178), we set $|J| = 1$ and $\text{tr}[\bar{L}] = 0$ to obtain Jaumann stress rates of orders $1, 2, \dots$

$$[{}^{(k)}\bar{T}^J] = \frac{D}{Dt}[{}^{(k-1)}\bar{T}^J] - [\bar{W}][{}^{(k-1)}\bar{T}^J] + [{}^{(k-1)}\bar{T}^J][\bar{W}] \quad ; \quad k = 1, 2, \dots \tag{5.187}$$

The first convected time derivative of Jaumann stress, i.e., Jaumann stress rate for incompressible matter can be written as

$$\frac{{}^J D}{Dt}[T] = \frac{D}{Dt}[\bar{T}] - [\bar{W}][\bar{T}] + [\bar{T}][\bar{W}] \tag{5.188}$$

5.13.4 Remarks

(1) Contravariant Cauchy stress measure $([\bar{T}^{(0)}])$ is derived using faces of the deformed tetrahedron in the current configuration. Whereas the covariant Cauchy stress tensors $([\bar{T}_{(0)}])$ uses a new tetrahedron in the deformed configuration such that covariant directions are perpendicular to the faces of this tetrahedron. Thus, the difference in the convected time derivatives of $([\bar{T}^{(0)}])$ and $([\bar{T}_{(0)}])$ is obviously expected. The deformation behavior obtained using contravariant and covariant convected time derivatives will obviously differ also.

(2) With increasing strain rates, the convected time derivatives of the covariant Cauchy stress measure will yield progressively more non-physical behaviors. For finite deformation and finite strain rates, only the convected time derivatives of the contravariant Cauchy stress tensor remain valid.

(3) Since the covariant Cauchy strain measures use a tetrahedron different than actual deformed tetrahedron, perhaps it is possible to adjust convected (upper convected) time derivatives of the contravariant Cauchy stress tensor $([\bar{T}^{(0)}])$ to obtain the first convected time derivatives (lower convected) of the covariant Cauchy stress tensor $([\bar{T}_{(0)}])$. Consider incompressible matter. If we consider the upper convected stress rate (5.184) and add $2[\bar{D}][T] + 2[T][\bar{D}]$ to the right side and change the meaning of $[T]$ to covariant measure we can write

$$\frac{\overset{\Delta}{D}}{Dt}[T] = \frac{D}{Dt}[T] - [\bar{L}][T] - [T][\bar{L}]^T + 2[\bar{D}][T] + 2[T][\bar{D}] \tag{5.189}$$

substituting $[\bar{D}] = \frac{1}{2}([\bar{L}] + [\bar{L}]^T)$ in (5.189) yields

$$\frac{\overset{\Delta}{D}}{Dt}[T] = \frac{D}{Dt}[T] + [\bar{L}]^T[T] + [T][\bar{L}] \tag{5.190}$$

which is the same as (5.185). Thus, the lower convected stress rate requires further deformation or distortion of the deformed tetrahedron. We note that the term added to the right side of (5.184) only contains $[\bar{D}]$ and not $[\bar{W}]$, hence the rotation of the actual deformed tetrahedron is precluded. This deformed configuration of the tetrahedron used in describing the lower convected stress rate is of course non-physical.

(4) We note that convected time derivatives are in the contravariant or covariant direction that are continuously changing during the evolution of the deforming matter. Thus, the convected time derivatives are in a continuously translating and rotating coordinate system. This suggests that these can also be derived by considering a coordinate transformation on a tensor of rank two in which the coordinate transformation matrix $\boldsymbol{Q}$ is a function of time, i.e using non-Galilean transformation. Derivation of first convected time derivatives of a symmetric contravariant or covariant tensor and Jaumann tensor using non-Galilean transformation is presented in Section 5.15.3.

(5) In Section 5.15.3, it is shown that the first convected time derivatives are objective. The objectivity proofs for first order convected time derivatives given in Section 5.15.3 can be easily extended to show that convected time derivatives of all orders (of contravariant, covariant and Jaumann tensors) are objective.

(6) In the derivations presented above, the second rank tensors are related to the physics of compressibility or incompressibility, thus, we see differences in the convected time derivatives in the two cases. Whereas in Section 5.15.3, the derivation is mathematical ($\bar{\boldsymbol{T}}$ is simply a tensor of rank two with no physics associated with it). Hence, this only holds for incompressible matter.

5.13.5 Convected time derivatives of the strain tensors

5.13.5.1 Covariant measure

Consider Green's strain tensor $[\varepsilon_{[0]}]$

$$[\varepsilon_{[0]}] = \frac{1}{2}\left([J]^T[J] - [I]\right) \tag{5.191}$$

Taking material derivative

$$\frac{D}{Dt}[\varepsilon_{[0]}] = \frac{1}{2}\Big(\frac{D}{Dt}([J]^T)[J] + [J]^T\frac{D}{Dt}[J]\Big) \tag{5.192}$$

Substituting from (5.154) and defining $[\gamma_{[1]}]$

$$[\gamma_{[1]}] = \frac{D}{Dt}[\varepsilon_{[0]}] = \frac{1}{2}\left([J]^T[\bar{L}]^T[J] + [J]^T[\bar{L}][J]\right) \tag{5.193}$$

$$\text{or} \quad [\gamma_{[1]}] = \frac{D}{Dt}[\varepsilon_{[0]}] = [J]^T\frac{1}{2}([\bar{L}]^T + [\bar{L}])[J] = [J]^T[\bar{D}][J] \tag{5.194}$$

$$\therefore \quad [\gamma_{(1)}] = \frac{1}{2}([\bar{L}]^T + [\bar{L}]) = [\bar{D}] \qquad \text{(Def.)} \tag{5.195}$$

The term $[\gamma_{(1)}]$ is the first convected time derivative of the covariant strain tensor $[\varepsilon_{[0]}]$. We can also obtain higher order convected time derivatives of $[\varepsilon_{[0]}]$ using a procedure similar to that used earlier in Section 5.13.2 for stresses, using $[\gamma_{[1]}]$. For example,

$$[\gamma_{[2]}] = \frac{D}{Dt}[\gamma_{[1]}] = \frac{D}{Dt}([J]^T[\gamma_{(1)}][J]) \tag{5.196}$$

$$\begin{aligned}[\gamma_{[2]}] =& \frac{D}{Dt}[\gamma_{[1]}] = [J]^T\frac{D}{Dt}([\gamma_{(1)}])[J] + \frac{D}{Dt}([J]^T)[\gamma_{(1)}][J] \\ &+ [J^T][\gamma_{(1)}]\frac{D}{Dt}[J]\end{aligned} \tag{5.197}$$

Substituting from (5.154) into (5.197), rearranging and grouping terms

$$[\gamma_{[2]}] = \frac{D}{Dt}[\gamma_{[1]}] = [J]^T\Big(\frac{D}{Dt}[\gamma_{(1)}] + [\bar{L}]^T[\gamma_{(1)}] + [\gamma_{(1)}][\bar{L}]\Big)[J] \tag{5.198}$$

If we define

$$[\gamma_{(2)}] = \frac{D}{Dt}[\gamma_{(1)}] + [\bar{L}]^T[\gamma_{(1)}] + [\gamma_{(1)}][\bar{L}] \qquad \text{(Def.)} \tag{5.199}$$

then we obtain the following:

$$[\gamma_{[2]}] = [J]^T[\gamma_{(2)}][J] \tag{5.200}$$

where $[\gamma_{(2)}]$ is the second convected time derivative of the covariant strain tensor $[\varepsilon_{[0]}]$. This procedure can be used to obtain convected time derivatives of the covariant strain tensor $[\varepsilon_{[0]}]$ of any desired order. In general we can write the following recursive relations that can be used to obtain convected time derivatives of any desired order k of the Green's strain tensor $[\varepsilon_{[0]}]$:

$$\left.\begin{aligned}&\frac{D}{Dt}[\gamma_{[k-1]}] = [\gamma_{[k]}] \\ &[\gamma_{[k]}] = [J]^T[\gamma_{(k)}][J] \\ &[\gamma_{(k)}] = \frac{D}{Dt}[\gamma_{(k-1)}] + [\bar{L}]^T[\gamma_{(k-1)}] + [\gamma_{(k-1)}][\bar{L}]\end{aligned}\right\} \quad k = 2, 3, \dots \tag{5.201}$$

where

$$[\gamma_{[1]}] = \frac{D}{Dt}[\varepsilon_{[0]}] = [J]^T[\gamma_{(1)}][J] \tag{5.202}$$

$$[\gamma_{(1)}] = \frac{1}{2}\big([\bar{L}] + [\bar{L}]^T\big) = [\bar{D}] \tag{5.203}$$

5.13.5.2 Contravariant measure

Next, we consider contravariant strain measure (5.153), i.e., Almansi strain tensor.

$$[\bar{\varepsilon}^{[0]}] = \frac{1}{2}\left([I] - [\bar{J}][\bar{J}]^T\right)$$

Taking its material derivative

$$\frac{D}{Dt}[\bar{\varepsilon}^{[0]}] = -\frac{1}{2}\Big(\frac{D}{Dt}([\bar{J}])[\bar{J}]^T + [\bar{J}]\frac{D}{Dt}[\bar{J}]^T\Big) \tag{5.204}$$

Substituting from (5.155) in (5.204) and defining $[\gamma^{[1]}]$

$$[\gamma^{[1]}] = \frac{D}{Dt}[\bar{\varepsilon}^{[0]}] = \frac{1}{2}\left([\bar{J}][\bar{L}][\bar{J}]^T + [\bar{J}][\bar{L}]^T[\bar{J}]^T\right) \tag{5.205}$$

$$\text{or} \quad [\gamma^{[1]}] = \frac{D}{Dt}[\bar{\varepsilon}^{[0]}] = [\bar{J}]\frac{1}{2}([\bar{L}] + [\bar{L}]^T)[\bar{J}]^T = [\bar{J}][\bar{D}][\bar{J}]^T \tag{5.206}$$

$$[\gamma^{(1)}] = \frac{1}{2}([\bar{L}] + [\bar{L}]^T) = [\bar{D}] \qquad \text{(Def.)} \tag{5.207}$$

The term $[\gamma^{(1)}]$ is the first convected time derivative of the contravariant strain tensor $[\bar{\varepsilon}^{[0]}]$. We note that $[\gamma^{(1)}] = [\gamma_{(1)}]$. We can also define higher order convected time derivatives of $[\bar{\varepsilon}^{[0]}]$ using a procedure similar to that used for the covariant case. For example,

$$[\gamma^{[2]}] = \frac{D}{Dt}[\gamma^{[1]}] = \frac{D}{Dt}([\bar{J}][\gamma^{(1)}][\bar{J}]^T) \tag{5.208}$$

$$\begin{aligned}[\gamma^{[2]}] = \frac{D}{Dt}[\gamma^{[1]}] =& [\bar{J}]\frac{D}{Dt}([\gamma^{(1)}])[\bar{J}]^T + \frac{D}{Dt}([\bar{J}])[\gamma^{(1)}][\bar{J}]^T \\ &+ [\bar{J}][\gamma^{(1)}]\frac{D}{Dt}[\bar{J}]^T\end{aligned} \tag{5.209}$$

Substituting from (5.155) into (5.209), rearranging and regrouping terms

$$[\gamma^{[2]}] = \frac{D}{Dt}[\gamma^{[1]}] = [\bar{J}]\left(\frac{D}{Dt}[\gamma^{(1)}] - [\bar{L}][\gamma^{(1)}] - [\gamma^{(1)}][\bar{L}]^T\right)[\bar{J}]^T \tag{5.210}$$

If we define

$$[\gamma^{(2)}] = \frac{D}{Dt}[\gamma^{(1)}] - [\bar{L}][\gamma^{(1)}] - [\gamma^{(1)}][\bar{L}]^T \qquad \text{(Def.)} \tag{5.211}$$

then we obtain

$$[\gamma^{[2]}] = [\bar{J}][\gamma^{(2)}][\bar{J}]^T \tag{5.212}$$

where $[\gamma^{(2)}]$ is the second convected time derivative of the contravariant strain tensor $[\bar{\varepsilon}^{[0]}]$. This procedure can be used to obtain contravariant convected time derivatives of tensor $[\bar{\varepsilon}^{[0]}]$ of any desired order. In general we can write the following recursive relations that can be used to obtain convected time derivatives of any desired order k of the Almansi strain tensor $[\bar{\varepsilon}^{[0]}]$:

$$\left.\begin{aligned}&\frac{D}{Dt}[\gamma^{[k-1]}] = [\gamma^{[k]}] \\ &[\gamma^{[k]}] = [\bar{J}][\gamma^{(k)}][\bar{J}]^T \\ &[\gamma^{(k)}] = \frac{D}{Dt}[\gamma^{(k-1)}] - [\bar{L}][\gamma^{(k-1)}] - [\gamma^{(k-1)}][\bar{L}]^T\end{aligned}\right\} k = 2, 3, \dots \tag{5.213}$$

with

$$[\gamma^{[1]}] = \frac{D}{Dt}[\bar{\varepsilon}^{[0]}] = [\bar{J}][\gamma^{(1)}][\bar{J}]^T \tag{5.214}$$

$$[\gamma^{(1)}] = \frac{1}{2}([\bar{L}] + [\bar{L}]^T) = [\bar{D}] \tag{5.215}$$

where $[\gamma_{(i)}]$ and $[\gamma^{(i)}]$; $i = 1, 2, \dots$ are fundamental kinematic tensors in contra- and co-variant bases. These are convected time derivatives of orders one, two and so on of the Green's strain and Almansi strain tensors in co- and contra-variant bases. These fundamental kinematic tensors (and others defined later) form the basis for deriving the rate constitutive theories for both incompressible and compressible matter in the Eulerian descriptions.

5.13.5.3 Jaumann strain rates

The strain rates conjugate to the Jaumann stress rates of various orders are the average of the convected rates of $[\varepsilon_{[0]}]$ and $[\bar{\varepsilon}^{[0]}]$. This gives us the following for Jaumann rates (only valid for small deformation).

$$[^{(k)}\gamma^J] = \frac{1}{2}([\gamma^{(k)}] + [\gamma_{(k)}]) \quad ; \quad [^{(k)}\gamma^J] = [\gamma^{(k)}] = [\gamma_{(k)}] \quad k = 1, 2, \dots \tag{5.216}$$

Substituting for $[\gamma^{(k)}]$ and $[\gamma_{(k)}]$ from (5.201) and (5.213) in (5.216) we obtain the following:

$$[^{(k)}\gamma^J] = \frac{D}{Dt}[^{(k-1)}\gamma^J] - [\bar{W}][^{(k-1)}\gamma^J] + [^{(k-1)}\gamma^J][\bar{W}] \; ; \; k = 2, 3, \dots \tag{5.217}$$

with

$$[^{(1)}\gamma^J] = \frac{1}{2}\left([\gamma^{(1)}] + [\gamma_{(1)}]\right) = \frac{1}{2}([\bar{D}] + [\bar{D}]^T) = [\bar{D}] \tag{5.218}$$

In (5.217), $[^{(k)}\gamma^J]$; $k = 1, 2, \dots$ are the Jaumann strain rates compatible with Jaumann stress rates.

5.14 Conjugate pairs of convected time derivatives of stress and strain tensors

Based on the undeformed and deformed tetrahedra and the definitions of various stress and strain measures we can conclude that covariant strain measures are conjugate to contravariant stress measures and vice-versa. The same holds in the case of convected time derivatives of the stress and strain tensors in contra- and co-variant bases. We summarize these in the following.

(1) $[\bar{T}^{(j)}]$; $j = 0, 1, \dots, m$ and $[\gamma_{(k)}]$; $k = 1, 2, \dots, n$ are conjugate. $[\bar{T}^{(j)}]$ are in contravariant basis and $[\gamma_{(k)}]$ are in covariant basis.

(2) $[\bar{T}_{(j)}]$; $j = 0, 1, \dots, m$ and $[\gamma^{(k)}]$; $k = 1, 2, \dots, n$ are conjugate. $[\bar{T}_{(j)}]$ are in covariant basis and $[\gamma^{(k)}]$ are in contravariant basis.

(3) $[^{(k)}\bar{T}^J]$; $k = 0, 1, \dots, m$ and $[^{(k)}\gamma^J]$; $k = 1, 2, \dots, n$ are conjugate Jaumann rates corresponding to stress and strain measures.

Conjugate convected rates of stress and strain tensors play an important role in the development of the rate constitutive theories.

5.15 Objective tensors and objective rates

Objective tensors are independent of the observer. If the quantities observed by an observer O in the current configuration and the same quantities observed by a different observer O^* are identical, then the quantities observed are objective, i.e., they are independent of the observer. An observer in the x-frame views coordinates of a material point in the current configuration as $\bar{x}_i$. Another observer O^* in the frame obtained by rigid translation and rigid rotation of x-frame observes the coordinates of the same material point as $\bar{x}_i^*$. As shown earlier (Chapter 3) $\bar{\boldsymbol{x}}$ and $\bar{\boldsymbol{x}}^*$ are related by the following in the non-inertial frame (Euclidean transformation or non-Galilean transformation).

$$\{\bar{x}^*\} = \{c(t)\} + [Q(t)]\{\bar{x}(t)\}] \tag{5.219}$$

In inertial frame (Galilean transformations)

$$\{\bar{x}^*\} = \{c\} + [Q]\{\bar{x}\} \tag{5.220}$$

in which $[Q(t)]$ and $[Q]$ are orthogonal rotation tensors and $\{c(t)\}$ and $\{c\}$ are translation vectors.

Definition of objective tensors

A tensor of rank zero $\bar{f}$, a tensor of rank one $\bar{\boldsymbol{v}}$ and a second rank tensor $\bar{\boldsymbol{T}}$ are objective if for the reference frames $(\bar{O}, \bar{\boldsymbol{x}})$ and $(\bar{O}^*, \bar{\boldsymbol{x}}^*)$ related by (5.219) or (5.220), the corresponding tensor of rank zero $\bar{f}^*$, tensor of rank one $\bar{\boldsymbol{v}}^*$ and the tensor of rank two $\bar{\boldsymbol{T}}^*$ are related by

$$\bar{f}^* = \bar{f} \quad \text{(scalar or tensor of rank zero)} \tag{5.221}$$

$$\bar{\boldsymbol{v}}^* = \bar{\boldsymbol{Q}} \cdot \bar{\boldsymbol{v}} \quad \text{(vector or tensor of rank one)} \tag{5.222}$$

$$\bar{\boldsymbol{T}}^* = \boldsymbol{Q} \cdot \bar{\boldsymbol{T}} \cdot \boldsymbol{Q}^T \quad \text{(tensor of rank two)} \tag{5.223}$$

In the following we consider transformation of tensors $\bar{\boldsymbol{u}}$, $\bar{\boldsymbol{v}}$, $\boldsymbol{J}$, $\bar{\boldsymbol{L}}$, $\bar{\boldsymbol{D}}$, $\bar{\boldsymbol{W}}$, $\boldsymbol{C}_{[0]}$, $\boldsymbol{F}_{[0]}$, $\boldsymbol{\varepsilon}_{[0]}$, $\boldsymbol{B}^{[0]}$, $\boldsymbol{S}_r$, $\boldsymbol{S}_l$, $\boldsymbol{R}$ etc. due to Galilean and non-Galilean transformations defined by (5.219) and (5.220). We recall that in Chapter 2, Section 2.10.2 transformations of tensors of rank $0, 1, \ldots$ due to rigid rotation of the frame have been presented to illustrate how the tensors of various ranks transform due to rigid rotation $[Q]$ of the frame. In Chapter 3, the influence of rigid rotation of the frame on strain tensors have been derived from the first principles. It was shown that $\boldsymbol{C}_{[0]}$, $\boldsymbol{F}_{[0]}$, $\boldsymbol{\varepsilon}_{[0]}$, tensors of rank two in Lagrangian description are effected if $\boldsymbol{x}$-coordinates are changed to $\boldsymbol{x}'$-coordinates by rigid rotation, but $\bar{\boldsymbol{C}}^{[0]}$, $\bar{\boldsymbol{F}}^{[0]}$, $\bar{\boldsymbol{\varepsilon}}^{[0]}$ remain unaffected. On the other hand if $\bar{\boldsymbol{x}}$ coordinates are changed to $\bar{\boldsymbol{x}}'$ through a rigid rotation, then $\bar{\boldsymbol{C}}^{[0]}$, $\bar{\boldsymbol{F}}^{[0]}$, $\bar{\boldsymbol{\varepsilon}}^{[0]}$ measures are affected but $\boldsymbol{C}_{[0]}$, $\boldsymbol{F}_{[0]}$, $\boldsymbol{\varepsilon}_{[0]}$ remain unaffected. Some of the derivations presented in the following result in the same conclusions as in Chapters 2 and 3, but the approach used in the following is different and is more compact and general.

5.15.1 Galilean transformation: tensors

Without loss of generality, we can assume $\{c\} = 0$ in (5.220) giving us

$$\{\bar{x}^*\} = [Q]\{\bar{x}\} \tag{5.224}$$

The bases in the two frames are related by

$$\{\mathbf{e}^*\} = [Q]\{\mathbf{e}\} \tag{5.225}$$

5.15.1.1 Displacement vector: $\boldsymbol{u}$

Displacement vector $\{\bar{u}\}$ is a tensor of rank one. Let $\{\bar{u}(\bar{\boldsymbol{x}})\}$ and $\{\bar{u}^*(\bar{\boldsymbol{x}})\}$ be displacement vectors in two frames then

$$\bar{\boldsymbol{u}} = \{\mathbf{e}\}^T\{\bar{u}\} \quad ; \quad \bar{\boldsymbol{u}}^* = \{\mathbf{e}^*\}^T\{\bar{u}^*\} \tag{5.226}$$

$$\therefore \quad \bar{\boldsymbol{u}}^* = (\{\mathbf{e}\}^T[Q]^T)\{\bar{u}^*\} = \{\mathbf{e}\}^T([Q]^T\{\bar{u}^*\}) = \{\mathbf{e}\}^T\{\bar{u}\} = \bar{\boldsymbol{u}} \tag{5.227}$$

Thus, the components of $\bar{\boldsymbol{u}}$ in $\bar{x}$- and $\bar{\boldsymbol{u}}^*$ in $\bar{x}^*$-frames are the same. Hence $\bar{\boldsymbol{u}}$ is objective, i.e., observer independent.

5.15.1.2 Velocity vector: $\bar{\boldsymbol{v}}$

Velocity vector $\{\bar{v}\}$ is also a tensor of rank one. Let $\{\bar{v}(\bar{\boldsymbol{x}})\}$ and $\{\bar{v}^*(\bar{\boldsymbol{x}}^*)\}$ be velocity tensors in $\bar{x}$-frame and $\bar{x}^*$-frame.

Using (5.227) and taking material derivative

$$\{\dot{\bar{x}}^*\} = [Q]\{\dot{\bar{x}}\} \quad \text{or} \quad \{\bar{v}^*\} = [Q]\{\bar{v}\} \tag{5.228}$$

Thus, velocity tensor $\{\bar{v}\}$ is frame invariant. Alternatively,

$$\bar{\boldsymbol{v}}^* = \{\mathbf{e}^*\}^T\{\bar{v}^*\} \quad ; \quad \bar{\boldsymbol{v}} = \{\mathbf{e}\}^T\{\bar{v}\} \tag{5.229}$$

$$\therefore \quad \bar{\boldsymbol{v}}^* = (\{\mathbf{e}\}^T[Q]^T)\{\bar{v}^*\} = \{\mathbf{e}\}^T([Q]^T\{\bar{v}^*\}) = \{\mathbf{e}\}^T\{\bar{v}\} = \bar{\boldsymbol{v}} \tag{5.230}$$

Thus, $\bar{\boldsymbol{v}}$ in $\bar{\boldsymbol{v}}^*$ in $\bar{x}$-frame and $\bar{x}^*$-frame are the same, i.e., $\bar{\boldsymbol{v}}$ is objective.

5.15.1.3 Deformation gradient tensor: $\boldsymbol{J}$

Let $[J^*]$ and $[J]$ be deformation gradient tensors in x- and x^*-frames (both are tensors of rank two).

$$[J^*] = \left[\frac{\partial \bar{\boldsymbol{x}}^*}{\partial \boldsymbol{x}}\right] = \left[\frac{\partial \bar{\boldsymbol{x}}^*}{\partial \bar{\boldsymbol{x}}}\right]\left[\frac{\partial \bar{\boldsymbol{x}}}{\partial \boldsymbol{x}}\right] = [Q][J] \tag{5.231}$$

Thus, the deformation gradient tensor, a tensor of rank two, transforms like a tensor of rank one, hence is not objective as $[J^*] \neq [Q][J][Q]^T$.

5.15.1.4 Velocity gradient tensor: $\bar{\boldsymbol{L}}$

Consider $[\bar{L}(\bar{\boldsymbol{x}})]$ and $[\bar{L}^*(\bar{\boldsymbol{x}}^*)]$ in $\bar{x}$- and $\bar{x}^*$-frames, tensors of rank two.

$$[\bar{L}^*] = \left[\frac{\partial \bar{\boldsymbol{v}}^*}{\partial \bar{\boldsymbol{x}}^*}\right] \quad ; \quad [\bar{L}] = \left[\frac{\partial \bar{\boldsymbol{v}}}{\partial \bar{\boldsymbol{x}}}\right] \tag{5.232}$$

$$[\bar{L}^*] = \left[\frac{\partial \bar{\boldsymbol{v}}^*}{\partial \bar{\boldsymbol{x}}^*}\right] = \left[\frac{\partial \bar{\boldsymbol{v}}^*}{\partial \bar{\boldsymbol{v}}}\right]\left[\frac{\partial \bar{\boldsymbol{v}}}{\partial \bar{\boldsymbol{x}}}\right]\left[\frac{\partial \bar{\boldsymbol{x}}}{\partial \bar{\boldsymbol{x}}^*}\right] = [Q][\bar{L}][Q]^T \tag{5.233}$$

$[\bar{L}]$ is a tensor of rank two and it transforms like a tensor of rank two. Alternatively, we note

$$\bar{\boldsymbol{L}} = \{\mathbf{e}_i\}^T[\bar{L}]\{\mathbf{e}_j\} \quad ; \quad \bar{\boldsymbol{L}}^* = \{\mathbf{e}_i^*\}^T[\bar{L}^*]\{\mathbf{e}_j^*\} \tag{5.234}$$

$$\therefore \quad \bar{\boldsymbol{L}}^* = (\{\mathbf{e}_i\}^T[Q]^T)([Q][\bar{L}][Q]^T)([Q]\{\mathbf{e}_j\}]) \tag{5.235}$$

$$= \{\mathbf{e}_i\}^T[\bar{L}]\{\mathbf{e}_j\} = \bar{\boldsymbol{L}} \tag{5.236}$$

That is $\bar{\boldsymbol{L}}$ and $\bar{\boldsymbol{L}}^*$ in $\bar{x}$-frame and $\bar{x}^*$-frame are the same implying that $\bar{\boldsymbol{L}}$ is objective.

5.15.1.5 Symmetric part of velocity gradient tensor: $\bar{\boldsymbol{D}}$

Let $\bar{\boldsymbol{D}}^*(\bar{\boldsymbol{x}}^*)$ and $\bar{\boldsymbol{D}}(\bar{\boldsymbol{x}})$ be symmetric parts of velocity gradient tensors $\bar{\boldsymbol{L}}^*$ and $\bar{\boldsymbol{L}}$, both symmetric tensors of rank two, then

$$\begin{aligned} [\bar{D}^*] &= \frac{1}{2}\left([\bar{L}^*] + [\bar{L}^*]^T\right) = \frac{1}{2}\left([Q][\bar{L}][Q]^T + [Q][\bar{L}]^T[Q]^T\right) \\ &= [Q]\left(\frac{1}{2}\left([\bar{L}] + [\bar{L}]^T\right)\right)[Q]^T = [Q][\bar{D}][Q]^T \end{aligned} \tag{5.237}$$

Thus, $[\bar{D}]$ transforms like a tensor of rank two. This is sufficient to conclude that $[\bar{D}]$ is objective, but we provide further proof.

$$\begin{aligned} \bar{\boldsymbol{D}} &= \{\mathbf{e}_i\}^T[\bar{D}]\{\mathbf{e}_j\} \quad ; \quad \bar{\boldsymbol{D}}^* = \{\mathbf{e}_i^*\}^T[\bar{D}^*]\{\mathbf{e}_j^*\} \\ \bar{\boldsymbol{D}}^* &= \{\mathbf{e}_i^*\}^T[\bar{D}^*]\{\mathbf{e}_j^*\} = \left(\{\mathbf{e}_i\}^T[Q]^T\right)\left([Q][\bar{D}][Q]^T\right)([Q]\{\mathbf{e}_j\}) \\ &= \{\mathbf{e}_i\}^T[\bar{D}]\{\mathbf{e}_j\} = \bar{\boldsymbol{D}} \end{aligned} \tag{5.238}$$

Thus, $\bar{\boldsymbol{D}}$ in $\bar{x}$-frame and $\bar{\boldsymbol{D}}^*$ in $\bar{x}^*$-frame are the same. Thus, to an observer in $\bar{x}$-frame and to another observer in $\bar{x}^*$-frame, $\bar{\boldsymbol{D}}$ appears the same.

5.15.1.6 Skew-symmetric part of velocity gradient tensor $\bar{\boldsymbol{L}}$: $\bar{\boldsymbol{W}}$

Let $\bar{\boldsymbol{W}}^*(\bar{\boldsymbol{x}}^*)$ and $\bar{\boldsymbol{W}}(\bar{\boldsymbol{x}})$ be skew-symmetric parts of the velocity gradient tensors $\bar{\boldsymbol{L}}^*$ and $\bar{\boldsymbol{L}}$, both antisymmetric tensors of rank two in $\bar{x}$- and $\bar{x}^*$-frames, then

$$\begin{aligned} [\bar{W}^*] &= \frac{1}{2}\left([\bar{L}^*] - [\bar{L}^*]^T\right) = \frac{1}{2}\left([Q][\bar{L}][Q]^T - [Q][\bar{L}]^T[Q]^T\right) \\ &= [Q]\left(\frac{1}{2}\left([\bar{L}] - [\bar{L}]^T\right)\right)[Q]^T = [Q][\bar{W}][Q]^T \end{aligned} \tag{5.239}$$

Thus, $[\bar{W}]$ transforms like a tensor of rank two. Alternatively, we note

$$\bar{\boldsymbol{W}} = \{\mathbf{e}_i\}^T[\bar{W}]\{\mathbf{e}_j\} \quad ; \quad \bar{\boldsymbol{W}}^* = \{\mathbf{e}_i^*\}^T[\bar{W}^*]\{\mathbf{e}_j^*\} \tag{5.240}$$

$$\begin{aligned} \bar{\boldsymbol{W}}^* &= \left(\{\mathbf{e}_i\}^T[Q]^T\right)\left([Q][\bar{W}][Q]^T\right)([Q]\{\mathbf{e}_j\}) \\ &= \{\mathbf{e}_i\}^T[\bar{W}]\{\mathbf{e}_j\} = \bar{\boldsymbol{W}} \end{aligned} \tag{5.241}$$

Thus, $\bar{\boldsymbol{W}}$ in $\bar{x}$-frame and $\bar{\boldsymbol{W}}^*$ in $\bar{x}^*$-frame are the same, hence $\bar{\boldsymbol{W}}$ is objective.

5.15.1.7 Cauchy strain tensor: $\boldsymbol{C}_{[0]}$

Let $\boldsymbol{C}^*_{[0]}$ and $\boldsymbol{C}_{[0]}$ be Cauchy strain tensors in x- and x^*-frames (symmetric tensor of rank two), then $[C^*_{[0]}] = [Q][C_{[0]}][Q]^T$. Alternatively,

$$\begin{aligned} &[C_{[0]}] = [J]^T[J] \quad ; \quad [C^*_{[0]}] = [J^*]^T[J^*] \\ \text{using} \quad &[J^*] = [Q][J] \end{aligned} \tag{5.242}$$

$$[C^*_{[0]}] = [J]^T[Q]^T[Q][J] = [J]^T[J] = [C_{[0]}] \tag{5.243}$$

Thus, $[C_{[0]}]$ in x-frame and $[C^*_{[0]}]$ in x^*-frame are the same, hence $[C_{[0]}]$ is objective. Thus, to observers in x- and x^*-frames, $[C_{[0]}]$ appears the same.

5.15.1.8 Finger strain tensor: $\boldsymbol{F}_{[0]}$

$$[F_{[0]}] = [C_{[0]}]^{-1} \quad ; \quad [F^*_{[0]}] = [C^*_{[0]}]^{-1} \tag{5.244}$$

Since $\boldsymbol{C}^*_{[0]} = \boldsymbol{C}_{[0]}$ we have

$$\boldsymbol{F}^*_{[0]} = \boldsymbol{F}_{[0]} \tag{5.245}$$

That is, $\boldsymbol{F}_{[0]}$ in x-frame and $\boldsymbol{F}^*_{[0]}$ in x^*-frame are the same, hence $\boldsymbol{F}_{[0]}$ is objective.

5.15.1.9 Green's strain tensor: $\boldsymbol{\varepsilon}_{[0]}$

Let $\boldsymbol{\varepsilon}^*_{[0]}$ and $\boldsymbol{\varepsilon}_{[0]}$ be Green's strain tensors in x^*- and x-frames, then $[\varepsilon^*_{[0]}] = [Q][\varepsilon_{[0]}][Q]^T$. Alternatively,

$$[\varepsilon^*_{[0]}] = \frac{1}{2}\left([J^*]^T[J^*] - [I]\right) = \frac{1}{2}\left([J]^T[Q]^T[Q][J] - [I]\right) \tag{5.246}$$

$$= \frac{1}{2}\left([J]^T[J] - [I]\right) = [\varepsilon_{[0]}] \tag{5.247}$$

Thus, $\boldsymbol{\varepsilon}_{[0]}$ in x-frame and $\boldsymbol{\varepsilon}^*_{[0]}$ in x^*-frame are the same, hence $\boldsymbol{\varepsilon}_{[0]}$ is objective. To an observer in x- and x^*-frames, $\boldsymbol{\varepsilon}_{[0]}$ appears the same.

5.15.1.10 Strain tensor $\boldsymbol{B}^{[0]}$

Consider $\boldsymbol{B}^{[0]}$ and $(\boldsymbol{B}^{[0]})^*$ in x- and x^*-frames, both symmetric tensors of rank two, then

$$\begin{aligned} &[(B^{[0]})^*] = [J^*][J^*]^T \quad ; \quad [B^{[0]}] = [J][J]^T \\ \text{using} \quad &[J^*] = [Q][J] \end{aligned} \tag{5.248}$$

$$[(B^{[0]})^*] = [J^*][J^*]^T = [Q][J][J]^T[Q]^T = [Q][B^{[0]}][Q]^T \tag{5.249}$$

Also

$$\boldsymbol{B}^{[0]} = \{\boldsymbol{e}_i\}^T[B^{[0]}]\{\boldsymbol{e}_j\} \quad ; \quad (\boldsymbol{B}^{[0]})^* = \{\boldsymbol{e}^*_i\}^T[(B^{[0]})^*]\{\boldsymbol{e}^*_j\} \tag{5.250}$$

$$(\boldsymbol{B}^{[0]})^* = \left([Q]^T\{\boldsymbol{e}_i\}^T\right)\left([Q][B^{[0]}][Q]^T\right)\left([Q]\{\boldsymbol{e}_j\}^T\right) \tag{5.251}$$

$$= \{\boldsymbol{e}_i\}^T[B^{[0]}]\{\boldsymbol{e}_j\} = \boldsymbol{B}^{[0]} \tag{5.252}$$

Thus, $\boldsymbol{B}^{[0]}$ in x-frame and $(\boldsymbol{B}^{[0]})^*$ in x^*-frame are the same, hence $\boldsymbol{B}^{[0]}$ is objective.

5.15.1.11 Right stretch $[S_r]$ and rotation $[R]$

$$[J] = [R][S_r] \quad ; \quad [J^*] = [R^*][S_r^*] \tag{5.253}$$
$$[J^*] = [Q][J] \tag{5.254}$$
$$\therefore \quad [J^*] = [R^*][S_r^*] = [Q][J] = [Q][R][S_r] \tag{5.255}$$
$$\therefore \quad [R^*][S_r^*] = [Q][R][S_r] \tag{5.256}$$

Since the polar decomposition is unique it follows that $[R^*]$ and $[R]$ are rotation tensors, therefore the following holds

$$[R^*] = [Q][R] \tag{5.257}$$

Hence, from (5.256) we have

$$[S_r^*] = [S_r] \tag{5.258}$$

Thus, right stretch $[S_r]$ is not objective as there is no transformation in (5.258).

5.15.1.12 Left stretch $[S_l]$ and rotation $[R]$

$$[J] = [S_l][R] \quad ; \quad [J^*] = [S_l^*][R^*] \tag{5.259}$$
$$\therefore \quad [J^*] = [Q][J] \tag{5.260}$$
$$[J^*] = [Q][J] = [Q][S_l][R] = \left([Q][S_l][Q]^T\right)[R^*] = [S_l^*][R^*] \tag{5.261}$$
$$\text{Thus,} \quad [S_l^*] = [Q][S_l][Q]^T \tag{5.262}$$

That is, left stretch tensor $[S_l]$ is objective. It transforms like a tensor of rank two.

5.15.2 Euclidean or Non-Galilean transformations: tensors

In this section, we consider objectivity of various tensors and their rates under non-Galilean transformation (5.219).

5.15.2.1 Physical significance of spin tensor

In Section 5.2, the spin tensor $\bar{\boldsymbol{W}}$ was obtained by decomposition of $\bar{\boldsymbol{L}}$ into symmetric tensor $\bar{\boldsymbol{D}}$ and skew-symmetric tensor $\bar{\boldsymbol{W}}$. In this section, we present another aspect of the spin tensor. Consider a transformation of x-frame into $\bar{x}$-frame by

$$\{\bar{x}\} = [Q(t)]\{x\} \tag{5.263}$$

Transformation (5.263) is a non-Galilean transformation in which the transformation matrix or tensor is a function of time. This transformation suggests that $\bar{\boldsymbol{x}}$ is obtained by the rigid rotation of $\boldsymbol{x}$.

$$\frac{D\{\bar{x}\}}{Dt} = \{\dot{\bar{x}}\} = \{\bar{v}\} = [\dot{Q}(t)]\{x\} \tag{5.264}$$

Since $[Q(t)]$ in (5.264) is orthogonal

$$[Q(t)][Q(t)]^T = [I] \tag{5.265}$$

Taking material derivative of both sides in (5.265)

$$[\dot{Q}(t)][Q(t)]^T + [Q(t)][\dot{Q}(t)]^T = 0 \tag{5.266}$$

$$\text{or} \quad [\dot{Q}(t)][Q(t)]^T = -[Q(t)][\dot{Q}(t)]^T \tag{5.267}$$

Let us define $[\Omega]$ as

$$[\Omega] = [\dot{Q}(t)][Q(t)]^T = -[Q(t)][\dot{Q}(t)]^T \tag{5.268}$$

From (5.268) we note that $[\Omega]$ is skew-symmetric. Substituting for $\{x\}$ from (5.263) into (5.264)

$$\{\bar{v}\} = [\dot{Q}(t)][Q(t)]^T\{\bar{x}\} \tag{5.269}$$

using (5.268) in (5.269) we obtain

$$\{\bar{v}\} = [\Omega]\{\bar{x}\} \tag{5.270}$$

$[\Omega]$ is the skew-symmetric spin tensor of angular velocities

$$[\Omega] = \begin{bmatrix} 0 & -\omega_3 & \omega_2 \\ \omega_3 & 0 & -\omega_1 \\ -\omega_2 & \omega_1 & 0 \end{bmatrix} \tag{5.271}$$

Recall equation (5.9)

$$\frac{D}{Dt}\{d\bar{x}\} = [\bar{L}]\{d\bar{x}\} \quad ; \quad \{d\bar{v}\} = [\bar{L}]\{d\bar{x}\} \tag{5.272}$$

using decomposition $[\bar{L}] = [\bar{D}] + [\bar{W}]$ (equation (5.15))

$$\{d\bar{v}\} = [\bar{D}]\{d\bar{x}\} + [\bar{W}]\{d\bar{x}\} \tag{5.273}$$

When $[\bar{D}]$ is zero we have pure rotation, hence (5.273) reduces to

$$\{d\bar{v}\} = [\bar{W}]\{d\bar{x}\} \tag{5.274}$$

$$\text{or} \quad \{\bar{v}\} = [\bar{W}]\{\bar{x}\} \tag{5.275}$$

Comparing (5.270) and (5.275), we have

$$[\Omega] = [\bar{W}] \tag{5.276}$$

That is skew-symmetric part of $[\bar{L}]$ is in fact angular velocity tensor associated with time dependent rigid rotation defined in (5.270).

5.15.2.2 Velocity: $\bar{\boldsymbol{v}}$

Consider (5.219) without $\{c(t)\}$, as the pure translation is of no consequence.

$$\{\bar{x}^*(t)\} = [Q(t)]\{\bar{x}(t)\} \tag{5.277}$$

Velocities are obtained by taking material derivative of (5.277)

$$\frac{D\{\bar{x}^*(t)\}}{Dt} = \{\bar{v}^*\} = [\dot{Q}(t)]\{\bar{x}(t)\} + [Q(t)]\{\bar{v}(t)\} \tag{5.278}$$

$$\text{or} \quad \{\bar{v}^*\} = [Q(t)]\{\bar{v}(t)\} + [\dot{Q}(t)]\{\bar{x}(t)\} \tag{5.279}$$

using (5.277) in (5.279)

$$\{\bar{v}^*\} = [Q(t)]\{\bar{v}(t)\} + [\dot{Q}(t)][Q(t)]^T\{\bar{x}^*(t)\} \tag{5.280}$$

$\{\bar{v}\}$ is a tensor of rank one but under the rigid rotation it is not objective. If it were objective we would have $\{\bar{v}^*\} = [Q(t)]\{\bar{v}(t)\}$ for any $[Q(t)]$, a pure rotation.

5.15.2.3 Velocity gradient tensor: $\bar{\boldsymbol{L}}$

Consider $\{\bar{v}^*\}$ in (5.280)

$$\{\bar{v}^*\} = [Q(t)]\{\bar{v}(t)\} + [\dot{Q}(t)][Q(t)]^T\{\bar{x}^*(t)\} \tag{5.281}$$

differentiate with respect to $\{\bar{x}^*\}$

$$\frac{\partial\{\bar{v}^*\}}{\partial\{\bar{x}^*\}} = [\bar{L}^*] = [Q(t)]\left[\frac{\partial\{\bar{v}(t)\}}{\partial\{\bar{x}(t)\}}\right]\left[\frac{\partial\{\bar{x}(t)\}}{\partial\{\bar{x}^*\}}\right] + [\dot{Q}(t)][Q(t)]^T \tag{5.282}$$

$$\text{or} \quad [\bar{L}^*] = [Q(t)][\bar{L}][Q(t)]^T + [\dot{Q}(t)][Q(t)]^T \tag{5.283}$$

Thus, $\bar{\boldsymbol{L}}$ is not objective under rigid rotation $\boldsymbol{Q}(t)$, i.e., under non-Galilean transformation. When $\boldsymbol{Q}(t)$ is not a function of time (Galilean transformation) then $\dot{\boldsymbol{Q}} = 0$ and we have

$$[\bar{L}^*] = [Q(t)][\bar{L}][Q(t)]^T \tag{5.284}$$

That is, under non-Galilean transformation $\bar{\boldsymbol{L}}$ transforms like a second rank tensor. Hence, $\bar{\boldsymbol{L}}$ is objective under non-Galilean transformation.

5.15.2.4 Symmetric part of velocity gradient tensor: $\bar{\boldsymbol{D}}$

$$[\bar{D}^*] = \frac{1}{2}\left([\bar{L}^*] + [\bar{L}^*]^T\right) \quad ; \quad [\bar{D}] = \frac{1}{2}\left([\bar{L}] + [\bar{L}]^T\right) \tag{5.285}$$

using (5.283) in $[\bar{D}^*]$

$$\begin{aligned}[\bar{D}^*] = \frac{1}{2}\Big[&\left([Q(t)][\bar{L}][\bar{Q}(t)]^T + [\dot{Q}(t)][Q(t)]^T\right) \\ &+ \left([Q(t)][\bar{L}]^T[Q(t)]^T + [Q(t)][\dot{Q}(t)]^T\right)\Big]\end{aligned} \tag{5.286}$$

regrouping terms

$$\begin{aligned}[\bar{D}^*] = [Q(t)]&\left(\frac{1}{2}([\bar{L}] + [\bar{L}]^T)\right)[Q(t)]^T \\ &+ \frac{1}{2}\left([\dot{Q}(t)][Q(t)]^T + [Q(t)][\dot{Q}(t)]^T\right)\end{aligned} \tag{5.287}$$

using (5.267), the second term in (5.287) is zero, hence we have (using $[\bar{D}] = \frac{1}{2}\left([\bar{L}] + [\bar{L}]^T\right)$

$$[\bar{D}^*] = [Q(t)][\bar{D}][Q(t)]^T \tag{5.288}$$

Hence, $\bar{\boldsymbol{D}}$ is objective under non-Galilean transformation we well as Galilean transformation.

5.15.2.5 Skew-symmetric part of velocity gradient tensor $\bar{\boldsymbol{L}}$: $\bar{\boldsymbol{W}}$

$$[\bar{W}^*] = \frac{1}{2}\left([\bar{L}^*] - [\bar{L}^*]^T\right) \quad \text{and} \quad [\bar{W}] = \frac{1}{2}\left([\bar{L}] - [\bar{L}]^T\right) \tag{5.289}$$

using $[\bar{L}^*]$ from (5.283) in $[\bar{W}^*]$

$$\begin{aligned}[\bar{W}^*] = \frac{1}{2}\Big[&\left([Q(t)[\bar{L}][Q(t)]^T + [\dot{Q}(t)][Q(t)]^T\right) \\ &- \left([Q(t)][\bar{L}]^T[Q(t)]^T + [Q(t)][\dot{Q}(t)]^T\right)\Big]\end{aligned} \tag{5.290}$$

regrouping terms

$$\begin{aligned}[\bar{W}^*] = {} & [Q(t)]\left(\frac{1}{2}\left([\bar{L}] - [\bar{L}]^T\right)\right)[Q(t)]^T \\ & + \frac{1}{2}\left([\dot{Q}(t)][Q(t)]^T - [Q(t)][\dot{Q}(t)]^T\right)\end{aligned} \tag{5.291}$$

using (5.267) in (5.291) and $[\bar{W}]$ from (5.289)

$$[\bar{W}^*] = [Q(t)][\bar{W}][Q(t)]^T + [\dot{Q}(t)][Q(t)]^T \tag{5.292}$$

Thus, $\bar{\boldsymbol{W}}$ is not objective under non-Galilean transformation. When $\boldsymbol{Q}$ is not a function of time (Galilean transformation), i.e., $\dot{\boldsymbol{Q}} = 0$, then $\bar{\boldsymbol{W}}$ is objective.

5.15.3 Objective rates

Consider an objective symmetric tensor $\bar{\boldsymbol{T}}$ under non-Galilean transformation

$$[\bar{T}^*] = [Q(t)][\bar{T}][Q(t)]^T \tag{5.293}$$

We inquire if $\frac{D[\bar{T}]}{Dt}$ is objective. Taking material derivative of (5.293)

$$\frac{D[\bar{T}^*]}{Dt} = [\dot{\bar{T}}^*] = [\dot{Q}(t)][\bar{T}][Q(t)]^T + [Q(t)][\dot{\bar{T}}][Q(t)]^T + [Q(t)][\bar{T}][\dot{Q}(t)]^T \tag{5.294}$$

Presence of $[\dot{Q}(t)][\bar{T}][Q(t)]^T$ and $[Q(t)][\bar{T}][\dot{Q}(t)]^T$ shows that $\frac{D[\bar{T}]}{Dt} = [\dot{\bar{T}}]$ is not objective. Clearly if $[Q]$ is not a function of time (Galilean transformation), then $[\dot{Q}(t)] = 0$ and (5.293) reduces to

$$[\dot{\bar{T}}^*] = [Q(t)][\dot{\bar{T}}][Q(t)]^T \tag{5.295}$$

Confirming that $[\dot{\bar{T}}^*]$ is objective under Galilean transformation. From the material derivative of $\bar{\boldsymbol{T}}^*$ we derive convected time derivatives, contravariant, covariant as well as Jaumann rates that are objective under non-Galilean as well as Galilean transformations. This requires that we eliminate $[\dot{Q}(t)]$ from (5.294) and bring in other desired measures.

5.15.3.1 Convected time derivative of a covariant tensor of rank two

Recall (equation (5.283)) $[\bar{L}^*]$

$$[\bar{L}^*] = [Q(t)][\bar{L}][Q(t)]^T + [\dot{Q}(t)][Q(t)]^T \tag{5.296}$$

Post multiply by $[Q(t)]$ and use $[Q(t)]^T[Q(t)] = [I]$ to obtain

$$[\bar{L}^*][Q(t)] = [Q(t)][\bar{L}] + [\dot{Q}(t)] \tag{5.297}$$

$$\therefore \quad [\dot{Q}(t)] = [\bar{L}^*][Q(t)] - [Q(t)][\bar{L}] \tag{5.298}$$

$$\text{and} \quad [\dot{Q}(t)]^T = [Q(t)]^T[\bar{L}^*]^T - [\bar{L}]^T[Q(t)]^T \tag{5.299}$$

substituting in (5.294)

$$\begin{aligned}\frac{D[\bar{T}^*]}{Dt} &= \left([\bar{L}^*][Q(t)] - [Q(t)][\bar{L}]\right)[\bar{T}][Q(t)]^T + [Q(t)][\dot{\bar{T}}][Q(t)]^T \\ &+ [Q(t)][\bar{T}]\left([Q(t)]^T[\bar{L}^*]^T - [\bar{L}]^T[Q(t)]^T\right)\end{aligned} \tag{5.300}$$

Grouping and rearranging terms

$$\begin{aligned}\frac{D[\bar{T}^*]}{Dt} &- [\bar{L}^*]\left([Q(t)][\bar{T}][Q(t)]^T\right) - [Q(t)][\bar{T}][Q(t)]^T[\bar{L}^*]^T \\ &= [Q(t)]\left([\dot{\bar{T}}] - [\bar{L}][\bar{T}] - [\bar{T}][\bar{L}]^T\right)[Q(t)]^T\end{aligned} \tag{5.301}$$

using $[\bar{T}^*] = [Q(t)[\bar{T}][Q(t)]^T$

$$\frac{D[\bar{T}^*]}{Dt} - [\bar{L}^*][\bar{T}^*] - [\bar{T}^*][\bar{L}^*]^T = [Q(t)]\left(\frac{D[\bar{T}]}{Dt} - [\bar{L}][\bar{T}] - [\bar{T}][\bar{L}]^T\right)[Q(t)]^T \tag{5.302}$$

From (5.302) we conclude that

$$[\bar{T}_{(1)}] = \frac{D[\bar{T}]}{Dt} - [\bar{L}][\bar{T}] - [\bar{T}][\bar{L}]^T \tag{5.303}$$

Convected time derivative $[\bar{T}_{(1)}]$ is objective under non-Galilean transformation. It is referred to as first convected time derivative of the covariant tensor $\bar{\boldsymbol{T}}$ (or $\bar{\boldsymbol{T}}_{(0)}$) (also see Sections 5.10.2 and 5.10.3).

5.15.3.2 Convected time derivative of a contravariant tensor of rank two

Recall (equation (5.283)) $\bar{\boldsymbol{L}}^*$

$$[\bar{L}^*] = [Q(t)][\bar{L}][Q(t)]^T + [\dot{Q}(t)][Q(t)]^T \tag{5.304}$$

using (5.267), i.e.

$$[\dot{Q}(t)][Q(t)]^T = -[Q(t)][\dot{Q}(t)]^T \tag{5.305}$$

in (5.304) we obtain

$$[\bar{L}^*] = [Q(t)][\bar{L}][Q(t)]^T - [Q(t)][\dot{Q}(t)]^T \tag{5.306}$$

Premultiply by $[Q(t)]^T$ and solve for $[\dot{Q}(t)]^T$

$$[\dot{Q}(t)]^T = [\bar{L}][Q(t)]^T - [Q(t)]^T[\bar{L}^*] \tag{5.307}$$

$$\therefore \quad [\dot{Q}(t)] = [Q(t)][\bar{L}]^T - [\bar{L}^*]^T[Q(t)] \tag{5.308}$$

Substituting from (5.307) and (5.308) into (5.294)

$$\begin{aligned}\frac{D[\bar{T}^*]}{Dt} &= \left([Q(t)][\bar{L}]^T - [\bar{L}^*]^T[Q(t)]\right)[\bar{T}][Q(t)]^T + [Q(t)][\dot{\bar{T}}][Q(t)]^T \\ &+ [Q(t)][\bar{T}]\left([\bar{L}][Q(t)]^T - [Q(t)]^T[\bar{L}^*]\right)\end{aligned} \tag{5.309}$$

Grouping and rearranging terms

$$\begin{aligned}&\frac{D[\bar{T}^*]}{Dt} + [\bar{L}^*]^T\left([Q(t)][\bar{T}][Q(t)]^T\right) + \left([Q(t)][\bar{T}][Q(t)]^T\right)[\bar{L}^*] \\ &\quad = [Q(t)]\left(\frac{D[\bar{T}]}{Dt} + [\bar{L}]^T[\bar{T}] + [\bar{T}][\bar{L}]\right)[Q(t)]^T\end{aligned} \tag{5.310}$$

$$\text{or} \quad \frac{D[\bar{T}^*]}{Dt} + [\bar{L}^*]^T[\bar{T}^*] + [\bar{T}^*][\bar{L}^*] = [Q(t)]\left(\frac{D[\bar{T}]}{Dt} + [\bar{L}]^T[\bar{T}] + [\bar{T}][\bar{L}]\right) \tag{5.311}$$

From (5.310), we conclude that

$$[\bar{T}^{(1)}] = \frac{D[\bar{T}]}{Dt} + [\bar{L}]^T[\bar{T}] + [\bar{T}][\bar{L}] \tag{5.312}$$

is the objective convected time derivative under non-Galilean transformation. It is referred to as the first convected time derivative of the contravariant tensor $\bar{\boldsymbol{T}}$ (or $\bar{\boldsymbol{T}}^{(0)}$) (also see Sections 5.10.2 and 5.10.3).

5.15.3.3 Jaumann convected time derivative

Jaumann convected time derivative of $[\bar{T}]$ defined as $[{}^{(1)}\bar{T}^J]$ is the average of the first convected time derivatives of contravariant and covariant tensors. Thus

$$[{}^{(1)}\bar{T}^J] = \frac{1}{2}\left([\bar{T}^{(1)}] + [\bar{T}_{(1)}]\right) \tag{5.313}$$

substituting from (5.303) and (5.312) and using $[\bar{W}] = \frac{1}{2}\left([\bar{L}] - [\bar{L}]^T\right)$

$$[{}^{(1)}\bar{T}^J] = \frac{[\bar{T}]}{Dt} - [\bar{W}][\bar{T}] + [\bar{T}][\bar{W}] \tag{5.314}$$

$[{}^{(1)}\bar{T}^J]$ is naturally objective as $[\bar{T}^{(1)}]$ and $[\bar{T}_{(1)}]$ are objective.

An independent and alternate derivation of the objectivity of Jaumann rate is given in the following. We begin with (5.292)

$$[\bar{W}^*] = [Q(t)][\bar{W}][Q(t)]^T + [\dot{Q}(t)][Q(t)]^T \tag{5.315}$$

Postmultiplying by $[Q(t)]$ and solving for $[\dot{Q}(t)]$

$$[\dot{Q}(t)] = [\bar{W}^*][Q(t)] - [Q(t)][\bar{W}] \tag{5.316}$$

$$\therefore \quad [\dot{Q}(t)]^T = [Q(t)]^T[\bar{W}^*]^T - [\bar{W}]^T[Q(t)]^T \tag{5.317}$$

substituting from (5.316) and (5.317) into (5.294)

$$\frac{D[\bar{T}^*]}{Dt} = \left([\bar{W}^*][Q(t)] - [Q(t)][\bar{W}]\right)[\bar{T}][Q(t)]^T + [Q(t)][\dot{\bar{T}}][Q(t)]^T \tag{5.318}$$

$$+ [Q(t)][\bar{T}]\left([Q(t)]^T[\bar{W}^*]^T - [\bar{W}]^T[Q(t)]^T\right)$$

Grouping and collecting terms

$$\begin{aligned}\frac{D[\bar{T}^*]}{Dt} &- [\bar{W}^*]\left([Q(t)][\bar{T}][Q(t)]^T\right) - [Q(t)][\bar{T}][Q(t)]^T[\bar{W}^*]^T \\ &= [Q(t)]\left(\frac{D[\bar{T}]}{Dt} - [\bar{W}][\bar{T}] - [\bar{T}][\bar{W}]^T\right)[Q(t)]^T\end{aligned} \tag{5.319}$$

using $[\bar{W}^*]^T = -[\bar{W}^*]$, $[\bar{W}]^T = -[\bar{W}]$ and $[\bar{T}^*] = [Q(t)][\bar{T}][Q(t)]^T$ in (5.319), we obtain

$$\frac{D[\bar{T}^*]}{Dt} - [\bar{W}^*][\bar{T}^*] + [\bar{T}^*][\bar{W}^*] = [Q(t)]\left(\frac{D[\bar{T}]}{Dt} - [\bar{W}][\bar{T}] + [\bar{T}][\bar{W}]\right)[Q(t)]^T \tag{5.320}$$

From (5.320) we conclude that

$$[{}^{(1)}\bar{T}^J] = \frac{D[\bar{T}]}{Dt} - [\bar{W}][\bar{T}] + [\bar{T}][\bar{W}] \tag{5.321}$$

is objective under non-Galilean transformation. ${}^{(1)}\bar{\boldsymbol{T}}^J$ is called the first convected time derivative of the Jaumann stress tensor, average of the convected time derivatives of contravariant and covariant Cauchy stress tensors (also see Sections 5.13.2 and 5.13.3)

5.15.4 Remarks

(1) In this section, criteria for objectivity of tensors of ranks zero, one and two are presented. Objectivity of $\bar{\boldsymbol{u}}$, $\bar{\boldsymbol{v}}$, $\boldsymbol{J}$, $\bar{\boldsymbol{L}}$, $\bar{\boldsymbol{D}}$, $\bar{\boldsymbol{W}}$, $\boldsymbol{C}_{[0]}$, $\boldsymbol{F}_{[0]}$, $\boldsymbol{\varepsilon}_{[0]}$, $\boldsymbol{B}^{[0]}$, $\boldsymbol{S}_r$, $\boldsymbol{S}_l$, $\boldsymbol{R}$, etc. is determined under Galilean as well as Euclidean or non-Galilean transformations.

(2) Convected time derivatives of order one of second rank contravariant, covariant and Jaumann tensors are derived using Euclidean or non-Galilean transformation and their objectivity is established.

(3) It is shown that the convected time derivatives derived here are same as those in Sections 5.13.2 and 5.13.3 for incompressible matter. In the derivations of the convected time derivatives presented in this section, $\bar{\boldsymbol{T}}$ is just a tensor of rank two without any physics associated with it. Thus, physics of compressibility cannot be accounted for in their derivations.

5.16 Summary

In this chapter, rate of deformation of length, area, volume, strain rate tensors, spin tensor and convected time derivatives of the stress tensors (for compressible and incompressible matter) and strain rate tensors in contra- and co-variant bases are presented. Objectivity of tensors of rank zero, one and two are established under Galilean as well as Euclidean or non-Galilean transformations. Convected time derivatives of various orders of contravariant, covariant and Jaumann stress tensors are derived for compressible as well as incompressible matter. Convected time derivatives of the Green and Almansi strain tensors are derived. Convected time derivatives of the co- and contra-variant and Jaumman tensors are also derived using non-Galilean transformation. This material plays a significant role in the derivations of the mathematical models for fluids based on the conservation and balance laws (Chapter 6) as well as in the derivations of the constitutive theories.

Problems

5.1 Let the velocity field in a deforming volume of matter be given by

$$\bar{v}_1 = 3\bar{x}_2 \quad , \quad \bar{v}_2 = 0 \quad , \quad \bar{v}_3 = 0$$

Find

(a) strain rate and spin tensors;
(b) the rate of extensions per unit length of the line segment $\{\bar{P}\bar{Q}\} = \delta[1,3,1]^T$, δ being an infinitesimal real number;
(c) maximum and minimum stretchings, i.e., rates extension per unit length.

5.2 Let the velocity field in a deforming matter be given by

$$\bar{v}_1 = 1.5\bar{x}_2^2 \quad , \quad \bar{v}_2 = 0 \quad , \quad \bar{v}_3 = 0$$

Find

(a) the rate of extension per unit length of a material line element, which in the present configuration passes through the point $(2.5, 1.5, 0)$ and is along the vector $(2, 2, 0)$;
(b) the strain rate tensor;
(c) the spin tensor;
(d) maximum and minimum stretchings and their directions.

5.3 Let the following

$$x_1 = \bar{x}_1$$
$$x_2 = (e^{-t} - 1)\bar{x}_1 + e^{-t}\bar{x}_2$$
$$x_3 = (e^{-t} - e^{t})\bar{x}_1 + \bar{x}_3$$

describe the deformation of a continuum.

(a) Determine the velocity field $\bar{v}_i = \bar{v}_i(\bar{x}_j, t)$.
(b) Show that for this deformation, $[\bar{L}] = [\dot{J}][J]^{-1}$ holds.
(c) Find the strain rate tensor $[\bar{D}]$ and spin tensor $[\bar{W}]$.

5.4 The motion of a continuous medium is given by

$$\bar{x}_1 = \frac{1}{2}(e^t + e^{-t})x_1 + \frac{1}{2}(e^t - e^{-t})x_2$$
$$\bar{x}_2 = \frac{1}{2}(e^t - e^{-t})x_1 + \frac{1}{2}(e^t + e^{-t})x_2$$
$$\bar{x}_3 = x_3$$

Determine

(a) the velocity components in the material description;
(b) the velocity components in the spatial description;
(c) spatial velocity gradient tensor and spin tensor;
(d) material velocity gradient tensor.

5.5 Consider the convected time derivative of strain tensors in contravariant and covariant bases and the Jaumann rates.

(a) Determine explicit expressions for $[\gamma^{(1)}]$, $[\gamma^{(2)}]$ and $[\gamma^{(3)}]$ in contravariant basis.
(b) Determine explicit expressions for $[\gamma_{(1)}]$, $[\gamma_{(2)}]$ and $[\gamma_{(3)}]$ in covariant basis.
(c) Also determine explicit expressions for $[{}^{(1)}\gamma^J]$, $[{}^{(2)}\gamma^J]$ and $[{}^{(3)}\gamma^J]$, Jaumann rates.

5.6 Consider $[\bar{\sigma}^{(0)}]$, $[\bar{\sigma}_{(0)}]$ and $[{}^{(0)}\bar{\sigma}^J]$, symmetric Cauchy stress tensors in Eulerian description in contravariant and covariant bases and Jaumann stress tensor.

(a) Determine explicit expressions for $[\bar{\sigma}^{(0)}]$, $[\bar{\sigma}^{(1)}]$ and $[\bar{\sigma}^{(3)}]$ in contravariant basis.
(b) Determine explicit expressions for $[\bar{\sigma}_{(0)}]$, $[\bar{\sigma}_{(1)}]$ and $[\bar{\sigma}_{(3)}]$ in covariant basis.
(c) Also determine explicit expressions for $[{}^{(1)}\bar{\sigma}^J]$, $[{}^{(2)}\bar{\sigma}^J]$ and $[{}^{(3)}\bar{\sigma}^J]$, Jaumann rates.

5.7 let $\{d\bar{x}_1\}$ and $\{d\bar{x}_2\}$ be two material lines at point $\underset{\sim}{\bar{Q}}$ in the current configuration. A surface element defined by these material lines is given by

$$d\bar{\boldsymbol{A}} = d\bar{\boldsymbol{x}}_1 \times d\bar{\boldsymbol{x}}_2 \tag{1}$$

Show that for (1) the following holds

$$\frac{D(d\bar{A}_i)}{Dt} = \text{tr}[\bar{D}]d\bar{A}_i - \bar{L}_{ji}d\bar{A}_j$$

5.8 Consider Polar decomposition of the deformation gradient tensor $[J]$.

$$[J] = [R][S_r]$$

Show that

(a) $[\bar{D}] = \frac{1}{2}\left([\dot{S}_r][S_r]^{-1} + [S_r]^{-1}[\dot{S}_r]\right)[R]^T$

(b) $[\bar{W}] = [\dot{R}][R]^T + \frac{1}{2}\left([R][\dot{S}_r][S_r]^{-1}[R]^T - \left([R][\dot{S}_r][S_r]^{-1}[R]^T\right)^T\right)$

5.9 Consider polar decomposition of the deformation gradient tensor $[J]$.

$$[J] = [S_l][R]$$

Derive expressions for

(a) $\bar{\boldsymbol{D}} = \bar{\boldsymbol{D}}(\boldsymbol{S}_l, \boldsymbol{R}, \dot{\boldsymbol{S}}_l, \dot{\boldsymbol{R}}, \dots)$

(a) $\bar{\boldsymbol{W}} = \bar{\boldsymbol{W}}(\boldsymbol{S}_l, \boldsymbol{R}, \dot{\boldsymbol{S}}_l, \dot{\boldsymbol{R}}, \dots)$

6 CONSERVATION AND BALANCE LAWS IN EULERIAN DESCRIPTION

6.1 Introduction

In this chapter, we derive mathematical description of a deforming volume of matter in Eulerian or spatial description using conservation and balance laws (CBL) of classical thermodynamics. In general, all processes can be classified into two categories: those that exhibit thermodynamic equilibrium and those in which there is lack of thermodynamic equilibrium, called non-equilibrium thermodynamic processes. Those processes in which the time scale of change in the process is much larger than the time scale of thermodynamic equilibrium are classified as *thermodynamic equilibrium processes.* On the other hand, if the time scale of change in the process is much smaller than the time scale of thermodynamic equilibrium, then the process is classified as *non-equilibrium thermodynamic process.* The development of the mathematical models for processes that exhibit thermodynamic equilibrium is obviously an important area of study.

The mathematical models for thermodynamic equilibrium processes can be derived using *principles of thermodynamics* commonly referred to as *conservation and balance laws.* There are four basic conservation and balance laws that are employed in deriving mathematical descriptions of continuous deforming matter that exhibits thermodynamic equilibrium: conservation of mass, kinematics of continuous matter, first law of thermodynamics (energy equation) and second law of thermodynamics (entropy of Clausius-Duhem inequality). We begin with the integral statements of the conservation and balance laws for a volume of matter $\bar{V}(t)$ with a closed boundary $\partial\bar{V}(t)$ in the current configuration. We present CBL in integral as well as differential form. The integral form permits isolated discontinuities of the integrand over the volume $\bar{V}(t)$. The differential or local form strictly requires the integrand to be continuous everywhere in the deforming volume which holds when the deforming matter is isotropic and homogeneous. In this case, the integral form of the CBL can be reduced to differential form at a material point. The differential form of CBL are also referred as local form of the conservation and the balance laws. The CBL are independent of constitution of the matter, hence are applicable to all solid and fluent continua. The CBL must be augmented by matter specific physics through the constitutive theories.

(1) ***Conservation of mass: CM***

For the volume or body of matter of constant mass in the reference (undeformed) configuration, the mass is conserved during deformation or motion. This leads to the law or principle of conservation of mass for the continuum.

DOI: 10.1201/9781003105336-6

(2) *Kinematics of continuum matter*

(a) Balance of linear momenta: BLM

Newton's second law of motion in classical mechanics can be reformulated for volume of continuous matter or continuum. This leads to the principle that the rate of change of linear momentum of a volume of matter or mass must be equal to the resultant forces acting on it. This is referred to as balance of linear momenta. It is a balance law and not a conservation law.

(b) Balance of angular momenta: BAM

If the deformed volume during evolution is in stable equilibrium, then the time rate of change of the total moment of momenta for the continuum is equal to the vector sum of the moments of external forces acting on the continuum. This is often referred to as the balance of angular momenta. This is obviously also a balance law and not a conservation law.

(3) *The first law of thermodynamics (FLT): balance of energy*

A volume or mass of matter must satisfy the first law of thermodynamics. Based on this law the heat added plus the total work done on a volume or mass of matter must be equal to the change in the energy of the volume or mass of matter. This is often referred to as the law of balance of energy. This is obviously a balance law and not a conservation law.

(4) *The second law of thermodynamics (SLT): entropy inequality or Clausius-Duhem inequality*

Every thermodynamic equilibrium process must satisfy the second law of thermodynamics or entropy inequality. The entropy imparted to a volume of matter by contacting and non-contacting sources must result in rate of change of entropy density of the volume of matter. The entropy inequality is a statement of irreversibility in most natural processes, especially those involving dissipation of energy. Analogous to the assumption of contact forces and body forces in the kinematics of continuous matter, and the heat flux and a source of energy in the balance of energy, entropy inequality is a similar postulate. Its importance in connection with conservation and balance laws and the development of the constitutive theory is shown in this chapter as well as in the subsequent chapters. The second law of thermodynamics resulting in entropy inequality is also a balance law.

We consider details of the development of differential forms of the mathematical models based on the conservation and balance laws in Eulerian description for compressible and incompressible matter in this chapter.

6.2 Localization theorem

If $\bar{\boldsymbol{f}}(\bar{\boldsymbol{x}})$ is a tensor of some rank that is continuous for all $\bar{\boldsymbol{x}} \in \bar{V}$ and if

$$\int_{\bar{V}} \bar{\boldsymbol{f}}(\bar{\boldsymbol{x}}) d\bar{V} = 0 \tag{6.1}$$

for all admissible subdomains $\underset{\sim}{\bar{V}} \subset \bar{V}$, then

$$\bar{\boldsymbol{f}}(\bar{\boldsymbol{x}}) = 0 \tag{6.2}$$

for all $\bar{\boldsymbol{x}} \in \bar{V}$.

Conversely if $\bar{\boldsymbol{f}}(\bar{\boldsymbol{x}}) = 0$ and if $\bar{\boldsymbol{f}}(\bar{\boldsymbol{x}})$ is continuous for all $\bar{\boldsymbol{x}} \in \bar{V}$, then (6.1) holds. In this case (6.1) is Riemann integral of (6.2).

Integral and differential forms of conservation and balance laws

In this chapter, we derive integral and differential forms of conservation and balance laws in Eulerian description. In this derivation based on CM, BLM, BAM, FLT and SLT, we begin with rate considerations in the current configuration that eventually leads to the following form for each conservation and balance law.

$$\int_{\bar{V}} \bar{\boldsymbol{G}}(\bar{\boldsymbol{x}}, \dots) d\bar{V} \tag{6.3}$$

The dots in the arguments of $\bar{\boldsymbol{G}}$ represent dependent variables and/or their derivatives. For example, we shall see that in case of BLM, (6.3) takes the following form (integral form of BLM)

$$\int_{\bar{V}} \bar{\boldsymbol{G}}(\bar{\boldsymbol{x}}, \bar{\boldsymbol{v}}, \bar{\boldsymbol{\nabla}} \cdot \bar{\boldsymbol{\sigma}}, \bar{\boldsymbol{F}}^b) d\bar{V} = 0 \tag{6.4}$$

Based on Localization theorem, if $\bar{\boldsymbol{G}}$ is continuous in all its arguments for all values of $\bar{\boldsymbol{x}} \in \bar{V}$, then

$$\bar{\boldsymbol{G}}(\bar{\boldsymbol{x}}, \bar{\boldsymbol{v}}, \bar{\boldsymbol{\nabla}} \cdot \bar{\boldsymbol{\sigma}}, \bar{\boldsymbol{F}}^b) = 0 \tag{6.5}$$

holds, which is the differential or local form of the BLM (shown later), and in this case (6.4) is the Riemann integral of (6.5).

The question of when is $\bar{\boldsymbol{G}}(\cdot)$ in (6.4) continuous for every $\bar{\boldsymbol{x}} \in \bar{V}$ is important to investigate further.

(1) When the mater in $\bar{V}$ is homogeneous and isotropic and when all arguments of $\bar{\boldsymbol{G}}(\cdot)$ in (6.4) belong to minimally conforming or higher order spaces, then $\bar{\boldsymbol{G}}(\cdot)$ is undoubtedly continuous in all of its arguments for all $\bar{\boldsymbol{x}} \in \bar{V}$, hence in this case, the differential form or local form of BLM in (6.5) is always valid from the integral form (6.4).

(2) When the matter is inhomogeneous and anisotropic, the volume $\bar{V}$ undoubtedly contains interfaces between material points consisting of bi-material interfaces that will either exhibit discontinuities of the gradients of displacements or of stresses. Thus, in this case, continuity of $\bar{\boldsymbol{G}}(\cdot)$ in all of its arguments for all $\bar{\boldsymbol{x}} \in \bar{V}$ cannot be ensured. Therefore, derivation of differential form (6.5) from (6.4) is not possible when the matter is not homogeneous and not isotropic.

(3) In case of energy equation, rate of work $\boldsymbol{\sigma} : \dot{\boldsymbol{\varepsilon}}$ requires continuity of both $\boldsymbol{\sigma}$ and $\boldsymbol{\varepsilon}$ everywhere in $\bar{V}$ which is not possible when bimaterial interfaces exist

in $\bar{V}$. Thus, in this case, differential forms of energy equation is not possible if the matter in $\bar{V}$ is not isotropic and homogeneous.

(4) Based on Remarks (2) and (3), we must consider the integral form of the conservation and balance laws when the matter is non-homogeneous and non-isotropic as differential or local form may not always be possible or may not be possible at all. But, when the matter is homogeneous and isotropic, (6.5) from (6.4) is always valid, i.e in this case, integral or differential forms are synonymous.

(5) Most writings on continuum mechanics [11,40,42,44,59,99,128,129] give differential form of the CBL based on localization theorem that requires continuity of the integrand without any explanation regarding the constitution of the matter in the volume for which this is possible.

(6) The reasoning based on physics and material considerations is transparent and clearly explains why and when localization theorem is valid. When the matter is isotropic and homogeneous, each material point in the volume $\bar{V}$ is identical, hence at all material point interfaces, all quantities are continuous in the volume $\bar{V}$. Thus, in this case $\bar{\boldsymbol{G}}(\cdot)$ is always continuous in all of its arguments for any $\bar{\boldsymbol{x}} \in \bar{V}$. Said in a different way, for isotropic and homogeneous matter, each material point in $\bar{V}$ is identical, hence how many material points are in $\bar{V}$ is of no consequence. Thus, the integral over $\bar{V}$ holds for a single material point. Hence, we have the differential form of the CBL (6.5) from (6.4).

(7) In the derivation of the constitutive theories, we shall observe that when using entropy inequality and representation theorem, the constitutive theories can only be derived for isotropic and homogeneous matter, further confirming the validity of isotropy and homogeneity in deriving differential forms of the conservation and balance laws.

(8) Throughout the book we cannot over emphasize the significance of the assumption of homogeneous and isotropic matter in deriving differential form or local form of the CBL as opposed to using localization theorem. Localization theorem is a mathematical fact, but it fails to explain or illustrate the physics in $\bar{V}$ for which it remains valid. In the rest of the book, we do not take any further issue with this, but simply use, "*based on localization theorem, the volume $\bar{V}(t)$ is arbitrary, hence the integrand can be set to zero,*" to obtain differential forms of the conservation and balance laws.

6.3 Mass density

Classical continuum mechanics is based on the fundamental idea that all material volumes or bodies possess continuous mass densities and the laws of motion and the constitution of the matter are valid for every part of the volume or body no matter how small. Based on this, completely enclosed volume dV by a closed surface with total mass dm contained in ΔV possesses a mass density ρ defined by

$$\rho = \lim_{\Delta V \to 0} \frac{\Delta m}{\Delta V} \tag{6.6}$$

In (6.6), ρ is independent of the size of ΔV and depends only on position vector $\boldsymbol{x}$ of a point in ΔV and time t. If we plot $\frac{\Delta m}{\Delta V}$ for progressively reducing ΔV, then we find when $\Delta V > \Delta\widetilde{V}$ (a threshold or critical value), ρ is nearly constant.

For $\Delta\widetilde{V} < \Delta V$, ρ begins to show dependence on ΔV. As ΔV approaches zero the dependence of ρ on ΔV becomes extreme. Thus, CCM is not a good theory in the range $\Delta V < \Delta\widetilde{V}$. For $\Delta V < \Delta\widetilde{V}$, molecular, atomic, granular physics become progressively more significant with progressively reducing ΔV.

In continuum mechanics, we assume that the limiting process (6.6) always exist at each material point. Both Δm and ΔV are always positive therefore the mass density $\rho = \rho(\boldsymbol{x}, t)$ is always positive.

6.4 Conservation of mass: CM

We assume that the mass of the matter contained in every small volume surrounding the material point $\underset{\sim}{Q}$ at x_i is conserved. Let ρ_0 be the density of the mass of matter contained in volume V surrounding the material point $\underset{\sim}{Q}$ in the reference configuration. Upon deformation the point $\underset{\sim}{Q}$ occupies position $\underset{\sim}{\bar{Q}}$ in the

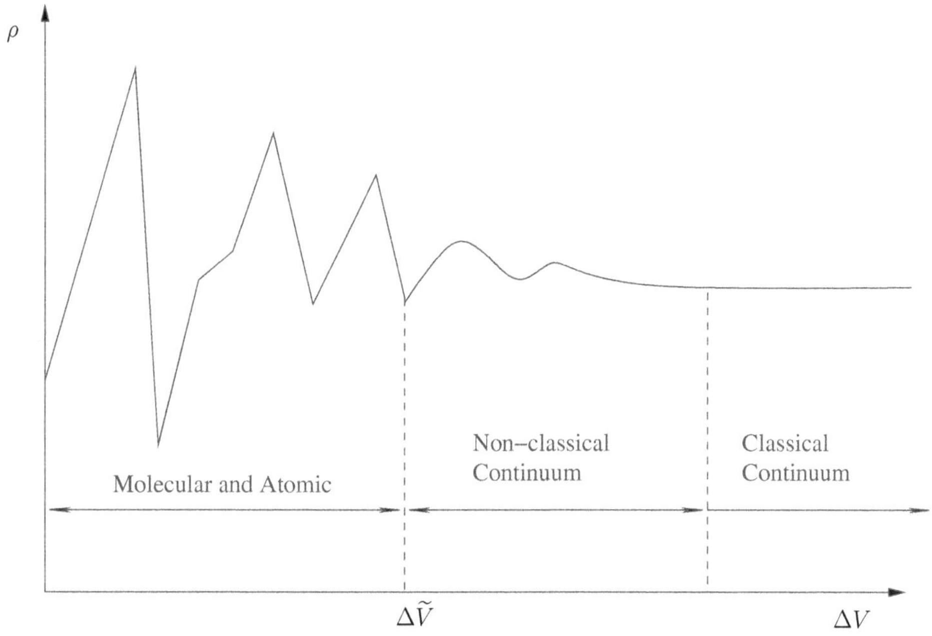

Figure 6.1: Density change with volume

current configuration at location $\bar{x}_i$ and the volume V changes to $\bar{V}$ with density $\bar{\rho}(\bar{x}_i, t) = \bar{\rho}$. Since the mass is conserved during deformation, the masses in volumes V and $\bar{V}$ are the same, i.e.

$$\int_V \rho_0 \, dV = \int_{\bar{V}(t)} \bar{\rho} d\bar{V} \tag{6.7}$$

At this point we could, for example, use $d\bar{V} = |J| \, dV$ in (6.7) to obtain

$$\int_V \rho_0 \, dV = \int_V |J| \, \rho dV \tag{6.8}$$

We begin with (6.7) to derive the continuity equation in Eulerian description that ensures conservation of mass in volume $\bar{V}$ surrounding material point $\bar{\underset{\sim}{Q}}$ in the current configuration during evolution. Conservation of mass implies that material derivative of (6.7) must be zero.

$$\frac{D}{Dt}\Big(\int_V \rho_0 \, dV\Big) = \frac{D}{Dt}\Big(\int_{\bar{V}(t)} \bar{\rho} d\bar{V}\Big) = 0 \tag{6.9}$$

Since V is not a function of time and $\rho_0 = \rho_0(\boldsymbol{x})$, the integral on left side of (6.9) is not a function of time, hence its material derivative is zero. Thus, (6.9) reduces to the following.

$$\frac{D}{Dt}\Big(\int_{\bar{V}(t)} \bar{\rho} d\bar{V}\Big) = 0 \tag{6.10}$$

Since the deforming volume $\bar{V}(t)$ is a function of time, material derivative $\frac{D}{Dt}$ cannot be taken inside the integral. We need special means to accomplish this, which is through the use of the transport theorem. We leave the derivation of conservation of mass temporarily at this stage and consider details of the transport theorem first.

6.5 Transport theorem

In continuum mechanics we often have the need to obtain the rate of change of integrals (over length, area, volume) that arise from conservation of mass, rate of change of linear momentum, etc. in which the domain of definition of the integrals is a function of time. The transport theorem is very helpful in these instances. Consider the following:

$$\bar{f} = \int_{\bar{V}(t)} \bar{g}(\bar{x}_1, \bar{x}_2, \bar{x}_3, t) d\bar{V} \tag{6.11}$$

where $\bar{V}(t)$ is an arbitrary volume in the current configuration. Consider the material derivative of $\bar{f}$ in (6.11)).

$$\frac{D\bar{f}}{Dt} = \frac{D}{Dt}\Big(\int_{\bar{V}(t)} \bar{g}(\bar{x}_1, \bar{x}_2, \bar{x}_3, t) d\bar{V}\Big) \tag{6.12}$$

Remarks

(1) Since the volume $\bar{V} = \bar{V}(t)$ is changing in time, the material derivative $\frac{D}{Dt}$ cannot be brought inside the integral.

(2) The volume of integration always contains the same material particles.

(3) In (6.11), the material derivative $\frac{D}{Dt}$ can be brought inside the integral if somehow we are able to change the volume of integration from $\bar{V}(t)$ to V,

the volume in the reference configuration (independent of time). This can be accomplished using either of two approaches described in the following.

6.5.1 Approach I

This derivation is much simpler than approach II in Section 6.5.2. We note that

$$\frac{D\bar{f}}{Dt} = \frac{D}{Dt}\Big(\int\limits_{\bar{V}(t)} \bar{g}(\bar{x}_i, t) d\bar{V} \Big) = \frac{D}{Dt}\Big(\int\limits_{V} g\,|J|\,dV \Big) \tag{6.13}$$

$$\therefore \quad \frac{D\bar{f}}{Dt} = \int\limits_{V} \frac{D}{Dt}(g\,|J|\,)dV \tag{6.14}$$

6.5.2 Approach II

This is an alternate approach to approach I in Section 6.5.1.

Multiply and divide $gd\bar{V}$ by $\bar{\rho}$ and note that

$$\bar{g}\,d\bar{V} = \frac{\bar{g}}{\bar{\rho}}\bar{\rho}\,d\bar{V} = \frac{\bar{g}}{\bar{\rho}}d\bar{m} = \frac{\bar{g}}{\bar{\rho}}dm \tag{6.15}$$

Substituting from (6.15) into (6.12)

$$\frac{D\bar{f}}{Dt} = \frac{D}{Dt}\Big(\int\limits_{\bar{V}(t)} \bar{g}(\bar{x}_1, \bar{x}_2, \bar{x}_3, t) d\bar{V} \Big) = \frac{D}{Dt}\Big(\int\limits_{m} \frac{\bar{g}}{\bar{\rho}} dm \Big) \tag{6.16}$$

Since m, mass of volume V or $\bar{V}(t)$, is independent of time, $\frac{D}{Dt}$ can be brought inside the integral in (6.16). Therefore

$$\frac{D\bar{f}}{Dt} = \frac{D}{Dt}\Big(\int\limits_{\bar{V}(t)} \bar{g}(\bar{x}_1, \bar{x}_2, \bar{x}_3, t) d\bar{V} \Big) = \int\limits_{m} \frac{D}{Dt}\Big(\frac{\bar{g}}{\bar{\rho}}\Big) dm \tag{6.17}$$

$$\therefore \quad \frac{D\bar{f}}{Dt} = \frac{D}{Dt}\Big(\int\limits_{\bar{V}(t)} \bar{g}(\bar{x}_1, \bar{x}_2, \bar{x}_3, t) d\bar{V} \Big) = \int\limits_{m} \Big(\frac{1}{\bar{\rho}}\frac{D\bar{g}}{Dt} - \frac{\bar{g}}{\bar{\rho}^2}\frac{D\bar{\rho}}{Dt} \Big) dm \tag{6.18}$$

We note from conservation of mass (in Lagrangian description, see Chapter 7) $dm = \rho_0 dV = \bar{\rho} d\bar{V}$. Since $d\bar{V} = |J|\,dV$ then $\rho_0 dV = \rho\,|J|\,dV$. Therefore, $\rho_0 = \rho\,|J|$ and we can write

$$\rho(x_i, t) = \frac{\rho_0(x_i)}{|J|} = \bar{\rho}(\bar{x}_i, t) \tag{6.19}$$

$$\therefore \quad \frac{D}{Dt}(\bar{\rho}(\bar{x}_i, t)) = \frac{D}{Dt}\Big(\frac{\rho_0(x_i)}{|J|}\Big) = \frac{1}{|J|}\frac{D}{Dt}(\rho_0(x_i)) + \rho_0(x_i)\frac{D}{Dt}\Big(\frac{1}{|J|}\Big) \tag{6.20}$$

$$= \rho_0(x_i)\frac{D}{Dt}\Big(\frac{1}{|J|}\Big) \quad ; \quad \text{as} \quad \frac{D}{Dt}(\rho_0(x_i)) = 0$$

$$= -\frac{\rho_0(x_i)}{|J|^2}\frac{D\,|J|}{Dt} \tag{6.21}$$

Substituting (6.21) into (6.18)

$$\begin{aligned}\frac{D\bar{f}}{Dt} &= \int_m \left(\frac{1}{\bar{\rho}} \frac{D\bar{g}}{Dt} - \frac{\bar{g}}{\bar{\rho}^2} \left(- \frac{\rho_0}{|J|^2} \frac{D|J|}{Dt} \right) \right) dm \\ &= \int_m \left(\frac{1}{\bar{\rho}} \frac{D\bar{g}}{Dt} + \bar{g} \frac{\rho_0}{\bar{\rho}^2 |J|^2} \frac{D|J|}{Dt} \right) dm\end{aligned} \tag{6.22}$$

Using (6.19), we can write the second term in the integrand of (6.22) as follows:

$$\bar{g} \frac{\rho_0}{\bar{\rho}^2 |J|^2} \frac{D|J|}{Dt} = g \frac{\rho_0}{(\rho_0 / |J|)^2 |J|^2} \frac{D|J|}{Dt} = \frac{g}{\rho_0} \frac{D|J|}{Dt} \tag{6.23}$$

Substituting from (6.23) into (6.22) and using $\bar{\rho}(\bar{x}_i, t) = \rho_0(x_i)/|J|$ from (6.19), we obtain

$$\begin{aligned}\frac{D\bar{f}}{Dt} &= \frac{D}{Dt} \left(\int_{\bar{V}(t)} \bar{g}(\bar{x}_i, t) d\bar{V} \right) = \int_m \left(\frac{|J|}{\rho_0} \frac{Dg}{Dt} + \frac{g}{\rho_0} \frac{D|J|}{Dt} \right) dm \qquad (6.24) \\ &= \int_m \left(|J| \frac{Dg}{Dt} + g \frac{D|J|}{Dt} \right) \frac{dm}{\rho_0} = \int_V \left(|J| \frac{Dg}{Dt} + g \frac{D|J|}{Dt} \right) dV\end{aligned}$$

$$\therefore \quad \frac{D\bar{f}}{Dt} = \frac{D}{Dt} \left(\int_{\bar{V}(t)} \bar{g}(\bar{x}_i, t) d\bar{V} \right) = \int_V \frac{D}{Dt} (g |J|) dV \tag{6.25}$$

where is same as (6.14)

6.5.3 Continued development of transport theorem

Using (6.14) or (6.25), we continue the development of transport theorem. Expanding the material derivative term in the integrand in (6.14) we can write from Chapter 5

$$\frac{D\bar{f}}{Dt} = \int_V \left(|J| \frac{Dg}{Dt} + g \frac{D|J|}{Dt} \right) dV \tag{6.26}$$

$$\text{Recall} \quad \frac{D|J|}{Dt} = |\dot{J}| = |J| \frac{\partial \bar{v}_i}{\partial \bar{x}_i} = (\bar{\boldsymbol{\nabla}} \cdot \bar{\boldsymbol{v}}) |J| \tag{6.27}$$

Substitute (6.27) in (6.26) and noting that $|J| \, dV = d\bar{V}$, we obtain

$$\begin{aligned}\therefore \quad \frac{D\bar{f}}{Dt} &= \int_V \left(\frac{D\bar{g}}{Dt} + \bar{g} \bar{\boldsymbol{\nabla}} \cdot \bar{\boldsymbol{v}} \right) d\bar{V} \\ &= \int_{\bar{V}} \left(\frac{\partial \bar{g}}{\partial t} + (\bar{\boldsymbol{v}} \cdot \bar{\boldsymbol{\nabla}}) \bar{g} + \bar{g} (\bar{\boldsymbol{\nabla}} \cdot \bar{\boldsymbol{v}}) \right) d\bar{V} \qquad (6.28) \\ &= \int_{\bar{V}} \left(\frac{\partial \bar{g}}{\partial t} + \bar{\boldsymbol{v}} \cdot (\bar{\boldsymbol{\nabla}} \bar{g}) + \bar{g} (\bar{\boldsymbol{\nabla}} \cdot \bar{\boldsymbol{v}}) \right) d\bar{V} = \int_{\bar{V}} \left(\frac{\partial \bar{g}}{\partial t} + \bar{\boldsymbol{\nabla}} \cdot (\bar{g} \bar{\boldsymbol{v}}) \right) d\bar{V}\end{aligned}$$

$$\therefore \quad \frac{D\bar{f}}{Dt} = \frac{D}{Dt}\Big(\int\limits_{\bar{V}(t)} \bar{g}(\bar{x}_i, t) d\bar{V} \Big) = \int\limits_{\bar{V}(t)} \Big(\frac{\partial \bar{g}}{\partial t} + \bar{\boldsymbol{\nabla}} \cdot (\bar{g}\, \bar{\boldsymbol{v}}) \Big) d\bar{V} \tag{6.29}$$

Equation (6.29) is the final form of the transport theorem. This theorem allows us to bring $\frac{D}{Dt}$ inside the integral when the volume of the integral is a function of time, i.e., $\bar{V}(t)$. This theorem is helpful in deriving mathematical models or governing differential equations based on conservation and balance laws in Eulerian description.

6.6 Conservation of mass: CM

6.6.1 Integral form of CM

Recall Equation (6.10)

$$\frac{D}{Dt}\Big(\int\limits_{\bar{V}(t)} \bar{\rho}(\bar{x}_i, t) d\bar{V} \Big) = 0 \tag{6.30}$$

Using the transport theorem, (6.30) can be written as

$$\int\limits_{\bar{V}(t)} \Big(\frac{\partial \bar{\rho}}{\partial t} + \bar{\boldsymbol{\nabla}} \cdot (\bar{\rho} \bar{\boldsymbol{v}}) \Big) d\bar{V} = 0 \quad \text{or} \quad \int\limits_{\bar{V}} \left(\frac{\partial \bar{\rho}}{\partial t} + \frac{\partial}{\partial \bar{x}_j} (\bar{\rho} \bar{v}_j) \right) d\bar{V} = 0 \tag{6.31}$$

Equation (6.31) is the integral form of the conservation of mass. The integral in (6.31) holds in Riemann sense when the integrand is continuous everywhere in $\bar{V}(t)$. If there are isolated discontinuities of the integrand in (6.31) in volume $\bar{V}(t)$, but the discontinuities are square integrable, then the integral in (6.31) holds in Lebesgue sense. This integral form must be used if the integrand is not continuous everywhere in $\bar{V}(t)$.

6.6.2 Differential form of CM

If the localization theorem holds for (6.31), then the volume $\bar{V}(t)$ in (6.31) is arbitrary and the integrand in (6.31) can be set to zero and we have

$$\frac{\partial \bar{\rho}}{\partial t} + \bar{\boldsymbol{\nabla}} \cdot (\bar{\rho} \bar{\boldsymbol{v}}) = 0 \quad \text{or} \quad \frac{\partial \bar{\rho}}{\partial t} + \frac{\partial}{\partial \bar{x}_j} (\bar{\rho} \bar{v}_j) = 0 \tag{6.32}$$

Equation (6.32) is the differential form of the *continuity equation* representing conservation of mass. It ensures local conservation of mass in a volume $\bar{V}$ surrounding a material point.

Remarks (differential form of CM)

(1) Equation (6.32) is valid for compressible matter.

(2) When the deforming matter is incompressible $\bar{\rho} = \rho_0 =$ constant, hence $\frac{\partial \bar{\rho}}{\partial t} = \frac{\partial \rho_0}{\partial t} = 0$, therefore (6.32) reduces to

$$\bar{\boldsymbol{\nabla}} \cdot (\rho_0 \bar{\boldsymbol{v}}) = \rho_0 (\bar{\boldsymbol{\nabla}} \cdot \bar{\boldsymbol{v}}) = 0 \qquad \text{or} \qquad \bar{\boldsymbol{\nabla}} \cdot \bar{\boldsymbol{v}} = 0 \tag{6.33}$$

That is, in this case *the velocity field is divergence free.*

(3) We note that (6.33) holds for evolutions as well as stationary processes when the matter is incompressible.

(4) If the matter is compressible but time evolution of the deformation reaches a stationary state, (6.32) reduces to

$$\bar{\boldsymbol{\nabla}} \cdot (\bar{\rho}\,\bar{\boldsymbol{v}}) = 0 \tag{6.34}$$

(5) We can also derive an alternate form of the continuity equation (6.32) for the compressible continua. Consider (6.32)

$$\frac{\partial \bar{\rho}}{\partial t} + \bar{\boldsymbol{\nabla}} \cdot (\bar{\rho}\,\bar{\boldsymbol{v}}) = 0 \tag{6.35}$$

Expanding the second term in (6.35) and regrouping,

$$\left(\frac{\partial \bar{\rho}}{\partial t} + \bar{\boldsymbol{v}} \cdot \bar{\boldsymbol{\nabla}}\bar{\rho}\right) + \bar{\rho}(\bar{\boldsymbol{\nabla}} \cdot \bar{\boldsymbol{v}}) = 0 \tag{6.36}$$

$$\text{or} \quad \frac{D\bar{\rho}}{Dt} + \bar{\rho}(\bar{\boldsymbol{\nabla}} \cdot \bar{\boldsymbol{v}}) = 0 \tag{6.37}$$

$$\text{or} \quad \frac{D\bar{\rho}}{Dt} + \bar{\rho}\,\mathrm{div}(\bar{\boldsymbol{v}}) = 0 \tag{6.38}$$

6.7 Kinematics of continuous media: BLM

In this section, we consider Newton's second law of motion for particle mechanics and reformulate it for a deforming volume of matter. Newton's laws of motion for particle mechanics are given in the following.

(1) *Newton's first law of motion*

In an inertial frame of reference, a free particle continues its state of rest or motion.

(2) *Newton's second law of motion*

In an inertial frame, the rate of change of linear momentum of a particle equals the resultant forces acting on it, that is

$$\frac{d}{dt}(m\,\boldsymbol{v}) = m\,\boldsymbol{a} = \sum \boldsymbol{F}$$

where $\boldsymbol{a} = \frac{d\boldsymbol{v}}{dt}$ is acceleration. Here, we have assumed that the mass m of the particle does not change. The term $m\boldsymbol{v}$ is the linear momentum and $\sum \boldsymbol{F}$ is the resultant force.

(3) *Newton's third law of motion*

To every action there is an equal and opposite reaction.

Remarks

(1) Newton's laws of motion are only valid in inertial frame in which the translation and rotation of the observer or of the frame of measurement are not functions of time.

(2) Newton's laws are for particle physics. Newton's second law of motion stated for particle mechanics must be reformulated for continuum mechanics when using it for a deforming volume of matter.

6.7.1 Preliminary considerations

Consider a volume V in the reference configuration with closed boundary ∂V (Figure 6.2). Let $\bar{V}(t)$ be the corresponding volume in the current configuration with closed boundary $\partial\bar{V}$. We generally consider two types of forces in the mechanics of non-polar continuum.

Body forces

Body forces are functions of the volume or mass of the matter. Forces such as gravity forces that act on all particles in the body (or matter) as a result of some external body not in direct contact with the one being studied are referred to as body forces. The body force is measured as a force per unit mass or per unit volume at the present location at a point in the continuum. Thus, if the body force per unit mass is F_i^b or $F_{x_i}^b$; $i = 1,2,3$ or $\boldsymbol{F}^b$, then the body force on an element of material enclosed in volume $d\bar{V}$ with mass $d\bar{m}$ or dm is dmF_i^b ; $i = 1,2,3$ or $dm\boldsymbol{F}^b$.

Surface force

Surface forces are the contact forces that exist across a surface of a body or volume or mass, which may be internal or external. It is generally assumed that the action of that part of the body exterior to the shaded region (Figure 6.2) on the part enclosed in the shaded region is equivalent to a system of forces acting on the bounding surface of the shaded region. This is due to cut principle of Cauchy (see Chapter 4).

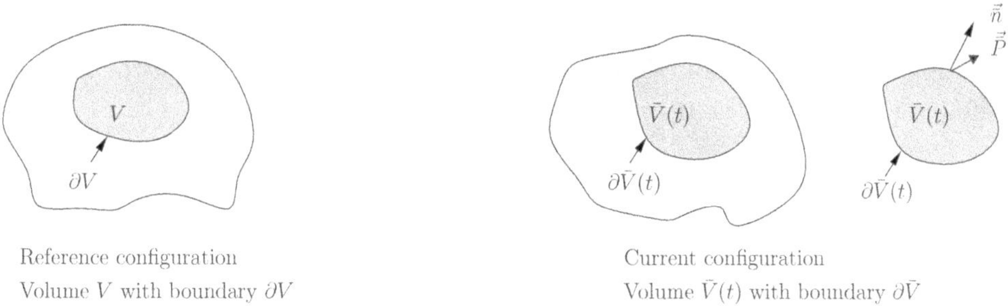

Figure 6.2: Reference and current configurations

6.7.2 Derivation of equations of BLM

We consider application of Newton's second law to a deforming volume of continua. Consider the undeformed or reference configuration (Figure 6.2) with an isolated volume of matter V with a closed boundary ∂V. Let x-frame be the fixed frame in which all material points are identified by their position coordinates. Consider an elementary tetrahedron (Figure 6.3(a)) whose oblique face $A_1A_2A_3$ is part of the boundary ∂V and with faces parallel to the planes of the x-frame. Upon finite deformation, the reference configuration deforms into the current configuration at time t (Figure 6.3(b)). The volume V changes to $\bar{V}(t)$ with its closed boundary $\partial\bar{V}(t)$. The material lines containing the edges of the elementary tetrahedron in the reference configuration deform into non-orthogonal curvilinear lines ($\tilde{o}$-$\tilde{x}_1$, $\tilde{o}$-$\tilde{x}_2$ and $\tilde{o}$-$\tilde{x}_3$). The edges of the deformed tetrahedron in the current configuration are formed by the covariant base vectors $\tilde{\boldsymbol{g}}_i$ (Figure 6.3(b)) that are tangent to the deformed material lines in the current configuration at point $\bar{\underset{\sim}{Q}}$. We must consider the deformed tetrahedron for applying Newton's second law. First, we note several important points regarding the deformed tetrahedron in the current configuration. (1) All of its faces $\bar{O}\bar{A}_2\bar{A}_3$, $\bar{O}\bar{A}_3\bar{A}_1$ and $\bar{O}\bar{A}_1\bar{A}_2$ are flat planes. (2) Faces $\bar{O}\bar{A}_2\bar{A}_3$, $\bar{O}\bar{A}_3\bar{A}_1$ and $\bar{O}\bar{A}_1\bar{A}_2$ are not orthogonal to each other (unlike their maps in the reference configuration). (3) The vectors normal to the faces of the deformed tetrahedron are contravariant base vectors $\tilde{\boldsymbol{g}}^i$. (4) The oblique plane $\bar{A}_1\bar{A}_2\bar{A}_3$ is part of boundary $\partial\bar{V}$. The deformed tetrahedron in Figure 6.3(b) represents true deformation of the tetrahedron in Figure 6.3(a). We must consider application of Newton's second law to the deformed tetrahedron in the current configuration.

Let $\bar{\boldsymbol{P}}$, given by $\{\bar{P}\} = [\bar{P}_{x_1}, \bar{P}_{x_2}, \bar{P}_{x_3}]^T$, with $\bar{P}_{x_i}$ being its components in the x-frame, be the resultant force per unit area acting on the oblique plane $\bar{A}_1\bar{A}_2\bar{A}_3$ with unit exterior normal $\bar{\boldsymbol{n}}$. The faces $\bar{O}\bar{A}_2\bar{A}_3$, $\bar{O}\bar{A}_2\bar{A}_1$ and $\bar{O}\bar{A}_1\bar{A}_2$ have tractions acting on them that result in average stress distribution over these areas. Details are shown in Figure 6.3. Let $\bar{V}(t)$ be the volume of the deformed tetrahedron. Let the oblique surface $\bar{A}_1\bar{A}_2\bar{A}_3$ be represented by $\partial\bar{V}$ and $\bar{\boldsymbol{F}}_{\partial\bar{V}}$ be the resultant force acting on $\partial\bar{V}$. Application of Newton's second law of motion to the deformed volume $\bar{V}(t)$ of the deformed tetrahedron (balance of linear momenta) can be written as:

$$\begin{pmatrix}\textbf{Rate of change of}\\ \textbf{linear momenta of}\\ \textbf{the mass in}\\ \textbf{volume } \bar{V}(t)\end{pmatrix} = \begin{pmatrix}\bar{\boldsymbol{F}}^b_{\bar{V}}\ \textbf{; Body}\\ \textbf{forces acting}\\ \textbf{on the mass in}\\ \textbf{volume } \bar{V}(t)\end{pmatrix} + \begin{pmatrix}\bar{\boldsymbol{F}}_{\partial\bar{V}}\ \textbf{; Re-}\\ \textbf{sultant force}\\ \textbf{acting on}\\ \textbf{surface } \partial\bar{V}\end{pmatrix} \tag{6.39}$$

Let $\bar{\boldsymbol{F}}^b$ (or $\bar{F}^b_i$) be the body force per unit mass in x-frame acting on the mass of volume $\bar{V}(t)$. Let $\bar{\boldsymbol{F}}^b_{\bar{V}}$ be the total body force acting on $\bar{V}(t)$, then

$$\bar{\boldsymbol{F}}^b_{\bar{V}} = \int\limits_{\bar{V}(t)} \bar{\boldsymbol{F}}^b \bar{\rho} d\bar{V} \tag{6.40}$$

Next we consider $\bar{\boldsymbol{F}}_{\partial\bar{V}}$. This requires careful considerations and physical interpretation of the stress tensor acting on the faces of the deformed tetrahedron. Since $\bar{\boldsymbol{F}}_{\partial\bar{V}}$ is the total resultant force acting on surface $\partial\bar{V}$ and $\bar{\boldsymbol{P}}$ is the total resultant force per unit area acting on elemental area $d\bar{A}$ of $\partial\bar{V}$, we can write

$$\bar{\boldsymbol{F}}_{\partial\bar{V}} = \int\limits_{\partial\bar{V}} \bar{\boldsymbol{P}} d\bar{A} \tag{6.41}$$

We recall from Chapter 4 that contravariant stress tensor $\underset{\sim}{\boldsymbol{\sigma}}^{(0)}$ acting on the faces of the deformed tetrahedron of Figure 6.3(b) is a natural way to define the stress tensor. Using dyads $\tilde{\boldsymbol{g}}_i \otimes \tilde{\boldsymbol{g}}_j$.

$$\boldsymbol{\sigma}^{(0)} = \tilde{\boldsymbol{g}}_i \otimes \tilde{\boldsymbol{g}}_j \underset{\sim}{\sigma}^{(0)}_{ij} = \tilde{\boldsymbol{g}}_i \tilde{\boldsymbol{g}}_j \underset{\sim}{\sigma}^{(0)}_{ij} \tag{6.42}$$

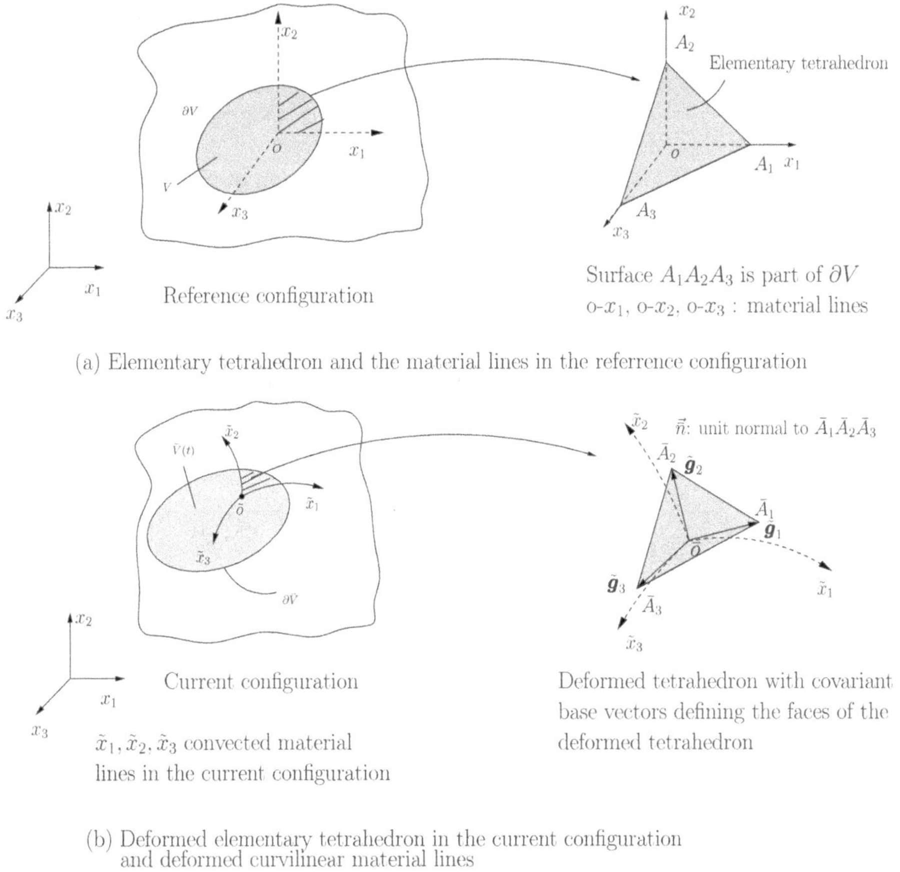

Figure 6.3: Elementary tetrahedra in the reference and current configurations

in which $\boldsymbol{\sigma}^{(0)}$ is the contravariant Cauchy stress tensor. Upon substituting definitions of $\tilde{\boldsymbol{g}}_i$ and $\tilde{\boldsymbol{g}}_j$ in (6.42) we can obtain explicit expressions for $\sigma_{ij}^{(0)}$ of $\boldsymbol{\sigma}^{(0)}$ with dyads $\boldsymbol{e}_i \otimes \boldsymbol{e}_j$. Thus, we can view $\boldsymbol{\sigma}^{(0)}$ as a Cauchy stress tensor acting on a tetrahedron at the location $\bar{\boldsymbol{x}}$ whose faces are orthogonal to each other and parallel to the planes of x-frame but with the same oblique plane $\bar{A}_1\bar{A}_2\bar{A}_3$ as that of the deformed tetrahedron of Figure 6.3(b). The tensor $\boldsymbol{\sigma}^{(0)}$ in (6.42) is symmetric (shown later) and is a Lagrangian description of the Cauchy stress tensor from which we can define Eulerian description $\bar{\boldsymbol{\sigma}}^{(0)}$; i.e., $\bar{\boldsymbol{\sigma}}^{(0)}$ is Eulerian description of contravariant Cauchy stress tensor as a function of $\bar{\boldsymbol{x}}$. With $\bar{\boldsymbol{\sigma}}^{(0)}$ and $\bar{\boldsymbol{P}}$ defined, we can now use the Cauchy principle (see Chapter 4).

$$\bar{\boldsymbol{P}} = (\bar{\boldsymbol{\sigma}}^{(0)})^T \cdot \bar{\boldsymbol{n}} \tag{6.43}$$

$$\bar{\boldsymbol{F}}_{\partial \bar{V}} = \int_{\partial \bar{V}} (\bar{\boldsymbol{\sigma}}^{(0)})^T \cdot \bar{\boldsymbol{n}} \, d\bar{A} = \int_{\partial \bar{V}} (\bar{\boldsymbol{\sigma}}^{(0)})^T \cdot d\bar{\boldsymbol{A}} \tag{6.44}$$

The rate of change of linear momentum for volume $\bar{V}(t)$ is given by

$$\frac{D}{Dt}\int_{\bar{V}(t)} \bar{\rho}\,\bar{\boldsymbol{v}}\,d\bar{V} \tag{6.45}$$

Using (6.40), (6.44) and (6.45) in (6.39) (i.e., application of Newton's second law to the volume $\bar{V}(t)$) gives the following in x-frame.

$$\frac{D}{Dt}\int_{\bar{V}(t)} \bar{\rho}\,\bar{\boldsymbol{v}}\,d\bar{V} = \int_{\bar{V}(t)} \bar{\boldsymbol{F}}^b \bar{\rho}\,d\bar{V} + \int_{\partial\bar{V}(t)} (\bar{\boldsymbol{\sigma}}^{(0)})^T \cdot d\bar{\boldsymbol{A}} \tag{6.46}$$

We note that $\bar{\boldsymbol{\sigma}}^{(0)}$ is the symmetric Cauchy stress tensor in the x-frame, but in the following derivation we treat it to be non-symmetric before we present a more general proof of its symmetry in this chapter.

Application of Gauss's divergence theorem to the integral over $\partial\bar{V}$ yields

$$\frac{D}{Dt}\int_{\bar{V}(t)} \bar{\rho}\,\bar{\boldsymbol{v}}\,d\bar{V} = \int_{\bar{V}(t)} \bar{\boldsymbol{F}}^b \bar{\rho}\,d\bar{V} + \int_{\bar{V}(t)} \bar{\boldsymbol{\nabla}} \cdot \bar{\boldsymbol{\sigma}}^{(0)} d\bar{V} \tag{6.47}$$

or

$$\frac{D}{Dt}\int_{\bar{V}(t)} \bar{\rho}\,\bar{v}_i\,d\bar{V} = \int_{\bar{V}(t)} \bar{F}_i^b \bar{\rho}\,d\bar{V} + \int_{\bar{V}(t)} \frac{\partial\bar{\sigma}_{ji}^{(0)}}{\partial\bar{x}_j} d\bar{V} \tag{6.48}$$

Equation (6.47) or (6.48) is a result of direct application of Newton's second law of motion to the volume $\bar{V}(t)$ in the current configuration, i.e., balance of linear momenta. Equation (6.48) are as the *most basic integral form of balance of linear momenta equations*. The first term in (6.48) contains the material derivative of the volume integral, in which the volume is a function of time t that needs to be expanded. In the following, we begin with (6.48) and present two alternate approaches of deriving a more explicit simplified forms of (6.48).

6.7.3 Approach I: Integral and differential forms of BLM

Consider (6.48) and apply the transport theorem to the left side by treating $\bar{\rho}\bar{v}_i$ as g, to obtain

$$\int_{\bar{V}(t)} \Big(\frac{\partial(\bar{\rho}\bar{v}_i)}{\partial t} + \bar{\boldsymbol{\nabla}} \cdot (\bar{\rho}\,\bar{v}_i\,\bar{\boldsymbol{v}})\Big) d\bar{V} = \int_{\bar{V}(t)} \bar{F}_i^b \bar{\rho} d\bar{V} + \int_{\bar{V}(t)} \frac{\partial\bar{\sigma}_{ji}^{(0)}}{\partial\bar{x}_j} d\bar{V} \tag{6.49}$$

or

$$\int_{\bar{V}(t)} \Big(\frac{\partial(\bar{\rho}\bar{v}_i)}{\partial t} + \frac{\partial}{\partial\bar{x}_j}(\bar{v}_i(\bar{\rho}\bar{v}_j)) - \bar{\rho}\bar{F}_i^b - \frac{\partial\bar{\sigma}_{ji}^{(0)}}{\partial\bar{x}_j}\Big) d\bar{V} = 0 \tag{6.50}$$

Equation (6.50) in the integral form of the balance of linear momenta. The integral in (6.50) is Riemann if the integrand is continuous everywhere in the volume $V(t)$. If there are isolated discontinuities of the integrand in $\bar{V}(t)$, but the discontinuities are square integrable, then the integrand (6.50) holds in the Lebesgue sense. This integral form must be used if the integrand is not continuous everywhere in $\bar{V}(t)$. If the localization theorem holds for (6.50), then volume $V(t)$ is arbitrary and we can set the integrand equal to zero.

$$\frac{\partial(\bar{\rho}\bar{v}_i)}{\partial t} + \bar{\boldsymbol{\nabla}} \cdot (\bar{\rho}\,\bar{v}_i\,\bar{\boldsymbol{v}}) - \bar{\rho}\bar{F}_i^b - \frac{\partial \bar{\sigma}_{ji}^{(0)}}{\partial \bar{x}_j} = 0 \tag{6.51}$$

$$\text{or} \quad \frac{\partial(\bar{\rho}\bar{v}_i)}{\partial t} + \frac{\partial(\bar{v}_i(\bar{\rho}\bar{v}_j))}{\partial \bar{x}_j} - \bar{\rho}\bar{F}_i^b - \frac{\partial \bar{\sigma}_{ji}^{(0)}}{\partial \bar{x}_j} = 0 \tag{6.52}$$

Equations (6.51) and (6.52) are differential form or local form of equation for BLM. In the following we consider if the integral and the differential form of BLM ((6.50) and (6.51) or (6.52)) can be further simplified.

6.7.3.1 Simplified form of integral form of BLM

Consider (6.50), expand first two terms in the integrand

$$\int_{\bar{V}(t)} \left(\bar{v}_i \frac{\partial \bar{\rho}}{\partial t} + \bar{\rho}\frac{\partial \bar{v}_i}{\partial t} + \bar{v}_i \frac{\partial(\bar{\rho}\bar{v}_j)}{\partial \bar{x}_j} + \bar{\rho}\bar{v}_j \frac{\partial \bar{v}_i}{\partial \bar{x}_j} - \bar{\rho}\bar{F}_i^b - \frac{\partial \bar{\sigma}_{ji}^{(0)}}{\partial \bar{x}_j} \right) d\bar{V} = 0 \tag{6.53}$$

Group the first and the third terms in the integrand.

$$\int_{\bar{V}(t)} \left(\bar{v}_i \left(\frac{\partial \bar{\rho}}{\partial t} + \frac{\partial(\bar{\rho}\bar{v}_j)}{\partial \bar{x}_j} \right) + \bar{\rho}\frac{\partial \bar{v}_i}{\partial t} + \bar{\rho}\bar{v}_j \frac{\partial \bar{v}_i}{\partial \bar{x}_j} - \bar{\rho}\bar{F}_i^b - \frac{\partial \bar{\sigma}_{ji}^{(0)}}{\partial \bar{x}_j} \right) d\bar{V} = 0 \tag{6.54}$$

From integral form of continuity equation (CM) (6.31) we have

$$\int_{\bar{V}(t)} \left(\frac{\partial \bar{\rho}}{\partial t} + \frac{\partial}{\partial \bar{x}_j} (\bar{\rho}\bar{v}_j) \right) d\bar{V} = 0 \tag{6.55}$$

Clearly (6.55) cannot be substituted for the first term in the integrand of (6.54). Thus, further simplification of the integral form of BLM (6.50) is not possible. Hence, we must maintain it in the form stated in (6.50) or (6.54).

6.7.3.2 Simplified form of differential form of BLM

We begin with the differential form of (6.53).

$$\bar{v}_i \frac{\partial \bar{\rho}}{\partial t} + \bar{\rho}\frac{\partial \bar{v}_i}{\partial t} + \bar{v}_i \frac{\partial(\bar{\rho}\bar{v}_j)}{\partial \bar{x}_j} + \bar{\rho}\bar{v}_j \frac{\partial \bar{v}_i}{\partial \bar{x}_j} - \bar{\rho}\bar{F}_i^b - \frac{\partial \bar{\sigma}_{ji}^{(0)}}{\partial \bar{x}_j} = 0 \tag{6.56}$$

group the first and the third terms in (6.56)

$$\bar{v}_i \left(\frac{\partial \bar{\rho}}{\partial t} + \frac{\partial(\bar{\rho}\bar{v}_i)}{\partial \bar{x}_j} \right) + \bar{\rho}\frac{\partial \bar{v}_i}{\partial t} + \bar{\rho}\bar{v}_j \frac{\partial \bar{v}_j}{\partial \bar{x}_j} - \bar{\rho}\bar{F}_i^b - \frac{\partial \bar{\sigma}_{ji}^{(0)}}{\partial \bar{x}_j} = 0 \tag{6.57}$$

From the differential form of the continuity equation (6.32)

$$\frac{\partial \bar{\rho}}{\partial t} + \frac{\partial(\bar{\rho}\bar{v}_j)}{\partial \bar{x}_j} = 0 \tag{6.58}$$

Substituting (6.58) in (6.57) we obtain

$$\bar{\rho}\frac{\partial \bar{v}_i}{\partial t} + \bar{\rho}\bar{v}_j\frac{\partial \bar{v}_i}{\partial \bar{x}_j} - \bar{\rho}\bar{F}_i^b - \frac{\partial \bar{\sigma}_{ji}^{(0)}}{\partial \bar{x}_j} = 0 \tag{6.59}$$

This is the simplified form of the differential or local form of BLM equations.

6.7.4 Approach II

6.7.4.1 Integral form of BLM

In this section, we consider an alternate approach of deriving integral form of BLM equations. We begin with (6.47)

$$\frac{D}{Dt}\int_{\bar{V}(t)} \bar{\rho}\bar{\boldsymbol{v}}d\bar{V} = \int_{\bar{V}(t)} \bar{\boldsymbol{F}}^b\bar{\rho}d\bar{V} + \int_{\bar{V}(t)} \bar{\boldsymbol{\nabla}}\cdot\bar{\boldsymbol{\sigma}}^{(0)}d\bar{V} \tag{6.60}$$

In approach I, we used transport theorem. Here we consider if we can use CM in integral form to simplify (6.60)

$$\int_V \rho_0 dV = \int_{\bar{V}} \bar{\rho}d\bar{V} \tag{6.61}$$

Consider (6.60) and rewrite the term on the left side

$$\frac{D}{Dt}\int_{\bar{V}(t)} \bar{\boldsymbol{v}}(\bar{\rho}d\bar{V}) = \int_{\bar{V}(t)} \bar{\boldsymbol{F}}^b\bar{\rho}d\bar{V} + \int_{\bar{V}(t)} \bar{\boldsymbol{\nabla}}\cdot\bar{\boldsymbol{\sigma}}^{(0)}d\bar{V} \tag{6.62}$$

Clearly (6.61) cannot be substituted in (6.62), thus, in this approach material derivative in (6.60) cannot be taken inside the integral without using transport theorem. This approach fails to yield integral form of BLM derived in approach I. Hence, simplified form is not possible either in this approach. Thus, the integral form of the BLM remains as (6.50) or (6.54) regardless of approach I or II.

6.7.4.2 Differential form of BLM

We begin with (6.47) or (6.60)

$$\frac{D}{Dt}\int_{\bar{V}(t)} \bar{\rho}\bar{\boldsymbol{v}}d\bar{V} = \int_{\bar{V}(t)} \bar{\boldsymbol{F}}^b\bar{\rho}d\bar{V} + \int_{\bar{V}(t)} \bar{\boldsymbol{\nabla}}\cdot\bar{\boldsymbol{\sigma}}^{(0)}d\bar{V} \tag{6.63}$$

We consider the term on the left side of (6.63) and use CM in differential form (in Lagrangian description).

$$\begin{aligned}\frac{D}{Dt}\int_{\bar{V}} \bar{\boldsymbol{v}}\,(\bar{\rho}d\bar{V}) &= \frac{D}{Dt}\int_V \boldsymbol{v}(\rho_0 dV) = \int_V \frac{D}{Dt}(\boldsymbol{v}\rho_0)dV \\ &= \int_V \frac{D\boldsymbol{v}}{Dt}\rho_0 dV = \int_{\bar{V}(t)} \frac{D\bar{\boldsymbol{v}}}{Dt}\bar{\rho}d\bar{V}\end{aligned} \tag{6.64}$$

substituting for (6.64) into (6.63) and transferring all terms to the left side

$$\int_{\bar{V}(t)} \left(\bar{\rho}\frac{D\bar{\boldsymbol{v}}}{Dt} - \bar{\rho}\bar{\boldsymbol{F}}^b - \bar{\boldsymbol{\nabla}}\cdot\bar{\boldsymbol{\sigma}}^{(0)} \right) d\bar{V} = 0 \tag{6.65}$$

If the localization theorem holds for (6.65), then we can set the integrand to zero.

$$\bar{\rho}\frac{D\bar{\boldsymbol{v}}}{Dt} - \bar{\rho}\bar{\boldsymbol{F}}^b - \bar{\boldsymbol{\nabla}}\cdot\bar{\boldsymbol{\sigma}}^{(0)} = 0 \tag{6.66}$$

This is the differential form of BLM. This form is particularly useful in deriving the differential form of the energy equation.

Expanding the material derivative in (6.66)

$$\bar{\rho}\left(\frac{\partial \bar{v}_i}{\partial t} + (\bar{\boldsymbol{v}}\cdot\bar{\boldsymbol{\nabla}})\bar{v}_i\right) - \bar{\rho}\bar{F}_i^b - \frac{\partial \bar{\sigma}_{ji}^{(0)}}{\partial \bar{x}_j} = 0 \tag{6.67}$$

$$\text{or} \quad \bar{\rho}\frac{\partial \bar{v}_i}{\partial t} + \bar{\rho}\bar{v}_j\frac{\partial \bar{v}_i}{\partial \bar{x}_j} - \bar{\rho}\bar{F}_i^b - \frac{\partial \bar{\sigma}_{ji}^{(0)}}{\partial \bar{x}_j} = 0 \tag{6.68}$$

Equation (6.68) are the same as derived in approach I, equations (6.59).

Remarks

(1) The BLM equations (6.67) or (6.68) are also called *Navier-Stokes equations* in fluid mechanics writings (also see Remark (2) in Section 6.11.1).

(2) Equations (6.67) and (6.68) are valid for compressible matter and hence $\bar{\rho} = \bar{\rho}(\bar{x}_i, t)$.

(3) When the matter is incompressible, density is constant, hence $\bar{\rho}(\bar{x}_i, t) = \rho_0(x_i) = \rho_0 =$ constant. Equations (6.67) and (6.68) remain valid for this case as well, except that we replace $\bar{\rho}(\bar{x}_i, t)$ by $\rho_0(x_i) = \rho_0$ (constant).

(4) When the motion is independent of time, (6.67) and (6.68) reduce to

$$\bar{\rho}\left(\bar{\boldsymbol{v}}\cdot\bar{\boldsymbol{\nabla}}\right)\bar{\boldsymbol{v}} - \bar{\rho}\bar{\boldsymbol{F}}^b - \bar{\boldsymbol{\nabla}}\cdot\bar{\boldsymbol{\sigma}}^{(0)} = 0 \tag{6.69}$$

For incompressible matter $\bar{\rho}(\bar{x}_i, t) = \rho_0(x_i) = \rho_0$ (constant) in (6.69).

(5) For incompressible matter with infinitesimally small time dependent motion and deformation, $\bar{\rho}\bar{v}_j\frac{\partial \bar{v}_i}{\partial \bar{x}_j}$ term can be neglected in (6.67) and (6.68) and the distinction between contravariant and covariant measures of stress disappears. In this case, (6.67) and (6.68) reduce to

$$\rho_0\frac{\partial \bar{\boldsymbol{v}}}{\partial t} - \rho_0\bar{\boldsymbol{F}}^b - \bar{\boldsymbol{\nabla}}\cdot\bar{\boldsymbol{\sigma}}^{(0)} = 0 \tag{6.70}$$

In this case $\bar{x}_i \simeq x_i$ and $\rho(\bar{x}_i, t) = \rho_0(x_i) = \rho_0$.

(6) In Remark (5), if the motion is independent of time, then (6.70) reduces to

$$\rho_0\bar{\boldsymbol{F}}^b + \bar{\boldsymbol{\nabla}}\cdot\bar{\boldsymbol{\sigma}}^{(0)} = 0 \tag{6.71}$$

Obviously, (6.71) also holds in Lagrangian description due to the fact that $\bar{x}_i \simeq x_i$. The distinction between Lagrangian and Eulerian description disappears. These are the well known *equations of equilibrium in linear theory of elasticity.* Based on Remark (5), $\bar{\boldsymbol{x}}$ can be replaced by $\boldsymbol{x}$ and the overbar can be eliminated from all quantities in (6.71).

(7) *Momentum equations for compressible matter using covariant Cauchy stress tensor* $\bar{\boldsymbol{\sigma}}_{(0)}$

In the derivation of the momentum equations, use of $\bar{\boldsymbol{\sigma}}^{(0)}$ is most appropriate, as this measure of stress is derived using the stress components acting on the faces of the actual deformed tetrahedron. Instead of using $\bar{\boldsymbol{\sigma}}^{(0)}$, if we decide to use $\bar{\boldsymbol{\sigma}}_{(0)}$, the covariant Cauchy stress tensor in Eulerian description, then we can obtain the following of balance of linear moment

$$\bar{\rho}\left(\frac{\partial \bar{\boldsymbol{v}}}{\partial t}+\bar{\boldsymbol{v}}\cdot\bar{\boldsymbol{\nabla}}\right)\bar{\boldsymbol{v}}-\bar{\rho}\bar{\boldsymbol{F}}^b-\bar{\boldsymbol{\nabla}}\cdot\bar{\boldsymbol{\sigma}}_{(0)}=0 \tag{6.72}$$

As explained in Chapter 4, the stress tensor $\bar{\boldsymbol{\sigma}}_{(0)}$ is non-physical, as it is derived using covariant stress components acting on a tetrahedron that is different than the true deformed tetrahedron.

(8) *Balance of linear momenta for compressible matter using the Jaumann stress tensor* ${}^{(0)}\bar{\boldsymbol{\sigma}}^J$

If we choose Jaumann stress tensor ${}^{(0)}\bar{\boldsymbol{\sigma}}^J$ instead of $\bar{\boldsymbol{\sigma}}^{(0)}$ in (6.68), then the momentum equations become

$$\bar{\rho}\left(\frac{\partial \bar{\boldsymbol{v}}}{\partial t}+(\bar{\boldsymbol{\nabla}}\cdot\bar{\boldsymbol{v}})\bar{\boldsymbol{v}}\right)-\bar{\rho}\bar{\boldsymbol{F}}^b-\bar{\boldsymbol{\nabla}}\cdot{}^{(0)}\bar{\boldsymbol{\sigma}}^J=0 \tag{6.73}$$

We recall from Chapter 4 from the definition of ${}^{(0)}\bar{\boldsymbol{\sigma}}^J$ that the stress measure ${}^{(0)}\bar{\boldsymbol{\sigma}}^J$ is also non-physical as it is average of $\bar{\boldsymbol{\sigma}}^{(0)}$ and $\bar{\boldsymbol{\sigma}}_{(0)}$.

6.8 Kinematics of continuous media: BAM

In this section, we derive integral as well as differential forms of BAM.

6.8.1 Integral form of BAM

The principle of balance of angular momenta for non-polar or monopolar continuous media can be stated as follows: *The time rate of change of the total moment of momentum for a continuum is equal to the vector sum of the moments of external forces acting on the continuum.* Thus, due to surface stress $\bar{\boldsymbol{P}}$, body force $\bar{\boldsymbol{F}}^b$ (per unit mass) and the momentum $\bar{\rho}\bar{v}d\bar{V}$ for an elemental mass $\bar{\rho}d\bar{V}$, we can write for a volume $\bar{V}$ with boundary $\partial\bar{V}$ in the current configuration.

$$\frac{D}{Dt}\int\limits_{\bar{V}(t)}\bar{\boldsymbol{x}}\times\bar{\rho}\,\bar{\boldsymbol{v}}\,d\bar{V}=\int\limits_{\partial\bar{V}(t)}\bar{\boldsymbol{x}}\times\bar{\boldsymbol{P}}\,d\bar{A}+\int\limits_{\bar{V}(t)}\bar{\boldsymbol{x}}\times\bar{\rho}\,\bar{\boldsymbol{F}}^b\,d\bar{V} \tag{6.74}$$

using Cauchy principle $\bar{\boldsymbol{P}}=(\bar{\boldsymbol{\sigma}}^{(0)})^T\cdot\bar{\boldsymbol{n}}$ or $\bar{P}_j=\bar{\sigma}^{(0)}_{mj}\,\bar{n}_m$ and expressing cross products using the permutation tensor.

$$\frac{D}{Dt}\int\limits_{\bar{V}(t)}\bar{\rho}\,\epsilon_{ijk}\,\bar{x}_i\,\bar{v}_j\,d\bar{V}=\int\limits_{\partial\bar{V}(t)}\epsilon_{ijk}\,\bar{x}_i\,\bar{\sigma}^{(0)}_{mj}\,\bar{n}_m\,d\bar{A}+\int\limits_{\bar{V}(t)}\bar{\rho}\,\epsilon_{ijk}\,\bar{x}_i\,\bar{F}^b_j\,d\bar{V} \tag{6.75}$$

Since $\bar{\rho} d\bar{V} = d\bar{m} = dm$

$$\begin{aligned}\frac{D}{Dt}\int_{\bar{V}} \epsilon_{ijk}\bar{x}_i\bar{v}_j\bar{\rho}d\bar{V} &= \frac{D}{Dt}\int_{m} \epsilon_{ijk}\bar{x}_i\bar{v}_j dm = \int_{m} \frac{D}{Dt}(\varepsilon_{ijk}x_i)v_j)dm \\ &= \int_{\bar{V}} \frac{D(\varepsilon_{ijk}\bar{x}_i\bar{v}_j)}{Dt}\bar{\rho}d\bar{V} = \int_{\bar{V}} \bar{\rho}\epsilon_{ijk}\left(\bar{v}_i\bar{v}_j + \bar{x}_i\frac{D\bar{v}_j}{Dt}\right)d\bar{V}\end{aligned} \tag{6.76}$$

Substituting (6.76) in (6.75) and using Gauss's divergence theorem for the first term on the right side of (6.75) and noting that $\frac{D\bar{x}_i}{Dt} = \bar{v}_i$, we can obtain

$$\begin{aligned}\int_{\bar{V}(t)} \bar{\rho}\,\epsilon_{ijk}\Big(\bar{v}_i\,\bar{v}_j + \bar{x}_i\frac{D\bar{v}_j}{Dt}\Big)d\bar{V} &= \int_{\bar{V}(t)} \epsilon_{ijk}\,\left(\bar{x}_i\,\bar{\sigma}^{(0)}_{mj}\right)_{,m}d\bar{V} \\ &\quad + \int_{\bar{V}(t)} \bar{\rho}\,\epsilon_{ijk}\,\bar{x}_i\,\bar{F}^b_j\,d\bar{V}\end{aligned} \tag{6.77}$$

We note that

$$\epsilon_{ijk}\,\bar{v}_i\,\bar{v}_j = 0 \tag{6.78}$$

hence, (6.77) reduces to the following

$$\int_{\bar{V}(t)} \bar{\rho}\,\epsilon_{ijk}\,\bar{x}_i\frac{D\bar{v}_j}{Dt}d\bar{V} = \int_{\bar{V}(t)} \epsilon_{ijk}\,\left(\bar{x}_i\,\bar{\sigma}^{(0)}_{mj}\right)_{,m}d\bar{V} + \int_{\bar{V}(t)} \bar{\rho}\,\epsilon_{ijk}\,\bar{x}_i\,\bar{F}^b_j\,d\bar{V} \tag{6.79}$$

We also note that

$$\begin{aligned}\left(\bar{x}_i\,\bar{\sigma}^{(0)}_{mj}\right)_{,m} &= \bar{x}_{i,m}\bar{\sigma}^{(0)}_{mj} + \bar{x}_i\,\bar{\sigma}^{(0)}_{mj,m} \\ &= \delta_{im}\,\bar{\sigma}^{(0)}_{mj} + \bar{x}_i\,\bar{\sigma}^{(0)}_{mj,m} \\ &= \bar{\sigma}^{(0)}_{ij} + \bar{x}_i\,\bar{\sigma}^{(0)}_{mj,m}\end{aligned} \tag{6.80}$$

Substituting from (6.80) into (6.79) and regrouping

$$\int_{\bar{V}(t)} \left(\epsilon_{ijk}\Big(\bar{x}_i\Big(\bar{\rho}\frac{D\bar{v}_j}{Dt} - \bar{\rho}\,\bar{F}^b_j - \bar{\sigma}^{(0)}_{mj,m}\Big)\Big) - \epsilon_{ijk}\,\bar{\sigma}^{(0)}_{ij}\right)d\bar{V} = 0 \tag{6.81}$$

The integral form of the BLM given by (6.50) is obviously not helpful in simplifying (6.81) further. Thus, (6.81) are the balance of angular momenta equations in the integral form. Based on (6.81), it is not possible to establish symmetry of Cauchy stress tensor $\bar{\boldsymbol{\sigma}}^{(0)}$. As in case of BLM, (6.81) holds in Riemann sense when the integrand is continuous everywhere in the volume $\bar{V}(t)$. If there are isolated discontinuities of the integrand in $\bar{V}(t)$, but the discontinuities are square integrable, then (6.81) holds in Lebesgue sense. This integral form must be used if the integrand is strictly not continuous everywhere in the volume $\bar{V}(t)$.

6.8.2 Differential form of BAM

We begin with (6.81) and use differential form of balance of linear momenta

$$\bar{\rho}\frac{D\bar{v}_j}{Dt} - \bar{\rho}\bar{F}_j^b - \frac{\partial\bar{\sigma}_{mj}^{(0)}}{\partial\bar{x}_m} = 0 \tag{6.82}$$

in (6.81) to obtain

$$\int\limits_{\bar{V}(t)} \epsilon_{ijk}\,\bar{\sigma}_{ij}^{(0)}\,d\bar{V} = 0 \tag{6.83}$$

If the localization theorem holds for (6.83), the volume $\bar{V}(t)$ is arbitrary, hence we can set the integrand in (6.83) to zero.

$$\epsilon_{ijk}\,\bar{\sigma}_{ij}^{(0)} = 0 \tag{6.84}$$

This is the differential form of BAM which can be written as

$$\bar{\sigma}_{12}^{(0)} - \bar{\sigma}_{21}^{(0)} = 0 \quad ; \quad \bar{\sigma}_{23}^{(0)} - \bar{\sigma}_{32}^{(0)} = 0 \quad ; \quad \bar{\sigma}_{31}^{(0)} - \bar{\sigma}_{13}^{(0)} = 0 \tag{6.85}$$

From (6.85) we conclude that Cauchy stress tensor $\bar{\boldsymbol{\sigma}}^{(0)}$ is symmetric, i.e., $\bar{\boldsymbol{\sigma}}^{(0)} = (\bar{\boldsymbol{\sigma}}^{(0)})^T$, hence $\boldsymbol{\sigma}^{(0)} = (\boldsymbol{\sigma}^{(0)})^T$ holds as well. A derivation establishing symmetry of $\bar{\boldsymbol{\sigma}}^{(0)}$ for small deformation case, i.e., for $\bar{\boldsymbol{\sigma}}$, has been presented in Chapter 4 but in the absence of body forces and momentum forces.

6.9 First law of thermodynamics: FLT

A deforming matter exhibiting thermodynamic equilibrium must satisfy the first law of thermodynamics. The first law of thermodynamics is a statement of balance of energy, hence yields the *energy equation.* Based on the first law of thermodynamics, the sum of the rate of work and the rate of heat added to a deforming volume (or mass) of matter will result in the rate of increase of the total energy of the system. Thus, in the current configuration at time t we can write the following

$$\frac{D\bar{E}_t}{Dt} = \frac{D\bar{Q}}{Dt} + \frac{D\bar{W}}{Dt} \tag{6.86}$$

In which $\frac{D\bar{E}_t}{Dt}$ is the rate of total energy, $\frac{D\bar{Q}}{Dt}$ is the rate of heat and $\frac{D\bar{W}}{Dt}$ is the rate of mechanical work. In (6.86) we have assumed that the rate of work results in entropy production and hence influences the thermal field and internal energy. This is referred to as the dissipation mechanism. This is the most fundamental form of the first law of thermodynamics. We integrate (6.86) over $\bar{V}(t)$ with boundary $\partial\bar{V}(t)$, then expand each term. First, the *total energy* $\bar{E}_t$ *of a volume of matter* $\bar{V}(t)$ can be written as

$$\bar{E}_t = \int\limits_{\bar{V}(t)} \bar{\rho}\Big(\bar{e} + \frac{1}{2}\bar{\boldsymbol{v}}\cdot\bar{\boldsymbol{v}} - \bar{\boldsymbol{F}}^b\cdot\bar{\boldsymbol{u}}\Big)d\bar{V} \tag{6.87}$$

where $\bar{e}$ is the specific internal energy, i.e., internal energy per unit mass, $\bar{\boldsymbol{F}}^b$ are the body force/unit mass and $\bar{\boldsymbol{u}}$ are displacement in x-frame. Consider the reference and the current configurations shown in Figures 6.1 and 6.2. Recall that $\bar{\boldsymbol{n}}$ is the outward unit normal to $d\bar{s}$ (oblique deformed plane $\bar{A}_1\bar{A}_2\bar{A}_3$ of $\partial\bar{V}(t)$ in Figure 6.3).

Let $\bar{\boldsymbol{q}}$ be the amount of heat per unit area and per unit time in the direction $\bar{\boldsymbol{n}}$. Then, the *rate of heat added to the volume of matter* is given by

$$\frac{D\bar{Q}}{Dt} = -\int_{\partial\bar{V}(t)} \bar{\boldsymbol{q}} \cdot \bar{\boldsymbol{n}}\, d\bar{A} \tag{6.88}$$

The negative sign is due to the fact that positive $\bar{\boldsymbol{q}}$ is in the direction of the outward normal which results in heat removal from the volume, hence a decrease in the rate of change of heat. We also note that $\bar{\boldsymbol{q}}$ has components in x-frame. Applying Gauss's divergence theorem, (6.88) can be written as

$$\frac{D\bar{Q}}{Dt} = -\int_{\partial\bar{V}(t)} \bar{\boldsymbol{q}} \cdot \bar{\boldsymbol{n}}\, d\bar{A} = -\int_{\bar{V}(t)} \bar{\boldsymbol{\nabla}} \cdot \bar{\boldsymbol{q}}\, d\bar{V} \tag{6.89}$$

To determine $\frac{D\bar{W}}{Dt}$, i.e., material derivative of work done $\bar{W}$, we proceed as follows. The resultant force in x-frame on the plane $\bar{A}_1\bar{A}_2\bar{A}_3$ with unit exterior normal $\bar{\boldsymbol{n}}$ is $\bar{\boldsymbol{P}}$ with velocities $\bar{\boldsymbol{v}}$, hence

$$\frac{D\bar{W}}{Dt} = \int_{\partial\bar{V}(t)} \bar{\boldsymbol{P}} \cdot \bar{\boldsymbol{v}}\, d\bar{A} = \int_{\partial\bar{V}(t)} \bar{P}_j \bar{v}_j\, d\bar{A} \tag{6.90}$$

The contravariant Cauchy stress tensor $\bar{\boldsymbol{\sigma}}^{(0)}$ and $\bar{\boldsymbol{P}}$ are related (Chapter 4). The reasons for choosing $\bar{\boldsymbol{\sigma}}^{(0)}$ are the same as those for choosing $\bar{\boldsymbol{\sigma}}^{(0)}$ in deriving momentum equations. We can write the following using the Cauchy principle:

$$\bar{\boldsymbol{P}} = (\bar{\boldsymbol{\sigma}}^{(0)})^T \cdot \bar{\boldsymbol{n}} \qquad \text{or} \qquad \bar{P}_j = \bar{\sigma}_{ij}^{(0)} \bar{n}_i \tag{6.91}$$

Substituting from (6.91) into (6.90)

$$\frac{D\bar{W}}{Dt} = \int_{\partial\bar{V}(t)} ((\bar{\boldsymbol{\sigma}}^{(0)})^T \cdot \bar{\boldsymbol{n}}) \cdot \bar{\boldsymbol{v}}\, d\bar{A} = \int_{\partial\bar{V}(t)} \bar{\boldsymbol{v}} \cdot ((\bar{\boldsymbol{\sigma}}^{(0)})^T \cdot \bar{\boldsymbol{n}}) d\bar{A} \tag{6.92}$$

$$= \int_{\partial\bar{V}(t)} (\bar{\boldsymbol{v}} \cdot (\bar{\boldsymbol{\sigma}}^{(0)})^T) \cdot \bar{\boldsymbol{n}}\, d\bar{A} = \int_{\partial\bar{V}(t)} (\bar{\boldsymbol{v}} \cdot (\bar{\boldsymbol{\sigma}}^{(0)})^T) \cdot d\bar{\boldsymbol{A}} \tag{6.93}$$

Using Gauss's divergence theorem, (6.93) can be written as (keeping in mind that $\bar{\boldsymbol{v}} \cdot (\bar{\boldsymbol{\sigma}}^{(0)})^T$ is a tensor of rank one)

$$\frac{D\bar{W}}{Dt} = \int_{\bar{V}(t)} \bar{\boldsymbol{\nabla}} \cdot (\bar{\boldsymbol{v}} \cdot (\bar{\boldsymbol{\sigma}}^{(0)})^T) d\bar{V} \tag{6.94}$$

Consider the integrand in (6.94).

$$\begin{aligned}
\bar{\boldsymbol{\nabla}} \cdot (\bar{\boldsymbol{v}} \cdot (\bar{\boldsymbol{\sigma}}^{(0)})^T) &= \boldsymbol{e}_i \frac{\partial}{\partial \bar{x}_i} \cdot (\boldsymbol{e}_j \bar{v}_j \cdot \boldsymbol{e}_k \boldsymbol{e}_l \bar{\sigma}_{lk}^{(0)}) \\
&= \boldsymbol{e}_i \frac{\partial}{\partial \bar{x}_i} \cdot (\boldsymbol{e}_j \cdot \boldsymbol{e}_k \boldsymbol{e}_l \bar{v}_j \bar{\sigma}_{lk}^{(0)})
\end{aligned}$$

$$
\begin{aligned}
&= \boldsymbol{e}_i \frac{\partial}{\partial \bar{x}_i} \cdot (\delta_{jk} \boldsymbol{e}_l \bar{v}_j \bar{\sigma}_{lk}^{(0)}) \\
&= \boldsymbol{e}_i \frac{\partial}{\partial \bar{x}_i} \cdot (\boldsymbol{e}_l \bar{v}_k \bar{\sigma}_{lk}^{(0)}) \\
&= \boldsymbol{e}_i \cdot \boldsymbol{e}_l \frac{\partial}{\partial \bar{x}_i} (\bar{v}_k \bar{\sigma}_{lk}^{(0)}) \\
&= \delta_{il} \frac{\partial}{\partial \bar{x}_i} (\bar{v}_k \bar{\sigma}_{lk}^{(0)}) = \frac{\partial}{\partial \bar{x}_i} (\bar{v}_k \bar{\sigma}_{ik}^{(0)}) \qquad (6.95) \\
&= \frac{\partial}{\partial \bar{x}_i} (\bar{v}_j \bar{\sigma}_{ij}^{(0)}) \\
&= \bar{v}_j \frac{\partial \bar{\sigma}_{ij}^{(0)}}{\partial \bar{x}_i} + \bar{\sigma}_{ij}^{(0)} \frac{\partial \bar{v}_j}{\partial \bar{x}_i} = \bar{v}_j \frac{\partial \bar{\sigma}_{ij}^{(0)}}{\partial \bar{x}_i} + \bar{\boldsymbol{\sigma}}^{(0)} : \bar{\boldsymbol{L}} \qquad (6.96)
\end{aligned}
$$

In the following we show that

$$
\bar{v}_j \frac{\partial \bar{\sigma}_{ij}^{(0)}}{\partial \bar{x}_i} = \bar{\boldsymbol{v}} \cdot (\bar{\boldsymbol{\nabla}} \cdot \bar{\boldsymbol{\sigma}}^{(0)}) \tag{6.97}
$$

$$
\begin{aligned}
\bar{\boldsymbol{v}} \cdot (\bar{\boldsymbol{\nabla}} \cdot \bar{\boldsymbol{\sigma}}^{(0)}) &= \bar{v}_j \boldsymbol{e}_j \cdot \left(\boldsymbol{e}_i \frac{\partial}{\partial \bar{x}_i} \cdot \boldsymbol{e}_k \boldsymbol{e}_l \bar{\sigma}_{kl}^{(0)} \right) \\
&= \bar{v}_j \boldsymbol{e}_j \cdot \left((\boldsymbol{e}_i \cdot \boldsymbol{e}_k) \boldsymbol{e}_l \frac{\partial \bar{\sigma}_{kl}^{(0)}}{\partial \bar{x}_i} \right) \\
&= \bar{v}_j \boldsymbol{e}_j \cdot \left(\delta_{ik} \boldsymbol{e}_l \frac{\partial \bar{\sigma}_{kl}^{(0)}}{\partial \bar{x}_i} \right) \\
&= \bar{v}_j \boldsymbol{e}_j \cdot \left(\boldsymbol{e}_l \frac{\partial \bar{\sigma}_{il}^{(0)}}{\partial \bar{x}_i} \right) \\
&= \bar{v}_j \frac{\partial \bar{\sigma}_{il}^{(0)}}{\partial \bar{x}_i} (\boldsymbol{e}_j \cdot \boldsymbol{e}_l) \\
&= \bar{v}_j \frac{\partial \bar{\sigma}_{il}^{(0)}}{\partial \bar{x}_i} \delta_{jl} = \bar{v}_j \frac{\partial \bar{\sigma}_{ij}^{(0)}}{\partial \bar{x}_i} = \bar{v}_j \frac{\partial \bar{\sigma}_{ij}^{(0)}}{\partial \bar{x}_i} \qquad (6.98)
\end{aligned}
$$

Substituting from (6.97) in (6.96)

$$
\bar{\boldsymbol{\nabla}} \cdot \left(\bar{\boldsymbol{v}} \cdot (\bar{\boldsymbol{\sigma}}^{(0)})^T \right) = \bar{\boldsymbol{v}} \cdot (\bar{\boldsymbol{\nabla}} \cdot \bar{\boldsymbol{\sigma}}^{(0)}) + \bar{\boldsymbol{\sigma}}^{(0)} : \bar{\boldsymbol{L}} \tag{6.99}
$$

Substituting from (6.99) in (6.94)

$$
\frac{D\bar{W}}{Dt} = \int\limits_{\bar{V}(t)} \left(\bar{\boldsymbol{v}} \cdot (\bar{\boldsymbol{\nabla}} \cdot \bar{\boldsymbol{\sigma}}^{(0)}) + \bar{\boldsymbol{\sigma}}^{(0)} : \bar{\boldsymbol{L}} \right) d\bar{V} \tag{6.100}
$$

Energy equation

Substituting from (6.87), (6.89) and (6.100) for $\bar{E}_t$, $\frac{D\bar{Q}}{Dt}$ and $\frac{D\bar{W}}{Dt}$ into (6.86), we obtain

$$\frac{D}{Dt}\int\limits_{\bar{V}(t)} \bar{\rho}(\bar{e}+\frac{1}{2}\bar{\boldsymbol{v}}\cdot\bar{\boldsymbol{v}}-\bar{\boldsymbol{F}}^b\cdot\bar{\boldsymbol{u}})d\bar{V} = -\int\limits_{\bar{V}(t)} \bar{\boldsymbol{\nabla}}\cdot\bar{\boldsymbol{q}}\,d\bar{V} + \int\limits_{\bar{V}(t)} \left(\bar{\boldsymbol{v}}\cdot(\bar{\boldsymbol{\nabla}}\cdot\bar{\boldsymbol{\sigma}}^{(0)})+\bar{\boldsymbol{\sigma}}^{(0)}:\bar{\boldsymbol{L}}\right)d\bar{V} \tag{6.101}$$

Equation (6.101) is the result of the first law of thermodynamics. This is known as the energy equation. This is the most fundamental form of the energy equation resulting from the first law of thermodynamics or balance of energy for a deformed volume of matter.

6.9.1 Integral form of FLT

The material derivative in (6.101) can be brought inside the integral only by using transport theorem. Another possible approach would have been to use differential form of CM in Lagrangian description, but this cannot be done in the integral form of the FLT. Using transport theorem we can write

$$\int\limits_{\bar{V}(t)} \left(\frac{\partial \bar{e}_t}{\partial t}+\bar{\boldsymbol{\nabla}}\cdot(\bar{e}_t\bar{\boldsymbol{v}})\right)d\bar{V} = -\int\limits_{\bar{V}(t)} \bar{\boldsymbol{\nabla}}\cdot\bar{\boldsymbol{q}}d\bar{V} + \int\limits_{\bar{V}(t)} \left(\bar{\boldsymbol{v}}\cdot(\bar{\boldsymbol{\nabla}}\cdot\bar{\boldsymbol{\sigma}}^{(0)})+\bar{\boldsymbol{\sigma}}^{(0)}:\bar{\boldsymbol{L}}\right)d\bar{V} \tag{6.102}$$

$$\text{where} \quad \bar{e}_t = \bar{\rho}\left(\bar{e}+\frac{1}{2}\bar{\boldsymbol{v}}\cdot\bar{\boldsymbol{v}}-\bar{\boldsymbol{F}}^b\cdot\bar{\boldsymbol{u}}\right) \tag{6.103}$$

$$\int\limits_{\bar{V}(t)} \left(\frac{\partial \bar{e}_t}{\partial t}+\bar{\boldsymbol{\nabla}}\cdot(\bar{e}_t\bar{\boldsymbol{v}})+\bar{\boldsymbol{\nabla}}\cdot\bar{\boldsymbol{q}}-\bar{\boldsymbol{v}}\cdot(\bar{\boldsymbol{\nabla}}\cdot\bar{\boldsymbol{\sigma}}^{(0)})-\bar{\boldsymbol{\sigma}}^{(0)}:\bar{\boldsymbol{L}}\right)d\bar{V} = 0 \tag{6.104}$$

Equation (6.104) is the integral form of the first law of thermodynamics. FLT (6.104) holds in Riemann sense when the integrand is continuous everywhere in the volume $\bar{V}(t)$. If there are isolated discontinuities of the integrand in (6.103) in volume $\bar{V}(t)$, but the discontinuities are square integrable, then (6.104) holds in Lebesgue sense. This integral form must be used if the integrand cannot be ensured to be continuous everywhere in the volume $\bar{V}(t)$.

6.9.2 Differential form of FLT

In this derivation, we can begin with (6.101) and use differential forms of previously derived CBL to obtain desired differential form. Consider the term on left side of (6.101) and using $\rho_0\,dV = \bar{\rho}d\bar{V}$ we have

$$\frac{D}{Dt}\int\limits_{\bar{V}(t)} \left(\bar{e}+\frac{1}{2}\bar{\boldsymbol{v}}\cdot\bar{\boldsymbol{v}}-\bar{\boldsymbol{F}}^b\cdot\bar{\boldsymbol{u}}\right)\bar{\rho}d\bar{V} = \frac{D}{Dt}\int\limits_{V} \left(e+\frac{1}{2}\boldsymbol{v}\cdot\boldsymbol{v}-\boldsymbol{F}^b\cdot\boldsymbol{u}\right)\rho_0\,dV \tag{6.105}$$

$$
\begin{aligned}
&= \int_V \frac{D}{Dt}\left(e + \frac{1}{2}\boldsymbol{v}\cdot\boldsymbol{v} - \boldsymbol{F}^b\cdot\boldsymbol{u}\right)\rho_0\, dV \quad , \quad \text{as } \frac{D\rho_0}{Dt} = 0 \\
&= \int_{\bar{V}(t)} \frac{D}{Dt}\left(\bar{e} + \frac{1}{2}\bar{\boldsymbol{v}}\cdot\bar{\boldsymbol{v}} - \bar{\boldsymbol{F}}^b\cdot\bar{\boldsymbol{u}}\right)\bar{\rho}\, d\bar{V} \\
&= \int_{\bar{V}(t)} \left(\frac{D\bar{e}}{Dt} + \frac{1}{2}\frac{D\bar{\boldsymbol{v}}}{Dt}\cdot\bar{\boldsymbol{v}} + \frac{1}{2}\bar{\boldsymbol{v}}\cdot\frac{D\bar{\boldsymbol{v}}}{Dt} - \bar{\boldsymbol{F}}^b\cdot\frac{D\bar{\boldsymbol{u}}}{Dt}\right)\bar{\rho}\, d\bar{V} \\
&= \int_{\bar{V}(t)} \left(\frac{D\bar{e}}{Dt} + \bar{\boldsymbol{v}}\cdot\frac{D\bar{\boldsymbol{v}}}{Dt} - \bar{\boldsymbol{F}}^b\cdot\frac{D\bar{\boldsymbol{u}}}{Dt}\right)\bar{\rho}\, d\bar{V}
\end{aligned}
\tag{6.106}
$$

substituting (6.106) into (6.101)

$$
\begin{aligned}
\int_{\bar{V}(t)} \bar{\rho}\left(\frac{D\bar{e}}{Dt} + \bar{\boldsymbol{v}}\cdot\frac{D\bar{\boldsymbol{v}}}{Dt} - \bar{\boldsymbol{F}}^b\cdot\bar{\boldsymbol{v}}\right)d\bar{V} = &-\int_{\bar{V}(t)} \bar{\boldsymbol{\nabla}}\cdot\bar{\boldsymbol{q}}\, d\bar{V} \\
&+ \int_{\bar{V}(t)} \left(\bar{\boldsymbol{v}}\cdot(\bar{\boldsymbol{\nabla}}\cdot\bar{\boldsymbol{\sigma}}^{(0)}) + \bar{\boldsymbol{\sigma}}^{(0)} : \bar{\boldsymbol{L}}\right)d\bar{V}
\end{aligned}
\tag{6.107}
$$

Based on the localization theorem if the integrand is continuous everywhere in volume $\bar{V}(t)$, then volume $\bar{V}(t)$ is arbitrary and we can set the integrand to zero.

$$
\bar{\rho}\left(\frac{D\bar{e}}{Dt} + \bar{\boldsymbol{v}}\cdot\frac{D\bar{\boldsymbol{v}}}{Dt} - \bar{\boldsymbol{F}}^b\cdot\bar{\boldsymbol{v}}\right) + \bar{\boldsymbol{\nabla}}\cdot\bar{\boldsymbol{q}} - \bar{\boldsymbol{v}}\cdot\left(\bar{\boldsymbol{\nabla}}\cdot\bar{\boldsymbol{\sigma}}^{(0)}\right) - \bar{\boldsymbol{\sigma}}^{(0)} : \bar{\boldsymbol{L}} = 0 \tag{6.108}
$$

Equation (6.108) can be further simplified by using (6.66) (differential form of momentum equations derived using approach II). Recall

$$
\bar{\rho}\frac{D\bar{\boldsymbol{v}}}{Dt} - \bar{\rho}\bar{\boldsymbol{F}}^b - \bar{\boldsymbol{\nabla}}\cdot\bar{\boldsymbol{\sigma}}^{(0)} = 0 \tag{6.109}
$$

Substituting $\bar{\rho}\frac{D\bar{\boldsymbol{v}}}{Dt}$ from (6.109) into (6.108)

$$
\begin{aligned}
\bar{\rho}\frac{D\bar{e}}{Dt} + \bar{\boldsymbol{v}}\cdot(\bar{\rho}\bar{\boldsymbol{F}}^b + \bar{\boldsymbol{\nabla}}\cdot\bar{\boldsymbol{\sigma}}^{(0)}) - \bar{\rho}\bar{\boldsymbol{F}}^b\cdot\bar{\boldsymbol{v}} + \bar{\boldsymbol{\nabla}}\cdot\bar{\boldsymbol{q}} \\
-\bar{\boldsymbol{v}}\cdot\left(\bar{\boldsymbol{\nabla}}\cdot\bar{\boldsymbol{\sigma}}^{(0)}\right) - \bar{\boldsymbol{\sigma}}^{(0)} : \bar{\boldsymbol{L}} = 0
\end{aligned}
\tag{6.110}
$$

Noting $\bar{\boldsymbol{v}}\cdot\bar{\boldsymbol{F}}^b\bar{\rho} = \bar{\rho}\bar{\boldsymbol{F}}^b\cdot\bar{\boldsymbol{v}}$ and simplifying (6.110), we obtain

$$
\bar{\rho}\frac{D\bar{e}}{Dt} + \bar{\boldsymbol{\nabla}}\cdot\bar{\boldsymbol{q}} - \bar{\boldsymbol{\sigma}}^{(0)} : \bar{\boldsymbol{L}} = 0 \tag{6.111}
$$

This is fundamental form of the differential form of *energy equation* from FLT. Details of $\bar{e}$ will be considered later. The last term in (6.111) can be expressed in terms of the symmetric and skew-symmetric parts of the velocity gradient tensor. We note that

$$
\bar{\boldsymbol{\sigma}}^{(0)} : \bar{\boldsymbol{L}} = \bar{\boldsymbol{\sigma}}^{(0)} : (\bar{\boldsymbol{D}} + \bar{\boldsymbol{W}}) = \bar{\boldsymbol{\sigma}}^{(0)} : \bar{\boldsymbol{D}} + \bar{\boldsymbol{\sigma}}^{(0)} : \bar{\boldsymbol{W}} \tag{6.112}
$$

$$
\text{Since} \quad \bar{\boldsymbol{\sigma}}^{(0)} : \bar{\boldsymbol{W}} = 0 \quad ; \quad \therefore \quad \bar{\boldsymbol{\sigma}}^{(0)} : \bar{\boldsymbol{L}} = \bar{\boldsymbol{\sigma}}^{(0)} : \bar{\boldsymbol{D}} \tag{6.113}
$$

Hence, (6.111) can be written as

$$\bar{\rho}\frac{D\bar{e}}{Dt} + \bar{\boldsymbol{\nabla}} \cdot \bar{\boldsymbol{q}} - \bar{\boldsymbol{\sigma}}^{(0)} : \bar{\boldsymbol{D}} = 0 \tag{6.114}$$

Equation (6.114) is sometimes more convenient as it uses the symmetric part of the velocity gradient tensor, i.e., $[\bar{D}]$, and is generally preferred over (6.111). In (6.114) we observe that $\bar{\boldsymbol{\sigma}}^{(0)}$ and $\bar{\boldsymbol{D}}$ as rate of work conjugate pair. $\bar{\boldsymbol{\sigma}}^{(0)} : \bar{\boldsymbol{D}}$ is the rate of work that may represent recoverable energy upon unloading or the rate of work that is converted into entropy (heat) or both depending upon the constitution of the matter. We can uniquely denote this rate of work for volume $\bar{V}$ by

$$\bar{W}_r = \int\limits_{\bar{V}(t)} \bar{\boldsymbol{\sigma}}^{(0)} : \bar{\boldsymbol{L}} d\bar{V} = \int\limits_{\bar{V}(t)} \bar{\boldsymbol{\sigma}}^{(0)} : \bar{\boldsymbol{D}} d\bar{V} \tag{6.115}$$

and the rate of work per unit volume by

$$\bar{w}_r = \bar{\boldsymbol{\sigma}}^{(0)} : \bar{\boldsymbol{L}} = \bar{\boldsymbol{\sigma}}^{(0)} : \bar{\boldsymbol{D}} \tag{6.116}$$

Remarks

As in the case of momentum equations, here also, we can use $\bar{\boldsymbol{\sigma}}_{(0)}$ or ${}^{(0)}\bar{\boldsymbol{\sigma}}^J$ as stress measures instead of $\bar{\boldsymbol{\sigma}}^{(0)}$ in (6.114), resulting in the following forms of the energy equations:

$$\bar{\rho}\frac{D\bar{e}}{Dt} + \bar{\boldsymbol{\nabla}} \cdot \bar{\boldsymbol{q}} - \bar{\boldsymbol{\sigma}}^{(0)} : \bar{\boldsymbol{D}} = 0 \tag{6.117}$$

$$\bar{\rho}\frac{D\bar{e}}{Dt} + \bar{\boldsymbol{\nabla}} \cdot \bar{\boldsymbol{q}} - {}^{(0)}\bar{\boldsymbol{\sigma}}^J : \bar{\boldsymbol{D}} = 0 \tag{6.118}$$

If we choose (6.117) or (6.118) as energy equation, then the conjugate pairs are $\bar{\boldsymbol{\sigma}}_{(0)}$, $\bar{\boldsymbol{D}}$ and ${}^{(0)}\bar{\boldsymbol{\sigma}}^J$, $\bar{\boldsymbol{D}}$. However, we note that $\bar{\boldsymbol{\sigma}}_{(0)}$ and ${}^{(0)}\bar{\boldsymbol{\sigma}}^J$ are non-physical.

6.10 Second law of thermodynamics: SLT

The fundamental principles of continuum mechanics: conservation of mass, balance of linear and angular momenta and balance of energy lead to continuity equation, balance of linear momenta equations, symmetry of Cauchy stress tensor and energy equation. These have already been established using a deformed tetrahedron with $\bar{\boldsymbol{x}}$ and t as independent variables. In addition to these, for all deforming matter to be in thermodynamic equilibrium, the second law of thermodynamics, i.e., *entropy inequality or Clausius-Duhem inequality*, must be satisfied. The second law of thermodynamics can be derived in terms of Helmholtz free energy density $\bar{\Phi}$ or Gibbs potential $\bar{\Psi}$. Since $\bar{\Phi}$ and $\bar{\Psi}$ are related, the two forms of the entropy inequality are equivalent. Both forms of entropy inequality are a fundamental starting point in the derivations of the constitutive theories. In this section, we consider derivations of entropy inequality in terms of Helmholtz free energy density $\bar{\Phi}$.

Consider a volume V of matter with closed boundary ∂V in the reference configuration. Upon deformation, V occupies $\bar{V}(t)$ and ∂V occupies $\partial\bar{V}(t)$ in the current configuration at time t.

6.10.1 Integral form of SLT

Consider volume $\bar{V}(t)$. Let $\bar{h}$ be the entropy flux between $\bar{V}(t)$ and the volume of the matter surrounding it, $\bar{s}$ be the source of entropy in $\bar{V}(t)$ due to non-contacting bodies (considered per unit mass). Let there exist $\bar{\eta}$, the specific entropy (entropy per unit mass) for $\bar{V}(t)$ bounded by $\partial\bar{V}(t)$ such that its rate of increase is at least equal to that supplied to $\bar{V}(t)$ from all sources (containing or non-contacting). Thus

$$\frac{D}{Dt}\int_{\bar{V}(t)} \bar{\eta}\bar{\rho}\, d\bar{V} \geq \int_{\partial\bar{V}(t)} \bar{h}\, d\bar{A} + \int_{\bar{V}(t)} \bar{s}\bar{\rho}\, d\bar{V} \tag{6.119}$$

We adopt Cauchy's principle for entropy flux $\bar{h}$, i.e., $\bar{h}$ at a point $\bar{x}_i$ on $\partial\bar{V}(t)$ depends on the orientation $\bar{\boldsymbol{n}}$ of $\partial\bar{V}(t)$ at $\bar{x}_i$, i.e.

$$\bar{h} = -\bar{\boldsymbol{\Psi}} \cdot \bar{\boldsymbol{n}} \tag{6.120}$$

in which $\bar{\boldsymbol{\Psi}}$ is similar to heat flux. Substituting from (6.120) into (6.119)

$$\frac{D}{Dt}\int_{\bar{V}(t)} \bar{\eta}\bar{\rho}\, d\bar{V} \geq -\int_{\partial\bar{V}(t)} \bar{\boldsymbol{\Psi}} \cdot \bar{\boldsymbol{n}}\, d\bar{A} + \int_{\bar{V}(t)} \bar{s}\bar{\rho}\, d\bar{V} \tag{6.121}$$

Using the divergence theorem for the first term on the right side of (6.121),

$$\frac{D}{Dt}\int_{\bar{V}(t)} \bar{\eta}\bar{\rho}\, d\bar{V} \geq -\int_{\bar{V}(t)} \bar{\Psi}_{i,i}\, d\bar{V} + \int_{\bar{V}(t)} \bar{s}\bar{\rho}\, d\bar{V} \tag{6.122}$$

Consider the left side of the inequality in (6.122) and use $\bar{\rho} d\bar{V} = \rho_0\, dV$ (CM).

$$\begin{aligned}\frac{D}{Dt}\int_{\bar{V}(t)} \bar{\eta}\bar{\rho}\, d\bar{V} &= \frac{D}{Dt}\int_{V} \eta\rho_0\, dV = \int_{V} \frac{D}{Dt}(\eta\rho_0)\, dV = \int_{V} \rho_0 \frac{D\eta}{Dt}\, dV \\ &= \int_{V} \frac{D\eta}{Dt}\rho_0\, dV = \int_{\bar{V}} \frac{D\bar{\eta}}{Dt}\bar{\rho} d\bar{V} = \int_{\bar{V}(t)} \bar{\rho}\frac{D\bar{\eta}}{Dt}\, d\bar{V}\end{aligned} \tag{6.123}$$

Substituting from (6.123) into (6.122)

$$\int_{\bar{V}(t)} \bar{\rho}\frac{D\bar{\eta}}{Dt}\, d\bar{V} \geq -\int_{\bar{V}(t)} \bar{\Psi}_{i,i}\, d\bar{V} + \int_{\bar{V}(t)} \bar{s}\bar{\rho}\, d\bar{V} \tag{6.124}$$

or

$$\int_{\bar{V}(t)} \left(\bar{\rho}\frac{D\bar{\eta}}{Dt} + \bar{\Psi}_{i,i} - \bar{s}\bar{\rho}\right) d\bar{V} \geq 0 \tag{6.125}$$

Equation (6.125) is the *most general form of the entropy inequality* also known as the *Clausius-Duhem inequality.* In continuum mechanics a different form of (6.125) is often more meaningful as well as useful. Based on the fact that the entropy due to contacting sources must naturally depend on $\bar{\boldsymbol{q}}$ (directly proportional) and $\bar{\theta}$

(inversely proportional), we can write the following

$$\bar{\boldsymbol{\Psi}} = \frac{\bar{\boldsymbol{q}}}{\bar{\theta}} \quad ; \quad \bar{s} = \frac{\bar{r}}{\bar{\theta}} \tag{6.126}$$

where $\bar{\theta}$ is absolute temperature (assumed to be greater than zero), $\bar{\boldsymbol{q}}$ is the heat vector and $\bar{r}$ is a suitable potential. Let

$$\bar{g}_i = \bar{\theta}_{,i} \tag{6.127}$$

Using the first equation in (6.126)

$$\bar{\Psi}_{i,i} = \frac{\bar{q}_{i,i}}{\bar{\theta}} - \frac{\bar{q}_i}{(\bar{\theta})^2}\bar{\theta}_{,i} = \frac{\bar{q}_{i,i}}{\bar{\theta}} - \frac{\bar{q}_i}{(\bar{\theta})^2}\bar{g}_i \tag{6.128}$$

Substituting from (6.128) and the second expression in (6.126) into (6.125)

$$\int\limits_{\bar{V}(t)} \left(\bar{\rho}\frac{D\bar{\eta}}{Dt} + \left(\frac{\bar{q}_{i,i}}{\bar{\theta}} - \frac{\bar{q}_i}{(\bar{\theta})^2}\bar{g}_i\right) - \frac{\bar{\rho}\bar{r}}{\bar{\theta}}\right) d\bar{V} \geq 0 \tag{6.129}$$

$$\text{or} \quad \int\limits_{\bar{V}(t)} \left(\bar{\rho}\bar{\theta}\frac{D\bar{\eta}}{Dt} + (\bar{q}_{i,i} - \bar{\rho}\bar{r}) - \frac{\bar{q}_i\bar{g}_i}{\bar{\theta}}\right) \geq 0 \tag{6.130}$$

Equation (6.130) is the *most common form the of integral form of Clausius-Duhem inequality.* If the integrand in (6.130) is continuous everywhere in $\bar{V}(t)$, then (6.130) holds in the Riemann sense. If the integrand in (6.130) has isolated discontinuities, but the discontinuities are square integrable, then (6.130) holds in Lebesgue sense. Thus, integral form of the SLT must be used if the integrand in (6.130) cannot be ensured to be continuous everywhere in the volume $\bar{V}(t)$.

6.10.2 Differential form of SLT

We begin with (6.130). Based on localization theorem if the integrand in (6.130) is continuous everywhere in volume $\bar{V}(t)$, then the inequality holds for the integrand.

$$\bar{\rho}\bar{\theta}\frac{D\bar{\eta}}{Dt} + (\bar{q}_{i,i} - \bar{\rho}\bar{r}) - \frac{\bar{q}_i\bar{g}_i}{\bar{\theta}} \geq 0 \tag{6.131}$$

Equation (6.131) can be further simplified using differential form of the energy equation from Section 6.9.2 (equation (6.111) with $\bar{\rho}\bar{r}$ added).

$$\bar{\rho}\frac{D\bar{e}}{Dt} + (\bar{\boldsymbol{\nabla}} \cdot \bar{\boldsymbol{q}} - \bar{\rho}\bar{r}) - \bar{\boldsymbol{\sigma}}^{(0)} : \bar{\boldsymbol{L}} = 0 \tag{6.132}$$

$$\therefore \quad \bar{\boldsymbol{\nabla}} \cdot \bar{\boldsymbol{q}} - \bar{\rho}\bar{r} = \bar{q}_{i,i} - \bar{\rho}\bar{r} = -\bar{\rho}\frac{D\bar{e}}{Dt} + \bar{\boldsymbol{\sigma}}^{(0)} : \bar{\boldsymbol{L}} \tag{6.133}$$

and $\bar{\sigma}_{ij}^{(0)}$ being the contravariant Cauchy stress tensor. Substituting from (6.133) into (6.131)

$$\bar{\rho}\bar{\theta}\frac{D\bar{\eta}}{Dt} - \bar{\rho}\frac{D\bar{e}}{Dt} + \bar{\boldsymbol{\sigma}}^{(0)} : \bar{\boldsymbol{L}} - \frac{\bar{q}_i\bar{g}_i}{\bar{\theta}} \geq 0 \tag{6.134}$$

$$\text{or} \quad \bar{\rho}\left(\bar{\theta}\frac{D\bar{\eta}}{Dt} - \frac{D\bar{e}}{Dt}\right) + \bar{\boldsymbol{\sigma}}^{(0)} : \bar{\boldsymbol{L}} - \frac{\bar{q}_i\bar{g}_i}{\bar{\theta}} \geq 0 \tag{6.135}$$

$$\text{or} \quad \bar{\rho}\Big(\frac{D\bar{e}}{Dt} - \bar{\theta}\frac{D\bar{\eta}}{Dt}\Big) - \bar{\boldsymbol{\sigma}}^{(0)} : \bar{\boldsymbol{L}} + \frac{\bar{q}_i \bar{g}_i}{\bar{\theta}} \leq 0 \tag{6.136}$$

Let $\bar{\Phi}$ be the Helmholtz free energy density (specific Helmholtz free energy) defined by

$$\bar{\Phi} = \bar{e} - \bar{\eta}\bar{\theta} \tag{6.137}$$

$$\therefore \quad \frac{D\bar{\Phi}}{Dt} = \frac{D\bar{e}}{Dt} - \bar{\eta}\frac{D\bar{\theta}}{Dt} - \bar{\theta}\frac{D\bar{\eta}}{Dt} \tag{6.138}$$

$$\text{or} \quad \frac{D\bar{e}}{Dt} - \bar{\theta}\frac{D\bar{\eta}}{Dt} = \frac{D\bar{\Phi}}{Dt} + \bar{\eta}\frac{D\bar{\theta}}{Dt} \tag{6.139}$$

Substituting from (6.139) into (6.136)

$$\bar{\rho}\Big(\frac{D\bar{\Phi}}{Dt} + \bar{\eta}\frac{D\bar{\theta}}{Dt}\Big) + \frac{\bar{\boldsymbol{q}} \cdot \bar{\boldsymbol{g}}}{\bar{\theta}} - \bar{\boldsymbol{\sigma}}^{(0)} : \bar{\boldsymbol{L}} \leq 0 \tag{6.140}$$

This is known as the *reduced form of Clausius-Duhem inequality.* Entropy inequality (6.140) plays an important role in the development of the constitutive theory. We note that the rate of work $\bar{W}_r$ (equation (6.116)) appears in entropy inequality (due to substitution from energy equation) also. Following the derivation of the energy equation in Section 6.9, we can replace $\bar{\boldsymbol{L}}$ in (6.140) with $\bar{\boldsymbol{D}}$. To obtain the following for the entropy inequality from (6.140):

$$\bar{\rho}\Big(\frac{D\bar{\Phi}}{Dt} + \bar{\eta}\frac{D\bar{\theta}}{Dt}\Big) + \frac{\bar{\boldsymbol{q}} \cdot \bar{\boldsymbol{g}}}{\bar{\theta}} - \bar{\boldsymbol{\sigma}}^{(0)} : \bar{\boldsymbol{D}} \leq 0 \tag{6.141}$$

Remarks

If we use covariant Cauchy stress tensor $\bar{\boldsymbol{\sigma}}_{(0)}$ or Jaumann stress tensor ${}^{(0)}\bar{\boldsymbol{\sigma}}^J$ as stress measures in momentum and energy equations, then $\bar{\boldsymbol{\sigma}}^{(0)}$ in (6.141) must be replaced by $\bar{\boldsymbol{\sigma}}_{(0)}$ or ${}^{(0)}\bar{\boldsymbol{\sigma}}^J$. The energy equation modifies accordingly.

$$\bar{\rho}\Big(\frac{D\bar{\Phi}}{Dt} + \bar{\eta}\frac{D\bar{\theta}}{Dt}\Big) + \frac{\bar{\boldsymbol{q}} \cdot \bar{\boldsymbol{g}}}{\bar{\theta}} - \bar{\boldsymbol{\sigma}}_{(0)} : \bar{\boldsymbol{D}} \leq 0 \tag{6.142}$$

$$\bar{\rho}\Big(\frac{D\bar{\Phi}}{Dt} + \bar{\eta}\frac{D\bar{\theta}}{Dt}\Big) + \frac{\bar{\boldsymbol{q}} \cdot \bar{\boldsymbol{g}}}{\bar{\theta}} - {}^{(0)}\bar{\boldsymbol{\sigma}}^J : \bar{\boldsymbol{D}} \leq 0 \tag{6.143}$$

6.11 Summary of mathematical model from CBL

6.11.1 Differential form of the CBL

In this section, we summarize the differential form of the mathematical model derived using conservation and balance laws utilizing stress measure $\bar{\boldsymbol{\sigma}}^{(0)}$ and Helmholtz free energy density $\bar{\Phi}$. The mathematical model holds for homogeneous, isotropic and compressible matter with finite deformation, finite strain rates and non-isothermal physics. The continuity equation, BLM, BAM, energy equation, entropy inequality resulting from the conservation and balance laws are summarized in the following using contavariant Cauchy stress tensor $\bar{\boldsymbol{\sigma}}^{(0)}$.

(1) Conservation of mass (CM): Continuity Equation

$$\frac{\partial \bar{\rho}}{\partial t} + \frac{\partial}{\partial \bar{x}_i}(\bar{\rho}\, \bar{v}_i) = 0 \tag{6.144}$$

(2) Balance of linear momenta (BLM): momentum equations

$$\bar{\rho}\frac{\partial \bar{v}_i}{\partial t} + \bar{\rho}\bar{v}_j \frac{\partial \bar{v}_i}{\partial \bar{x}_j} - \bar{\rho}\bar{F}_i^b - \frac{\partial \bar{\sigma}_{ji}^{(0)}}{\partial \bar{x}_j} = 0 \tag{6.145}$$

(3) Balance of angular momenta (BAM)

$$[\bar{\sigma}^{(0)}] = [\bar{\sigma}^{(0)}]^T \tag{6.146}$$

(4) First law of thermodynamics (FLT): energy equation

$$\bar{\rho}\frac{D\bar{e}}{Dt} + \frac{\partial \bar{q}_i}{\partial \bar{x}_i} - \bar{\sigma}_{ij}^{(0)} \bar{L}_{ji} = 0 \tag{6.147}$$

(5) Second law of thermodynamics (SLT): entropy inequality

$$\bar{\rho}\Big(\frac{D\bar{\Phi}}{Dt} + \bar{\eta}\frac{D\bar{\theta}}{Dt}\Big) + \frac{\bar{q}_i \bar{g}_i}{\bar{\theta}} - \bar{\sigma}_{ij}^{(0)} \bar{L}_{ji} \leq 0 \tag{6.148}$$

(6.144)–(6.148) can be written in an alternate form given in the following

(1) Conservation of mass (CM): continuity equation

$$\frac{\partial \bar{\rho}}{\partial t} + \bar{\boldsymbol{\nabla}} \cdot (\bar{\rho}\, \bar{\boldsymbol{v}}) = 0 \tag{6.149}$$

(2) Balance of linear momenta (BLM): momentum equations

$$\bar{\rho}\frac{D\bar{\boldsymbol{v}}}{Dt} - \bar{\rho}\, \bar{\boldsymbol{F}}^b - \bar{\boldsymbol{\nabla}} \cdot \bar{\boldsymbol{\sigma}}^{(0)} = 0 \tag{6.150}$$

(3) Balance of angular momenta (BAM)

$$\bar{\boldsymbol{\sigma}}^{(0)} = (\bar{\boldsymbol{\sigma}}^{(0)})^T \tag{6.151}$$

(4) First law of thermodynamics (FLT): energy equation

$$\bar{\rho}\frac{D\bar{e}}{Dt} + \bar{\boldsymbol{\nabla}} \cdot \bar{\boldsymbol{q}} - \bar{\boldsymbol{\sigma}}^{(0)} : \bar{\boldsymbol{L}} = 0 \tag{6.152}$$

(5) Second law of thermodynamics (SLT): entropy inequality

$$\bar{\rho}\Big(\frac{D\bar{\Phi}}{Dt} + \bar{\eta}\frac{D\bar{\theta}}{Dt}\Big) + \frac{\bar{\boldsymbol{q}} \cdot \bar{\boldsymbol{g}}}{\bar{\theta}} - \bar{\boldsymbol{\sigma}}^{(0)} : \bar{\boldsymbol{L}} \leq 0 \tag{6.153}$$

By replacing $\bar{\boldsymbol{\sigma}}^{(0)}$ with $\bar{\boldsymbol{\sigma}}_{(0)}$ and ${}^{(0)}\bar{\boldsymbol{\sigma}}^J$ in (6.145)–(6.148) (or (6.150)–(6.153)) we can obtain their forms in $\bar{\boldsymbol{\sigma}}_{(0)}$ and ${}^{(0)}\bar{\boldsymbol{\sigma}}^J$ stress measures.

Remarks

(1) This mathematical model consists of five partial differential equations: CM (1), BLM (3), FLT (1) in fourteen dependent variables: $\bar{\rho}$ (1), $\bar{\boldsymbol{v}}$ (3), $\bar{\boldsymbol{\sigma}}^{(0)}$

(6), $\bar{\boldsymbol{q}}$ (3), $\bar{\theta}$ (1), hence requires nine additional equations for closure. These are obtained from the constitutive theories for $\bar{\boldsymbol{\sigma}}^{(0)}$ (6) and $\bar{\boldsymbol{q}}$ (3). We note that in general, for compressible continua, $\bar{e} = \bar{e}(\bar{\rho}, \bar{\theta})$ and $\bar{\Phi} = \bar{\Phi}(\bar{\rho}, \bar{\theta}, \dots)$, $\bar{\eta} = \bar{\eta}(\bar{\rho}, \bar{\theta}, \dots)$, hence these are known functions. Thus, $\bar{e}$, $\bar{\Phi}$, $\bar{\eta}$ are not considered dependent variables in the mathematical model. The system of partial differential equations constituting the mathematical model is valid for isotropic and homogeneous compressible continuous matter. As stated earlier, the CBL are independent of the constitution of the matter, hence hold for all continua, solid or fluent (liquids and gases). Derivations of the constitutive theories consisting of matter specific physics are considered in Chapters 8–14.

(2) The balance of linear momenta equations (6.145) are also referred to as Navier-Stokes equations in the fluid mechanics writings. In some writings, the equations resulting after substituting constitutive theory for $\bar{\boldsymbol{\sigma}}^{(0)}$ for Newtonian fluids (compressible or incompressible) in (6.145) are called Navier-Stokes equations. In recent writings in fluid mechanics, it is well recognized that (6.145) without continuity and energy equations is incomplete description of the physics, hence conservation and balance laws ((6.144)-(6.148)) collectively are referred to as Navier-Stokes equations.

(3) We note that the CBL contain velocity $\bar{\boldsymbol{v}}$ and its spatial and temporal derivatives. Even though $\bar{\boldsymbol{v}} = \frac{D\bar{\boldsymbol{u}}}{Dt}$, unless we make this substitution in the CBL, displacements do not appear in them explicitly.

(4) If the constitutive theories can also be derived purely in terms of $\bar{\boldsymbol{v}}$ and its spatial derivatives (and temporal derivatives), then the complete mathematical model consisting of CBL and the constitutive theories only contains $\bar{\boldsymbol{v}}$ and its spatial and temporal derivatives. In this case we could treat velocities $\bar{\boldsymbol{v}}$ as observable quantities in which case $\bar{\boldsymbol{v}} \neq \frac{D\bar{\boldsymbol{u}}}{Dt}$ as $\bar{\boldsymbol{u}}$ is not observable, hence not monitored. The CBL described above and constitutive theories purely in terms of $\bar{\boldsymbol{v}}$ and its spatial and temporal derivatives are in fact what constitute the mathematical model for fluent continua in which $\bar{\boldsymbol{v}}$ are observable quantities.

(5) Even though one could consider the view point expressed in Remark (3), but the fact that $\bar{\boldsymbol{x}} = \bar{\boldsymbol{u}} + \boldsymbol{x}$ is essential for existence of Eulerian descriptions must be kept in mind. And the existence of deformed coordinates $\bar{\boldsymbol{x}}$ necessitates displacements $\boldsymbol{u}$ or $\bar{\boldsymbol{u}}$.

(6) Even though strain measures do not directly appear in the CBL (and constitutive theories for fluent continua), but the convected time derivatives of $\boldsymbol{\varepsilon}_{[0]}$ and $\bar{\boldsymbol{\varepsilon}}^{[0]}$ are essential in the constitutive theories for fluent continua.

6.11.2 Integral form of the CBL

The integral form of the equations resulting from CM, BLM, BAM, FLT and SLT are (6.30), (6.50), (6.81), (6.104) and (6.130), respectively. These are given in the following.

(1) Conservation of mass (CM): continuity equation

$$\int_{\bar{V}(t)} \left(\frac{\partial \bar{\rho}}{\partial t} + \bar{\boldsymbol{\nabla}} \cdot (\bar{\rho}\bar{\boldsymbol{v}})\right) d\bar{V} = 0 \tag{6.154}$$

(2) Balance of linear momenta (BLM): momentum equations

$$\int\limits_{\bar{V}(t)} \left(\frac{\partial(\bar{\rho}\bar{v}_i)}{\partial t} + \frac{\partial(\bar{v}_i(\bar{\rho}\bar{v}_j))}{\partial \bar{x}_j} - \bar{\rho}\bar{F}_i^b - \frac{\partial \bar{\sigma}_{ji}^{(0)}}{\partial \bar{x}_j} \right) d\bar{V} = 0 \tag{6.155}$$

(3) Balance of angular momenta (BAM)

$$\int\limits_{\bar{V}(t)} \left(\epsilon_{ijk} \left(\bar{x}_i \left(\bar{\rho} \frac{D\bar{v}_j}{Dt} - \bar{\rho}\, \bar{F}_j^b - \bar{\sigma}_{mj,m}^{(0)} \right) \right) - \epsilon_{ijk}\, \bar{\sigma}_{ij}^{(0)} \right) d\bar{V} = 0 \tag{6.156}$$

(4) First law of thermodynamics (FLT): energy equation

$$\int\limits_{\bar{V}(t)} \left(\frac{\partial \bar{e}_t}{\partial t} + \bar{\boldsymbol{\nabla}} \cdot (\bar{e}_t \bar{\boldsymbol{v}}) + \bar{\boldsymbol{\nabla}} \cdot \bar{\boldsymbol{q}} - \bar{\boldsymbol{v}} \cdot (\bar{\boldsymbol{\nabla}} \cdot \bar{\boldsymbol{\sigma}}^{(0)}) - \bar{\boldsymbol{\sigma}}^{(0)} : \bar{\boldsymbol{L}} \right) d\bar{V} = 0 \tag{6.157}$$

in which $\bar{e}_t = \bar{\rho} \left(\bar{e} + \frac{1}{2} \bar{\boldsymbol{v}} \cdot \bar{\boldsymbol{v}} - \bar{\boldsymbol{F}}^b \cdot \bar{\boldsymbol{u}} \right)$

(5) Second law of thermodynamics (SLT): entropy inequality

$$\int\limits_{\bar{V}(t)} \left(\bar{\rho}\bar{\theta} \frac{D\bar{\eta}}{Dt} + (\bar{q}_{i,i} - \bar{\rho}\bar{r}) - \frac{\bar{q}_i \bar{g}_i}{\bar{\theta}} \right) \geq 0 \tag{6.158}$$

These conservation and balance laws permit isolated discontinuities of the integrand provided the discontinuities are square integrable. When the integrand is not continuous everywhere in volume $\bar{V}(t)$, these conservation and balance laws must be used.

6.12 Summary

In this chapter, the conservation and balance laws of classical thermodynamics: conservation of mass, kinematics of continuous media (balance of linear and angular momenta), first law of thermodynamics and the second law of thermodynamics are formulated using integral forms in the current configuration in Eulerian description. The CBL have been presented in integral as well as differential forms. The assumptions of isotropic and homogeneous matter (in which case localization theorem always holds) allows us to derive differential form of the mathematical descriptions resulting from these at a material point, also called local form of the conservation and balance laws. These result in a system of partial differential equations: continuity equation, balance of linear momenta (BLM), balance of angular momenta (BAM) which implies that Cauchy stress tensor is symmetric, energy equation (FLT) and the entropy inequality (SLT). This mathematical model is independent of the constitution of the matter, i.e., matter specific physics, hence is valid for all deforming solid and fluent continua. The integral form of the CBL must be used if the continuity of the integrand in the CBL cannot be ensured everywhere in the volume of deforming matter.

Since the mathematical model resulting from the CBL does not consider matter specific physics, the mathematical model does not have closure. Additional nine equations needed for closure are provided by the constitutive theories of Cauchy stress tensor and heat vector. It is shown that the mathematical model derived here from the CBL in Eulerian description is naturally in velocities $\bar{\boldsymbol{v}}$ and that the displacments only appear in PDEs if we define $\bar{\boldsymbol{v}} = \frac{D\bar{\boldsymbol{u}}}{Dt}$ which is obviously true. However, if we assume velocities to be observable quantities, then $\bar{\boldsymbol{v}} \neq \frac{D\bar{\boldsymbol{u}}}{Dt}$. In this case, the mathematical model only contains velocities $\bar{\boldsymbol{v}}$ (observable) and does not have $\bar{\boldsymbol{u}}$ at all. This form of the mathematical model is used in fluent continua. Absence of $\bar{\boldsymbol{u}}$, in turn, allows displacements not to be monitored, thus, absence of strain also. Both of these aspects are suitable for fluent continua.

Problems

6.1 The velocity distribution between two parallel plates separated by a distance h is given by

$$\bar{v}_1 = \frac{\bar{x}_2}{h} v_0 - c\frac{\bar{x}_2}{h}\Big(1 - \frac{\bar{x}_2}{h}\Big) \quad ; \quad \bar{v}_2 = 0 \quad , \quad \bar{v}_3 = 0 \quad , \quad 0 < \bar{x}_2 < h$$

where $\bar{x}_2$ is measured from the normal to the bottom plate, $\bar{x}_1$ is taken along the plates, $\bar{v}_1$ is the velocity component parallel to the plates, v_0 is the velocity of the top plate in the $\bar{x}_1$ direction and c is a constant. Assume the fluid to be incompressible. Determine whether this velocity field satisfies the continuity equation. Find the volume rate of flow and average velocity.

6.2 Let the velocity field for a deforming matter be given by

$$\bar{v}_i = \frac{\bar{x}_i}{1+t} \qquad ; \qquad t \geq 0$$

Find the mass density of the material particle as a function of time. Assume ρ_0 to be density of the deforming matter in the reference configuration.

6.3 Consider the following velocity field:

$$\bar{v}_1 = \alpha\bar{x}_1 - \beta\bar{x}_2$$
$$\bar{v}_2 = \beta\bar{x}_1 + \alpha\bar{x}_2$$
$$\bar{v}_3 = \gamma\sqrt{\bar{x}_1^2 + \bar{x}_2^2}$$

where α, β and γ are constants. Assume that ρ_0 is the density in the reference configuration.

(a) For this velocity field, determine density $\bar{\rho}(t)$ in the current configuration.
(b) Is this deformation isochoric? If not, then under what conditions would it be isochoric.

6.4 The velocity field in a deforming matter is given by

$$\bar{v}_1 = 2\bar{x}_1\bar{x}_3 \qquad ; \qquad \bar{v}_2 = 2\bar{x}_2^2 t \qquad ; \qquad \bar{v}_3 = 3\bar{x}_2\bar{x}_3 t$$

and the Cauchy stress field $[\bar{\sigma}]$ is given by

$$[\bar{\sigma}] = \alpha[\bar{D}]$$

where α is a constant and $[\bar{D}]$ is the symmetric part of the velocity gradient tensor.

(a) Is the velocity field divergence free? If not, then under what condition would it be divergence free? What does divergence free velocity field mean?
(b) Find the body forces required to satisfy momentum equations.

6.5 Consider the deformation of a continuum resulting in Cauchy stresses given by $(\bar{\sigma}^{(0)})_{ij} = (\bar{\sigma}_{(0)})_{ij} = \bar{\sigma}_{ij} = -p\delta_{ij}$ in which p is constant. Show that

$$\bar{\sigma}_{ij}\bar{D}_{ij} = \frac{p}{\bar{\rho}}\frac{D\bar{\rho}}{Dt}$$

in which $\bar{D}_{ij}$ is the symmetric part of the velocity gradient tensor and $\bar{\rho}$ is the material density in the current configuration.

6.6 Consider a continuum for which contravariant Cauchy stresses are given by

$$\bar{\sigma}_{ij}^{(0)} = -p\,\delta_{ij} \tag{a}$$

and the heat conduction law is given by

$$\bar{q}_i = -k\,\frac{\partial\bar{\theta}}{\partial\bar{x}_i} \tag{b}$$

in which $\bar{\theta}$ is temperature, $\partial\bar{\theta}/\partial\bar{x}_i$ is gradient of temperature, p and k are constants. $\bar{q}_i$ is heat flux.

(a) Determine the energy equation using (a) and (b) assuming that specific internal energy $\bar{e} = \bar{e}(\bar{\theta})$.
(b) Also, determine the momentum equations for the stress field defined by (a).

6.7 Consider the following deformation field:

$$\bar{x}_1 = \frac{x_1}{1 + t^2 x_1} \quad , \quad \bar{x}_2 = x_2 \quad , \quad \bar{x}_3 = x_3$$

Determine material density as a function of $\boldsymbol{x}$ and t using continuity equation in Eulerian description. Use ρ_0 as constant material density in the reference configuration.

6.8 Consider the one-dimensional fully developed non-isothermal flow of a compressible fluid with ${}_d\bar{\sigma}^{(0)}_{x_1x_2} \neq 0$, $\partial\bar{v}_1(\bar{x}_2)/\partial\bar{x}_2 \neq 0$, where x_1 is direction of the flow with velocity $\bar{v}_1(\bar{x}_2)$ and ${}_d\bar{\sigma}^{(0)}_{x_1x_2}$ is the deviatoric shear stress in x_1x_2 plane. The flow is pressure driven in the positive x_1 direction, i.e., $\partial\bar{p}/\partial\bar{x}_1$ (negative) is specified.

(a) Obtain the expanded form of the equations resulting from the conservation and balance laws that constitute the mathematical model.
(b) Simplify the mathematical model derived in (a) by assuming the fluid to be incompressible.
(c) Determine how many additional equations are needed for these mathematical models in (a) and (b) to have closure.

6.9 Consider unsteady two-dimensional non-isothermal flow of a compressible fluid with $\bar{v}_1 = \bar{v}_1(\bar{x}_1, \bar{x}_2, t)$, $\bar{v}_2 = \bar{v}_2(\bar{x}_1, \bar{x}_2, t)$ as velocities in the x_1 and x_2 directions. Consider contravariant deviatoric Cauchy stress tensor ${}_d\bar{\sigma}_{ij}^{(0)}$; $i, j = 1, 2$ and thermodynamic pressure $\bar{p} = \bar{p}(\bar{\rho}, \bar{\theta})$.

(a) Obtain expanded form of the equations resulting from the conservation and balance laws that constitute the mathematical model.
(b) Simplify the mathematical model derived in (a) for incompressible fluid.
(c) Determine how many additional equations are needed for these mathematical models in (a) and (b) to have closure.

6.10 Consider evolution of a volume of compressible fluid in $\mathbb{R}^3$ using $\bar{v}_i(\bar{\boldsymbol{x}}, t)$ as velocities, ${}_d\bar{\sigma}_{ij}^{(0)}(\bar{\boldsymbol{x}}, t)$ as deviatoric Cauchy stresses and $\bar{p}(\bar{\rho}, \bar{\theta})$ as thermodynamic pressure.

(a) Obtain expanded form of the equations resulting from the conservation and balance laws.
(b) Simplify the mathematical model derived in (a) assuming the fluid to be incompressible.
(c) Determine how many additional equations are needed for these models in (a) and (b) to have closure.

7
CONSERVATION AND BALANCE LAWS IN LAGRANGIAN DESCRIPTION

7.1 Introduction

In this chapter, we consider conservation and balance laws (CBL) of classical thermodynamics in Lagrangian (or material or referential) description to derive mathematical description of compressible and incompressible deforming continua. Mathematical models derived in Lagrangian description are ideally suited to follow material point motion. Hence, are preferred for solid continua. Since the actions and reactions equilibrate in the current configuration, we always begin the derivation of the CBL in the current configuration by expressing each conservation and balance law for a deformed volume $\bar{V}(t)$ with closed boundary $\partial\bar{V}(t)$. These forms of the CBL are naturally integral forms and are in Eulerian description for a deformed volume $\bar{V}(t)$ with boundary $\partial\bar{V}(t)$. These are then converted to Lagrangian description from which we attempt to derive integral as well as differential forms of the CBL. We show that integral form of the CBL in Lagrangian description are only possible for CM. This is primarily due to the fact that the integral forms of all other balance laws in Lagrangian description require use of the differential form of the continuity equation resulting from CM in Lagrangian description. Differential form of the CBL in Lagrangian description are derived using localization theorem which always holds when the volume of deforming matter is isotropic and homogeneous.

We consider CBL: conservation of mass (CM), kinematics of continuous media (BLM and BAM), first law of thermodynamics (FLT) and second law of thermodynamics (SLT). These yield continuity equation (CM), force balance equations (BLM), establish symmetry of Cauchy stress tensor (BAM in differential form), energy equation (FLT) and the entropy or Clausius Duhem inequality (SLT).

Since the conservation and balance laws are independent of the constitution of the matter, the mathematical model derived here from the CBL hold for all deforming continua, solid as well as fluent. These mathematical models require additional equations related to the constitution of the matter, constitutive theories that provide closure to these mathematical models.

7.2 Mathematical model for deforming continua in Lagrangian description

In Lagrangian description, all quantities in the current configuration must be expressed as functions of undeformed material point coordinates $\boldsymbol{x}$, areas dA, $d\boldsymbol{A}$, volume V, all in the reference configuration, i.e., at time $t = t_0 = 0$ and current time t. Whereas in Eulerian description, all quantities in the current configuration are expressed as functions of $\bar{\boldsymbol{x}}$, $d\bar{A}$, $d\bar{\boldsymbol{A}}$, $\bar{V}(t)$, all in the current configuration and time

DOI: 10.1201/9781003105336-7

t. Consider a volume V with boundary ∂V in the reference configuration. Let $\bar{V}(t)$ and $\partial\bar{V}(t)$ be the deformed volume and its boundary in the current configuration. In the derivations presented here we consider finite deformation and finite strain. The conservation and balance laws in Eulerian description have already been presented in Chapter 6. In the following we consider CBL in Lagrangian description.

7.3 Conservation of mass: (CM)

7.3.1 Integral form of CM

We assume that the mass of the matter contained in every small volume surrounding the material point $\underset{\sim}{Q}$ is conserved. Let ρ_0 be the density of the mass of matter contained in volume V surrounding the material point $\underset{\sim}{Q}$ in the reference configuration. Upon deformation the point $\underset{\sim}{Q}$ occupies position $\underset{\sim}{\bar{Q}}$ in the current configuration and the volume V changes to $\bar{V}$ with density $\bar{\rho} = \bar{\rho}(\bar{\boldsymbol{x}}, t) = \rho(\boldsymbol{x}, t) = \rho$. Since the mass is conserved during deformation, the mass in volume V is same as the mass in volume $\bar{V}$, i.e.

$$\int_V \rho_0 \, dV = \int_{\bar{V}(t)} \bar{\rho} \, d\bar{V} \tag{7.1}$$

But

$$d\bar{V} = |J| dV \tag{7.2}$$

$$\therefore \quad \int_V \rho_0 \, dV = \int_V \rho(\boldsymbol{x}, t) |J| dV \tag{7.3}$$

This is the integral form of the equation resulting from the conservation of mass. In (7.3), when the integrands are continuous everywhere in V, then the integrals in (7.3) hold in the Riemann sense. If the integrands in (7.3) have isolated discontinuities, but the discontinuities are square integrable, then (7.3) holds in the Lebesgue sense. Equation (7.3) must be used if the continuity of the integrand in (7.3) cannot be ensured everywhere in volume V.

7.3.2 Differential form of CM

Based on localization theorem, if the integrand in (7.3) is continuous everywhere in volume V, then volume V is arbitrary, hence we can equate the integrands in (7.3).

$$\rho_0(\boldsymbol{x}) = |J(\boldsymbol{x}, t)| \rho(\boldsymbol{x}, t) \tag{7.4}$$

where $|J|$ is the determinant of the deformation gradient tensor $[J]$.

Equation (7.4) gives a relationship between the mass density ρ at time t in the current configuration and the mass density ρ_0 in the reference configuration at (at time $t_0 = 0$) material points $\underset{\sim}{\bar{Q}}$ and $\underset{\sim}{Q}$. Equation (7.4) is called the *continuity equation in Lagrangian description.* It ensures local conservation of mass during the deformation process at each material point.

Remarks

(1) Equation (7.4) is the differential form of the continuity equation. It ensures continuity of matter through local conservation of mass at each material point, hence for the entire volume of matter when the matter is isotropic and homogeneous.

(2) When the deforming matter is incompressible, density remains constant, i.e., $\rho_0(\boldsymbol{x}) = \rho(\boldsymbol{x}, t)$, in which case the continuity equation (7.4) in Lagrangian description reduces to

$$|J(\boldsymbol{x}, t)| = 1 \tag{7.5}$$

for all values of time.

(3) We recall that

$$d\bar{V} = |J|\, dV \tag{7.6}$$

Hence, for incompressible continua, we have

$$d\bar{V} = dV \tag{7.7}$$

Such deformation is *volume preserving* and is often referred to as *isochoric* deformation.

(4) Thus, for isochoric deformation, $|J(x_1, x_2, x_3, t)| = 1$ at each material point for all values of time.

(5) When the deformation gradient tensor $[J]$ is known, determination of $\rho(\boldsymbol{x}, t)$, density in the current configuration at a material point $\boldsymbol{x}$, is a simple calculation using (7.4). Thus, continuity equation is not part of the mathematical model and density $\rho(\boldsymbol{x}, t)$ is not a dependent variable in Lagrangian description of CBL.

7.4 Balance of linear momenta: (BLM)

Newton's laws of motion, definition of body forces, cut principle of Cauchy etc. described in Chapter 6 naturally hold here as well. We consider application of Newton's second law to the deformed volume of matter $\bar{V}(t)$ with boundary $\partial\bar{V}(t)$ in the current configuration, i.e., we consider balance of linear momenta for volume $\bar{V}(t)$.

Consider the undeformed or reference configuration (Figure 7.1(a)) with an isolated volume of matter V with a closed boundary ∂V. All material points are identified by their position coordinates in the fixed x-frame. Consider an elementary tetrahedron (Figure 7.1(a)) whose oblique face $A_1A_2A_3$ is part of the boundary ∂V and its faces are parallel to the planes of the x-frame. Upon finite deformation, the reference configuration deforms into the current configuration at time t (Figure 7.1(b)). The volume V changes to $\bar{V}(t)$ with its closed boundary ∂V changing into $\partial\bar{V}(t)$. The material lines containing the edges of the elementary tetrahedron in the reference configuration deform into non-orthogonal curvilinear lines ($\tilde{o}$-$\tilde{x}_1$, $\tilde{o}$-$\tilde{x}_2$ and $\tilde{o}$-$\tilde{x}_3$). The edges of the deformed tetrahedron in the current configuration are formed by the covariant base vectors $\tilde{\boldsymbol{g}}_i$ (Figure 7.1(b)). We must consider the deformed tetrahedron in applying BLM using Newton's second law. First, we note several important points regarding the deformed tetrahedron in the current configuration: (1) All of its faces are flat planes. (2) Faces $\bar{O}\bar{A}_2\bar{A}_3$, $\bar{O}\bar{A}_3\bar{A}_1$ and $\bar{O}\bar{A}_1\bar{A}_2$ are not orthogonal to each other (but their maps in the reference configuration are orthogonal to each other). (3) The vectors normal to the faces of the deformed tetrahedron are contravariant base vectors $\tilde{\boldsymbol{g}}^i$. (4) The oblique plane $\bar{A}_1\bar{A}_2\bar{A}_3$ is part of boundary $\partial\bar{V}$. The deformed tetrahedron in Figure 7.1(b) represents true deformation of the tetrahedron in Figure 7.1(a). We must consider application of Newton's second law to this deformed tetrahedron of figure 7.1(b).

Let $\bar{\boldsymbol{P}}$ given by $\{\bar{P}\} = [\bar{P}_{x_1}, \bar{P}_{x_2}, \bar{P}_{x_3}]^T$, with $\bar{P}_{x_i}$ being its components in the x-frame, be the resultant force per unit area acting on the oblique plane $\bar{A}_1\bar{A}_2\bar{A}_3$ with unit exterior normal $\bar{\boldsymbol{n}}$. The faces $\bar{O}\bar{A}_2\bar{A}_3$, $\bar{O}\bar{A}_3\bar{A}_1$ and $\bar{O}\bar{A}_1\bar{A}_2$ have tractions acting on them that result in average stress distribution over these areas. Details are shown in Figure 6.2. Let $\bar{V}(t)$ be the volume of the deformed tetrahedron. Let the oblique surface $\bar{A}_1\bar{A}_2\bar{A}_3$ be represented by $\partial\bar{V}$ and $\bar{\boldsymbol{F}}_{\partial\bar{V}}$ be the resultant force acting on $\partial\bar{V}$. We must begin with the deformed volume $\bar{V}(t)$ with its boundary $\partial\bar{V}(t)$, as the balance of linear momenta must hold for this deformed volume. Application of Newton's second law of motion to the deformed volume $\bar{V}(t)$ of the deformed tetrahedron yields

$$\begin{pmatrix} \textbf{Rate of change of} \\ \textbf{linear momenta of} \\ \textbf{the mass in} \\ \textbf{volume } \bar{V}(t) \end{pmatrix} = \begin{pmatrix} \bar{\boldsymbol{F}}^b_{\bar{V}}\textbf{ ; Body} \\ \textbf{forces acting} \\ \textbf{on the mass in} \\ \textbf{volume } \bar{V}(t) \end{pmatrix} + \begin{pmatrix} \bar{\boldsymbol{F}}_{\partial\bar{V}}\textbf{ ; Re-} \\ \textbf{sultant force} \\ \textbf{acting on} \\ \textbf{surface } \partial\bar{V} \end{pmatrix} \tag{7.8}$$

Following Section 6.7.2 in Chapter 6 we can write

$$\bar{\boldsymbol{F}}^b_{\bar{V}} = \int_{\bar{V}(t)} \bar{\boldsymbol{F}}^b \bar{\rho}\, d\bar{V} \quad ; \quad \bar{\boldsymbol{F}}_{\partial\bar{V}} = \int_{\partial\bar{V}} \bar{\boldsymbol{P}} d\bar{A} \tag{7.9}$$

in which $\bar{\boldsymbol{F}}^b_{\bar{V}}$ is the body force per unit mass in x-frame acting on the mass in volume $\bar{V}(t)$. $\bar{\boldsymbol{P}}$ is the average force per unit deformed area $d\bar{A}$ of $\partial\bar{V}(t)$. Rate of change of linear momenta for volume $\bar{V}(t)$ is given by

$$\frac{D}{Dt}\int_{\bar{V}(t)} \bar{\rho}\,\bar{\boldsymbol{v}}\, d\bar{V} \tag{7.10}$$

Using (7.9) and (7.10) in (7.8) we obtain

$$\frac{D}{Dt}\int_{\bar{V}(t)} \bar{\rho}\,\bar{\boldsymbol{v}}\, d\bar{V} = \int_{\bar{V}(t)} \bar{\boldsymbol{F}}^b \bar{\rho}\, d\bar{V} + \int_{\partial\bar{V}} \bar{\boldsymbol{P}} d\bar{A} \tag{7.11}$$

Following Section 6.7.2 in Chapter 6 or Section 4.12 in Chapter 4 we can write (using Cauchy principle)

$$\bar{\boldsymbol{P}} = (\bar{\boldsymbol{\sigma}}^{(0)})^T \cdot \bar{\boldsymbol{n}} \tag{7.12}$$

in which $\bar{\boldsymbol{\sigma}}^{(0)}$ is the contravariant Cauchy stress tensor and $\bar{\boldsymbol{n}}$ is the unit exterior normal to the area $d\bar{\boldsymbol{A}}$. Substituting from (7.12) into (7.11) gives

$$\frac{D}{Dt}\int\limits_{\bar{V}(t)} \bar{\rho}\,\bar{\boldsymbol{v}}\,d\bar{V} = \int\limits_{\bar{V}(t)} \bar{\boldsymbol{F}}^b\bar{\rho}\,d\bar{V} + \int\limits_{\partial\bar{V}} (\bar{\boldsymbol{\sigma}}^{(0)})^T \cdot \bar{\boldsymbol{n}}\,d\bar{A} \tag{7.13}$$

7.4.1 Differential form of BLM

Derivation of the momentum equations in Lagrangian description requires that we express all quantities of interest as a function of x_1, x_2, x_3 and time t. In specific, in (7.13) all quantities in the integrand must become a function of x_i and time t, $d\bar{V}$ must be expressed in terms of dV and then of course $\bar{V}(t)$ will automatically change into V. Using the differential form of CM

$$\rho_0 = \rho(\boldsymbol{x},t)|\boldsymbol{J}| \quad \text{and} \quad d\bar{V} = |\boldsymbol{J}|dV| \tag{7.14}$$

$$\rho_0 dV = \rho(\boldsymbol{x},t)|\boldsymbol{J}|dV = \rho(\boldsymbol{x},t)d\bar{V} = \bar{\rho}d\bar{V} \tag{7.15}$$

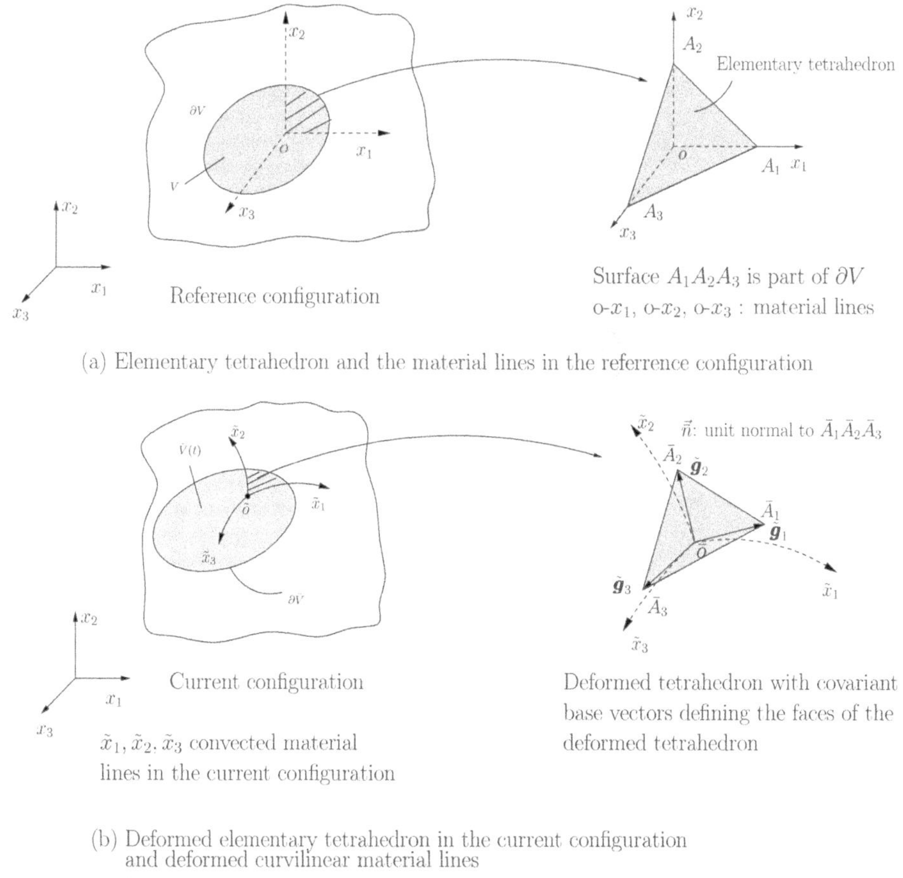

(a) Elementary tetrahedron and the material lines in the referrence configuration

(b) Deformed elementary tetrahedron in the current configuration and deformed curvilinear material lines

Figure 7.1: Elementary tetrahedron in the reference and current configurations

We note that these cannot be used if we wish to derive the integral form of BLM in Lagrangian description.

Following Section 4.12 in Chapter 4 we can write the following using $\boldsymbol{\sigma}^*$, the first Piola-Kirchhoff stress tensor.

$$(\bar{\boldsymbol{\sigma}}^{(0)})^T \cdot \bar{\boldsymbol{n}}\, d\bar{A} = (\boldsymbol{\sigma}^*)^T \cdot \boldsymbol{n}\, dA = (\boldsymbol{\sigma}^*)^T \cdot d\boldsymbol{A} \tag{7.16}$$

Using (7.14)–(7.16), (7.13) can be written as

$$\frac{D}{Dt}\int_V \rho_0\, \boldsymbol{v}\, dV = \int_V \boldsymbol{F}^b \rho_0\, dV + \int_{\partial V} (\boldsymbol{\sigma}^*)^T \cdot d\boldsymbol{A} \tag{7.17}$$

and applying Gauss's divergence theorem to the integral over ∂V (keeping in mind that $\boldsymbol{\sigma}^*$ is a non-symmetric tensor of rank two)

$$\frac{D}{Dt}\int_V \rho_0\, \boldsymbol{v}\, dV = \int_V \boldsymbol{F}^b \rho_0\, dV + \int_V \boldsymbol{\nabla} \cdot \boldsymbol{\sigma}^*\, dV \tag{7.18}$$

$$\text{or} \quad \int_V \Big(\rho_0 \frac{D\boldsymbol{v}}{Dt} - \rho_0 \boldsymbol{F}^b - \boldsymbol{\nabla} \cdot \boldsymbol{\sigma}^*\Big) dV = 0 \tag{7.19}$$

Based on localization theorem if the integrand in (7.19) is continuous everywhere in volume V, then the volume V is arbitrary, hence we can set the integrand to zero.

$$\rho_0 \frac{D\boldsymbol{v}}{Dt} - \rho_0 \boldsymbol{F}^b - \boldsymbol{\nabla} \cdot \boldsymbol{\sigma}^* = 0 \tag{7.20}$$

Equation (7.20) *is the fundamental form resulting from the balance of linear momenta in Lagrangian description.* In matrix and vector notation (using Murnaghan's notation [64]), we can write (7.20) as

$$\rho_0 \frac{\partial\{v\}}{\partial t} - \rho_0 \{F^b\} - [\sigma^*]^T \{\nabla\} = 0 \tag{7.21}$$

since the stress measures $\boldsymbol{\sigma}^*$, $\boldsymbol{\sigma}^{(0)}$ and $\boldsymbol{\sigma}^{[0]}$ are related, using

$$[\sigma^*]^T = |J| [\sigma^{(0)}]^T [J^{-1}]^T \tag{7.22}$$

$$[\sigma^*]^T = [J] [\sigma^{[0]}]^T \tag{7.23}$$

we can express (7.21) in terms of $\boldsymbol{\sigma}^{(0)}$ or $\boldsymbol{\sigma}^{[0]}$.

$$\rho_0 \frac{\partial\{v\}}{\partial t} - \rho_0 \{F^b\} - (|J| [\sigma^{(0)}]^T [J^{-1}]^T)\{\nabla\} = 0 \tag{7.24}$$

$$\rho_0 \frac{\partial\{v\}}{\partial t} - \rho_0 \{F^b\} - ([J] [\sigma^{[0]}]^T)\{\nabla\} = 0 \tag{7.25}$$

Equations (7.21), (7.24) and (7.25) using $\boldsymbol{\sigma}^*$ and $\boldsymbol{\sigma}^{[0]}$ are three alternate forms of the force balance resulting from BLM.

Remarks

(1) First, we note that the integral form of BLM in Lagrangian description is not possible as it requires use of the differential form of CM.

(2) We can also use the covariant Cauchy stress tensor $\boldsymbol{\sigma}_{(0)}$ or Jaumann stress tensor ${}^{(0)}\boldsymbol{\sigma}^J$ in (7.24) (with corresponding modification in (7.25)). Since $\boldsymbol{\sigma}_{(0)}$ and ${}^{(0)}\boldsymbol{\sigma}^J$ stress measures are non-physical, the BLM equations resulting from their use are non-physical as well.

(3) The force balance equations (7.21), (7.24) and (7.25) resulting from BLM can also be written in terms of the displacements $\boldsymbol{u}$ (or $\{u\}$) by using

$$\boldsymbol{v} = \frac{D\boldsymbol{u}}{Dt} = \frac{\partial \boldsymbol{u}}{\partial t} \quad \text{or} \quad \{v\} = \frac{D\{u\}}{Dt} = \frac{\partial \{u\}}{\partial t} \tag{7.26}$$

(4) If the deformation is independent of time, i.e., if the evolution has reached *stationary state*, then for this state, the BLM equations remain valid but without the terms containing $\frac{\partial \boldsymbol{v}}{\partial t}$ or $\frac{\partial\{v\}}{\partial t}$ or alternatively, without $\frac{\partial^2 \boldsymbol{u}}{\partial t^2}$ or $\frac{\partial^2\{u\}}{\partial t^2}$ terms, hence for this case we have

$$\rho_0 \boldsymbol{F}^b + \boldsymbol{\nabla} \cdot \boldsymbol{\sigma} = 0 \tag{7.27}$$

These are the well known *equations of equilibrium in theory of elasticity.*

(5) If the deforming matter is *incompressible*, then BLM equations must be modified using

$$|J| = 1 \quad ; \quad |J|^{-1} = 1 \quad ; \quad \rho(\boldsymbol{x}, t) = \rho_0(\boldsymbol{x}) \tag{7.28}$$

If the *stationary state* of the evolution needs to be described for incompressible matter, then the terms containing $\frac{\partial \boldsymbol{v}}{\partial t}$ and $\frac{\partial\{v\}}{\partial t}$ must be removed from the BLM equations in addition to using (7.28).

(6) If the time-dependent deformation is *infinitesimal*, then $\bar{x}_i \approx x_i$ implying $[J] = [I]$, $|J| = 1$. For this case, all stress measures are the same, i.e.

$$\boldsymbol{\sigma}^* \approx \boldsymbol{\sigma}^{[0]} \approx \boldsymbol{\sigma}^{(0)} = \boldsymbol{\sigma} \tag{7.29}$$

Hence we can write the following using $\boldsymbol{\sigma}$, the Cauchy stress tensor.

$$\rho_0 \frac{\partial \boldsymbol{v}}{\partial t} - \rho_0 \boldsymbol{F}^b - \boldsymbol{\nabla} \cdot \boldsymbol{\sigma} = 0 \tag{7.30}$$

7.5 Balance of angular momenta: (BAM)

7.5.1 Differential form of BAM

If the deformed volume of matter in the current configuration is in equilibrium, then the time rate of change of the moment of linear momenta for the continuum must be equal to the vector sum of the moments of the forces acting on the continuum. When expressed in Lagrangian description this leads to (7.31). See Chapter 6 for the derivation of the differential form in Eulerian description, then convert the final result to Lagrangian description to obtain the following.

$$\epsilon_{ijk}\sigma_{ij}^{(0)} = 0 \quad \text{or} \quad \boldsymbol{\sigma}^{(0)} = (\boldsymbol{\sigma}^{(0)})^T \tag{7.31}$$

Thus, this balance law establishes symmetry of the Cauchy stress tensor $\boldsymbol{\sigma}^{(0)}$, but yields no additional equations.

7.6 First law of thermodynamics: (FLT)

7.6.1 Differential form of the FLT

For thermodynamic equilibrium during evolution of a deforming matter, the first law of thermodynamics must be satisfied (in addition to other conservation and balance laws). The first law of thermodynamics states that the sum of rate of work and the rate of heat added to a deforming volume or mass of matter will result in a rate of increase of energy of the system. The equation resulting from FLT is called the *energy equation* as it is a statement of the balance of energy. Eulerian description for deformed volume of matter in the current configuration is the most logical to begin consideration of the formulation of this balance law as in the current configuration actions and reactions equilibrate. Following Chapter 6, we can write the following in the current configuration in Eulerian description.

$$\frac{D\bar{E}_t}{Dt} = \frac{D\bar{Q}}{Dt} + \frac{D\bar{W}}{Dt} \tag{7.32}$$

in which $\frac{D\bar{E}}{Dt}$, $\frac{D\bar{Q}}{Dt}$ and $\frac{D\bar{W}}{Dt}$ are rate of total energy, rate of heat and the rate of mechanical work. We integrate (7.32) over $\bar{V}(t)$ with closed boundary $\partial\bar{V}(t)$, then expand each term. First, the *total energy $\bar{E}_t$ for a volume of matter $\bar{V}(t)$* can be written as

$$\bar{E}_t = \int\limits_{\bar{V}(t)} \bar{\rho}(\bar{e} + \frac{1}{2}\bar{\boldsymbol{v}} \cdot \bar{\boldsymbol{v}} - \bar{\boldsymbol{F}}^b \cdot \bar{\boldsymbol{u}})d\bar{V} \tag{7.33}$$

where $\bar{e}$ is the specific internal energy, i.e., internal energy per unit mass. $\bar{\boldsymbol{F}}^b$ are the body forces per unit mass and $\bar{\boldsymbol{u}}$ are displacements in x-frame. Consider the reference and the current configurations shown in Figures 6.1 and 6.2. Unit vector $\bar{\boldsymbol{n}}$ is the outward unit normal to $d\bar{s}$ (oblique deformed plane $\bar{A}_1\bar{A}_2\bar{A}_3$ of $\partial\bar{V}(t)$ in Figure 6.3). Let $\bar{\boldsymbol{q}}$ be the amount of heat per unit area and per unit time in the direction $\bar{\boldsymbol{n}}$, then the *rate of heat added to the volume of matter* is given by

$$\frac{D\bar{Q}}{Dt} = -\int\limits_{\partial\bar{V}(t)} \bar{\boldsymbol{q}} \cdot \bar{\boldsymbol{n}}\, d\bar{A} \tag{7.34}$$

The negative sign is due to the fact that positive $\bar{\boldsymbol{q}}$ is in the direction of the outward normal, which means heat removed from the volume and hence decrease in the rate of change of heat. Following Chapter 6, rate of work can be written as

$$\frac{D\bar{W}}{Dt} = \int\limits_{\partial\bar{V}(t)} \bar{\boldsymbol{P}} \cdot \bar{\boldsymbol{v}}\, d\bar{A} = \int\limits_{\partial\bar{V}(t)} \bar{P}_j \bar{v}_j\, d\bar{A} \tag{7.35}$$

Substituting from (7.33), (7.34) and (7.35) into (7.32)

$$\frac{D}{Dt}\int\limits_{\bar{V}(t)} \bar{\rho}\left(\bar{e} + \frac{1}{2}\bar{\boldsymbol{v}} \cdot \bar{\boldsymbol{v}} - \bar{\boldsymbol{F}}^b \cdot \bar{\boldsymbol{u}}\right)d\bar{V} = -\int\limits_{\partial\bar{V}(t)} \bar{\boldsymbol{q}} \cdot \bar{\boldsymbol{n}}\, d\bar{A} + \int\limits_{\partial\bar{V}(t)} \bar{\boldsymbol{P}} \cdot \bar{\boldsymbol{v}}\, d\bar{A} \tag{7.36}$$

In (7.36), we convert all quantities and integrals to the undeformed volume V with closed boundary ∂V. Using the differential form of CM

$$\rho_0 = \rho(\boldsymbol{x},t)|\boldsymbol{J}| \quad \text{and} \quad d\bar{V} = |\boldsymbol{J}|dV| \tag{7.37}$$

$$\rho_0 dV = \rho(\boldsymbol{x},t)|\boldsymbol{J}|dV = \rho(\boldsymbol{x},t)d\bar{V} = \bar{\rho}d\bar{V} \tag{7.38}$$

We note that these cannot be used if we wish to derive the integral form of the FLT in Lagrangian description

$$\int\limits_{\partial V} \bar{\boldsymbol{q}} \cdot \bar{\boldsymbol{n}}\, d\bar{A} = \int\limits_{\partial V} \boldsymbol{q} \cdot \boldsymbol{n}\, dA \tag{7.39}$$

From Section 4.12, Chapter 4 (Equation (4.113)) we can write

$$\begin{aligned} \int\limits_{\partial\bar{V}} \bar{\boldsymbol{P}} \cdot \bar{\boldsymbol{v}}\, d\bar{A} &= \int\limits_{\partial V} \left(\boldsymbol{v} \cdot (\boldsymbol{\sigma}^*)^T\right) \cdot \boldsymbol{n}\, dA \\ \text{or} \quad \int\limits_{\partial\bar{V}} \bar{\boldsymbol{P}} \cdot \bar{\boldsymbol{v}}\, d\bar{A} &= \int\limits_{\partial V} \left(\boldsymbol{v} \cdot (\boldsymbol{\sigma}^*)^T\right) \cdot d\boldsymbol{A} \end{aligned} \tag{7.40}$$

Using (7.39) and (7.40) in (7.36)

$$\frac{D}{Dt}\int\limits_V \rho_0 \left(e + \frac{1}{2}\boldsymbol{v}\cdot\boldsymbol{v} - \boldsymbol{F}^b\cdot\boldsymbol{u}\right)dV = -\int\limits_{\partial V} \boldsymbol{q}\cdot\boldsymbol{n}\, dA + \int\limits_{\partial V} \left(\boldsymbol{v}\cdot(\boldsymbol{\sigma}^*)^T\right)\cdot d\boldsymbol{A} \tag{7.41}$$

Applying Gauss's divergence theorem to the integrals over ∂V in (7.41) (keeping in mind that $\boldsymbol{v} \cdot (\boldsymbol{\sigma}^*)^T$ is a tensor of rank one)

$$\frac{D}{Dt}\int\limits_V \rho_0 \left(e + \frac{1}{2}\boldsymbol{v}\cdot\boldsymbol{v} - \boldsymbol{F}^b\cdot\boldsymbol{u}\right)dV = -\int\limits_V \boldsymbol{\nabla}\cdot\boldsymbol{q}\, dV + \int\limits_V \boldsymbol{\nabla}\cdot\left(\boldsymbol{v}\cdot(\boldsymbol{\sigma}^*)^T\right) dV \tag{7.42}$$

or

$$\int\limits_V \frac{D}{Dt}\left(\rho_0 \left(e + \frac{1}{2}\boldsymbol{v}\cdot\boldsymbol{v} - \boldsymbol{F}^b\cdot\boldsymbol{u}\right)\right)dV = -\int\limits_V \boldsymbol{\nabla}\cdot\boldsymbol{q}\, dV + \int\limits_V \boldsymbol{\nabla}\cdot\left(\boldsymbol{v}\cdot(\boldsymbol{\sigma}^*)^T\right) dV \tag{7.43}$$

Consider integral on the left (7.43). Expanding material derivative and noting $\frac{D\boldsymbol{u}}{Dt} = \boldsymbol{v}$, we can write

$$\int\limits_V \frac{D}{Dt}\left(\rho_0 \left(e + \frac{1}{2}\boldsymbol{v}\cdot\boldsymbol{v} - \boldsymbol{F}^b\cdot\boldsymbol{u}\right)\right)dV = \int\limits_V \left(\rho_0 \frac{De}{Dt} + \rho_0 \boldsymbol{v}\cdot\frac{D\boldsymbol{v}}{Dt} - \rho_0 \boldsymbol{F}^b\cdot\boldsymbol{v}\right) dV \tag{7.44}$$

Using (7.44) in (7.43) and combining all integrals

$$\int\limits_V \left(\rho_0 \frac{De}{Dt} + \rho_0 \boldsymbol{v}\cdot\frac{D\boldsymbol{v}}{Dt} - \rho_0 \boldsymbol{F}^b\cdot\boldsymbol{v} + \boldsymbol{\nabla}\cdot\boldsymbol{q} - \boldsymbol{\nabla}\cdot\left(\boldsymbol{v}\cdot(\boldsymbol{\sigma}^*)^T\right)\right) dV = 0 \tag{7.45}$$

Based on localization theorem if the integrand in (7.45) is continuous everywhere, then the volume V is arbitrary, hence the integrand in (7.45) can be set to zero.

$$\rho_0 \frac{De}{Dt} + \rho_0 \boldsymbol{v} \cdot \frac{D\boldsymbol{v}}{Dt} - \rho_0 \boldsymbol{F}^b \cdot \boldsymbol{v} + \boldsymbol{\nabla} \cdot \boldsymbol{q} - \boldsymbol{\nabla} \cdot \left(\boldsymbol{v} \cdot (\boldsymbol{\sigma}^*)^T\right) = 0 \tag{7.46}$$

The term $\boldsymbol{\nabla} \cdot \left(\boldsymbol{v} \cdot (\boldsymbol{\sigma}^*)^T\right)$ can be expanded as shown in the following.

$$\begin{aligned}
\boldsymbol{\nabla} \cdot \left(\boldsymbol{v} \cdot (\boldsymbol{\sigma}^*)^T\right) &= \boldsymbol{e}_i \frac{\partial}{\partial x_i} \cdot (\boldsymbol{e}_j v_j \cdot \boldsymbol{e}_k \boldsymbol{e}_l \sigma^*_{lk}) \\
&= \boldsymbol{e}_i \frac{\partial}{\partial x_i} \cdot (\boldsymbol{e}_j \cdot \boldsymbol{e}_k \boldsymbol{e}_l v_j \sigma^*_{lk}) \\
&= \boldsymbol{e}_i \frac{\partial}{\partial x_i} \cdot (\delta_{jk} \boldsymbol{e}_l v_j \sigma^*_{lk}) \\
&= \boldsymbol{e}_i \frac{\partial}{\partial x_i} \cdot (\boldsymbol{e}_l v_k \sigma^*_{lk}) \\
&= \boldsymbol{e}_i \cdot \boldsymbol{e}_l \frac{\partial}{\partial x_i} (v_k \sigma^*_{lk}) \\
&= \delta_{il} \frac{\partial}{\partial x_i} (v_k \sigma^*_{lk}) = \frac{\partial}{\partial x_i} (v_k \sigma^*_{ik}) = \frac{\partial}{\partial x_i} (v_j \sigma^*_{ij}) \\
&= v_j \frac{\partial \sigma^*_{ij}}{\partial x_i} + \sigma^*_{ij} \frac{\partial v_j}{\partial x_i}
\end{aligned}$$

Since $\sigma^*_{ij} \dfrac{\partial v_j}{\partial x_i} = \boldsymbol{\sigma}^* : \dot{\boldsymbol{J}}$

$$\therefore \quad \boldsymbol{\nabla} \cdot \left(\boldsymbol{v} \cdot (\boldsymbol{\sigma}^*)^T\right) = v_j \frac{\partial \sigma^*_{ij}}{\partial x_i} + \boldsymbol{\sigma}^* : \dot{\boldsymbol{J}} \tag{7.47}$$

In the following we show that

$$v_j \frac{\partial \sigma^*_{ij}}{\partial x_i} = \boldsymbol{v} \cdot (\boldsymbol{\nabla} \cdot \boldsymbol{\sigma}^*) \tag{7.48}$$

$$\begin{aligned}
\boldsymbol{v} \cdot (\boldsymbol{\nabla} \cdot \boldsymbol{\sigma}^*) &= v_j \boldsymbol{e}_j \cdot \left(\boldsymbol{e}_i \frac{\partial}{\partial x_i} \cdot \boldsymbol{e}_k \boldsymbol{e}_l \sigma^*_{kl}\right) \\
&= v_j \boldsymbol{e}_j \cdot \left((\boldsymbol{e}_i \cdot \boldsymbol{e}_k) \boldsymbol{e}_l \frac{\partial \sigma^*_{kl}}{\partial x_i}\right) \\
&= v_j \boldsymbol{e}_j \cdot \left(\delta_{ik} \boldsymbol{e}_l \frac{\partial \sigma^*_{kl}}{\partial x_i}\right) \\
&= v_j \boldsymbol{e}_j \cdot \left(\boldsymbol{e}_l \frac{\partial \sigma^*_{il}}{\partial x_i}\right) \\
&= v_j \frac{\partial \sigma^*_{il}}{\partial x_i} (\boldsymbol{e}_j \cdot \boldsymbol{e}_l) \\
&= v_j \frac{\partial \sigma^*_{il}}{\partial x_i} \delta_{jl} = v_j \frac{\partial \sigma^*_{ij}}{\partial x_i}
\end{aligned} \tag{7.49}$$

Hence we can write (using (7.47) and (7.49))

$$\boldsymbol{\nabla} \cdot \left(\boldsymbol{v} \cdot (\boldsymbol{\sigma}^*)^T\right) = \boldsymbol{v} \cdot (\boldsymbol{\nabla} \cdot \boldsymbol{\sigma}^*) + \boldsymbol{\sigma}^* : \dot{\boldsymbol{J}} \tag{7.50}$$

From the BLM, equations (7.20) we have

$$\rho_0 \frac{D\boldsymbol{v}}{Dt} = \rho_0 \boldsymbol{F}^b + \boldsymbol{\nabla} \cdot \boldsymbol{\sigma}^* \tag{7.51}$$

Substituting (7.51) into (7.46)

$$\rho_0 \frac{De}{Dt} + \boldsymbol{v} \cdot \left(\rho_0 \boldsymbol{F}^b + \boldsymbol{\nabla} \cdot \boldsymbol{\sigma}^*\right) - \rho_0 \boldsymbol{F}^b \cdot \boldsymbol{v} + \boldsymbol{\nabla} \cdot \boldsymbol{q} - \boldsymbol{v} \cdot \left(\boldsymbol{\nabla} \cdot \boldsymbol{\sigma}^*\right) - \boldsymbol{\sigma}^* : \dot{\boldsymbol{J}} = 0 \tag{7.52}$$

or

$$\rho_0 \frac{\partial e}{\partial t} + \boldsymbol{\nabla} \cdot \boldsymbol{q} - \boldsymbol{\sigma}^* : \dot{\boldsymbol{J}} = 0 \tag{7.53}$$

Equation (7.53) *is the the energy equation in Lagrangian description resulting from FLT*. From (7.53), we observe that W_r rate of mechanical work for volume V is given by

$$W_r = \int_V \boldsymbol{\sigma}^* : \dot{\boldsymbol{J}} dV \tag{7.54}$$

and w_r, rate of mechanical work per unit volume is given by

$$w_r = \boldsymbol{\sigma}^* : \dot{\boldsymbol{J}} \tag{7.55}$$

7.6.2 Rate of mechanical work conjugate pairs in the energy equation (differential form)

In the derivation of the energy equation presented in Chapter 6 in Eulerian description we note that the rate of work resulting in rate of energy stored, or rate of entropy production (rate of dissipation) or both depending upon the constitution of the matter for a volume $\bar{V}(t)$ is given by (using contravariant Cauchy stress tensor $\bar{\boldsymbol{\sigma}}^{(0)}$)

$$\bar{W}_r = \int_{\bar{V}(t)} \bar{\boldsymbol{\sigma}}^{(0)} : \bar{\boldsymbol{L}} \, d\bar{V} \tag{7.56}$$

and the same rate of work per unit volume is given by

$$\bar{w}_r = \bar{\boldsymbol{\sigma}}^{(0)} : \bar{\boldsymbol{L}} \tag{7.57}$$

$\bar{W}_r$ in (7.56) and $\bar{w}_r$ in (7.57) can be converted to Lagrangian description using $\boldsymbol{\sigma}^*$ or $\boldsymbol{\sigma}^{[0]}$ as stress measures to obtain W_r and w_r. When $\boldsymbol{\sigma}^*$ is used as stress measure we expect W_r and w_r in (7.54) and (7.55) are the same as $\bar{W}_r$ and $\bar{w}_r$ in (7.56) and (7.57). Details are given in the following.

7.6.2.1 Using $\boldsymbol{\sigma}^*$ as a stress measure

We recall (Chapter 4) that

$$[\bar{\sigma}^{(0)}]^T = |J|^{-1} [\sigma^*]^T [J]^T \tag{7.58}$$

$$[\dot{J}] = [\bar{L}][J] \quad ; \quad \therefore \quad [\bar{L}] = [\dot{J}][J]^{-1} \tag{7.59}$$

We note that (treating $[\bar{\sigma}^{(0)}]$ as non-symmetric)

$$\bar{W}_r = \int_{\bar{V}(t)} \operatorname{tr}\left([\bar{\sigma}^{(0)}][\bar{L}]\right) d\bar{V} = \int_{\bar{V}(t)} \operatorname{tr}\left([\bar{\sigma}^{(0)}]^T[\bar{L}]^T\right) d\bar{V} \tag{7.60}$$

Substituting from (7.58) and (7.59) in (7.60) and changing $\bar{V}(t)$ to V and using $d\bar{V} = |J|dV$, we can write

$$W_r = \int_V \operatorname{tr}\left(\left(|J|^{-1}[\sigma^*]^T[J]^T\right)\left([J^T]^{-1}[\dot{J}]^T\right)\right)|J|dV \tag{7.61}$$

$$= \int_V |J|^{-1}\operatorname{tr}[\sigma^*]^T[\dot{J}]^T|J|dV \tag{7.62}$$

$$W_r = \int_V \operatorname{tr}\left([\sigma^*]^T[\dot{J}]^T\right) dV = \int_V \operatorname{tr}\left([\sigma^*][\dot{J}]\right) dV = \int_V \boldsymbol{\sigma}^* : \dot{\boldsymbol{J}} dV \tag{7.63}$$

Therefore

$$w_r = \operatorname{tr}\left([\boldsymbol{\sigma}^*][\dot{J}]\right) = \boldsymbol{\sigma}^* : \dot{\boldsymbol{J}} \tag{7.64}$$

W_r and w_r in (7.63) and (7.64) are same as those in (7.54) and (7.55).

7.6.2.2 Using $\boldsymbol{\sigma}^{[0]}$ as a stress measure

Recall

$$[\sigma^*]^T = [J][\sigma^{[0]}]^T \tag{7.65}$$

$$[\dot{J}]^T = [J]^T[\bar{L}]^T \tag{7.66}$$

using (7.63) and substituting (7.65) and (7.66)

$$W_r = \int_V \operatorname{tr}\left([\sigma^*]^T[\dot{J}]^T\right) dV = \int_V \operatorname{tr}\left([J][\sigma^{[0]}]^T[J]^T[\bar{L}]^T\right) dV \tag{7.67}$$

In CCM, $\boldsymbol{\sigma}^{(0)}$ is symmetric, hence $\boldsymbol{\sigma}^{[0]}$ is also symmetric, thus, $([J][\sigma^{[0]}]^T[J]^T)$ is symmetric. Using additive decomposition of $[\bar{L}]$, $[\bar{L}] = [\bar{D}] + [\bar{W}]$ in (7.67)

$$W_r = \int_V \operatorname{tr}\left([J][\sigma^{[0]}]^t[J]^T\right)\left([\bar{D}] + [\bar{W}]^T\right) dV \tag{7.68}$$

Since $[\bar{W}]$ is skew-symmetric, we have

$$\operatorname{tr}\left([J][\sigma^{[0]}]^T[J]^T\right)[\bar{W}]^T = 0 \tag{7.69}$$

Thus, (7.68) reduces to

$$W_r = \int_V \operatorname{tr}\left([J][\sigma^{[0]}]^T[J]^T[\bar{D}]\right) dV \tag{7.70}$$

The order of multiplication in trace does not matter, (7.70) can be written as

$$W_r = \int_V \operatorname{tr}\left([\sigma^{[0]}]^T \left([J]^T[\bar{D}][J]\right)\right) dV \tag{7.71}$$

Recall, from Section 5.13.5

$$\frac{D[\varepsilon_{[0]}]}{Dt} = [\dot{\varepsilon}_{[0]}] = [\dot{\varepsilon}_{[0]}]^T = [J]^T[\bar{D}][J] \tag{7.72}$$

$$\begin{aligned} \therefore \quad W_r &= \int_V \operatorname{tr}\left([\sigma^{[0]}]^T[\dot{\varepsilon}_{[0]}]^T\right) dV = \int_V \operatorname{tr}\left([\sigma^{[0]}][\dot{\varepsilon}_{[0]}]\right) dV \\ &= \int_V \boldsymbol{\sigma}^{[0]} : \dot{\boldsymbol{\varepsilon}}_{[0]} dV \end{aligned} \tag{7.73}$$

$$\text{and} \quad w_r = \operatorname{tr}\left([\sigma^{[0]}]^T[\dot{\varepsilon}_{[0]}]^T\right) = \operatorname{tr}\left([\sigma^{[0]}][\dot{\varepsilon}_{[0]}]\right) = \boldsymbol{\sigma}^{[0]} : \dot{\boldsymbol{\varepsilon}}_{[0]} \tag{7.74}$$

Remarks

(1) In CCM $\boldsymbol{\sigma}^{(0)}$ and $\boldsymbol{\sigma}^{[0]}$ both are symmetric stress tensors. When $\boldsymbol{\sigma}^{(0)}$ is not symmetric as in NCCM, then $\boldsymbol{\sigma}^{[0]}$ is not symmetric either. In this case, (7.69) does not hold, hence $\boldsymbol{\sigma}^{[0]}$, $\dot{\boldsymbol{\varepsilon}}_{[0]}$ are not conjugate pairs in W_r and w_r, i.e., (7.73) and (7.74) do not hold.

(2) However, (7.67) holds when $[\sigma^*]$ is not symmetric. Rearranging tensor in the product in (7.67) gives

$$W_r = \int_V \operatorname{tr}\left([\sigma^{[0]}]\left([J]^T[\bar{L}]^T[J]\right)\right) dV = \int_V \operatorname{tr}\left([\sigma^{[0]}]\left([J]^T[\bar{L}][J]\right)\right) dV \tag{7.75}$$

Thus, in this case, $[\sigma^{[0]}]$ and $[J]^T[\bar{L}][J]$ are conjugate. Also,

$$w_r = \operatorname{tr}\left([\sigma^{[0]}]\left([J]^T[\bar{L}][J]\right)\right) \tag{7.76}$$

7.6.3 Energy equation in equivalent rate of work conjugate measures

Using

$$w_r = \operatorname{tr}\left([\sigma^*][\dot{J}]\right) = \operatorname{tr}\left([\sigma^{[0]}][\dot{\varepsilon}_{[0]}]\right) = |J| \operatorname{tr}\left([J]^{-1}[\bar{\sigma}^{(0)}][\dot{J}]\right) \tag{7.77}$$

the energy equation can be expressed in the following forms:

$$\rho_0 \frac{\partial e}{\partial t} + \boldsymbol{\nabla} \cdot \boldsymbol{q} - \boldsymbol{\sigma}^* : \dot{\boldsymbol{J}} = 0 \tag{7.78}$$

$$\rho_0 \frac{\partial e}{\partial t} + \boldsymbol{\nabla} \cdot \boldsymbol{q} - \boldsymbol{\sigma}^{[0]} : \dot{\boldsymbol{\varepsilon}}_{[0]} = 0 \tag{7.79}$$

$$\rho_0 \frac{\partial e}{\partial t} + \boldsymbol{\nabla} \cdot \boldsymbol{q} - |J| \operatorname{tr}\left([J]^{-1}[\bar{\sigma}^{(0)}][\dot{J}]\right) = 0 \tag{7.80}$$

Remarks

(1) If the matter is incompressible, then $\rho(\boldsymbol{x},t) = \rho_0(\boldsymbol{x})$ for all values of time, hence $\rho_0 = |J|\rho$ implies that $|J| = 1$. For this deformation physics, the energy equation can be modified using $|J| = 1$.

(2) When deformation and strains are infinitesimal, $\bar{\boldsymbol{x}} \simeq \boldsymbol{x}$, $[J] = [J]^{-1} = [\bar{J}] = [I]$ and all stress measures reduce to the symmetric Cauchy stress tensor $\boldsymbol{\sigma}$.

$$\boldsymbol{\sigma}^* \simeq \boldsymbol{\sigma}^{[0]} \simeq \bar{\boldsymbol{\sigma}}^{(0)} \simeq \boldsymbol{\sigma} = \boldsymbol{\sigma}^T \tag{7.81}$$

and

$$[\dot{\varepsilon}_{[0]}] = \frac{1}{2}\left([\dot{J}]^T[J] + [J]^T[\dot{J}]\right) \simeq \frac{1}{2}\left([\dot{J}] + [\dot{J}]^T\right) = [{}_s^d\dot{J}] \tag{7.82}$$

on the other hand

$$[\dot{J}] = [{}^d\dot{J}] = [{}_s^d\dot{J}] + [{}_a^d\dot{J}] \tag{7.83}$$

$$\therefore \quad \boldsymbol{\sigma}^* : \dot{\boldsymbol{J}} = \boldsymbol{\sigma}^* : {}^d\dot{\boldsymbol{J}} = \boldsymbol{\sigma}^* : \left({}_s^d\dot{\boldsymbol{J}} + {}_a^d\dot{\boldsymbol{J}}\right) \tag{7.84}$$

$$\boldsymbol{\sigma}^* : \dot{\boldsymbol{J}} = \boldsymbol{\sigma} : {}_s^d\dot{\boldsymbol{J}} \quad ; \quad \boldsymbol{\sigma} : {}_a^d\dot{\boldsymbol{J}} = 0 \tag{7.85}$$

Also, using (7.79)

$$\boldsymbol{\sigma}^{[0]} : \dot{\boldsymbol{\varepsilon}}_{[0]} = \boldsymbol{\sigma} : \dot{\boldsymbol{\varepsilon}}_{[0]} = \boldsymbol{\sigma} : {}_s^d\dot{\boldsymbol{J}} \tag{7.86}$$

Thus, for this deformation physics

$$w_r = \boldsymbol{\sigma}^* : \dot{\boldsymbol{J}} = \boldsymbol{\sigma}^{[0]} : \dot{\boldsymbol{\varepsilon}}_{[0]} = \boldsymbol{\sigma} : {}_s^d\dot{\boldsymbol{J}} \tag{7.87}$$

Thus, all three forms of energy equations (7.78)-(7.80) reduce to

$$\rho_0 \frac{\partial e}{\partial t} + \boldsymbol{\nabla} \cdot \boldsymbol{q} - \boldsymbol{\sigma} : {}_s^d\dot{\boldsymbol{J}} = 0 \tag{7.88}$$

7.7 Second law of thermodynamics: (SLT)

7.7.1 Differential form of SLT

In this section, we derive the *entropy inequality or Clausius-Duhem inequality* in Lagrangian description using SLT. We begin with SLT in Eulerian description using current configuration of the deforming volume of matter.

Consider a volume of matter $\bar{V}(t)$ with closed boundary $\partial\bar{V}(t)$ in the current configuration. Let $\bar{h}$ be the entropy flux between $\bar{V}(t)$ and the volume of the matter surrounding it, $\bar{s}$ be the source of entropy in $\bar{V}(t)$ due to non-contacting bodies (considered per unit mass). Let there exist $\bar{\eta}$, the specific entropy (entropy per unit mass) for $\bar{V}(t)$ bounded by $\partial\bar{V}(t)$ such that its rate of increase is at least equal to that supplied to $\bar{V}(t)$ from all sources (containing or non-contacting). Thus

$$\frac{D}{Dt}\int\limits_{\bar{V}(t)} \bar{\eta}\bar{\rho}\, d\bar{V} \geq \int\limits_{\partial\bar{V}(t)} \bar{h}\, d\bar{A} + \int\limits_{\bar{V}(t)} \bar{s}\bar{\rho}\, d\bar{V} \tag{7.89}$$

We adopt Cauchy principle for entropy flux $\bar{h}$, i.e., $\bar{h}$ at a point $\bar{x}_i$ on $\partial\bar{V}(t)$ depends on the orientation $\bar{\boldsymbol{n}}$ of $\partial\bar{V}(t)$ at $\bar{x}_i$

$$\bar{h} = -\bar{\boldsymbol{\Psi}} \cdot \bar{\boldsymbol{n}} \tag{7.90}$$

in which $\bar{\boldsymbol{\Psi}}$ is similar to heat flux. Substituting from (7.90) into (7.89)

$$\frac{D}{Dt}\int_{\bar{V}(t)} \bar{\eta}\bar{\rho}\, d\bar{V} \geq -\int_{\partial\bar{V}(t)} \bar{\boldsymbol{\Psi}} \cdot \bar{\boldsymbol{n}}\, d\bar{A} + \int_{\bar{V}(t)} \bar{s}\bar{\rho}\, d\bar{V} \tag{7.91}$$

We need to transform (7.91) to Lagrangian description. Using differential form of CM

$$\begin{gathered} \rho_0 = \rho(\boldsymbol{x},t)|\boldsymbol{J}| \quad \text{and} \quad d\bar{V} = |\boldsymbol{J}|dV \\ \rho_0\, dV = \rho(\boldsymbol{x},t)|\boldsymbol{J}|dV = \rho(\boldsymbol{x},t)d\bar{V} = \bar{\rho}d\bar{V} \\ \int_{\partial\bar{V}} \bar{\boldsymbol{\Psi}} \cdot \bar{\boldsymbol{n}}\, d\bar{A} = \int_{\partial V} \boldsymbol{\Psi} \cdot \boldsymbol{n}\, dA \end{gathered} \tag{7.92}$$

Using (7.92) in (7.91) and noting that $\bar{\eta}$ and $\bar{s}$ in Lagrangian description become η and s we can obtain

$$\frac{D}{Dt}\int_V \eta\rho_0\, dV \geq -\int_{\partial V} \boldsymbol{\Psi} \cdot \boldsymbol{n}\, dA + \int_V s\rho_0\, dV \tag{7.93}$$

Using Gauss's divergence theorem for the integral over ∂V gives (noting that $\boldsymbol{\psi}$ is a tensor of rank one)

$$\frac{D}{Dt}\int_V \eta\rho_0\, dV \geq -\int_V \boldsymbol{\nabla} \cdot \boldsymbol{\Psi}\, dV + \int_V s\rho_0\, dV \tag{7.94}$$

$$\text{or} \quad \int_V \left(\rho_0 \frac{D\eta}{Dt} + \boldsymbol{\nabla} \cdot \boldsymbol{\Psi} - \rho_0 s\right) dV \geq 0 \tag{7.95}$$

Based on localization theorem if the integrand in (7.95) is continuous everywhere in volume V, then the volume V is arbitrary and the inequality applies to the integrand.

$$\rho_0 \frac{D\eta}{Dt} + \boldsymbol{\nabla} \cdot \boldsymbol{\Psi} - \rho_0 s \geq 0 \tag{7.96}$$

Equation (7.96) *is the most general form of the entropy inequality, also known as Clausius-Duhem inequality.* In continuum mechanics, a different form of (7.96) is often more meaningful as well as useful. Based on the fact that the entropy due to contacting surfaces must naturally depend upon $\boldsymbol{q}$ (directly proportional) and temperature θ (inversely proportional), we can write

$$\boldsymbol{\Psi} = \frac{\boldsymbol{q}}{\theta} \quad ; \quad s = \frac{r}{\theta} \tag{7.97}$$

where θ is absolute temperature (assumed to be greater than zero), $\boldsymbol{q}$ is the heat vector and r is a suitable potential, then

$$\boldsymbol{\nabla}\cdot\boldsymbol{\Psi} = \Psi_{i,i} = \frac{q_{i,i}}{\theta} - \frac{q_i}{\theta^2}\theta_{,i} = \frac{q_{i,i}}{\theta} - \frac{q_i g_i}{\theta^2} = \frac{\boldsymbol{\nabla}\cdot\boldsymbol{q}}{\theta} - \frac{\boldsymbol{q}\cdot\boldsymbol{g}}{\theta^2} \tag{7.98}$$

Substituting (7.97) and (7.98) into (7.96)

$$\rho_0 \frac{D\eta}{Dt} + \frac{\boldsymbol{\nabla}\cdot\boldsymbol{q}}{\theta} - \frac{\boldsymbol{q}\cdot\boldsymbol{g}}{\theta^2} - \rho_0 \frac{r}{\theta} \geq 0 \tag{7.99}$$

Multiplying throughout by θ (noting that $\theta > 0$)

$$\rho_0 \theta \frac{D\eta}{Dt} + \boldsymbol{\nabla}\cdot\boldsymbol{q} - \rho_0 r - \frac{\boldsymbol{q}\cdot\boldsymbol{g}}{\theta} \geq 0 \tag{7.100}$$

Equation (7.100) *is the most commonly used form of Clausius-Duhem inequality.* Equation (7.100) can be further given a different form using the energy equation. From the energy equation (after inserting the $\rho_0 r$ term) we have

$$\rho_0 \frac{De}{Dt} + \boldsymbol{\nabla}\cdot\boldsymbol{q} - \rho_0 r - \boldsymbol{\sigma}^* : \dot{\boldsymbol{J}} = 0 \tag{7.101}$$

$$\therefore \quad \boldsymbol{\nabla}\cdot\boldsymbol{q} - \rho_0 r = -\rho_0 \frac{De}{Dt} + \boldsymbol{\sigma}^* : \dot{\boldsymbol{J}} \tag{7.102}$$

Substituting from (7.102) into (7.100)

$$\rho_0 \theta \frac{D\eta}{Dt} - \rho_0 \frac{De}{Dt} - \frac{\boldsymbol{q}\cdot\boldsymbol{g}}{\theta} + \sigma^*_{ji}\dot{J}_{ij} \geq 0 \tag{7.103}$$

$$\text{or} \quad \rho_0 \left(\frac{De}{Dt} - \theta \frac{D\eta}{Dt} \right) + \frac{\boldsymbol{q}\cdot\boldsymbol{g}}{\theta} - \boldsymbol{\sigma}^* : \dot{\boldsymbol{J}} \leq 0 \tag{7.104}$$

Let Φ be the Helmholtz free energy density (specific Helmholtz free energy) defined by

$$\Phi = e - \eta\theta \tag{7.105}$$

$$\therefore \quad \frac{De}{Dt} - \theta \frac{D\eta}{Dt} = \frac{D\Phi}{Dt} + \eta \frac{D\theta}{Dt} \tag{7.106}$$

Substituting from (7.106) into (7.104)

$$\rho_0 \left(\frac{D\Phi}{Dt} + \eta \frac{D\theta}{Dt} \right) + \frac{\boldsymbol{q}\cdot\boldsymbol{g}}{\theta} - \boldsymbol{\sigma}^* : \dot{\boldsymbol{J}} \leq 0 \tag{7.107}$$

Equation (7.107) *is the final form of entropy inequality in Lagrangian description,* which can also be recast into alternate forms by noting that

$$\boldsymbol{\sigma}^* : \dot{\boldsymbol{J}} = \boldsymbol{\sigma}^{[0]} : \dot{\boldsymbol{\varepsilon}}_{[0]} = |J| \operatorname{tr}\left([J]^{-1}[\sigma^{(0)}][\dot{J}] \right) \tag{7.108}$$

Hence we obtain the following from (7.107) using (7.108)

$$\rho_0 \left(\frac{D\Phi}{Dt} + \eta \frac{D\theta}{Dt} \right) + \frac{\boldsymbol{q}\cdot\boldsymbol{g}}{\theta} - \boldsymbol{\sigma}^* : \dot{\boldsymbol{J}} \leq 0 \tag{7.109}$$

$$\rho_0 \left(\frac{D\Phi}{Dt} + \eta \frac{D\theta}{Dt} \right) + \frac{\boldsymbol{q}\cdot\boldsymbol{g}}{\theta} - \boldsymbol{\sigma}^{[0]} : \dot{\boldsymbol{\varepsilon}}_{[0]} \leq 0 \tag{7.110}$$

$$\text{or} \quad \rho_0 \left(\frac{\partial \Phi}{\partial t} + \eta \frac{\partial \theta}{\partial t} \right) + \frac{\boldsymbol{q} \cdot \boldsymbol{g}}{\theta} - |J| \operatorname{tr} \left([J]^{-1} [\sigma^{(0)}][\dot{J}] \right) \leq 0 \tag{7.111}$$

Equations (7.109)–(7.111) are *three alternate forms of the entropy inequality in Lagrangian description.* We remark that we can also use $\boldsymbol{\sigma}_{(0)}$ or ${}^{(0)}\boldsymbol{\sigma}^J$ stress measures and obtain similar form of entropy inequality. Limitations of the stress measures $\boldsymbol{\sigma}_{(0)}$ and ${}^{(0)}\boldsymbol{\sigma}^J$ have already been discussed.

Remarks

(1) Based on the fact that w_r is same in (7.108), it is natural to conclude that any one of (7.109)–(7.111) entropy inequalities could be chosen. A closer examination of the physics in w_r term is necessary.

(2) We recall that $\boldsymbol{\sigma}^*$ measure is based on the correspondence rule $d\boldsymbol{F} = d\bar{\boldsymbol{F}}$, implying small deformation, small strain. Furthermore, $[J]$ is not a measure of finite strain. Thus, w_r term in (7.109), though it represents correct rate of dissipation, but is only valid to represent small deformation, small strain physics. This suggests that (7.109) is only appropriate for small deformation, small strain physics. Similar arguments hold for (7.111) as it does not have measure of finite strain either.

(3) In the derivation of $[\sigma^{[0]}]$ we use $\{d\bar{F}\} = [J]\{dF\}$ as the correspondence rule and $[\varepsilon_{[0]}]$ is a measure of finite strain, hence (7.110) is most appropriate for finite deformation, finite strain physics.

7.8 Second law of thermodynamics using Gibbs potential (differential form)

Gibbs potential or Gibbs thermodynamic potential or Gibbs free energy is used to determine the maximum amount of reversible work that may be performed by a thermodynamic system at constant temperature and pressure.

7.8.1 Using $\boldsymbol{\sigma}^{[0]}$ and $\boldsymbol{\varepsilon}_{[0]}$ as conjugate pair

In this section, we derive entropy inequality in Lagrangian description in terms of Gibbs potential Ψ using $\boldsymbol{\sigma}^{[0]}$ and covariant Green's strain tensor $\boldsymbol{\varepsilon}_{[0]}$. Consider entropy inequality (7.110) in terms of Helmholtz free energy density Φ.

$$\rho_0 \left(\dot{\Phi} + \eta \dot{\theta} \right) + \frac{\boldsymbol{q} \cdot \boldsymbol{q}}{\theta} - \boldsymbol{\sigma}^{[0]} : \dot{\boldsymbol{\varepsilon}}_{[0]} \leq 0 \tag{7.112}$$

We recall that Φ and Ψ are related [30, 31] through

$$\Psi = \Phi - \frac{1}{\rho_0} \boldsymbol{\sigma}^{[0]} : \boldsymbol{\varepsilon}_{[0]} \tag{7.113}$$

Hence

$$\dot{\Psi} = \dot{\Phi} - \frac{1}{\rho_0} \dot{\boldsymbol{\sigma}}^{[0]} : \boldsymbol{\varepsilon}_{[0]} - \frac{1}{\rho_0} \boldsymbol{\sigma}^{[0]} : \dot{\boldsymbol{\varepsilon}}_{[0]} \tag{7.114}$$

Substituting for $\dot{\Phi}$ from (7.114) into (7.112)

$$\rho_0 \left(\dot{\Psi} + \frac{1}{\rho_0}\dot{\boldsymbol{\sigma}}^{[0]} : \boldsymbol{\varepsilon}_{[0]} + \frac{1}{\rho_0}\boldsymbol{\sigma}^{[0]} : \dot{\boldsymbol{\varepsilon}}_{[0]} + \eta\dot{\theta} \right) + \frac{1}{\theta}\boldsymbol{q}\cdot\boldsymbol{g} - \boldsymbol{\sigma}^{[0]} : \dot{\boldsymbol{\varepsilon}}_{[0]} \leq 0 \tag{7.115}$$

Simplifying (7.115)

$$\rho_0 \left(\dot{\Psi} + \eta\dot{\theta} \right) + \frac{\boldsymbol{q}\cdot\boldsymbol{g}}{\theta} + \dot{\boldsymbol{\sigma}}^{[0]} : \boldsymbol{\varepsilon}_{[0]} \leq 0 \tag{7.116}$$

Equation (7.116) is *the most fundamental form of the entropy inequality in terms of Gibbs potential* Ψ *and conjugate measures* $\boldsymbol{\sigma}^{[0]}$ *and* $\boldsymbol{\varepsilon}_{[0]}$.

7.8.2 Using $\boldsymbol{\sigma}^{[0]}$ and $\boldsymbol{C}_{[0]}$ as conjugate pair

In this section, we present an alternate form of (7.116) using $\boldsymbol{\sigma}^{[0]}$ and covariant Cauchy strain tensor $\boldsymbol{C}_{[0]}$ in Lagrangian description. This form is sometimes more convenient. Recall

$$\boldsymbol{\varepsilon}_{[0]} = \frac{1}{2}\left(\boldsymbol{C}_{[0]} - \boldsymbol{I}\right) \tag{7.117}$$

$$\therefore \quad \dot{\boldsymbol{\varepsilon}}_{[0]} = \frac{1}{2}\dot{\boldsymbol{C}}_{[0]} \tag{7.118}$$

Substituting from (7.118) into (7.112)

$$\rho_0 \left(\dot{\Phi} + \eta\dot{\theta} \right) + \frac{\boldsymbol{q}\cdot\boldsymbol{g}}{\theta} - \frac{1}{2}\boldsymbol{\sigma}^{[0]} : \dot{\boldsymbol{C}}_{[0]} \leq 0 \tag{7.119}$$

Also using (7.117) in (7.113)

$$\Psi = \Phi - \frac{1}{2\rho_0}\boldsymbol{\sigma}^{[0]} : \boldsymbol{C}_{[0]} - \frac{1}{2\rho_0}\boldsymbol{\sigma}^{[0]} : \boldsymbol{I} \tag{7.120}$$

$$\dot{\Psi} = \dot{\Phi} - \frac{1}{2\rho_0}\dot{\boldsymbol{\sigma}}^{[0]} : \boldsymbol{C}_{[0]} - \frac{1}{2\rho_0}\boldsymbol{\sigma}^{[0]} : \dot{\boldsymbol{C}}_{[0]} - \frac{1}{2\rho_0}\dot{\boldsymbol{\sigma}}^{[0]} : \boldsymbol{I} \tag{7.121}$$

Substituting for $\dot{\Phi}$ from (7.121) into (7.119) and simplifying

$$\rho_0 \left(\dot{\Psi} + \eta\dot{\theta} \right) + \frac{\boldsymbol{q}\cdot\boldsymbol{g}}{\theta} + \frac{1}{2}\dot{\boldsymbol{\sigma}}^{[0]} : \left(\boldsymbol{C}_{[0]} - \boldsymbol{I}\right) \leq 0 \tag{7.122}$$

The easier way to derive (7.122) is to substitute (7.117) into (7.116). Equation (7.122) is also the *fundamental form of the entropy inequality in* Ψ, $\dot{\boldsymbol{\sigma}}^{[0]}$ *and* $\boldsymbol{C}_{[0]}$. We note that even though (7.122) contains $\boldsymbol{C}_{[0]}$ instead of $\boldsymbol{\varepsilon}_{[0]}$, the conjugate pair remains as $\dot{\boldsymbol{\sigma}}^{[0]}$ and $\boldsymbol{\varepsilon}_{[0]}$. This is rather obvious.

Gibbs potential provides alternate means of deriving constitutive theories [112].

7.9 Summary of differential form of conservation and balance laws

Conservation and balance laws derived here in Lagrangian description are summarized in the following (using Φ as well as ψ).

Conservation of mass: CM

$$\rho_0 = |J|\rho(\boldsymbol{x},t) \quad ; \quad \text{continuity equation} \tag{7.123}$$

Balance of linear momenta: BLM

$$\left.\begin{aligned}
&\rho_0 \frac{\partial\{v\}}{\partial t} - \rho_0\{F^b\} - \big(|J|[\sigma^{(0)}][J^{-1}]^T\big)\{\nabla\} = 0\\
\text{or}\\
&\rho_0 \frac{\partial\{v\}}{\partial t} - \rho_0\{F^b\} - \big([J][\sigma^{[0]}]\big)\{\nabla\} = 0\\
\text{or}\\
&\rho_0 \frac{\partial\{v\}}{\partial t} - \rho_0\{F^b\} - [\sigma^*]^T\{\nabla\} = 0\\
&\{v\} = \frac{D}{Dt}\{u\}
\end{aligned}\right\} ; \ \text{momentum equations} \tag{7.124}$$

Balance of angular momenta: BAM

$$[\sigma^{(0)}] = [\sigma^{(0)}]^T \tag{7.125}$$

First law of thermodynamics: FLT

$$\left.\begin{aligned}
&\rho_0 \frac{\partial e}{\partial t} + \boldsymbol{\nabla}\cdot\boldsymbol{q} - |J|\text{tr}\big([J^T]^{-1}[\sigma^{(0)}]^T[\dot{J}^T]\big) = 0\\
\text{or}\\
&\rho_0 \frac{\partial e}{\partial t} + \boldsymbol{\nabla}\cdot\boldsymbol{q} - \boldsymbol{\sigma}^{[0]} : \dot{\boldsymbol{\varepsilon}}_{[0]} = 0\\
\text{or}\\
&\rho_0 \frac{\partial e}{\partial t} + \boldsymbol{\nabla}\cdot\boldsymbol{q} - \boldsymbol{\sigma}^* : \dot{\boldsymbol{J}} = 0
\end{aligned}\right\} ; \ \text{energy equation} \tag{7.126}$$

Second law of thermodynamics

$$\left.\begin{aligned}
&\rho_0\left(\frac{\partial\Phi}{\partial t} + \eta\frac{\partial\theta}{\partial t}\right) + \frac{\boldsymbol{q}\cdot\boldsymbol{g}}{\theta} - |J|\text{tr}\big([J^T]^{-1}[\sigma^{(0)}]^T[\dot{J}]^T\big) \leq 0\\
\text{or}\\
&\rho_0\left(\frac{\partial\Phi}{\partial t} + \eta\frac{\partial\theta}{\partial t}\right) + \frac{\boldsymbol{q}\cdot\boldsymbol{g}}{\theta} - \boldsymbol{\sigma}^{[0]} : \dot{\boldsymbol{\varepsilon}}_{[0]} \leq 0\\
\text{or}\\
&\rho_0\left(\frac{\partial\Phi}{\partial t} + \eta\frac{\partial\theta}{\partial t}\right) + \frac{\boldsymbol{q}\cdot\boldsymbol{g}}{\theta} - \boldsymbol{\sigma}^* : \dot{\boldsymbol{J}} \leq 0
\end{aligned}\right\} ; \ \text{entropy inequality in } \Phi \tag{7.127}$$

$$\left.\begin{aligned}
&\rho_0\left(\frac{\partial\Psi}{\partial t} + \eta\frac{\partial\theta}{\partial t}\right) + \frac{\boldsymbol{q}\cdot\boldsymbol{g}}{\theta} + \dot{\boldsymbol{\sigma}}^{[0]} : \boldsymbol{\varepsilon}_{[0]} \leq 0\\
\text{or}\\
&\rho_0\left(\frac{\partial\Psi}{\partial t} + \eta\frac{\partial\theta}{\partial t}\right) + \frac{\boldsymbol{q}\cdot\boldsymbol{g}}{\theta} + \frac{1}{2}\dot{\boldsymbol{\sigma}}^{[0]} : (\boldsymbol{C}_{[0]} - \boldsymbol{I}) \leq 0
\end{aligned}\right\} ; \ \text{entropy inequality in } \Psi \tag{7.128}$$

Remarks

(1) The CBL in Lagrangian description in the differential form derived here always hold for all isotropic, homogeneous continuous matter.

(2) From the conservation of mass, we note that $\rho(\boldsymbol{x},t) = \dfrac{\rho_0}{|J|}$, i.e., $\rho(\boldsymbol{x},t)$ at time

t in the current configuration is deterministic once the deformation, i.e., when $\boldsymbol{J}$ is known.

(3) The mathematical models from CBL consists of four partial differential equations: BLM (3), FLT (1) in thirteen dependent variables: $\boldsymbol{v}$ (3), $\boldsymbol{\sigma}^{[0]}$ (6), $\boldsymbol{q}$ (3), θ (1), hence nine more equations are required for closure. These are obtained from the constitutive theories for $\boldsymbol{\sigma}^{[0]}$ and $\boldsymbol{q}$. We note that in general for compressible continua $e = e(\rho, \theta)$, $\Phi = \Phi(\rho, \theta, \dots)$, $\eta = \eta(\rho, \theta, \dots)$, hence known, thus, e, Φ and η are not considered as dependent variables in the mathematical model.

(4) We can also substitute $\boldsymbol{v} = \frac{D\boldsymbol{u}}{Dt} = \frac{\partial \boldsymbol{u}}{\partial t}$ in the mathematical model which gives us $\boldsymbol{u}$ as dependent variables in the model opposed to velocities $\boldsymbol{v}$.

(5) The constitutive theories contain matter specific physics. When the CBL are augmented with the equations resulting from the constitutive theory, the resulting mathematical model has closure.

(6) The balance of linear momenta equations (7.124) representing force balance in x-frame are also referred to as *equations of equilibrium* in the theory of elasticity and writings on solid continua.

7.10 Summary

The differential forms of the mathematical models have been based on localization theorem for compressible and incompressible matter in Lagrangian description using conservation and balance laws. The integral form of the CBL in Lagrangian description cannot be derived except for CM due to the fact that their derivation requires use of the differential form of CM. The momentum equations, energy equation and entropy inequality are presented using different conjugate stress and strain pairs. The entropy inequality has been derived using Helmholtz free energy density as well as Gibbs potential in Lagrangian description. These two forms of the entropy inequality permit alternate derivations of constitutive theories. The general derivations of the mathematical models are valid for finite deformation of solids in which the rate of mechanical work results in entropy production and strain energy influencing the thermal field and internal energy. In the case of thermoelastic solids, the rate of mechanical work does not contribute to entropy production. For solid matter, Lagrangian descriptions are the most meaningful and easier to use in applications.

Problems

7.1 If $[J] = [\frac{\partial\{\bar{x}\}}{\partial\{x\}}]$ in which $\{\bar{x}\}$ and $\{x\}$ are coordinates of a material point in the current and reference configuration, then, show that for infinitesimal deformation in which infinitesimals of order two or higher of $\frac{\partial u_i}{\partial x_j}$ are neglected, we can obtain

(a)
$$|J| = 1 + \frac{\partial u_i}{\partial x_i}$$

(b)
$$\rho = \rho_0\left(1 - \frac{\partial u_i}{\partial x_i}\right)$$

in which u_i are displacements along o-x_1, o-x_2 and o-x_3 axes of o-$x_1x_2x_3$ frame.

7.2 Consider the motion of a body described by

$$\bar{x}_1 = \frac{x_1}{1+tx_1} \quad , \quad \bar{x}_2 = x_2 \quad , \quad , \bar{x}_3 = x_3 \tag{7.129}$$

Determine the material density as a function of $(\boldsymbol{x}, t)$ and $(\bar{\boldsymbol{x}}, t)$, i.e., $\rho(\boldsymbol{x}, t)$ and $\bar{\rho}(\bar{\boldsymbol{x}}, t)$ in the current configuration using

(a) conservation of mass in Lagrangian description;
(b) conservation of mass in Eulerian description.

Use ρ_0 as density in the reference configuration at time $t = 0$.

7.3 Consider the following deformation field:

$$\bar{x}_1 = \frac{x_1}{1+t^2x_1} \quad , \quad \bar{x}_2 = x_2 \quad , \quad \bar{x}_3 = x_3 \tag{7.130}$$

Determine the material density as a function of $(\boldsymbol{x}, t)$ and $(\bar{\boldsymbol{x}}, t)$, i.e., $\rho(\boldsymbol{x}, t)$ and $\bar{\rho}(\bar{\boldsymbol{x}}, t)$ in the current configuration using

(a) conservation of mass in Lagrangian description;
(b) conservation of mass in Eulerian description.

Use ρ_0 as density in the reference configuration at time $t = 0$.

7.4 Consider one-dimensional deformation of an isotropic, homogeneous, compressible thermoelastic solid continua. Use second Piola-Kirchhoff stress $\boldsymbol{\sigma}^{[0]}$ and Green's strain $\boldsymbol{\varepsilon}_{[0]}$ as conjugate measures of stress and finite strain in Lagrangian description.

(a) Obtain expanded form of the equations resulting from the conservation and the balance laws that constitute the mathematical model. Use x_1 direction of x-frame as the direction of deformation or displacement.
(b) Simplify the mathematical model derived in (a) by assuming the deformation to be infinitesimal and the solid continua to be incompressible.
(c) Determine how many additional equations are needed for the mathematical models in (a) and (b) to have closure.

7.5 Consider one-dimensional deformation of an isotropic, homogeneous compressible thermoviscoelastic solid continua (with and without memory). Consider second Piola-Kirchhoff stress $\boldsymbol{\sigma}^{[0]}$ and Green's strain tensor $\boldsymbol{\varepsilon}_{[0]}$ as conjugate measures of stress and finite strain in Lagrangian description.

(a) Obtain the expanded form of the equations resulting from the conservation and the balance laws that constitute the mathematical model. Use x_1 direction of x-frame as the direction of deformation or displacement.
(b) Simplify the mathematical model derived in (a) by assuming the deformation to be infinitesimal and the solid continua to be incompressible.
(c) Determine how many additional equations are needed for the mathematical models in (a) and (b) to have closure.

7.6 Consider two-dimensional deformation in x_1x_2 space of x-frame of an isotropic, homogeneous, compressible thermoelastic solid continua. Choose second Piola-Kirchhoff stress $\boldsymbol{\sigma}^{[0]}$ and Green's strain $\boldsymbol{\varepsilon}_{[0]}$ as conjugate measures of stress and finite strain in Lagrangian description.

(a) Obtain expanded form of the equations resulting from the conservation and the balance laws.
(b) Simplify the mathematical model derived in (a) for infinitesimal deformation of an incompressible thermoelastic solid.
(c) Determine how many additional equations are needed for the mathematical models in (a) and (b) to have closure.

7.7 Consider two-dimensional deformation in x_1x_2 space of x-frame of an isotropic, homogeneous, compressible thermoviscoelastic solid continua with and without memory. Consider $\boldsymbol{\sigma}^{[0]}$, second Piola-Kirchhoff stress and $\boldsymbol{\varepsilon}_{[0]}$, Green's strain tensor as conjugate measures of stress and finite strain in Lagrangian description.

(a) Obtain expanded form of the equations resulting from the conservation and the balance laws.
(b) Simplify the mathematical model derived in (a) for infinitesimal deformation of an incompressible thermoviscoelastic solid.
(c) Determine how many additional equations are needed for the mathematical models in (a) and (b) to have closure.

7.8 Consider three-dimensional deformation in x-frame of an isotropic, homogeneous, compressible thermoelastic solid continua. Using $\boldsymbol{\sigma}^{[0]}$, second Piola-Kirchhoff stress and $\boldsymbol{\varepsilon}_{[0]}$, Green's strain tensor as conjugate measures of stress and finite strain in Lagrangian description,

(a) obtain the expanded form of the equations resulting from the conservation and the balance laws;
(b) simplify the mathematical model derived in (a) for infinitesimal deformation of an incompressible thermoelastic solid;
(c) determine how many additional equations are needed for the mathematical models in (a) and (b) to have closure.

7.9 Consider three-dimensional deformation in x-frame of an isotropic, homogeneous, compressible thermoviscoelastic solid continua (with and without memory). Choose $\boldsymbol{\sigma}^{[0]}$, second Piola-Kirchhoff stress tensor and $\boldsymbol{\varepsilon}_{[0]}$, Green's strain tensor as conjugate measures of stress and finite strain in Lagrangian description.

(a) Obtain the expanded form of the equations resulting from the conservation and the balance laws.
(b) Simplify the mathematical model derived in (a) for infinitesimal deformation of an incompressible thermoviscoelastic solids.
(c) Determine how many additional equations are needed for the mathematical models in (a) and (b) to have closure.

7.10 Let $\{d\bar{A}\}$ and $\{dA\}$ be deformed and undeformed areas in the current and reference configurations, then

(a) Show that

$$\{d\bar{A}\} = \frac{\rho_0}{\rho}\left[[J]^{-1}\right]^T \{dA\} \tag{1}$$

(b) Using (1) show that

$$\text{div}\left(|J|\left[[J]^{-1}\right]^T\right) = 0 \tag{2}$$

Hint: Use details given in Section 3.19.3.

8
CONSTITUTIVE THEORIES

8.1 Introduction

The constitutive theories are intended to provide relationships between the kinematics of deformation and material constitution with the objective of giving mathematical forms to those quantities that are simply assumed to exist in a deforming continua when deriving laws of thermodynamics. Dependence of stress on strain and the material constitution, dependence of heat vector on temperature gradients and the material constitution are simple but illustrative examples. In deriving constitutive theory, we must ensure that thermodynamic equilibrium is not violated by the constitutive theories. That is, all constitutive theories must satisfy CBL, in particular entropy inequality (SLT) so that thermodynamic equilibrium is ensured during the evolution.

Historically, the development of the constitutive theories began with *phenomenological approaches* in which the observed physics is described using various combinations of springs and dash-pots in series and/or parallel. In the early and subsequent developments, these models enjoyed a large degree of success in giving mathematical form to various observed physical behaviors for one-dimensional cases with infinitesimal deformation. This approach, though simple, has many serious drawbacks: (i) the extensions of these one-dimensional constitutive models to continuous media in two and three dimensions are generally not possible and (ii) these models have no *thermodynamic basis*, i.e., in their derivations, entropy inequality (CBL in general) is not considered. Thus, when such constitutive models are used in conjunction with the mathematical models derived using conservation and balance laws, the resulting mathematical models may not ensure *thermodynamic equilibrium* during evolution. For viscoelastic fluids and solids, the *integral constitutive models* using the phenomenological approach have also been derived and are used. These constitutive models also have similar shortcomings. In this book, we do not address phenomenological constitutive models. Interested readers are encouraged to see references [20, 21, 26].

In this chapter, we consider general concepts, guidelines, axioms of constitutive theory and methodologies that are helpful in the derivation of the constitutive theories for continuous matter, both solid and fluent continua. In case of solid continua consideration of: finite deformation, finite strain, compressibility, incompressibility, small strain, small deformation, thermoelastic behavior and thermoviscoelastic behavior must be addressed generally in Lagrangian description in terms of physics as well as the influence of the choice of measures on the constitutive variables and their argument tensors. In case of fluent continua, in principle we have similar considerations but from the point of view of Eulerian description. Choice of basis, consistent stress measures and their convected time derivatives, conjugate convected time derivatives of strain measures and their considerations in compressible as well as incompressible thermoviscous and thermoviscoelastic fluent continua are of significant importance in deriving the constitutive theories for the fluent continua. The

DOI: 10.1201/9781003105336-8

material presented in this chapter in principle applies to all solid as well as fluent continua. Distinctions whenever necessary are made in explaining the concepts that are only applicable to solid continua or fluent continua.

The mathematical models based on CBL are independent of the constitution of the matter, hence the CBL in Eulerian well as Lagrangian description are applicable to all continua (solid or fluent). In the derivation of the CBL, the stress tensor and heat vector are assumed to be present in the deforming matter without regard to the specific constitution of the matter. However, we know that the same disturbance applied to different continuous media will undoubtedly produce different deformation, i.e., response. We have shown in Chapters 6 and 7 that the mathematical models resulting from the conservation and balance laws in Eulerian as well as Lagrangian descriptions do not have closure due to missing information related to constitution of the matter, i.e., matter specific physics. Thus, without matter specific physics, the mathematical models resulting from the conservation and balance laws cannot be used in applications. We refer to this problem as *lack of closure of the mathematical model from CBL*. We have seen that the mathematical model from the conservation and balance laws contain more dependent variables than the number of equations. The additional equations needed to provide closure to the mathematical model derived using CBL are obtained from the constitutive theories derived using kinematics of deformation and constitution of the matter.

For simple solids and fluids, relationships between the stress field and deformation, heat vector and temperature gradients and the properties of the matter are referred to as *constitutive equations or theories*. Considerations for the development of general constitutive theories that contain simpler theories as subset is the subject matter of study in this chapter and the subsequent chapters. *For a deforming matter to be in thermodynamic equilibrium, conservation law and balance laws must be satisfied.* The conservation of mass (CM), balance of linear momenta (BLM), balance of angular momenta (BAM), first law of thermodynamics (FLT) and second law of thermodynamics (SLT). While CM, BLM and FLT yield partial differential equations, BAM only establishes symmetry of the Cauchy stress tensor, and SLT is an inequality that must also be satisfied in addition to the other conservation and balance laws to ensure thermodynamic equilibrium in the deforming continua.

In any deforming continua the total deformation in volume $\bar{V}(t)$ can be additively decomposed into volumetric $\bar{V}_v(t)$ and distortional $\bar{V}_d(t)$ physics, i.e., $\bar{V}(t) = \bar{V}_v(t) + \bar{V}_d(t)$. This holds true regardless of the type of continua or the type of deformation. The volumetric aspects are related to compressibility, incompressibility or the influence of non-isothermal physics on the deforming volume. The distortional aspects on the other hand are the physics that result in change of shape or distortion of the deforming volume of matter as well as dissipation and memory mechanisms. The volumetric and distortional aspects are mutually exclusive, hence cannot be described by a single constitutive theory for the stress tensor. Volumetric aspects, when quantified mathematically in the current configuration using the Cauchy stress tensor result in the same mathematical description regardless of the type of continua (solid or fluent), type of deformation or the constitution of the matter. This is not the case for the distortional aspects of the deformation physics. Thus, the constitutive theories for the volumetric deformation physics are derived in this chapter and referenced in subsequent chapters when needed. The constitutive theories for distortional physics are highly dependent on the type of deformation and the constitution of the matter, thus, require specific consideration (presented in Chapters 9-13).

In the following we present axioms of constitutive theory that serve as a guide in deriving the constitutive theories. This is followed by various considerations that are also helpful in the derivation of the constitutive theories. The thermodynamic approach of deriving constitutive theory is presented next, beginning with entropy inequality and followed by the derivation of constitutive theories for volumetric physics and lastly the final form of the entropy inequality called *the reduced form of the entropy inequality* which is suitable for deriving constitutive theories for distortional aspects of the physics. Details of representation theorem are given with two illustrative examples, using a constitutive tensor of rank two and a constitutive tensor of rank one. Material coefficients are derived in each case. Various non-thermodynamic approaches of deriving constitutive theories are also discussed.

In deriving constitutive theories for the continuous matter, determination of constitutive variables (tensors) and their argument tensors is the first important task. Generally, examination of dependent variable in the SLT and the conjugate pairs in conjunction with axiom of causality (see Section 8.2) is sufficient in the initial determination of the constitutive variables for which the constitutive theories are sought. Determination of the argument tensors of the constitutive variable is facilitated by the conjugate pairs in the SLT and the axiom of equipresence (see Section 8.2).

8.2 Axioms of constitutive theory

In the following we list and discuss axioms of constitutive theory that provide helpful guidelines in determining constitutive variables, their argument tensors as well as guide in deriving constitutive theories in classical continuum mechanics. In non-classical continuum mechanics [116, 117, 123, 124] additional balance laws(s) over and beyond CCM may be needed due to new physics to ensure thermodynamic equilibrium of the deforming continua. These additional balance law(s) may require additional dependent variable(s) (over and beyond CCM) that may lead to additional constitutive variables. Thus, in using the following axioms (established for CCM) in NCCM, care must be exercised. The axioms of constitutive theory for CCM given in the following are fundamental in the development of the constitutive theory.

(i) Axiom of causality
(ii) Axiom of determinism
(iii) Axiom of equipresence
(iv) Axiom of objectivity
(v) Axiom of material invariance
(vi) Axiom of neighborhood
(vii) Axiom of smooth neighborhood
(viii) Axiom of memory
(ix) Axiom of admissibility
(x) Axiom of frame invariance
(xi) Additional axiom(s)

Axiom of causality

Based on this axiom, we consider motion (i.e., displacements) of the material points of a body and their temperatures (θ) as self evident observable effects in every thermomechanical behavior of matter. The remaining quantities, other than those that can be derived by simple integration and differentiation using motion and temperature of material points that enter the expression of entropy generation or production, are causes or dependent variables in the development of constitutive theory.

In the case of thermomechanical behavior of deforming matter, the following cannot be constitutive variables based on this principle.

$$\begin{aligned} \bar{\boldsymbol{x}} = \bar{\boldsymbol{x}}(\boldsymbol{x},t) \quad ; \quad \boldsymbol{x} = \boldsymbol{x}(\bar{\boldsymbol{x}},t) \quad ; \quad \bar{\theta} = \bar{\theta}(\bar{\boldsymbol{x}},t) \quad ; \quad \theta = \theta(\boldsymbol{x},t) \\ \bar{\boldsymbol{u}} = \bar{\boldsymbol{u}}(\bar{\boldsymbol{x}},t) \quad ; \quad \boldsymbol{u} = \boldsymbol{u}(\boldsymbol{x},t) \end{aligned} \tag{8.1}$$

Based on (8.1), the velocity can be derived using material derivative of $\boldsymbol{u}$ or $\bar{\boldsymbol{u}}$, and the density in the current configuration is deterministic from the conservation of mass or continuity equation. Thus, in describing the entropy production, the quantities that remain to be prescribed are stress tensor $\boldsymbol{\sigma}$, heat vector $\bar{\boldsymbol{q}}$, Helmholtz free energy density $\bar{\Phi}$ and entropy density $\bar{\eta}$. Therefore, these are constitutive variables, hence can be expressed in terms of (8.1) and/or their simple differentiations as well as in terms of some additional quantities.

Axiom of determinism

The values of the thermomechanical functions $\bar{\boldsymbol{\sigma}}$, $\bar{\boldsymbol{q}}$, $\bar{\Phi}$ and $\bar{\eta}$ at a material point $\bar{\boldsymbol{x}}$ in the current configuration at time t are determined by the history of motion and temperature of all material points. Thus, for material points $\bar{\boldsymbol{x}}'$ at time $t' \leq t$, we have

$$\bar{\boldsymbol{\sigma}}(\bar{\boldsymbol{x}},t) = \bar{\boldsymbol{\sigma}}(\bar{\boldsymbol{x}}(\bar{\boldsymbol{x}}',t'),\bar{\theta}(\bar{\boldsymbol{x}}',t'),\bar{\boldsymbol{x}},t) \tag{8.2}$$

$$\bar{\boldsymbol{q}}(\bar{\boldsymbol{x}},t) = \bar{\boldsymbol{q}}(\bar{\boldsymbol{x}}(\bar{\boldsymbol{x}}',t'),\bar{\theta}(\bar{\boldsymbol{x}}',t'),\bar{\boldsymbol{x}},t) \tag{8.3}$$

$$\bar{\Phi}(\bar{\boldsymbol{x}},t) = \bar{\Phi}(\bar{\boldsymbol{x}}(\bar{\boldsymbol{x}}',t'),\bar{\theta}(\bar{\boldsymbol{x}}',t'),\bar{\boldsymbol{x}},t) \tag{8.4}$$

$$\bar{\eta}(\bar{\boldsymbol{x}},t) = \bar{\eta}(\bar{\boldsymbol{x}}(\bar{\boldsymbol{x}}',t'),\bar{\theta}(\bar{\boldsymbol{x}}',t'),\bar{\boldsymbol{x}},t) \tag{8.5}$$

We note that $\bar{\boldsymbol{x}}$ and t are functions of $\bar{\boldsymbol{x}}'$ and t'.

Axiom of equipresence

At the onset of establishing the argument tensors of the constitutive variables, all constitutive variables can contain the same list of argument tensors. This is called *principle of equipresence*, i.e all argument tensors have equal presence in all constitutive variables. If $\bar{\boldsymbol{\sigma}}$, $\bar{\boldsymbol{q}}$, $\bar{\Phi}$, $\bar{\eta}$ are the constitutive variables, then instead of

$$\bar{\boldsymbol{\sigma}} = \bar{\boldsymbol{\sigma}}(\bar{\boldsymbol{D}},\bar{\theta}) \quad ; \quad \bar{\boldsymbol{q}} = \bar{\boldsymbol{q}}(\bar{\boldsymbol{g}},\bar{\theta}) \quad ; \quad \bar{\Phi} = \bar{\Phi}(\bar{\theta}) \quad ; \quad \bar{\eta} = \bar{\eta}(\bar{\theta}) \tag{8.6}$$

We can consider

$$\bar{\boldsymbol{\sigma}} = \bar{\boldsymbol{\sigma}}(\bar{\boldsymbol{D}},\bar{\boldsymbol{g}},\bar{\theta}) \quad ; \quad \bar{\boldsymbol{q}} = \bar{\boldsymbol{q}}(\bar{\boldsymbol{D}},\bar{\boldsymbol{g}},\bar{\theta}) \quad ; \quad \bar{\Phi} = \bar{\Phi}(\bar{\boldsymbol{D}},\bar{\boldsymbol{g}},\bar{\theta}) \quad ; \quad \bar{\eta} = \bar{\eta}(\bar{\boldsymbol{D}},\bar{\boldsymbol{g}},\bar{\theta}) \tag{8.7}$$

That is, all constitutive variables $\bar{\boldsymbol{\sigma}}$, $\bar{\boldsymbol{q}}$, $\bar{\Phi}$ and $\bar{\eta}$ can contain the same argument tensors $\bar{\boldsymbol{D}}$, $\bar{\boldsymbol{g}}$, $\bar{\theta}$.

Axiom of objectivity

The constitutive equations must be *form invariant or objective* with respect to rigid motion (translation and rotation) of the spatial frame of reference. Alternatively, the constitutive equations must be independent of the observer translation and rotation. In the inertial frame considered here, translations and rotations (orthogonal) are not functions of time. That is, due to rigid rotation of the frame or the rotation of the observer in which the constitutive theory is derived, the form of the constitutive theory should not change.

As an example, consider $[\sigma]$, $[\varepsilon]$ and $[I]$, symmetric tensors of rank two in x-frame.

Let these tensors be related by (constitutive theory for $[\sigma]$)

$$[\sigma] = 2\mu[\varepsilon] + \lambda(\text{tr}[\varepsilon])[I] \tag{8.8}$$

in which μ and λ are known coefficients.

Consider a coordinate transformation T given by

$$T \; : \; \boldsymbol{x}' = \boldsymbol{x}'(\boldsymbol{x}, t) \tag{8.9}$$

If T is due to pure rotation independent of time, then

$$\{x'\} = [Q]\{x\} \tag{8.10}$$

and

$$[\sigma'] = [Q][\sigma][Q]^T \quad \text{and} \quad [\varepsilon'] = [Q][\varepsilon][Q]^T \tag{8.11}$$

$[\sigma']$ and $[\varepsilon']$ are maps of $[\sigma]$ and $[\varepsilon]$ in x'-frame. If we premultiply (8.8) by $[Q]$ and postmultiply by $[Q]^T$, then we can write

$$[Q][\sigma][Q]^T = 2\mu[Q][\varepsilon][Q]^T + \lambda(\text{tr}([Q][\varepsilon][Q]^T))[I] \tag{8.12}$$

Substituting from (8.11) into (8.12)

$$[\sigma'] = 2\mu[\varepsilon'] + \lambda(\text{tr}[\varepsilon'])[I] \tag{8.13}$$

Comparing (8.8) with (8.13), we note that both have the same form, (8.8) is in x-frame and (8.13) is in x'-frame, where x'-frame is obtained by rigid rotation of x-frame. Thus, to an observer after rotation, the constitutive theory (8.8) appears as (8.13), i.e., the same. Thus, we can conclude that (8.8) is *form invariant or objective* under rigid rotation of the frame.

Constitutive tensors and their argument tensors must always be objective tensors (also see Chapter 5, Section 5.15)

Axiom of material invariance

If the constitutive equations do not change when $(\bar{x}_1, \bar{x}_2, \bar{x}_3)$ are changed to $(\bar{x}_1, \bar{x}_2, -\bar{x}_3)$, then this represents reflection of the material frame of reference with respect to the plane $\bar{x}_3 = 0$. This may be due to crystallographic orientations of the material points in the matter. Thus, according to this principle, the constitutive equations must be form invariant under a group of orthogonal transformations representing planes of symmetry in the matter. The constitutive theory must incorporate or reflect this property, and orthogonal transformations representing form invariance. According to this principle, the constitutive equations must be form invariant under a group of orthogonal transformations representing planes of symmetry in the matter.

Axiom of neighborhood

According to this axiom, the value of a constitutive variable Q at $\bar{\boldsymbol{x}}$ is not effected appreciably by constitutive variable Q at distant material points i.e., the values of Q at material points that are not in the immediate neighborhood of the material point under consideration. Based on this axiom, *all constitutive theories in CCM are local to each material point under consideration.*

Axiom of smooth neighborhood

If the constitutive functionals in the derivation of constitutive theories are sufficiently smooth, then they can be approximated by the functionals in the field of real functions. Based on this axiom, Taylor series expansions hold in a neighborhood of a material point. This axiom is fundamental in establishing material coefficients in the constitutive theories. We consider an example in the following. Let

$$\alpha = \alpha(I_1, I_2, \theta) \tag{8.14}$$

be a constitutive functional at a material point $\bar{\boldsymbol{x}}$ such that $\alpha(I_1, I_2, \theta)$ is smooth in the neighborhood of the map of $\bar{\boldsymbol{x}}$ in a known configuration $\underline{\Omega}$, then we can expand $\alpha(I_1, I_2, \theta)$ in Taylor series in I_1, I_2, θ about a known configuration $\underline{\Omega}$ at the map of $\bar{\boldsymbol{x}}$. We generally retain only up to linear terms in I_1, I_2 and θ for the sake of simplicity of the resulting constitutive theory.

$$\alpha(I_1, I_2, \theta) = \alpha\big|_{\underline{\Omega}} + \left.\frac{\partial \alpha}{\partial I_1}\right|_{\underline{\Omega}} (I_1 - I_1\big|_{\underline{\Omega}}) + \left.\frac{\partial \alpha}{\partial I_2}\right|_{\underline{\Omega}} (I_2 - I_2\big|_{\underline{\Omega}}) + \left.\frac{\partial \alpha}{\partial \theta}\right|_{\underline{\Omega}} (\theta - \theta\big|_{\underline{\Omega}}) \tag{8.15}$$

Axiom of memory

The values of the constitutive variables in the current configuration (time t) are not affected appreciably by the constitutive variables at a distant past. That is $Q(t)$ in the current configuration at $\bar{\boldsymbol{x}}$ is not affected significantly by $Q(t')$, $t' <<< t$. Based on this axiom, the constitutive variables values are only affected by the immediate past. This is often referred to as the *axiom of memory* or *axiom of fading memory.* The material only remembers the events in the immediate past but not those in the distant past.

Axiom of admissibility

The constitutive equations must be consistent with the principles of CCM i.e., the constitutive equations must satisfy all conservation and balance laws (CM, BLM, BAM, FLT, SLT). While all axioms are important in deriving constitutive theories, axiom of causality, axiom of objectivity, axiom of material invariance, axiom of smooth neighborhood and axiom of admissibility are of special significance.

In the following sections we consider some other important aspects such as compressibility, incompressibility, choices of stress and strain measures, finite deformation and finite strain, small deformation, small strain, finite strain rates, stress decomposition, etc. for solid and fluent continua. Clear understanding of associated physics and consistent choices of measures are vital elements in establishing successful and physically meaningful constitutive theories.

Axiom of frame invariance

In the derivation of the constitutive theories we use functionals such as Helmholtz free energy density $\Phi = \Phi(\boldsymbol{\varepsilon}_{[0]})$ or strain energy density $\pi = \pi(\boldsymbol{\varepsilon}_{[0]})$, then use derivatives of $\Phi(\boldsymbol{\varepsilon}_{[0]})$ or $\pi(\boldsymbol{\varepsilon}_{[0]})$ with respect to $\boldsymbol{\varepsilon}_{[0]}$ to derive constitutive theory for $\boldsymbol{\sigma}^{[0]}$.

$$\boldsymbol{\sigma}^{[0]} = \rho_0 \frac{\partial \Phi(\boldsymbol{\varepsilon}_{[0]})}{\partial \boldsymbol{\varepsilon}_{[0]}} \quad \text{or} \quad \boldsymbol{\sigma}^{[0]} = \rho_0 \frac{\partial \pi(\boldsymbol{\varepsilon}_{[0]})}{\partial \boldsymbol{\varepsilon}_{[0]}} \tag{8.16}$$

In order for the material coefficients in the resulting constitutive theories to be frame invariant, we must replace dependence of Φ and π on $\boldsymbol{\varepsilon}_{[0]}$ with dependence on invariants of $\boldsymbol{\varepsilon}_{[0]}$, i.e., instead of (8.16) we must consider

$$\boldsymbol{\sigma}^{[0]} = \rho_0 \frac{\partial \Phi(I_{\varepsilon_{[0]}}, II_{\varepsilon_{[0]}}, III_{\varepsilon_{[0]}})}{\partial \boldsymbol{\varepsilon}_{[0]}} \quad \text{or} \quad \rho_0 \frac{\partial \pi(I_{\varepsilon_{[0]}}, II_{\varepsilon_{[0]}}, III_{\varepsilon_{[0]}})}{\partial \boldsymbol{\varepsilon}_{[0]}} \tag{8.17}$$

Using (8.17) ensures that the material coefficients in the resulting constitutive theories will be functions of $I_{\varepsilon_{[0]}}$, $II_{\varepsilon_{[0]}}$, $III_{\varepsilon_{[0]}}$, hence frame invariant. *Thus, as a general rule, the arguments of the functionals used in deriving constitutive theories must be frame invariant.*

Additional Axioms

The quantities that are not dependent variables in the conservation and balance laws can neither be constitutive tensors nor the argument tensors of the constitutive tensors. For example, density $\rho(\boldsymbol{x}, t)$ is not a dependent variable in the CBL in Lagrangian description as it is deterministic from the CM, thus, $\rho(\boldsymbol{x}, t)$ can neither be a constitutive variable nor can it be the argument tensor of the constitutive variables in Lagrangian description. On the other hand, $\bar{\rho}(\bar{\boldsymbol{x}}, t)$ is a dependent variable in the CBL in Eulerian description. Choice of $\bar{\rho}$ as a constitutive variable is precluded as it is observable, but $\bar{\rho}$ can be an argument tensor of the constitutive variables in the Eulerian description.

The constitutive tensors and their argument tensors must be objective. This applies to the constitutive tensors and their argument tensors of rank zero, one and higher.

8.3 Approaches of deriving constitutive theories

There are many different methodologies and approaches used for deriving constitutive theories in the published works. Broadly, these can be grouped in two categories: thermodynamic approach and non-thermodynamic approaches. We consider both approaches of deriving constitutive theories in the following.

8.3.1 Thermodynamic approach

In this approach we adhere to laws of thermodynamics as far as possible in deriving the constitutive theories so that the derived constitutive theories ensure thermodynamic equilibrium. The starting point in this approach is always the entropy inequality in Eulerian description. CBL, but more specifically entropy inequality and the conjugate pairs in it facilitate determination of constitutive variables and their argument tensors. The constitutive theories for the volumetric deformation physics are derived using conservation of mass, the Helmholtz free energy and the entropy inequality in Eulerian description. This derivation also yields reduced form of the entropy inequality in Eulerian description from which constitutive theories in Eulerian and Lagrangian descriptions can be derived for solid and fluent continua using representation theorem followed by determination of material coefficients.

8.3.2 Other approaches (not strictly thermodynamic)

These approaches do not strictly adhere to laws of thermodynamics in the derivation of the constitutive theories. Some theories use some aspects of the laws of thermodynamics whereas others are purely based on phenomenological, empirical or considerations based on experimental data. The constitutive theories so derived almost always lead to lack of thermodynamic equilibrium (if admissible in the CBL). Many constitutive theories derived using non-thermodynamic approach for non-homogeneous, non-isotropic matter (such as composites) are not admissible in CBL of CCM as these only hold for isotropic and homogeneous matter. We list some of these approaches in the following:

(1) Using energy functionals: strain energy density and complementary strain energy density functionals
(2) Using Taylor series expansion of strain energy or complementary strain energy functionals
(3) Using phenomenologically constructed potentials
(4) Using experimentally measured data
(5) Purely phenomenological approaches based on 1D springs and dashpots.

8.4 Considerations in the constitutive theories

The conservation and balance laws in Eulerian description derived in the current configuration are most fundamental as the actions and reactions only equilibrate in this configuration. Thus, this is a natural setting to give mathematical form to the deformation physics. The Cauchy stress tensor, convected time derivatives of Green's and Almansi strain tensors are examples of the definitions that require consideration of the current configuration. Many measures used in Lagrangian description such as first and second Piola-Kirchhoff stress tensors and Jaumann stress tensor are based on assumed correspondence rule and the true physics of deformation in the current configuration. Such measures though useful, but they obscure the real physics, hence their direct use in the derivation of the constitutive theories may lead to erroneous results or may not even permit derivation of constitutive theories. We note that the conservation and balance laws in Eulerian description are derived for continuous, isotropic and homogeneous matter without regard to the constitution of the matter, hence these are equally valid and applicable to solid as well as fluent continua. Initiating derivations of constitutive theories in Eulerian description is not only natural, but in many instances necessitated due to the fact that the required physics in the desired form may only be realized in Eulerian description.

In any deforming volume of matter, i.e., solid or fluent continua, there are two major aspects of the deformation physics, i.e., *volumetric and distortional*. Volumetric aspects are related to compressibility, incompressibility and non-isothermal physics that may result in change of volume or incompressibility. Volumetric aspects of the deformation physics do not influence distortional aspects. Distortional aspects on the other hand are related to the change in shape of the deforming volume of matter without change in volume, dissipation and memory mechanisms. Clearly, distortional aspect of the physics exclude volumetric aspects and vice versa. Thus, volumetric and distortional aspects of the physics of deformation are mutually exclusive, yet constitutive theory of the stress tensor must describe both aspects of the deformation physics. Therein lies the challenge in developing constitutive theories for stress tensor. We discuss both of these aspects of deformation physics in the following sections.

8.4.1 Common deformation physics

There may be many different aspects of the deformation physics that must be accounted for in deriving constitutive theories for solid and fluent continua. However, there are some aspects of the deformation physics that are fundamental and common in all deforming solid or fluent continua. These aspects of the deformation physics exist in a deforming continua regardless of whether the continua is solid or fluent and regardless of the specific constitution of the matter. It is perhaps prudent to consider constitutive theories for such physics at one place so that they can be referenced whenever needed without unnecessary repetition. We consider these aspects in the three following sections.

8.4.1.1 Compressibility

In a deforming volume of matter of constant mass, a change in volume due to compressibility must result in change in density if the mass is conserved. Thus, conservation of mass must be used to address compressibility physics of deforming continua (solid or fluent). In Lagrangian description, conservation of mass simply states

$$\rho_0(\boldsymbol{x}) = |\boldsymbol{J}|\rho(\boldsymbol{x}, t) \tag{8.18}$$

Continuity equation (8.18) states that if $|\boldsymbol{J}|$ is known, then $\rho(\boldsymbol{x}, t) = \rho_0/|\boldsymbol{J}|$, i.e., $\rho(\boldsymbol{x}, t)$ is deterministic from (8.18) once the deformation is known. Thus, in Lagrangian description, density $\rho(\boldsymbol{x}, t)$ is not a dependent variable in the mathematical model (CBL), hence cannot be used as an argument tensor of a constitutive tensor (axioms of constitutive theory). Thus, the continuity equation (8.18), though it describes density change due to compressibility, but cannot be used to address compressibility physics in the derivation of the constitutive theories.

In Eulerian description, conservation of mass at a material point (local conservation of mass) is given by

$$\frac{D\bar{\rho}}{Dt} + \bar{\rho}(\bar{\boldsymbol{\nabla}} \cdot \bar{\boldsymbol{v}}) = 0 \tag{8.19}$$

The density $\bar{\rho}(\bar{\boldsymbol{x}}, t)$ is not deterministic from (8.19). Hence, $\bar{\rho}(\bar{\boldsymbol{x}}, t)$ must be considered as a dependent variable in the conservation and balance laws. This allows us to include density $\bar{\rho}(\bar{\boldsymbol{x}}, t)$ as an argument tensor of the constitutive variables which, in turn, allows us to use (8.19) in incorporating compressibility physics in the constitutive theories. Thus, it is rather straight forward to conclude that to address compressibility physics (for solid or fluent continua) in the constitutive theories we must consider the entropy inequality in Eulerian description as well as conservation of mass in Eulerian description. Dependence of compressibility physics on density $\bar{\rho}(\bar{\boldsymbol{x}}, t)$ is obvious. Non-isothermal physics may also influence density. An unconstrained volume of matter will experience increase or decrease in density with uniform temperature decrease or rise. Thus, at the onset we observe that the compressibility physics when incorporated in the constitutive theories must show dependence on $\bar{\rho}(\bar{\boldsymbol{x}}, t)$ and $\bar{\theta}(\bar{\boldsymbol{x}}, t)$ regardless of the constitution of the matter and the type of deformation physics (as long as the continuum is compressible). We shall see subsequently that this is indeed the case when compressibility is considered in Eulerian description in the current configuration using the Cauchy stress tensor ${}^{(0)}\bar{\boldsymbol{\sigma}}$ (see Section 8.5.1).

8.4.1.2 Incompressibility

In an incompressible or nearly incompressible continua there is no change of volume during deformation. Hence, for a fixed mass, the density remains constant (unchanged). Thus, from the conservation of mass in Lagrangian description (8.18), we have the following as the incompressibility condition.

$$|\boldsymbol{J}| = 1 \quad \text{as} \quad \rho_0(\boldsymbol{x}) = \rho_0(\boldsymbol{x}, t) \tag{8.20}$$

As in the case of compressible matter, (8.20) cannot be used to incorporate the incompressibility condition in the constitutive theories. From the conservation of mass in Eulerian description (8.19), incompressibility condition ($\bar{\rho}(\bar{\boldsymbol{x}}, t) = \rho_0(\boldsymbol{x}) =$ constant) implies that

$$\bar{\rho}(\bar{\boldsymbol{\nabla}} \cdot \bar{\boldsymbol{v}}) = 0 \quad \text{or} \quad \bar{\boldsymbol{\nabla}} \cdot \bar{\boldsymbol{v}} = 0 \tag{8.21}$$

Equation (8.21) describing the incompressibility condition can be used in the derivation of the constitutive theories representing incompressibility. In case of the non-isothermal physics, a completely constrained volume of matter when subjected to uniform increase or decrease in temperature will result in uniform compressive or tensile pressure field on the boundary of the volume without the change of volume, hence incompressible physics.

Thus, the incompressibility physics in Eulerian description is described by (8.21) and dependence on temperature $\bar{\theta}(\bar{\boldsymbol{x}},t)$. Hence, these must be used in deriving the constitutive theory for non-isothermal incompressible continua.

8.4.1.3 Distortional physics, dissipation and memory mechanisms

Distortion or change of shape of the deforming volume of constant mass without change in volume is the distortional aspect of the physics that is present in all deforming continua, solid or fluent. Thus, distortional physics is not dependent on $\bar{\rho}(\bar{\boldsymbol{x}},t)$ and $\bar{\theta}(\bar{\boldsymbol{x}},t)$ as these describe purely volumetric aspects of the deformation physics. Additionally, since dissipation and memory mechanisms are non-volumetric aspects, they must be considered together with distortional physics.

Remarks

(1) Volumetric aspects: compressibility, incompressibility, non-isothermal physics and the distortional aspects of the physics are mutually exclusive.

(2) Both volumetric and distortional aspects must be incorporated in the constitutive theories for ${}^{(0)}\bar{\boldsymbol{\sigma}}$. However, a single constitutive theory for ${}^{(0)}\bar{\boldsymbol{\sigma}}$ cannot possibly describe both aspects of the deformation physics as they are mutually exclusive.

(3) Volumetric and distortional aspects are present in virtually all types of deformation physics of solid and fluent continua in some form or the other.

(4) Volumetric aspects when expressed through constitutive theories of ${}^{(0)}\bar{\boldsymbol{\sigma}}$ remain invariant of specific deformation physics. That is, physics of compressibility, incompressibility and non-isothermal physics when expressed through the constitutive theory for ${}^{(0)}\bar{\boldsymbol{\sigma}}$ remain the same regardless of type of deformation physics. Thus, it is prudent to consider derivation(s) of constitutive theories related to the volumetric aspects of the deformation physics in this chapter so that its repetition in the subsequent chapters can be avoided.

8.5 Thermodynamic approach of deriving constitutive theories

In the thermodynamic approach, we always begin with the entropy inequality in Eulerian description.

8.5.1 Entropy inequality in Eulerian description

When considering the entropy inequality in Eulerian description, the choice of basis for the Cauchy stress tensor is important. $\bar{\boldsymbol{\sigma}}^{(0)}$ or $\bar{\boldsymbol{\sigma}}_{(0)}$ are contra- and co-variant Cauchy stress tensors and ${}^{(0)}\bar{\boldsymbol{\sigma}}^J$ is the Jaumann stress tensor. Conjugate to these are convected time derivatives $\boldsymbol{\gamma}_{(i)}$, $\boldsymbol{\gamma}^{(i)}$ and ${}^{(i)}\boldsymbol{\gamma}^J$; $i = 1,2,\dots,n$ of Green's strain tensor $\boldsymbol{\varepsilon}_{[0]}$, the Almansi tensor $\bar{\boldsymbol{\varepsilon}}^{[0]}$ and the Jaumann rates that are the average of the two. To make the derivation of constitutive theories basis independent, we choose ${}^{(0)}\bar{\boldsymbol{\sigma}}$ as Cauchy stress tensor (that could be $\bar{\boldsymbol{\sigma}}^{(0)}$, $\bar{\boldsymbol{\sigma}}_{(0)}$ or ${}^{(0)}\bar{\boldsymbol{\sigma}}^J$) and ${}^{(j)}\boldsymbol{\gamma}$; $j = 1,2,\dots,n$ as conjugate convected time derivatives of the strain tensors (that could be $\boldsymbol{\gamma}_{(i)}$, $\boldsymbol{\gamma}^{(i)}$ or ${}^{(i)}\boldsymbol{\gamma}^J$). Noting that ${}^{(1)}\boldsymbol{\gamma} = \boldsymbol{\gamma}^{(1)} = \boldsymbol{\gamma}_{(1)} = {}^{(1)}\boldsymbol{\gamma}^J = \bar{\boldsymbol{D}}$, $\bar{\boldsymbol{D}}$ being symmetric part of the velocity gradient tensor $\bar{\boldsymbol{L}}$. Using the basis independent Cauchy stress tensor and the strain rate measures we can write the entropy inequality in the Eulerian description as follows (Chapter 6).

$$\bar{\rho}\left(\frac{D\bar{\Phi}}{Dt}+\bar{\eta}\frac{D\bar{\theta}}{Dt}\right)-{}^{(0)}\bar{\boldsymbol{\sigma}}:\bar{\boldsymbol{D}}+\frac{\bar{\boldsymbol{q}}\cdot\bar{\boldsymbol{g}}}{\bar{\theta}}\leq 0 \tag{8.22}$$

We consider isotropic, homogeneous, compressible continua with non-isothermal physics.

8.5.2 Additive decomposition of Cauchy stress tensor ${}^{(0)}\bar{\boldsymbol{\sigma}}$

A single constitutive theory for ${}^{(0)}\bar{\boldsymbol{\sigma}}$ cannot possibly describe both volumetric and distortional physics as these are mutually exclusive. Thus, we must consider additive decomposition of the Cauchy stress tensor ${}^{(0)}\bar{\boldsymbol{\sigma}}$.

$${}^{(0)}\bar{\boldsymbol{\sigma}}={}^{(0)}_{\ e}\bar{\boldsymbol{\sigma}}+{}^{(0)}_{\ d}\bar{\boldsymbol{\sigma}} \tag{8.23}$$

in which ${}^{(0)}_{\ e}\bar{\boldsymbol{\sigma}}$ is called the equilibrium Cauchy stress tensor and ${}^{(0)}_{\ d}\bar{\boldsymbol{\sigma}}$ is the deviatoric Cauchy stress tensor. The constitutive theory for ${}^{(0)}_{\ e}\bar{\boldsymbol{\sigma}}$ must consider volumetric aspects of the deformation physics and the distortional aspects of the deformation physics are due to the constitutive theory of ${}^{(0)}_{\ d}\bar{\boldsymbol{\sigma}}$. Thus, at the onset ${}^{(0)}_{\ e}\bar{\boldsymbol{\sigma}}={}^{(0)}_{\ e}\bar{\boldsymbol{\sigma}}(\bar{\rho},\bar{\theta})$ must hold as the volumetric aspects are purely due to $\bar{\rho}$ and $\bar{\theta}$. We do not substitute additive decomposition (8.23) in the entropy inequality (8.22) yet. Instead, we proceed with the Cauchy stress ${}^{(0)}\bar{\boldsymbol{\sigma}}$ in the entropy inequality (8.22) to observe the consequences of doing so.

8.5.3 Constitutive tensors, their argument tensors and SLT

Based on axiom of objectivity, we must ensure that the constitutive tensors and their argument tensors are *form invariant or objective* (see Chapter 5, Section 5.15). We consider the entropy inequality (8.22) in Cauchy stress tensor ${}^{(0)}\bar{\boldsymbol{\sigma}}$. From the entropy inequality (8.22) (as well as other conservation and balance laws) it is straight forward to conclude that ${}^{(0)}\bar{\boldsymbol{\sigma}}$, $\bar{\boldsymbol{q}}$, $\bar{\Phi}$ and $\bar{\eta}$ are possible choice of constitutive tensors. Since $\bar{\Phi}$, $\bar{\eta}$ and $\bar{e}$ are related, $\bar{e}$ is not considered as constitutive tensor. The conjugate pairs ${}^{(0)}\bar{\boldsymbol{\sigma}}:\bar{\boldsymbol{D}}$ and $\frac{\bar{\boldsymbol{q}}\cdot\bar{\boldsymbol{g}}}{\bar{\theta}}$ in the entropy inequality (8.22) suggest that $\bar{\boldsymbol{D}}$ can be an argument tensor of ${}^{(0)}\bar{\boldsymbol{\sigma}}$ and $\bar{\boldsymbol{g}}$ can be an argument tensor of $\bar{\boldsymbol{q}}$. Compressibility (or incompressibility) and non-isothermal physics suggest that $\bar{\rho}$ and $\bar{\theta}$ must also be argument tensors of ${}^{(0)}\bar{\boldsymbol{\sigma}}$ and $\bar{\boldsymbol{q}}$. Thus, we have

$${}^{(0)}\bar{\boldsymbol{\sigma}}={}^{(0)}\bar{\boldsymbol{\sigma}}(\bar{\rho},\bar{\boldsymbol{D}},\bar{\theta}) \tag{8.24}$$

$$\bar{\boldsymbol{q}}=\bar{\boldsymbol{q}}(\bar{\rho},\bar{\boldsymbol{g}},\bar{\theta}) \tag{8.25}$$

Argument tensor of $\bar{\Phi}$ and $\bar{\eta}$ can be chosen based on principle of equipresence.

$$\bar{\Phi}=\bar{\Phi}(\bar{\rho},\bar{\boldsymbol{D}},\bar{\boldsymbol{g}},\bar{\theta}) \tag{8.26}$$

$$\bar{\eta}=\bar{\eta}(\bar{\rho},\bar{\boldsymbol{D}},\bar{\boldsymbol{g}},\bar{\theta}) \tag{8.27}$$

We could have also used the principle of equipresence to change argument tensors of ${}^{(0)}\bar{\boldsymbol{\sigma}}$ and $\bar{\boldsymbol{q}}$ of (8.24) and (8.25), but we have not done so due to the fact that the conjugate pairs in the term ${}^{(0)}\bar{\boldsymbol{\sigma}}:\bar{\boldsymbol{D}}$ and $\frac{\bar{\boldsymbol{q}}\cdot\bar{\boldsymbol{g}}}{\bar{\theta}}$ suggest against it. The argument tensors of the constitutive variables in (8.24)–(8.27) can be augmented by additional argument tensor if the physics requires so. We shall see in Chapters 10–13 that this is indeed the case in the derivations of the ordered rate constitutive theories. Using (8.26) we can write

$$\frac{D\bar{\Phi}}{Dt} = \frac{\partial\bar{\Phi}}{\partial\bar{\rho}}\dot{\bar{\rho}} + \frac{\partial\bar{\Phi}}{\partial\bar{\boldsymbol{D}}} : \dot{\bar{\boldsymbol{D}}} + \frac{\partial\bar{\Phi}}{\partial\bar{\boldsymbol{g}}} \cdot \dot{\bar{\boldsymbol{g}}} + \frac{\partial\bar{\Phi}}{\partial\bar{\theta}}\dot{\bar{\theta}} \tag{8.28}$$

Substituting (8.28) in the entropy inequality (8.22)

$$\bar{\rho}\left(\frac{\partial\bar{\Phi}}{\partial\bar{\rho}}\dot{\bar{\rho}} + \frac{\partial\bar{\Phi}}{\partial\bar{\boldsymbol{D}}} : \dot{\bar{\boldsymbol{D}}} + \frac{\partial\bar{\Phi}}{\partial\bar{\boldsymbol{g}}} \cdot \dot{\bar{\boldsymbol{g}}} + \frac{\partial\bar{\Phi}}{\partial\bar{\theta}}\dot{\bar{\theta}} + \bar{\eta}\dot{\bar{\theta}}\right) - {}^{(0)}\bar{\boldsymbol{\sigma}} : \bar{\boldsymbol{D}} + \frac{\bar{\boldsymbol{q}} \cdot \bar{\boldsymbol{g}}}{\bar{\theta}} \leq 0 \tag{8.29}$$

From continuity equation in Eulerian description we have (compressibility condition)

$$\dot{\bar{\rho}} = -\bar{\rho}(\bar{\boldsymbol{\nabla}} \cdot \bar{\boldsymbol{v}}) = -\bar{\rho}\bar{D}_{kk} = -\bar{\rho}\bar{D}_{ki}\delta_{ik} = -\bar{\rho}\boldsymbol{\delta} : \bar{\boldsymbol{D}} \tag{8.30}$$

Substituting (8.30) in (8.29) and regrouping terms

$$\left(-\bar{\rho}^2\frac{\partial\bar{\Phi}}{\partial\bar{\rho}}\boldsymbol{\delta} - {}^{(0)}\bar{\boldsymbol{\sigma}}\right) : \bar{\boldsymbol{D}} + \bar{\rho}\frac{\partial\bar{\Phi}}{\partial\bar{\boldsymbol{D}}} : \dot{\bar{\boldsymbol{D}}} + \bar{\rho}\frac{\partial\bar{\Phi}}{\partial\bar{\boldsymbol{g}}} \cdot \dot{\bar{\boldsymbol{g}}} + \bar{\rho}\left(\bar{\eta} + \frac{\partial\bar{\Phi}}{\partial\bar{\theta}}\right)\dot{\bar{\theta}} + \frac{\bar{\boldsymbol{q}} \cdot \bar{\boldsymbol{g}}}{\bar{\theta}} \leq 0 \tag{8.31}$$

The entropy inequality (8.31) will hold for arbitrary but admissible choices of $\dot{\bar{\boldsymbol{D}}}$, $\dot{\bar{\boldsymbol{g}}}$ and $\dot{\bar{\theta}}$ if the following conditions are satisfied

$$\bar{\rho}\frac{\partial\bar{\Phi}}{\partial\bar{\boldsymbol{D}}} = 0 \implies \bar{\Phi} \neq \bar{\Phi}(\bar{\boldsymbol{D}}) \tag{8.32}$$

$$\bar{\rho}\frac{\partial\bar{\Phi}}{\partial\bar{\boldsymbol{g}}} = 0 \implies \bar{\Phi} \neq \bar{\Phi}(\bar{\boldsymbol{g}}) \tag{8.33}$$

$$\bar{\rho}\left(\bar{\eta} + \frac{\partial\bar{\Phi}}{\partial\bar{\theta}}\right) = 0 \implies \bar{\eta} = -\frac{\partial\bar{\Phi}}{\partial\bar{\theta}} \tag{8.34}$$

Equation (8.32) and (8.33) imply that $\bar{\Phi}$ is not a function of $\bar{\boldsymbol{D}}$ and $\bar{\boldsymbol{g}}$, and based on equation (8.34), $\bar{\eta}$ is not a constitutive variable as it is deterministic from $\bar{\Phi}$. The constitutive tensors and their argument tensor in (8.24)–(8.27) can be modified using (8.32)–(8.34)

$$\bar{\Phi} = \bar{\Phi}(\bar{\rho}, \bar{\theta}) \tag{8.35}$$

$${}^{(0)}\bar{\boldsymbol{\sigma}} = {}^{(0)}\bar{\boldsymbol{\sigma}}(\bar{\rho}, \bar{\boldsymbol{D}}, \bar{\theta}) \tag{8.36}$$

$$\bar{\boldsymbol{q}} = \bar{\boldsymbol{q}}(\bar{\rho}, \bar{\boldsymbol{g}}, \bar{\theta}) \tag{8.37}$$

and the entropy inequality (8.31) reduces to

$$\left(-\bar{\rho}^2\frac{\partial\bar{\Phi}(\bar{\rho},\bar{\theta})}{\partial\bar{\rho}}\boldsymbol{\delta} - {}^{(0)}\bar{\boldsymbol{\sigma}}\right) : \bar{\boldsymbol{D}} + \frac{\bar{\boldsymbol{q}} \cdot \bar{\boldsymbol{g}}}{\bar{\theta}} \leq 0 \tag{8.38}$$

The entropy inequality as stated in (8.38) must be kept in this form at this stage. For example, setting coefficient of $\bar{\boldsymbol{D}}$ to zero (for arbitrary but admissible $\bar{\boldsymbol{D}}$) would imply that

$$
{}^{(0)}\bar{\boldsymbol{\sigma}} = -\bar{\rho}^2 \frac{\partial \bar{\Phi}(\bar{\rho}, \bar{\theta})}{\partial \bar{\rho}} = {}^{(0)}\bar{\boldsymbol{\sigma}}(\bar{\rho}, \bar{\theta}) \tag{8.39}
$$

which states that ${}^{(0)}\bar{\boldsymbol{\sigma}}$ only depends upon $\bar{\rho}$ and $\bar{\theta}$ which is incorrect based on (8.36). In (8.38), we must introduce additive stress decomposition of ${}^{(0)}\bar{\boldsymbol{\sigma}}$ (equation (8.23)) before we can proceed further. We note from this derivation that a single constitutive theory for ${}^{(0)}\bar{\boldsymbol{\sigma}}$ to describe volumetric and distortional deformation physics is not possible.

8.5.4 Constitutive theory for equilibrium Cauchy stress tensor ${}^{(0)}_{\ e}\bar{\boldsymbol{\sigma}}$ (volumetric deformation physics): Eulerian description, Helmholtz free energy density

8.5.4.1 Compressible, non-isothermal (Eulerian)

Substituting additive stress decomposition (8.23) in (8.38) and noting that

$$
{}^{(0)}_{\ e}\bar{\boldsymbol{\sigma}} = {}^{(0)}_{\ e}\bar{\boldsymbol{\sigma}}(\bar{\rho}, \bar{\theta}) \tag{8.40}
$$

$$
{}^{(0)}_{\ d}\bar{\boldsymbol{\sigma}} = {}^{(0)}_{\ d}\bar{\boldsymbol{\sigma}}(\bar{\rho}, \bar{\boldsymbol{D}}, \bar{\theta}) \tag{8.41}
$$

$$
\text{and} \quad {}^{(0)}_{\ d}\bar{\boldsymbol{\sigma}}(\bar{\rho}, 0, \bar{\theta}) = 0 \tag{8.42}
$$

we can write (after regrouping terms)

$$
\left(-\bar{\rho}^2 \frac{\partial \bar{\Phi}(\bar{\rho}, \bar{\theta})}{\partial \bar{\rho}} \boldsymbol{\delta} - {}^{(0)}_{\ e}\bar{\boldsymbol{\sigma}} \right) : \bar{\boldsymbol{D}} - {}^{(0)}_{\ d}\bar{\boldsymbol{\sigma}} : \bar{\boldsymbol{D}} + \frac{\bar{\boldsymbol{q}} \cdot \bar{\boldsymbol{g}}}{\bar{\theta}} \leq 0 \tag{8.43}
$$

Entropy inequality (8.43) holds for arbitrary but admissible $\bar{\boldsymbol{D}}$ if the coefficient $\bar{\boldsymbol{D}}$ in the first term is set to zero.

$$
{}^{(0)}_{\ e}\bar{\boldsymbol{\sigma}} = -\bar{\rho}^2 \frac{\partial \bar{\Phi}(\bar{\rho}, \bar{\theta})}{\partial \bar{\rho}} \boldsymbol{\delta} = \bar{p}(\bar{\rho}, \bar{\theta}) \boldsymbol{\delta} \quad ; \quad \text{(compressive, non-isothermal)} \tag{8.44}
$$

$$
\text{where} \quad \bar{p}(\bar{\rho}, \bar{\theta}) = -\bar{\rho}^2 \frac{\partial \bar{\Phi}(\bar{\rho}, \bar{\theta})}{\partial \bar{\rho}} \tag{8.45}
$$

Equation (8.44) is the constitutive theory of equilibrium Cauchy stress tensor for compressible, non-isothermal case in Eulerian description. $\bar{p}(\bar{\rho}, \bar{\theta})$ is thermodynamic pressure, generally referred to as the equation of state. $\bar{p}(\bar{\rho}, \bar{\theta})$ can be established analytically, experimentally or by any other means to describe the compressibility deformation physics as long as it is continuous and differentiable in its arguments.

8.5.4.2 Incompressible, non-isothermal (Eulerian)

When the deforming matter is incompressible, there is no change of volume. Therefore, for fixed mass, we have $\bar{\rho}(\bar{\boldsymbol{x}}, t) = \rho(\boldsymbol{x}, t) = \rho_0$, i.e., density is constant.

For this case

$$\dot{\bar{\rho}} = -\bar{\rho}(\bar{\boldsymbol{\nabla}} \cdot \bar{\boldsymbol{v}}) = 0 \quad \text{(CM)}$$
$$\text{and} \quad \frac{\partial \bar{\Phi}(\bar{\rho}, \bar{\theta})}{\partial \bar{\rho}} = \frac{\partial \bar{\Phi}(\rho_0, \bar{\theta})}{\partial \bar{\rho}} = 0 \tag{8.46}$$

Hence, the constitutive theory for ${}^{(0)}_{\;e}\bar{\boldsymbol{\sigma}}$ cannot be derived using (8.44). Using (8.46), the entropy inequality (8.43) reduces to

$$-{}^{(0)}_{\;e}\bar{\boldsymbol{\sigma}} : \bar{\boldsymbol{D}} - {}^{(0)}_{\;d}\bar{\boldsymbol{\sigma}} : \bar{\boldsymbol{D}} + \frac{\bar{\boldsymbol{q}} \cdot \bar{\boldsymbol{g}}}{\bar{\theta}} \leq 0 \tag{8.47}$$

In order to derive the constitutive theory for ${}^{(0)}_{\;e}\bar{\boldsymbol{\sigma}}$ for incompressible matter we must introduce the incompressibility condition using conservation of mass in Eulerian description (8.46) in the entropy inequality (8.47).

$$\bar{\boldsymbol{\nabla}} \cdot \bar{\boldsymbol{v}} = \bar{D}_{kk} = \bar{D}_{ki}\delta_{ik} = \boldsymbol{\delta} : \bar{\boldsymbol{D}} = 0 \tag{8.48}$$

We note that when (8.48) holds, the following also holds.

$$\bar{p}(\bar{\theta})\boldsymbol{\delta} : \bar{\boldsymbol{D}} = 0 \tag{8.49}$$

where $\bar{p}(\bar{\theta})$ is the Lagrange multiplier. Adding (8.49) to (8.47) and regrouping terms.

$$\left(\bar{p}(\bar{\theta})\boldsymbol{\delta} - {}^{(0)}_{\;e}\bar{\boldsymbol{\sigma}}\right) : \bar{\boldsymbol{D}} - {}^{(0)}_{\;d}\bar{\boldsymbol{\sigma}} : \bar{\boldsymbol{D}} + \frac{\bar{\boldsymbol{q}} \cdot \bar{\boldsymbol{g}}}{\bar{\theta}} \leq 0 \tag{8.50}$$

Entropy inequality (8.50) holds for arbitrary but admissible $\bar{\boldsymbol{D}}$ if the coefficient of $\bar{\boldsymbol{D}}$ in the first term is set to zero.

$${}^{(0)}_{\;e}\bar{\boldsymbol{\sigma}} = \bar{p}(\bar{\theta})\boldsymbol{\delta} \quad ; \quad \text{(incompressible, non-isothermal)} \tag{8.51}$$

Equation (8.51) is the constitutive theory for the equilibrium Cauchy stress tensor for incompressible, non-isothermal case in Eulerian description.

8.5.4.3 Compressible, isothermal (Eulerian)

In this case, $\bar{\Phi} = \bar{\Phi}(\bar{\rho})$, hence the derivation presented in Section 8.5.4.1 holds by replacing $\bar{\Phi}(\bar{\rho}, \bar{\theta})$ with $\bar{\Phi}(\bar{\rho})$, and we have

$${}^{(0)}_{\;e}\bar{\boldsymbol{\sigma}} = -\bar{\rho}^2 \frac{\partial \bar{\Phi}(\bar{\rho})}{\partial \bar{\rho}}\boldsymbol{\delta} = \bar{p}(\bar{\rho})\boldsymbol{\delta} \quad ; \quad \text{(compressible, isothermal)} \tag{8.52}$$

$$\bar{p}(\bar{\rho}) = -\bar{\rho}^2 \frac{\partial \bar{\Phi}(\bar{\rho})}{\partial \bar{\rho}} \tag{8.53}$$

Equation (8.52) is the constitutive theory for the equilibrium Cauchy stress tensor ${}^{(0)}_{\;e}\bar{\boldsymbol{\sigma}}$ for the compressible, isothermal case (Eulerian description).

8.5.4.4 Incompressible, isothermal (Eulerian)

For this case the derivation presented in Section 8.5.4.2 holds, but the Lagrange multiplier $\bar{p}(\bar{\theta})$ must be replaced with $\bar{p}$, thus, the constitutive theory is given by

(Eulerian description (8.51) with $\bar{p}$ in place of $\bar{p}(\bar{\theta})$).

$$ {}^{(0)}_{\;e}\bar{\boldsymbol{\sigma}} = \bar{p}\boldsymbol{\delta} \quad ; \quad \text{(incompressible, isothermal)} \tag{8.54} $$

Remarks

(1) In Sections 8.5.4.1–8.5.4.4, constitutive theories for equilibrium Cauchy stress tensor ${}^{(0)}_{\;e}\bar{\boldsymbol{\sigma}}$ have been derived for different deformation physics.

(2) These constitutive theories are in Eulerian description in the current configuration describing volumetric deformation physics.

(3) The constitutive theories hold for fluent as well as solid continua as no such distinction has been made in their derivation.

(4) We observe that equilibrium Cauchy stress tensor ${}^{(0)}_{\;e}\bar{\boldsymbol{\sigma}}$ is basis independent, as it is a pressure field, but we continue with this notation as the corresponding deviatoric part is basis dependent.

(5) In case of solid continua, CBL in Lagrangian description are preferred due to convenience of following material points during deformation.

(6) In the following section the constitutive theories derived in Sections 8.5.4.1–8.5.4.4 are expressed in Lagrangian description for various deformation physics.

8.5.5 Constitutive theories for equilibrium stress ${}^{(0)}_{\;e}\boldsymbol{\sigma}$ (volumetric deformation): Lagrangian description

As is well known in Lagrangian description, the CBL and the measures used in them are expressed as functions of undeformed coordinates $\boldsymbol{x}$ of the material points and time t. This presents no problem in expressing constitutive theory for equilibrium Cauchy stress tensor ${}^{(0)}_{\;e}\bar{\boldsymbol{\sigma}}$ derived in Eulerian description in the current configuration as ${}^{(0)}_{\;e}\boldsymbol{\sigma}$, the equilibrium Cauchy stress tensor in the current configuration in Lagrangian description. The finite deformation and finite strain aspects of the deformation in solid continua necessitate the use of first and second Piola-Kirchhoff stress tensors in the reference configuration that are derived using Cauchy stress tensors in Lagranigan description and the correspondence rules. The constitutive theories for these for volumetric deformation can be obtained using the constitutive theory for ${}^{(0)}_{\;e}\boldsymbol{\sigma}$. We present the details in the following.

8.5.5.1 Equilibrium Cauchy stress tensor ${}^{(0)}_{\;e}\boldsymbol{\sigma}$ (Lagrangian description)

Compressible, non-isothermal

Using (8.44) and (8.45) we can write

$$ {}^{(0)}_{\;e}\boldsymbol{\sigma} = p(\rho,\theta)\boldsymbol{\delta} \tag{8.55} $$

$$ p(\rho,\theta) = -\rho^2 \frac{\partial \Phi(\rho,\theta)}{\partial \rho} \tag{8.56} $$

Incompressible, non-isothermal

Using (8.51) we can write

$$ {}^{(0)}_{\;e}\boldsymbol{\sigma} = p(\theta)\boldsymbol{\delta} \tag{8.57} $$

Compressible, isothermal

From (8.52) and (8.53), we can obtain

$$ {}^{(0)}_{\;e}\boldsymbol{\sigma} = p(\rho)\boldsymbol{\delta} \tag{8.58} $$

$$ p(\rho) = -\rho^2 \frac{\partial \Phi(\rho)}{\partial \rho} \tag{8.59} $$

8.5.5.2 Incompressible, isothermal

$$ {}^{(0)}_{\;e}\boldsymbol{\sigma} = p\boldsymbol{\delta} \tag{8.60} $$

8.5.6 Constitutive theories for equilibrium contravariant second Piola-Kirchhoff stress tensor ${}_e\boldsymbol{\sigma}^{[0]}$

For the CBL in Lagrangian description, $\boldsymbol{\sigma}^{[0]}$ and $\dot{\boldsymbol{\varepsilon}}_{[0]}$ are the rate of work conjugate pair that are meaningful in finite deformation, finite strain deformation physics. Therefore, in this section we only consider the contravariant Cauchy stress tensor $\boldsymbol{\sigma}^{(0)}$ (instead of basis independent measure ${}^{(0)}\boldsymbol{\sigma}$) that is used to define the contravariant second Piola-Kirchhoff stress tensor. Regardless, as stated in remarks, the equilibrium Cauchy stress is basis independent. Recall that $\boldsymbol{\sigma}^{(0)}$ and $\boldsymbol{\sigma}^{[0]}$ are related by (see Chapter 4).

$$ \boldsymbol{\sigma}^{[0]} = |\boldsymbol{J}|(\boldsymbol{J}^{-1}) \cdot \boldsymbol{\sigma}^{(0)} \cdot (\boldsymbol{J}^{-1})^T \tag{8.61} $$

By using ${}_e\boldsymbol{\sigma}^{(0)}$ in (8.61) we can obtain the constitutive theory for ${}_e\boldsymbol{\sigma}^{[0]}$. We consider various deformation physics in the following.

8.5.6.1 Finite deformation, finite strain, compressible, non-isothermal

$$ {}_e\boldsymbol{\sigma}^{(0)} = p(\rho,\theta)\boldsymbol{\delta} \tag{8.62} $$

$$ p(\rho,\theta) = -\rho^2 \frac{\partial \Phi(\rho,\theta)}{\partial \rho} \tag{8.63} $$

$$ \begin{aligned} \therefore \quad {}_e\boldsymbol{\sigma}^{[0]} &= |\boldsymbol{J}|(\boldsymbol{J}^{-1}) \cdot {}_e\boldsymbol{\sigma}^{(0)} \cdot (\boldsymbol{J}^{-1})^T \\ &= |\boldsymbol{J}|(\boldsymbol{J}^{-1}) \cdot p(\rho,\theta)\boldsymbol{\delta} \cdot (\boldsymbol{J}^{-1})^T \\ \text{or} \quad {}_e\boldsymbol{\sigma}^{[0]} &= |\boldsymbol{J}| p(\rho,\theta)(\boldsymbol{J}^T \cdot \boldsymbol{J})^{-1} \end{aligned} \tag{8.64} $$

Equation (8.64) is the constitutive theory for the second Piola-Kirchhoff equilibrium stress tensor.

8.5.6.2 Finite deformation, finite strain, compressible, isothermal

$$ {}_e\boldsymbol{\sigma}^{(0)} = p(\rho)\boldsymbol{\delta} \tag{8.65} $$

$$ p(\rho) = -\rho^2 \frac{\partial \Phi(\rho)}{\partial \rho} \tag{8.66} $$

$$ \therefore \quad {}_e\boldsymbol{\sigma}^{[0]} = |\boldsymbol{J}|(\boldsymbol{J}^{-1}) \cdot {}_e\boldsymbol{\sigma}^{(0)} \cdot (\boldsymbol{J}^{-1})^T \tag{8.67} $$

$$ \text{or} \quad {}_e\boldsymbol{\sigma}^{[0]} = |\boldsymbol{J}| p(\rho)(\boldsymbol{J}^T \cdot \boldsymbol{J})^{-1} \tag{8.68} $$

Equation (8.68) is the desired constitutive theory for ${}_e\boldsymbol{\sigma}^{[0]}$.

8.5.6.3 Finite deformation, finite strain, incompressible, non-isothermal

$$ {}_e\boldsymbol{\sigma}^{(0)} = p(\theta)\boldsymbol{\delta} \tag{8.69}$$

$$\therefore \quad {}_e\boldsymbol{\sigma}^{[0]} = |\boldsymbol{J}|(\boldsymbol{J}^{-1}) \cdot p(\theta)\boldsymbol{\delta} \cdot (\boldsymbol{J}^{-1})^T \tag{8.70}$$

$$\text{or} \quad {}_e\boldsymbol{\sigma}^{[0]} = |\boldsymbol{J}|p(\theta)(\boldsymbol{J}^T \cdot \boldsymbol{J})^{-1} \tag{8.71}$$

Equation (8.71) is the desired constitutive theory.

8.5.6.4 Finite deformation, finite strain, incompressible, isothermal

$$ {}_e\boldsymbol{\sigma}^{(0)} = p\boldsymbol{\delta} \tag{8.72}$$

$$\therefore \quad {}_e\boldsymbol{\sigma}^{[0]} = |\boldsymbol{J}|(\boldsymbol{J}^{-1}) \cdot p\boldsymbol{\delta} \cdot (\boldsymbol{J}^{-1})^T \tag{8.73}$$

$$\text{or} \quad {}_e\boldsymbol{\sigma}^{[0]} = |\boldsymbol{J}|p(\boldsymbol{J}^T \cdot \boldsymbol{J})^{-1} \tag{8.74}$$

8.5.6.5 Constitutive theory for equilibrium stress tensor: Small deformation, small strain

In this case, the Cauchy stress measure is $\boldsymbol{\sigma}$ (basis independent) and the strain measure is $\boldsymbol{\varepsilon}$, the linear part of Green's strain tensor. Since the deformed and undeformed configurations are virtually the same, the first and second Piola-Kirchhoff stress tensors reduce to the basis independent Cauchy stress tensor. We then have the following

$$\left.\begin{aligned}
{}_e\boldsymbol{\sigma} &= p(\rho,\theta)\boldsymbol{\delta} \quad ; \quad \text{compressible, non-isothermal}\\
{}_e\boldsymbol{\sigma} &= p(\rho)\boldsymbol{\delta} \quad ; \quad \text{compressible, isothermal}\\
{}_e\boldsymbol{\sigma} &= p(\theta)\boldsymbol{\delta} \quad ; \quad \text{incompressible, non-isothermal}\\
{}_e\boldsymbol{\sigma} &= p\boldsymbol{\delta} \quad ; \quad \text{incompressible, isothermal}
\end{aligned}\right. \tag{8.75}$$

8.5.6.6 Reduced form of entropy inequality

In this section, we consider the reduced form of the entropy inequality in Eulerian and Lagrangian descriptions. This is the final form of the entropy inequality after the constitutive theory for the equilibrium stress tensor has been derived.

8.5.6.7 Eulerian description

After deriving the constitutive theory for ${}^{(0)}_{\ e}\bar{\boldsymbol{\sigma}}$, the entropy inequality (8.43) reduces to

$$ -{}^{(0)}_{\ d}\bar{\boldsymbol{\sigma}} : \bar{\boldsymbol{D}} + \frac{\bar{\boldsymbol{q}} \cdot \bar{\boldsymbol{g}}}{\bar{\theta}} \leq 0 \tag{8.76}$$

We refer to (8.76) as the *reduced form of the entropy inequality.* In (8.76), ${}^{(0)}_{\ d}\bar{\boldsymbol{\sigma}} : \bar{\boldsymbol{D}}$ represents the rate of mechanical work that is expanded in distorting the volume of the deforming matter and in dissipation and memory mechanisms. The entropy inequality (8.76) is satisfied if

$$ {}^{(0)}_{\ d}\bar{\boldsymbol{\sigma}} : \bar{\boldsymbol{D}} > 0 \tag{8.77}$$

$$\text{and} \quad \frac{\bar{\boldsymbol{q}} \cdot \bar{\boldsymbol{g}}}{\bar{\theta}} \leq 0 \tag{8.78}$$

Inequality (8.77) states that constitutive theory for ${}^{(0)}_{d}\bar{\boldsymbol{\sigma}}$ must result in positive rate of work and the constitutive theory for $\bar{\boldsymbol{q}}$ must satisfy (8.78). Inequalities (8.77) and (8.78) form the basis for deriving constitutive theories for ${}^{(0)}_{d}\bar{\boldsymbol{\sigma}}$ and $\bar{\boldsymbol{q}}$ in Eulerian description (generally for fluent continua). From the conjugate pairs in (8.76), we can write

$$ {}^{(0)}_{d}\bar{\boldsymbol{\sigma}} = {}^{(0)}_{d}\bar{\boldsymbol{\sigma}}(\bar{\rho}, \bar{\boldsymbol{D}}, \bar{\theta}) \tag{8.79} $$

$$ \text{and} \quad \bar{\boldsymbol{q}} = \bar{\boldsymbol{q}}(\bar{\rho}, \bar{\boldsymbol{g}}, \bar{\theta}) \tag{8.80} $$

$\bar{\rho}$ and $\bar{\theta}$ are always argument tensors of the constitutive tensors in fluent continua. Argument tensors of ${}^{(0)}_{d}\bar{\boldsymbol{\sigma}}$ and $\bar{\boldsymbol{q}}$ in (8.79) and (8.80) can be augmented with additional argument tensors based on desired physics as done in Chapter 12 and 13 when deriving ordered rate constitutive theories for ${}^{(0)}_{d}\bar{\boldsymbol{\sigma}}$. Once the argument tensors and their constitutive tensors are established, constitutive theories can be derived using representation theorem, shown in the following sections.

8.5.6.8 Lagrangian description

In order to be able to derive the constitutive theory for the deviatoric stress tensor and heat vector in Lagrangian description, we need the reduced form of (8.76) in Lagrangian description. We recall from Chapter 7 that the rate of work conjugate pairs in Lagrangian description yield the same rate of work per unit volume as (8.77) are

$$ \boldsymbol{\sigma}^{*} : \dot{\boldsymbol{J}} \quad ; \quad \boldsymbol{\sigma}^{[0]} : \dot{\boldsymbol{\varepsilon}}_{[0]} \tag{8.81} $$

The conjugate pair $\boldsymbol{\sigma}^{[0]} : \dot{\boldsymbol{\varepsilon}}_{[0]}$ needs to be considered for finite deformation, finite strain deformation physics, as $\boldsymbol{\varepsilon}_{[0]}$ is the measure of finite strain but $\boldsymbol{J}$ is not. Furthermore, $\boldsymbol{\sigma}^{*}$ is based on the correspondence rule that is only valid for small deformation physics. We also note that, since the rate of work is due to $\int_{\partial\bar{V}} \bar{\boldsymbol{P}} \cdot \bar{\boldsymbol{v}} d\bar{A}$, which is the same in Eulerian as well as Lagrangian description. Thus,

$$ {}^{(0)}_{d}\bar{\boldsymbol{\sigma}} : \bar{\boldsymbol{D}} = \boldsymbol{\sigma}^{*} : \dot{\boldsymbol{J}} = \boldsymbol{\sigma}^{[0]} : \dot{\boldsymbol{\varepsilon}}_{[0]} \tag{8.82} $$

Finite deformation, finite strain

For finite deformation, finite strain, the reduced form of the entropy inequality in Lagrangian description can be written as (using (8.76))

$$ -{}_{d}\boldsymbol{\sigma}^{[0]} : \dot{\boldsymbol{\varepsilon}}_{[0]} + \frac{\boldsymbol{q} \cdot \boldsymbol{g}}{\theta} \leq 0 \tag{8.83} $$

From which we can conclude (similar to (8.77) and (8.78))

$$ {}_{d}\boldsymbol{\sigma}^{[0]} : \boldsymbol{\varepsilon}_{[0]} > 0 $$

$$ \frac{\boldsymbol{q} \cdot \boldsymbol{g}}{\theta} \leq 0 $$

$$ {}_{d}\boldsymbol{\sigma}^{[0]} = {}_{d}\boldsymbol{\sigma}^{[0]}(\boldsymbol{\varepsilon}_{[0]}, \theta) \tag{8.84} $$

$$ \boldsymbol{q} = \boldsymbol{q}(\boldsymbol{g}, \theta) \tag{8.85} $$

Argument tensors of ${}_d\boldsymbol{\sigma}^{[0]}$ and $\boldsymbol{q}$ can be further augmented by additional argument tensors if the physics of the deformation necessitates so (as done in Chapters 10-11). Constitutive theories for ${}_d\boldsymbol{\sigma}^{[0]}$ and $\boldsymbol{q}$ are derived using representation theorem (presented in the following sections). Dependence on θ and the term $\frac{\boldsymbol{q}\cdot\boldsymbol{g}}{\theta}$ can be removed in (8.79) for isothermal case.

Small deformation, small strain

The deviatoric stress measure is ${}_d\boldsymbol{\sigma}$, the basis independent Cauchy stress tensor and the strain measure is $\boldsymbol{\varepsilon}$, the linear part of Green's strain tensor. Hence, the reduced form of the entropy inequality in Lagrangian description becomes

$$-{}_d\boldsymbol{\sigma} : \dot{\boldsymbol{\varepsilon}} + \frac{\boldsymbol{q}\cdot\boldsymbol{g}}{\theta} \leq 0 \tag{8.86}$$

$$\text{and} \quad {}_d\boldsymbol{\sigma} = {}_d\boldsymbol{\sigma}(\boldsymbol{\varepsilon}, \theta, \dots) \tag{8.87}$$

$$\boldsymbol{q} = \boldsymbol{q}(\boldsymbol{g}, \theta) \tag{8.88}$$

Equations (8.87) and (8.88) are used to derive constitutive theories for ${}_d\boldsymbol{\sigma}$ and $\boldsymbol{q}$ using representation theorem (presented in the following). For isothermal case, dependence on θ and the term $\frac{\boldsymbol{q}\cdot\boldsymbol{g}}{\theta}$ can be removed in (8.86)–(8.88).

8.6 Representation Theorem

Representation theorem is based on pioneering works of Spencer, Wang and Zheng on the theory of isotropic tensors [19, 30, 31, 73, 84, 87, 90, 93, 94, 96–98, 126, 134–137, 144, 145]. The essence of their work in simple terms is stated in this section. A valid constitutive tensor must be in a finite dimensional linear space. The finite dimensional space naturally has dimension and must have a unique basis. Their work shows that for a valid constitutive tensor with known valid argument tensors, the dimension of the space of the constitutive tensor and the corresponding minimum basis of the space (integrity) can be determined using the argument tensors of the constitutive tensor. The tensors constituting the basis are called the *combined generators* of the argument tensors. Once we know the basis (generators) of the space of the constitutive tensor, the constitutive tensor can be expressed as a linear combination of the combined generators (basis). The coefficients in the linear combination can be functions of the combined invariants of the argument tensors of the constitutive variables as well as functions of the argument tensors of rank zero. This approach of deriving constitutive theories is called *representation theorem*. The works of Spencer, Wang and Zheng establish the following.

(1) A constitutive tensor can be a tensor of rank one and its argument tensors can be tensors of rank zero, one, two, etc.

(2) A constitutive tensor can also be a symmetric tensor of rank two. Its argument tensors can be symmetric and skew-symmetric tensors of rank two, tensors of rank one and tensors of rank zero.

(3) A constitutive tensor can also be a skew-symmetric tensor of rank two. Its argument tensors can be symmetric and skew-symmetric tensors of rank two, tensors of rank one and tensors of rank zero.

(4) A constitutive tensor cannot be a non-symmetric tensor. In this case, the dimension of the space of the constitutive tensor and the basis of this space

cannot be established using the theory of generators for tensor valued isotropic functions presented by Spencer, Wang and Zheng. Thus, a constitutive tensor cannot be a non-symmetric as it is not possible to establish the dimension and the basis of its space.

Remarks

(1) Based on remarks (1)–(3), when non-symmetric tensors appear in rate of work conjugate pairs, they must be decomposed into symmetric and skew-symmetric tensors using additive decomposition to establish conjugate pairs.

(2) In this book, we only consider CCM in which case all stress tensors are tensors of rank two and are symmetric except $\boldsymbol{\sigma}^*$. All strain measures and their convected time derivatives are symmetric tensors of rank two as well.

(3) In NCCM, the Cauchy stress tensor and the Cauchy moment tensor both can be non-symmetric. Hence, based on Spencer, Wang and Zheng's work, care must be taken in the consideration of constitutive tensors and their argument tensors.

(4) Dependence of the coefficients α^i ; $i = 0, 1, \dots, N$ in the linear combination (8.89) on the combined invariants ${}^{\sigma}\underset{\sim}{I}^j$; $j = 1, 2 \dots, M$ of the argument tensors of the constitutive tensor $\boldsymbol{T}$ and the tensors of rank zero, r_k ; $k = 1, 2 \dots$ in the arguments of $\boldsymbol{T}$ (shown in (8.90)) has been established in the works of Spencer, Wang, and Zheng et. al. In Chapter 9, Section 9.3.1 (under some assumptions) in the derivation of the constitutive theory for ${}_d\boldsymbol{\sigma}^{[0]}$ for incompressible elastic solid continua using Helmholtz free energy density, we establish exactly the same form of the constitutive theory with the same combined generators in the linear combination as those from the representation theorem. The derivation also establishes dependence of the coefficients in the linear combination on the combined invariants of the argument tensor of the constitutive tensor ${}_d\boldsymbol{\sigma}^{[0]}$ in the current configuration. This is an independent derivation (independent of representation theorem) that provides the proof of dependence of the coefficients in the linear combination on the combined invariants of the argument tensors of the constitutive tensor in the representation theorem.

8.6.1 Def: Representation theorem

Let $\boldsymbol{T} \in S$ be a valid constitutive tensor with valid argument tensors and let $\underset{\sim}{\boldsymbol{G}}^i \in S$; $i = 1, 2, \dots, N$ be the combined generators of the argument tensors of $\boldsymbol{T}$ that are of the same type and rank as the constitutive tensor $\boldsymbol{T}$. S is the linear space of the constitutive tensor $\boldsymbol{T}$. Also, let $\underset{\sim}{I}^j$; $j = 1, 2, \dots, M$ be combined invariants of the same argument tensors of $\boldsymbol{T}$. Then, based on representation theorem we can express tensor $\boldsymbol{T}$ as follows in the current configuration. We have included $\boldsymbol{I}$ as a generator for the sake of generality, but it can be omitted if not admissible.

$$\boldsymbol{T} = \alpha^0 \boldsymbol{I} + \sum_{i=1}^{N} \alpha^i \underset{\sim}{\boldsymbol{G}}^i \tag{8.89}$$

Since $\boldsymbol{I} \in S$, ${}^{\sigma}\underset{\sim}{\boldsymbol{G}}^i \in S$; $i = 1, 2, \dots, N$; $\boldsymbol{T} \in S$

$$\alpha^i = \alpha^i(\underset{\sim}{I}^j, r_k) \quad ; \quad i = 0, 1, \dots, N \quad ; \quad j = 1, 2, \dots, M \quad ; \quad k = 1, 2, \dots \tag{8.90}$$

In (8.90), r_k ; $k = 1, 2, \ldots$ are tensors of rank zero in the arguments of tensor $\boldsymbol{T}$. Tensors $\boldsymbol{I}$, $\underset{\sim}{\boldsymbol{G}}^i$; $i = 1, 2, \ldots, N$ form the basis of the space of tensor $\boldsymbol{T}$ and are referred to as integrity or complete basis (complete irreducible set). When $\boldsymbol{T}$ is a skew-symmetric tensor $\alpha^0 = 0$, i.e., the first term on the right side of (8.89) is not included as in this case $\boldsymbol{I}$ cannot be a generator.

The key element in representation theorem or the theory of generators and invariants is the determination of the basis, i.e., integrity using the combined generators of the argument tensors and of course, determination of the combined invariants.

For example, if we consider $[T([S])]$, where $[T]$ and $[S]$ are symmetric tensors of rank two, which obey the following invariance due to orthogonal transformation $[Q]$

$$[T([Q][S][Q]^T)] = [Q][T([S])][Q]^T \tag{8.91}$$

then the tensor $[T]$ has the following form:

$$[T] = \alpha_0[I] + \alpha_1[S] + \alpha_2[S]^2 \tag{8.92}$$

where α_0, α_1 and α_2 are functions of the invariants of $[S]$, i.e., $i_s = \text{tr}([S])$, $ii_s = \text{tr}([S]^2)$ and $iii_s = \text{tr}([S]^3)$ or I_s, $I\!I_s$ and $I\!I\!I_s$. The tensors $[I]$, $[S]$, $[S]^2$ are generators of the constitutive tensor $[T]$ and form the basis or integrity. If the arguments of $[T]$ consist of more than one tensor (can be of different rank), then a linear combination like (8.92) would contain all combined generators (tensors of the same rank as $[T]$) of the argument tensors of $[T]$. Likewise, the coefficients in the linear combination would be functions of the argument tensors of rank zero and the combined invariants of the argument tensors of rank one and two of $\boldsymbol{T}$. For combined generators and invariants of the argument tensors (can be of different ranks) see references [19, 30, 31, 73, 84, 87, 90, 93, 94, 96–98, 126, 134–137, 144, 145] and Appendix A.

In the following, we consider two examples in which the constitutive variables are a symmetric tensor of rank two and a tensor of rank one and their argument tensors are tensors of rank two, one and zero.

8.6.2 Constitutive theory for a symmetric tensor of rank two

Let $\boldsymbol{T}$ be a symmetric constitutive tensor of rank two with argument tensors $\boldsymbol{A}_1$, $\boldsymbol{A}_2$, $\boldsymbol{a}_1$, $\boldsymbol{a}_2$ and θ in which $\boldsymbol{A}_1$ and $\boldsymbol{A}_2$ are symmetric tensors of rank two, $\boldsymbol{a}_1$ and $\boldsymbol{a}_2$ are tensors of rank one and θ is a tensor of rank zero. Then,

$$\boldsymbol{T} = \boldsymbol{T}(\boldsymbol{A}_1, \boldsymbol{A}_2, \boldsymbol{a}_1, \boldsymbol{a}_2, \theta) \tag{8.93}$$

Our objective is to derive a constitutive theory for $\boldsymbol{T}$ using representation theorem. Let $\underset{\sim}{\boldsymbol{G}}^i$; $i = 1, 2, \ldots, N$ be the combined generators of the argument tensors of $\boldsymbol{T}$ that are symmetric tensors of rank two and let $\underset{\sim}{I}^j$; $j = 1, 2, \ldots, M$ be the combined invariants of the same argument tensors, all in the current configuration. Then $\boldsymbol{I}$, $\underset{\sim}{\boldsymbol{G}}^i$; $i = 1, 2, \ldots, N$ constitute the complete basis (integrity) of the space containing tensor $\boldsymbol{T}$. Hence, $\boldsymbol{T}$ can be expressed as a linear combination of $\boldsymbol{I}$, $\underset{\sim}{\boldsymbol{G}}^i$; $i = 1, 2, \ldots, N$ in the current configuration

$$\boldsymbol{T} = \underset{\sim}{\alpha}^0 \boldsymbol{I} + \sum_{i=1}^{N} \underset{\sim}{\alpha}^i \underset{\sim}{\boldsymbol{G}}^i \tag{8.94}$$

$$\text{in which} \quad \underset{\sim}{\alpha}^i = \underset{\sim}{\alpha}^i(\underset{\sim}{I}^j, \theta) \quad ; \quad j = 1, 2, \ldots, M \quad ; \quad i = 0, 1, \ldots, N \tag{8.95}$$

The coefficient $\underset{\sim}{\alpha}^i$; $i = 0, 1, \ldots, N$ in the linear combination are also in the current configuration. We note that in the constitutive theory (8.94), $\underset{\sim}{\alpha}^i$; $i = 0, 1, \ldots, N$ are not material coefficients. These are simply coefficients in the linear combination in (8.94) that show dependence on invariants $\underset{\sim}{I}^j$; $j = 1, 2, \ldots, M$ and θ. Derivation of material coefficients is given in the following section.

8.6.2.1 Material coefficients

The material coefficients in the constitutive theory (8.94) and (8.95) are derived using the axiom of smooth neighborhood. Based on this axiom, when the constitutive functionals are sufficiently smooth, they can be approximated by the functional in the field of real functions. Thus, if α^i are sufficiently smooth in their arguments in a known configuration $\underline{\Omega}$, then α^i can be expanded in Taylor series in their arguments about the known configuration $\underline{\Omega}$. Clearly the known configuration refers to a configuration at time $t_1 < t$ in which t refers to the current configuration.

Once the constitutive theory has been expressed as a linear combination of the combined generators constituting integrity (basis) and the dependence of the coefficients in the linear combination has been established on combined invariants and tensors of rank zero, the process for deriving the material coefficient remains the same regardless of the type of constitutive tensor (i.e., symmetric tensor of rank two or tensor of rank one or others).

Consider constitutive theory (8.94) and (8.95). We expand each $\underset{\sim}{\alpha}^i$; $i = 0, 1, \ldots, N$ in Taylor series in $\underset{\sim}{I}^j$; $j = 1, 2 \ldots, M$ and θ about a known configuration $\underline{\Omega}$ and retain only up to linear terms in $\underset{\sim}{I}^j$; $j = 1, 2, \ldots, M$ and θ (for simplicity).

$$\underset{\sim}{\alpha}^i = \underset{\sim}{\alpha}^i\Big|_{\underline{\Omega}} + \sum_{j=1}^{M} \frac{\partial \underset{\sim}{\alpha}^i}{\partial (\underset{\sim}{I}^j)}\Bigg|_{\underline{\Omega}} \left(\underset{\sim}{I}^j - \underset{\sim}{I}^j\Big|_{\underline{\Omega}}\right) + \frac{\partial \underset{\sim}{\alpha}^i}{\partial \theta}\Bigg|_{\underline{\Omega}} (\theta - \theta|_{\underline{\Omega}}) \quad ; \quad i = 0, 1, \ldots, N \tag{8.96}$$

Substitute $\underset{\sim}{\alpha}^i$; $i = 0, 1, \ldots, N$ from (8.96) in (8.94)

$$\begin{aligned} \boldsymbol{T} = &\left(\underset{\sim}{\alpha}^0\Big|_{\underline{\Omega}} + \sum_{j=1}^{M} \frac{\partial \underset{\sim}{\alpha}^0}{\partial (\underset{\sim}{I}^j)}\Bigg|_{\underline{\Omega}} \left(\underset{\sim}{I}^j - \underset{\sim}{I}^j\Big|_{\underline{\Omega}}\right) + \frac{\partial \underset{\sim}{\alpha}^0}{\partial \theta}\Bigg|_{\underline{\Omega}} (\theta - \theta|_{\underline{\Omega}}) \right) \boldsymbol{I} \\ &+ \sum_{i=1}^{N} \left(\underset{\sim}{\alpha}^i\Big|_{\underline{\Omega}} + \sum_{j=1}^{M} \frac{\partial \underset{\sim}{\alpha}^i}{\partial (\underset{\sim}{I}^j)}\Bigg|_{\underline{\Omega}} \left(\underset{\sim}{I}^j - \underset{\sim}{I}^j\Big|_{\underline{\Omega}}\right) + \frac{\partial \underset{\sim}{\alpha}^i}{\partial \theta}\Bigg|_{\underline{\Omega}} (\theta - \theta|_{\underline{\Omega}}) \right) \underset{\sim}{\boldsymbol{G}}^i \end{aligned} \tag{8.97}$$

collecting coefficients (those terms defined in known configuration $\underline{\Omega}$) of $\boldsymbol{I}, \underset{\sim}{I}^j\boldsymbol{I}, \underset{\sim}{\boldsymbol{G}}^i$, $\underset{\sim}{I}^j\underset{\sim}{\boldsymbol{G}}^i$, $(\theta - \theta|_{\underline{\Omega}})\underset{\sim}{\boldsymbol{G}}^i$ and $(\theta - \theta|_{\underline{\Omega}})\boldsymbol{I}$ and defining new coefficients we can write

$$\begin{aligned} T^0\Big|_{\underline{\Omega}} &= \underset{\sim}{\alpha}^0\Big|_{\underline{\Omega}} + \sum_{j=1}^{M} \frac{\partial \underset{\sim}{\alpha}^0}{\partial (\underset{\sim}{I}^j)}\Bigg|_{\underline{\Omega}} \left(-\underset{\sim}{I}^j\Big|_{\underline{\Omega}}\right) \quad ; \quad \underset{\sim}{a}_j = \frac{\partial \underset{\sim}{\alpha}^0}{\partial (\underset{\sim}{I}^j)}\Bigg|_{\underline{\Omega}} \\ \underset{\sim}{b}_i &= \underset{\sim}{\alpha}^i\Big|_{\underline{\Omega}} + \sum_{j=1}^{M} \frac{\partial \underset{\sim}{\alpha}^i}{\partial (\underset{\sim}{I}^j)}\Bigg|_{\underline{\Omega}} \left(-\underset{\sim}{I}^j\Big|_{\underline{\Omega}}\right) \quad ; \quad \underset{\sim}{c}_{ij} = \frac{\partial \underset{\sim}{\alpha}^i}{\partial (\underset{\sim}{I}^j)}\Bigg|_{\underline{\Omega}} \end{aligned} \tag{8.98}$$

$$
\underset{\sim}{d}_i = \left.\frac{\partial \underset{\sim}{\alpha}^i}{\partial \theta}\right|_{\underline{\Omega}} \quad ; \quad \alpha_{tm} = - \left.\frac{\partial \underset{\sim}{\alpha}^0}{\partial \theta}\right|_{\underline{\Omega}}
$$
$$
i = 1, 2 \ldots, N \quad ; \quad j = 1, 2, \ldots, M
$$

$$
\begin{aligned}
\boldsymbol{T} = {} & T^0\big|_{\underline{\Omega}} \boldsymbol{I} + \sum_{j=1}^{M} \underset{\sim}{a}_j \underset{\sim}{I}^j \boldsymbol{I} + \sum_{i=1}^{N} \underset{\sim}{b}_i \underset{\sim}{\boldsymbol{G}}^i + \sum_{i=1}^{N} \sum_{j=1}^{M} \underset{\sim}{c}_{ij} \underset{\sim}{I}^j \underset{\sim}{\boldsymbol{G}}^i \\
& + \sum_{i=1}^{N} \underset{\sim}{d}_i (\theta - \theta|_{\underline{\Omega}}) \underset{\sim}{\boldsymbol{G}}^i - \alpha_{tm} (\theta - \theta|_{\underline{\Omega}}) \boldsymbol{I}
\end{aligned}
\tag{8.99}
$$

This is the final form of the constitutive theory for constitutive tensor $\boldsymbol{T}$. $\underset{\sim}{a}_j$, $\underset{\sim}{b}_i$, $\underset{\sim}{c}_{ij}$, $\underset{\sim}{d}_i$ and α_{tm} ; $i = 1, 2, \ldots, N$; $j = 1, 2, \ldots, M$ are material coefficients. These are functions of $\underset{\sim}{I}^j\big|_{\underline{\Omega}}$; $j = 1, 2, \ldots, M$ and $\theta|_{\underline{\Omega}}$.

Remarks

(1) This constitutive theory requires $(M + 2N + MN + 1)$ material coefficients. We note that $T^0\big|_{\underline{\Omega}}$ is a known field.

(2) This constitutive theory is nonlinear in the argument tensors of $\boldsymbol{T}$ as some generators and and invariants are nonlinear functions of the argument tensors. Secondly, terms containing products of generators and the invariants may be nonlinear.

(3) The constitutive theory is based on integrity, hence, uses a complete basis of the space of the constitutive tensor $\boldsymbol{T}$.

(4) Simplified linear as well as nonlinear constitutive theories for $\boldsymbol{T}$ can be extracted from (8.99) depending upon the desired physics.

(5) The constitutive theory permits material coefficients to be functions of deformation and temperature through their dependence on the invariants and θ.

(6) We shall see in the subsequent chapters that the constitutive theories for the stress tensor follow the derivation presented here for the constitutive theory for tensor $\boldsymbol{T}$.

(7) A significant aspect of the general approach of deriving the constitutive theory presented in this section is that it is based on the entropy inequality in conjunction with the theory of isotropic tensors, i.e., representation theorem. The constitutive theories derived using this approach are more likely to ensure thermodynamic equilibrium during evolution of the deforming solid continua.

8.6.2.2 Combined generators and combined invariants

Combined generators

The combined generators $\underset{\sim}{\boldsymbol{G}}^i$; $i = 1, 2, \ldots, N$ and the combined invariants ${}^{\sigma}\underset{\sim}{I}^j$; $j = 1, 2, \ldots, M$ of the argument tensors $\boldsymbol{A}_1$, $\boldsymbol{A}_2$, $\boldsymbol{a}_1$, $\boldsymbol{a}_2$, $\bar{\theta}$ of $\boldsymbol{T}$ in (8.93) can be determined using the tables in Appendix A. Since $\boldsymbol{T}$ is a symmetric tensor of rank two, we need to determine the combined generators of tensors $\boldsymbol{A}_1$, $\boldsymbol{A}_2$, $\boldsymbol{a}_1$, $\boldsymbol{a}_2$ and $\bar{\theta}$ that are symmetric tensors of rank two. Using Table A.3 of Appendix A, we can determine the number of generators for various combinations of the argument tensors listed in Table 8.1. Explicit expression of $\underset{\sim}{\boldsymbol{G}}^i$; $i = 1, 2, \dots, 20$ are given in Table 8.2.

Table 8.1: Number of combined generators: $\underset{\sim}{\boldsymbol{G}}^i$; $i = 1, 2, \dots, N$ $(N = 20)$ of tensor $\boldsymbol{T}$ in (8.93)

One at a time	$\boldsymbol{A}_1$	$\boldsymbol{A}_2$	$\boldsymbol{a}_1$	$\boldsymbol{a}_2$	θ	
$(\underset{\sim}{\boldsymbol{G}}^i$; $i = 1, 2, \dots, 6)$	2	2	1	1	0	
Two at a time	$\boldsymbol{A}_1, \boldsymbol{A}_2$	$\boldsymbol{A}_1, \boldsymbol{a}_1$	$\boldsymbol{A}_1, \boldsymbol{a}_2$	$\boldsymbol{A}_2, \boldsymbol{a}_1$	$\boldsymbol{A}_2, \boldsymbol{a}_2$	$\boldsymbol{a}_1, \boldsymbol{a}_2$
$(\underset{\sim}{\boldsymbol{G}}^i$; $i = 7, 8, \dots, 18)$	3	2	2	2	2	1
Three at a time	$\boldsymbol{A}_1, \boldsymbol{A}_2, \boldsymbol{a}_1$	$\boldsymbol{A}_1, \boldsymbol{A}_2, \boldsymbol{a}_2$	$\boldsymbol{A}_1, \boldsymbol{a}_1, \boldsymbol{a}_2$	$\boldsymbol{A}_2, \boldsymbol{a}_1, \boldsymbol{a}_2$		
$(\underset{\sim}{\boldsymbol{G}}^{19}, \underset{\sim}{\boldsymbol{G}}^{20})$	0	0	1	1		
Four at a time	$\boldsymbol{A}_1, \boldsymbol{A}_2, \boldsymbol{a}_1, \boldsymbol{a}_2$					
(none)	0					

Table 8.2: Combined generators $\underset{\sim}{\boldsymbol{G}}^i$; $i = 1, 2 \dots, 20$ of the argument tensor of $\boldsymbol{T}$ in (8.93): symmetric tensors of rank two

Arguments	Generators
(1) None	$[I]$
(2) One at a time	
$[A_1]$	$[\underset{\sim}{G}^1] = [A_1]$, $[\underset{\sim}{G}^2] = [A_1]^2$
$[A_2]$	$[\underset{\sim}{G}^3] = [A_2]$, $[\underset{\sim}{G}^4] = [A_2]^2$
$\{a_1\}$	$[\underset{\sim}{G}^5] = \{a_1\}\{a_1\}^T$
$\{a_2\}$	$[\underset{\sim}{G}^6] = \{a_2\}\{a_2\}^T$

Table 8.2: (Contd.) Combined generators $\underset{\sim}{\boldsymbol{G}}^i$; $i = 1, 2 \dots, 20$ of the argument tensor of $\boldsymbol{T}$ in (8.93): symmetric tensors of rank two

Arguments	Generators
(3) Two at a time	
$[A_1]$, $[A_2]$	$[\underset{\sim}{G}^7] = [A_1][A_2] + [A_2][A_1]$, $[\underset{\sim}{G}^8] = [A_1]^2[A_2] + [A_1][A_2]^2$, $[\underset{\sim}{G}^9] = [A_1][A_2]^2 + [A_2]^2[A_1]$
$[A_1]$, $\{a_1\}$	$[\underset{\sim}{G}^{10}] = \{a_1\}\{[A_1]\{a_1\}\}^T + \{[A_1]\{a_1\}\}\{a_1\}^T$, $[\underset{\sim}{G}^{11}] = \{a_1\}\{[A_1]^2\{a_1\}\}^T + \{[A_1]^2\{a_1\}\}\{a_1\}^T$
$[A_1]$, $\{a_2\}$	$[\underset{\sim}{G}^{12}] = \{a_2\}\{[A_1]\{a_2\}\}^T + \{[A_1]\{a_2\}\}\{a_2\}^T$, $[\underset{\sim}{G}^{13}] = \{a_2\}\{[A_1]^2\{a_2\}\}^T + \{[A_1]^2\{a_2\}\}\{a_2\}^T$
$[A_2]$, $\{a_1\}$	$[\underset{\sim}{G}^{14}] = \{a_1\}\{[A_2]\{a_1\}\}^T + \{[A_2]\{a_1\}\}\{a_1\}^T$, $[\underset{\sim}{G}^{15}] = \{a_1\}\{[A_2]^2\{a_1\}\}^T + \{[A_2]^2\{a_1\}\}\{a_1\}^T$
$[A_2]$, $\{a_2\}$	$[\underset{\sim}{G}^{16}] = \{a_2\}\{[A_2]\{a_2\}\}^T + \{[A_2]\{a_2\}\}\{a_2\}^T$, $[\underset{\sim}{G}^{17}] = \{a_2\}\{[A_2]^2\{a_2\}\}^T + \{[A_2]^2\{a_2\}\}\{a_2\}^T$
$\{a_1\}$, $\{a_2\}$	$[\underset{\sim}{G}^{18}] = \{a_1\}\{a_2\}^T + \{a_2\}\{a_1\}^T$
(4) Three at a time	
$[A_1]$, $[A_2]$, $\{a_1\}$	none
$[A_1]$, $[A_2]$, $\{a_2\}$	none
$[A_1]$, $\{a_1\}$, $\{a_2\}$	$[\underset{\sim}{G}^{19}] = [A_1]\left[\{a_1\}\{a_2\}^T - \{a_2\}\{a_1\}^T\right] - \left[\{a_1\}\{a_2\}^T - \{a_2\}\{a_1\}^T\right][A_1]$
$[A_2]$, $\{a_1\}$, $\{a_2\}$	$[\underset{\sim}{G}^{20}] = [A_2]\left[\{a_1\}\{a_2\}^T - \{a_2\}\{a_1\}^T\right] - \left[\{a_1\}\{a_2\}^T - \{a_2\}\{a_1\}^T\right][A_2]$
(5) Four at a time	
$[A_1]$, $[A_2]$, $\{a_1\}$, $\{a_2\}$	none

Combined invariants

Using Table A.1 of Appendix A, we can obtain the following number of invariants for various combinations of the argument tensors of $\boldsymbol{T}$ in (8.93). Explicit expressions of $\underset{\sim}{I}^j$; $j = 1, 2, \dots, 30$ are given in Table 8.4.

Table 8.3: Number of combined invariants: $\underset{\sim}{I}^j$; $j = 1, 2, \dots, M$ $(M = 30)$ of the argument tensors of $\boldsymbol{T}$ in (8.93)

One at a time	$\boldsymbol{A}_1$	$\boldsymbol{A}_2$	$\boldsymbol{a}_1$	$\boldsymbol{a}_2$	θ	
$(\underset{\sim}{I}^j\ ;\ j = 1, 2, \dots, 8)$	3	3	1	1	0	
Two at a time	$\boldsymbol{A}_1, \boldsymbol{A}_2$	$\boldsymbol{A}_1, \boldsymbol{a}_1$	$\boldsymbol{A}_1, \boldsymbol{a}_2$	$\boldsymbol{A}_2, \boldsymbol{a}_1$	$\boldsymbol{A}_2, \boldsymbol{a}_2$	$\boldsymbol{a}_1, \boldsymbol{a}_2$
$(\underset{\sim}{I}^j\ ;\ j = 9, 10, \dots, 23)$	6	2	2	2	2	1
Three at a time	$\boldsymbol{A}_1, \boldsymbol{A}_2, \boldsymbol{a}_1$	$\boldsymbol{A}_1, \boldsymbol{A}_2, \boldsymbol{a}_2$	$\boldsymbol{A}_1, \boldsymbol{a}_1, \boldsymbol{a}_2$	$\boldsymbol{A}_2, \boldsymbol{a}_1, \boldsymbol{a}_2$		
$(\underset{\sim}{I}^j\ ;\ j = 24, 25, \dots, 29)$	1	1	2	2		
Four at a time	$\boldsymbol{A}_1, \boldsymbol{A}_2, \boldsymbol{a}_1, \boldsymbol{a}_2$					
$(\underset{\sim}{I}^{30})$	1					

Table 8.4: Combined invariants $\underset{\sim}{I}^j$; $j = 1, 2, \dots, 30$ of argument tensors of $\boldsymbol{T}$ in (8.93)

Arguments	Invariants
(1) One at a time	
$[A_1]$	$\underset{\sim}{I}^1 = \text{tr}[A_1]\ ,\quad \underset{\sim}{I}^2 = \text{tr}[A_1]^2\ ,\quad \underset{\sim}{I}^3 = \text{tr}[A_1]^3$
$[A_2]$	$\underset{\sim}{I}^4 = \text{tr}[A_2]\ ,\quad \underset{\sim}{I}^5 = \text{tr}[A_2]^2\ ,\quad \underset{\sim}{I}^6 = \text{tr}[A_2]^3$
$\{a_1\}$	$\underset{\sim}{I}^7 = \{a_1\}^T\{a_1\}$
$\{a_2\}$	$\underset{\sim}{I}^8 = \{a_2\}T\{a_2\}$

Table 8.4: (Contd.) Combined invariants $\underset{\sim}{I}^j$; $j = 1, 2, \dots, 30$ of argument tensors of $\boldsymbol{T}$ in (8.93)

Arguments	Invariants
(2) Two at a time	
$[A_1]$, $[A_2]$	$\underset{\sim}{I}^9 = \text{tr}([A_1][A_2])$, $\underset{\sim}{I}^{10} = \text{tr}([A_1]^2[A_2])$ $\underset{\sim}{I}^{11} = \text{tr}([A_1][A_2]^2)$, $\underset{\sim}{I}^{12} = \text{tr}([A_1]^2[A_2]^2)$ $\underset{\sim}{I}^{13} = \text{tr}([A_1][A_2] + [A_2][A_1])$ $\underset{\sim}{I}^{14} = \text{tr}([A_1][A_2] - [A_2][A_1])$
$[A_1]$, $\{a_1\}$	$\underset{\sim}{I}^{15} = \{a_1\}^T([A_1]\{a_1\})$, $\underset{\sim}{I}^{16} = \{a_1\}^T([A_1]^2\{a_1\})$
$[A_1]$, $\{a_2\}$	$\underset{\sim}{I}^{17} = \{a_2\}^T([A_1]\{a_2\})$, $\underset{\sim}{I}^{18} = \{a_2\}^T([A_1]^2\{a_2\})$
$[A_2]$, $\{a_1\}$	$\underset{\sim}{I}^{19} = \{a_1\}^T([A_2]\{a_1\})$, $\underset{\sim}{I}^{20} = \{a_1\}^T([A_2]^2\{a_1\})$
$[A_2]$, $\{a_2\}$	$\underset{\sim}{I}^{21} = \{a_2\}^T([A_2]\{a_2\})$, $\underset{\sim}{I}^{22} = \{a_2\}^T([A_2]^2\{a_2\})$
$\{a_1\}$, $\{a_2\}$	$\underset{\sim}{I}^{23} = \{a_1\}^T\{a_2\}$
(4) Three at a time	
$[A_1]$, $[A_2]$, $\{a_1\}$	$\underset{\sim}{I}^{24} = \{a_1\}^T([A_1][A_2]\{a_1\})$
$[A_1]$, $[A_2]$, $\{a_2\}$	$\underset{\sim}{I}^{25} = \{a_2\}^T([A_1][A_2]\{a_1\})$
$[A_1]$, $\{a_1\}$, $\{a_2\}$	$\underset{\sim}{I}^{26} = \{a_1\}^T([A_1]\{a_2\})$ $\underset{\sim}{I}^{27} = \{a_1\}^T([A_1]^2\{a_2\})$
$[A_2]$, $\{a_1\}$, $\{a_2\}$	$\underset{\sim}{I}^{28} = \{a_1\}^T([A_2]\{a_2\})$ $\underset{\sim}{I}^{29} = \{a_1\}^T([A_2]^2\{a_2\})$
(5) Four at a time	
$[A_1]$, $[A_2]$, $\{a_1\}$, $\{a_2\}$	$\underset{\sim}{I}^{30} = \{a_1\}^T([A_1][A_2] - [A_2][A_1])\{a_2\}$

8.6.3 Constitutive variable is a tensor of rank one

Let $\boldsymbol{q}$ be a constitutive variable, a tensor of rank one with the same argument tensors $\boldsymbol{A}_1$, $\boldsymbol{A}_2$, $\boldsymbol{a}_1$, $\boldsymbol{a}_2$ and θ as used in case of $\boldsymbol{T}$ in (8.93).

$$\boldsymbol{q} = \boldsymbol{q}(\boldsymbol{A}_1, \boldsymbol{A}_2, \boldsymbol{a}_1, \boldsymbol{a}_2, \theta) \tag{8.100}$$

Here also our objective is to derive a constitutive theory for $\boldsymbol{q}$ using representation theorem in the current configuration. Let $\underset{\sim}{\boldsymbol{g}}^i$; $i = 1, 2, \dots, \underset{\sim}{N}$ be the combined generators of the argument tensors of $\boldsymbol{q}$ in (8.100) that are tensors of rank one and let ${}^q\underset{\sim}{I}^j$; $j = 1, 2, \dots, \underset{\sim}{M}$ be the combined invariants of the same argument tensors of $\boldsymbol{q}$ as in (8.100). We note that $\underset{\sim}{\boldsymbol{g}}^i$ and $\underset{\sim}{I}^j$ are all in the current configuration. We remark that if $\boldsymbol{q}$ is heat flux, then a unit vector cannot be a generator as uniform temperature field does not result in $\boldsymbol{q}$. Thus, we only consider $\underset{\sim}{\boldsymbol{g}}^i$; $i = 1, 2, \dots, \underset{\sim}{N}$ to constitute the basis of the space containing $\boldsymbol{q}$, hence $\boldsymbol{q}$ can be expressed as a linear combination of $\underset{\sim}{\boldsymbol{g}}^i$; $i = 1, 2, \dots, \underset{\sim}{N}$.

$$\boldsymbol{q} = \sum_{i=1}^{\underset{\sim}{N}} \alpha^i \underset{\sim}{\boldsymbol{g}}^i \tag{8.101}$$

$$\text{in which} \quad \alpha^i = \alpha^i({}^q\underset{\sim}{I}^j, \theta) \quad ; \quad j = 1, 2, \dots, \underset{\sim}{M} \quad ; \quad i = 1, 2, \dots, \underset{\sim}{N} \tag{8.102}$$

If $\boldsymbol{q}$ is heat flux, then we would have used negative sign for the sum in (8.101). In this case also, the coefficients α^i are the linear combination of $\underset{\sim}{\boldsymbol{g}}^i$ in (8.102), and are also defined in the current configuration. In the constitutive theory (8.101), α^i are not material coefficients. They are simply coefficients in the linear combination that are dependent on ${}^q\underset{\sim}{I}^j$; $j = 1, 2, \dots, \underset{\sim}{M}$ and θ in the current configuration.

8.6.3.1 Material coefficients

Based on the axiom of smooth neighborhood, if α^i are sufficiently smooth functions of their arguments then α^i can be expanded in Taylor series in the arguments of α^i about a known configuration $\underline{\Omega}$. Consider the constitutive theory (8.101) and (8.102). We expand each α^i ; $i = 0, 1, 2 \dots, \underset{\sim}{N}$ in Taylor series in ${}^q\underset{\sim}{I}^j$; $j = 1, 2, \dots, \underset{\sim}{M}$ and θ about a known configuration $\underline{\Omega}$ and retain only up to linear terms in $\underset{\sim}{I}^j$; $j = 1, 2, \dots, \underset{\sim}{M}$ and θ (for simplicity).

$$\alpha^i = \alpha^i\big|_{\underline{\Omega}} + \sum_{j=1}^{\underset{\sim}{M}} \frac{\partial \alpha^i}{\partial ({}^q\underset{\sim}{I}^j)}\bigg|_{\underline{\Omega}} \left({}^q\underset{\sim}{I}^j - {}^q\underset{\sim}{I}^j\big|_{\underline{\Omega}}\right) + \frac{\partial \alpha^i}{\partial \theta}\bigg|_{\underline{\Omega}} (\theta - \theta|_{\underline{\Omega}}) \quad ; \quad i = 1, 2, \dots \tag{8.103}$$

Substituting α^i from (8.103) into (8.101)

$$\boldsymbol{q} = \sum_{i=1}^{\underset{\sim}{N}} \left(\alpha^i\big|_{\underline{\Omega}} + \sum_{j=1}^{\underset{\sim}{M}} \frac{\partial \alpha^i}{\partial ({}^q\underset{\sim}{I}^j)}\bigg|_{\underline{\Omega}} \left({}^q\underset{\sim}{I}^j - \underset{\sim}{I}^j\big|_{\underline{\Omega}}\right) + \frac{\partial \alpha^i}{\partial \theta}\bigg|_{\underline{\Omega}} \left(\theta - \theta|_{\underline{\Omega}}\right) \right) \underset{\sim}{\boldsymbol{g}}^i \tag{8.104}$$

Collecting coefficients (those defined in the known configuration $\underline{\Omega}$) of (8.104) and defining new coefficients

$$\tilde{b}_i = \alpha^i\Big|_{\Omega} + \sum_{j=1}^{\underset{\sim}{M}} \frac{\partial \alpha^i}{\partial({}^q\underset{\sim}{I}^j)}\Bigg|_{\Omega} \left(- {}^q\underset{\sim}{I}^j\Big|_{\Omega}\right) \tag{8.105}$$

$$\tilde{c}_{ij} = \frac{\partial \alpha^i}{\partial({}^q\underset{\sim}{I}^j)}\Bigg|_{\Omega} \quad ; \quad i = 1, 2, \dots, \underset{\sim}{N} \quad ; \quad j = 1, 2, \dots, \underset{\sim}{M} \tag{8.106}$$

$$\tilde{d}_i = \frac{\partial \alpha^i}{\partial \theta}\Bigg|_{\Omega} \quad ; \quad i = 1, 2, \dots, \underset{\sim}{N} \tag{8.107}$$

We can write

$$\boldsymbol{q} = \sum_{i=1}^{\underset{\sim}{N}} \tilde{b}_i \underset{\sim}{\boldsymbol{g}}_i + \sum_{i=1}^{\underset{\sim}{N}} \sum_{j=1}^{\underset{\sim}{M}} \tilde{c}_{ij} ({}^q\underset{\sim}{I}^j) \underset{\sim}{\boldsymbol{g}}^i + \sum_{i=1}^{\underset{\sim}{N}} \tilde{d}_i \left(\theta - \theta\Big|_{\Omega}\right) \underset{\sim}{\boldsymbol{g}}^i \tag{8.108}$$

8.6.3.2 Combined generators and combined invariants

Combined generators

The combined generators $\underset{\sim}{\boldsymbol{g}}^i$; $i = 1, 2, \dots, \underset{\sim}{N}$ and the combined invariants ${}^q\underset{\sim}{I}^j$; $j = 1, 2, \dots, \underset{\sim}{M}$ of the argument tensors $\boldsymbol{A}_1$, $\boldsymbol{A}_2$, $\boldsymbol{a}_1$, $\boldsymbol{a}_2$ and θ of $\boldsymbol{q}$ in (8.101) can be determined using the Tables A.1 and A.2 of Appendix A. Since $\boldsymbol{q}$ is a tensor of rank one, we need to determine combined generators of tensors $\boldsymbol{A}_1$, $\boldsymbol{A}_2$, $\boldsymbol{a}_1$, $\boldsymbol{a}_2$ and θ that are tensors of rank one. Using Table A.2 of Appendix A, we can obtain the following number of generators for various combination of the argument tensors of $\boldsymbol{q}$ in (8.100). These are listed in Table 8.5. Explicit expressions of $\underset{\sim}{\boldsymbol{g}}^i$; $i = 1, 2, \dots, \boldsymbol{N}$ ($N = 14$) are given in Table 8.6.

Table 8.5: Number of combined generators $\tilde{\boldsymbol{g}}^i$; $i = 1, 2, \dots, \underset{\sim}{N}$ ($\underset{\sim}{N} = 14$) of the argument tensors of $\boldsymbol{q}$ in (8.100)

One at a time	$\boldsymbol{A}_1$	$\boldsymbol{A}_2$	$\boldsymbol{a}_1$	$\boldsymbol{a}_2$	θ	
($\underset{\sim}{\boldsymbol{g}}^1$,$\underset{\sim}{\boldsymbol{g}}^2$)	0	0	1	1	0	
Two at a time	$\boldsymbol{A}_1, \boldsymbol{A}_2$	$\boldsymbol{A}_1, \boldsymbol{a}_1$	$\boldsymbol{A}_1, \boldsymbol{a}_2$	$\boldsymbol{A}_2, \boldsymbol{a}_1$	$\boldsymbol{A}_2, \boldsymbol{a}_2$	$\boldsymbol{a}_1, \boldsymbol{a}_2$
($\underset{\sim}{\boldsymbol{g}}^i$; $i = 3, 4, \dots, 10$)	0	2	2	2	2	0
Three at a time	$\boldsymbol{A}_1, \boldsymbol{A}_2, \boldsymbol{a}_1$	$\boldsymbol{A}_1, \boldsymbol{A}_2, \boldsymbol{a}_2$	$\boldsymbol{A}_1, \boldsymbol{a}_1, \boldsymbol{a}_2$	$\boldsymbol{A}_2, \boldsymbol{a}_1, \boldsymbol{a}_2$		
($\underset{\sim}{\boldsymbol{g}}^i$; $i = 11, 12, \dots, 14$)	2	2	0	0		
Four at a time	$\boldsymbol{A}_1, \boldsymbol{A}_2, \boldsymbol{a}_1, \boldsymbol{a}_2$					
(none)	0					

Table 8.6: Combined generators $\underset{\sim}{\boldsymbol{g}}^i$; $i = 1, 2, \ldots, \widetilde{N}$, of the argument tensors of $\boldsymbol{q}$ in (8.100): tensors of rank one

Arguments	Generators
(1) None	None
(2) One at a time	
$[A_1]$	0
$[A_2]$	0
$\{a_1\}$	$\{\underset{\sim}{g}^1\} = \{a_1\}$
$\{a_2\}$	$\{\underset{\sim}{g}^2\} = \{a_2\}$
(3) Two at a time	
$[A_1]$, $[A_2]$	0
$[A_1]$, $\{a_1\}$	$\{\underset{\sim}{g}^3\} = [A_1]\{a_1\}$, $\{\underset{\sim}{g}^4\} = [A_1]^2\{a_1\}$
$[A_1]$, $\{a_2\}$	$\{\underset{\sim}{g}^5\} = [A_1]\{a_2\}$, $\{\underset{\sim}{g}^6\} = [A_1]^2\{a_2\}$
$[A_2]$, $\{a_1\}$	$\{\underset{\sim}{g}^7\} = [A_2]\{a_1\}$, $\{\underset{\sim}{g}^8\} = [A_2]^2\{a_1\}$
$[A_2]$, $\{a_2\}$	$\{\underset{\sim}{g}^9\} = [A_2]\{a_2\}$, $\{\underset{\sim}{g}^{10}\} = [A_2]^2\{a_2\}$
$\{a_1\}$, $\{a_2\}$	0
(4) Three at a time	
$[A_1]$, $[A_2]$, $\{a_1\}$	$\{\underset{\sim}{g}^{11}\} = ([A_1][A_2] + [A_2][A_1])\{a_1\}$ $\{\underset{\sim}{g}^{12}\} = ([A_1][A_2] - [A_2][A_1])\{a_1\}$
$[A_1]$, $[A_2]$, $\{a_2\}$	$\{\underset{\sim}{g}^{13}\} = ([A_1][A_2] + [A_2][A_1])\{a_2\}$ $\{\underset{\sim}{g}^{14}\} = ([A_1][A_2] - [A_2][A_1])\{a_2\}$
$[A_1]$, $\{a_1\}$, $\{a_2\}$	0
$[A_2]$, $\{a_1\}$, $\{a_2\}$	0

Combined invariants

Since the argument tensors of $\boldsymbol{q}$ in (8.100) and those of $\boldsymbol{T}$ in (8.93) are same, the combined invariants ${}^q\underset{\sim}{I}^j$; $j = 1, 2, \dots, \boldsymbol{M}$ of the argument tensors of $\boldsymbol{q}$ in (8.100) are identical to those of the argument tensors of $\boldsymbol{T}$ in (8.93) listed in Table 8.3.

8.7 Other approaches of deriving constitutive theories (not strictly thermodynamic)

In these approaches of deriving constitutive theories one does not strictly adhere to the laws of thermodynamics. We discuss a few of these approaches in the following.

For thermoelastic solids with incompressible, isothermal deformation physics constitutive theories for the deviatoric stress tensor have been derived using Helmholtz free energy density in many published works. Constitutive theories for this physics have also been considered in the published works using strain energy and complementary strain energy density functionals.

There are many phenomenological approaches currently used for establishing constitutive theories particularly for non-isotropic, non-homogeneous continua. As for such continua, thermodynamic approach of deriving constitutive theories is not possible. The most commonly used approach is to consider strain energy density functional π, a function of strain tensor and temperature. π is expanded in Taylor series in the strain tensor about a known configuration and then differentiating π with respect to strain tensor to obtain a general constitutive theory for the stress tensor. Reduction in the material coefficients is carried out by using symmetry of stress and strain tensors as well as consideration of a group of symmetry transformations. Constitutive theories for the stress tensor for generally anisotropic and orthotropic solid continua are derived using this approach, but such constitutive theories are not supported by CCM. There are some approaches in which experimental measurements of material behavior are used to construct empirical relationships describing the constitutive theories for the stress tensor and heat vector. Obviously, these approaches are not based on principles of thermodynamics.

Since such approaches are not based on SLT, the resulting constitutive theories are not ensured to satisfy SLT, thus, could result in violation of thermodynamic equilibrium in the deforming continua.

Constitutive theories have also been derived using phenomenologically constructed potentials similar to Φ or energy functionals. This approach is quite common in NCCM where non-symmetric constitutive tensors are often used. Such constitutive theories are likely to be in violation of SLT, hence may exhibit lack of thermodynamic equilibrium during deformation.

For solid continua, specially for thermoviscous and thermoviscoelastic solids, various combinations of 1D springs and 1D dashpots have been considered to match the observed or expected behavior. This approach is purely phenomenological as in this approach one tries to mimic the observed phenomenon using 1D spring and 1D dashpots. It is well known that this approach cannot be extended to $\mathbb{R}^2$ and $\mathbb{R}^3$, hence is rather not useful for continuous matter.

There are many other published works in which the complexity of physics necessitates phenomenological approaches in deriving constitutive theories.

Remarks

(1) Regardless of the approach used for establishing constitutive theories, the constitutive theories must always satisfy the entropy inequality, otherwise the deformation established using these constitutive theories is in violation of thermodynamic equilibrium.
(2) The thermodynamic approach of deriving constitutive theories for solid and fluent continua in conjunction with representation theorem always ensure that the resulting constitutive theories satisfy the entropy inequality, hence thermodynamic equilibrium is always ensured during deformation.

8.8 Summary

This chapter contains axioms of constitutive theory and basic concepts and principles of deriving constitutive theories for solid continua in Lagrangian description and for fluent continua in Eulerian description, strictly based on the entropy inequality in conjunction with representation theorem. The constitutive theories derived using entropy inequality in conjunction with representation theorem always satisfy entropy inequality, hence conservation and balance laws. Thus, these theories always ensure thermodynamic equilibrium in the deforming continua.

It is shown that constitutive theories can be initiated using entropy inequality. Constitutive tensors are determined using SLT (as well as other balance laws). The argument tensors of the constitutive tensors are established using conjugate pairs in the entropy inequality, the principle of equipresence and based on consideration of additional physics that is either not present in the entropy inequality or is not obvious from the entropy inequality. Substitution of $\frac{D\Phi}{Dt}$ (or $\frac{D\bar{\Phi}}{Dt}$) in the entropy inequality and using considerations of arbitrary but admissible rate quantities, the entropy inequality is satisfied if the coefficients of the rate terms are set to zero. This results in elimination of some quantities as constitutive tensors as well as modification of the argument tensors of the remaining constitutive tensors. It is shown that additive decomposition of the stress tensor into equilibrium and deviatoric tensors describing pure volume change and pure shape change of a volume of matter facilitates derivations of the constitutive theories. While the constitutive theory for the equilibrium stress tensor addresses compressibility, i.e volume change or incompressibility as well as non-isothermal physics, the constitutive theory for deviatoric stress tensor addresses distortion of the volume of matter, dissipation and memory mechanisms. Constitutive theory for equilibrium stress tensor is derived using $\bar{\Phi}$ for compressible matter. In the case of incompressible matter, the incompressibility condition when incorporated in the entropy inequality permits derivation of the constitutive theory for the equilibrium stress. Constitutive theories for the deviatoric stress tensor and the heat vector can be derived using representation theorem. Material coefficients are established using Taylor series expansion of the coefficients used in the linear combination in using representation theorem about a known configuration.

Using decomposition of the deformation into volumetric and distortional physics and through further considerations regarding volumetric deformation, it is shown that derivation of the constitutive theory for equilibrium Cauchy stress tensor ${}^{(0)}_{\ e}\bar{\boldsymbol{\sigma}}$ addressing volumetric deformation is only possible by considering entropy inequality and conservation of mass, both in Eulerian description. It is shown that constitutive theory for ${}^{(0)}_{e}\bar{\boldsymbol{\sigma}}$ describing volumetric deformation physics is invariant of the type of continua, type of deformation and constitution of the matter. Constitutive theories for ${}^{(0)}_{e}\bar{\boldsymbol{\sigma}}$ for compressible, incompressible, isothermal and non-isothermal physics are derived. These are then utilized to obtain their Lagrangian form such as $\boldsymbol{\sigma}^{(0)}$, $\boldsymbol{\sigma}^{[0]}$ etc. The reduced form of the entropy inequality (obtained after deriving the constitutive theory for ${}^{(0)}_{e}\bar{\boldsymbol{\sigma}}$) in Eulerian description is presented from which the reduced form of the entropy inequality is obtained in the Lagrangian description. The conjugate pairs in the reduced form of the entropy inequalities in conjunction with representation theorem can be used to derive constitutive theories for fluent and solid continua in Eulerian and Lagrangian descriptions (Chapters 9-13).

We have also discussed derivations of the constitutive theories for thermoelastic solid continua based on energy functionals. Strain energy density and complementary strain energy density functionals are possible choices. Limitations of this approach are also presented. We have also discussed the derivation of constitutive theories using phenomenological approaches. These are necessitated when the continua is not isotropic and homogeneous (as in case of composites). Here we have discussed the approach based on Taylor series expansion of strain energy density functional in its arguments. The reduction process of material coefficients based on symmetry of stress and strain tensors as well as based on material symmetry group transformation is used to derive constitutive theories for generally anisotropic solid continua as well as orthotropic solid continua. In this category empirical constitutive theories based on experimentally determined material response are also considered. The constitutive theory in this category may result in violation of SLT, hence may lead to lack of thermodynamic equilibrium in the deforming solid continua.

A brief discussion of phenomenological approaches of deriving constitutive theories in $\mathbb{R}^1$ using 1D springs and 1D dashpots is also discussed. Such constitutive theories cannot be extended to $\mathbb{R}^2$ and $\mathbb{R}^3$. Hence, this approach cannot be used for continuous matter. The entropy inequality is rarely considered in these approaches. Hence, these constitutive theories almost always results in violation of thermodynamic equilibrium.

9 CONSTITUTIVE THEORIES FOR THERMOELASTIC SOLIDS

9.1 Introduction

In thermoelastic solids, the mechanical deformation physics can be linear or non-linear. In such solids, the rate of mechanical work does not result in the rate of entropy production. Hence, no heat is generated due to rate of mechanical work. In general, the deformation physics can result in finite deformation, finite strain, compressible or incompressible, non-isothermal or isothermal behavior as well as small strain, small deformation, non-isothermal, or isothermal behavior. All these aspects require careful considerations in deriving constitutive theories. In a deforming solid continua the deformation consists of two major aspects: change in volume without change in shape or distortion of volume and change in shape, i.e., distortion of volume without change in volume, i.e., volumetric deformation and distortional deformation. Incompressibility condition in which there is no change of volume is a special case of volumetric deformation. These two aspects of the deformation physics, volumetric and distortional are mutually exclusive. Thus, they must be addressed separately in deriving constitutive theories for the stress tensor. Stress tensor $\boldsymbol{\sigma}$ (appropriately chosen measure) is additively decomposed into equilibrium ${}_e\boldsymbol{\sigma}$ and deviatoric ${}_d\boldsymbol{\sigma}$ stress tensors. The change in volume in compressible and incompressible matter and non-isothermal physics, i.e., volumetric aspects, are considered in the constitutive theory for equilibrium stress ${}_e\boldsymbol{\sigma}$. Pure distortion of the volume without change in volume, i.e., the distortional aspect, is due to the constitutive theory for ${}_d\boldsymbol{\sigma}$. It is perhaps more meaningful to think of thermoelastic solid continua from the point of view described above as opposed to merely reversible or irreversible deformation physics.

Compressibility, incompressibility, finite deformation, finite strain, small deformation, small strain in conjunction with isothermal or non-isothermal considerations necessitate that we consider appropriate choices of stress and strain measures as well as suitable approaches of deriving constitutive theories. We present details of the stress and strain measure requirements in the constitutive theories for different types of physics and the approaches considered in deriving them.

In this chapter, we consider thermodynamic as well as some other approaches that may be strictly thermodynamic approaches of deriving constitutive theories for solid continua for various type of distortional physics. It has been shown in Chapter 8, that the volumetric aspects of deformation describing compressibility, incompressibility and non-isothermal effects remain the same in various type of deformations when described by the constitutive theories for the equilibrium Cauchy stress tensor ${}^{(0)}_{e}\bar{\boldsymbol{\sigma}}$ (basis independent) in Eulerian descriptions. ${}^{(0)}_{e}\bar{\boldsymbol{\sigma}}$ then can be converted to Lagrangian description and finally to the second Piola-Kirchhoff equilibrium stress tensor ${}_e\boldsymbol{\sigma}^{[0]}$ and the entropy inequality reduces to what has been referred to as the *reduced form of the entropy inequality*. The reduced form of the

DOI: 10.1201/9781003105336-9

entropy inequality can be used to derive the constitutive theories for the deviatoric stress tensor and heat vector. Furthermore, since the constitutive theory for heat vector remains unaffected by the type of mechanical deformation, we present the details of the constitutive theory for heat vector $\boldsymbol{q}$ after we have considered various constitutive theories for the deviatoric stress tensor.

9.2 Thermodynamic approach

It has been shown in Chapter 8 that regardless of the type of continua (solid or fluent) we must always begin the derivation of constitutive theories with the entropy inequality in Eulerian description in the current configuration and: (1) derive the constitutive theory for equilibrium Cauchy stress ${}^{(0)}_{e}\bar{\boldsymbol{\sigma}}$, then convert it into Lagrangian description and finally transform to the desired type of measure (second Piola-Kirchhoff stress or any other) (2) obtain the reduced form of the entropy inequality from which constitutive theories for derviatoric stress and heat vector can be derived. We follow this approach in the following sections for deriving constitutive theories for the deviatoric stress tensor for various types of deformations.

9.2.1 Finite deformation, finite strain, compressible, non-isothermal

For this deformation physics, CBL in Lagrangian description using $\boldsymbol{\sigma}^{[0]}$ and $\boldsymbol{\varepsilon}_{[0]}$ as stress and strain measures are ideally suited. Deformation physics consists of finite motion of material points, finite strain, hence change in volume (compressibility), change in shape, i.e., distortion of volume as well as non-isothermal effect that may result in volume change and/or thermal stress field. We consider additive decomposition of $\boldsymbol{\sigma}^{[0]}$ into equilibrium stress ${}_{e}\boldsymbol{\sigma}^{[0]}$ and deviatoric stress ${}_{d}\boldsymbol{\sigma}^{[0]}$. The constitutive theory for ${}_{e}\boldsymbol{\sigma}^{[0]}$ addresses compressibility or incompressibility and the thermal aspect whereas the constitutive theory for ${}_{d}\boldsymbol{\sigma}^{[0]}$ incorporates the physics of change in shape or distortion.

Following Chapter 8, the constitutive theory for ${}_{e}\boldsymbol{\sigma}^{[0]}$ for this deformation physics is given by ((8.64) and (8.63))

$$ {}_{e}\boldsymbol{\sigma}^{[0]} = |\boldsymbol{J}| p(\rho,\theta)(\boldsymbol{J}^T \cdot \boldsymbol{J})^{-1} \tag{9.1} $$

$$ \text{where} \quad p(\rho,\theta) = -\rho^2 \frac{\partial \Phi(\rho,\theta)}{\partial \rho} \tag{9.2} $$

and the reduced form of the entropy inequality is given by (8.85)

$$ -{}_{d}\boldsymbol{\sigma}^{[0]} : \dot{\boldsymbol{\varepsilon}}_{[0]} + \frac{\boldsymbol{q}\cdot\boldsymbol{g}}{\theta} \leq 0 \tag{9.3} $$

in which

$$ {}_{d}\boldsymbol{\sigma}^{[0]} = {}_{d}\boldsymbol{\sigma}^{[0]}(\boldsymbol{\varepsilon}_{[0]},\theta) \tag{9.4} $$

$$ \boldsymbol{q} = \boldsymbol{q}(\boldsymbol{g},\theta) \tag{9.5} $$

In Lagrangian description, density $\rho(\boldsymbol{x},t)$ is deterministic from the conservation of mass, hence is not a dependent variable in the CBL, thus, cannot be an argument tensor in (9.4) and (9.5).

9.2.1.1 Constitutive theory for ${}_d\boldsymbol{\sigma}^{[0]}$

We derive the constitutive theory for ${}_d\boldsymbol{\sigma}^{[0]}$ using (9.4) and representation theorem. Let ${}^\sigma\underset{\sim}{\boldsymbol{G}}^i$; $i = 1, 2, \dots, N$ be the combined generators of the argument tensors $\boldsymbol{\varepsilon}_{[0]}$ and θ (symmetric tensor of rank two and tensor of rank zero) that are symmetric tensors of rank two. Let ${}^\sigma\underset{\sim}{I}^j$; $j = 1, 2, \dots, M$ be the combined invariants of the tensors of $\boldsymbol{\varepsilon}_{[0]}$ and θ. Then, we can express ${}_d\boldsymbol{\sigma}^{[0]}$ as a linear combination of $\boldsymbol{I}$ and ${}^\sigma\underset{\sim}{\boldsymbol{G}}^i$; $i = 1, 2, \dots, N$ in the current configuration.

$$
{}_d\boldsymbol{\sigma}^{[0]} = {}^\sigma\underset{\sim}{\alpha}^0 \boldsymbol{I} + \sum_{i=1}^{N} {}^\sigma\underset{\sim}{\alpha}^i \, {}^\sigma\underset{\sim}{\boldsymbol{G}}^i \tag{9.6}
$$

$$
\text{in which} \quad {}^\sigma\underset{\sim}{\alpha}^i = {}^\sigma\underset{\sim}{\alpha}^i({}^\sigma\underset{\sim}{I}^j, \theta) \quad ; \quad j = 1, 2, \dots, M \quad ; \quad i = 0, 1, \dots, N \tag{9.7}
$$

Based on the argument tensors in (9.4) we have the following for the combined generators ($N = 2$) and invariants ($N = 3$).

$$
{}^\sigma\underset{\sim}{\boldsymbol{G}}^1 = \boldsymbol{\varepsilon}_{[0]} \quad ; \quad {}^\sigma\underset{\sim}{\boldsymbol{G}}^2 = (\boldsymbol{\varepsilon}_{[0]})^2 \tag{9.8}
$$

$$
{}^\sigma\underset{\sim}{I}^1 = I_{\varepsilon_{[0]}} \quad ; \quad {}^\sigma\underset{\sim}{I}^2 = II_{\varepsilon_{[0]}} \quad ; \quad {}^\sigma\underset{\sim}{I}^3 = III_{\varepsilon_{[0]}} \tag{9.9}
$$

Material coefficients

Using (9.6) and (9.7) we derive material coefficients in the constitutive theory (9.6) using the procedure described in Chapter 8. We present a compact derivation in the following. We expand ${}^\sigma\underset{\sim}{\alpha}^i$; $i = 1, 2, \dots, N$ in Taylor series in ${}^\sigma\underset{\sim}{I}^j$; $j = 1, 2, \dots, M$ about a known configuration $\underline{\Omega}$ and retain only up to linear terms in ${}^\sigma\underset{\sim}{I}^j$; $j = 1, 2, \dots, M$. Taylor series expansion in θ is not considered as the stress field due to non-isothermal physical has already been considered in ${}_e\boldsymbol{\sigma}^{[0]}$.

$$
{}^\sigma\underset{\sim}{\alpha}^i = {}^\sigma\underset{\sim}{\alpha}^i\Big|_{\underline{\Omega}} + \sum_{j=1}^{M} \left.\frac{\partial\, {}^\sigma\underset{\sim}{\alpha}^i}{\partial({}^\sigma\underset{\sim}{I}^j)}\right|_{\underline{\Omega}} \left({}^\sigma\underset{\sim}{I}^j - {}^\sigma\underset{\sim}{I}^j\Big|_{\underline{\Omega}} \right) \tag{9.10}
$$

substituting (9.10) in (9.6) and collecting coefficients of $\boldsymbol{I}$, ${}^\sigma\underset{\sim}{I}^j\boldsymbol{I}$, ${}^\sigma\underset{\sim}{\boldsymbol{G}}^i$ and ${}^\sigma\underset{\sim}{I}^j\,{}^\sigma\underset{\sim}{\boldsymbol{G}}^i$ and defining new coefficients using

$$
\begin{aligned}
\underset{\sim}{\sigma}^0\Big|_{\underline{\Omega}} &= {}^\sigma\underset{\sim}{\alpha}^0\Big|_{\underline{\Omega}} + \sum_{j=1}^{M} \left.\frac{\partial\, {}^\sigma\underset{\sim}{\alpha}^0}{\partial({}^\sigma\underset{\sim}{I}^j)}\right|_{\underline{\Omega}} \left(-\, {}^\sigma\underset{\sim}{I}^j\Big|_{\underline{\Omega}}\right) \quad ; \quad {}^\sigma\underset{\sim}{a}_j = \left.\frac{\partial\, {}^\sigma\underset{\sim}{\alpha}^0}{\partial({}^\sigma\underset{\sim}{I}^j)}\right|_{\underline{\Omega}} \\
{}^\sigma\underset{\sim}{b}_i &= {}^\sigma\underset{\sim}{\alpha}^i\Big|_{\underline{\Omega}} + \sum_{j=1}^{M} \left.\frac{\partial\, {}^\sigma\underset{\sim}{\alpha}^i}{\partial({}^\sigma\underset{\sim}{I}^j)}\right|_{\underline{\Omega}} \left(-\, {}^\sigma\underset{\sim}{I}^j\Big|_{\underline{\Omega}}\right) \quad ; \quad {}^\sigma\underset{\sim}{c}_{ij} = \left.\frac{\partial\, {}^\sigma\underset{\sim}{\alpha}^i}{\partial({}^\sigma\underset{\sim}{I}^j)}\right|_{\underline{\Omega}} \\
i &= 1, 2 \dots, N \quad ; \quad j = 1, 2, \dots, M
\end{aligned} \tag{9.11}
$$

we can write the following for (9.6)

$$
{}_d\boldsymbol{\sigma}^{[0]} = \underset{\sim}{\sigma}^0\Big|_{\underline{\Omega}} \boldsymbol{I} + \sum_{j=1}^{M} {}^\sigma\underset{\sim}{a}_j ({}^\sigma\underset{\sim}{I}^j) \boldsymbol{I} + \sum_{i=1}^{N} {}^\sigma\underset{\sim}{b}_i \, {}^\sigma\underset{\sim}{\boldsymbol{G}}^i + \sum_{i=1}^{N} \sum_{j=1}^{M} {}^\sigma\underset{\sim}{c}_{ij} ({}^\sigma\underset{\sim}{I}^j) \, {}^\sigma\underset{\sim}{\boldsymbol{G}}^i \tag{9.12}
$$

$\underset{\sim}{\sigma}^0\big|_{\underline{\Omega}}$ is a known stress in the configuration $\underline{\Omega}$, hence is not a material coefficient. The material coefficients ${}^{\sigma}\underset{\sim}{a}_j$, ${}^{\sigma}\underset{\sim}{b}_i$ and ${}^{\sigma}\underset{\sim}{c}_{ij}$ that can be functions of ${}^{\sigma}\underset{\sim}{I}^j\big|_{\underline{\Omega}}$; $j = 1, 2, \dots, M$ and $\theta|_{\underline{\Omega}}$. This constitutive theory requires $(M + NM + N)$ material coefficients.

9.2.1.2 Example 9.1: Explicit form of the constitutive theory for ${}_d\boldsymbol{\sigma}^{[0]}$ in $\boldsymbol{\varepsilon}_{[0]}$

We can write explicit form of the constitutive theory for ${}_d\boldsymbol{\sigma}^{[0]}$ in terms of $\boldsymbol{\varepsilon}_{[0]}$ using generators and invariants defined by (9.8) and (9.9). In this case (9.6) reduces to

$$ {}_d\boldsymbol{\sigma}^{[0]} = {}^{\sigma}\underset{\sim}{\alpha}^0 \boldsymbol{I} + {}^{\sigma}\underset{\sim}{\alpha}^1 \boldsymbol{\varepsilon}_{[0]} + {}^{\sigma}\underset{\sim}{\alpha}^2 \boldsymbol{\varepsilon}^2_{[0]} \tag{9.13} $$

and the constitutive theory ${}_d\boldsymbol{\sigma}^{[0]}$ in (9.12) becomes the following

$$ \begin{aligned} {}_d\boldsymbol{\sigma}^{[0]} = {} & \underset{\sim}{\sigma}^0\big|_{\underline{\Omega}} \boldsymbol{I} + \sum_{j=1}^{3} {}^{\sigma}\underset{\sim}{a}_j ({}^{\sigma}\underset{\sim}{I}^j) \boldsymbol{I} + {}^{\sigma}\underset{\sim}{b}_1 \boldsymbol{\varepsilon}_{[0]} + {}^{\sigma}\underset{\sim}{b}_2 \boldsymbol{\varepsilon}^2_{[0]} \\ & + \sum_{j=1}^{3} {}^{\sigma}\underset{\sim}{I}^j \left({}^{\sigma}\underset{\sim}{c}_{1j} \boldsymbol{\varepsilon}_{[0]} + {}^{\sigma}\underset{\sim}{c}_{2j} \boldsymbol{\varepsilon}^2_{[0]} \right) \end{aligned} \tag{9.14} $$

Remarks

(1) We note that the constitutive theory (9.14) contains up to fifth degree terms of the components of $\boldsymbol{\varepsilon}_{[0]}$ and up to tenth degree terms of displacement gradients.

(2) This constitutive theory is based on integrity (complete basis).

(3) This constitutive theory requires eleven material coefficients ($N = 2$, $M = 3$) that must generally be determined experimentally. Secondly, all material coefficients in the constitutive theory may not be of equal significance. This of course depends upon application.

(4) In the following we derive a simplified form of the constitutive theory for ${}_d\boldsymbol{\sigma}^{[0]}$ using (9.14).

9.2.1.3 Example 9.2: Simplified linear form of the constitutive theory for ${}_d\boldsymbol{\sigma}^{[0]}$

If we neglect second and higher degree terms in the components of $\boldsymbol{\varepsilon}_{[0]}$, i.e., $\boldsymbol{\varepsilon}^2_{[0]}$, $II_{\varepsilon_{[0]}}$, $III_{\varepsilon_{[0]}}$, $I_{\varepsilon_{[0]}}\boldsymbol{\varepsilon}_{[0]}, \dots$, etc., we can obtain a much simplified constitutive theory for ${}_d\boldsymbol{\sigma}^{[0]}$. Based on the above assumption, (9.14) reduces to the following

$$ {}_d\boldsymbol{\sigma}^{[0]} = \underset{\sim}{\sigma}^0\big|_{\underline{\Omega}} \boldsymbol{I} + {}^{\sigma}\underset{\sim}{a}_1 (\operatorname{tr} \boldsymbol{\varepsilon}_{[0]}) \boldsymbol{I} + {}^{\sigma}\underset{\sim}{b}_1 \boldsymbol{\varepsilon}_{[0]} \tag{9.15} $$

If we choose ${}^{\sigma}\underset{\sim}{b}_1 = 2\mu$ and ${}^{\sigma}\underset{\sim}{a}_1 = \lambda$, new notation for the material coefficients (to conform to the published works), then we can write (9.15) as

$$ {}_d\boldsymbol{\sigma}^{[0]} = \underset{\sim}{\sigma}^0\big|_{\underline{\Omega}} \boldsymbol{I} + 2\mu \boldsymbol{\varepsilon}_{[0]} + \lambda (\operatorname{tr} \boldsymbol{\varepsilon}_{[0]}) \boldsymbol{I} \tag{9.16} $$

In (9.16), μ, λ are material coefficients which can be functions of the invariants of $\boldsymbol{\varepsilon}_{[0]}$ and θ in a known configuration $\underline{\Omega}$.

Remarks

(1) Since $\boldsymbol{\varepsilon}_{[0]} = \frac{1}{2}(\boldsymbol{C}_{[0]} - \boldsymbol{I})$, we can use (9.16) to express ${}_d\boldsymbol{\sigma}^{[0]}$ as a function of $\boldsymbol{C}_{[0]}$, giving us a constitutive theory for ${}_d\boldsymbol{\sigma}^{[0]}$ in terms of $\boldsymbol{C}_{[0]}$, but with new material coefficients

$$ {}_d\boldsymbol{\sigma}^{[0]} = \underset{\sim}{\sigma}^0\big|_{\underline{\Omega}} \boldsymbol{I} + \underset{\sim}{\mu}(\boldsymbol{C}_{[0]} - \boldsymbol{I}) + \underset{\sim}{\lambda}(\operatorname{tr}\boldsymbol{C}_{[0]} - \operatorname{tr}\boldsymbol{I})\boldsymbol{I} \tag{9.17} $$

in which $\underset{\sim}{\mu} = \mu$ and $\underset{\sim}{\lambda} = \dfrac{\lambda}{2}$

(2) If we assume that the material coefficients are only functions of $(I_{\varepsilon_{[0]}})\big|_{\underline{\Omega}}$ and $\theta\big|_{\underline{\Omega}}$, then the material coefficients in (9.16) defined by (9.11) reduce to the following

$$ \underset{\sim}{\sigma}^0\big|_{\underline{\Omega}} = {}^{\sigma}\underset{\sim}{\alpha}^0\big|_{\underline{\Omega}} - \frac{\partial\, {}^{\sigma}\underset{\sim}{\alpha}^0}{\partial({}^{\sigma}\underset{\sim}{I}^0)}\bigg|_{\underline{\Omega}} (\operatorname{tr}[\varepsilon_{[0]}])\big|_{\underline{\Omega}} \quad ; \quad 2\mu = {}^{\sigma}\underset{\sim}{\alpha}^1\big|_{\underline{\Omega}} - \frac{\partial\, {}^{\sigma}\underset{\sim}{\alpha}^1}{\partial({}^{\sigma}\underset{\sim}{I}^1)}\bigg|_{\underline{\Omega}} (\operatorname{tr}[\varepsilon_{[0]}])\big|_{\underline{\Omega}} $$
$$ \lambda = \frac{\partial\, {}^{\sigma}\underset{\sim}{\alpha}^0}{\partial({}^{\sigma}\underset{\sim}{I}^1)}\bigg|_{\underline{\Omega}} \tag{9.18} $$

(3) If we consider the material coefficients to be functions of $I_{\varepsilon_{[0]}}$ and θ (as in Remark (2)) but choose $\underline{\Omega}$ to be reference configuration, then we have

$$ \begin{aligned} [\varepsilon_{[0]}]_{\underline{\Omega}} &= [\varepsilon_{[0]}]_0 = 0 \quad ; \quad (I_{\varepsilon_{[0]}})_{\underline{\Omega}} = (I_{\varepsilon_{[0]}})_0 = 0 \\ (II_{\varepsilon_{[0]}})_{\underline{\Omega}} &= (II_{\varepsilon_{[0]}})_0 = 0 \quad ; \quad (III_{\varepsilon_{[0]}})_{\underline{\Omega}} = (III_{\varepsilon_{[0]}}) = 0 \end{aligned} \tag{9.19} $$

and the coefficient in (9.11) reduce to the following

$$ \underset{\sim}{\sigma}^0\big|_{\underline{\Omega}} = {}^{\sigma}\underset{\sim}{\alpha}^0\big|_0 \quad ; \quad 2\mu_0 = {}^{\sigma}\underset{\sim}{\alpha}^1\big|_0 \quad ; \quad \lambda_0 = \frac{\partial\, {}^{\sigma}\underset{\sim}{\alpha}^0}{\partial(I_{\varepsilon_{[0]}})}\bigg|_0 \tag{9.20} $$

using (9.20) in (9.16), we can write

$$ {}_d\boldsymbol{\sigma}^{[0]} = \underset{\sim}{\sigma}^0\big|_{\underline{\Omega}} \boldsymbol{I} + 2\mu_0\boldsymbol{\varepsilon}_{[0]} + \lambda_0(\operatorname{tr}\boldsymbol{\varepsilon}_{[0]})\boldsymbol{I} \tag{9.21} $$

Material coefficients μ_0, λ_0 can be function of θ_0. Clearly (9.20) is a special case of constitutive theory (9.16).

(4) Constitutive theory (9.21) does not permit deformation dependent material coefficients during the evolution. Constitutive theory (9.21) is commonly used in published writings.

(5) It is instructive to represent constitutive theory (9.16) in matrix and vector form using Voigt's notation. Let

$$ \left\{{}_d\sigma^{[0]}\right\}^T = \left[{}_d\sigma_{11}^{[0]}, {}_d\sigma_{22}^{[0]}, {}_d\sigma_{33}^{[0]}, {}_d\sigma_{23}^{[0]}, {}_d\sigma_{31}^{[0]}, {}_d\sigma_{12}^{[0]}\right] \tag{9.22} $$

$$ \text{and} \quad \left\{\varepsilon_{[0]}\right\}^T = \left[(\varepsilon_{[0]})_{11}, (\varepsilon_{[0]})_{22}, (\varepsilon_{[0]})_{33}, (\varepsilon_{[0]})_{23}, (\varepsilon_{[0]})_{31}, (\varepsilon_{[0]})_{12}\right] \tag{9.23} $$

Then using (9.22) and (9.23), we can write (9.16) as

$$\left\{{}_d\sigma^{[0]}\right\} = \{\sigma^0\} + [D]\{\varepsilon_{[0]}\} \tag{9.24}$$

in which

$$\{\sigma^0\} = \underset{\sim}{\sigma}^0\Big|_{\underline{\Omega}} \begin{Bmatrix} 1 \\ 1 \\ 1 \\ 0 \\ 0 \\ 0 \end{Bmatrix} \quad ; \quad [D] = \begin{bmatrix} 2\mu+\lambda & \lambda & \lambda & 0 & 0 & 0 \\ \lambda & 2\mu+\lambda & \lambda & 0 & 0 & 0 \\ \lambda & \lambda & 2\mu+\lambda & 0 & 0 & 0 \\ 0 & 0 & 0 & 2\mu & 0 & 0 \\ 0 & 0 & 0 & 0 & 2\mu & 0 \\ 0 & 0 & 0 & 0 & 0 & 2\mu \end{bmatrix} \tag{9.25}$$

Constitutive theories (9.16) and (9.24) are identical. In applications requiring computations, perhaps (9.24) is more convenient.

Remarks

The equilibrium Cauchy stress tensor ${}_e\bar{\boldsymbol{\sigma}}^{(0)}$ is a diagonal tensor, but ${}_e\boldsymbol{\sigma}^{[0]}$ is not a diagonal tensor. Physics of pure volume change is in the constitutive theory for ${}_e\bar{\boldsymbol{\sigma}}^{(0)}$. This volume change is not obvious from the constitutive theory of ${}_e\boldsymbol{\sigma}^{[0]}$ as $\boldsymbol{\sigma}^{[0]}$ stress tensor is a hypothetical tensor based on assumed correspondence rule. Thus, ${}_e\boldsymbol{\sigma}^{[0]}$ and ${}_d\boldsymbol{\sigma}^{[0]}$ are hypothetical tensors too. Pure distortion of the volume in the current configuration is due to ${}_d\bar{\boldsymbol{\sigma}}^{(0)}$.

9.2.2 Finite deformation, finite strain, compressible, isothermal: reversible deformation

This deformation physics is similar to that defined in Section 9.2.1 except that the deformation process is isothermal, i.e., change in volume and/or stress due to non-isothermal effects are not considered. Here also we consider additive decomposition of $\boldsymbol{\sigma}^{[0]}$ into ${}_e\boldsymbol{\sigma}^{[0]}$ and ${}_d\boldsymbol{\sigma}^{[0]}$. Constitutive theory for ${}_e\boldsymbol{\sigma}^{[0]}$ accounts for change in volume (compressibility) and the constitutive theory for ${}_d\boldsymbol{\sigma}^{[0]}$ describes change in shape of the volume or distortion of the volume.

In this case, temperature θ is not an argument tensor of the constitutive variable and does not appear in the entropy inequality either. Following Chapter 8, the constitutive theory for ${}_e\boldsymbol{\sigma}^{[0]}$ is given by ((8.70) and (8.68))

$${}_e\boldsymbol{\sigma}^{[0]} = |\boldsymbol{J}| p(\rho)(\boldsymbol{J}^T \cdot \boldsymbol{J})^{-1} \tag{9.26}$$

$$p(\rho) = -\rho^2 \frac{\partial \Phi(\rho)}{\partial \rho} \tag{9.27}$$

and the reduced form of the entropy inequality is given by (8.83) in the absence of $\frac{\boldsymbol{q}\cdot\boldsymbol{g}}{\theta}$ term.

$$-{}_d\boldsymbol{\sigma}^{[0]} : \dot{\boldsymbol{\varepsilon}}_{[0]} \leq 0 \tag{9.28}$$

$$\text{in which} \quad {}_d\boldsymbol{\sigma}^{[0]} = {}_d\boldsymbol{\sigma}^{[0]}(\boldsymbol{\varepsilon}_{[0]}) \tag{9.29}$$

9.2.2.1 Constitutive theory for ${}_d\boldsymbol{\sigma}^{[0]}$

We derive constitutive theory for ${}_d\boldsymbol{\sigma}^{[0]}$ using (9.29) and representation theorem. The combined generators and invariants of the arguments of ${}_d\boldsymbol{\sigma}^{[0]}$ are given by (9.8) and (9.9) ($N = 2$, $M = 3$). Using representation theorem, we can write

$$ {}_d\boldsymbol{\sigma}^{[0]} = {}^{\sigma}\underset{\sim}{\alpha}^0 \boldsymbol{I} + \sum_{i=1}^{N} {}^{\sigma}\underset{\sim}{\alpha}^i ({}^{\sigma}\underset{\sim}{\boldsymbol{G}}^i) \tag{9.30} $$

Following Section 9.2.1.1, we can write

$$ {}_d\boldsymbol{\sigma}^{[0]} = \underset{\sim}{\sigma}^0\Big|_{\underset{\sim}{\Omega}} \boldsymbol{I} + \sum_{j=1}^{M} {}^{\sigma}\underset{\sim}{a}_j ({}^{\sigma}\underset{\sim}{I}^j)\boldsymbol{I} + \sum_{i=1}^{N} {}^{\sigma}\underset{\sim}{b}_i {}^{\sigma}\underset{\sim}{\boldsymbol{G}}^i + \sum_{i=1}^{N}\sum_{j=1}^{M} {}^{\sigma}\underset{\sim}{c}_{ij} ({}^{\sigma}\underset{\sim}{I}^j) {}^{\sigma}\underset{\sim}{\boldsymbol{G}}^i \tag{9.31} $$

in which ${}^{\sigma}\underset{\sim}{\alpha}^i = {}^{\sigma}\underset{\sim}{\alpha}^i({}^{\sigma}\underset{\sim}{I}^j)$; $j = 1, 2, \dots, M$. We note that ${}^{\sigma}\underset{\sim}{\alpha}^i$ are not function of temperature θ. The remaining details of deriving material coefficients and simplified theories follow Section 9.2.1.1 (omitted here).

9.2.3 Finite deformation, finite strain, incompressible, non-isothermal

Due to finite deformation, finite strain, valid stress and strain measures remain $\boldsymbol{\sigma}^{[0]}$ and $\boldsymbol{\varepsilon}_{[0]}$. The incompressibility condition implies no change in volume. We perform stress decomposition of $\boldsymbol{\sigma}^{[0]}$ into ${}_e\boldsymbol{\sigma}^{[0]}$ and ${}_d\boldsymbol{\sigma}^{[0]}$. As usual volumetric and thermal physics is incorporated in ${}_e\boldsymbol{\sigma}^{[0]}$ and ${}_d\boldsymbol{\sigma}^{[0]}$ describes distortional physics. There are many applications in which the motion of the material points is finite, strain may or may not be finite but the continua may be treated incompressible due to an insignificant change in volume. A slender cantilever beam subjected to moment at the free end bending into an arc of a circle and eventually deforming into a full circle with progressively increasing moments is a good example. The material points in this example experience a finite motion but the strains may be moderate with virtually little or no compressibility. This implies that the constitutive theory for equilibrium stress ${}_e\boldsymbol{\sigma}^{[0]}$ must enforce incompressibility and non-isothermal physics and the constitutive theory for deviatoric stress ${}_d\boldsymbol{\sigma}^{[0]}$ should result in distortion of volume.

Following Chapter 8, the constitutive theory for ${}_e\boldsymbol{\sigma}^{[0]}$ is given by (equation (8.73))

$$ {}_e\boldsymbol{\sigma}^{[0]} = |\boldsymbol{J}| p(\theta) (\boldsymbol{J}^T \cdot \boldsymbol{J})^{-1} \tag{9.32} $$

and the reduced form of the entropy inequality is given by (equation (8.83))

$$ -{}_d\boldsymbol{\sigma}^{[0]} : \dot{\boldsymbol{\varepsilon}}_{[0]} + \frac{\boldsymbol{q} \cdot \boldsymbol{g}}{\theta} \leq 0 \tag{9.33} $$

in which

$$ {}_d\boldsymbol{\sigma}^{[0]} = {}_d\boldsymbol{\sigma}^{[0]}(\boldsymbol{\varepsilon}_{[0]}, \theta) \tag{9.34} $$
$$ \boldsymbol{q} = \boldsymbol{q}(\boldsymbol{g}, \theta) \tag{9.35} $$

Constitutive theory for ${}_d\boldsymbol{\sigma}^{[0]}$ is derived using (9.34) and the representation theorem.

9.2.3.1 Constitutive theory for ${}_d\boldsymbol{\sigma}^{[0]}$

Since the argument tensors of ${}_d\boldsymbol{\sigma}^{[0]}$ in (9.34) are the same as those in (9.4), the constitutive theory for ${}_d\boldsymbol{\sigma}^{[0]}$ and its simplified forms are exactly the same as those in Section 9.2.1.1, hence, it is not repeated here.

9.2.4 Finite deformation, finite strain, incompressible, isothermal: reversible deformation

The stress measure $\boldsymbol{\sigma}^{[0]}$ and strain measure $\boldsymbol{\varepsilon}_{[0]}$ remain valid here as well. We decompose $\boldsymbol{\sigma}^{[0]}$ into ${}_e\boldsymbol{\sigma}^{[0]}$ and ${}_d\boldsymbol{\sigma}^{[0]}$. The constitutive theory for ${}_e\boldsymbol{\sigma}^{[0]}$ is given by (equation (8.74))

$$ {}_e\boldsymbol{\sigma}^{[0]} = |\boldsymbol{J}|p(\boldsymbol{J}^T \cdot \boldsymbol{J})^{-1} \tag{9.36}$$

Where p is mechanical pressure. The reduced form of the entropy inequality is given by (equation (8.83) in the absence of $\frac{\boldsymbol{q}\cdot\boldsymbol{g}}{\theta}$ term).

$$ -{}_d\boldsymbol{\sigma}^{[0]} : \dot{\boldsymbol{\varepsilon}}_{[0]} \leq 0 \tag{9.37}$$

$$ \text{in which} \quad {}_d\boldsymbol{\sigma}^{[0]} = {}_d\boldsymbol{\sigma}^{[0]}(\boldsymbol{\varepsilon}_{[0]}) \tag{9.38}$$

The constitutive theory for ${}_d\boldsymbol{\sigma}^{[0]}$ is derived using (9.38) and representation theorem.

9.2.4.1 Constitutive theory for ${}_d\boldsymbol{\sigma}^{[0]}$

Since (9.38) is same as (9.29), the derivation of the constitutive theory for ${}_d\boldsymbol{\sigma}^{[0]}$ remains same as in Section 9.2.2.1, hence not repeated here.

9.2.5 Small deformation, small strain

In this deformation physics, the Cauchy stress tensor $\boldsymbol{\sigma}$ (basis independent) and linear part $\boldsymbol{\varepsilon}$ of the Green's strain tensor ($\boldsymbol{\varepsilon}_{[0]}$) are work conjugate measures. We consider non-isothermal and isothermal deformation physics.

9.2.5.1 Non-isothermal case

We consider additive decomposition of the Cauchy stress tensor $\boldsymbol{\sigma}$ into the equilibrium Cauchy stress tensor ${}_e\boldsymbol{\sigma}$ and deviatoric Cauchy stress tensor ${}_d\boldsymbol{\sigma}$. The constitutive theory for ${}_e\boldsymbol{\sigma}$ describes volumetric deformation whereas ${}_d\boldsymbol{\sigma}$ addresses distortional deformation physics. We consider non-isothermal and isothermal deformation physics in the following sections.

Following details in Chapter 8, for this deformation physics, the constitutive theories for the equilibrium Cauchy stress tensor ${}_e\boldsymbol{\sigma}$ are given by (equation (8.75)):

$$ {}_e\boldsymbol{\sigma} = p(\rho,\theta)\boldsymbol{\delta} \quad ; \quad \text{compressible} \tag{9.39}$$

$$ {}_e\boldsymbol{\sigma} = p(\theta)\boldsymbol{\delta} \quad ; \quad \text{incompressible} \tag{9.40}$$

and the reduced form of the entropy inequality is given by (equation (8.86))

$$ -{}_d\boldsymbol{\sigma} : \dot{\boldsymbol{\varepsilon}} + \frac{\boldsymbol{q}\cdot\boldsymbol{g}}{\theta} \leq 0 \tag{9.41}$$

$$ \text{in which} \quad {}_d\boldsymbol{\sigma} = {}_d\boldsymbol{\sigma}(\boldsymbol{\varepsilon},\theta) \tag{9.42}$$

The constitutive theory for ${}_d\boldsymbol{\sigma}$ is derived using (9.42) and representation theorem.

We can write

$$ {}_d\boldsymbol{\sigma} = {}^{\sigma}\underset{\sim}{\alpha}^0 \boldsymbol{I} + {}^{\sigma}\underset{\sim}{\alpha}^1 \boldsymbol{\varepsilon} + {}^{\sigma}\underset{\sim}{\alpha}^2 \boldsymbol{\varepsilon}^2 \tag{9.43} $$

$$ \text{in which} \quad \boldsymbol{\varepsilon} = {}^{\sigma}\underset{\sim}{\boldsymbol{G}}^1 \quad \text{and} \quad \boldsymbol{\varepsilon}^2 = {}^{\sigma}\underset{\sim}{\boldsymbol{G}}^2 \; ; \; N = 2 \tag{9.44} $$

are the combined generators (symmetric tensors of rank two) of the argument tensors $\boldsymbol{\varepsilon}$ and θ and

$$ {}^{\sigma}\underset{\sim}{\alpha}^i = {}^{\sigma}\underset{\sim}{\alpha}^i \left({}^{\sigma}\underset{\sim}{I}^j, \theta \right) \quad ; \quad j = 1, 2, 3 \tag{9.45} $$

$$ \text{in which} \quad {}^{\sigma}\underset{\sim}{I}^1 = I_{\varepsilon} \quad , \quad {}^{\sigma}\underset{\sim}{I}^2 = II_{\varepsilon} \quad , \quad {}^{\sigma}\underset{\sim}{I}^3 = III_{\varepsilon} \quad ; \quad M = 3 \tag{9.46} $$

are the combined invariants of the argument tensors $\boldsymbol{\varepsilon}$ and θ. The material coefficients are determined in the usual way by considering Taylor series expansion of each ${}^{\sigma}\underset{\sim}{\alpha}^i$ in ${}^{\sigma}\underset{\sim}{I}^j$ about a known configuration $\underline{\Omega}$ and retaining only up to linear terms in ${}^{\sigma}\underset{\sim}{I}^j$. Taylor series expansion in θ is not considered as ${}_e\boldsymbol{\sigma}$ already accounts for it.

$$ {}^{\sigma}\underset{\sim}{\alpha}^i = {}^{\sigma}\underset{\sim}{\alpha}^i \Big|_{\underline{\Omega}} + \sum_{j=1}^{M} \left. \frac{\partial {}^{\sigma}\underset{\sim}{\alpha}^i}{\partial {}^{\sigma}\underset{\sim}{I}^j} \right|_{\underline{\Omega}} \left({}^{\sigma}\underset{\sim}{I}^j - {}^{\sigma}\underset{\sim}{I}^j \Big|_{\underline{\Omega}} \right) \quad ; \quad i = 0, 1, \dots, N \quad ; \quad j = 1, 2, \dots, M \tag{9.47} $$

substituting ${}^{\sigma}\underset{\sim}{\alpha}^i$ from (9.47) in (9.43) and collecting coefficients of the terms defined in the current configuration (and introducing new notation, see Section 9.2.1.1)

$$ {}_d\boldsymbol{\sigma} = \underset{\sim}{\sigma}^0 \Big|_{\underline{\Omega}} \boldsymbol{I} + \sum_{j=1}^{M} {}^{\sigma}\underset{\sim}{a}_j ({}^{\sigma}\underset{\sim}{I}^j) \boldsymbol{I} + \sum_{i=1}^{N} {}^{\sigma}\underset{\sim}{b}_i {}^{\sigma}\underset{\sim}{\boldsymbol{G}}^i + \sum_{i=1}^{N} \sum_{j=1}^{M} {}^{\sigma}\underset{\sim}{c}_{ij} ({}^{\sigma}\underset{\sim}{I}^j) {}^{\sigma}\underset{\sim}{\boldsymbol{G}}^i \tag{9.48} $$

Equations (9.39), (9.40) and (9.48) are the constitutive theories for ${}_e\boldsymbol{\sigma}$ and ${}_d\boldsymbol{\sigma}$. ${}^{\sigma}\underset{\sim}{a}_j$, ${}^{\sigma}\underset{\sim}{b}_i$ and ${}^{\sigma}\underset{\sim}{c}_{ij}$ are material coefficients that can be functions of ${}^{\sigma}\underset{\sim}{I}^j \Big|_{\underline{\Omega}}$; $j = 0, 1, \dots, M$ and $\theta|_{\underline{\Omega}}$.

Simplified linear theory

Simplified form of linear constitutive theory for the deviatoric Cauchy stress tensor ${}_d\boldsymbol{\sigma}$ in which non-linear terms are neglected is easily obtained using (9.48) and is given by

$$ {}_d\boldsymbol{\sigma} = \underset{\sim}{\sigma}^0 \Big|_{\underline{\Omega}} \boldsymbol{I} + 2\mu\boldsymbol{\varepsilon} + \lambda(\operatorname{tr}\boldsymbol{\varepsilon})\boldsymbol{I} \tag{9.49} $$

$$ \text{in which} \quad 2\mu = {}^{\sigma}\underset{\sim}{b}_1 \quad , \quad \lambda = {}^{\sigma}\underset{\sim}{a}_1 \tag{9.50} $$

We note that μ and λ in this small deformation, small strain theory are not the same as μ and λ in (9.16) for finite deformation, finite strain. Using Voigt's notation, we can write (9.49) as

$$ \{{}_d\sigma\} = \{\sigma^0\} + [D]\{\varepsilon\} \tag{9.51} $$

in which

$$ \{\sigma^0\} = \underset{\sim}{\sigma}^0 \Big|_{\underline{\Omega}} [1, 1, 1, 0, 0, 0]^T \tag{9.52} $$

and

$$[D] = \begin{bmatrix} 2\mu+\lambda & \lambda & \lambda & 0 & 0 & 0 \\ \lambda & 2\mu+\lambda & \lambda & 0 & 0 & 0 \\ \lambda & \lambda & 2\mu+\lambda & 0 & 0 & 0 \\ 0 & 0 & 0 & 2\mu & 0 & 0 \\ 0 & 0 & 0 & 0 & 2\mu & 0 \\ 0 & 0 & 0 & 0 & 0 & 2\mu \end{bmatrix} \tag{9.53}$$

μ and λ are called Lamé's constants and are related to the modulus of elasticity E and Poisson's ratio ν by:

$$\lambda = \frac{\nu E}{(1+\nu)(1-2\nu)} \quad ; \quad \mu = \frac{E}{2(1+\nu)} \tag{9.54}$$

μ is the shear modulus. $\{\sigma^0\}$ is initial stress field. The constitutive theory ((9.49) or (9.51)) is called Hooke's law in which stresses are linearly related to strains.

9.2.5.2 Isothermal case: reversible deformation

Following the derivations in Chapter 8, for this deformation physics, the constitutive theories for equilibrium Cauchy stress ${}_e\boldsymbol{\sigma}$ are given by (equation (8.75))

$${}_e\boldsymbol{\sigma} = p(\rho)\boldsymbol{\delta} \quad ; \quad \text{compressible} \tag{9.55}$$

$${}_e\boldsymbol{\sigma} = p\boldsymbol{\delta} \quad ; \quad \text{incompressible} \tag{9.56}$$

Where p is mechanical pressure. The reduced form of the entropy inequality is given by ((8.88) without the $\frac{\boldsymbol{q}\cdot\boldsymbol{g}}{\theta}$ term)

$$-{}_d\boldsymbol{\sigma} : \dot{\boldsymbol{\varepsilon}} \leq 0 \tag{9.57}$$

$$\text{in which} \quad {}_d\boldsymbol{\sigma} = {}_d\boldsymbol{\sigma}(\boldsymbol{\varepsilon}) \tag{9.58}$$

The constitutive theory for ${}_d\boldsymbol{\sigma}$ is derived using (9.58) and representation theorem. The derivations are exactly the same as in Section 9.2.5.1 except θ is removed from all expressions in Section 9.2.5.1. The final constitutive theory is given by (9.48), but the material coefficients in this case are only dependent on invariants ${}^{\sigma}\underset{\sim}{I}^j\big|_{\underline{\Omega}}$. The simplified linear theory in Section 9.2.5.1, hold here as well.

9.3 Other approaches of deriving constitutive theories (not necessarily thermodynamic)

The approaches of deriving constitutive theories considered in this section do not necessarily make use of CBL, in particular the entropy inequality. Thus, the constitutive theories derived using these approaches may violate thermodynamic equilibrium. However, in many instances such constitutive theories are necessitated due to complexity of material behavior that is beyond the scope of CCM thermodynamic approach of deriving constitutive theories. We keep in mind that the constitutive theories derived using this approach may or may not be admissible in the CBL of CCM derived in Chapter 7. We consider some of the commonly used approaches in the following.

9.3.1 Constitutive theory for $\boldsymbol{\sigma}^{[0]}$ using Φ: reversible

The purpose of providing details of this derivation is multifold. First, we want to show that within certain assumptions it is possible to derive a constitutive theory for ${}_d\boldsymbol{\sigma}^{[0]}$ using Helmholtz free energy density. The resulting constitutive theory is a linear combination of the generators $\boldsymbol{I}$, $\boldsymbol{\varepsilon}_{[0]}$ and $\boldsymbol{\varepsilon}_{[0]}^2$ and the coefficients in the linear combination are functions of the invariants of $\boldsymbol{\varepsilon}_{[0]}$ (i.e., $I_{\varepsilon_{[0]}}$, $II_{\varepsilon_{[0]}}$, $III_{\varepsilon_{[0]}}$). Secondly, when this constitutive theory is compared with one obtained using ${}_d\boldsymbol{\sigma}^{[0]} = {}_d\boldsymbol{\sigma}^{[0]}(\boldsymbol{\varepsilon}_{[0]})$ and the representation theorem, we find that the combined generators in the linear combination in two approaches are the same ($\boldsymbol{I}$, $\boldsymbol{\varepsilon}_{[0]}$, $\boldsymbol{\varepsilon}_{[0]}^2$). Thus, the coefficients in the linear combination must be same as well, confirming that the coefficients in the linear combination in the derivation using representation theorem are functions of the invariants of $\boldsymbol{\varepsilon}_{[0]}$ (i.e., $I_{\varepsilon_{[0]}}$, $II_{\varepsilon_{[0]}}$, $III_{\varepsilon_{[0]}}$). This is rather simple, but important independent proof of the dependence of the coefficients in linear combination of the generator on the invariants of the argument tensors of the constitutive tensor in the derivation using representation theorem.

If we consider finite deformation, finite strain but the matter to be incompressible and the deformation physics to be isothermal, then we have seen that (from equation (8.80))

$$ {}_e\boldsymbol{\sigma}^{(0)} = p\boldsymbol{\delta} \tag{9.59} $$

$$ \text{Hence} \quad {}_e\boldsymbol{\sigma}^{[0]} = |\boldsymbol{J}|p(\boldsymbol{J}^T \cdot \boldsymbol{J})^{-1} \tag{9.60} $$

and the reduced form of the entropy inequality is given by

$$ -{}_d\boldsymbol{\sigma}^{[0]} : \dot{\boldsymbol{\varepsilon}}_{[0]} \leq 0 \tag{9.61} $$

$$ \text{in which} \quad {}_d\boldsymbol{\sigma}^{[0]} = {}_d\boldsymbol{\sigma}^{[0]}(\boldsymbol{\varepsilon}_{[0]}) \tag{9.62} $$

The constitutive theory for ${}_d\boldsymbol{\sigma}^{[0]}$ can be derived using (9.62) and the representation theorem as shown earlier. We have also seen that under no circumstances the constitutive theory for $\boldsymbol{\sigma}^{[0]}$ is possible using Φ due to the fact that $\bar{\Phi} = \bar{\Phi}(\bar{\rho}, \bar{\theta})$ and both, volumetric deformation and distortional deformation, cannot be described by a single constitutive theory for $\boldsymbol{\sigma}^{[0]}$. *Nonetheless, if we ignore ${}_e\boldsymbol{\sigma}^{[0]}$ in this case, then ${}_d\boldsymbol{\sigma}^{[0]} = \boldsymbol{\sigma}^{[0]}$ and we can obtain the following from the entropy inequality.*

$$ \left(\rho_0 \frac{\partial \Phi}{\partial \boldsymbol{\varepsilon}_{[0]}} - \boldsymbol{\sigma}^{[0]} \right) : \dot{\boldsymbol{\varepsilon}}_{[0]} \leq 0 \tag{9.63} $$

which suggests that

$$ \boldsymbol{\sigma}^{[0]} = \rho_0 \frac{\partial \Phi(\boldsymbol{\varepsilon}_{[0]})}{\partial \boldsymbol{\varepsilon}_{[0]}} \tag{9.64} $$

We can derive the constitutive theory for $\boldsymbol{\sigma}^{[0]}$ using (9.64). The principle of frame invariance of the constitutive theories require that Φ in (9.63) be a function of the invariants of $\boldsymbol{\varepsilon}_{[0]}$ rather than $\boldsymbol{\varepsilon}_{[0]}$. Thus, we have (using principal invariants).

$$ \boldsymbol{\sigma}^{[0]} = \rho_0 \frac{\partial \Phi(I_{\varepsilon_{[0]}}, II_{\varepsilon_{[0]}}, III_{\varepsilon_{[0]}})}{\partial \boldsymbol{\varepsilon}_{[0]}} \tag{9.65} $$

We derive the constitutive theory for $\boldsymbol{\sigma}^{[0]}$ using (9.65). Recall the principal invariants of $\boldsymbol{\varepsilon}_{[0]}$.

$$\begin{aligned} I_{\varepsilon_{[0]}} &= \text{tr}([\varepsilon_{[0]}]) = (\varepsilon_{[0]})_{ll} \\ II_{\varepsilon_{[0]}} &= \frac{1}{2}\Big(\big(\text{tr}([\varepsilon_{[0]}])\big)^2 - \text{tr}([\varepsilon_{[0]}]^2)\Big) = \frac{1}{2}\big((\varepsilon_{[0]})_{ll}(\varepsilon_{[0]})_{kk} - (\varepsilon_{[0]})_{kl}(\varepsilon_{[0]})_{lk}\big) \\ III_{\varepsilon_{[0]}} &= \det([\varepsilon_{[0]}]) \end{aligned} \tag{9.66}$$

Using (9.65)

$$[{}_d\sigma^{[0]}] = \rho_0 \left(\frac{\partial \Phi}{\partial I_{\varepsilon_{[0]}}} \frac{\partial I_{\varepsilon_{[0]}}}{\partial [\varepsilon_{[0]}]} + \frac{\partial \Phi}{\partial II_{\varepsilon_{[0]}}} \frac{\partial II_{\varepsilon_{[0]}}}{\partial [\varepsilon_{[0]}]} + \frac{\partial \Phi}{\partial III_{\varepsilon_{[0]}}} \frac{\partial III_{\varepsilon_{[0]}}}{\partial [\varepsilon_{[0]}]} \right) \tag{9.67}$$

We determine $\dfrac{\partial I_{\varepsilon_{[0]}}}{\partial [\varepsilon_{[0]}]}$, $\dfrac{\partial II_{\varepsilon_{[0]}}}{\partial [\varepsilon_{[0]}]}$ and $\dfrac{\partial III_{\varepsilon_{[0]}}}{\partial [\varepsilon_{[0]}]}$ in the following.

Consider $\dfrac{\partial I_{\varepsilon_{[0]}}}{\partial [\varepsilon_{[0]}]}$:

$$\frac{\partial I_{\varepsilon_{[0]}}}{\partial (\varepsilon_{[0]})_{ij}} = \frac{\partial (\varepsilon_{[0]})_{ll}}{\partial (\varepsilon_{[0]})_{ij}} = \delta_{ij} \tag{9.68}$$

$$\text{or} \quad \frac{\partial I_{\varepsilon_{[0]}}}{\partial [\varepsilon_{[0]}]} = [I] \tag{9.69}$$

Consider $\dfrac{\partial II_{\varepsilon_{[0]}}}{\partial (\varepsilon_{[0]})_{ij}}$:

$$\begin{aligned} \frac{\partial II_{\varepsilon_{[0]}}}{\partial (\varepsilon_{[0]})_{ij}} &= \frac{1}{2}\Big(- \frac{\partial (\varepsilon_{[0]})_{kl}}{\partial (\varepsilon_{[0]})_{ij}} (\varepsilon_{[0]})_{lk} - (\varepsilon_{[0]})_{kl} \frac{\partial (\varepsilon_{[0]})_{lk}}{\partial (\varepsilon_{[0]})_{ij}} \\ &\quad + \frac{\partial (\varepsilon_{[0]})_{ll}}{\partial (\varepsilon_{[0]})_{ij}} (\varepsilon_{[0]})_{kk} + (\varepsilon_{[0]})_{ll} \frac{\partial (\varepsilon_{[0]})_{kk}}{\partial (\varepsilon_{[0]})_{ij}} \Big) \\ &= \frac{1}{2}\left(-(\varepsilon_{[0]})_{ij} - (\varepsilon_{[0]})_{ij} + (\varepsilon_{[0]})_{kk}\delta_{ij} + (\varepsilon_{[0]})_{ll}\delta_{ij} \right) \\ &= -(\varepsilon_{[0]})_{ij} + (\varepsilon_{[0]})_{kk}\delta_{ij} \end{aligned} \tag{9.70}$$

$$\text{or} \quad \frac{\partial II_{\varepsilon_{[0]}}}{\partial [\varepsilon_{[0]}]} = -[\varepsilon_{[0]}] + I_{\varepsilon_{[0]}}[I] \tag{9.71}$$

Consider $\dfrac{\partial III_{\varepsilon_{[0]}}}{\partial (\varepsilon_{[0]})_{ij}}$: Recall that $III_{\varepsilon_{[0]}} = \det[\varepsilon_{[0]}]$. Thus, we can write

$$\frac{\partial III_{\varepsilon_{[0]}}}{\partial [\varepsilon_{[0]}]} = (\det[\varepsilon_{[0]}])[\varepsilon_{[0]}]^{-1} = III_{\varepsilon_{[0]}}[\varepsilon_{[0]}]^{-1} \tag{9.72}$$

Substituting from (9.69), (9.71) and (9.72) into (9.67)

$$[{}_d\sigma^{[0]}] = \rho_0 \left(\frac{\partial \Phi}{\partial I_{\varepsilon_{[0]}}}[I] + \frac{\partial \Phi}{\partial II_{\varepsilon_{[0]}}}(-[\varepsilon_{[0]}] + I_{\varepsilon_{[0]}}[I]) + \frac{\partial \Phi}{\partial III_{\varepsilon_{[0]}}} III_{\varepsilon_{[0]}}[\varepsilon_{[0]}]^{-1} \right) \tag{9.73}$$

Collecting the coefficients of $[I]$, $[\varepsilon_{[0]}]$ and $[\varepsilon_{[0]}]^{-1}$

$$\begin{aligned}[{}_d\sigma^{[0]}] = {}&\rho_0 \Big(\frac{\partial \Phi}{\partial I_{\varepsilon_{[0]}}} + \frac{\partial \Phi}{\partial II_{\varepsilon_{[0]}}} I_{\varepsilon_{[0]}} \Big) [I] \\ &+ \Big(-\rho_0 \frac{\partial \Phi}{\partial II_{\varepsilon_{[0]}}} \Big) [\varepsilon_{[0]}] + \Big(\rho_0 \frac{\partial \Phi}{\partial III_{\varepsilon_{[0]}}} III_{\varepsilon_{[0]}} \Big) [\varepsilon_{[0]}]^{-1}\end{aligned} \tag{9.74}$$

Let

$$\begin{aligned}{}^{\sigma}\alpha_0 &= \rho_0 \Big(\frac{\partial \Phi}{\partial I_{\varepsilon_{[0]}}} + \frac{\partial \Phi}{\partial II_{\varepsilon_{[0]}}} I_{\varepsilon_{[0]}} \Big) \\ {}^{\sigma}\alpha_1 &= -\rho_0 \frac{\partial \Phi}{\partial II_{\varepsilon_{[0]}}} \\ {}^{\sigma}\alpha_{-1} &= \rho_0 \frac{\partial \Phi}{\partial III_{\varepsilon_{[0]}}} III_{\varepsilon_{[0]}}\end{aligned} \tag{9.75}$$

Then

$$[{}_d\sigma^{[0]}] = {}^{\sigma}\alpha_0 [I] + {}^{\sigma}\alpha_1 [\varepsilon_{[0]}] + {}^{\sigma}\alpha_{-1} [\varepsilon_{[0]}]^{-1} \tag{9.76}$$

Recall, Hamilton-Cayley theorem

$$[\varepsilon_{[0]}]^3 - I_{\varepsilon_{[0]}} [\varepsilon_{[0]}]^2 + II_{\varepsilon_{[0]}} [\varepsilon_{[0]}] - III_{\varepsilon_{[0]}} [I] = 0 \tag{9.77}$$

For a non-singular $[\varepsilon_{[0]}]$, i.e., $III_{\varepsilon_{[0]}} \neq 0$, we can solve for $[\varepsilon_{[0]}]^{-1}$ from (9.77) to obtain

$$[\varepsilon_{[0]}]^{-1} = \frac{1}{III_{\varepsilon_{[0]}}} \big([\varepsilon_{[0]}]^2 - I_{\varepsilon_{[0]}} [\varepsilon_{[0]}] + II_{\varepsilon_{[0]}} [I] \big) \tag{9.78}$$

Substituting from (9.78) into (9.76)

$$[{}_d\sigma^{[0]}] = {}^{\sigma}\alpha_0 [I] + {}^{\sigma}\alpha_1 [\varepsilon_{[0]}] + \frac{{}^{\sigma}\alpha_{-1}}{III_{\varepsilon_{[0]}}} \big([\varepsilon_{[0]}]^2 - I_{\varepsilon_{[0]}} [\varepsilon_{[0]}] + II_{\varepsilon_{[0]}} [I] \big) \tag{9.79}$$

Collecting coefficients of $[I]$, $[\varepsilon_{[0]}]$ and $[\varepsilon_{[0]}]^2$ in (9.79)

$$\begin{aligned}[{}_d\sigma^{[0]}] = {}&\Big({}^{\sigma}\alpha_0 + \frac{{}^{\sigma}\alpha_{-1}}{III_{\varepsilon_{[0]}}} II_{\varepsilon_{[0]}} \Big) [I] \\ &+ \Big({}^{\sigma}\alpha_1 - \frac{{}^{\sigma}\alpha_{-1}}{III_{\varepsilon_{[0]}}} I_{\varepsilon_{[0]}} \Big) [\varepsilon_{[0]}] + \Big(\frac{{}^{\sigma}\alpha_{-1}}{III_{\varepsilon_{[0]}}} \Big) [\varepsilon_{[0]}]^2\end{aligned} \tag{9.80}$$

Let

$${}^{\sigma}\tilde{\alpha}_0 = {}^{\sigma}\alpha_0 + \frac{{}^{\sigma}\alpha_{-1}}{III_{\varepsilon_{[0]}}} II_{\varepsilon_{[0]}} \quad ; \quad {}^{\sigma}\tilde{\alpha}_1 = {}^{\sigma}\alpha_1 - \frac{{}^{\sigma}\alpha_{-1}}{III_{\varepsilon_{[0]}}} I_{\varepsilon_{[0]}} \quad ; \quad {}^{\sigma}\tilde{\alpha}_2 = \frac{{}^{\sigma}\alpha_{-1}}{III_{\varepsilon_{[0]}}} \tag{9.81}$$

Using (9.81) in (9.80) we can write

$$\boldsymbol{\sigma}^{[0]} = {}^{\sigma}\tilde{\alpha}_0 \boldsymbol{I} + {}^{\sigma}\tilde{\alpha}_1 \boldsymbol{\varepsilon}_{[0]} + {}^{\sigma}\tilde{\alpha}_2 (\boldsymbol{\varepsilon}_{[0]})^2 \tag{9.82}$$

Since ${}^{\sigma}\alpha_i = {}^{\sigma}\alpha_i(\rho_0\,,\, I_{\varepsilon_{[0]}}\,,\, II_{\varepsilon_{[0]}}\,,\, III_{\varepsilon_{[0]}})$; $i = 0, 1, -1$ (from (9.75)) we can conclude from (9.81) that ${}^{\sigma}\tilde{\alpha}_i = {}^{\sigma}\tilde{\alpha}_i(\rho_0\,,\, I_{\varepsilon_{[0]}}\,,\, II_{\varepsilon_{[0]}}\,,\, III_{\varepsilon_{[0]}})$; $i = 0, 1, 2$. Thus, we have a constitutive theory for ${}_d\boldsymbol{\sigma}^{[0]}$ in terms of $\boldsymbol{\varepsilon}_{[0]}$ in (9.76) or (9.82). Equation (9.82) is preferred over (9.76) as it does not involve the inverse of $\boldsymbol{\varepsilon}_{[0]}$.

Both (9.76) and (9.82) hold in the current configuration. Thus, the coefficients in (9.76) and (9.82) are functions of the invariants of $\boldsymbol{\varepsilon}_{[0]}$ in the current configuration. These are not material coefficients as the material coefficients must be defined in a known configuration. Material coefficients can be derived using the usual approach (Chapter 8).

Remarks

(1) If we consider (9.60) and use representation theorem, then we have

$$ {}_d\boldsymbol{\sigma}^{[0]} = {}^\sigma\underset{\sim}{\alpha}^0 \boldsymbol{I} + \sum_{i=1}^{N} {}^\sigma\underset{\sim}{\alpha}^i \, {}^\sigma\underset{\sim}{\boldsymbol{G}}^i \tag{9.83} $$

In this case, $N = 2$ and ${}^\sigma\underset{\sim}{\boldsymbol{G}}^1 = \boldsymbol{\varepsilon}_{[0]}$ and ${}^\sigma\underset{\sim}{\boldsymbol{G}}^2 = \boldsymbol{\varepsilon}^2_{[0]}$. Thus, (9.83) becomes

$$ {}_d\boldsymbol{\sigma}^{[0]} = {}^\sigma\underset{\sim}{\alpha}^0 \boldsymbol{I} + {}^\sigma\underset{\sim}{\alpha}^1 \boldsymbol{\varepsilon}_{[0]} + {}^\sigma\underset{\sim}{\alpha}^1 \boldsymbol{\varepsilon}^2_{[0]} \tag{9.84} $$

Comparing (9.84) with (9.82), we conclude that

$$ {}^\sigma\underset{\sim}{\alpha}^i = {}^\sigma\underset{\sim}{\alpha}^i (I_{\varepsilon_{[0]}}, II_{\varepsilon_{[0]}}, III_{\varepsilon_{[0]}}) \quad ; \quad i = 0, 1, 2 \tag{9.85} $$

That is, ${}^\sigma\underset{\sim}{\alpha}^i$ in (9.84) are exactly the same as ${}^\sigma\underset{\sim}{\alpha}^i$ in (9.82). Hence, ${}^\sigma\underset{\sim}{\alpha}^i$ in (9.84) must be functions of the invariants of $\boldsymbol{\varepsilon}_{[0]}$.

(2) Though derivation is based on Helmholtz free energy density, we are able to establish dependence of the coefficients in the linear combination of the combined generators in the representation theorem on the combined invariants of the argument tensors of ${}_d\boldsymbol{\sigma}^{[0]}$.

9.3.2 Constitutive theory for $\boldsymbol{\sigma}^{[0]}$ using strain energy density π

If we consider *finite deformation, finite strain but incompressible and isothermal deformation* and neglect ${}_e\boldsymbol{\sigma}^{[0]}$ (as in Section 9.3.1), then ${}_d\boldsymbol{\sigma}^{[0]} = \boldsymbol{\sigma}^{[0]}$ and we can write the strain energy density function $\pi = \pi(\boldsymbol{\sigma}^{[0]}, \boldsymbol{\varepsilon}_{[0]})$. For a volume of matter V we have

$$ \text{the rate of strain energy} = \int_V \boldsymbol{\sigma}^{[0]} : \dot{\boldsymbol{\varepsilon}}_{[0]} dV \tag{9.86} $$

If π is the strain energy density, then the rate of strain energy for volume V is also given by

$$ \text{rate of strain energy} = \frac{D}{Dt} \int_V \pi \rho_0 \, dV = \int_V \frac{D\pi}{Dt} \rho_0 \, dV \tag{9.87} $$

Equating (9.86) and (9.87) we obtain

$$ \int_V \left(\rho_0 \frac{D\pi}{Dt} - \boldsymbol{\sigma}^{[0]} : \boldsymbol{\varepsilon}_{[0]} \right) dV = 0 \tag{9.88} $$

For isotropic, homogeneous matter, volume V is arbitrary, hence we have

$$\rho_0 \frac{D\pi}{Dt} - \boldsymbol{\sigma}^{[0]} : \dot{\boldsymbol{\varepsilon}}_{[0]} = 0 \tag{9.89}$$

If we consider $\pi = \pi(\boldsymbol{\varepsilon}_{[0]})$, then

$$\frac{D\pi}{Dt} = \frac{\partial \pi}{\partial \boldsymbol{\varepsilon}_{[0]}} : \dot{\boldsymbol{\varepsilon}}_{[0]} \tag{9.90}$$

using (9.90) in (9.89)

$$\left(\rho_0 \frac{\partial \pi}{\partial \boldsymbol{\varepsilon}_{[0]}} - \boldsymbol{\sigma}^{[0]} \right) : \dot{\boldsymbol{\varepsilon}}_{[0]} \tag{9.91}$$

Clearly $\dot{\boldsymbol{\varepsilon}}_{[0]} \neq 0$, hence we have

$$\boldsymbol{\sigma}^{[0]} = \rho_0 \frac{\partial \pi(\boldsymbol{\varepsilon}_{[0]})}{\partial \boldsymbol{\varepsilon}_{[0]}} \tag{9.92}$$

Since constitutive theories must be frame invariant, instead of (9.92), we must consider π as a function of the invariants of $\boldsymbol{\varepsilon}_{[0]}$ (using principal invariants).

$$\boldsymbol{\sigma}^{[0]} = \rho_0 \frac{\partial \pi(I_{\boldsymbol{\varepsilon}_{[0]}}, II_{\boldsymbol{\varepsilon}_{[0]}}, III_{\boldsymbol{\varepsilon}_{[0]}})}{\partial \boldsymbol{\varepsilon}_{[0]}} \tag{9.93}$$

Derivation of the constitutive theory for $\boldsymbol{\sigma}^{[0]}$ using (9.93) follows exactly the same procedure as presented in Section 9.3.1 using (9.65), i.e., using

$$\boldsymbol{\sigma}^{[0]} = \rho_0 \frac{\partial \Phi(I_{\boldsymbol{\varepsilon}_{[0]}}, II_{\boldsymbol{\varepsilon}_{[0]}}, III_{\boldsymbol{\varepsilon}_{[0]}})}{\partial \boldsymbol{\varepsilon}_{[0]}} \tag{9.94}$$

Thus, using derivation of Section 9.3.1 and replacing Φ with π we can obtain the constitutive theory. The final outcome is same as (9.82).

$$\boldsymbol{\sigma}^{[0]} = {}^{\sigma}\underset{\sim}{\alpha}^{0} \boldsymbol{I} + {}^{\sigma}\underset{\sim}{\alpha}^{1} \boldsymbol{\varepsilon}_{[0]} + {}^{\sigma}\underset{\sim}{\alpha}^{2} (\boldsymbol{\varepsilon}_{[0]})^2 \tag{9.95}$$

Derivation of material coefficients and the final expression for the constitutive theory for $\boldsymbol{\sigma}^{[0]}$ follows details presented in Chapter 8.

9.3.3 Constitutive theory for $\boldsymbol{\sigma}^{[0]}$ using $\pi = \pi(\boldsymbol{\varepsilon}_{[0]})$ and Taylor series expansion: non-isotropic, non-homogeneous solid continua

Non-homogeneous and non-isotropic solid continua exist in nature. For example, wood as well as those synthesized for engineering applications, plywood, ceramics, laminated composites, etc. Thermodynamic framework of CCM is well founded for homogeneous, isotropic continua, but the extension of CCM to accommodate non-homogeneous and non-isotropic continua is generally not possible.

In this section, we only address the constitutive theory considerations for non-isotropic solid continua. It is rather obvious that the constitutive theories for such behavior is not possible using entropy inequality as it only holds for isotropic, homogeneous matter. Thus, the constitutive theories for such matter have to consider non-thermodynamic approaches. Secondly, these constitutive theories cannot be used in the differential form of the conservation and balance laws that are only valid for homogeneous, isotropic continua. The mathematical models describing equilibrium (not thermodynamic equilibrium), static or dynamic behavior of such deforming matter are generally derived using energy methods. We keep in mind that the mathematical models derived using energy methods almost always exhibit lack of thermodynamic equilibrium as these are not derived using the conservation and balance laws of CCM. There are also many other issues in this approach (see Chapter 16). Nonetheless, this approach provides means of solving engineering problems with complex material behavior that cannot be considered within the framework of CCM or NCCM.

We consider $\pi = \pi(\boldsymbol{\varepsilon}_{[0]})$ and expand π in Taylor series in $\boldsymbol{\varepsilon}_{[0]}$ about a known configuration $\underline{\Omega}$. This is valid based on the axiom of smooth neighborhood.

$$
\begin{aligned}
\pi = \pi\big|_{\underline{\Omega}} &+ \frac{\partial \pi}{\partial (\varepsilon_{[0]})_{ij}}\bigg|_{\underline{\Omega}} \Big((\varepsilon_{[0]})_{ij} - ((\varepsilon_{[0]})_{ij})_{\underline{\Omega}} \Big) \\
&+ \frac{1}{2!} \frac{\partial^2 \pi}{\partial (\varepsilon_{[0]})_{ij} \partial (\varepsilon_{[0]})_{kl}}\bigg|_{\underline{\Omega}} \Big((\varepsilon_{[0]})_{ij} - ((\varepsilon_{[0]})_{ij})_{\underline{\Omega}} \Big)\Big((\varepsilon_{[0]})_{kl} - ((\varepsilon_{[0]})_{kl})_{\underline{\Omega}} \Big) \\
&+ \frac{1}{3!} \frac{\partial^3 \pi}{\partial (\varepsilon_{[0]})_{ij} \partial (\varepsilon_{[0]})_{kl} \partial (\varepsilon_{[0]})_{pq}}\bigg|_{\underline{\Omega}} \\
&\Big((\varepsilon_{[0]})_{ij} - ((\varepsilon_{[0]})_{ij})_{\underline{\Omega}} \Big)\Big((\varepsilon_{[0]})_{kl} - ((\varepsilon_{[0]})_{kl})_{\underline{\Omega}} \Big)\Big((\varepsilon_{[0]})_{pq} - ((\varepsilon_{[0]})_{pq})_{\underline{\Omega}} \Big) + \dots
\end{aligned}
\tag{9.96}
$$

Let

$$
\begin{aligned}
\rho_0 \pi\big|_{\underline{\Omega}} &= \widetilde{C} \\
\rho_0 \frac{\partial \pi}{\partial (\varepsilon_{[0]})_{ij}}\bigg|_{\underline{\Omega}} &= \widetilde{C}_{ij} \\
\frac{\rho_0}{2!} \frac{\partial^2 \pi}{\partial (\varepsilon_{[0]})_{ij} \partial (\varepsilon_{[0]})_{kl}}\bigg|_{\underline{\Omega}} &= \widetilde{C}_{ijkl} \\
\frac{\rho_0}{3!} \frac{\partial^3 \pi}{\partial (\varepsilon_{[0]})_{ij} \partial (\varepsilon_{[0]})_{kl} \partial (\varepsilon_{[0]})_{pq}}\bigg|_{\underline{\Omega}} &= \widetilde{C}_{ijklpq}
\end{aligned}
\tag{9.97}
$$

Substituting (9.97) in (9.96)

$$
\begin{aligned}
\rho_0 \pi = \widetilde{C} &+ \widetilde{C}_{ij} \Big((\varepsilon_{[0]})_{ij} - ((\varepsilon_{[0]})_{ij})_{\underline{\Omega}} \Big) \\
&+ \widetilde{C}_{ijkl} \Big((\varepsilon_{[0]})_{ij} - ((\varepsilon_{[0]})_{ij})_{\underline{\Omega}} \Big)\Big((\varepsilon_{[0]})_{kl} - ((\varepsilon_{[0]})_{kl})_{\underline{\Omega}} \Big) \\
&+ \widetilde{C}_{ijklpq} \Big((\varepsilon_{[0]})_{ij} - ((\varepsilon_{[0]})_{ij})_{\underline{\Omega}} \Big)\Big((\varepsilon_{[0]})_{kl} - ((\varepsilon_{[0]})_{kl})_{\underline{\Omega}} \Big)\Big((\varepsilon_{[0]})_{pq} - ((\varepsilon_{[0]})_{pk})_{\underline{\Omega}} \Big) \\
&+ \dots
\end{aligned}
\tag{9.98}
$$

Recall, from (9.92)

$$
\sigma_{mn}^{[0]} = \rho_0 \frac{\partial \pi}{\partial (\varepsilon_{[0]})_{mn}} = \frac{\partial (\rho_0 \pi)}{\partial (\varepsilon_{[0]})_{mn}}
\tag{9.99}
$$

Differentiating $\rho_0 \pi$ in (9.98) with respect to $\boldsymbol{\varepsilon}_{[0]}$ and noting that

$$
\begin{aligned}
&\frac{\partial \widetilde{C}}{\partial (\varepsilon_{[0]})_{mn}} = 0 \quad ; \quad \frac{\partial \widetilde{C}_{ij}}{\partial (\varepsilon_{[0]})_{mn}} = 0 \quad ; \quad \frac{\partial \widetilde{C}_{ijkl}}{\partial (\varepsilon_{[0]})_{mn}} = 0 \quad ; \quad \frac{\partial \widetilde{C}_{ijklpq}}{\partial (\varepsilon_{[0]})_{mn}} = 0 \\
&\frac{\partial}{\partial (\varepsilon_{[0]})_{mn}} \Big((\varepsilon_{[0]})_{ij} - ((\varepsilon_{[0]})_{ij})_{\underline{\Omega}} \Big) = \delta_{im}\delta_{jn} \\
&\frac{\partial}{\partial (\varepsilon_{[0]})_{mn}} \Big((\varepsilon_{[0]})_{ij} - ((\varepsilon_{[0]})_{ij})_{\underline{\Omega}} \Big)\Big((\varepsilon_{[0]})_{kl} - ((\varepsilon_{[0]})_{kl})_{\underline{\Omega}} \Big) \\
&\quad = \delta_{im}\delta_{jn}\Big((\varepsilon_{[0]})_{kl} - ((\varepsilon_{[0]})_{kl})_{\underline{\Omega}} \Big) + \delta_{km}\delta_{ln}\Big((\varepsilon_{[0]})_{ij} - ((\varepsilon_{[0]})_{ij})_{\underline{\Omega}} \Big) \\
&\frac{\partial}{\partial (\varepsilon_{[0]})_{mn}} \Big((\varepsilon_{[0]})_{ij} - ((\varepsilon_{[0]})_{ij})_{\underline{\Omega}} \Big)\Big((\varepsilon_{[0]})_{kl} - ((\varepsilon_{[0]})_{kl})_{\underline{\Omega}} \Big)\Big((\varepsilon_{[0]})_{pq} - ((\varepsilon_{[0]})_{pq})_{\underline{\Omega}} \Big) \\
&\quad = \delta_{im}\delta_{jn}\Big((\varepsilon_{[0]})_{kl} - ((\varepsilon_{[0]})_{kl})_{\underline{\Omega}} \Big)\Big((\varepsilon_{[0]})_{pq} - ((\varepsilon_{[0]})_{pq})_{\underline{\Omega}} \Big) \\
&\qquad + \delta_{km}\delta_{ln}\Big((\varepsilon_{[0]})_{ij} - ((\varepsilon_{[0]})_{ij})_{\underline{\Omega}} \Big)\Big((\varepsilon_{[0]})_{pq} - ((\varepsilon_{[0]})_{pq})_{\underline{\Omega}} \Big) \\
&\qquad + \delta_{pm}\delta_{qn}\Big((\varepsilon_{[0]})_{ij} - ((\varepsilon_{[0]})_{ij})_{\underline{\Omega}} \Big)\Big((\varepsilon_{[0]})_{kl} - ((\varepsilon_{[0]})_{kl})_{\underline{\Omega}} \Big)
\end{aligned}
\tag{9.100}
$$

Substituting (9.100) in (9.99), we can write

$$
\begin{aligned}
\sigma_{mn}^{[0]} &= \widetilde{C}_{ij}\delta_{im}\delta_{jn} \\
&\quad + \widetilde{C}_{ijkl}\left(\delta_{im}\delta_{jn}\Big((\varepsilon_{[0]})_{kl} - ((\varepsilon_{[0]})_{kl})_{\underline{\Omega}} \Big) + \delta_{km}\delta_{ln}\Big((\varepsilon_{[0]})_{ij} - ((\varepsilon_{[0]})_{ij})_{\underline{\Omega}} \Big)\right) \\
&\quad + \widetilde{C}_{ijklpq}\left(\delta_{im}\delta_{jn}\Big((\varepsilon_{[0]})_{kl} - ((\varepsilon_{[0]})_{kl})_{\underline{\Omega}} \Big)\Big((\varepsilon_{[0]})_{pq} - ((\varepsilon_{[0]})_{pq})_{\underline{\Omega}} \Big) \right. \\
&\qquad + \delta_{km}\delta_{ln}\Big((\varepsilon_{[0]})_{ij} - ((\varepsilon_{[0]})_{ij})_{\underline{\Omega}} \Big)\Big((\varepsilon_{[0]})_{pq} - ((\varepsilon_{[0]})_{pq})_{\underline{\Omega}} \Big) \\
&\qquad \left. + \delta_{pm}\delta_{qn}\Big((\varepsilon_{[0]})_{ij} - ((\varepsilon_{[0]})_{ij})_{\underline{\Omega}} \Big)\Big((\varepsilon_{[0]})_{kl} - ((\varepsilon_{[0]})_{kl})_{\underline{\Omega}} \Big)\right)
\end{aligned}
\tag{9.101}
$$

Using properties of Kronecker delta

$$
\begin{aligned}
\sigma_{mn}^{[0]} &= \widetilde{C}_{mn} \\
&\quad + \widetilde{C}_{mnkl}\Big((\varepsilon_{[0]})_{kl} - ((\varepsilon_{[0]})_{kl})_{\underline{\Omega}} \Big) + \widetilde{C}_{ijmn}\Big((\varepsilon_{[0]})_{ij} - ((\varepsilon_{[0]})_{ij})_{\underline{\Omega}} \Big) \\
&\quad + \widetilde{C}_{mnklpq}\Big((\varepsilon_{[0]})_{kl} - ((\varepsilon_{[0]})_{kl})_{\underline{\Omega}} \Big)\Big((\varepsilon_{[0]})_{pq} - ((\varepsilon_{[0]})_{pq})_{\underline{\Omega}} \Big) \\
&\quad + \widetilde{C}_{ijmnpq}\Big((\varepsilon_{[0]})_{ij} - ((\varepsilon_{[0]})_{ij})_{\underline{\Omega}} \Big)\Big((\varepsilon_{[0]})_{pq} - ((\varepsilon_{[0]})_{pq})_{\underline{\Omega}} \Big) \\
&\quad + \widetilde{C}_{ijklmn}\Big((\varepsilon_{[0]})_{ij} - ((\varepsilon_{[0]})_{ij})_{\underline{\Omega}} \Big)\Big((\varepsilon_{[0]})_{kl} - ((\varepsilon_{[0]})_{kl})_{\underline{\Omega}} \Big) + \dots
\end{aligned}
\tag{9.102}
$$

The constitutive theory can be further simplified by noting that sum $\boldsymbol{\sigma}^{[0]}$ and $\boldsymbol{\varepsilon}_{[0]}$ are symmetric tensors (of rank two). Hence, $\widetilde{C}_{mnkl} = \widetilde{C}_{klmn}$ and the pair of indices mn, kl, pq in tensor C_{mnklpq} can be interchanged. Thus, we can write (9.102) as

$$
\begin{aligned}
\sigma_{mn}^{[0]} &= \widehat{C}_{mn} + \widehat{C}_{mnij}\Big((\varepsilon_{[0]})_{ij} - ((\varepsilon_{[0]})_{ij})_{\underline{\Omega}} \Big) \\
&\quad + \widehat{C}_{mnijkl}\Big((\varepsilon_{[0]})_{ij} - ((\varepsilon_{[0]})_{ij})_{\underline{\Omega}} \Big)\Big((\varepsilon_{[0]})_{kl} - ((\varepsilon_{[0]})_{kl})_{\underline{\Omega}} \Big) + \dots
\end{aligned}
\tag{9.103}
$$

in which $\widehat{C}_{mn} = \widetilde{C}_{mn}$, $\widehat{C}_{mnij} = 2\widetilde{C}_{mnij}$, $\widehat{C}_{mnijkl} = 3\widetilde{C}_{mnijkl}$. $\widehat{C}_{mn}$, $\widehat{C}_{mnij}$ and $\widehat{C}_{mnijkl}$ are tensors of rank two, four and six respectively. This constitutive theory (9.103) for $\boldsymbol{\sigma}^{[0]}$ is quadratic in the components of $\boldsymbol{\varepsilon}_{[0]}$.

A linear constitutive theory can be obtained from (9.103)

$$\sigma_{mn}^{[0]} = \widehat{C}_{mn} + \widehat{C}_{mnij}(\varepsilon_{[0]})_{ij} - \widehat{C}_{mnij}((\varepsilon_{[0]})_{ij})_{\underline{\Omega}} \tag{9.104}$$

We also note that the pair of indices mn and ij are also interchangeable in $\widetilde{C}_{mn}$ and $\widetilde{C}_{mnij}$ due to symmetry of tensors $\widehat{C}_{mn}$, $\widehat{C}_{mnij}$

Remarks

(1) $\widehat{C}_{mn}$ in (9.103) and (9.104) is initial stress field (known) and the last term on right side of (9.104) are stresses due to thermal strains.

(2) This constitutive theory (9.104) requires 81 material coefficients defined by the fourth rank tensors $\widehat{C}_{mnij}$. Since $\widehat{C}_{mnij}$ is symmetric, only 21 material coefficients are independent. Whereas the linear constitutive theory for $\boldsymbol{\sigma}^{[0]}$ derived using entropy inequality and the representation theorem (for isotropic homogeneous matter) requires only two material coefficients.

(3) The constitutive theories in (9.103) and (9.104): (i) do not have a thermodynamic basis (ii) cannot ensure thermodynamic equilibrium during deformation (iii) cannot be used in the conservation and the balance laws derived for isotropic, homogeneous matter, i.e., CBL of CCM.

(4) Mathematical models that utilize such constitutive theories can possibly be derived using energy methods.

(5) We note that in this approach material coefficients $\widehat{C}_{mnij}$ are functions of $(\boldsymbol{\varepsilon}_{[0]})_{\underline{\Omega}}$, hence are not frame invariant. Thus, this approach can only support constant material coefficients.

(6) This constitutive theory holds for anisotropic materials but with symmetry of $\widehat{C}_{mnij}$, it requires 21 independent material coefficients.

(7) Since $\boldsymbol{\sigma}^{[0]}$ and $\boldsymbol{\varepsilon}_{[0]}$ are symmetric tensors of rank two and $\boldsymbol{\varepsilon}_{[0]}$ is argument tensor of $\boldsymbol{\sigma}^{[0]}$, we can express $\boldsymbol{\sigma}^{[0]}$ in terms of $\boldsymbol{\varepsilon}_{[0]}$ through a tensor of rank four. Thus, we can write (neglecting thermal field)

$$\sigma_{mn}^{[0]} = \widehat{C}_{mn} + \widehat{C}_{mnij}(\varepsilon_{[0]})_{ij} \tag{9.105}$$

which is same as (9.104) (in the absence of thermal term).

9.3.3.1 Simplified form of the constitutive theories for $\boldsymbol{\sigma}^{[0]}$

In the following we derive simplified forms of the constitutive theories (9.104) or (9.105). We have seen that $\widehat{C}_{ij}$, $\widehat{C}_{ijkl}$ are symmetric tensors. The material tensor $\widehat{C}_{ijkl}$ has the following properties.

$$\begin{aligned} \widehat{C}_{ijkl} &= \widehat{C}_{jikl} \quad ; \quad \widehat{C}_{ijkl} = \widehat{C}_{ijlk} \\ \widehat{C}_{ijkl} &= \widehat{C}_{jilk} \quad ; \quad \widehat{C}_{ijkl} = \widehat{C}_{klij} \end{aligned} \tag{9.106}$$

That is, the tensor $\widehat{C}_{ijkl}$ is symmetric, thus, the number of independent constants in the tensor $\widehat{C}_{ijkl}$ reduces to 21. For simplicity, *if we assume the initial stress and initial strain in* (9.104) *to be zero*, then (9.104) in view of (9.106) can be written in the matrix and vector form as follows:

$$\begin{Bmatrix} \sigma_{11}^{[0]} \\ \sigma_{22}^{[0]} \\ \sigma_{33}^{[0]} \\ \sigma_{23}^{[0]} \\ \sigma_{31}^{[0]} \\ \sigma_{12}^{[0]} \end{Bmatrix} = \begin{bmatrix} \widehat{C}_{1111} & \widehat{C}_{1122} & \widehat{C}_{1133} & \widehat{C}_{1123} & \widehat{C}_{1113} & \widehat{C}_{1112} \\ & \widehat{C}_{2222} & \widehat{C}_{2233} & \widehat{C}_{2223} & \widehat{C}_{2213} & \widehat{C}_{2212} \\ & & \widehat{C}_{3333} & \widehat{C}_{3323} & \widehat{C}_{3313} & \widehat{C}_{3312} \\ & \text{SYMM.} & & \widehat{C}_{2323} & \widehat{C}_{2313} & \widehat{C}_{2312} \\ & & & & \widehat{C}_{1313} & \widehat{C}_{1312} \\ & & & & & \widehat{C}_{1212} \end{bmatrix} \begin{Bmatrix} (\varepsilon_{[0]})_{11} \\ (\varepsilon_{[0]})_{22} \\ (\varepsilon_{[0]})_{33} \\ (\varepsilon_{[0]})_{23} \\ (\varepsilon_{[0]})_{31} \\ (\varepsilon_{[0]})_{12} \end{Bmatrix} \tag{9.107}$$

or

$$\{\sigma^{[0]}\} = [\widehat{C}]\{\varepsilon_{[0]}\} \tag{9.108}$$

If we introduce the following new notation for $\widehat{C}_{ijkl}$

$$\begin{aligned} 11 \to 1 \quad , \quad 22 \to 2 \quad , \quad 33 \to 3 \\ 23 \to 4 \quad , \quad 13 \to 5 \quad , \quad 12 \to 6 \end{aligned} \tag{9.109}$$

then $[\widehat{C}]$ in (9.108) can be written as

$$[\widehat{C}] = \begin{bmatrix} \widehat{C}_{11} & \widehat{C}_{12} & \widehat{C}_{13} & \widehat{C}_{14} & \widehat{C}_{15} & \widehat{C}_{16} \\ & \widehat{C}_{22} & \widehat{C}_{23} & \widehat{C}_{24} & \widehat{C}_{25} & \widehat{C}_{26} \\ & & \widehat{C}_{33} & \widehat{C}_{34} & \widehat{C}_{35} & \widehat{C}_{36} \\ & \text{SYMM.} & & \widehat{C}_{44} & \widehat{C}_{45} & \widehat{C}_{46} \\ & & & & \widehat{C}_{55} & \widehat{C}_{56} \\ & & & & & \widehat{C}_{66} \end{bmatrix} \tag{9.110}$$

If we assume that $[\widehat{C}]$ is invertible, then

$$\{\varepsilon_{[0]}\} = [\widehat{C}]^{-1}\{\sigma^{[0]}\} = [\hat{S}]\{\sigma^{[0]}\} \tag{9.111}$$

(a) *Influence of change of frame on* $\widehat{C}_{ijkl}$

We recall that $\boldsymbol{\sigma}^{[0]}$ and $\boldsymbol{\varepsilon}_{[0]}$ are symmetric tensors of rank two defined in fixed x-frame. Consider a change of frame from x-frame to x'-frame due to pure rotation, i.e., consider rigid rotation of the reference configuration due to $[Q]$ (not a function of time).

$$\boldsymbol{e}_i' = Q_{ij}\boldsymbol{e}_j \tag{9.112}$$

in which $\boldsymbol{e}_j$ and $\boldsymbol{e}_i'$ are base vectors in x- and x'-frames and Q_{ij} is an orthogonal rotation matrix. Then, $\boldsymbol{\sigma}^{[0]}$ and $\boldsymbol{\varepsilon}_{[0]}$ in x'-frame are obtained using

$$[(\sigma^{[0]})'] = [Q][\sigma^{[0]}][Q]^T \quad \text{or} \quad (\sigma_{ij}^{[0]})' = Q_{ip}Q_{jq}\sigma_{pq}^{[0]} \tag{9.113}$$

$$[(\varepsilon_{[0]})'] = [Q][\varepsilon_{[0]}][Q]^T \quad \text{or} \quad (\varepsilon_{[0]})_{ij}' = Q_{ip}Q_{jq}(\varepsilon_{[0]})_{pq} \tag{9.114}$$

and

$$[\sigma^{[0]}] = [Q]^T[(\sigma^{[0]})'][Q] \quad \text{or} \quad \sigma^{[0]}_{ij} = Q_{pi}Q_{qj}(\sigma^{[0]}_{pq})' \tag{9.115}$$

$$[\varepsilon_{[0]}] = [Q]^T[(\varepsilon_{[0]})'][Q] \quad \text{or} \quad (\varepsilon_{[0]})_{ij} = Q_{pi}Q_{qj}(\varepsilon_{[0]})'_{pq} \tag{9.116}$$

Consider (9.104) in the absence of initial stress and initial strain fields, i.e., neglecting first and the last terms in right hand side of (9.104).

$$\sigma^{[0]}_{ij} = \widehat{C}_{ijkl}(\varepsilon_{[0]})_{kl} \tag{9.117}$$

Substitute for $(\varepsilon_{[0]})_{kl}$ from (9.116)

$$\sigma^{[0]}_{ij} = \widehat{C}_{ijkl}Q_{pk}Q_{ql}(\varepsilon_{[0]})'_{pq} \tag{9.118}$$

Premultiply (9.118) by $Q_{ri}Q_{sj}$

$$Q_{ri}Q_{sj}\sigma^{[0]}_{ij} = Q_{ri}Q_{sj}\widehat{C}_{ijkl}Q_{pk}Q_{ql}(\varepsilon_{[0]})'_{pq} \tag{9.119}$$

Using (9.113)

$$(\sigma^{[0]}_{rs})' = Q_{ri}Q_{sj}\widehat{C}_{ijkl}Q_{pk}Q_{ql}(\varepsilon_{[0]})'_{pq} \tag{9.120}$$

$$\text{or} \quad (\sigma^{[0]}_{rs})' = \widehat{C}'_{rspq}(\varepsilon_{[0]})'_{pq} \tag{9.121}$$

where

$$\widehat{C}'_{rspq} = Q_{ri}Q_{sj}\widehat{C}_{ijkl}Q_{pk}Q_{ql} \tag{9.122}$$

Thus, due to rigid rotation of the reference configuration, $\widehat{C}_{ijkl}$ in the x-frame transforms into $\widehat{C}'_{rspq}$ according to (9.122)). Hence, $\widehat{C}_{ijkl}$ *is a tensor of rank four in x-frame. Likewise, $\widehat{C}'_{ijkl}$ is also a tensor of rank four but in the x'-frame.*

Remarks

(1) We note that 81 material coefficients in (9.104) provide the most general relationship of $\boldsymbol{\sigma}^{[0]}$ to $\boldsymbol{\varepsilon}_{[0]}$ purely based on Taylor series expansion. This is in violation of the axiom of frame invariance as $\boldsymbol{\varepsilon}_{[0]}$ is not frame invariant.

(2) Due to symmetry considerations, reduction of 81 material coefficients to 21 in (9.107) or (9.111) is mathematically justified. Proceeding further with (9.107) or (9.111) is a matter of choice knowing full well its limitation. This theory is non-physical for isotropic, homogeneous matter as it lacks thermodynamic foundation.

(3) We realize that constitutive theories (9.107) or (9.111) exhibit directional dependence at a material point, which is not possible for homogeneous, isotropic solid matter.

(4) In spite of limitations in (1)–(3), we proceed further using (9.111) due to two reasons: first, simplifications of (9.111) to various limited material behaviors is commonly used in engineering, particularly for small deformation small strain applications and secondly, as a matter of curiosity to find out under

what assumptions these theories will be in agreement with those derived using principles of continuum mechanics.

(b) *Simplification of* $[\widehat{C}]$ *in* (9.110)

The material stiffness coefficient matrix $[\widehat{C}]$ in (9.110) can be further simplified if we consider some specific types of materials.

Material symmetry

Many materials exhibit planes of symmetry. Consider x-frame to be the original frame and x'-frame to represent planes of symmetry. If

$$x'_1 = -x_1 \quad , \quad x'_2 = -x_2 \quad \text{and} \quad x'_3 = -x_3 \tag{9.123}$$

$$\text{that is} \quad Q_{ij} = -\delta_{ij} \quad \text{in} \quad x'_i = Q_{ij} x_j \tag{9.124}$$

then, for this special case, using (9.124) in (9.122), we can show that

$$\widehat{C}'_{ijkl} = \widehat{C}_{ijkl} \tag{9.125}$$

Monoclinic materials

When the stiffness or elastic coefficients at a material point have the same value for every possible set of coordinate directions that are mirror images of each other, the material is called monoclinic material. Consider x'-frame to be a mirror image of x-frame under the conditions

$$x'_1 = x_1 \quad , \quad x'_2 = x_2 \quad \text{and} \quad x'_3 = -x_3 \tag{9.126}$$

that is, $x_1 x_2$ plane is parallel to the plane of symmetry. Equations (9.126) imply that $[Q]$ in $x'_i = Q_{ij} x_j$ is given by

$$[Q] = \begin{bmatrix} 1 & 0 & 0 \\ 0 & 1 & 0 \\ 0 & 0 & -1 \end{bmatrix} \tag{9.127}$$

Using (9.127) in (9.122) we obtain the following:

$$\widehat{C}'_{ijkl} = \widehat{C}_{ijkl} \tag{9.128}$$

If we use (9.127) in (9.113) then we obtain

$$[(\sigma^{[0]})'] = \begin{bmatrix} 1 & 0 & 0 \\ 0 & 1 & 0 \\ 0 & 0 & -1 \end{bmatrix} [\sigma^{[0]}] \begin{bmatrix} 1 & 0 & 0 \\ 0 & 1 & 0 \\ 0 & 0 & -1 \end{bmatrix} \tag{9.129}$$

In the following: $x_1 \to 1$, $x_2 \to 2$, $x_3 \to 3$ which gives

$$\begin{bmatrix} (\sigma^{[0]}_{11})' & (\sigma^{[0]}_{12})' & (\sigma^{[0]}_{13})' \\ & (\sigma^{[0]}_{22})' & (\sigma^{[0]}_{23})' \\ \text{SYMM.} & & (\sigma^{[0]}_{33})' \end{bmatrix} = \begin{bmatrix} \sigma^{[0]}_{11} & \sigma^{[0]}_{12} & -\sigma^{[0]}_{13} \\ & \sigma^{[0]}_{22} & -\sigma^{[0]}_{23} \\ \text{SYMM.} & & \sigma^{[0]}_{33} \end{bmatrix} \tag{9.130}$$

Equality (9.130) implies

$$
\begin{aligned}
(\sigma^{[0]})'_{11} &= \sigma^{[0]}_{11} \quad ; \quad (\sigma^{[0]})'_{12} = \sigma^{[0]}_{12} \\
(\sigma^{[0]})'_{22} &= \sigma^{[0]}_{22} \quad ; \quad (\sigma^{[0]})'_{13} = -\sigma^{[0]}_{13} \\
(\sigma^{[0]})'_{33} &= \sigma^{[0]}_{33} \quad ; \quad (\sigma^{[0]})'_{23} = -\sigma^{[0]}_{23}
\end{aligned}
\tag{9.131}
$$

Similarly, if we use (9.127) in (9.114) and follow the same procedure as used above for stress we obtain

$$
\begin{aligned}
(\varepsilon_{[0]})'_{11} &= (\varepsilon_{[0]})_{11} \quad ; \quad (\varepsilon_{[0]})'_{12} = (\varepsilon_{[0]})_{12} \\
(\varepsilon_{[0]})'_{22} &= (\varepsilon_{[0]})_{22} \quad ; \quad (\varepsilon_{[0]})'_{13} = -(\varepsilon_{[0]})_{13} \\
(\varepsilon_{[0]})'_{33} &= (\varepsilon_{[0]})_{33} \quad ; \quad (\varepsilon_{[0]})'_{23} = -(\varepsilon_{[0]})_{23}
\end{aligned}
\tag{9.132}
$$

Consider

$$\{\sigma^{[0]}\} = [\widehat{C}]\{\varepsilon_{[0]}\} \tag{9.133}$$

$$\{(\sigma^{[0]})'\} = [\widehat{C}']\{(\varepsilon_{[0]})'\} = [\widehat{C}]\{(\varepsilon_{[0]})'\} \quad \text{as} \quad [\widehat{C}] = [\widehat{C}'] \tag{9.134}$$

Using (9.134)

$$
\begin{aligned}
(\sigma^{[0]}_{11})' =& \widehat{C}_{11}(\varepsilon_{[0]})'_{11} + \widehat{C}_{12}(\varepsilon_{[0]})'_{22} + \widehat{C}_{13}(\varepsilon_{[0]})'_{33} \\
&+ \widehat{C}_{14}(\varepsilon_{[0]})'_{23} + \widehat{C}_{15}(\varepsilon_{[0]})'_{31} + \widehat{C}_{16}(\varepsilon_{[0]})'_{12}
\end{aligned}
\tag{9.135}
$$

Substituting from (9.132) into (9.135)

$$
\begin{aligned}
(\sigma^{[0]}_{11})' =& \widehat{C}_{11}(\varepsilon_{[0]})_{11} + \widehat{C}_{12}(\varepsilon_{[0]})_{22} + \widehat{C}_{13}(\varepsilon_{[0]})_{33} \\
&- \widehat{C}_{14}(\varepsilon_{[0]})_{23} - \widehat{C}_{15}(\varepsilon_{[0]})_{31} + \widehat{C}_{16}(\varepsilon_{[0]})_{12}
\end{aligned}
\tag{9.136}
$$

Also from (9.133)

$$
\begin{aligned}
\sigma^{[0]}_{11} =& \widehat{C}_{11}(\varepsilon_{[0]})_{11} + \widehat{C}_{12}(\varepsilon_{[0]})_{22} + \widehat{C}_{13}(\varepsilon_{[0]})_{33} \\
&+ \widehat{C}_{14}(\varepsilon_{[0]})_{23} + \widehat{C}_{15}(\varepsilon_{[0]})_{31} + \widehat{C}_{16}(\varepsilon_{[0]})_{12}
\end{aligned}
\tag{9.137}
$$

Since $\sigma^{[0]}_{11} = (\sigma^{[0]}_{11})'$, subtracting (9.136) from (9.137) we obtain

$$\widehat{C}_{14}(\varepsilon_{[0]})_{23} + \widehat{C}_{15}(\varepsilon_{[0]})_{31} = 0 \quad \forall (\varepsilon_{[0]})_{23} \text{ and } (\varepsilon_{[0]})_{31} \tag{9.138}$$

Equation (9.138) implies that

$$
\begin{aligned}
\widehat{C}_{14} = 0 \quad &\text{and} \quad \widehat{C}_{15} = 0 \\
\text{or} \quad \widehat{C}_{1123} = 0 \quad &\text{and} \quad \widehat{C}_{1113} = 0
\end{aligned}
$$

Similarly, by considering $(\sigma^{[0]}_{22})'$, $\sigma^{[0]}_{22}$, and $(\sigma^{[0]}_{33})'$, $\sigma^{[0]}_{33}$, and $(\sigma^{[0]}_{12})'$, $\sigma^{[0]}_{12}$, and following the same procedure, we obtain

$$
\begin{aligned}
&\widehat{C}_{24} = 0 \ , \ \widehat{C}_{25} = 0 \quad \text{or} \quad \widehat{C}_{2223} = 0 \ , \ \widehat{C}_{2213} = 0 \quad \text{for} \quad (\sigma^{[0]}_{22})' \ , \ \sigma^{[0]}_{22} \\
&\widehat{C}_{34} = 0 \ , \ \widehat{C}_{35} = 0 \quad \text{or} \quad \widehat{C}_{3323} = 0 \ , \ \widehat{C}_{3313} = 0 \quad \text{for} \quad (\sigma^{[0]}_{33})' \ , \ \sigma^{[0]}_{33} \\
&\widehat{C}_{46} = 0 \ , \ \widehat{C}_{56} = 0 \quad \text{or} \quad \widehat{C}_{2312} = 0 \ , \ \widehat{C}_{1312} = 0 \quad \text{for} \quad (\sigma^{[0]}_{12})' \ , \ \sigma^{[0]}_{12}
\end{aligned}
$$

Thus, the 21 material coefficients in $[\widehat{C}]$ of (9.110)) are now reduced to $21-8=13$. The new $[\widehat{C}]$ has the following form (for monoclinic materials):

$$[\widehat{C}] = \begin{bmatrix} \widehat{C}_{11} & \widehat{C}_{12} & \widehat{C}_{13} & 0 & 0 & \widehat{C}_{16} \\ \widehat{C}_{12} & \widehat{C}_{22} & \widehat{C}_{23} & 0 & 0 & \widehat{C}_{26} \\ \widehat{C}_{13} & \widehat{C}_{23} & \widehat{C}_{33} & 0 & 0 & \widehat{C}_{36} \\ 0 & 0 & 0 & \widehat{C}_{44} & \widehat{C}_{45} & 0 \\ 0 & 0 & 0 & \widehat{C}_{45} & \widehat{C}_{55} & 0 \\ \widehat{C}_{16} & \widehat{C}_{26} & \widehat{C}_{36} & 0 & 0 & \widehat{C}_{66} \end{bmatrix} \tag{9.139}$$

We note that for monoclinic materials, shear strain $((\varepsilon_{[0]})_{12}$ in this case) can produce normal stress, and normal stresses ($\sigma_{11}^{[0]}$, $\sigma_{22}^{[0]}$, $\sigma_{33}^{[0]}$ in this case) can produce shear strain $((\varepsilon_{[0]})_{12}$ in this case).

Orthotropic materials

We call materials in which three orthogonal planes of symmetry exist *orthotropic materials*. If we consider x_1x_2, x_2x_3 and x_3x_1 as planes of symmetry, then the transformation matrices associated with them are

$$[{}^1Q] = \begin{bmatrix} 1 & 0 & 0 \\ 0 & 1 & 0 \\ 0 & 0 & -1 \end{bmatrix} \quad , \quad [{}^2Q] = \begin{bmatrix} -1 & 0 & 0 \\ 0 & 1 & 0 \\ 0 & 0 & 1 \end{bmatrix} \quad , \quad [{}^3Q] = \begin{bmatrix} 1 & 0 & 0 \\ 0 & -1 & 0 \\ 0 & 0 & 1 \end{bmatrix} \tag{9.140}$$

The term $[{}^1Q]$ has already been considered in the case of monoclinic materials, thus we only need to investigate the influence of $[{}^2Q]$ and $[{}^3Q]$ on $[\widehat{C}]$ defined by (9.139). We can show that under these two transformations, the invariance of $[\widehat{C}]$ requires

$$\widehat{C}_{1112} = \widehat{C}_{16} = 0 \ , \ \widehat{C}_{2212} = \widehat{C}_{26} = 0 \ , \ \widehat{C}_{3312} = \widehat{C}_{36} = 0 \ , \ \widehat{C}_{2313} = \widehat{C}_{45} = 0 \tag{9.141}$$

Thus, for orthotropic materials the coefficients in $[\widehat{C}]$ reduce to $13-4=9$ and we can write the following for $[\widehat{C}]$ based on (9.139) and (9.141):

$$[\widehat{C}] = \begin{bmatrix} \widehat{C}_{11} & \widehat{C}_{12} & \widehat{C}_{13} & 0 & 0 & 0 \\ \widehat{C}_{12} & \widehat{C}_{22} & \widehat{C}_{23} & 0 & 0 & 0 \\ \widehat{C}_{13} & \widehat{C}_{23} & \widehat{C}_{33} & 0 & 0 & 0 \\ 0 & 0 & 0 & \widehat{C}_{44} & 0 & 0 \\ 0 & 0 & 0 & 0 & \widehat{C}_{55} & 0 \\ 0 & 0 & 0 & 0 & 0 & \widehat{C}_{66} \end{bmatrix} \tag{9.142}$$

Since $[\widehat{C}]$ is symmetric, only nine material coefficients are required in (9.142). In this derivation $\boldsymbol{\sigma}^{[0]}$ and $\boldsymbol{\varepsilon}_{[0]}$ are rate of work conjugate. Hence, the constitutive theory is valid for finite deformation finite strain. Physical meaning of the material coefficients in (9.142) becomes clear in small deformation, small strain constitutive theory given in a following section.

9.3.4 Constitutive theory for $\boldsymbol{\sigma}$ using $\pi(\boldsymbol{\varepsilon})$: small deformation, small strain

We recall that based on the strain energy density approach (Section 9.3.2), we have the following for small deformation, small strain incompressible, non-isothermal deformation physics.

$$[\sigma] = \rho_0 \frac{\partial \pi(\boldsymbol{\varepsilon})}{\partial \boldsymbol{\varepsilon}} \tag{9.143}$$

If we consider $\pi(\boldsymbol{\varepsilon})$ and expand π in Taylor series in $\boldsymbol{\varepsilon}$, then we can derive the following linear constitutive theory for $\boldsymbol{\sigma}$ for orthotropic material (similar to (9.142)).

$$\{\sigma\} = \{\sigma^0\} + [D]\{\varepsilon\} \tag{9.144}$$

In the absence of initial stress field we can write (9.144) as

$$[\sigma] = [D]\{\varepsilon\} \tag{9.145}$$

$[D]$ is symmetric and contains only nine independent material coefficients

$$[D] = \begin{bmatrix} D_{11} & D_{12} & D_{13} & 0 & 0 & 0 \\ & D_{12} & D_{13} & 0 & 0 & 0 \\ & & D_{13} & 0 & 0 & 0 \\ & \text{SYMM.} & & D_{44} & 0 & 0 \\ & & & & D_{55} & 0 \\ & & & & & D_{66} \end{bmatrix} \tag{9.146}$$

$$\text{or} \quad \{\varepsilon\} = [S]\{\sigma\} \quad ; \quad [S] = [D]^{-1} \tag{9.147}$$

For small deformation, small strain and linear elastic material behavior, superposition holds, i.e., the response for a combination of loads applied together is the sum of the responses due to individual loads. From elementary mechanics of materials we assume that for such materials the components of matrix $[S]$ can be expressed in terms of Young's moduli E_1, E_2, E_3, Poisson's ratios ν_{21}, ν_{31}, ν_{12}, ν_{32}, ν_{13}, ν_{23} and shear moduli G_{23}, G_{31} and G_{12}. Hence, we can write (using the matrix and vector notations)

$$\begin{Bmatrix} \varepsilon_{11} \\ \varepsilon_{22} \\ \varepsilon_{33} \\ \varepsilon_{23} \\ \varepsilon_{31} \\ \varepsilon_{12} \end{Bmatrix} = \begin{bmatrix} \frac{1}{E_1} & -\frac{\nu_{21}}{E_2} & -\frac{\nu_{31}}{E_3} & 0 & 0 & 0 \\ -\frac{\nu_{12}}{E_1} & \frac{1}{E_2} & -\frac{\nu_{32}}{E_3} & 0 & 0 & 0 \\ -\frac{\nu_{13}}{E_1} & -\frac{\nu_{23}}{E_2} & \frac{1}{E_3} & 0 & 0 & 0 \\ 0 & 0 & 0 & \frac{1}{G_{23}} & 0 & 0 \\ 0 & 0 & 0 & 0 & \frac{1}{G_{31}} & 0 \\ 0 & 0 & 0 & 0 & 0 & \frac{1}{G_{12}} \end{bmatrix} \begin{Bmatrix} \sigma_{11} \\ \sigma_{22} \\ \sigma_{33} \\ \sigma_{23} \\ \sigma_{31} \\ \sigma_{12} \end{Bmatrix} \tag{9.148}$$

Since $[D]$ and $[S]$ are symmetric, the following must hold:

$$\frac{\nu_{12}}{E_1} = \frac{\nu_{21}}{E_2} \quad ; \quad \frac{\nu_{13}}{E_1} = \frac{\nu_{31}}{E_3} \quad ; \quad \frac{\nu_{23}}{E_2} = \frac{\nu_{32}}{E_3} \tag{9.149}$$

That is, only three of the six Poisson's ratios are independent. If we choose ν_{21}, ν_{31} and ν_{32}, then ν_{12}, ν_{13} and ν_{23} are deterministic from (9.149). Thus, for such materials E_1, E_2, E_3, ν_{21}, ν_{31}, ν_{32}, G_{23}, G_{31} and G_{12} are (nine) independent material constants that must be determined experimentally. The constitutive theory (9.148) is often called Generalized Hooke's law for orthotropic materials.

Isotropic materials

For isotropic materials, material properties are independent of direction at every material point. For such materials

$$E_1 = E_2 = E_3 = E \tag{9.150}$$

$$\nu_{21} = \nu_{31} = \nu_{32} = \nu \tag{9.151}$$

$$G_{23} = G_{31} = G_{12} = G = \frac{E}{2(1+\nu)} \tag{9.152}$$

For this case, we can write (9.148) as follows using Einstein notations.

$$\varepsilon_{ij} = \frac{1+\nu}{E}\sigma_{ij} - \frac{\nu}{E}\sigma_{kk}\delta_{ij} \tag{9.153}$$

Inverse of (9.153) gives

$$\sigma_{ij} = \frac{E}{1+\nu}(\varepsilon_{ij}) + \frac{E}{(1+\nu)(1-2\nu)}(\varepsilon_{kk})\delta_{ij} \tag{9.154}$$

$$\text{or} \quad \sigma_{ij} = 2\mu(\varepsilon_{ij}) + \lambda(\varepsilon_{kk})\delta_{ij} \tag{9.155}$$

in which

$$\mu = \frac{E}{2(1+\nu)} \quad \text{and} \quad \lambda = \frac{\nu E}{(1+\nu)(1-2\nu)} \tag{9.156}$$

μ and λ are usual Lamé's constants and E, ν are modulus of elasticity and Poisson's ratio. As usual $[\varepsilon]$ is given by

$$[\varepsilon] = \begin{bmatrix} u_{1,1} & \frac{1}{2}(u_{1,2}+u_{2,1}) & \frac{1}{2}(u_{1,3}+u_{3,1}) \\ & u_{2,2} & \frac{1}{2}(u_{2,3}+u_{3,2}) \\ \text{SYMM.} & & u_{3,3} \end{bmatrix} \tag{9.157}$$

9.4 Constitutive theories for the heat vector $\boldsymbol{q}$: Lagrangian description

The constitutive theory for heat vector $\boldsymbol{q}$ remain the same regardless of the type of mechanical deformation. Hence, the following derivations hold for deformation physics considered in the previous sections. The conditions resulting from the entropy inequality require that (as $\theta > 0$)

$$\boldsymbol{q} \cdot \boldsymbol{g} \leq 0 \tag{9.158}$$

be satisfied by the constitutive theories for $\boldsymbol{q}$ regardless of how they are derived. We can take two approaches to derive constitutive theories for $\boldsymbol{q}$. In the first approach, we strictly use (9.158) to derive the constitutive theory for $\boldsymbol{q}$. This constitutive theory for $\boldsymbol{q}$ will naturally satisfy the entropy inequality as it is derived using the conditions resulting from it. In the second approach we use $\boldsymbol{g}$ and θ as the argument tensors of $\boldsymbol{q}$ and then use representation theorem [19,73,84,87,90,93,94,96–98,126, 134–137,144,145]. The constitutive theories derived using this approach must satisfy (9.158) so that the deforming matter will be in thermodynamic equilibrium during evolution. We present the derivation of the constitutive theories for $\boldsymbol{q}$ using both approaches and present comparisons of the resulting constitutive theories, discuss assumptions, and make some remarks regarding their merits and shortcomings.

9.4.1 Constitutive theory for $\boldsymbol{q}$ using entropy inequality

This derivation based on (9.158) is fundamental and can be found in any textbook on continuum mechanics [30, 31]. We present details in the following to point out the assumptions used in the derivation as they play a significant role when comparing this constitutive theory with the theories resulting from representation theorem. We begin with (9.158). Equation (9.158) implies that

$$\boldsymbol{q} \cdot \boldsymbol{g} = \beta \leq 0 \tag{9.159}$$

Using the inequality, we obtain

$$\frac{\partial \beta}{\partial \boldsymbol{g}} = \boldsymbol{q} \tag{9.160}$$

β has a maximum value at $\boldsymbol{g} = 0$ (based on equation (9.159)), hence

$$\left.\frac{\partial \beta}{\partial \boldsymbol{g}}\right|_{\boldsymbol{g}=0} = \boldsymbol{q}\Big|_{\boldsymbol{g}=0} = 0 \tag{9.161}$$

That is, *heat flux vanishes in the absence of temperature gradient.* Thus, the constitutive theory for $\boldsymbol{q}$ must be a function of $\boldsymbol{g}$. At this stage, many possibilities exist. The simplest of course is assuming that $\boldsymbol{q}$ is proportional to $-\boldsymbol{g}$, i.e., $\boldsymbol{q}$ is a linear function of $-\boldsymbol{g}$. Using $\boldsymbol{k}(\theta)$ as constant of proportionality we can write

$$\boldsymbol{q} = -\boldsymbol{k}(\theta) \cdot \boldsymbol{g} \quad \text{or} \quad q_i = -k_{ij}(\theta) g_j \tag{9.162}$$

from which we define

$$\frac{\partial \boldsymbol{q}}{\partial \boldsymbol{g}} = -\boldsymbol{k}(\theta) \quad \text{or} \quad \frac{\partial q_i}{\partial g_j} = -k_{ij}(\theta) \tag{9.163}$$

Also, from (9.160)

$$\frac{\partial^2 \beta}{\partial^2 \boldsymbol{g}} = \frac{\partial \boldsymbol{q}}{\partial \boldsymbol{g}} = -\boldsymbol{k}(\theta) \leq 0 \quad \text{or} \quad \frac{\partial^2 \beta}{\partial g_j \partial g_i} = \frac{\partial q_i}{\partial g_j} = -k_{ij}(\theta) \leq 0 \tag{9.164}$$

From (9.164), we conclude that $[k]$ is positive-semidefinite and all its eigenvalues are non-negative. Equation (9.162) is the *Fourier heat conduction law* in Lagrangian description. The thermal conductivity matrix $[k]$ does not have to be symmetric but is often assumed to be. In general, in this constitutive theory for $\boldsymbol{q}$, the coefficients of $[k]$ can be functions of temperature θ. This constitutive theory is based on the assumption that $\boldsymbol{q}$ is a linear function of $\boldsymbol{g}$. *While mathematically* (9.162) *is justified, we note that for isotropic, homogeneous matter, $\boldsymbol{q}$ cannot exhibit directional dependence.* Hence, (9.162) must be reduced to

$$\boldsymbol{q} = -k(\theta)\boldsymbol{g} \quad \text{or} \quad q_i = k(\theta)\, g_i \tag{9.165}$$

This constitutive theory requires only one material coefficient, thermal conductivity $k(\theta)$ that can be temperature dependent (Power law, Sutherland law, etc.). The constitutive theory (9.165) is the standard Fourier heat conduction law.

9.4.2 Constitutive theories for $\boldsymbol{q}$ using representation theorem

In this approach, the heat vector $\boldsymbol{q}$, a tensor of rank one, is expressed as a linear combination of the combined generators (only tensors of rank one) of its argument tensors. The material coefficients in the linear combination are assumed to be functions of the combined invariants of the argument tensors and temperature θ. The material coefficients are derived by expanding each coefficient in the linear combination in Taylor series about a known configuration. In this approach it is obvious that the explicit form of the constitutive theory for $\boldsymbol{q}$ depends on the argument tensors of $\boldsymbol{q}$ and the terms retained in the Taylor series expansion of the coefficients in the linear combination. We present this derivation in the following.

9.4.2.1 Using argument tensors strictly based on conjugate pair

In this derivation, we consider (based on conjugate pairs in the reduced entropy inequality)

$$\boldsymbol{q} = \boldsymbol{q}(\boldsymbol{g}, \theta) \tag{9.166}$$

Tensors $\boldsymbol{q}$ and $\boldsymbol{g}$ are tensors of rank one and θ is a tensor of rank zero. The only combined generator of rank one of the argument tensors $\boldsymbol{g}$ and θ is $\boldsymbol{g}$, hence based on representation theorem, we can write

$$\boldsymbol{q} = -{}^q\!\alpha \boldsymbol{g} \tag{9.167}$$

Material coefficients

The coefficient ${}^q\!\alpha$ is a function of the combined invariants of $\boldsymbol{g}$, θ, i.e., $\{g\}^T\{g\}$ and temperature θ. Let us define ${}^q\underset{\sim}{I} = \{g\}^T\{g\}$ to simplify the details of further derivation. We note that (9.167) holds in the current configuration in which the deformation is not known. Hence, in (9.167), ${}^q\!\alpha = {}^q\!\alpha({}^q\underset{\sim}{I}, \theta)$ is not yet deterministic and it not a material coefficient. To determine material coefficients in (9.167), we expand ${}^q\!\alpha({}^q\underset{\sim}{I}, \theta)$ in Taylor series about a known configuration $\underline{\Omega}$ in ${}^q\underset{\sim}{I}$ and θ and retain only up to linear terms in ${}^q\underset{\sim}{I}$ and θ (for simplicity).

$${}^q\!\alpha = {}^q\!\alpha\Big|_{\underline{\Omega}} + \frac{\partial\, {}^q\!\alpha}{\partial\, {}^q\underset{\sim}{I}}\bigg|_{\underline{\Omega}} \left({}^q\underset{\sim}{I} - ({}^q\underset{\sim}{I})_{\underline{\Omega}}\right) + \frac{\partial\, {}^q\!\alpha}{\partial \theta}\bigg|_{\underline{\Omega}} (\theta - \theta_{\underline{\Omega}}) \tag{9.168}$$

Substituting from (9.168) into (9.167)

$$\boldsymbol{q} = -\left({}^q\!\alpha\Big|_{\underline{\Omega}} + \frac{\partial\, {}^q\!\alpha}{\partial\, {}^q\underset{\sim}{I}}\bigg|_{\underline{\Omega}} \left({}^q\underset{\sim}{I} - ({}^q\underset{\sim}{I})_{\underline{\Omega}}\right) + \frac{\partial\, {}^q\!\alpha}{\partial \theta}\bigg|_{\underline{\Omega}} (\theta - \theta_{\underline{\Omega}})\right)\boldsymbol{g} \tag{9.169}$$

We note that ${}^q\alpha\big|_{\underline{\Omega}}$, $\dfrac{\partial\, {}^q\alpha}{\partial\, {}^q\underset{\sim}{I}}\Big|_{\underline{\Omega}}$ and $\dfrac{\partial\, {}^q\alpha}{\partial \theta}\Big|_{\underline{\Omega}}$ are functions of $({}^q\underset{\sim}{I})_{\underline{\Omega}}$ and $\theta_{\underline{\Omega}}$, whereas ${}^q\alpha$ in (9.167) is a function of ${}^q\underset{\sim}{I}$ and θ in the current configuration. From (9.169) we can write the following, noting that ${}^q\underset{\sim}{I} = \{g\}^T\{g\}$

$$\boldsymbol{q} = -{}^q\alpha\big|_{\underline{\Omega}}\boldsymbol{g} - \frac{\partial {}^q\alpha}{\partial {}^q\underset{\sim}{I}}\Big|_{\underline{\Omega}}(\{g\}^T\{g\})\boldsymbol{g} + \frac{\partial {}^q\alpha}{\partial {}^q\underset{\sim}{I}}\Big|_{\underline{\Omega}}(\{g\}^T\{g\})_{\underline{\Omega}}\boldsymbol{g} - \frac{\partial {}^q\alpha}{\partial \theta}\Big|_{\underline{\Omega}}(\theta - \theta_{\underline{\Omega}})\boldsymbol{g} \tag{9.170}$$

or

$$\boldsymbol{q} = -\left({}^q\alpha\big|_{\underline{\Omega}} - \frac{\partial {}^q\alpha}{\partial {}^q\underset{\sim}{I}}\Big|_{\underline{\Omega}}(\{g\}^T\{g\})_{\underline{\Omega}}\right)\boldsymbol{g} - \frac{\partial {}^q\alpha}{\partial {}^q\underset{\sim}{I}}\Big|_{\underline{\Omega}}(\{g\}^T\{g\})\boldsymbol{g} - \frac{\partial {}^q\alpha}{\partial \theta}\Big|_{\underline{\Omega}}(\theta - \theta_{\underline{\Omega}})\boldsymbol{g} \tag{9.171}$$

Let

$$\begin{aligned} k(\theta_{\underline{\Omega}}, ({}^q\underset{\sim}{I})_{\underline{\Omega}}) &= {}^q\alpha\big|_{\underline{\Omega}} - \frac{\partial {}^q\alpha}{\partial {}^q\underset{\sim}{I}}\Big|_{\underline{\Omega}}(\{g\}^T\{g\})_{\underline{\Omega}} \\ k_1(\theta_{\underline{\Omega}}, ({}^q\underset{\sim}{I})_{\underline{\Omega}}) &= \frac{\partial {}^q\alpha}{\partial {}^q\underset{\sim}{I}}\Big|_{\underline{\Omega}} \\ k_2(\theta_{\underline{\Omega}}, ({}^q\underset{\sim}{I})_{\underline{\Omega}}) &= \frac{\partial {}^q\alpha}{\partial \theta}\Big|_{\underline{\Omega}} \end{aligned} \tag{9.172}$$

Then

$$\boldsymbol{q} = -k\boldsymbol{g} - k_1(\{g\}^T\{g\})\boldsymbol{g} - k_2(\theta - \theta_{\underline{\Omega}})\boldsymbol{g} \tag{9.173}$$

This is the simplest possible constitutive theory based on conjugate pairs in the entropy inequality, representation theorem and (9.166). This constitutive theory uses integrity, the complete basis of the space of $\boldsymbol{q}$. The only assumption in this theory beyond (9.166) is the truncation of the Taylor series in (9.168) beyond linear terms in ${}^q\underset{\sim}{I}$ and θ. The constitutive theory for $\boldsymbol{q}$ in (9.173) is cubic in $\boldsymbol{q}$. It contains linear and cubic terms in $\boldsymbol{g}$, but does not contain a quadratic term in $\boldsymbol{g}$.

9.5 General remarks

In this section, we make general remarks regarding the use of mathematical models (CBL) utilizing the constitutive theories presented in this chapter.

(1) The mechanical work is completely recoverable upon unloading. That is no part of the rate of mechanical work is converted into entropy.

(2) These materials have elasticity regardless of whether the deformation physics is linear or non-linear.

  (a) In the case of non-linear deformation $\boldsymbol{\sigma}^{[0]}$ and $\boldsymbol{\varepsilon}_{[0]}$ are appropriate measures of stress and strain.

  (b) When the deformation and strains are small, the Cauchy stress tensor $\boldsymbol{\sigma}$ and strain tensor $\boldsymbol{\varepsilon}$, the linear part of Green's strain are appropriate measures of stress and strain.

(3) Due to elasticity (E, ν) and mass (ρ), such materials have stiffness (a load, displacement relationship). Thus, the following holds in such materials.

(a) Finite speed of sound ($\sqrt{E/\rho}$). This permits propagation of stress waves with finite speed without shape change for linear case and with shape distortion for non-linear case.
(b) These materials exhibit linear and non-linear vibrational characteristics upon application of frequency and time dependent disturbances.
(c) These materials have the concept of natural vibrations, natural frequencies of vibrations and mode shapes (at least in the linear case).

(4) In such materials, the coupling between non-isothermal behavior and mechanical deformation is only due to equilibrium stress if the material coefficients are not dependent on temperature.

9.6 Summary

In this chapter, constitutive theories have been derived for thermoelastic solid continua in Lagrangian description using thermodynamic as well as non-thermo dynamic approaches. Total deformation is decomposed into volumetric and distortional deformation through additive decomposition of the stress tensor into equilibrium and deviatoric stress tensors. Constitutive theories for the equilibrium stress tensor addressing volumetric deformation are not rederived, but references are made to their derivation for finite deformation, finite strain, compressible, incompressible, isothermal and non-isothermal deformation physics as well as small deformation, small strain, isothermal and non-isothermal deformation in Chapter 8. The reduced form of the entropy inequality derived in Chapter 8 in Lagrangian description is used to derive constitutive theories for the deviatoric stress tensor and the heat vector using conjugate pairs in the entropy inequality and the representation theorem for the various deformation physics considered in case of constitutive theories for the equilibrium stress tensor.

Separate derivations are presented for the constitutive theory for the deviatoric stress tensor for small deformation, small strain deformation physics. Material coefficients are derived in all cases. For small deformation, small strain physics, a linear constitutive theory is also presented for deviatoric stress tensor in which the material coefficients are identified as Lamé's constants and their relationship to modulus of elasticity and Poisson's ratio is presented.

Non-thermodynamic approaches of constitutive theories considered in this chapter include: those based on Helmholtz free energy density, strain energy density, Taylor series expansion of strain energy density $\pi(\boldsymbol{\varepsilon}_{[0]})$ and $\pi(\boldsymbol{\varepsilon})$ in its argument, purely phenomenological approach based on 1D springs and 1D dashpots, and empirically derived theories based on experimental data. In the non-thermodynamic approach based on Taylor series expansion of $\pi(\boldsymbol{\varepsilon}_{[0]})$ or $\pi(\boldsymbol{\varepsilon})$, most general anisotropic constitutive theories requiring 81 material coefficients and simplified orthotropic constitutive theory requiring nine material coefficients (called orthotropic materials that are used widely in composites) are presented. These constitutive theories obviously are not admissible in the differential forms of CBL of CCM which are only valid for isotropic and homogeneous matter. All constitutive theories that are not based on thermodynamic approach are likely to be in violation of thermodynamic equilibrium during evolution of the deformation.

Problems

9.1 Derive the constitutive theory for finite deformation of thermoelastic solids from first principles using the theory of generators and invariants in which

(a) the second Piola-Kirchhoff stress $\boldsymbol{\sigma}^{[0]}$ is a linear function of $\boldsymbol{\varepsilon}_{[0]}$, Green's strain tensor and temperature θ;

(b) the second Piola-Kirchhoff stress $\boldsymbol{\sigma}^{[0]}$ is a linear function of Cauchy strain tensor $\boldsymbol{C}_{[0]}$ and temperature θ.

In both cases, derive the material coefficients. Determine the relationship between the material coefficients in the theories derived in (a) and (b) if one exists. Give explicit forms of the equations for $\mathbb{R}^1$ and $\mathbb{R}^2$ for both (a) and (b). Also, represent these using Voigt's notation.

9.2 Derive the constitutive theory for finite deformation of thermoelastic solids from first principles using the strain energy density as a function of the invariants of the Green's strain tensor and temperature in which

(a) the second Piola-Kirchhoff stress $\boldsymbol{\sigma}^{[0]}$ is a linear function of $\boldsymbol{\varepsilon}_{[0]}$, Green's strain tensor and temperature θ;

(b) the second Piola-Kirchhoff stress $\boldsymbol{\sigma}^{[0]}$ is a linear function of the Cauchy strain tensor $\boldsymbol{C}_{[0]}$ and temperature θ.

In both cases, derive the material coefficients. Determine the relationship between the material coefficients in the theories derived in (a) and (b) if one exists. Give explicit forms of the equations for $\mathbb{R}^1$ and $\mathbb{R}^2$ for both (a) and (b). Also, represent these using Voigt's notation.

9.3 Derive an expression for the strain energy density function for an isotropic, homogeneous elastic solid undergoing finite deformation by assuming that the stress tensor is up to a quadratic function of the conjugate strain tensor.

9.4 Consider an orthogonal coordinate transformation in which x-frame changes to $\tilde{x}$-frame through

$$\{\tilde{x}\} = \begin{bmatrix} \cos\theta & \sin\theta & 0 \\ -\sin\theta & \cos\theta & 0 \\ 0 & 0 & 1 \end{bmatrix} \{x\}$$

Consider second Piola-Kirchhoff stress $\boldsymbol{\sigma}^{[0]}$ and Green's strain $\boldsymbol{\varepsilon}_{[0]}$ in x-frame and $\tilde{\boldsymbol{\sigma}}^{[0]}$ and Green's strain $\tilde{\boldsymbol{\varepsilon}}_{[0]}$ in $\tilde{x}$-frame. Using Voigt's vector notation for the stress and strain tensors in the two frames, determine the matrix (or matrices) that describe the transformation of the two groups of stress and strain tensors in the two frames.

9.5 Derive the constitutive theory for finite deformation of thermoelastic solids from first principles using the theory of generators and invariants in which

(a) the Green's strain tensor $\boldsymbol{\varepsilon}_{[0]}$ is a linear function of the second Piola-Kirchhoff stress tensor $\boldsymbol{\sigma}^{[0]}$ and temperature θ;

(b) the Cauchy strain tensor $\boldsymbol{C}_{[0]}$ is a linear function of the second Piola-Kirchhoff stress tensor $\boldsymbol{\sigma}^{[0]}$ and temperature θ.

In both cases, derive the material coefficients. Determine the relationship between the material coefficients in the theories derived in (a) and (b) if one exists. Derive the explicit form of the equations for $\mathbb{R}^1$ and $\mathbb{R}^2$ for both (a) and (b). Express these equations in matrix and vector form using Voigt's notation.

9.6 Derive the constitutive theory for finite deformation of thermoelastic solids from first principles using the complementary strain energy density function as a function of the invariants of the second Piola-Kirchhoff stress tensor and temperature in which

(a) the Green's strain tensor $\boldsymbol{\varepsilon}_{[0]}$ is a linear function of the second Piola-Kirchhoff stress tensor $\boldsymbol{\sigma}^{[0]}$ and temperature θ;

(b) the Cauchy strain tensor $\boldsymbol{C}_{[0]}$ is a linear function of the second Piola-Kirchhoff stress tensor $\boldsymbol{\sigma}^{[0]}$ and temperature θ.

In both cases, derive the material coefficients. Determine the relationship between the material coefficients in the theories derived in (a) and (b) if one exists. Derive the explicit form of the equations for $\mathbb{R}^1$ and $\mathbb{R}^2$ for both (a) and (b). Express these equations in matrix and vector form using Voigt's notation.

9.7 Consider isotropic, homogeneous continuous matter with finite deformation, finite strain. Let $[\sigma^{[0]}]$ and $[\dot{\varepsilon}_{[0]}]$ be the rate of work conjugate pair. Let $\pi = \pi([\varepsilon_{[0]}])$ be the strain energy density functional. Derive the constitutive theory for $[\sigma^{[0]}]$ using

$$[\sigma^{[0]}] = \rho_0 \frac{\partial \pi([\varepsilon_{[0]}])}{\partial [\varepsilon_{[0]}]} \tag{1}$$

Consider $\pi = \pi(i_{\varepsilon_{[0]}}, ii_{\varepsilon_{[0]}}, iii_{\varepsilon_{[0]}})$ due to frame invariance requirement of the constitutive theory. In (1), $i_{\varepsilon_{[0]}} = \text{tr}[\varepsilon_{[0]}]$, $ii_{\varepsilon_{[0]}} = \text{tr}([\varepsilon_{[0]}]^2)$, $iii_{\varepsilon_{[0]}} = \text{tr}([\varepsilon_{[0]}]^3)$.

10
CONSTITUTIVE THEORIES FOR THERMOVISCOELASTIC SOLIDS WITHOUT MEMORY

10.1 Introduction

In this chapter, we consider constitutive theories for thermoviscoelastic solid continua in Lagrangian description that possess elasticity and mechanism of dissipation but have no memory. In such solid continua, a part of mechanical work results in rate of entropy generation through dissipation mechanism. This part of the rate of work is obviously not recoverable. In this chapter, we consider constitutive theories for stress tensor that describe this physics for finite deformation, finite strain, compressible, incompressible and non-isothermal deformation as well as for small deformation, small strain and non-isothermal deformation with dissipation mechanism but without memory.

In finite deformation, finite strain deformation physics, the second Piola-Kirchhoff stress tensor $\boldsymbol{\sigma}^{[0]}$ and Green's strain tensor $\boldsymbol{\varepsilon}_{[0]}$ are appropriate measures of stress and strain. The dissipation mechanism considered here is due to constitution of the material (more pronounced in polymeric solids). From the physics of viscous fluent continua we recall that the dissipation mechanism is due to viscosity of the fluid (material property), the deviatoric stress tensor and the strain rate tensor. The deviatoric stresses are produced in the fluid due to the resistance offered by the fluid particles to the strain rates due to viscosity. A parallel mechanism also exists in thermoviscoelastic solids in which elastic behavior is due to $\boldsymbol{\varepsilon}_{[0]}$, while dissipation and additional stress mechanisms are due to strain rate $\dot{\boldsymbol{\varepsilon}}_{[0]}$. Since $\boldsymbol{\varepsilon}_{[0]}$ is a nonlinear measure of strain in displacement gradients, both elasticity and dissipation mechanisms are nonlinear mechanism in thermoviscoelastic solid continua when the deformation and the strains are finite. Compressibility or incompressibiltiy and thermal physics merely influence volumetric deformation physics, hence are addressed by the constitutive theory for equilibrium stress. Distortion of volume (change of shape) and dissipation mechanisms are addressed by the constitutive theory of the deviatoric stress tensor.

In the case of small deformation, small strain deformation physics, distinction between co- and contra-variant bases disappears and we only have Cauchy stress tensor $\boldsymbol{\sigma}$ (independent of basis) and $\boldsymbol{\varepsilon}$, linear part of the Green's strain tensor as appropriate measures of stress and strain. As in case of finite deformation, finite strain, dissipation mechanism and additional stress is due to strain rate $\dot{\boldsymbol{\varepsilon}}$ in addition to the stresses due to $\boldsymbol{\varepsilon}$.

Constitutive theories for heat vector $\boldsymbol{q}$ are also considered. These remain same regardless of mechanical deformation physics. In all cases complete constitutive theories based on integrity including material coefficients are presented first, followed by simplified non-linear and linear constitutive theories. Derivation of all constitutive

DOI: 10.1201/9781003105336-10

theories are initiated with the reduced form of the entropy inequality and the conjugate pairs in it. Constitutive variables and their argument tensors are established (SLT suffices for this purpose). Argument tensors of the constitutive variables are augmented to include additional physics that is either not obvious in SLT or has not been considered in the derivation of SLT.

As discussed in Chapter 8, the total deformation of a volume of matter consists of volumetric and distortional deformation. Volumetric deformation considers compressibility, incompressibility and non-isothermal aspects of the physics of deformation, whereas distortional aspects are due to change of shape of the volume. Volumetric and distortional physics are mutually exclusive. Volumetric aspects when quantified in terms of the Cauchy stress tensor in the current configuration remain the same regardless of the type deformation or constitution of the matter. Thus, the constitutive theories for the equilibrium stress tensor for the deformation physics considered in this chapter remain the same as those derived in Chapter 8. What remains to be considered in this chapter are the constitutive theories for the deviatoric stress tensor and the heat vector. We consider these constitutive theories for different types of deformation physics. We primarily consider thermodynamic approaches with some discussion of non-thermodynamic and phenomenological approaches.

10.2 Finite deformation, finite strain

For this deformation physics, the second Piola-Kirchoff stress $\boldsymbol{\sigma}^{[0]}$ and the Green's strain $\boldsymbol{\varepsilon}_{[0]}$ are appropriate measures of stress and strain. We perform additive stress decomposition of $\boldsymbol{\sigma}^{[0]}$ into equilibrium second Piola-Kirchhoff stress tensor ${}_{e}\boldsymbol{\sigma}^{[0]}$ and deviatoric second Piola-Kirchhoff stress tensor ${}_{d}\boldsymbol{\sigma}^{[0]}$ to facilitate derivation of constitutive theories for volumetric and distortion deformation physics.

$$\boldsymbol{\sigma}^{[0]} = {}_{e}\boldsymbol{\sigma}^{[0]} + {}_{d}\boldsymbol{\sigma}^{[0]} \tag{10.1}$$

Due to dissipation, there is entropy production. Hence, the deformation physics is always is non-isothermal.

10.2.1 Equilibrium stress tensor ${}_{e}\boldsymbol{\sigma}^{[0]}$

The constitutive theory for the equilibrium second Piola-Kirchoff stress tensor has been derived in Chapter 8. We have the following.

Compressible matter

$$ {}_{e}\boldsymbol{\sigma}^{[0]} = |\boldsymbol{J}| p(\rho,\theta)(\boldsymbol{J}^T \cdot \boldsymbol{J})^{-1} \tag{10.2}$$

$$ p(\rho,\theta) = -\rho^2 \frac{\partial \Phi(\rho,\theta)}{\partial \rho} \tag{10.3}$$

Incompressible matter

$$ {}_{e}\boldsymbol{\sigma}^{[0]} = |\boldsymbol{J}| p(\theta)(\boldsymbol{J}^T \cdot \boldsymbol{J})^{-1} \tag{10.4}$$

and the reduced form of the entropy inequality can be written as (Chapter 8).

$$ -{}_{d}\boldsymbol{\sigma}^{[0]} : \dot{\boldsymbol{\varepsilon}}_{[0]} + \frac{\boldsymbol{q}\cdot\boldsymbol{g}}{\theta} \leq 0 \tag{10.5}$$

When ${}_d\boldsymbol{\sigma}^{[0]} : \boldsymbol{\varepsilon}_{[0]} > 0$ and $\frac{\boldsymbol{q}\cdot\boldsymbol{g}}{\theta} \leq 0$, (10.5) is satisfied. These inequalities serve as restrictions in the constitutive theories for ${}_d\boldsymbol{\sigma}^{[0]}$ and $\boldsymbol{q}$. Based on the conjugate pairs in (10.5), we can write (including θ as an argument tensor).

$$ {}_d\boldsymbol{\sigma}^{[0]} = {}_d\boldsymbol{\sigma}^{[0]}(\boldsymbol{\varepsilon}_{[0]}, \theta) \tag{10.6} $$

$$ \boldsymbol{q} = \boldsymbol{q}(\boldsymbol{g}, \theta) \tag{10.7} $$

10.2.2 Deviatoric stress ${}_d\boldsymbol{\sigma}^{[0]}$

Due to the presence of dissipation mechanism $\dot{\boldsymbol{\varepsilon}}_{[0]}$ or $\boldsymbol{\varepsilon}_{[1]}$ needs to be an argument tensor of ${}_d\boldsymbol{\sigma}^{[0]}$. If we further consider that strain rates of up to orders n, contribute to dissipation and additional stresses, then in addition to $\boldsymbol{\varepsilon}_{[1]}$ as an argument tensor of ${}_d\boldsymbol{\sigma}^{[0]}$, we also need to consider $\boldsymbol{\varepsilon}_{[i]}$; $i = 2, 3, \ldots, n$ as additional argument tensors of ${}_d\boldsymbol{\sigma}^{[0]}$ in (10.6). Thus, the argument tensors of ${}_d\boldsymbol{\sigma}^{[0]}$ in (10.6) can be augmented with $\boldsymbol{\varepsilon}_{[i]}$; $i = 1, 2, \ldots, n$.

$$ {}_d\boldsymbol{\sigma}^{[0]} = {}_d\boldsymbol{\sigma}^{[0]}(\boldsymbol{\varepsilon}_{[0]}, \boldsymbol{\varepsilon}_{[i]}, \theta) \quad ; \quad i = 1, 2, \ldots, n \tag{10.8} $$

We consider ${}_d\boldsymbol{\sigma}^{[0]}$ with its argument tensors defined in (10.8) to derive a constitutive theory for ${}_d\boldsymbol{\sigma}^{[0]}$ using representation theorem. Let ${}^{\sigma}\underset{\sim}{\boldsymbol{G}}^i$; $i = 1, 2, \ldots, N$ be the combined generators of the argument tensors of ${}_d\boldsymbol{\sigma}^{[0]}$ in (10.8) that are symmetric tensors of rank two. Then, ${}_d\boldsymbol{\sigma}^{[0]}$ in the current configuration can be expressed as a linear combination of $\boldsymbol{I}$ and ${}^{\sigma}\underset{\sim}{\boldsymbol{G}}^i$; $i = 1, 2, \ldots, N$

$$ {}_d\boldsymbol{\sigma}^{[0]} = {}^{\sigma}\underset{\sim}{\alpha}^0 \boldsymbol{I} + \sum_{i=1}^{N} {}^{\sigma}\underset{\sim}{\alpha}^i ({}^{\sigma}\underset{\sim}{\boldsymbol{G}}^i) \tag{10.9} $$

$$ \text{in which} \quad {}^{\sigma}\underset{\sim}{\alpha}^i = {}^{\sigma}\underset{\sim}{\alpha}^i({}^{\sigma}\underset{\sim}{I}^j, \theta) \quad ; \quad j = 1, 2, \ldots, M \quad ; \quad i = 0, 1, \ldots, N \tag{10.10} $$

${}^{\sigma}\underset{\sim}{I}^j$; $j = 1, 2, \ldots, M$ are the combined invariants of the same argument tensors of ${}_d\boldsymbol{\sigma}^{[0]}$ in (10.8). We note that ${}^{\sigma}\underset{\sim}{\alpha}^i$; $i = 1, 2, \ldots, M$ are not material coefficients. These are merely coefficients in the linear combination (10.9).

Material coefficients

To determine the material coefficients in (10.9), we consider Taylor series expansion of ${}^{\sigma}\underset{\sim}{\alpha}^i$; $i = 0, 1, \ldots, N$ in ${}^{\sigma}\underset{\sim}{I}^j$; $j = 1, 2, \ldots, M$ about a known configuration $\underline{\Omega}$ and retain only up to linear terms in ${}^{\sigma}\underset{\sim}{I}^j$ (for simplicity). We do not consider Taylor series expansion of ${}^{\sigma}\underset{\sim}{\alpha}^i$; $i = 0, 1, \ldots, N$ in θ as the thermal stress field has already been accounted for in the constitutive theory for ${}_e\boldsymbol{\sigma}^{[0]}$.

$$ {}^{\sigma}\underset{\sim}{\alpha}^i = {}^{\sigma}\underset{\sim}{\alpha}^i\Big|_{\underline{\Omega}} + \sum_{j=1}^{M} \frac{\partial\, {}^{\sigma}\underset{\sim}{\alpha}^i}{\partial\, {}^{\sigma}\underset{\sim}{I}^j}\Bigg|_{\underline{\Omega}} \left({}^{\sigma}\underset{\sim}{I}^j - {}^{\sigma}\underset{\sim}{I}^j\Big|_{\underline{\Omega}}\right) \quad ; \quad i = 0, 1, \ldots, N \quad ; \quad j = 1, 2 \ldots, M \tag{10.11} $$

substituting (10.11) in (10.9)

$$
\begin{aligned}
{}_d\boldsymbol{\sigma}^{[0]} = & \left({}^{\sigma}\underset{\sim}{\alpha}^0\Big|_{\underline{\Omega}} + \sum_{j=1}^{M} \frac{\partial\, {}^{\sigma}\underset{\sim}{\alpha}^0}{\partial\, {}^{\sigma}\underset{\sim}{I}^j}\Bigg|_{\underline{\Omega}} \left({}^{\sigma}\underset{\sim}{I}^j - {}^{\sigma}\underset{\sim}{I}^j\Big|_{\underline{\Omega}} \right) \right) \boldsymbol{I} \\
& + \sum_{i=1}^{N} \left({}^{\sigma}\underset{\sim}{\alpha}^i\Big|_{\underline{\Omega}} + \sum_{j=1}^{M} \frac{\partial\, {}^{\sigma}\underset{\sim}{\alpha}^i}{\partial\, {}^{\sigma}\underset{\sim}{I}^j}\Bigg|_{\underline{\Omega}} \left({}^{\sigma}\underset{\sim}{I}^j - {}^{\sigma}\underset{\sim}{I}^j\Big|_{\underline{\Omega}} \right) \right) {}^{\sigma}\boldsymbol{G}^i
\end{aligned}
\tag{10.12}
$$

Collecting coefficients in (10.12) (of the terms defined in the current configuration) of $\boldsymbol{I}$, ${}^{\sigma}\underset{\sim}{I}^j\boldsymbol{I}$, ${}^{\sigma}\underset{\sim}{\boldsymbol{G}}^i$, ${}^{\sigma}\underset{\sim}{I}^j({}^{\sigma}\underset{\sim}{\boldsymbol{G}}^i)$ and defining

$$
\begin{aligned}
\underset{\sim}{\sigma}^0\Big|_{\underline{\Omega}} &= {}^{\sigma}\underset{\sim}{\alpha}^0\Big|_{\underline{\Omega}} - \sum_{j=1}^{M} \frac{\partial\, {}^{\sigma}\underset{\sim}{\alpha}^0}{\partial\, {}^{\sigma}\underset{\sim}{I}^j}\Bigg|_{\underline{\Omega}} \left({}^{\sigma}\underset{\sim}{I}^j\Big|_{\underline{\Omega}} \right) \quad ; \quad {}^{\sigma}\underset{\sim}{a}_j = \frac{\partial\, {}^{\sigma}\underset{\sim}{\alpha}^0}{\partial\, {}^{\sigma}\underset{\sim}{I}^j}\Bigg|_{\underline{\Omega}} \\
{}^{\sigma}\underset{\sim}{b}_i &= {}^{\sigma}\underset{\sim}{\alpha}^i\Big|_{\underline{\Omega}} - \sum_{j=1}^{M} \frac{\partial\, {}^{\sigma}\underset{\sim}{\alpha}^i}{\partial\, {}^{\sigma}\underset{\sim}{I}^j}\Bigg|_{\underline{\Omega}} \left({}^{\sigma}\underset{\sim}{I}^j\Big|_{\underline{\Omega}} \right) \quad ; \quad {}^{\sigma}\underset{\sim}{c}_{ij} = \frac{\partial\, {}^{\sigma}\underset{\sim}{\alpha}^i}{\partial\, {}^{\sigma}\underset{\sim}{I}^j}\Bigg|_{\underline{\Omega}} \\
& i = 1, 2, \dots, N \quad ; \quad j = 1, 2, \dots, M
\end{aligned}
\tag{10.13}
$$

Equation (10.12) can be written as

$$
{}_d\boldsymbol{\sigma}^{[0]} = \underset{\sim}{\sigma}^0\Big|_{\underline{\Omega}} \boldsymbol{I} + \sum_{j=1}^{M} {}^{\sigma}\underset{\sim}{a}_j({}^{\sigma}\underset{\sim}{I}^j)\boldsymbol{I} + \sum_{i=1}^{N}\sum_{j=1}^{M} {}^{\sigma}\underset{\sim}{c}_{ij}({}^{\sigma}\underset{\sim}{I}^j){}^{\sigma}\underset{\sim}{\boldsymbol{G}}^i + \sum_{i=1}^{N} {}^{\sigma}\underset{\sim}{b}_i({}^{\sigma}\underset{\sim}{\boldsymbol{G}}^i) \tag{10.14}
$$

where ${}^{\sigma}\underset{\sim}{a}_j$, ${}^{\sigma}\underset{\sim}{b}_i$, ${}^{\sigma}\underset{\sim}{c}_{ij}$; $i = 1, 2, \dots, N$; $j = 1, 2, \dots, M$ are material coefficients in the known configuration $\underline{\Omega}$. In (10.14), $\underset{\sim}{\sigma}^0\Big|_{\underline{\Omega}}$ is the initial stress field in the known configuration $\underline{\Omega}$. This constitutive theory for ${}_d\boldsymbol{\sigma}^{[0]}$ requires $(M+N+MN)$ material coefficients. The constitutive theory is based on integrity (complete basis of the space containing ${}_d\boldsymbol{\sigma}^{[0]}$). The material coefficients ${}^{\sigma}\underset{\sim}{a}_j$, ${}^{\sigma}\underset{\sim}{c}_{ij}$, ${}^{\sigma}\underset{\sim}{b}_j$; $i = 1, 2, \dots, N$ and $j = 1, 2, \dots, M$ can be functions of ${}^{\sigma}\underset{\sim}{I}^j\Big|_{\underline{\Omega}}$; $j = 1, 2, \dots, M$ and $\theta\Big|_{\underline{\Omega}}$.

In this constitutive theory the dissipation mechanism is due to strain rates $\boldsymbol{\varepsilon}_{[i]}$; $i = 1, 2, \dots, n$, up to order n. *Thus, we refer to this constitutive theory as an ordered rate constitutive theory of up to order n.* By choosing a specific value of n, ordered rate constitutive theories of desired order can be obtained.

Simplified constitutive theories for ${}_d\boldsymbol{\sigma}^{[0]}$

The constitutive theory (10.14) uses complete basis, but requires too many material coefficients. Simplified constitutive theories with fewer material coefficients that are more suitable for specific applications can be derived by appropriate choice of the highest order of the strain rate n and retaining only desired generators and invariants. We consider some examples in the following.

10.2.2.1 A simplified linear constitutive theory of order n

We consider an ordered rate constitutive theory of order n that is linear in the components of the $\boldsymbol{\varepsilon}_{[0]}$ and $\boldsymbol{\varepsilon}_{[i]}$; $i = 1, 2, \dots, n$. In such a constitutive theory, the combine generators of the argument tensors of ${}_d\boldsymbol{\sigma}^{[0]}$ that are symmetric tensors of rank two are $\boldsymbol{\varepsilon}_{[0]}$, $\boldsymbol{\varepsilon}_{[i]}$; $i = 1, 2, \dots, n$. We can write the following using (10.14) (after redefining material coefficients).

$$ {}_d\boldsymbol{\sigma}^{[0]} = \underset{\sim}{\sigma}^0\Big|_{\underline{\Omega}} \boldsymbol{I} + a_1\boldsymbol{\varepsilon}_{[0]} + a_2(\operatorname{tr}\boldsymbol{\varepsilon}_{[0]})\boldsymbol{I} + \sum_{i=1}^{n} b_i^1\boldsymbol{\varepsilon}_{[i]} + \sum_{i=1}^{n} b_i^2(\operatorname{tr}\boldsymbol{\varepsilon}_{[i]})\boldsymbol{I} \tag{10.15} $$

The material coefficients a_1, a_2, b_i^1 and b_i^2 ; $i = 1, 2, \dots, n$ can be functions of ${}^{\sigma}\underset{\sim}{I}^j\Big|_{\underline{\Omega}}$; $j = 1, 2, \dots, M$ and $\theta|_{\underline{\Omega}}$. This constitutive theory (10.15) can also be written in matrix and vector notations if we employ Voigt's notation. Let

$$ \{{}_d\sigma^{[0]}\}^T = [{}_d\sigma_{11}^{[0]}, {}_d\sigma_{22}^{[0]}, {}_d\sigma_{33}^{[0]}, {}_d\sigma_{23}^{[0]}, {}_d\sigma_{31}^{[0]}, {}_d\sigma_{12}^{[0]}] \tag{10.16} $$

If we use the same arrangement of the components of $\boldsymbol{\varepsilon}_{[0]}$ and $\boldsymbol{\varepsilon}_{[i]}$; $i = 1, 2, \dots, n$ as in (10.16) for ${}_d\boldsymbol{\sigma}^{[0]}$, then we can write (10.15) in the following form.

$$ \{{}_d\sigma^{[0]}\} = \{\sigma^0\} + [\underset{\sim}{a}]\{\varepsilon_{[0]}\} + \sum_{i=1}^{n} [\underset{\sim}{b}_i]\{\varepsilon_{[i]}\} \tag{10.17} $$

in which the matrix $[\underset{\sim}{a}]$ is defined as

$$ [\underset{\sim}{a}] = \begin{bmatrix} a_1 + a_2 & a_2 & a_2 & 0 & 0 & 0 \\ a_2 & a_1 + a_2 & a_2 & 0 & 0 & 0 \\ a_2 & a_2 & a_1 + a_2 & 0 & 0 & 0 \\ 0 & 0 & 0 & a_1 & 0 & 0 \\ 0 & 0 & 0 & 0 & a_1 & 0 \\ 0 & 0 & 0 & 0 & 0 & a_1 \end{bmatrix} \tag{10.18} $$

and

$$ \{\sigma^0\}^T = [\underset{\sim}{\sigma}^0\Big|_{\underline{\Omega}}, \underset{\sim}{\sigma}^0\Big|_{\underline{\Omega}}, \underset{\sim}{\sigma}^0\Big|_{\underline{\Omega}}, 0, 0, 0] \tag{10.19} $$

Components of each matrix $[\underset{\sim}{b}_i]$ can be obtained from (10.18) by replacing a_1 and a_2 with b_i^1 and b_i^2.

10.2.2.2 Linear constitutive theory of order one ($n = 1$)

Using $n = 1$ in (10.17) and (10.15) we obtain the most simple constitutive theory for ${}_d\boldsymbol{\sigma}^{[0]}$ of order one that is linear in the components of $\boldsymbol{\varepsilon}_{[0]}$ and $\boldsymbol{\varepsilon}_{[1]}$.

$$ \{{}_d\sigma^{[0]}\} = \{\sigma^0\} + [\underset{\sim}{a}]\{\varepsilon_{[0]}\} + [\underset{\sim}{b}_1]\{\varepsilon_{[1]}\} \tag{10.20} $$

$$ \text{or} \quad {}_d\boldsymbol{\sigma}^{[0]} = \underset{\sim}{\sigma}^0\Big|_{\underline{\Omega}} \boldsymbol{I} + a_1\boldsymbol{\varepsilon}_{[0]} + a_2(\operatorname{tr}\boldsymbol{\varepsilon}_{[0]})\boldsymbol{I} + b_1^1\boldsymbol{\varepsilon}_{[1]} + b_1^2(\operatorname{tr}\boldsymbol{\varepsilon}_{[1]})\boldsymbol{I} \tag{10.21} $$

If we define

$$ a_1 = 2\underset{\sim}{\mu} \quad ; \quad a_2 = \underset{\sim}{\lambda} \quad ; \quad b_1^1 = 2\underset{\sim}{\eta} \quad ; \quad b_1^2 = \underset{\sim}{k} \tag{10.22} $$

then (10.21) can be written as

$$ {}_d\boldsymbol{\sigma}^{[0]} = \underset{\sim}{\sigma}^0\Big|_{\underline{\Omega}} \boldsymbol{I} + 2\underset{\sim}{\mu}\boldsymbol{\varepsilon}_{[0]} + \underset{\sim}{\lambda}(\operatorname{tr}\boldsymbol{\varepsilon}_{[0]})\boldsymbol{I} + 2\underset{\sim}{\eta}\boldsymbol{\varepsilon}_{[1]} + \underset{\sim}{k}(\operatorname{tr}\boldsymbol{\varepsilon}_{[1]})\boldsymbol{I} \tag{10.23} $$

Material coefficients $\underset{\sim}{\mu}$ and $\underset{\sim}{\lambda}$ are similar to Lamé's constants and $\underset{\sim}{\eta}$ and $\underset{\sim}{k}$ are similar to first and second viscosities.

10.2.2.3 Remarks

(1) Ordered rate constitutive theory has been derived for finite deformation, finite strain, compressible as well as incompressible matter using ${}_d\boldsymbol{\sigma}^{[0]}$ and $\boldsymbol{\varepsilon}_{[0]}$ as stress and strain measures as well as $\boldsymbol{\varepsilon}_{[i]}$; $i = 1, 2, \dots, n$ as strain rate measures.

(2) Since $\boldsymbol{\varepsilon}_{[0]}$ and $\boldsymbol{\varepsilon}_{[i]}$; $i = 1, 2, \dots, n$ are non-linear measures in terms of displacement gradients, elastic behavior as well as dissipation mechanism are non-linear in this constitutive theory.

(3) Inclusion of each strain rate in the dissipation mechanism requires two material coefficients in the constitutive theory that is linear in the components of $\boldsymbol{\varepsilon}_{[i]}$. This is similar to elastic response of solid continua.

(4) We note that in the absence of the first three terms on the right side of (10.21), the constitutive theory for the stress tensor appears quite similar to Newton's law of viscosity for compressible fluent continua in Lagrangian description.

(5) We note that

$$\boldsymbol{\varepsilon}_{[0]} = \frac{1}{2}\left(\boldsymbol{J}^T \cdot \boldsymbol{J} - \boldsymbol{I}\right) = \frac{1}{2}\left({}^d\boldsymbol{J} + {}^d\boldsymbol{J}^T + {}^d\boldsymbol{J}^T \cdot {}^d\boldsymbol{J}\right) \tag{10.24}$$

$$\boldsymbol{\varepsilon}_{[1]} = \frac{D\boldsymbol{\varepsilon}_{[0]}}{Dt} = \frac{1}{2}\left(\dot{\boldsymbol{J}}^T \cdot \boldsymbol{J} + \boldsymbol{J}^T \cdot \dot{\boldsymbol{J}}\right) \tag{10.25}$$

$$\text{or} \quad \boldsymbol{\varepsilon}_{[i]} = \frac{D\boldsymbol{\varepsilon}_{[i-1]}}{Dt} \quad ; \quad i = 1, 2, \dots, n \tag{10.26}$$

Thus, we see that strain rates are rather complex non-linear expressions in displacement gradients and their rates.

10.3 Small strain, small deformation

For small strain, small deformation, the Cauchy stress tensor $\boldsymbol{\sigma}$ (basis independent) and $\boldsymbol{\varepsilon}$, the linear part of the Green's strain tensor are appropriate stress and strain measures. We perform additive decomposition of Cauchy stress tensor $\boldsymbol{\sigma}$.

$$\boldsymbol{\sigma} = {}_e\boldsymbol{\sigma} + {}_d\boldsymbol{\sigma} \tag{10.27}$$

As usual, ${}_e\boldsymbol{\sigma}$ incorporates volumetric physics and ${}_d\boldsymbol{\sigma}$ describes distortional physics.

10.3.1 Equilibrium stress tensor ${}_e\boldsymbol{\sigma}$

The constitutive theory for ${}_e\boldsymbol{\sigma}$ has been derived in Chapter 8 (equation (8.77)). We can write the following.

$$ {}_e\boldsymbol{\sigma} = p(\rho, \theta)\boldsymbol{\delta} \quad ; \quad \text{compressible} \tag{10.28}$$

$$ {}_e\boldsymbol{\sigma} = p(\theta)\boldsymbol{\delta} \quad ; \quad \text{incompressible} \tag{10.29}$$

and the reduced form of the entropy inequality (Chapter 8) can be written as

$$-{}_d\boldsymbol{\sigma} : \dot{\boldsymbol{\varepsilon}} + \frac{\boldsymbol{q} \cdot \boldsymbol{g}}{\theta} \leq 0 \tag{10.30}$$

Based on the conjugate pairs in (10.30), we can write

$${}_d\boldsymbol{\sigma} = {}_d\boldsymbol{\sigma}(\boldsymbol{\varepsilon}, \theta) \tag{10.31}$$

$$\text{and} \quad \boldsymbol{q} = \boldsymbol{q}(\boldsymbol{g}, \theta) \tag{10.32}$$

10.3.2 Deviatoric stress ${}_d\boldsymbol{\sigma}$

Following Section 10.2.2, the argument tensors of ${}_d\boldsymbol{\sigma}$ in (10.31) are augmented by $\boldsymbol{\varepsilon}_{(i)}$; $i = 1, 2, \dots, n$. This implies that each strain rate contributes to ${}_d\boldsymbol{\sigma}$ and dissipation.

$${}_d\boldsymbol{\sigma} = {}_d\boldsymbol{\sigma}(\boldsymbol{\varepsilon}, \boldsymbol{\varepsilon}_{(i)}, \theta) \quad ; \quad i = 1, 2, \dots, n \tag{10.33}$$

in which $\boldsymbol{\varepsilon}_{(i)}$ are simply time derivatives of linear Green's strain $\boldsymbol{\varepsilon}$. The constitutive theory for ${}_d\boldsymbol{\sigma}$ is derived using (10.33) and representation theorem. Let ${}^\sigma\underset{\sim}{\boldsymbol{G}}^i$; $i = 1, 2, \dots, N$ be the combined generators of the argument tensors of ${}_d\boldsymbol{\sigma}$ in (10.33) that are symmetric tensors of rank two. Then, ${}_d\boldsymbol{\sigma}$ in the current configuration can be expressed by a linear combination of $\boldsymbol{I}$ and ${}^\sigma\underset{\sim}{\boldsymbol{G}}^i$; $i = 1, 2, \dots, N$

$${}_d\boldsymbol{\sigma} = {}^\sigma\underset{\sim}{\alpha}^0 \boldsymbol{I} + \sum_{i=1}^{N} {}^\sigma\underset{\sim}{\alpha}^i \, {}^\sigma\underset{\sim}{\boldsymbol{G}}^i \tag{10.34}$$

$$\text{in which} \quad {}^\sigma\underset{\sim}{\alpha}^i = {}^\sigma\underset{\sim}{\alpha}^i({}^\sigma\underset{\sim}{I}^j, \theta) \quad i = 0, 1, \dots, N \quad ; \quad j = 1, 2, \dots, M \tag{10.35}$$

${}^\sigma\underset{\sim}{I}^j$; $j = 1, 2, \dots, M$ are the combined invariants of the same argument tensors of ${}_d\boldsymbol{\sigma}$ in (10.33). To determine the material coefficients in (10.34), we consider Taylor series expansion of ${}^\sigma\underset{\sim}{\alpha}^i$; $i = 0, 1, \dots, N$ in ${}^\sigma\underset{\sim}{I}^j$; $j = 1, 2, \dots, M$ about a known configuration $\underline{\Omega}$ and retain only up to linear terms in ${}^\sigma\underset{\sim}{I}^j$; $j = 1, 2, \dots, M$. We do not consider Taylor series expansion θ, as the thermal stress field is already accounted for in the constitutive theory for ${}_e\boldsymbol{\sigma}$.

$${}^\sigma\underset{\sim}{\alpha}^i = {}^\sigma\underset{\sim}{\alpha}^i\Big|_{\underline{\Omega}} + \sum_{j=1}^{M} \left.\frac{\partial\, {}^\sigma\underset{\sim}{\alpha}^i}{\partial\, {}^\sigma\underset{\sim}{I}^j}\right|_{\underline{\Omega}} \left({}^\sigma\underset{\sim}{I}^j - {}^\sigma\underset{\sim}{I}^j\Big|_{\underline{\Omega}} \right) \tag{10.36}$$

substituting (10.36) in (10.34)

$$\begin{aligned} {}_d\boldsymbol{\sigma} = & \left({}^\sigma\underset{\sim}{\alpha}^0\Big|_{\underline{\Omega}} + \sum_{j=1}^{M} \left.\frac{\partial\, {}^\sigma\underset{\sim}{\alpha}^0}{\partial\, {}^\sigma\underset{\sim}{I}^j}\right|_{\underline{\Omega}} \left({}^\sigma\underset{\sim}{I}^j - {}^\sigma\underset{\sim}{I}^j\Big|_{\underline{\Omega}} \right) \right) \boldsymbol{I} \\ & + \sum_{i=1}^{N} \left({}^\sigma\underset{\sim}{\alpha}^i\Big|_{\underline{\Omega}} + \sum_{j=1}^{M} \left.\frac{\partial\, {}^\sigma\underset{\sim}{\alpha}^i}{\partial\, {}^\sigma\underset{\sim}{I}^j}\right|_{\underline{\Omega}} \left({}^\sigma\underset{\sim}{I}^j - {}^\sigma\underset{\sim}{I}^j\Big|_{\underline{\Omega}} \right) \right) {}^\sigma\underset{\sim}{\boldsymbol{G}}^i \end{aligned} \tag{10.37}$$

Collecting coefficients of $\boldsymbol{I}$, ${}^\sigma\underset{\sim}{I}^j \boldsymbol{I}$, ${}^\sigma\underset{\sim}{\boldsymbol{G}}^i$, ${}^\sigma\underset{\sim}{I}^j({}^\sigma\underset{\sim}{\boldsymbol{G}}^i)$ and defining

$$\begin{aligned} & \underset{\sim}{\sigma}^0\Big|_{\underline{\Omega}} = {}^\sigma\underset{\sim}{\alpha}^0\Big|_{\underline{\Omega}} - \sum_{j=1}^{M} \left.\frac{\partial\, {}^\sigma\underset{\sim}{\alpha}^0}{\partial\, {}^\sigma\underset{\sim}{I}^j}\right|_{\underline{\Omega}} \left({}^\sigma\underset{\sim}{I}^j\Big|_{\underline{\Omega}} \right) \quad ; \quad {}^\sigma\underset{\sim}{a}_j = \left.\frac{\partial\, {}^\sigma\underset{\sim}{\alpha}^0}{\partial\, {}^\sigma\underset{\sim}{I}^j}\right|_{\underline{\Omega}} \\ & {}^\sigma\underset{\sim}{b}_i = {}^\sigma\underset{\sim}{\alpha}^i\Big|_{\underline{\Omega}} - \sum_{j=1}^{M} \left.\frac{\partial\, {}^\sigma\underset{\sim}{\alpha}^i}{\partial\, {}^\sigma\underset{\sim}{I}^j}\right|_{\underline{\Omega}} \left({}^\sigma\underset{\sim}{I}^j\Big|_{\underline{\Omega}} \right) \quad ; \quad {}^\sigma\underset{\sim}{c}_{ij} = \left.\frac{\partial\, {}^\sigma\underset{\sim}{\alpha}^i}{\partial\, {}^\sigma\underset{\sim}{I}^j}\right|_{\underline{\Omega}} \\ & i = 1, 2, \dots, N \quad ; \quad j = 1, 2, \dots, M \end{aligned} \tag{10.38}$$

Equation (10.37) can be written as

$$_d\boldsymbol{\sigma} = \underset{\sim}{\sigma}^0\Big|_{\underline{\Omega}} \boldsymbol{I} + \sum_{j=1}^{M} {}^{\sigma}\underset{\sim}{a}_j({}^{\sigma}\underset{\sim}{I}^j)\boldsymbol{I} + \sum_{i=1}^{N}\sum_{j=1}^{M} {}^{\sigma}\underset{\sim}{c}_{ij}({}^{\sigma}\underset{\sim}{I}^j){}^{\sigma}\underset{\sim}{\boldsymbol{G}}^i + \sum_{i=1}^{N} {}^{\sigma}\underset{\sim}{b}_i {}^{\sigma}\underset{\sim}{\boldsymbol{G}}^i \tag{10.39}$$

In which $\underset{\sim}{\sigma}^0\Big|_{\underline{\Omega}}$ is the known initial stress field in the known configuration $\underline{\Omega}$. This constitutive theory requires $(M + N + MN)$ material coefficients. The constitutive theory is based on integrity (complete basis of the space $_d\boldsymbol{\sigma}$). ${}^{\sigma}\underset{\sim}{a}_j$, ${}^{\sigma}\underset{\sim}{b}_i$ and ${}^{\sigma}\underset{\sim}{c}_{ij}$; $i = 1, 2, \dots, N$; $j = 1, 2, \dots, M$ are material coefficients that can be functions of ${}^{\sigma}\underset{\sim}{I}^j\Big|_{\underline{\Omega}}$; $j = 1, 2, \dots, M$ and $\theta|_{\underline{\Omega}}$.

In this constitutive theory the dissipation mechanism is due to strain rates $\boldsymbol{\varepsilon}_{(i)}$; $i = 1, 2, \dots, n$. Thus, it is an ordered rate dissipation mechanism of up to order n. By choosing a specific value of n we can select the desired physics of dissipation.

Simplified constitutive theories for $_d\boldsymbol{\sigma}$

The constitutive theory (10.39) for $_d\boldsymbol{\sigma}$ uses complete basis of the space of $_d\boldsymbol{\sigma}$, but requires too many material coefficients. Simplified constitutive theories with fewer material coefficients can be derived by appropriate choice of n and retaining only desired generators and the invariants. We consider some simple theories in the following.

10.3.2.1 A simplified linear constitutive theory of order n

We consider an ordered rate constitutive theory for $_d\boldsymbol{\sigma}$ of order n that is linear in $\boldsymbol{\varepsilon}$ and $\boldsymbol{\varepsilon}_{(i)}$; $i = 1, 2, \dots, n$. For this constitutive theory the combined generators of the argument tensors that are symmetric tensors of rank two are $\boldsymbol{\varepsilon}$, $\boldsymbol{\varepsilon}_{(i)}$; $i = 1, 2, \dots, n$ and the combined invariants of the same argument tensors are $I_{\boldsymbol{\varepsilon}}$, $I_{\boldsymbol{\varepsilon}_{(i)}}$; $i = 1, 2, \dots, n$, i.e., traces of $\boldsymbol{\varepsilon}$ and $\boldsymbol{\varepsilon}_{(i)}$; $i = 1, 2, \dots, n$. Using (10.39), we can write the following for this choice of generators and invariants (redefining material coefficients).

$$_d\boldsymbol{\sigma} = \underset{\sim}{\sigma}^0\Big|_{\underline{\Omega}} \boldsymbol{I} + a_1\boldsymbol{\varepsilon} + a_2(\operatorname{tr}\boldsymbol{\varepsilon})\boldsymbol{I} + \sum_{i=1}^{n} b_i^1\boldsymbol{\varepsilon}_{(i)} + \sum_{i=1}^{n} b_i^2(\operatorname{tr}\boldsymbol{\varepsilon}_{(i)})\boldsymbol{I} \tag{10.40}$$

The material coefficients a_1, a_2, b_i^1, b_i^2 ; $i = 1, 2, \dots, n$ can be functions of ${}^{\sigma}\underset{\sim}{I}^j\Big|_{\underline{\Omega}}$; $j = 1, 2, \dots, M$ and $\theta|_{\underline{\Omega}}$. This constitutive theory (10.40) can also be written in matrix and vector form using Voigt's notation. Let

$$\{_d\sigma\}^T = [_d\sigma_{11}, {_d\sigma_{22}}, {_d\sigma_{33}}, {_d\sigma_{23}}, {_d\sigma_{31}}, {_d\sigma_{12}}] \tag{10.41}$$

If we use the same arrangement of the components of $\boldsymbol{\varepsilon}$ and $\boldsymbol{\varepsilon}_{(i)}$; $i = 1, 2, \dots, n$ as used for $_d\boldsymbol{\sigma}$ in (10.41), then we can write (10.41) in the following form.

$$\{_d\sigma\} = \{\sigma^0\} + [\underset{\sim}{a}]\{\varepsilon\} + \sum_{i=1}^{n} [\underset{\sim}{b}_i]\{\varepsilon_{(i)}\} \tag{10.42}$$

The coefficients of $[\underset{\sim}{a}]$ are given by (10.18) and $\{\sigma^0\}$ is defined by (10.19). Matrices $[\underset{\sim}{b}_i]$ are obtained from $[\underset{\sim}{a}]$ by replacing a_1 and a_2 with b_i^1 and b_i^2.

10.3.2.2 A simplified linear constitutive theory of order one ($n = 1$)

Using $n = 1$ in (10.42) and (10.40) we obtain the most simple constitutive theory for ${}_d\boldsymbol{\sigma}$ of order one that is linear in the components of $\boldsymbol{\varepsilon}$ and $\boldsymbol{\varepsilon}_{(1)}$

$$\{{}_d\sigma\} = \{\sigma^0\} + [\underset{\sim}{a}]\{\varepsilon\} + [\underset{\sim}{b}_1]\{\varepsilon_{(1)}\} \tag{10.43}$$

$${}_d\boldsymbol{\sigma} = \underset{\sim}{\sigma}^0\Big|_{\underset{\sim}{\Omega}} \boldsymbol{I} + a_1\boldsymbol{\varepsilon} + a_2(\operatorname{tr}\boldsymbol{\varepsilon})\boldsymbol{I} + b_1^1\boldsymbol{\varepsilon}_{(1)} + b_1^2(\operatorname{tr}\boldsymbol{\varepsilon}_{(1)})\boldsymbol{I} \tag{10.44}$$

$$\text{or} \quad {}_d\boldsymbol{\sigma} = \underset{\sim}{\sigma}^0\Big|_{\underset{\sim}{\Omega}} \boldsymbol{I} + 2\underset{\sim}{\mu}\boldsymbol{\varepsilon}_{[0]} + \underset{\sim}{\mu}(\operatorname{tr}\boldsymbol{\varepsilon}_{[0]})\boldsymbol{I} + 2\underset{\sim}{\eta}\boldsymbol{\varepsilon}_{[1]} + \underset{\sim}{k}(\operatorname{tr}\boldsymbol{\varepsilon}_{[1]})\boldsymbol{I} \tag{10.45}$$

Material coefficients $\underset{\sim}{\mu}$, $\underset{\sim}{\lambda}$, $\underset{\sim}{\eta}$ and $\underset{\sim}{k}$ are defined in (10.22). We remark that these coefficients are not necessarily the same as those in (10.23) for the finite deformation, finite strain case.

Remarks

(1) Since $\boldsymbol{\varepsilon}$ is a linear function of displacement gradient tensors, both elasticity as well as dissipation mechanism are linear in displacement gradients.

(2) Inclusion of each strain rate in the dissipation mechanism requires two material coefficients. This is similar and parallel to elastic deformation.

(3) In this constitutive theory (10.42) or (10.43), the material coefficients a_1 and a_2 are Lamé's constants 2μ and λ which can also be expressed in terms of modulus of elasticity E and Poisson's ratio ν (if we assume that the presence of dissipation does not influence elastic material coefficients).

(4) We note that

$$\boldsymbol{\varepsilon}_{(0)} = \frac{1}{2}\left({}^d\boldsymbol{J} + {}^d\boldsymbol{J}^T\right) \tag{10.46}$$

$$\boldsymbol{\varepsilon}_{(1)} = \frac{1}{2}\left({}^d\dot{\boldsymbol{J}} + {}^d\dot{\boldsymbol{J}}^T\right) = \frac{D\boldsymbol{\varepsilon}_{(0)}}{Dt} = \frac{\partial}{\partial t}\boldsymbol{\varepsilon}_{(0)} \tag{10.47}$$

Hence, $\boldsymbol{\varepsilon}_{(i)} = \frac{\partial^i \boldsymbol{\varepsilon}_{(0)}}{\partial^i t}$; $i = 1, 2, \dots, n$. Thus, we note that due to linear strains and Lagrangian description, the strain rates are simple partial derivatives with respect to time.

10.4 Small deformation, small strain, incompressible, isothermal: Kelvin-Voigt model

For this deformation physics ${}_e\boldsymbol{\sigma}$ is given by (Chapter 8).

$${}_e\boldsymbol{\sigma} = p\boldsymbol{\delta} \tag{10.48}$$

If we assume that ${}_e\boldsymbol{\sigma} = 0$, then there is no distinction between total Cauchy stress $\boldsymbol{\sigma}$ and deviatoric Cauchy ${}_d\boldsymbol{\sigma}$. Thus, we can use $\boldsymbol{\sigma}$ instead of ${}_d\boldsymbol{\sigma}$ in the constitutive theory for the stress tensor. This assumption is necessary to compare with Kelvin-Voigt model.

The classical *Kelvin-Voigt model* originates with the simple assumption that the mechanisms of elasticity and dissipation in incompressible viscoelastic solids without memory for the simple one-dimensional case can be viewed as being due to a spring and a dashpot arranged in parallel. The stress due to the spring is proportional to the strain, while the stress due to the dashpot is considered to be proportional to the strain rate. Thus, for the one-dimensional case (infinitesimal deformation), we can write

$$\sigma_{11} = c_1 \varepsilon_{11} + c_2 \frac{\partial(\varepsilon_{11})}{\partial t} \quad ; \quad \varepsilon_{11} = \frac{\partial u_1}{\partial x_1} \tag{10.49}$$

This model in $\mathbb{R}^1$ has been considered in published works [26]. Equation (10.49) is called the Kelvin-Voigt constitutive model for small strain, small deformation, incompressible isothermal, 1D viscoelastic solid continua without memory.

Next, we consider the constitutive theory derived here for the same physics, i.e., equation (10.44) in which ${}_d\boldsymbol{\sigma}$ is replaced by $\boldsymbol{\sigma}$ (and neglecting initial stress field $\{\sigma^0\}$).

$$\sigma_{11} = a_1 \varepsilon_{11} + a_2 \varepsilon_{11} + b_1^1 (\varepsilon_{(1)})_{11} + b_2^1 (\varepsilon_{(1)})_{11} \tag{10.50}$$

$$\text{or} \quad \sigma_{11} = (a_1 + a_2)\varepsilon_{11} + (b_1^1 + b_2^1)(\varepsilon_{(1)})_{11} \tag{10.51}$$

defining

$$c_1 = a_1 + a_2 \quad ; \quad c_2 = b_1^1 + b_2^1 \tag{10.52}$$

Equation (10.51) becomes

$$\sigma_{11} = c_1 \varepsilon_{11} + c_2 \frac{\partial(\varepsilon_{11})}{\partial t} \tag{10.53}$$

Remarks

(1) We note that the constitutive theory for stress tensor derived in Section 10.3 reduces to the Kelvin-Voigt model in $\mathbb{R}^1$.

(2) Since the Kelvin-Voigt model is phenomenological, it cannot be extended to $\mathbb{R}^2$ or $\mathbb{R}^3$ or in general for viscoelastic solid continua. This is a rather serious drawback of all phenomenological models.

(3) The dissipation mechanism is always accompanied by entropy production which requires stress decomposition $\boldsymbol{\sigma} = {}_e\boldsymbol{\sigma} + {}_d\boldsymbol{\sigma}$ in which case the constitutive theories derived in this chapter for ${}_e\boldsymbol{\sigma}$ and ${}_d\boldsymbol{\sigma}$ hold, but the Kelvin-Voigt model has no mechanism to account for this physics.

(4) We note that the Kelvin-Voigt model contains only a single material coefficient for dissipation (due to $\mathbb{R}^1$) and has no mechanism to extend this to two material coefficients that are required in $\mathbb{R}^2$ and $\mathbb{R}^3$ as shown here.

10.5 1D wave propagation in viscoelastic solid continua

We consider complete mathematical model for 1D wave propagation in viscoelastic solid continua (small deformation, small strain, incompressible, isothermal). Balance of linear momenta and constitutive theory are given by (in the absence of body forces).

$$\rho_0 \frac{\partial^2 u_1}{\partial t^2} - \frac{\partial \sigma_{11}}{\partial x_1} = 0 \quad \forall (x_1, t) \in \Omega_{x_1,t} = \Omega_{x_1} \times \Omega_t \tag{10.54}$$

$$\text{and} \quad \sigma_{11} = c_1 \frac{\partial u_1}{\partial x_1} + c_2 \frac{\partial}{\partial t}\left(\frac{\partial u_1}{\partial x_1}\right) \tag{10.55}$$

substituting (10.55) in (10.54)

$$\rho_0 \frac{\partial^2 u_1}{\partial t^2} - c_1 \frac{\partial^2 u_1}{\partial x_1^2} - c_2 \frac{\partial}{\partial t}\left(\frac{\partial^2 u_1}{\partial x_1^2}\right) = 0 \quad \forall (x_1, t) \in \Omega_{x_1,t} = \Omega_{x_1} \times \Omega_t \tag{10.56}$$

This is the final mathematical model for 1D wave propagation resulting from the theory presented here as well as due to the Kelvin-Voigt model.

10.5.1 Alternate phenomenological model for dissipation

In a large majority of published works in engineering [2–4, 6–10, 13], it is common to assume that dissipation is due to a dashpot like mechanism that provide a resisting force to the motion that is proportional to the velocity. This force is included in the balance of linear momenta giving rise to (10.58) force balance in x_1-direction.

Thus, in this case

$$\sigma_{11} = c_1 \frac{\partial u_1}{\partial x_1} \quad ; \quad F_1^d = -\underset{\sim}{c}_2 \frac{\partial u_1}{\partial t} \tag{10.57}$$

F_1^d is the resisting force due to dashpot. Substituting σ_{11} from (10.57) in BLM (10.54) and subtracting F_1^d, we obtain the final mathematical model for 1D wave propagation based on this assumed phenomenological dissipation mechanism.

$$\rho_0 \frac{\partial^2 u_1}{\partial t^2} - c_1 \frac{\partial^2 u_1}{\partial x_1^2} + \underset{\sim}{c}_2 \frac{\partial u_1}{\partial t} = 0 \quad \forall (x_1, t) \in \Omega_{x_1,t} \tag{10.58}$$

We remark that the concept of dissipation proportional to velocity as used in (10.58) has no physical basis and is purely ad hoc. It is worth noting that if we decouple space and time, i.e., if we consider separation of variables in (10.56) for $u_1(x_1, t)$, then the resulting ODEs in time will exhibit dissipation proportional to velocity. We consider details in the following. Let

$$u_1(x_1, t) = g(x_1) h(t) \quad \forall (x_1, t) \in \Omega_{x_1,t} = \Omega_{x_1} \times \Omega_t = (0, L) \times (0, \tau) \tag{10.59}$$

in which $g(x_1)$ is a known function, L is the spatial domain and τ is the final value of time t. Substituting (10.59) in (10.56) and integrating over $\bar{\Omega}_{x_1}$

$$\left(\int_{\bar{\Omega}_{x_1}} g(x_1) \rho_0 \, dx_1\right) \frac{\partial^2 h(t)}{\partial t^2} - \left(c_1 \int_{\bar{\Omega}_{x_1}} \frac{\partial^2 g(x_1)}{\partial x_1^2} dx_1\right) h(t) - \left(\underset{\sim}{c}_2 \int_{\bar{\Omega}_{x_1}} g(x_1) dx_1\right) \frac{dh(t)}{dt} = 0 \tag{10.60}$$

$$\text{or} \quad m \frac{d^2 h(t)}{dt^2} - k h(t) - c \frac{dh(t)}{dt} = 0 \tag{10.61}$$

This is similar to an ordinary differential equation (ODE in time) describing a spring, mass and damper system. We note that strain rate dependent dissipation mechanism in BLM results in velocity dependent dissipation in the resulting ODEs in time only upon decoupling of space and time. Thus, use of $\underset{\sim}{c}_2 \frac{\partial u_1}{\partial t}$ in (10.58) for damping force is invalid and is bound to result in erroneous evolution from the solution of (10.58).

10.5.2 Model Problem: Numerical Studies

10.5.2.1 Model A

We present numerical studies using the mathematical models (10.56) and (10.58) (referred to as Model A and Model B) realizing that the mathematical model (10.58) has an incorrect mechanism of dissipation, nonetheless the evolution from (10.58) is of interest in comparison to that from (10.56). We calculate evolution described by (10.56) and (10.58) numerically using the finite element method. In performing numerical studies using the finite element method (or any other method), we must non-dimensionalize the mathematical model. First, consider (10.56) and rewrite it using hat ($\wedge$) on all quantities signifying that all quantities have their usual dimensions.

$$\widehat{\rho}_0 \frac{\partial^2 \widehat{u}_1}{\partial \widehat{t}^2} - \widehat{c}_1 \frac{\partial^2 \widehat{u}_1}{\partial \widehat{x}_1^2} - \widehat{c}_2 \frac{\partial}{\partial \widehat{t}} \left(\frac{\partial^2 \widehat{u}_1}{\partial \widehat{x}_1^2} \right) = 0 \quad \forall (\widehat{x}_1, \widehat{t}) \in \Omega_{\widehat{x}_1, \widehat{t}} \tag{10.62}$$

We consider $\Omega_{\widehat{x}_1,\widehat{t}} = \Omega_{\widehat{x}_1} \times \Omega_{\widehat{t}} = [0, \widehat{L}] \times (0, \widehat{\tau})$. We choose reference quantities with subscript 0 (zero) or *ref* and define dimensionless quantities using these. Let

$$\rho_0 = \frac{\widehat{\rho}_0}{(\rho_0)_{ref}} \; ; \; u_1 = \frac{\widehat{u}_1}{u_0} \; ; \; x_1 = \frac{\widehat{x}_1}{L_0} \; ; \; v_1 = \frac{\widehat{v}_1}{v_0} \; ; \; t = \frac{\widehat{t}}{t_0} \; ; \; t_0 = \frac{L_0}{v_0} \; ; \; E = \frac{\widehat{E}}{E_0} \tag{10.63}$$

We note that $\widehat{c}_1$ is $\widehat{E}$, the modulus of elasticity and $u_0 = L_0$. We choose L_0 and v_0, thus $t_0 = L_0/v_0$. At this stage, we have two options: (1) either we use reference speed of sound as v_0, L_0 and reference of modulus of elasticity E_0 (thus, $\sigma_0 = E_0$, σ_0 being reference stress) (2) or we can choose $\sigma_0 = E_0 = (\rho_0)_{ref} v_0^2$, reference characteristic kinetic energy, L_0 and choose a suitable reference velocity v_0. We consider details of both choices.

First, we consider (10.62) with $\widehat{c}_1 = \widehat{E}$ and use (10.63)

$$\left((\rho_0)_{ref} \frac{L_0}{t_0^2} \right) \rho_0 \frac{\partial^2 u_1}{\partial t^2} - \left(\frac{E_0 L_0}{L_0^2} \right) E \frac{\partial^2 u_1}{\partial x_1^2} - \left(\frac{\widehat{c}_2 L_0}{t_0 L_0^2} \right) \frac{\partial}{\partial t} \left(\frac{\partial^2 u_1}{\partial x_1^2} \right) = 0 \tag{10.64}$$

Using $t_0 = L_0/v_0$ in (10.64), we have

$$\left(\frac{(\rho_0)_{ref} v_0^2}{L_0} \right) \rho_0 \frac{\partial^2 u_1}{\partial t^2} - \left(\frac{E_0}{L_0} \right) E \frac{\partial^2 u_1}{\partial x_1^2} - \left(\frac{\widehat{c}_2 v_0}{L_0^2} \right) \frac{\partial}{\partial t} \left(\frac{\partial^2 u_1}{\partial x_1^2} \right) = 0 \tag{10.65}$$

Dividing throughout by $((\rho_0)_{ref} v_0^2 / L_0)$

$$\rho_0 \frac{\partial^2 u_1}{\partial t^2} - \left(\frac{E_0}{(\rho_0)_{ref} v_0^2} \right) E \frac{\partial^2 u_1}{\partial x_1^2} - \left(\frac{\widehat{c}_2}{L_0 (\rho_0)_{ref} v_0} \right) \frac{\partial}{\partial t} \left(\frac{\partial^2 u_1}{\partial x_1^2} \right) = 0 \tag{10.66}$$

Using speed of sound

$$
\begin{aligned}
v_0 &= \sqrt{\frac{E_0}{(\rho_0)_{ref}}} \quad ; \quad E_0 = (\rho_0)_{ref} v_0^2 \\
\therefore \quad \frac{E_0}{(\rho_0)_{ref} v_0^2} &= 1 \text{ and we define } \frac{\widehat{c}_2}{L_0 (\rho_0)_{ref} v_0} = c_2^d
\end{aligned}
\tag{10.67}
$$

using (10.67) in (10.66), we have the following.

$$
\begin{aligned}
\text{Model A:} \quad & \rho_0 \frac{\partial^2 u_1}{\partial t^2} - E \frac{\partial^2 u_1}{\partial x_1^2} - c_2^d \frac{\partial}{\partial t}\left(\frac{\partial^2 u_1}{\partial x_1^2}\right) = 0 \\
& \forall (x_1, t) \in \Omega_{x_1,t} = (0, L) \times (0, \tau)
\end{aligned}
\tag{10.68}
$$

c_2^d is the dimensionless dissipation coefficient.

Using characteristic kinetic energy

Let $E_0 = \sigma_0 = (\rho_0)_{ref} v_0^2$

$$
\sigma_0 = E_0 = (\rho_0)_{\text{ref}} v_0^2 \tag{10.69}
$$

$$
\frac{E_0}{(\rho_0)_{ref} v_0^2} = 1 \text{ and defining } \frac{\widehat{c}_2}{L_0 (\rho_0)_{ref} v_0} = c_2^d \tag{10.70}
$$

Using (10.69) is the same as using (10.67), hence using (10.69) in (10.66) we obtain the same dimensionless mathematical model as in (10.68). Equation (10.68) is the final form of the dimensionless form of Model A.

10.5.2.2 Model B

Next, we nondimensionalize (10.58). First, we rewrite (10.58) using ($\wedge$) on all quantities, indicating that all quantities have their usual dimensions.

$$
\widehat{\rho}_{ref} \frac{\partial^2 \widehat{u}_1}{\partial \widehat{t}^2} - \widehat{E} \frac{\partial^2 \widehat{u}_1}{\partial \widehat{x}_1^2} + \widehat{\underset{\sim}{c}}_2 \frac{\partial \widehat{u}_1}{\partial \widehat{t}} = 0 \quad \forall (\widehat{x}_1, \widehat{t}) \in \Omega_{(\widehat{x}_1, \widehat{t})} = \Omega_{\widehat{x}_1} \times \Omega_{\widehat{t}} = (0, \widehat{L}) \times (0, \widehat{\tau})
\tag{10.71}
$$

Using reference speed of sound or reference kinetic energy we can obtain the following (using the same procedure as used for Model A).

$$
\begin{aligned}
\text{Model B:} \quad & \rho_0 \frac{\partial^2 u_1}{\partial t^2} - E \frac{\partial^2 u_1}{\partial x_1^2} + \underset{\sim}{c}_2^d \frac{\partial u_1}{\partial t} = 0 \quad \forall \quad \Omega_{x_1,t} = \Omega_{x_1} \times \Omega_t \\
& \underset{\sim}{c}_2^d = \frac{\widehat{\underset{\sim}{c}}_2 L_0}{(\rho_0)_{ref} v_0}
\end{aligned}
\tag{10.72}
$$

The choice of v_0 changes depending upon whether we consider reference speed of sound or reference kinetic energy. $\underset{\sim}{c}_2^d$ is the dimensionless damping coefficient.

In the numerical studies, we consider an axial rod of dimensionless length one unit and choose $(\rho_0)_{ref} = \widehat{\rho}_0$ and $E_0 = \widehat{E}$ so that $\rho_0 = 1$ and $E = 1$. The spatial domain $[0, 1]$ for an increment of time $\Delta t = 0.1$, i.e., the space-time strip $[0, L] \times [0, \Delta t]$, is discretized using a uniform mesh of eight nine-node, p-version, higher order global differentiability, space-time finite elements [101–103, 105, 106]. Figure

10.1 shows details of the space-time domain for a time increment $\Delta t = t_{n+1} - t_n$ and boundary conditions as well as the initial conditions.

The left end of the rod is clamped (impermeable boundary) and the right end is subjected to a compressive piecewise-cubic strain distribution, such that the strain is continuous and differentiable with respect to time, over a time period of $2\Delta t$. For $t \geq 2\Delta t$, the applied strain at the right end of the rod is zero. We choose a p-level of 9 in space and time and a local approximation of class $C^{11}(\bar{\Omega}^e_{x_1,t})$, i.e., of class C^1 in space and time. The finite element formulations used for computing numerical solutions for both models (models A and B) use the space-time least squares process constructed using residual functionals [101–103, 105, 106].

The resulting computational process is unconditionally stable. Evolutions are computed using a space-time strip with time marching [101–103, 105, 106]. For the choice of local approximation $(C^{11}(\bar{\Omega}^e_{x_1,t}))$, the integrals in the finite element processes are Lebesgue, but due to the smoothness of the evolution for the 8-element discretization with a p-level of 9, the residual functionals are on the order of $O(10^{-6})$ or lower for all space-time strips for both model problems, confirming good accuracy of evolution.

For both model problems, the evolutions are completed for $0 \leq t \leq 2.0$. Since c_2^d and $\underset{\sim}{c}_2^d$ in model problems A and B do not have the same meaning (i.e., physics), a direct comparison of the evolutions for the same values of c_2^d and $\underset{\sim}{c}_2^d$ is not meaningful. For this reason, we choose a range of values for c_2^d and $\underset{\sim}{c}_2^d$ to show behaviors of dissipation in models A and B. We note that for Model A, $\sigma_{11} = E\frac{\partial u_1}{\partial x_1} + c_2 \frac{\partial}{\partial t}\left(\frac{\partial u_1}{\partial x_1}\right)$ whereas for Model B, $\sigma_{11} = E\frac{\partial u_1}{\partial x_1}$. We monitor σ_{11}, $\frac{\partial u_1}{\partial x_1}$ for Model A and σ_{11} (same as $\frac{\partial u_1}{\partial x_1}$ as $E = 1$) for Model B.

Figures 10.2 and 10.3 show plots of stress σ_{11} versus x_1 for different values of time for $c_2^d = 0.0$, 0.1, and 0.5 for model A.

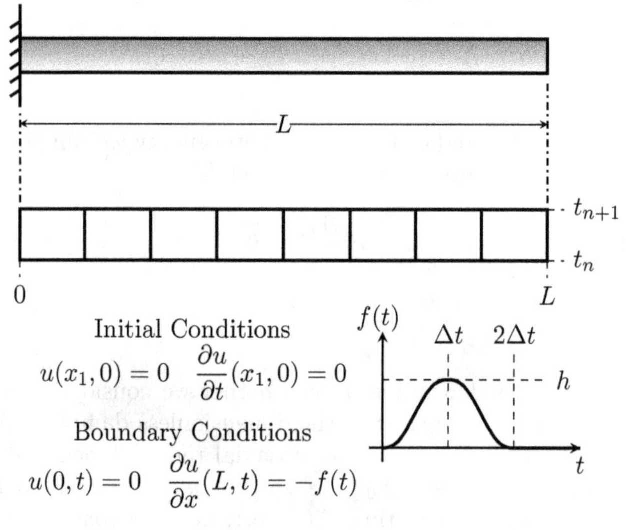

Figure 10.1: Schematic of the model problem

Due to dimensionless $E = 1$ and $\rho_0 = 1$, the wave speed is one. Thus, a stress pulse of base $2\Delta t = 0.2$ will travel 0.2 units of distance in 0.2 units of time. From Figure 10.2 (σ_{11} versus x_1) we note that for all three values of damping coefficient c_2^d the pulse is completely in the rod at $t = 0.2$. When $c_2^d = 0.0$, the amplitude and the support of stress wave are preserved during evolution $0.1 \leq t \leq 1.0$). When c_2^d is non-zero, amplitude decay and base elongation are observed. Higher value of c_2^d results in more amplitude decay and more pronounced base elongation. At $t = 1.0$, the stress wave has traveled one unit of distance and is incident at the impermeable boundary at $x_1 = 0.0$. At $t = 1.1$, due to reflection of the stress wave, its amplitude doubles as clearly seen for $c_2^d = 0.0$. At $t = 1.2$, 1.8 and 2.0, the reflected wave travels back towards $x_1 = 1.0$. At $t = 2.0$, the reflected stress wave is incident on the free boundary at $x = 1.0$, hence results in reflected tensile stress wave that travels back towards $x_1 = 0.0$ ($t = 2.2, 2.3$). It is interesting to note the reflection process that occurs between $2.0 \leq t \leq 2.2$.

Figures 10.4 and 10.5 show plots of $\frac{\partial u_1}{\partial x_1}$ versus x_1 for Model A for different values of time for $c_2^d = 0.0$, 0.1, 0.5. Differences in σ_{11} versus x_1 and $\frac{\partial u_1}{\partial x_1}$ (same as $E\frac{\partial u_1}{\partial x_1}$) versus x_1 are most significant for initial time steps as seen in Figures 10.2 and 10.4 for $t = 0.1$, 0.2 and 0.3. Evolution for subsequent values of time is quite similar to σ_{11} versus x_1 in Figures 10.2 and 10.3.

Figures 10.6 and 10.7 show plots of σ_{11} versus x_1 for model B for different values of time for $\underset{\sim}{c}_2^d = 0$, 1.0, and 2.0. From Figures 10.6 and 10.7, for model B, we note that for $\underset{\sim}{c}_2^d = 0$, we have exactly the same behavior as in Figures 10.2 and 10.3 as in this case, the two models are identical. For non-zero values of $\underset{\sim}{c}_2^d$, the evolutions in Figures 10.6 and 10.7 show amplitude decay of the stress wave, but the base of the wave is preserved during the evolution, regardless of the values of $\underset{\sim}{c}_2^d$.

Thus, this mechanism of dissipation in Model B is quite different compared to that in model A. From Figures 10.6 and 10.7, we note that progressively increasing values of $\underset{\sim}{c}_2^d$ result in progressively decaying amplitudes of the stress wave, indicating progressively increased dissipation, but the base of the stress wave remains unaltered during the entire evolution.

We note the reflection of the wave from the impermeable boundary (at $x_1 = 0.0$) at $t = 1.1$ doubles the amplitude of the reflected stress wave. Reflection of the stress wave from the free boundary at $x_1 = 1.0$ for $2 \leq t \leq 2.3$ resulting in a tensile wave is simulated correctly also.

10.6 Constitutive theories for heat vector $\boldsymbol{q}$

The constitutive theory for heat vector $\boldsymbol{q}$ remains the same as derived in Chapter 9, Section 9.4, hence not repeated here.

10.7 General remarks

In the following we make some general remarks regarding the use of mathematical models (CBL) utilizing the constitutive theories presented in this chapter.

(1) Such materials have mechanisms of elasticity as well as dissipation. A part of the rate of work is converted into rate of entropy production that alters the thermal field. The work expanded in entropy generation is not recoverable upon unloading.

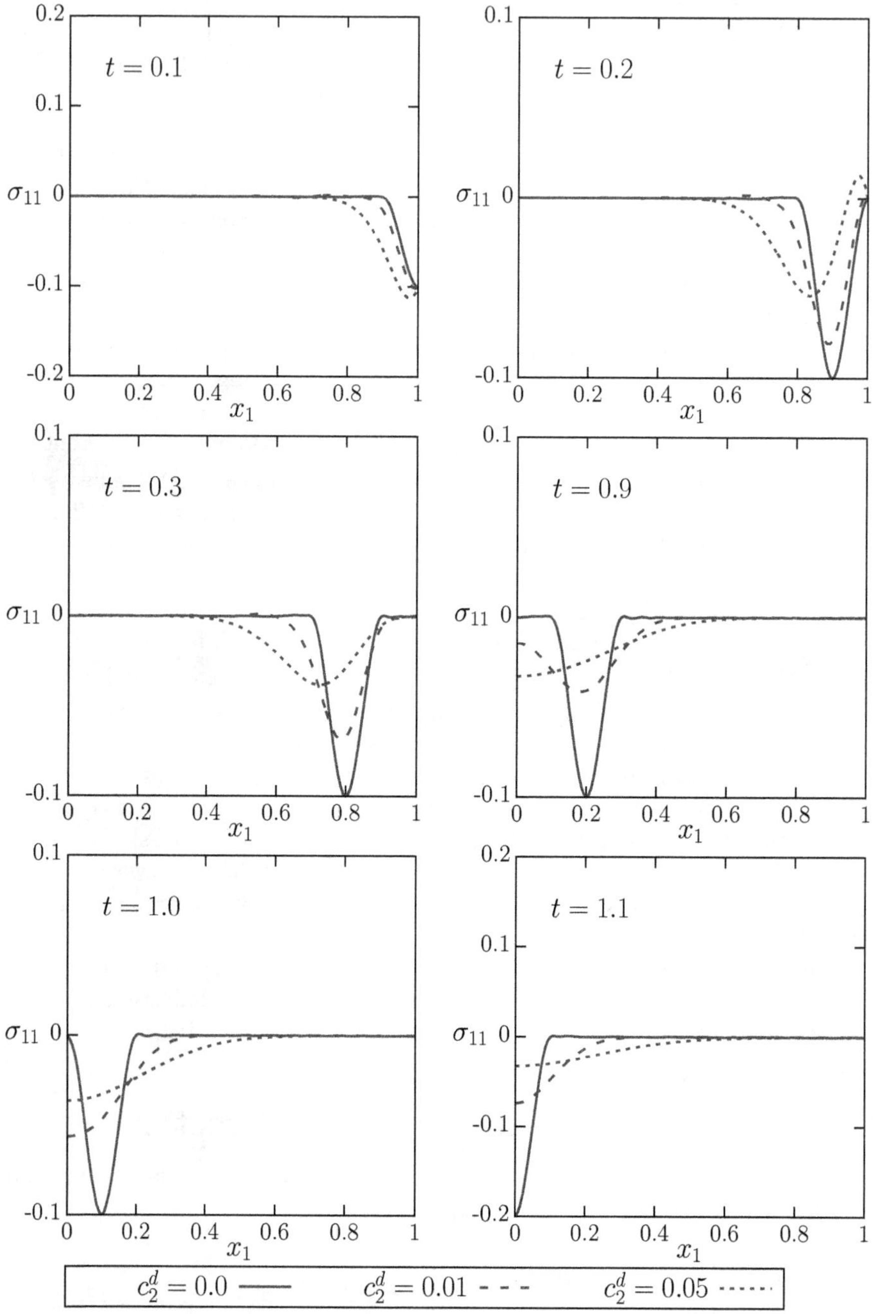

Figure 10.2: Time evolution of the propagation of an applied strain pulse in a one-dimensional axially deforming rod with dissipation based on strain rate (Model A: **strain rate-based damping**)

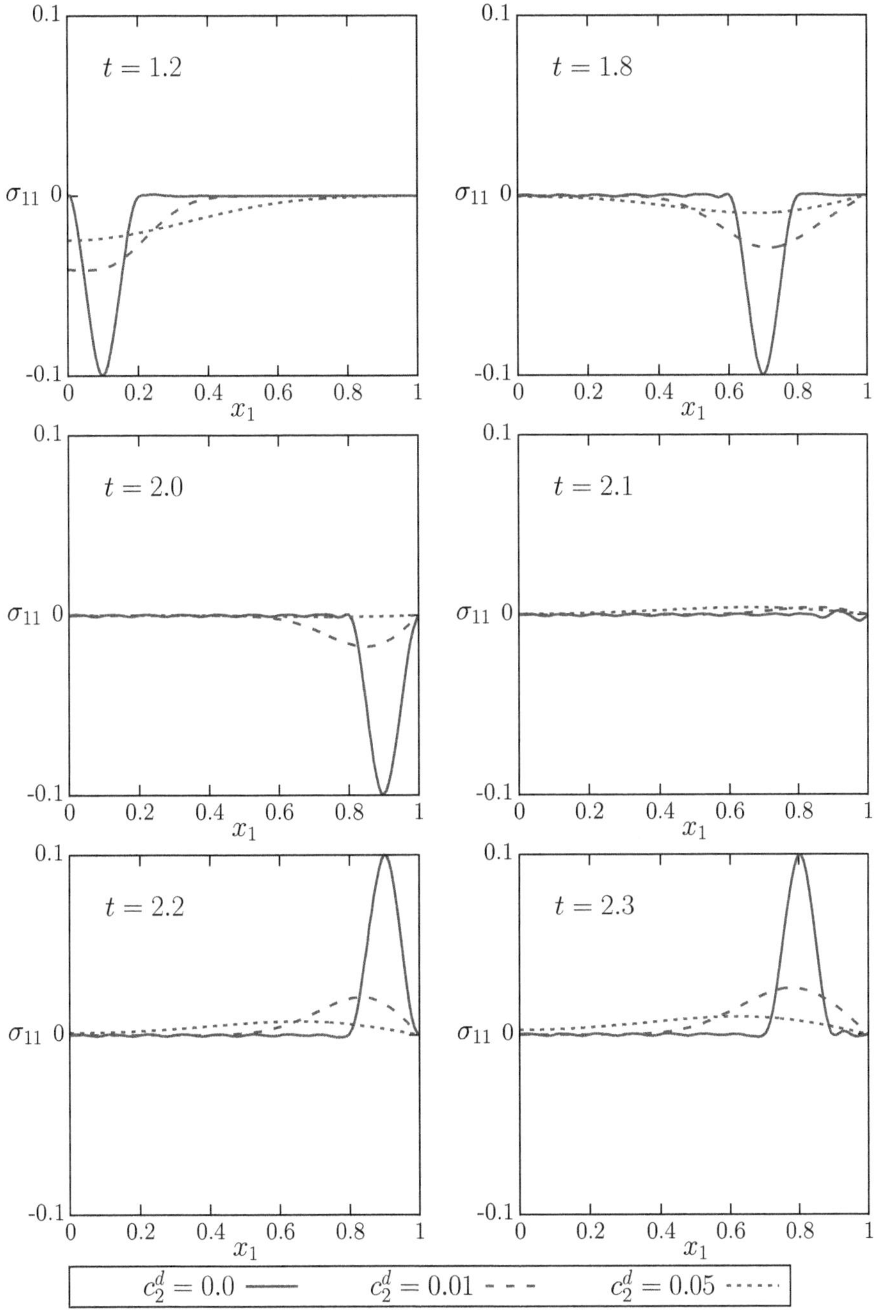

Figure 10.3: Continued time evolution of the propagation of an applied strain pulse in a one-dimensional axially deforming rod with dissipation based on strain rate (Model A: **strain rate-based damping**)

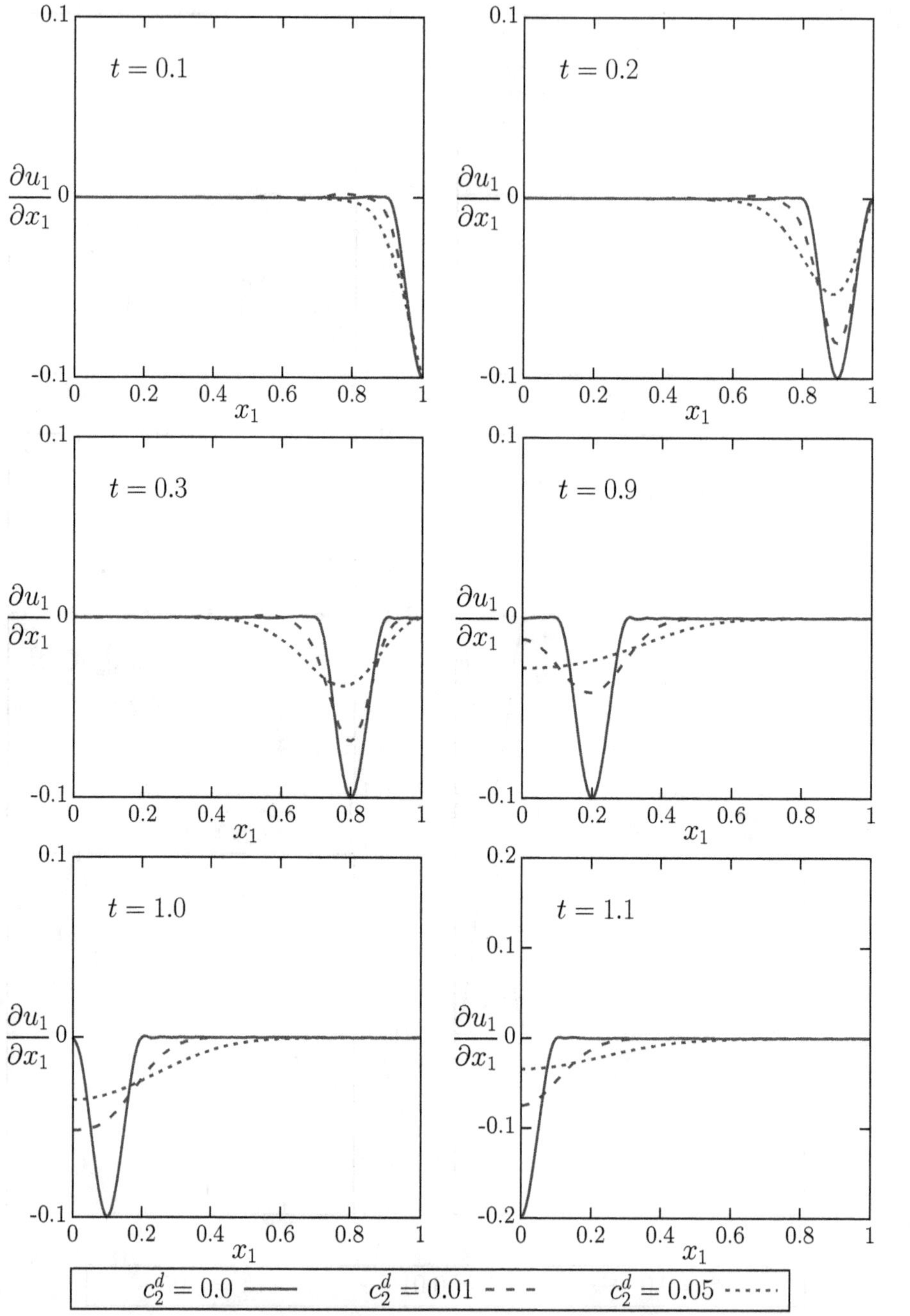

Figure 10.4: Time evolution of the propagation of an applied strain pulse in a one-dimensional axially deforming rod with dissipation based on strain rate (Model A: **strain rate-based damping**)

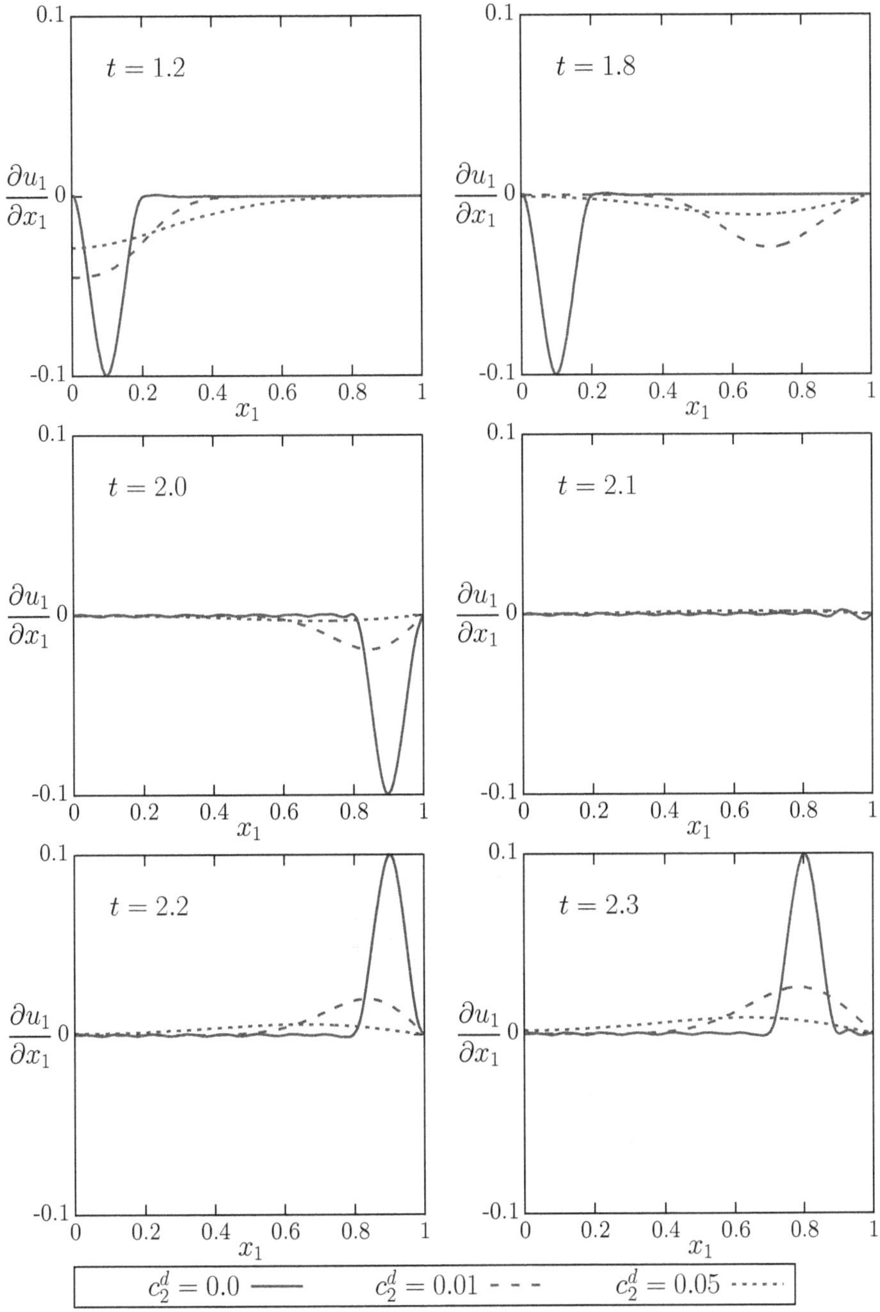

Figure 10.5: Continued time evolution of the propagation of an applied strain pulse in a one-dimensional axially deforming rod with dissipation based on strain rate (Model A: **strain rate-based damping**)

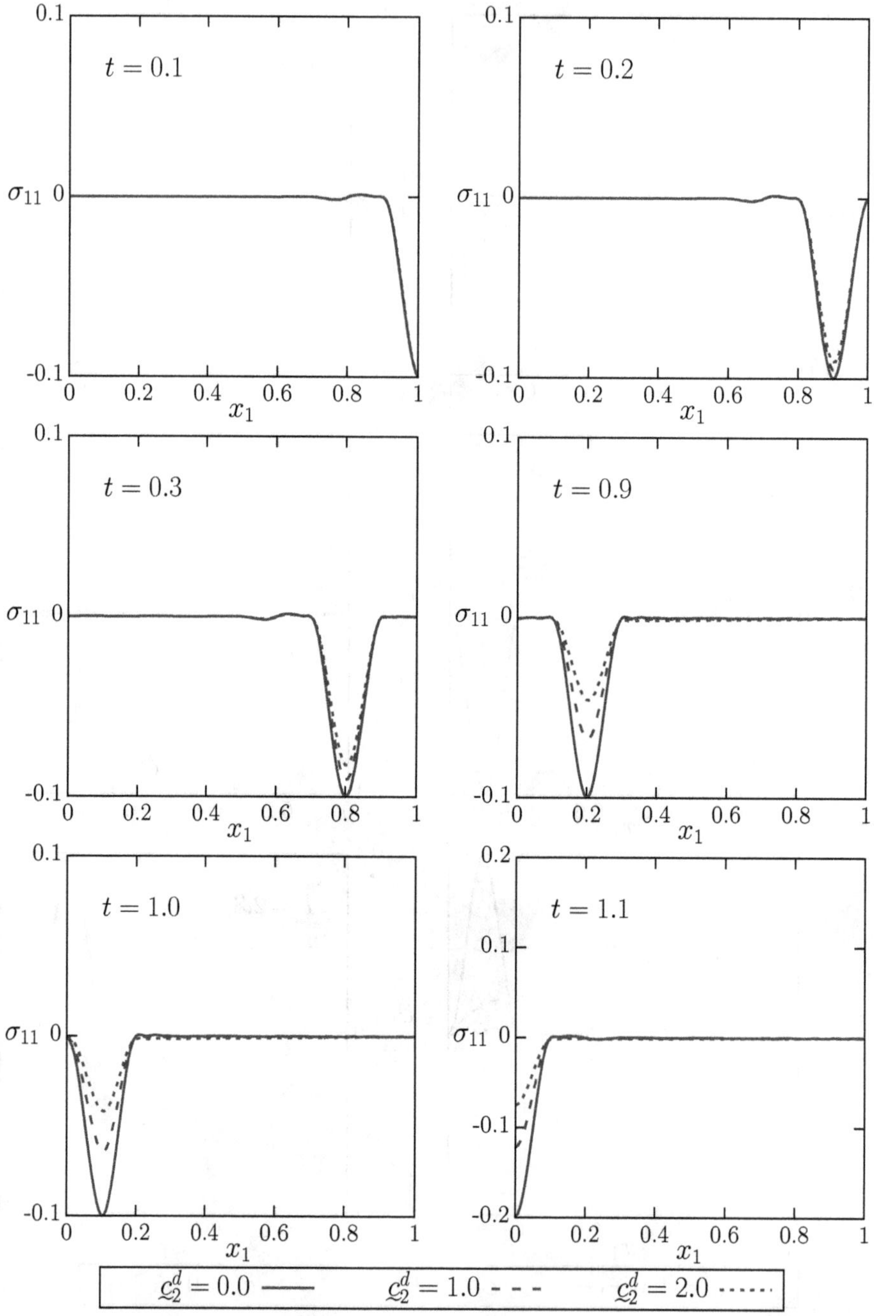

Figure 10.6: Continued time evolution of the propagation of an applied strain pulse in a one-dimensional axially deforming rod with dissipation based on velocity (Model B: **velocity-based damping**)

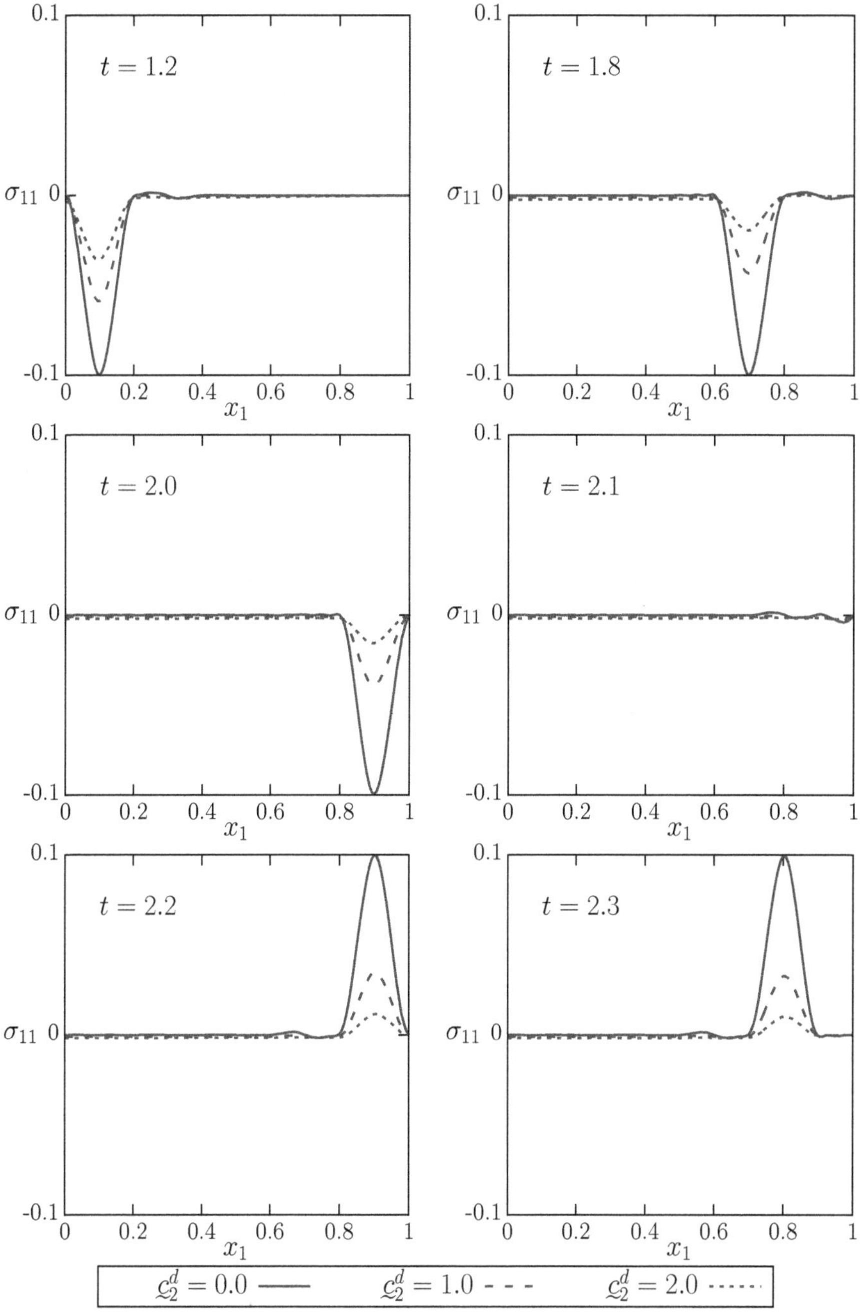

Figure 10.7: Continued time evolution of the propagation of an applied strain pulse in a one-dimensional axially deforming rod with dissipation based on velocity (Model B: **velocity-based damping**)

(2) For an insulated system, continued application of cyclic loading would result in increase in temperature of the volume of matter with increasing time.

(3) The choices of $\boldsymbol{\sigma}^{[0]}$, $\boldsymbol{\varepsilon}_{[0]}$ as work conjugate pair for finite deformation, finite strain and $\boldsymbol{\sigma}$, $\boldsymbol{\varepsilon}$ for small deformation, small strain holds here as well regardless of whether the material is compressible or incompressible.

(4) In case of finite deformation, finite strain, the dissipation mechanism is non-linear [125] due to the fact that strain rates are non-linear measure of displacement gradients. This is an active area of research at present.

(5) Due to elasticity (E, ν) and density (ρ), such materials have stiffness, thus, the following hold for such materials.

   (a) Such materials have finite speed of sound ($\sqrt{E/\rho}$). This permits propagation of stress waves for small deformation, small strain, but with amplitude decay and base elongation (if energy content of the wave is fixed) due to dissipation. In the case of non-linear deformation physics, the same holds as for linear case but the wave shapes get distorted.

   (b) Such materials exhibit damped linear and non-linear vibrational characteristics upon application of frequency and time dependent disturbance.

   (c) Concept of natural vibration exists only in undamped linear small deformation, small strain physics, but the physics of vibrations does exist in non-linear case as well.

10.8 Summary

In this chapter, the reduced form of the entropy inequality has been utilized in conjunction with representation theorem to derive constitutive theories for the stress tensor for: (1) finite deformation, finite strain, compressible and incompressible, non-isothermal thermoviscoelastic solid continua (without memory) in which the dissipation mechanism is dependent on strain rates $\boldsymbol{\varepsilon}_{[i]}$; $i = 1, 2, \ldots, n$ (2) small deformation, small strain, non-isothermal thermoviscoelastic solid continua using Cauchy stress $\boldsymbol{\sigma}$ as a constitutive variable with the linear part of Green's strain tensor and its rates up to order n as its argument tensors. These constitutive theories are based on integrity, hence use complete basis, but require too many material coefficients. Simplified form of the constitutive theories for the stress tensor that are linear in the argument tensors are also presented. In all constitutive theories material coefficients are derived and discussed. The constitutive theory for Cauchy stress tensor $\boldsymbol{\sigma}$ for small deformation, small strain, incompressible and isothermal physics in $\mathbb{R}^1$ is shown to be same as the Kelvin-Voigt phenomenological model for $\boldsymbol{\sigma}$ for viscoelastic solid continua if the equilibrium stress is neglected. It is shown that inclusion of each strain rate in the dissipation mechanism requires two additional material coefficients regardless of $\mathbb{R}^1$, $\mathbb{R}^2$ or $\mathbb{R}^3$, whereas the Kelvin-Voigt model needs on only one material coefficients due to its phenomenological origin in $\mathbb{R}^1$, hence cannot be extended to solid continua in $\mathbb{R}^2$ and $\mathbb{R}^3$. Dissipation force proportional to velocity added to BLM commonly used in engineering writings and applications has been considered in model problem studies. It is shown that addition of this force as resisting force to motion is in violation of physics and that the dependence of dissipation on velocity is only realized when separation of variables is used for balance of linear momenta containing strain rate dependent dissipation mechanism. A model problem consisting of stress wave propagation in $\mathbb{R}^1$ is used to illustrate the validity of dissipation mechanism presented in this chapter based on strain rate and is compared with dissipation mechanism proportional to velocity to further illustrate lack of validity of dissipation mechanism proportional to velocity.

Constitutive theories are also considered (referenced to the derivation in Chapter 9) for heat vector $\boldsymbol{q}$: (i) Strictly based on conjugate pairs $\boldsymbol{q}$, $\boldsymbol{g}$ in the entropy inequality (ii) using representation theorem. These hold regardless of the nature of mechanical deformation. Material coefficients in all constitutive theories can be functions of the combined invariants of the argument tensors of the constitutive variables and temperature θ in a known configuration $\underline{\Omega}$.

Problems

10.1 Derive a constitutive theory for a thermoviscoelastic solid without memory in which the dependent variables in the constitutive theory are functions of strain, strain rate and temperature gradient. Consider finite deformation.

10.2 Derive an ordered rate constitutive theory for finite deformation of thermoviscoelastic solid without memory from first principles using the theory of generators and invariants in which

(a) the deviatoric second Piola-Kirchhoff stress is a linear function of the Green's strain, Green's strain rate tensor and temperature θ;
(b) the deviatoric second Piola-Kirchhoff stress is a linear function of the Cauchy strain tensor, Cauchy strain rate tensor and temperature θ.

In both cases, derive the material coefficients. Determine the relationship between the material coefficients in the theories derived in (a) and (b) if one exists. Give explicit forms of the equations for $\mathbb{R}^1$ and $\mathbb{R}^2$ for both (a) and (b). Also, represent these in matrix and vector forms using Voigt's notation.

10.3 Derive an ordered rate constitutive theory for finite deformation of thermoviscoelastic solids without memory from first principles using the theory of generators and invariants in which

(a) the time derivative of order n of the Green's strain tensor is a linear function of the deviatoric second Piola-Kirchhoff stress tensor, Green's strain tensor, the time derivatives of up to order $(n-1)$ of the Green's strain tensor and temperature θ;
(b) the time derivative of order n of the Cauchy strain tensor is a linear function of the deviatoric second Piola-Kirchhoff stress tensor, Cauchy strain tensor, the time derivatives of up to orders $(n-1)$ of the Cauchy strain tensor and the temperature θ.

In both cases, derive the material coefficients. Determine the relationship between the material coefficients in the theories derived in (a) and (b) if one exists. Give explicit forms of the equations for $\mathbb{R}^1$ and $\mathbb{R}^2$ for both (a) and (b). Also, represent these in matrix and vector forms using Voigt's notation.

10.4 Derive a linear ordered rate constitutive theory for the heat vector that is consistent with the choice of argument tensors in Problem 10.2. Include temperature gradient as an additional argument. Give the explicit expanded form of the equations.

10.5 Derive a linear ordered rate constitutive theory for the heat vector that is consistent with the choice of argument tensors in Problem 10.3. Include temperature gradient as an additional argument. Derive the explicit expanded form of the equations.

11

CONSTITUTIVE THEORIES FOR THERMOVISCOELASTIC SOLIDS WITH MEMORY

11.1 Introduction

In this chapter, we consider constitutive theories for thermoviscoelastic solids with memory in Lagrangian description. Such solids possess elasticity, dissipation and memory mechanisms. Part of the rate of mechanical work results in entropy generation through dissipation mechanism that influences thermal field. Memory mechanism in such solids is incorporated through constitutive theories for the deviatoric stress tensor that are partial differential equations in space and time. In this chapter we consider constitutive theories for: (i) finite deformation, finite strain, compressible, incompressible and non-isothermal physics with dissipation and memory mechanisms (ii) small deformation, small strain, non-isothermal physics with dissipation and memory mechanisms.

In case of finite deformation, finite strain deformation physics, the second Piola-Kirchhoff stress tensor $\boldsymbol{\sigma}^{[0]}$ and Green's strain tensor $\boldsymbol{\varepsilon}_{[0]}$ are appropriate measures of stress and strains. As discussed in Chapter 10, the most basic form of dissipation is due to strain rate $\dot{\boldsymbol{\varepsilon}}_{[0]}$ or $\boldsymbol{\varepsilon}_{[1]}$. Since $\boldsymbol{\varepsilon}_{[0]}$ is a non-linear function of displacements gradients, in such solid continua elasticity, dissipation and memory mechanisms are non-linear functions of displacement gradient tensor. Compressibility, incompressibility and non-isothermal physics merely influences volumetric deformation described by the constitutive theory for the equilibrium stress tensor. Distortion of volume (change of shape), dissipation and memory are addressed by the constitutive theory for the deviatoric stress tensor.

In case of small deformation, small strain, the Cauchy stress tensor $\boldsymbol{\sigma}$ (basis independent) and $\boldsymbol{\varepsilon}$, the linear part of the Green's strain tensor are appropriate choices. As in the case of finite deformation, here also the dissipation mechanism is due to strain rate(s) and the memory is due to constitutive theory for the deviatoric stress tensor, a partial differential equation in space coordinates and time.

Constitutive theories for heat vector $\boldsymbol{q}$ are also considered. These remain the same regardless of the type of mechanical deformation. In all cases complete constitutive theories based on integrity including material coefficients are presented first, followed by simplified and linear constitutive theories.

Since the volumetric deformation physics when expressed in the Cauchy stress tensor in the current configuration is invariant of the deformation type and material constitution, the constitutive theories for equilibrium stress tensor describing volumetric deformation physics derived in Chapter 8 are used here (without rederivation). Constitutive theories for the deviatoric stress tensor addressing the distortional physics, dissipation and memory are derived using the reduced form of the entropy inequality and the representation theorem. We consider the constitutive

DOI: 10.1201/9781003105336-11

theory for deviatoric stress for finite deformation, finite strain deformation physics as well as for small strain, small deformation. We only consider thermodynamic approach in this chapter, but some references are made to the phenomenological approaches of deriving constitutive theories.

11.2 Finite deformation, finite strain

The second Piola-Kirchhoff stress tensor $\boldsymbol{\sigma}^{[0]}$ and Green's strain tensor $\boldsymbol{\varepsilon}_{[0]}$ are appropriate measures of stress and strain in Lagrangian description for this deformation physics. We perform additive decomposition of $\boldsymbol{\sigma}^{[0]}$ into equilibrium and deviatoric stress tensors ${}_e\boldsymbol{\sigma}^{[0]}$ and ${}_d\boldsymbol{\sigma}^{[0]}$.

$$\boldsymbol{\sigma}^{[0]} = {}_e\boldsymbol{\sigma}^{[0]} + {}_d\boldsymbol{\sigma}^{[0]} \tag{11.1}$$

Constitutive theory of ${}_e\boldsymbol{\sigma}^{[0]}$ addresses volumetric deformation physics. Distortion, dissipation and memory aspects of the deformation physics are described by the constitutive theory for ${}_d\boldsymbol{\sigma}^{[0]}$. Following Chapter 8, the constitutive theory for ${}_e\boldsymbol{\sigma}^{[0]}$ is given by ((8.44), (8.45) and (8.52), (8.53))

Compressible

$${}_e\boldsymbol{\sigma}^{[0]} = |\boldsymbol{J}| p(\rho,\theta)(\boldsymbol{J}^T \cdot \boldsymbol{J})^{-1} \tag{11.2}$$

$$p(\rho,\theta) = -\rho^2 \frac{\partial \Phi(\rho,\theta)}{\partial \rho} \tag{11.3}$$

Incompressible

$${}_e\boldsymbol{\sigma}^{[0]} = |\boldsymbol{J}| p(\theta)(\boldsymbol{J}^T \cdot \boldsymbol{J})^{-1} \tag{11.4}$$

The reduced form of the entropy inequality can be written as (equation (8.83)).

$$-{}_d\boldsymbol{\sigma}^{[0]} : \dot{\boldsymbol{\varepsilon}}_{[0]} + \frac{\boldsymbol{q} \cdot \boldsymbol{g}}{\theta} \leq 0 \tag{11.5}$$

The entropy inequality (11.5) is satisfied if ${}_d\boldsymbol{\sigma}^{[0]} : \dot{\boldsymbol{\varepsilon}}_{[0]} > 0$ and $\frac{\boldsymbol{q}\cdot\boldsymbol{g}}{\theta} \leq 0$, These serve as restrictions on the constitutive theory for ${}_d\boldsymbol{\sigma}^{[0]}$ and $\boldsymbol{q}$.

11.2.1 Constitutive theory for ${}_d\boldsymbol{\sigma}^{[0]}$

Based on the conjugate pairs in (11.5) and due to the dissipation mechanism, at the very least we must have the following argument tensors of ${}_d\boldsymbol{\sigma}^{[0]}$

$${}_d\boldsymbol{\sigma}^{[0]} = {}_d\boldsymbol{\sigma}^{[0]}(\boldsymbol{\varepsilon}_{[0]}, \boldsymbol{\varepsilon}_{[1]}, \theta) \tag{11.6}$$

$$\text{and} \quad \boldsymbol{q} = \boldsymbol{q}(\boldsymbol{g}, \theta) \tag{11.7}$$

in which $\boldsymbol{\varepsilon}_{[1]}$ is the strain rate ($\dot{\boldsymbol{\varepsilon}}_{[0]}$). If we assume that strain rates of orders higher than one contribute to dissipation mechanism, then $\boldsymbol{\varepsilon}_{[1]}$ must be replaced by $\boldsymbol{\varepsilon}_{[i]}$; $i = 1, 2, \dots, n$ in (11.6). From the development of the constitutive theories for thermoviscoelastic fluids (Chapter 13) and their simplifications resulting in Maxwell model, Oldroyd-B and Giesekus models [20, 21], we know that for such fluids to have memory, at the very least the constitutive theory must consider the first convected time derivative of the stress tensor as a constitutive variable with stress and strain rate as its argument tensor (in addition to others). The generalization of this concept leads to the stress rate of order m to be the constitutive variable with stress tensor and its rates up to orders $m-1$ as its argument tensors. Thus, the constitutive theories under consideration here, must consider strain tensor $\boldsymbol{\varepsilon}_{[0]}$, strain rate tensors $\boldsymbol{\varepsilon}_{[i]}$; $i=1,2,\dots,n$, stress tensor ${}_d\boldsymbol{\sigma}^{[0]}$ and stress rate tensors ${}_d\boldsymbol{\sigma}^{[j]}$; $j=1,2,\dots,m$, keeping in mind that the highest stress rate tensor (i.e., ${}_d\boldsymbol{\sigma}^{[m]}$) must be chosen as the constitutive tensor. Thus, we finally have the following constitutive tensors and their argument tensors.

$$
\begin{aligned}
{}_d\boldsymbol{\sigma}^{[m]} &= {}_d\boldsymbol{\sigma}^{[m]}({}_d\boldsymbol{\sigma}^{[j]}, \boldsymbol{\varepsilon}_{[0]}, \boldsymbol{\varepsilon}_{[i]}, \theta) && (11.8)\\
i &= 1,2,\dots,n \quad ; \quad j = 0,1,\dots,m-1 \\
\boldsymbol{q} &= \boldsymbol{q}(\boldsymbol{g},\theta) && (11.9)
\end{aligned}
$$

We consider ${}_d\boldsymbol{\sigma}^{[m]}$ with its argument tensors defined in (11.8) to derive the constitutive theory for ${}_d\boldsymbol{\sigma}^{[m]}$ using representation theorem. Let ${}^{\sigma}\underset{\sim}{\boldsymbol{G}}^i$; $i=1,2,\dots,N$ be the combined generators of the argument tensors of ${}_d\boldsymbol{\sigma}^{[m]}$ in (11.8) that are symmetric tensors of rank two. Then, ${}_d\boldsymbol{\sigma}^{[m]}$ in the current configuration can be expressed as a linear combination of $\boldsymbol{I}$ and ${}^{\sigma}\underset{\sim}{\boldsymbol{G}}^i$; $i=1,2,\dots,N$.

$$
{}_d\boldsymbol{\sigma}^{[m]} = {}^{\sigma}\underset{\sim}{\alpha}^0 \boldsymbol{I} + \sum_{i=1}^{N} {}^{\sigma}\underset{\sim}{\alpha}^i \, {}^{\sigma}\underset{\sim}{\boldsymbol{G}}^i \tag{11.10}
$$

$$
\text{in which} \quad {}^{\sigma}\underset{\sim}{\alpha}^i = {}^{\sigma}\underset{\sim}{\alpha}^i({}^{\sigma}\underset{\sim}{I}^j, \theta) \quad ; \quad i=0,1,\dots,N \quad ; \quad j=1,2,\dots,M \tag{11.11}
$$

${}^{\sigma}\underset{\sim}{I}^j$; $j=1,2,\dots,M$ are the combined invariants of the same argument tensors of ${}_d\boldsymbol{\sigma}^{[m]}$ in (11.8).

Material coefficients

To determine material coefficients in the constitutive theory (11.10), we consider Taylor series expansion of ${}^{\sigma}\underset{\sim}{\alpha}^i$; $i=0,1,\dots,N$ in ${}^{\sigma}\underset{\sim}{I}^j$ $j=1,2,\dots,M$ about a known configuration $\underline{\Omega}$ and retain only up to linear terms in ${}^{\sigma}\underset{\sim}{I}^j$; $j=1,2,\dots,M$ (for simplicity). We do not consider Taylor series expansion of ${}^{\sigma}\underset{\sim}{\alpha}^i$; $i=0,1,\dots,N$ in θ as the effect of θ on the stress field has already been accounted for in the constitutive theory for ${}_e\boldsymbol{\sigma}^{[0]}$.

$$
{}^{\sigma}\underset{\sim}{\alpha}^i = {}^{\sigma}\underset{\sim}{\alpha}^i\Big|_{\underline{\Omega}} + \sum_{j=1}^{M} \frac{\partial\, {}^{\sigma}\underset{\sim}{\alpha}^i}{\partial\, {}^{\sigma}\underset{\sim}{I}^j}\Bigg|_{\underline{\Omega}} \left({}^{\sigma}\underset{\sim}{I}^j - {}^{\sigma}\underset{\sim}{I}^j\Big|_{\underline{\Omega}} \right) \quad ; \quad i=0,1,\dots,N \tag{11.12}
$$

substituting (11.12) in (11.10)

$$
\begin{aligned}
{}_d\boldsymbol{\sigma}^{[m]} = & \left({}^{\sigma}\underset{\sim}{\alpha}^0\Big|_{\underline{\Omega}} + \sum_{j=1}^{M} \frac{\partial\, {}^{\sigma}\underset{\sim}{\alpha}^0}{\partial\, {}^{\sigma}\underset{\sim}{I}^j}\Bigg|_{\underline{\Omega}} \left({}^{\sigma}\underset{\sim}{I}^j - {}^{\sigma}\underset{\sim}{I}^j\Big|_{\underline{\Omega}} \right) \right) \boldsymbol{I} \\
& + \sum_{i=1}^{N} \left({}^{\sigma}\underset{\sim}{\alpha}^i\Big|_{\underline{\Omega}} + \sum_{j=1}^{M} \frac{\partial\, {}^{\sigma}\underset{\sim}{\alpha}^i}{\partial\, {}^{\sigma}\underset{\sim}{I}^j}\Bigg|_{\underline{\Omega}} \left({}^{\sigma}\underset{\sim}{I}^j - {}^{\sigma}\underset{\sim}{I}^j\Big|_{\underline{\Omega}} \right) \right) {}^{\sigma}\underset{\sim}{\boldsymbol{G}}^i
\end{aligned} \tag{11.13}
$$

Collecting coefficients of (terms defined in the current configuration) $\boldsymbol{I}$, ${}^{\sigma}\underset{\sim}{I}^j\boldsymbol{I}$, ${}^{\sigma}\underset{\sim}{\boldsymbol{G}}^i$, ${}^{\sigma}\underset{\sim}{I}^j({}^{\sigma}\underset{\sim}{\boldsymbol{G}}^i)$ and defining

$$
\begin{aligned}
\underset{\sim}{\sigma}^0\Big|_{\underline{\Omega}} &= {}^{\sigma}\underset{\sim}{\alpha}^0\Big|_{\underline{\Omega}} - \sum_{j=1}^{M} \frac{\partial^{\sigma}\underset{\sim}{\alpha}^0}{\partial^{\sigma}\underset{\sim}{I}^j}\Bigg|_{\underline{\Omega}} \left({}^{\sigma}\underset{\sim}{I}^j\Big|_{\underline{\Omega}}\right) \quad ; \quad {}^{\sigma}\underset{\sim}{a}_j = \frac{\partial^{\sigma}\underset{\sim}{\alpha}^0}{\partial^{\sigma}\underset{\sim}{I}^j}\Bigg|_{\underline{\Omega}} \\
{}^{\sigma}\underset{\sim}{b}_j &= {}^{\sigma}\underset{\sim}{\alpha}^i\Big|_{\underline{\Omega}} - \sum_{j=1}^{M} \frac{\partial^{\sigma}\underset{\sim}{\alpha}^i}{\partial^{\sigma}\underset{\sim}{I}^j}\Bigg|_{\underline{\Omega}} \left({}^{\sigma}\underset{\sim}{I}^j\Big|_{\underline{\Omega}}\right) \quad ; \quad {}^{\sigma}\underset{\sim}{c}_{ij} = \frac{\partial^{\sigma}\underset{\sim}{\alpha}^i}{\partial^{\sigma}\underset{\sim}{I}^j}\Bigg|_{\underline{\Omega}} \\
& i = 1,2,\dots,N \quad ; \quad j = 1,2,\dots,M
\end{aligned}
\tag{11.14}
$$

equation (11.13) can be written as

$$
{}_d\boldsymbol{\sigma}^{[m]} = \underset{\sim}{\sigma}^0\Big|_{\underline{\Omega}}\boldsymbol{I} + \sum_{j=1}^{M} {}^{\sigma}\underset{\sim}{a}_j({}^{\sigma}\underset{\sim}{I}^j)\boldsymbol{I} + \sum_{i=1}^{N}\sum_{j=1}^{M} {}^{\sigma}\underset{\sim}{c}_{ij}({}^{\sigma}\underset{\sim}{I}^j)\,{}^{\sigma}\underset{\sim}{\boldsymbol{G}}^i + \sum_{i=1}^{N} {}^{\sigma}\underset{\sim}{b}_i\,{}^{\sigma}\underset{\sim}{\boldsymbol{G}}^i \tag{11.15}
$$

${}^{\sigma}\underset{\sim}{a}_j$, ${}^{\sigma}\underset{\sim}{b}_i$, ${}^{\sigma}\underset{\sim}{c}_{ij}$; $i = 1,2,\dots,N$; $j = 1,2,\dots,M$ are material coefficients defined in a known configuration $\underline{\Omega}$. In equation (11.15), $\underset{\sim}{\sigma}^0\big|_{\underline{\Omega}}$ is the initial stress field in the known configuration $\underline{\Omega}$. This constitutive theory for ${}_d\boldsymbol{\sigma}^{[0]}$ requires $(M+N+MN)$ material coefficients and is based on integrity (complete basis) of the space containing ${}_d\boldsymbol{\sigma}^{[m]}$. The material coefficients can be functions of ${}^{\sigma}\underset{\sim}{I}^j\big|_{\underline{\Omega}}$; $j = 1,2,\dots,M$ and $\theta\big|_{\underline{\Omega}}$.

In this constitutive theory the dissipation mechanism is due to strain rates $\boldsymbol{\varepsilon}_{[i]}$; $i = 1,2,\dots,n$ and the memory mechanism is due to strain rates as well as stress rates of up to orders n and m respectively. *Thus, this constitutive theory is referred to as ordered rate constitutive theory of orders* (m,n). By approximate choices of m and n specific forms of the constitutive theory for the deviatoric second Piola-Kirchhoff stress tensor can be obtained.

Simplified form of the constitutive theories for ${}_d\boldsymbol{\sigma}^{[m]}$

Different choices of m and n and specific choices of combined generators and invariants will yield different constitutive theories. Choices of N and M, number of combined generators and number of combined invariants can be dependent on specific applications. In the following we consider some simple constitutive theories.

11.2.1.1 A simplified linear constitutive theory orders m and n

We consider an ordered rate constitutive theory of up to orders m and n that is linear in the components of $\boldsymbol{\varepsilon}_{[0]}$, $\boldsymbol{\varepsilon}_{[i]}$; $i = 1,2,\dots,n$ and ${}_d\boldsymbol{\sigma}^{[j]}$; $j = 0,1,\dots,m-1$. This constitutive theory will contain generators $\boldsymbol{I}$, $\boldsymbol{\varepsilon}_{[0]}$, $\boldsymbol{\varepsilon}_{[i]}$; $i = 1,2,\dots,n$, ${}_d\boldsymbol{\sigma}^{[j]}$; $j = 0,1,\dots,m-1$ and invariants $\mathrm{tr}\,\boldsymbol{\varepsilon}_{[0]}$, $\mathrm{tr}\,\boldsymbol{\varepsilon}_{[i]}$; $i = 1,2,\dots,n$; $\mathrm{tr}({}_d\boldsymbol{\sigma}^{[j]})$; $j = 0,1,\dots,m-1$. Using (11.15) for these generators and invariants, we can write (after redefining material coefficients)

$$
\begin{aligned}
{}_d\boldsymbol{\sigma}^{[m]} = {} & \underset{\sim}{\sigma}^0\Big|_{\underline{\Omega}}\boldsymbol{I} + a_1\boldsymbol{\varepsilon}_{[0]} + a_2\,\mathrm{tr}(\boldsymbol{\varepsilon}_{[0]})\boldsymbol{I} + \sum_{i=1}^{n} b_i^1\boldsymbol{\varepsilon}_{[i]} + \sum_{i=1}^{n} b_i^2\,\mathrm{tr}\,\boldsymbol{\varepsilon}_{[i]}\boldsymbol{I} \\
& + \sum_{j=0}^{m-1} c_j^1({}_d\boldsymbol{\sigma}^{[j]}) + \sum_{j=0}^{m-1} c_j^2\,\mathrm{tr}({}_d\boldsymbol{\sigma}^{[j]})\boldsymbol{I}
\end{aligned}
\tag{11.16}
$$

The material coefficients of a_1, a_2, b_i^1, b_i^2, c_j^1, c_j^2 ; $i = 1, 2, \ldots, n$; $j = 0, 1, \ldots, m-1$ can be function of invariants ${}^{\sigma}\underset{\sim}{I}^j\big|_{\Omega}$; $j = 1, 2, \ldots, M$ and $\theta|_{\Omega}$.

This constitutive theory can also be written in the matrix and the vector form using Voigt's notation.

Let

$$\{{}_d\sigma^{[j]}\}^T = [\ {}_d\sigma_{11}^{[j]}, {}_d\sigma_{22}^{[j]}, {}_d\sigma_{33}^{[j]}, {}_d\sigma_{23}^{[j]}, {}_d\sigma_{31}^{[j]}, {}_d\sigma_{12}^{[j]}\ ] \quad ; \quad j = 0, 1, \ldots, m \tag{11.17}$$

If we use the same arrangement of coefficients of $\boldsymbol{\varepsilon}_{[0]}$ and $\boldsymbol{\varepsilon}_{[i]}$; $i = 1, 2, \ldots, n$, Then we can write (11.16) in the following form

$$\{{}_d\sigma^{[m]}\} = \{\sigma^0\} + [\underset{\sim}{a}]\{\varepsilon_{[0]}\} + \sum_{i=1}^{n} [\underset{\sim}{b}_i]\{\varepsilon_{[i]}\} + \sum_{j=0}^{m-1} [\underset{\sim}{c}_j]\{{}_d\sigma^{[j]}\} \tag{11.18}$$

in which

$$[\underset{\sim}{a}] = \begin{bmatrix} a_1 + a_2 & a_2 & a_2 & 0 & 0 & 0 \\ a_2 & a_1 + a_2 & a_2 & 0 & 0 & 0 \\ a_2 & a_2 & a_1 + a_2 & 0 & 0 & 0 \\ 0 & 0 & 0 & a_1 & 0 & 0 \\ 0 & 0 & 0 & 0 & a_1 & 0 \\ 0 & 0 & 0 & 0 & 0 & a_1 \end{bmatrix} \tag{11.19}$$

$$\text{and} \quad \{\sigma^0\}^T = [\underset{\sim}{\sigma}^0|_{\Omega}, \underset{\sim}{\sigma}^0|_{\Omega}, \underset{\sim}{\sigma}^0|_{\Omega}, 0, 0, 0]$$

Coefficients of matrices $[\underset{\sim}{b}_i]$ and $[\underset{\sim}{c}_j]$ can be obtained using (11.19) and replacing a_1, a_2 with b_i^1, b_i^2 and c_j^1, c_j^2, respectively. We note that in the constitutive theory (11.18), each strain rate requires two material coefficients. Inclusion of each stress rate also requires two material coefficients.

11.2.1.2 A linear constitutive theory of orders $m = 1$ and n

In this case, we have

$${}_d\boldsymbol{\sigma}^{[1]} = {}_d\boldsymbol{\sigma}^{[1]}(\boldsymbol{\varepsilon}, \boldsymbol{\varepsilon}_{[i]}, {}_d\boldsymbol{\sigma}^{[0]}, \theta) \quad ; \quad i = 1, 2, \ldots, n \tag{11.20}$$

using $m = 1$ in (11.16) and (11.18) we have the corresponding constitutive theory

$$\begin{aligned} {}_d\boldsymbol{\sigma}^{[1]} &= \underset{\sim}{\sigma}^0\big|_{\Omega}\boldsymbol{I} + a_1\boldsymbol{\varepsilon}_{[0]} + a_2(\mathrm{tr}\,\boldsymbol{\varepsilon}_{[0]})\boldsymbol{I} \\ &+ \sum_{i=1}^{n} b_i^1\boldsymbol{\varepsilon}_{[i]} + \sum_{i=1}^{n} b_i^2(\mathrm{tr}\,\boldsymbol{\varepsilon}_{[i]})\boldsymbol{I} + c_0^1({}_d\boldsymbol{\sigma}^{[0]}) + c_0^2(\mathrm{tr}\,{}_d\boldsymbol{\sigma}^{[0]})\boldsymbol{I} \end{aligned} \tag{11.21}$$

or in matrix and vector notation

$$\{{}_d\sigma^{[1]}\} = \{\sigma^0\} + [\underset{\sim}{a}]\{\varepsilon_{[0]}\} + \sum_{i=1}^{n} [\underset{\sim}{b}_i]\{\varepsilon_{[i]}\} + [\underset{\sim}{c}_0]\{{}_d\sigma^{[0]}\} \tag{11.22}$$

The constitutive theory (11.21) can be written in a slightly different form using new definition of material coefficients that are in accordance with what is used currently in polymeric fluids. We divide each term in equation (11.21) by $-c_0^1$ and define (assuming $c_0^1 \neq 0$ and neglecting $\underset{\sim}{\sigma}^0\big|_{\underline{\Omega}}$)

$$\begin{aligned} \left(-\frac{1}{c_0^1}\right) &= \lambda \;,\quad \left(-\frac{a_1}{c_0^1}\right) = 2\underset{\sim}{\mu} \;,\quad \left(-\frac{a_2}{c_0^1}\right) = \underset{\sim}{\lambda} \\ \left(-\frac{b_i^1}{c_0^1}\right) &= 2\eta_i \;,\quad \left(-\frac{b_i^2}{c_0^1}\right) = k_i \;,\quad \left(-\frac{c_0^2}{c_0^1}\right) = \underset{\sim}{\beta} \end{aligned} \tag{11.23}$$

Then, (11.21) can be written as

$$_d\boldsymbol{\sigma}^{[0]} + \lambda({}_d\boldsymbol{\sigma}^{[1]}) = 2\underset{\sim}{\mu}\boldsymbol{\varepsilon}_{[0]} + \underset{\sim}{\lambda}(\operatorname{tr}\boldsymbol{\varepsilon}_{[0]})\boldsymbol{I} + \sum_{i=1}^{n} 2\eta_i\boldsymbol{\varepsilon}_{[i]} + \sum_{i=1}^{n} k_i(\operatorname{tr}\boldsymbol{\varepsilon}_{[i]})\boldsymbol{I} + \underset{\sim}{\beta}\operatorname{tr}({}_d\boldsymbol{\sigma}^{[0]})\boldsymbol{I} \tag{11.24}$$

λ is similar as to relaxation time, $\underset{\sim}{\mu}$ and $\underset{\sim}{\lambda}$ are similar to Lamé's constants, η_i and k_i are parallel to first and second viscosities (in viscous fluent continua) for strain rate $\boldsymbol{\varepsilon}_{[i]}$. The term containing $\underset{\sim}{\beta}$ is generally neglected.

11.2.1.3 A linear constitutive theory of orders $m = 1$, $n = 1$

In this particular case, the constitutive variable is ${}_d\boldsymbol{\sigma}^{[1]}$ and its argument tensors are given by

$$_d\boldsymbol{\sigma}^{[1]} = {}_d\boldsymbol{\sigma}^{[1]}(\boldsymbol{\varepsilon}_{[0]}, \boldsymbol{\varepsilon}_{[1]}, {}_d\boldsymbol{\sigma}^{[0]}, \theta) \tag{11.25}$$

The combined generators of the argument tensors in (11.25) are $\boldsymbol{I}$, $\boldsymbol{\varepsilon}_{[0]}$, $\boldsymbol{\varepsilon}_{[1]}$, ${}_d\boldsymbol{\sigma}^{[0]}$. The combined invariants of the argument tensors in (11.25) are $\operatorname{tr}(\boldsymbol{\varepsilon}_{[0]})$, $\operatorname{tr}(\boldsymbol{\varepsilon}_{[1]})$, $\operatorname{tr}({}_d\boldsymbol{\sigma}^{[0]})$ and the constitutive theories (11.18), (11.16) and (11.24) reduce to

$$\{{}_d\sigma^{[1]}\} = \{\sigma^0\} + [\underset{\sim}{a}]\{\varepsilon_{[0]}\} + [\underset{\sim}{b}_1]\{\varepsilon_{[1]}\} + [\underset{\sim}{c}_0]\{{}_d\sigma^{[0]}\} \tag{11.26}$$

$$\begin{aligned} {}_d\boldsymbol{\sigma}^{[1]} &= \underset{\sim}{\sigma}^0\big|_{\underline{\Omega}}\boldsymbol{I} + a_1\boldsymbol{\varepsilon}_{[0]} + a_2(\operatorname{tr}\boldsymbol{\varepsilon}_{[0]})\boldsymbol{I} \\ &\quad + b_1^1\boldsymbol{\varepsilon}_{[1]} + b_1^2(\operatorname{tr}\boldsymbol{\varepsilon}_{[1]})\boldsymbol{I} + \underset{\sim}{c}_0^1({}_d\boldsymbol{\sigma}^{[0]}) + \underset{\sim}{c}_0^2(\operatorname{tr}({}_d\boldsymbol{\sigma}^{[0]})) \end{aligned} \tag{11.27}$$

and

$$\begin{aligned} {}_d\boldsymbol{\sigma}^{[0]} + \lambda({}_d\boldsymbol{\sigma}^{[1]}) &= 2\underset{\sim}{\mu}\boldsymbol{\varepsilon}_{[0]} + \underset{\sim}{\lambda}(\operatorname{tr}\boldsymbol{\varepsilon}_{[0]})\boldsymbol{I} + \underset{\sim}{\eta}\boldsymbol{\varepsilon}_{[1]} \\ &\quad + \underset{\sim}{\kappa}(\operatorname{tr}\boldsymbol{\varepsilon}_{[1]})\boldsymbol{I} + \underset{\sim}{\beta}(\operatorname{tr}{}_d\boldsymbol{\sigma}^{[0]})\boldsymbol{I} \end{aligned} \tag{11.28}$$

Remarks

(1) The linear constitutive theory of orders one ($m = 1$, $n = 1$) is the simplest possible constitutive theory for thermoviscoelastic solids with memory.

(2) The linear constitutive theory of orders one requires six material coefficients, two for elasticity, two for dissipation, and two associated with the memory mechanism.

(3) In the general constitutive theory of orders m and n, inclusion of each strain rate for dissipation mechanism requires two additional material coefficients. Likewise, inclusion of each stress rate also requires two additional material coefficients.

(4) Both, dissipation and memory mechanism are non-linear in displacement gradients due to the fact that strain and strain rates are non-linear functions of displacement gradients.

(5) Deriving relaxation modulus for finite deformation, finite strain deformation physics is quite difficult. We consider derivation of memory modulus for small deformation, small strain case in a subsequent section.

11.3 Small deformation, small strain

For small deformation, small strain deformation physics, the Cauchy stress tensor $\boldsymbol{\sigma}$ (basis independent) and $\boldsymbol{\varepsilon}$, the linear part of the Green's strain tensor are appropriate measures of stress and strain. Due to dissipation, the deformation is non-isothermal. The additive decomposition of $\boldsymbol{\sigma}$ gives:

$$\boldsymbol{\sigma} = {}_e\boldsymbol{\sigma} + {}_d\boldsymbol{\sigma} \tag{11.29}$$

Constitutive theory for ${}_e\boldsymbol{\sigma}$ addresses volumetric deformation physics whereas distortional deformation, dissipation and memory aspects are incorporated in the constitutive theory for ${}_d\boldsymbol{\sigma}$. Following details in Chapter 8, the constitutive theory for ${}_e\boldsymbol{\sigma}$ is given by (equation (8.75))

$$ {}_e\boldsymbol{\sigma} = p(\rho,\theta)\boldsymbol{\delta} \quad ; \quad \text{compressible} \tag{11.30}$$

$$ {}_e\boldsymbol{\sigma} = p(\theta)\boldsymbol{\delta} \quad ; \quad \text{incompressible} \tag{11.31}$$

and the reduced form of the entropy inequality is given by

$$-{}_d\boldsymbol{\sigma} : \dot{\boldsymbol{\varepsilon}} + \frac{\boldsymbol{q}\cdot\boldsymbol{g}}{\theta} \leq 0 \tag{11.32}$$

Inequality (11.32) holds if ${}_d\boldsymbol{\sigma} : \dot{\boldsymbol{\varepsilon}} > 0$ and $\frac{\boldsymbol{q}\cdot\boldsymbol{g}}{\theta} \leq 0$. These serve as restrictions on the constitutive theories for ${}_d\boldsymbol{\sigma}$ and $\boldsymbol{q}$. Constitutive theory for ${}_d\boldsymbol{\sigma}$ is derived using (11.32) and the representation theorem.

11.3.1 Constitutive theory for ${}_d\boldsymbol{\sigma}$

Following Section 11.2.1, we consider $\boldsymbol{\varepsilon}_{(i)}$; $i = 1,2,\dots,n$ and stress rates ${}_d\boldsymbol{\sigma}^{[j]}$; $j = 1,2,\dots,m$ as well as stress tensor $\boldsymbol{\sigma}$ or $\boldsymbol{\sigma}^{(0)}$ and the strain tensor $\boldsymbol{\varepsilon}$ or $\boldsymbol{\varepsilon}_{(0)}$. The argument tensors of the constitutive tensor $\boldsymbol{\sigma}^{(m)}$ can be written as

$$\begin{gathered} {}_d\boldsymbol{\sigma}^{(m)} = {}_d\boldsymbol{\sigma}^{(m)}(\boldsymbol{\varepsilon}, \boldsymbol{\varepsilon}_{(i)}, {}_d\boldsymbol{\sigma}^{(j)}, \theta) \\ i = 1,2,\dots,n \quad ; \quad j = 0,1,\dots,m-1 \end{gathered} \tag{11.33}$$

We derive constitutive theory for ${}_d\boldsymbol{\sigma}^{(m)}$ using representation theorem. Let ${}^{\sigma}\underset{\sim}{\boldsymbol{G}}^i$; $i = 1,2,\dots,N$ be the combined generators of the argument tensors of ${}_d\boldsymbol{\sigma}^{(m)}$ in (11.33) that are symmetric tensors of rank two. Then, ${}_d\boldsymbol{\sigma}^{(m)}$ can be expressed as a linear combination of $\boldsymbol{I}$ and ${}^{\sigma}\underset{\sim}{\boldsymbol{G}}^i$; $i = 1,2\dots,N$ in the current configuration.

$$ {}_d\boldsymbol{\sigma}^{(m)} = {}^{\sigma}\underset{\sim}{\alpha}^0\boldsymbol{I} + \sum_{i=1}^{N} {}^{\sigma}\underset{\sim}{\alpha}^i\,{}^{\sigma}\underset{\sim}{\boldsymbol{G}}^i \tag{11.34}$$

$$\text{in which} \quad {}^{\sigma}\underset{\sim}{\alpha}^i = {}^{\sigma}\underset{\sim}{\alpha}^i({}^{\sigma}\underset{\sim}{I}^j, \theta) \quad ; \quad i = 0,1,\dots,N \quad ; \quad j = 1,2,\dots,M \tag{11.35}$$

${}^{\sigma}\underset{\sim}{I}^j$; $j = 1,2,\dots,M$ are the combined invariants of the argument tensors of ${}_d\boldsymbol{\sigma}^{(m)}$ in (11.33). To determine the material coefficients in the constitutive theory (11.34), we expand each ${}^{\sigma}\underset{\sim}{\alpha}^i \; ; i = 0, 1, \dots, N$ in Taylor series in ${}^{\sigma}\underset{\sim}{I}^j \; ; j = 1, 2, \dots, M$ about a known configuration $\underline{\Omega}$ and retain only up to linear terms in ${}^{\sigma}\underset{\sim}{I}^j \; ; j = 1, 2, \dots, M$. We do not consider Taylor series expansion in θ as the thermal stress field is already accounted for in the constitutive theory for ${}_e\boldsymbol{\sigma}$.

$$
{}^{\sigma}\underset{\sim}{\alpha}^i = {}^{\sigma}\underset{\sim}{\alpha}^i\Big|_{\underline{\Omega}} + \sum_{j=1}^{M} \frac{\partial\, {}^{\sigma}\underset{\sim}{\alpha}^i}{\partial\, {}^{\sigma}\underset{\sim}{I}^j}\Bigg|_{\underline{\Omega}} \left({}^{\sigma}\underset{\sim}{I}^j - {}^{\sigma}\underset{\sim}{I}^j\Big|_{\underline{\Omega}} \right) \quad ; \quad i = 0, 1, \dots, N \tag{11.36}
$$

substituting form (11.36) into (11.34)

$$
\begin{aligned}
{}_d\boldsymbol{\sigma}^{(m)} = & \left({}^{\sigma}\underset{\sim}{\alpha}^0\Big|_{\underline{\Omega}} + \sum_{j=1}^{M} \frac{\partial\, {}^{\sigma}\underset{\sim}{\alpha}^0}{\partial\, {}^{\sigma}\underset{\sim}{I}^j}\Bigg|_{\underline{\Omega}} \left({}^{\sigma}\underset{\sim}{I}^j - {}^{\sigma}\underset{\sim}{I}^j\Big|_{\underline{\Omega}} \right) \right) \boldsymbol{I} \\
& + \sum_{i=1}^{N} \left({}^{\sigma}\underset{\sim}{\alpha}^i\Big|_{\underline{\Omega}} + \sum_{j=1}^{M} \frac{\partial\, {}^{\sigma}\underset{\sim}{\alpha}^i}{\partial\, {}^{\sigma}\underset{\sim}{I}^j}\Bigg|_{\underline{\Omega}} \left({}^{\sigma}\underset{\sim}{I}^j - {}^{\sigma}\underset{\sim}{I}^j\Big|_{\underline{\Omega}} \right) \right) {}^{\sigma}\underset{\sim}{\boldsymbol{G}}^i
\end{aligned}
\tag{11.37}
$$

Collecting coefficients of $\boldsymbol{I}$, ${}^{\sigma}\underset{\sim}{I}^j\boldsymbol{I}$, ${}^{\sigma}\underset{\sim}{\boldsymbol{G}}^i$, ${}^{\sigma}\underset{\sim}{I}^j({}^{\sigma}\underset{\sim}{\boldsymbol{G}}^i)$ and defining

$$
\begin{aligned}
\underset{\sim}{\sigma}^0\Big|_{\underline{\Omega}} &= {}^{\sigma}\underset{\sim}{\alpha}^0\Big|_{\underline{\Omega}} - \sum_{j=1}^{M} \frac{\partial\, {}^{\sigma}\underset{\sim}{\alpha}^0}{\partial\, {}^{\sigma}\underset{\sim}{I}^j}\Bigg|_{\underline{\Omega}} \left({}^{\sigma}\underset{\sim}{I}^j\Big|_{\underline{\Omega}} \right) \quad ; \quad {}^{\sigma}\underset{\sim}{a}_j = \frac{\partial\, {}^{\sigma}\underset{\sim}{\alpha}^0}{\partial\, {}^{\sigma}\underset{\sim}{I}^j}\Bigg|_{\underline{\Omega}} \\
{}^{\sigma}\underset{\sim}{b}_i &= {}^{\sigma}\underset{\sim}{\alpha}^i\Big|_{\underline{\Omega}} - \sum_{j=1}^{M} \frac{\partial\, {}^{\sigma}\underset{\sim}{\alpha}^i}{\partial\, {}^{\sigma}\underset{\sim}{I}^j}\Bigg|_{\underline{\Omega}} \left({}^{\sigma}\underset{\sim}{I}^j\Big|_{\underline{\Omega}} \right) \quad ; \quad {}^{\sigma}\underset{\sim}{c}_{ij} = \frac{\partial\, {}^{\sigma}\underset{\sim}{\alpha}^i}{\partial\, {}^{\sigma}\underset{\sim}{I}^j}\Bigg|_{\underline{\Omega}} \\
& i = 1, 2, \dots, N \quad ; \quad j = 1, 2, \dots, M
\end{aligned}
\tag{11.38}
$$

equation (11.37) can be written as

$$
{}_d\boldsymbol{\sigma}^{(m)} = \underset{\sim}{\sigma}\Big|_{\underline{\Omega}} \boldsymbol{I} + \sum_{j=1}^{M} {}^{\sigma}\underset{\sim}{a}_j ({}^{\sigma}\underset{\sim}{I}^j) \boldsymbol{I} + \sum_{i=1}^{N} \sum_{j=1}^{M} {}^{\sigma}\underset{\sim}{c}_{ij} ({}^{\sigma}\underset{\sim}{I}^j) {}^{\sigma}\underset{\sim}{\boldsymbol{G}}^i + \sum_{i=1}^{N} {}^{\sigma}\underset{\sim}{b}_i {}^{\sigma}\underset{\sim}{\boldsymbol{G}}^i \tag{11.39}
$$

in which $\underset{\sim}{\sigma}^0\Big|_{\underline{\Omega}}$ is the known initial stress field in the known configuration $\underline{\Omega}$. This constitutive theory requires $(M + N + MN)$ material coefficients. The constitutive theory is based on integrity (complete basis of the space of ${}_d\boldsymbol{\sigma}^{(m)}$). ${}^{\sigma}\underset{\sim}{a}_j$, ${}^{\sigma}\underset{\sim}{c}_{ij}$ and ${}^{\sigma}\underset{\sim}{b}_i$; $i = 1, 2, \dots, N$; $j = 1, 2, \dots, M$ are material coefficients that can be functions of ${}^{\sigma}\underset{\sim}{I}^j\Big|_{\underline{\Omega}}$; $j = 1, 2, \dots, M$ and $\theta|_{\underline{\Omega}}$.

In this constitutive theory the dissipation mechanism is due to strain rates and the memory mechanism is due to strain rates as well as stress rates. *This is an ordered rate constitutive theory of orders up to m and n in stress and strain rates, respectively.*

Simplified constitutive theories

This constitutive theory (11.39) for ${}_d\boldsymbol{\sigma}^{(m)}$ is based on integrity. Though complete, it requires too many material coefficients. Simplified constitutive theories with fewer material coefficients can be derived by appropriate choices of n and m and by retaining only the desired generators and invariants. We consider simple linear constitutive theories in the following.

11.3.1.1 Simplified linear constitutive theory of orders m and n

We consider an ordered rate constitutive theory for ${}_d\boldsymbol{\sigma}^{(m)}$ of orders m and n that is linear in $\boldsymbol{\varepsilon}$, $\boldsymbol{\varepsilon}_{(i)}$; $i = 1, 2, \dots, n$ as well as linear in ${}_d\boldsymbol{\sigma}^{(j)}$; $j = 0, 1, \dots, m-1$. For this constitutive theory the combined generators of the argument tensors that are symmetric tensors of rank two are $\boldsymbol{\varepsilon}$, $\boldsymbol{\varepsilon}_{(i)}$; $i = 1, 2, \dots, n$, ${}_d\boldsymbol{\sigma}^{(j)}$; $j = 0, 1, \dots, m-1$. The combined invariants of the argument tensors are I_{ε}, $I_{\varepsilon_{(i)}}$; $i = 1, 2, \dots, n$, $I_{{}_d\sigma^{(j)}}$; $j = 0, 1, \dots, m-1$, i.e., traces of $\boldsymbol{\varepsilon}$, $\boldsymbol{\varepsilon}_{(i)}$; $i = 1, 2, \dots, n$ and ${}_d\boldsymbol{\sigma}^{(j)}$; $j = 0, 1, \dots, m-1$. We can write the following for this constitutive theory (using (11.39) after redefining the material coefficients).

$$
\begin{aligned}
{}_d\boldsymbol{\sigma}^{(m)} &= \underset{\sim}{\sigma}^0\Big|_{\underline{\Omega}} \boldsymbol{I} + a_1\boldsymbol{\varepsilon} + a_2(\operatorname{tr}\boldsymbol{\varepsilon})\boldsymbol{I} + \sum_{i=1}^{n} b_i^1 \boldsymbol{\varepsilon}_{(i)} + \sum_{i=1}^{n} b_i^2 (\operatorname{tr}\boldsymbol{\varepsilon}_{(i)})\boldsymbol{I} \\
&+ \sum_{j=0}^{m-1} c_j^1 ({}_d\boldsymbol{\sigma}^{(j)}) + \sum_{j=0}^{m-1} c_j^2 \operatorname{tr}({}_d\boldsymbol{\sigma}^{(j)})\boldsymbol{I}
\end{aligned}
\tag{11.40}
$$

The material coefficients a_1, a_2, b_i^1, b_i^2, c_j^1, c_j^2 ; $i = 1, 2, \dots, n$; $j = 0, 1, \dots, m-1$ can be functions of ${}^{\sigma}\underset{\sim}{I}^j\big|_{\underline{\Omega}}$; $j = 1, 2, \dots, m$ and $\theta\big|_{\underline{\Omega}}$. This constitutive theory (11.40) can also be written in matrix and vector form using Voigt's notation

$$
\{{}_d\sigma^{(m)}\} = \{\sigma^0\} + [\underset{\sim}{a}]\{\varepsilon\} + \sum_{i=1}^{n} [\underset{\sim}{b}_i]\{\varepsilon_{(i)}\} + \sum_{j=0}^{m-1} [\underset{\sim}{c}_j]\{{}_d\sigma^{(j)}\}
\tag{11.41}
$$

The arrangement of the coefficients of ${}_d\boldsymbol{\sigma}^{(m)}$ in $\{{}_d\sigma^{(m)}\}$ is same as (11.17). Same arrangement of the coefficients is also used for strain, strain rates and stress rates. Coefficients of matrix $[\underset{\sim}{a}]$ for $\{\sigma^0\}$ are defined in (11.19). Coefficients of matrices $[\underset{\sim}{b}_i]$ and $[\underset{\sim}{c}_j]$ are obtained from $[\underset{\sim}{a}]$ by replacing a_1, a_2 with b_i^1, b_i^2 and c_j^1, c_j^2, respectively.

11.3.1.2 Simplified linear constitutive theories of orders $m = 1, n = 1$

For this case, the constitutive variable is ${}_d\boldsymbol{\sigma}^{(1)}$ and its argument tensors are

$$
{}_d\boldsymbol{\sigma}^{(1)} = {}_d\boldsymbol{\sigma}^{(1)}(\boldsymbol{\varepsilon}, \boldsymbol{\varepsilon}_{(i)}, {}_d\boldsymbol{\sigma}^{(0)}, \theta) \quad ; \quad i = 1, 2, \dots, n
\tag{11.42}
$$

The constitutive theories (11.40) and (11.41) reduce to the following.

$$
\begin{aligned}
{}_d\boldsymbol{\sigma}^{(1)} &= \underset{\sim}{\sigma}^0\Big|_{\underline{\Omega}} \boldsymbol{I} + a_1\boldsymbol{\varepsilon} + a_2(\operatorname{tr}\boldsymbol{\varepsilon})\boldsymbol{I} + \sum_{i=1}^{n} b_i^1 \boldsymbol{\varepsilon}_{(i)} + \sum_{i=1}^{n} b_i^2 (\operatorname{tr}\boldsymbol{\varepsilon}_{(i)})\boldsymbol{I} \\
&+ c_0^1({}_d\boldsymbol{\sigma}^{(0)}) + c_0^2(\operatorname{tr}{}_d\boldsymbol{\sigma}^{(0)})\boldsymbol{I}
\end{aligned}
\tag{11.43}
$$

and in matrix and vector notation

$$
\{{}_d\sigma^{(1)}\} = \{\sigma^0\} + [\underset{\sim}{a}]\{\varepsilon\} + \sum_{i=1}^{n} [\underset{\sim}{b}_i]\{\varepsilon_{(i)}\} + [\underset{\sim}{c}_0]\{{}_d\sigma^{(0)}\}
\tag{11.44}
$$

The constitutive theory (11.43) can be written in a slightly different form using new definition of material coefficients that are in accordance with what is used in polymeric fluids. Dividing (11.43) by $-c_0^1$ and defining (assuming $c_0^1 \neq 0$ and neglecting $\underset{\sim}{\sigma}^0\big|_{\underline{\Omega}}$)

$$\left(-\frac{1}{c_0^1}\right) = \lambda \quad , \quad \left(-\frac{a_1}{c_0^1}\right) = 2\underset{\sim}{\mu} \quad , \quad \left(-\frac{a_2}{c_0^1}\right) = \underset{\sim}{\lambda}$$
$$\left(-\frac{b_i^1}{c_0^1}\right) = 2\eta_i \quad , \quad \left(-\frac{b_i^2}{c_0^1}\right) = k_i \quad , \quad \left(-\frac{c_0^2}{c_0^1}\right) = \underset{\sim}{\beta} \tag{11.45}$$

we can write (11.43) as

$$\boldsymbol{\sigma}^{(0)} + \lambda\frac{\partial}{\partial t}({}_d\boldsymbol{\sigma}^{(0)}) = 2\underset{\sim}{\mu}\boldsymbol{\varepsilon} + \underset{\sim}{\lambda}(\operatorname{tr}\boldsymbol{\varepsilon})\boldsymbol{I} + \sum_{i=1}^{n} 2\eta_i(\boldsymbol{\varepsilon}_{(i)}) + \sum_{i=1}^{n} \kappa_i(\operatorname{tr}\boldsymbol{\varepsilon}_{(i)})\boldsymbol{I} + \underset{\sim}{\beta}\operatorname{tr}({}_d\boldsymbol{\sigma})\boldsymbol{I} \tag{11.46}$$

λ is similar to relaxation time, $\underset{\sim}{\mu}$ and $\underset{\sim}{\lambda}$ are similar to Lamé's constants (assuming that the elastic constants are not effected by dissipation), η_i and κ_i are similar to first and second viscosity (as in thermo-viscous compressible fluids) for strain rate $\boldsymbol{\varepsilon}_{(i)}$. The term containing $\underset{\sim}{\beta}$ is generally neglected.

11.3.1.3 Linear constitutive theory of orders one ($m = 1, n = 1$)

This is the most simplified linear constitutive theory for deviatoric Cauchy stress tensor. In this case, ${}_d\boldsymbol{\sigma}^{(1)}$ is the constitutive variable with the following argument tensors

$${}_d\boldsymbol{\sigma}^{(1)} = {}_d\boldsymbol{\sigma}^{(1)}(\boldsymbol{\varepsilon}, \boldsymbol{\varepsilon}_{(1)}, {}_d\boldsymbol{\sigma}^{(0)}, \theta) \tag{11.47}$$

For this case, the constitutive theories (11.44) and (11.43) simplify to

$$\{{}_d\sigma^{(1)}\} = \{\sigma^0\} + [\underset{\sim}{a}]\{\varepsilon\} + [\underset{\sim}{b}_1]\{\varepsilon_{(1)}\} + [\underset{\sim}{c}]\{{}_d\sigma^{(0)}\} \tag{11.48}$$

$$\text{or} \quad {}_d\boldsymbol{\sigma}^{(1)} = \underset{\sim}{\sigma}^0\big|_{\underline{\Omega}}\boldsymbol{I} + a_1\boldsymbol{\varepsilon} + a_2(\operatorname{tr}\boldsymbol{\varepsilon})\boldsymbol{I} + b_1^1\boldsymbol{\varepsilon}_{(1)} + b_1^2(\operatorname{tr}\boldsymbol{\varepsilon}_{(1)})\boldsymbol{I} + c_0^1({}_d\boldsymbol{\sigma}^{(0)}) + c_0^2(\operatorname{tr}{}_d\boldsymbol{\sigma}^{(0)})\boldsymbol{I} \tag{11.49}$$

The constitutive theory (11.49) can be written in an alternate form using new definition of material coefficients (in (11.45)). We divide throughout by $-c_0^1$ and define new material coefficients as in (11.45) (defining $\underset{\sim}{\eta} = \eta_1$ and $\underset{\sim}{\kappa} = \kappa_1$).

$${}_d\boldsymbol{\sigma}^{(0)} + \lambda\frac{\partial}{\partial t}({}_d\boldsymbol{\sigma}^{(0)}) = 2\underset{\sim}{\mu}\boldsymbol{\varepsilon} + \underset{\sim}{\lambda}(\operatorname{tr}\boldsymbol{\varepsilon})\boldsymbol{I} + 2\underset{\sim}{\eta}\boldsymbol{\varepsilon}_{(1)} + \underset{\sim}{\kappa}(\operatorname{tr}\boldsymbol{\varepsilon}_{(1)})\boldsymbol{I} + \underset{\sim}{\beta}(\operatorname{tr}{}_d\boldsymbol{\sigma}^{(0)})\boldsymbol{I} \tag{11.50}$$

In (11.50), $2\underset{\sim}{\mu}$ and $\underset{\sim}{\lambda}$ are Lamé's constants (assuming dissipation does not influence elastic constants), $\underset{\sim}{\eta}$ and $\underset{\sim}{\kappa}$ are similar to first and second viscosities (as in compressible fluids). Generally, for small deformation, small strain, $\underset{\sim}{\kappa}(\operatorname{tr}\boldsymbol{\varepsilon}_{(1)})\boldsymbol{I}$ term may be small but we do not neglect it. The $\underset{\sim}{\beta}(\operatorname{tr}({}_d\boldsymbol{\sigma}^{(0)})\boldsymbol{I}$ term is often neglected. Thus, (11.50) reduces to the following

$${}_d\boldsymbol{\sigma}^{(0)} + \lambda\frac{\partial}{\partial t}({}_d\boldsymbol{\sigma}^{(0)}) = 2\underset{\sim}{\mu}\boldsymbol{\varepsilon} + \underset{\sim}{\lambda}(\operatorname{tr}\boldsymbol{\varepsilon})\boldsymbol{I} + 2\underset{\sim}{\eta}\boldsymbol{\varepsilon}_{(1)} + \underset{\sim}{\kappa}(\operatorname{tr}\boldsymbol{\varepsilon}_{(1)})\boldsymbol{I} \tag{11.51}$$

The constitutive theory (11.51) resembles Maxwell model for polymeric fluids (Chapter 13).

11.4 Memory modulus or relaxation modulus

The memory modulus or relaxation modulus establishes how the material relaxes, i.e., the stresses decay upon cesation of external disturbance. We consider one dimensional case (x_1-direction only) of the linear constitutive theory of orders $m = 1$ and n (equation (11.46) in the absence of term containing $\underset{\sim}{\beta}$). Defining ${}_d\sigma_{11} = {}_d\sigma_{11}^{(0)}$ and $\frac{\partial\, {}_d\sigma_{11}^{(0)}}{\partial t} = {}_d\sigma_{11}^{(1)}$, we can write

$$ {}_d\sigma_{11} + \lambda \frac{\partial}{\partial t}({}_d\sigma_{11}) = e_1 \varepsilon_{11} + \sum_{i=1}^{n} \underset{\sim}{\eta_i} \frac{\partial^i}{\partial t^i}(\varepsilon_{11}) \tag{11.52} $$

in which $e_1 = (2\mu + \lambda)$, $\underset{\sim}{\eta_i} = 2\eta_i + k_i$. We recall the differential equation

$$ \frac{dy}{dt} + P(t)y = Q(t) \tag{11.53} $$

has a solution

$$ y = e^{-\int P(t)dt}\Big[\int Q(t)e^{\int P(t)dt}dt + C\Big] \tag{11.54} $$

where C is the constant of integration. Comparing (11.52) and (11.53) we have

$$ \begin{aligned} y = {}_d\sigma_{11} \quad &; \quad P(t) = \frac{1}{\lambda} \\ Q(t) = \frac{1}{\lambda}\Big(e_1\varepsilon_{11} &+ \sum_{i=1}^{n} \underset{\sim}{\eta_i} \frac{\partial^i \varepsilon_{11}}{\partial t^i}\Big) \end{aligned} \tag{11.55} $$

Using (11.55) in (11.54) we obtain (using $\int p(t)dt = \int \frac{1}{\lambda}dt = \frac{t}{\lambda}$

$$ {}_d\sigma_{11} = e^{-\frac{t}{\lambda}}\Big[\int Q(t')e^{\frac{t'}{\lambda}}dt' + C\Big] \tag{11.56} $$

Using the integration limits $-\infty$ to t

$$ {}_d\sigma_{11} = \int_{-\infty}^{t} Q(t')e^{\frac{-(t-t')}{\lambda}}dt' + Ce^{-\frac{t}{\lambda}} \tag{11.57} $$

If $Q(t)$ is finite at $t = -\infty$, then ${}_d\sigma_{11}$ is also finite. Hence, the constant C must be zero. Thus, we have

$$ {}_d\sigma_{11} = \int_{-\infty}^{t} \Big(Q(t')e^{\frac{-(t-t')}{\lambda}}\Big)dt' \tag{11.58} $$

The quantity in parentheses is called the *relaxation modulus.* Based on (11.58), the stress at time t (current configuration) depends on strain and strain rates due to Q at time t, as well as the strain and strain rates at all past times t' with a weighting factor (the relaxation modulus) that decays exponentially as one goes backward in time. Thus, such materials have *fading memory* or *stress relaxation.* When λ is zero in (11.52) we have

$$_d\sigma_{11} = e_1 \varepsilon_{11} + \sum_{i=1}^{n} \underset{\sim}{\eta_i} \frac{\partial^i \varepsilon_{11}}{\partial t^i} \tag{11.59}$$

The material behavior described by (11.59) has no stress relaxation or fading memory due to non-existence of relaxation modulus as integrand of (11.58) is zero. Equation (11.59) is the same as that derived in reference [114] (also Chapter 10) for thermoviscoelastic solids without memory.

11.5 Zener constitutive model

The Zener constitutive model (small deformation, small strain, incompressible) for 1D viscoelastic behavior of solids with memory [26] is given by

$$\sigma_{11} + \lambda \frac{\partial}{\partial t}(\sigma_{11}) = e_1 \frac{\partial u_1}{\partial x_1} + \underset{\sim}{\eta_1} \frac{\partial}{\partial t}\left(\frac{\partial u_1}{\partial x_1}\right) \tag{11.60}$$

where σ_{11} is total Cauchy stress in x_1 direction. Using the constitutive theory derived here for $1D$ case ($m = 1$, $n = 1$, (11.52)) we can write the following for deviatoric Cauchy stress,

$$_d\sigma_{11} + \lambda \frac{\partial}{\partial t}(_d\sigma_{11}) = e_1 \frac{\partial u_1}{\partial x_1} + \underset{\sim}{\eta_1} \frac{\partial}{\partial t}\left(\frac{\partial u_1}{\partial x_1}\right) \tag{11.61}$$

The basic difference between the two constitutive theories ((11.60) and (11.61)) is that Zener model uses total stress, whereas the constitutive theory derived here (11.61) is in deviatoric stress. Based on the derivation presented in this chapter, Zener model cannot be derived using continuum mechanics principles of constitutive theory. This is due to the fact that presence of dissipation necessitates stress decomposition resulting in $_e\boldsymbol{\sigma}$ and $_d\boldsymbol{\sigma}$.

11.6 Model problem: numerical studies

In this section, we consider a model problem consisting of 1D wave propagation in thermoviscoelastic medium with memory. We consider small deformation, small strain deformation physics, hence the material is nearly incompressible. Furthermore, even though dissipation mechanism generates entropy that influences thermal field accounted for by the constitutive theory for equilibrium Cauchy stress tensor $_e\boldsymbol{\sigma}$. In the present study if we assume this to be negligible, then we can assume equilibrium Cauchy stress $_e\sigma_{11} = 0$. For this deformation physics $\sigma_{11} = {_d\sigma_{11}}$. Thus, if we neglect entropy production due to dissipation, then the 1D Zener constitutive model (11.60) and the constitutive theory derived here (11.61) are the same. This assumption is obviously not valid due to the presence of dissipation. If we assume $\sigma_{11} = {_d\sigma_{11}}$, then the balance of linear momenta in the x_1 direction and the constitutive theory are given by (in the absence of body forces) the following, constituting the complete mathematical model.

$$\rho_0 \frac{\partial^2 u_1}{\partial t^2} - \frac{\partial}{\partial x_1}(_d\sigma_{11}) = 0 \tag{11.62}$$

$$_d\sigma_{11} + \lambda \frac{\partial}{\partial t}(_d\sigma_{11}) = e_1 \frac{\partial u_1}{\partial x_1} + \underset{\sim}{\eta_1} \frac{\partial}{\partial t}\left(\frac{\partial u_1}{\partial x_1}\right) \tag{11.63}$$

In this mathematical model ${}_d\sigma_{11}$ must be maintained as a dependent variable as the substitution of ${}_d\sigma_{11}$ from (11.63) in (11.62) is not possible. We derive dimensionless form of (11.62) and (11.63). We rewrite (11.62) and (11.63) by introducing hat ($\wedge$) on each variable signifying that all quantities have their usual dimensions or units.

$$\widehat{\rho}_0 \frac{\partial^2 \widehat{u}_1}{\partial \widehat{t}^2} - \frac{\partial {}_d\widehat{\sigma}_{11}}{\partial \widehat{x}_1} = 0 \tag{11.64}$$

$${}_d\widehat{\sigma}_{11} + \widehat{\lambda}\frac{\partial}{\partial \widehat{t}}({}_d\widehat{\sigma}_{11}) = \widehat{e}_1 \frac{\partial \widehat{u}_1}{\partial \widehat{x}_1} + \widehat{\underset{\sim}{\eta}}_1 \frac{\partial}{\partial \widehat{t}}\left(\frac{\partial \widehat{u}_1}{\partial \widehat{x}_1}\right) \tag{11.65}$$

we choose reference quantities with subscript zero or "ref" and define the following dimensionless variables using them

$$\rho_0 = \frac{\widehat{\rho}_0}{(\rho_0)_{ref}} \quad ; \quad u_1 = \frac{\widehat{u}_1}{u_0} \quad ; \quad x_1 = \frac{\widehat{x}_1}{L_0} \quad ; \quad v_1 = \frac{v_1}{v_0} \quad ; \quad t = \frac{\widehat{t}}{t_0} \quad ; \quad {}_d\sigma_{11} = \frac{{}_d\widehat{\sigma}_{11}}{\tau_0} \tag{11.66}$$

where L_0 is reference length, hence $u_0 = L_0$. If we choose v_0 as the reference velocity (generally the reference speed of sound based on reference quantities), then $t_0 = \frac{u_0}{v_0} = \frac{L_0}{v_0}$. τ_0 is reference stress; hence, we do not need to define reference force. We choose reference stress $\tau_0 = (\rho_0)_{ref} v_0^2$, the characteristic kinetic energy (which has the same units as stress). The dimensionless form of (11.64) and (11.65) and become

$$\rho_0 \frac{\partial^2 u_1}{\partial t^2} - \frac{\partial}{\partial x_1}({}_d\sigma_{11}) = 0 \tag{11.67}$$

$${}_d\sigma_{11} + De\frac{\partial {}_d\sigma_{11}}{\partial t} = e_1^d \frac{\partial u_1}{\partial x_1} + \eta_1^d \frac{\partial}{\partial t}\left(\frac{\partial u_1}{\partial x_1}\right) \tag{11.68}$$

$$\text{where} \quad De = \frac{\widehat{\lambda} v_0}{L_0} \quad ; \quad e_1^d = \frac{\widehat{e}_1}{\tau_0} \quad ; \quad \eta_1^d = \frac{\widehat{\underset{\sim}{\eta}}_1}{(\rho_0)_{ref} v_0 L_0} \tag{11.69}$$

De is called Deborah number, the dimensionless relaxation time. e_1^d and $\underset{\sim}{\eta}_1^d$ are dimensionless elasticity and dissipation coefficients. We note that when $De = 0$, the constitutive theory reduces to that for viscoelastic solid without memory. We consider the numerical solution of (11.67) and (11.68) for $De \geq 0$. When $De = 0$, there is no rheology. Hence, the solutions for $De = 0$ and $De > 0$ can be compared to illustrate the influence of rheology.

In the numerical studies we consider an axial rod of dimensionless length one unit and choose $(\rho_0)_{\text{ref}} = \widehat{\rho}_0$ so that $\rho_0 = 1$. The spatial domain $[0, 1]$ for an increment of time $\Delta t = t_{n+1} - t_n$, i.e., the nth space-time strip $[0, 1] \times [t_n, t_{n+1}]$, is discretized using a uniform mesh of eight space-time p-version finite elements with higher order global differentiability [121, 122].

Figure 11.1 shows details of the space-time domain for an increment of time $\Delta t = t_{n+1} - t_n$, boundary conditions, as well as initial conditions. The left end of the rod is clamped (impermeable boundary) and the right end is subjected to a continuous and differentiable compressive piecewise-cubic strain distribution over a time period of $2\Delta t$. For $t \geq 2\Delta t$, the applied strain at the right end of the rod is zero. We choose p-level of 9 in space and time, and the local approximation is of

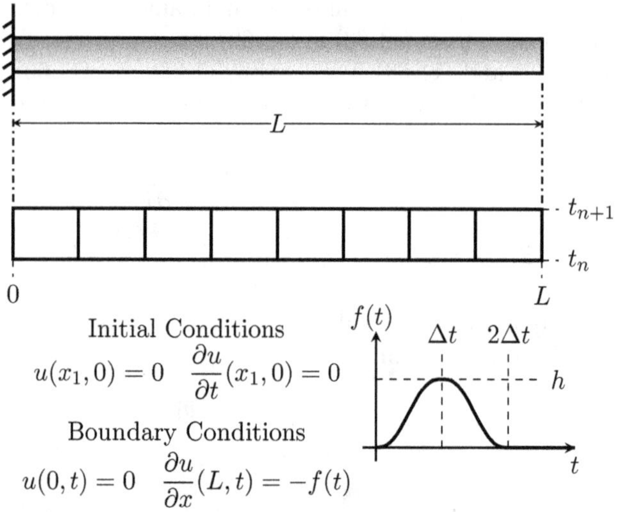

Figure 11.1: Schematic of the model problem and boundary conditions, initial conditions

class $C^{11}(\bar{\Omega}_{x_1t})$, i.e., C^1 in space as well as time. The finite element formulation used for computing numerical solutions is based on space-time least squares process constructed using the residual functional. The resulting computational processes are unconditionally stable. Evolutions are computed using a space-time strip with time marching [101–103, 105, 106]. For the choice of local approximation $C^{11}(\bar{\Omega}_{x_1t})$, the integrals in the finite element processes are Lebesgue, but due to the smoothness of the evolution, for the eight-element discretization with p-level of 9, the residual functionals are on the order of $O(10^{-8})$ for all space-time strips, confirming good accuracy of the computed evolution.

We note $e_1^d = 1$ is dimensionless modulus of elasticity and dimensionless density $\rho_0 = 1$; hence, the elastic wave speed is one. Figure 11.2 and 11.3 show plots of ${}_d\sigma_{11}$ versus x_1 for $0 \leq t \leq 2.3$ for $De = 0.0$, 0.004, 0.008. When $De = 0.0$, there is dissipation but no memory. The stress pulse of duration $2\Delta t = 0.2$ advances 0.2 units in negative x_1 direction as shown in Figure 11.2 for $t = 0.2$. Further propagation of stress pulse in the negative x_1 direction is shown at $t = 0.3$. Stress reflection from the impermeable boundary at $x_1 = 0.0$ can be observed in Figure 11.2 for $t = 0.9$, 1.0, 1.1 and at $t = 1.2$ in Figure 11.3. At $t = 1.8$ (Figure 11.3) reflected wave reaches free boundary at $x_1 = 1.0$ and reflects and propagates (to the left) as a tensile wave (Figure 11.3) for $1.8 \leq t \leq 2.3$.

In Figures 11.2 and 11.3 we observe that for $De = 0.004$, there is elasticity and dissipation but no memory. For $De = 0.004$, we observe higher values of stresses during evolution compared to $De = 0.0$ and for $De = 0.008$ even higher stresses than those for $De = 0.004$ throughout the evolution, demonstrating the progressively more pronounce memory with progressively increasing Deborah number.

Figure 11.4 and 11.5 show plots of strain $\varepsilon_{11} = \frac{\partial u_1}{\partial x_1}$ versus x_1 for $0.1 \leq t \leq 2.3$. We observe behavior similar to ${}_d\sigma_{11}$ versus x_1 shown in Figures 11.2 and 11.3 (but obviously not identical).

When De $= 0.0$, the behavior is thermoviscoelastic, but without memory. As time elapses, the stress wave experiences base elongation and amplitude decay due to dissipation as shown in the figures. At time $t = 2.0$, waves show almost complete amplitude decay when $De = 0.0$. For non-zero Deborah number, the stress amplitude is higher than for De = 0.0 throughout the evolution, due to rheology. Increasing Deborah numbers produce increasing values of stress during evolution compared to $De = 0.0$. Peak stress and strain values for De = 0.008 are higher than those for $De = 0.004$ throughout the evolution as expected, showing the more pronounced memory aspect with higher De.

11.7 Constitutive theories for heat vector $\boldsymbol{q}$

The constitutive theory for heat vector $\boldsymbol{q}$ remains the same as derived in Chapter 9, Section 9.4, hence not repeated here.

11.8 General remarks

We make some remarks in the following that may be helpful regarding the use of mathematical models (CBL) utilizing the constitutive theories presented in this chapter.

(1) Such materials have mechanisms of elasticity, dissipation and memory. Dissipation mechanism is due to strain rates, and memory (rheology) is due to long chain molecules of the polymer incorporated in the constitutive theory through convected stress rates.

(2) Due to dissipation, a part of the rate of mechanical work is expanded in entropy generation, which is not recoverable upon unloading.

(3) Upon cessation of loading, such materials take finite amount of time to return to unstressed (or relaxed) state. This time is controlled by relaxation time, a material coefficient. Relaxation modulus (function of relaxation time) can be used to determine the actual time it takes for the volume of matter to achieve unstressed, relaxed state upon cessation of loading.

(4) When $\boldsymbol{\sigma}^{[0]}$ and $\boldsymbol{\varepsilon}_{[0]}$ are choices of stress and strain measures (for finite deformation, finite strain), the dissipation and the memory mechanism are non-linear functions of displacement gradients [125]. This is also an active area of research at present.

(5) Due to elasticity (E, ν) and density (ρ_0) such materials have stiffness and mass, thus, the following hold for the materials.

   (a) Finite speed of sound as $\sqrt{E/\rho}$ is hard to establish theoretically due to dissipation and memory but can be used as a guide. Stress wave can form and propagate in such materials but with attenuation, i.e., amplitude decay and base elongation occurs (when the energy content of the wave is fixed). If $\boldsymbol{\varepsilon}_{[0]}$ is the strain measure (as opposed to $\boldsymbol{\varepsilon}$), then the distortion of wave shape occurs during propagation.

   (b) Even though frequency and time dependent excitations do cause vibrations in such material but not in as pure of a form as for thermoelastic solid continua.

   (c) Since the constitutive theories for such materials contain strain rates and stress rates, the mathematical models and the physics described by them does not lend to the concepts of natural vibrations, natural frequencies and mode shapes.

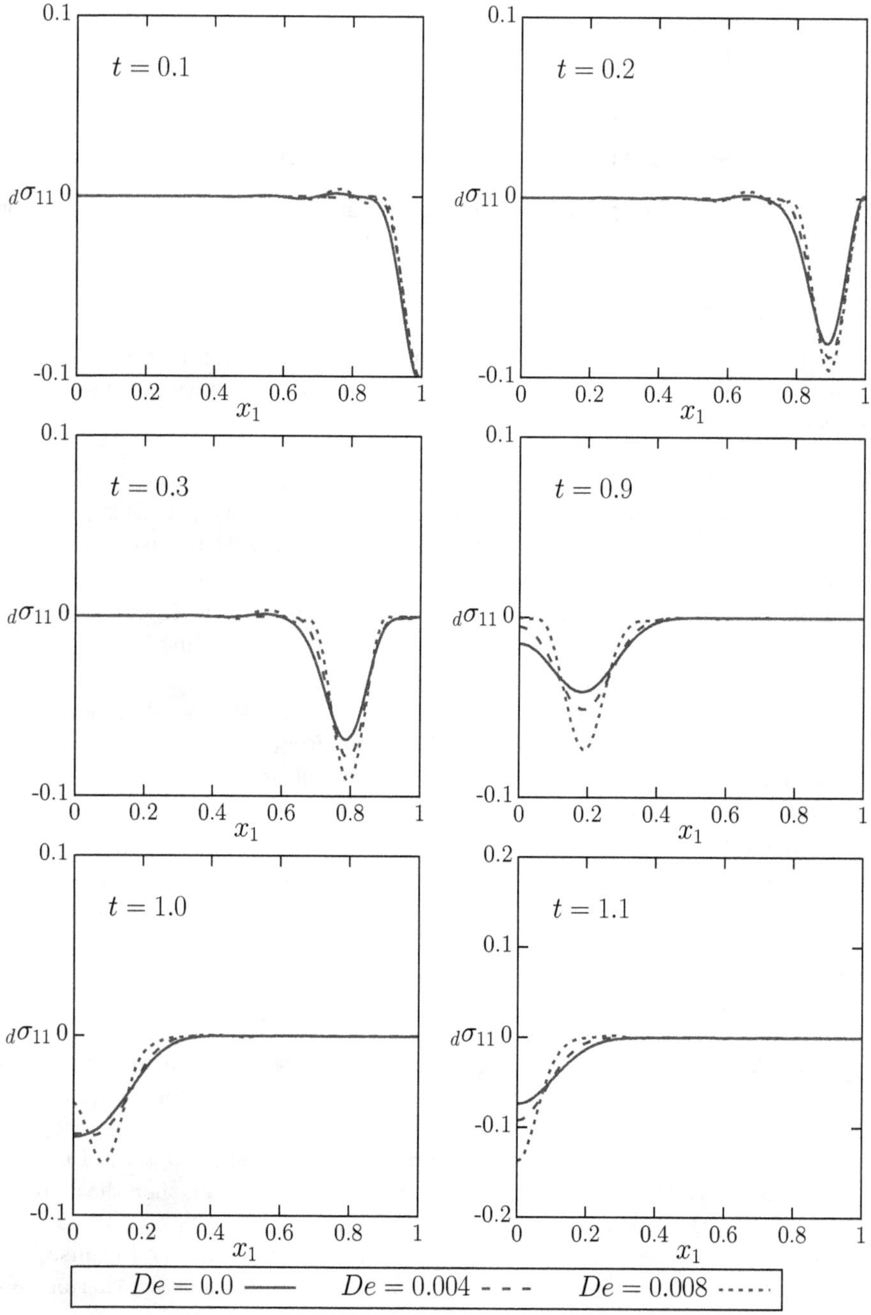

Figure 11.2: Evolutions of stress for $e_1^d = 1.0$, $\eta_1^d = 0.01$

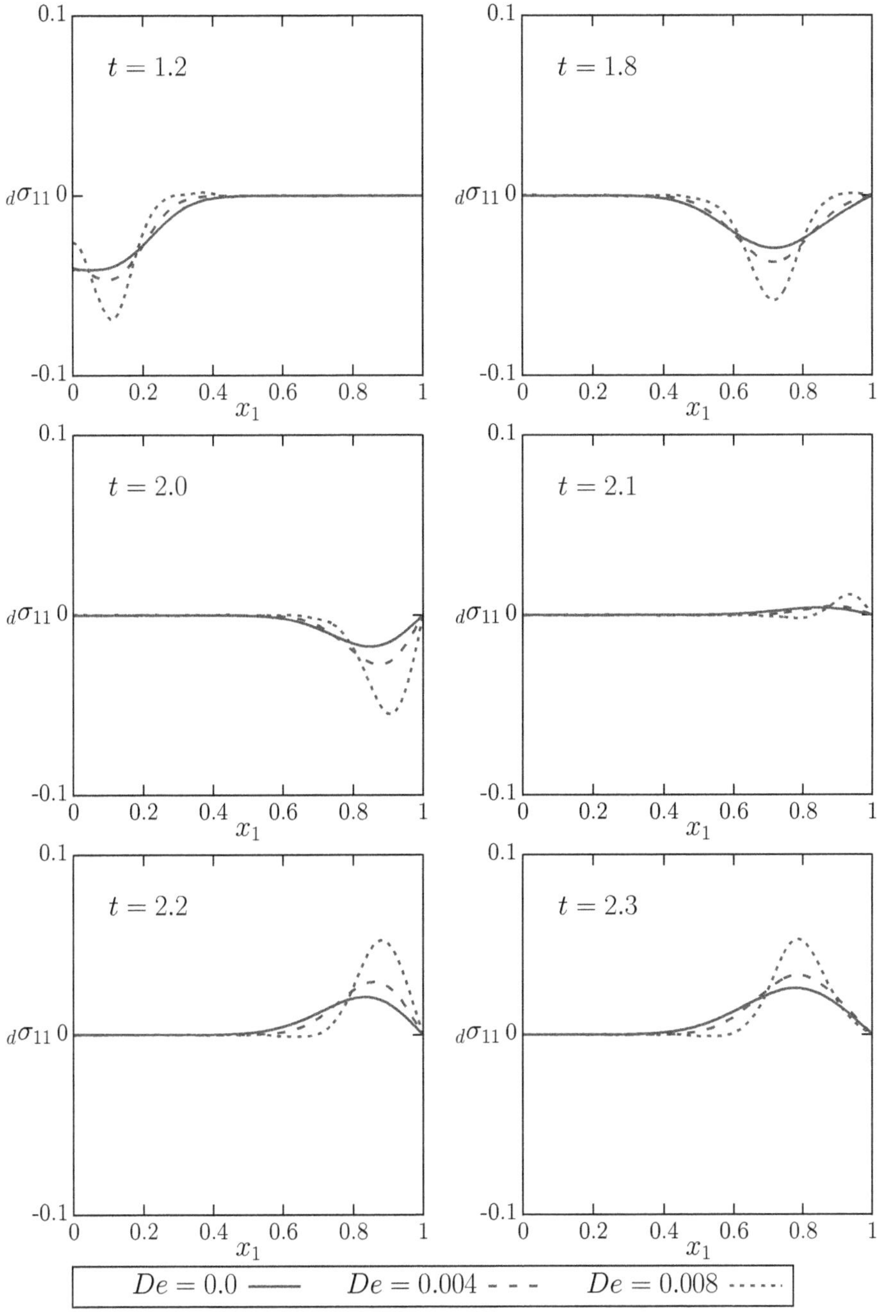

Figure 11.3: Evolution of stress for $e_1^d = 1.0$, $\eta_1^d = 0.01$

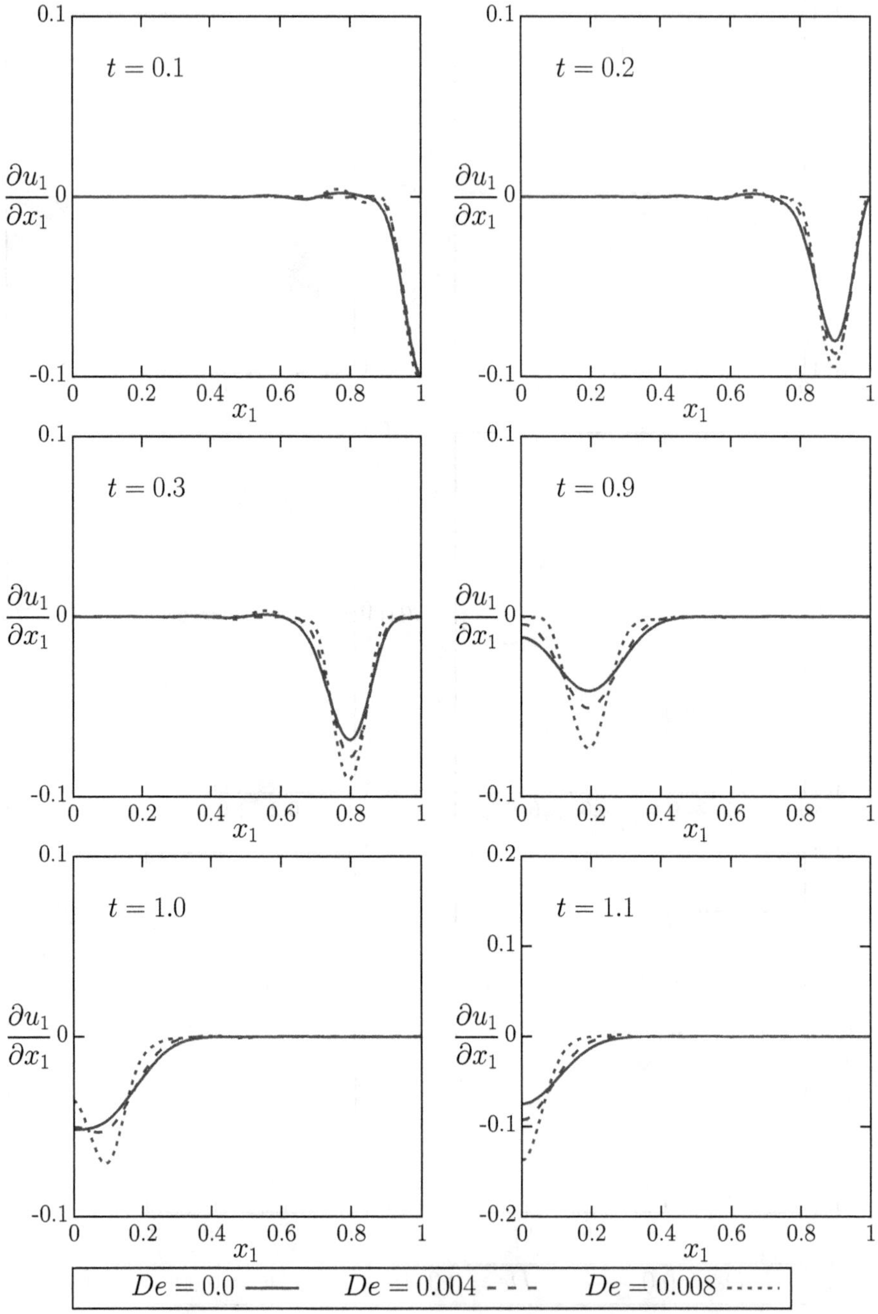

Figure 11.4: Evolution of strain for $e_1^d = 1.0$, $\eta_1^d = 0.01$

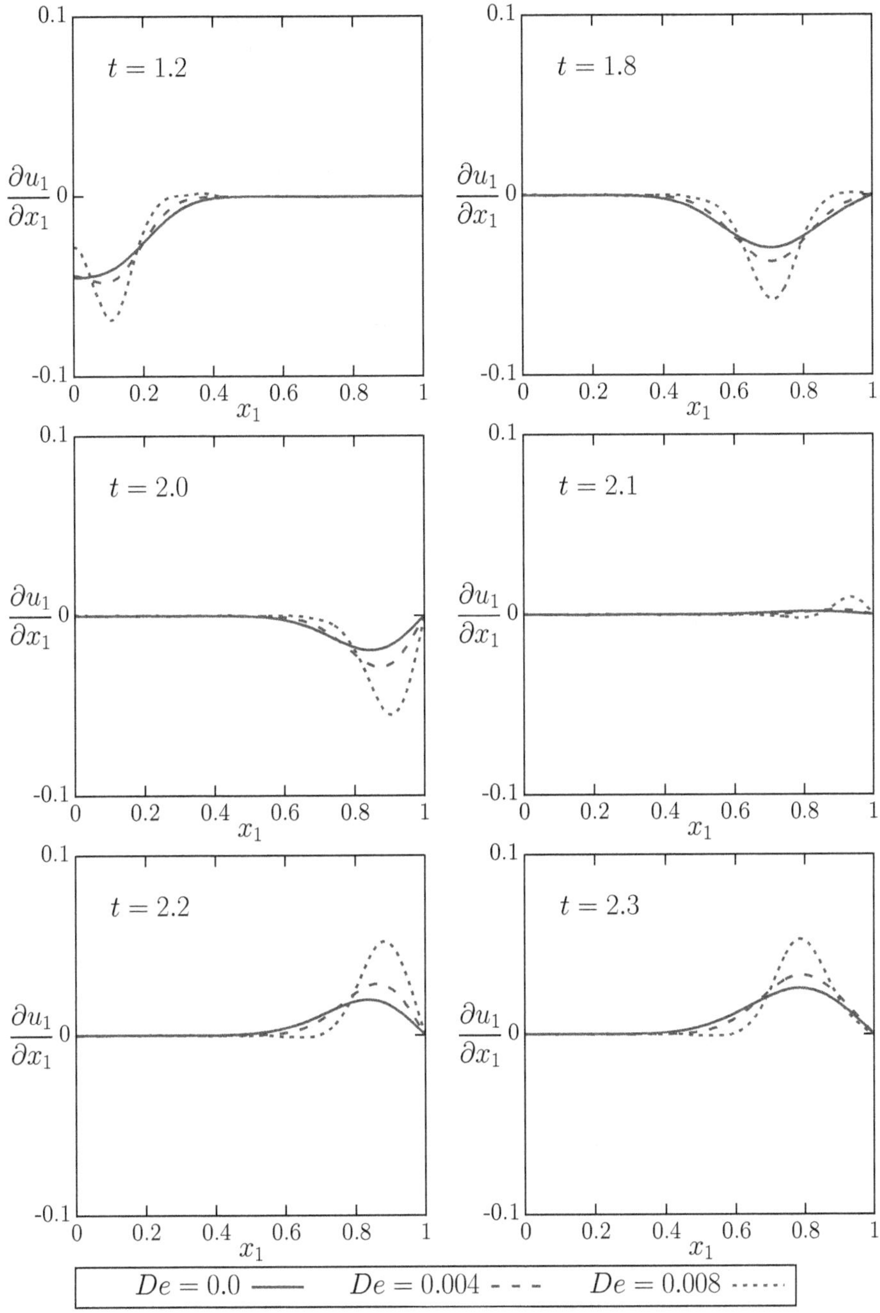

Figure 11.5: Evolution of strain for $e_1^d = 1.0$, $\eta_1^d = 0.01$

11.9 Summary

In this chapter, the reduced form of the entropy and representation theorem are utilized to derive constitutive theory for the deviatoric stress tensor for thermoviscoelastic solid continua with memory. Constitutive theories for the equilibrium stress tensor describing volumetric deformation, compressibility, incompressibility and non-isothermal physics presented in Chapter 8 is referenced here without repeating the derivation. Constitutive theories for deviatoric stress tensor addressing distortional physics, dissipation mechanism and memory are presented for finite deformation, finite strain, compressible, incompressible and non-isothermal physics as well as for small deformation, small strain physics.

In case of finite deformation, finite strain the ordered rate constitutive theories of orders m and n in deviatoric second Piola-Kirchhoff stress tensor ${}_d\boldsymbol{\sigma}^{[0]}$ and Green's strain tensor $\boldsymbol{\varepsilon}_{[0]}$ require ${}_d\boldsymbol{\sigma}^{[m]}$ to be the constitutive tensor with $\boldsymbol{\varepsilon}_{[0]}$, $\boldsymbol{\varepsilon}_{[i]}$; $i = 1, 2, \ldots, n$ and ${}_d\boldsymbol{\sigma}^{[j]}$; $j = 0, 1, \ldots, m-1$ and θ to be its argument tensors.

This constitutive theory is based on integrity (complete of the space of ${}_d\boldsymbol{\sigma}^{[m]}$). Simplified linear forms of the constitutive theory are presented for: (a) orders m and n (b) orders $m = 1$, n and (c) $m = 1$ and $n = 1$. Matrix and vector forms of these constitutive theories are also given. The constitutive theory of orders $m = 1$ and n is used to redefine material coefficients to illustrate relaxation time, Lamé's constants, first and second viscosities for finite deformation, finite strain, compressible and incompressible physics in the present work. These are similar to those used in published works.

Constitutive theories for small deformation, small strain, deformation physics with dissipation and memory are parallel to those for finite deformation, finite strain but the stress measure is basis independent Cauchy stress tensor and the strain measure is linear part of Green's strain tensor. Constitutive theories of orders m, n based on complete integrity as well as the linear constitutive theory of orders: m, n; $m = 1$, n and $m = 1$, $n = 1$ are presented. Material coefficients such as relaxation time, Lam e's constants, first and second viscosity are defined.

For small deformation, small strain deformation physics, linear constitutive theory of orders $m = 1$, $n = 1$ is used to derive relaxation modulus. Zener model is presented and compared with the present theory to demonstrate lack of validity of Zener model based on CCM when the deformation is non-isothermal (as always is the case in the presence of dissipation), and when the mathematical models are in $\mathbb{R}^2$ and $\mathbb{R}^3$. A model problem consisting of one dimensional stress wave propagation in thermoviscoelastic incompressible continua with memory is presented for small deformation, small strain physics (neglecting thermal effects) to illustrate rheology aspect of the constitutive theories derived in this chapter. Since the constitutive theories for heat vector remain the same as in Chapters 9 and 10, only the reference is made without derivation.

Problems

11.1 Derive a first order linear rate constitutive theory for the deviatoric second Piola-Kirchhoff stress tensor for a thermoviscoelastic solid with memory from first principles using the theory of generators and invariants in which

(a) the argument tensors are Green's strain tensor, Green's strain tensor rates, deviatoric second Piola-Kirchhoff stress tensor and temperature θ;
(b) the constitutive theory depends on the Cauchy strain tensor, rates of Cauchy strain tensor, second Piola-Kirchhoff stress tensor and temperature θ.

In both cases, derive material coefficients. Determine the relationship between the material coefficients in the theories derived in (a) and (b) if one exists. Give explicit forms of the equations for $\mathbb{R}^1$ and $\mathbb{R}^2$ for both (a) and (b). Also, represent these in matrix and vector forms using Voigt's notation.

11.2 Derive an ordered linear rate constitutive theory for the deviatoric second Piola-Kirchhoff stress tensor for finite deformation of thermoviscoelastic solids with memory from first principles using the theory of generators and invariants in which

(a) the time derivative of order n of the Green's strain tensor is a linear function of the deviatoric second Piola-Kirchhoff stress tensor and its rate, Green's strain tensor, rate of Green's strain tensors up to orders $(n-1)$ and temperature θ;
(b) the constitutive theory utilizes Cauchy strain tensor instead of Green's strain tensor and is based on the same approach as in (a).

Derive material coefficients in both (a) and (b). Give explicit forms of the equations for $\mathbb{R}^1$ and $\mathbb{R}^2$ for both (a) and (b). Also, represent these in matrix and vector forms using Voigt's notation.

11.3 Derive a linear ordered rate constitutive theory for the heat vector that is consistent with the choice of argument tensors in Problem 11.1. Include temperature gradient as an additional argument. Give explicit expanded forms of the equations.

11.4 Derive a linear ordered rate constitutive theory for the heat vector that is consistent with the choice of argument tensors in Problem 11.2. Include temperature gradient as an additional argument. Derive explicit expanded forms of the equations.

12 CONSTITUTIVE THEORIES FOR THERMOVISCOUS FLUIDS

12.1 Introduction

Thermoviscous fluids have mechanism of dissipation, i.e., some (or all) of the rate of mechanical work results in rate of entropy generation due to viscosity of the fluid that provides resisting force to the relative motion between the fluid particles. Rate of entropy production naturally influences thermal field in the fluid. Thermoviscous fluids can be compressible or nearly incompressible. The dissipation mechanism is present in both as this physics is due to the viscosity of the medium. Inviscid fluent continua are non-physical, as without viscosity the fluent continua cannot exist. The derivation of the constitutive theories for compressible and incompressible thermoviscous fluent continua is the subject of study in this chapter.

Newton's law of viscosity for compressible and incompressible thermoviscous fluids is well known and is widely used as a constitutive theory for deviatoric Cauchy stress tensor. Constitutive models for generalized Newtonian fluids such as power law, Carreau-Yasuda model, etc. are extensions of Newton's law of viscosity in which the viscosity of the medium is dependent on the deformation field. The constitutive theories for thermoviscous fluids in this chapter are derived using entropy inequality in conjunction with representation theorem. These constitutive theories are based on integrity; hence, utilize complete basis of the space of the constitutive variables. It is shown that all currently used constitutive theories for thermoviscous fluids are very limited subsets of the constitutive theories presented in this chapter.

We recall from Chapter 6 that for fluent continua the Eulerian description is the only possible way to consider the conservation and balance laws. This of course stems from the fact that complex motion of fluid particles rules out the Lagrangian description. Secondly, dissipation in the physics of fluid motion and deformation and measure of stress only requires strain rates, symmetric part of the velocity gradient tensor and not the strain tensor, thus, displacements of the fluid particles are not needed. Both of these requirements suggest that in the description of fluid motion and deformation, velocities can be observable quantities. Hence, the conservation and balance laws in Eulerian description in velocities are suitable. In Eulerian description, we consider deformed tetrahedron in the current configuration to define contravariant, covariant Cauchy and Jaumann stress tensors, $\bar{\boldsymbol{\sigma}}^{(0)}$, $\bar{\boldsymbol{\sigma}}_{(0)}$ and ${}^{(0)}\bar{\boldsymbol{\sigma}}^{J}$. In the derivation of constitutive theories, it is more general to consider a basis independent Cauchy stress measure ${}^{(0)}\bar{\boldsymbol{\sigma}}$ that can be $\bar{\boldsymbol{\sigma}}^{(0)}$ or $\bar{\boldsymbol{\sigma}}_{(0)}$ or ${}^{(0)}\bar{\boldsymbol{\sigma}}^{J}$. The measures of deformation rates in fluent continua are convected time derivatives of the strain tensors in covariant directions (tangent to the deformed material) and contravariant (reciprocal to the covariant directions) directions obtained using convected time derivatives of Green's strain tensor $\boldsymbol{\varepsilon}_{[0]}$ and Almansi strain tensors $\bar{\boldsymbol{\varepsilon}}^{[0]}$. We recall that the contravariant Cauchy stress measure is rate of work conjugate with covariant convected time derivatives of the strain tensor $\boldsymbol{\varepsilon}_{[0]}$. Likewise, the

DOI: 10.1201/9781003105336-12

covariant Cauchy stress tensor is rate of work conjugate to the contravariant convected time derivatives of the Almansi strain tensor $\bar{\boldsymbol{\varepsilon}}^{[0]}$. We consider details of the these measures and their role in the constitutive theories in the following sections in this chapter.

As discussed in Chapter 8, the deformed volume $\bar{V}(t)$ in the current configuration contains volumetric deformation $\bar{V}_v(t)$ and distortional deformation $\bar{V}_d(t)$. The volumetric deformation physics accounts for compressibility, incompressibility and non-isothermal effects whereas the distortion deformation physics for thermoviscous fluent continua accounts for distortion of the volume and the dissipation mechanism.

12.2 Preliminary considerations

In fluent continua we must consider CBL in Eulerian description in which velocities $\bar{\boldsymbol{v}}(\bar{\boldsymbol{x}},t)$ are observable quantities. The choice of appropriate stress measure(s) and convected time derivatives of the strain tensors is crucial as well. In the CBL and the derivation of the constitutive theories we choose basis independent Cauchy stress tensor ${}^{(0)}\bar{\boldsymbol{\sigma}}$ that could be $\bar{\boldsymbol{\sigma}}^{(0)}$, $\bar{\boldsymbol{\sigma}}_{(0)}$ or ${}^{(0)}\bar{\boldsymbol{\sigma}}^J$. We also consider basis independent convected time derivatives ${}^{(i)}\boldsymbol{\gamma}$; $i=1,2,\dots,n$ that could be covariant convected time derivatives $\boldsymbol{\gamma}_{(i)}$; $i=1,2,\dots,n$ of the Green's strain tensors $\boldsymbol{\varepsilon}_{[0]}$ or could be contravariant convected time derivatives $\boldsymbol{\gamma}^{(i)}$; $i=1,2,\dots,n$ of the Almansi strain tensor $\bar{\boldsymbol{\varepsilon}}^{[0]}$ or Jaumann rates ${}^{(i)}\boldsymbol{\gamma}^J$; $i=1,2,\dots,n$ which are average of $\boldsymbol{\gamma}_{(i)}$ and $\boldsymbol{\gamma}^{(i)}$; $i=1,2,\dots,n$ rates. We note that (see Chapter 5).

$$\boldsymbol{\gamma}^{(1)}=\boldsymbol{\gamma}_{(1)}=\bar{\boldsymbol{D}}={}^{(1)}\boldsymbol{\gamma}^J \tag{12.1}$$

where $\bar{\boldsymbol{D}}$ is the symmetric part of the velocity gradient tensor. Thus, ${}^{(0)}\bar{\boldsymbol{\sigma}}$, $\bar{\boldsymbol{D}}$ or $\bar{\boldsymbol{\sigma}}^{(0)}$, $\bar{\boldsymbol{D}}$ or $\bar{\boldsymbol{\sigma}}_{(0)}$, $\bar{\boldsymbol{D}}$ or ${}^{(0)}\bar{\boldsymbol{\sigma}}^J$, $\bar{\boldsymbol{D}}$ are rate of work conjugate pairs. If we assume that convected time derivatives ${}^{(i)}\boldsymbol{\gamma}$; $i=2,3,\dots,n$ also contribute to the dissipation mechanism, then we must consider

$${}^{(0)}\bar{\boldsymbol{\sigma}} \quad \text{and} \quad {}^{(i)}\boldsymbol{\gamma} \quad ; \quad i=1,2,\dots,n \tag{12.2}$$

as basis independent rate of work conjugate pairs in the derivation of the constitutive theories. Basis dependent constitutive theories can be obtained by choosing

$$\begin{aligned} &{}^{(0)}\bar{\boldsymbol{\sigma}},{}^{(i)}\boldsymbol{\gamma} \quad \text{as} \quad \bar{\boldsymbol{\sigma}}^{(0)},\boldsymbol{\gamma}_{(i)} \\ &{}^{(0)}\bar{\boldsymbol{\sigma}},{}^{(i)}\boldsymbol{\gamma} \quad \text{as} \quad \bar{\boldsymbol{\sigma}}_{(0)},\boldsymbol{\gamma}^{(i)} \\ &{}^{(0)}\bar{\boldsymbol{\sigma}},{}^{(i)}\boldsymbol{\gamma} \quad \text{as} \quad {}^{(0)}\bar{\boldsymbol{\sigma}}^J,{}^{(i)}\boldsymbol{\gamma}^J \\ &i=1,2,\dots,n \end{aligned} \tag{12.3}$$

We perform additive stress decomposition of Cauchy stress tensor ${}^{(0)}\bar{\boldsymbol{\sigma}}$ into equilibrium stress tensor ${}^{(0)}_{e}\bar{\boldsymbol{\sigma}}$ and deviatoric stress tensor ${}_{d}\bar{\boldsymbol{\sigma}}^{(0)}$. Volumetric deformation physics is described by the constitutive theory for ${}^{(0)}_{e}\bar{\boldsymbol{\sigma}}$. Distortional deformation physics and the dissipation mechanism are due to the constitutive theory for ${}^{(0)}_{d}\bar{\boldsymbol{\sigma}}$. Furthermore, volumetric deformation physics in Eulerian description is invariant of the type of deformation and the constitution of the matter.

12.3 Constitutive theory for equilibrium Cauchy stress ${}^{(0)}_{\ e}\bar{\boldsymbol{\sigma}}$

The derivation of the constitutive theory for ${}^{(0)}_{\ e}\bar{\boldsymbol{\sigma}}$ has been presented in Chapter 8, see Section 8.5.4. We can write

$$ {}^{(0)}_{\ e}\bar{\boldsymbol{\sigma}} = \bar{p}(\bar{\rho},\bar{\theta})\boldsymbol{\delta} \quad ; \quad \text{compressible} \tag{12.4} $$

$$ {}^{(0)}_{\ e}\bar{\boldsymbol{\sigma}} = \bar{p}(\bar{\theta})\boldsymbol{\delta} \quad ; \quad \text{incompressible} \tag{12.5} $$

$$ \text{in which} \quad \bar{p}(\bar{\rho},\bar{\theta}) = -\bar{\rho}^2 \frac{\partial \bar{\Phi}(\bar{\rho},\bar{\theta})}{\partial \bar{\rho}} \tag{12.6} $$

$\bar{p}(\bar{\rho},\bar{\theta})$ is thermodynamic pressure or the equation of state for the deforming matter. We note that the equilibrium Cauchy stress tensor constitutive theory in (12.4) and (12.5) is basis independent as it is pressure field. We continue using the notation in (12.4) and (12.5) for more clarity when it is added to the deviatoric Cauchy stress tensor. The reduced form of the entropy inequality is given by (equation (8.76))

$$ -{}^{(0)}_{\ d}\bar{\boldsymbol{\sigma}} : \bar{\boldsymbol{D}} + \frac{\bar{\boldsymbol{q}} \cdot \bar{\boldsymbol{g}}}{\bar{\theta}} \leq 0 \tag{12.7} $$

The entropy inequality (12.7) is satisfied if ${}^{(0)}_{\ d}\bar{\boldsymbol{\sigma}} : \bar{\boldsymbol{D}} > 0$ and $\frac{\bar{\boldsymbol{q}}\cdot\bar{\boldsymbol{g}}}{\bar{\theta}} \leq 0$. These serve as restrictions on the constitutive theories for ${}^{(0)}_{\ d}\bar{\boldsymbol{\sigma}}$ and $\bar{\boldsymbol{q}}$. In Eulerian descriptions density $\bar{\rho}$ is a dependent variable in CBL, thus, it can be included as an argument tensor of the constitutive tensors. Due to non-isothermal physics, $\bar{\theta}$ as argument tensor of the constitutive tensors is a natural choice. Thus, strictly based on the conjugate pairs in (12.7) we can write

$$ {}^{(0)}_{\ d}\bar{\boldsymbol{\sigma}} = {}^{(0)}_{\ d}\bar{\boldsymbol{\sigma}}(\bar{\rho},\bar{\boldsymbol{D}},\bar{\theta}) \tag{12.8} $$

$$ \text{and} \quad \bar{\boldsymbol{q}} = \bar{\boldsymbol{q}}(\bar{\rho},\bar{\boldsymbol{g}},\bar{\theta}) \tag{12.9} $$

12.4 Constitutive theory for deviatoric Cauchy stress ${}^{(0)}_{\ d}\bar{\boldsymbol{\sigma}}$

We recall that

$$ \bar{\boldsymbol{D}} = \boldsymbol{\gamma}^{(1)} = \boldsymbol{\gamma}_{(1)} = {}^{(1)}\boldsymbol{\gamma} \tag{12.10} $$

If we consider the dissipation mechanism to be dependent on ${}^{(i)}\boldsymbol{\gamma}$; $i = 1, 2, \dots, n$, then an ordered rate theory of dissipation up to orders n of the convected time derivatives of the strain tensor is possible. Thus, $\bar{\boldsymbol{D}}$ in (12.8) can be replaced by ${}^{(i)}\boldsymbol{\gamma}$; $i = 1, 2, \dots, n$ and we have

$$ {}^{(0)}_{\ d}\bar{\boldsymbol{\sigma}} = {}^{(0)}_{\ d}\bar{\boldsymbol{\sigma}}(\bar{\rho}, {}^{(i)}\boldsymbol{\gamma}, \bar{\theta}) \quad ; \quad i = 1, 2, \dots, n \tag{12.11} $$

We can derive a constitutive theory for ${}^{(0)}_{\ d}\bar{\boldsymbol{\sigma}}$ using (12.11) and representation theorem. Let ${}^{\sigma}\underset{\sim}{\boldsymbol{G}}^i$; $i = 1, 2, \dots, N$ be the combined generators of the argument tensors of ${}^{(0)}_{\ d}\bar{\boldsymbol{\sigma}}$ in (12.11) that are symmetric tensors of rank two. Then, $\boldsymbol{I}$, ${}^{\sigma}\underset{\sim}{\boldsymbol{G}}^i$; $i = 1, 2, \dots, N$ constitute the basis of the space of ${}^{(0)}_{\ d}\bar{\boldsymbol{\sigma}}$. Hence, we can express ${}^{(0)}_{\ d}\bar{\boldsymbol{\sigma}}$ as a linear combination of the generators $\boldsymbol{I}$, ${}^{\sigma}\underset{\sim}{\boldsymbol{G}}^{i}$; $i = 1, 2, \dots, N$ in the current configuration.

$$ {}^{(0)}_{\;d}\bar{\boldsymbol{\sigma}} = {}^{\sigma}\underset{\sim}{\alpha}^{0}\boldsymbol{I} + \sum_{i=1}^{N} {}^{\sigma}\underset{\sim}{\alpha}^{i}({}^{\sigma}\underset{\sim}{\boldsymbol{G}}^{i}) \tag{12.12} $$

in which the coefficients ${}^{\sigma}\underset{\sim}{\alpha}^{i}$; $i = 0, 1, \dots, N$ in the linear combination (12.12) are function of $\bar{\rho}$, $\bar{\theta}$ and combined invariants of the argument tensors of ${}^{(0)}_{\;d}\bar{\boldsymbol{\sigma}}$ in (12.11), ${}^{\sigma}\underset{\sim}{I}^{j}$; $j = 1, 2, \dots, M$ in the current configuration. That is

$$ {}^{\sigma}\underset{\sim}{\alpha}^{i} = {}^{\sigma}\underset{\sim}{\alpha}^{i}(\bar{\rho}, \bar{\theta}, {}^{\sigma}\underset{\sim}{I}^{j}) \quad ; \quad j = 1, 2, \dots, M \quad ; \quad i = 0, 1, \dots, N \tag{12.13} $$

To determine material coefficients in the constitutive theory (12.12), we expand ${}^{\sigma}\underset{\sim}{\alpha}^{i}$; $i = 0, 1, \dots, N$ in ${}^{\sigma}\underset{\sim}{I}^{j}$; $j = 1, 2, \dots, M$ about a known configuration $\underline{\Omega}$ and retain only up to linear terms in ${}^{\sigma}\underset{\sim}{I}^{j}$; $j = 1, 2, \dots, M$ (for simplicity). We do not consider Taylor series expansion of ${}^{\sigma}\underset{\sim}{\alpha}^{i}$; $i = 0, 1, \dots, N$ in $\bar{\rho}$ and $\bar{\theta}$ as their influence on the Cauchy stress tensor has already been accounted for in deriving the constitutive theory for ${}^{(0)}_{\;e}\bar{\boldsymbol{\sigma}}$.

$$ {}^{\sigma}\underset{\sim}{\alpha}^{i} = {}^{\sigma}\underset{\sim}{\alpha}^{i}\Big|_{\underline{\Omega}} + \sum_{j=1}^{M} \frac{\partial\, {}^{\sigma}\underset{\sim}{\alpha}^{i}}{\partial\, {}^{\sigma}\underset{\sim}{I}^{j}}\Bigg|_{\underline{\Omega}} \left({}^{\sigma}\underset{\sim}{I}^{j} - {}^{\sigma}\underset{\sim}{I}^{j}\Big|_{\underline{\Omega}}\right) \quad ; \quad i = 0, 1, \dots, N \tag{12.14} $$

Substituting ${}^{\sigma}\underset{\sim}{\alpha}^{i}$ from (12.14) in (12.12)

$$ \begin{aligned} {}^{(0)}_{\;d}\bar{\boldsymbol{\sigma}} = & \left({}^{\sigma}\underset{\sim}{\alpha}^{0}\Big|_{\underline{\Omega}} + \sum_{j=1}^{M} \frac{\partial\, {}^{\sigma}\underset{\sim}{\alpha}^{0}}{\partial\, {}^{\sigma}\underset{\sim}{I}^{j}}\Bigg|_{\underline{\Omega}} \left({}^{\sigma}\underset{\sim}{I}^{j} - {}^{\sigma}\underset{\sim}{I}^{j}\Big|_{\underline{\Omega}}\right)\right)\boldsymbol{I} \\ & + \sum_{i=1}^{N} \left({}^{\sigma}\underset{\sim}{\alpha}^{i}\Big|_{\underline{\Omega}} + \sum_{j=1}^{M} \frac{\partial\, {}^{\sigma}\underset{\sim}{\alpha}^{i}}{\partial\, {}^{\sigma}\underset{\sim}{I}^{j}}\Bigg|_{\underline{\Omega}} \left({}^{\sigma}\underset{\sim}{I}^{j} - {}^{\sigma}\underset{\sim}{I}^{j}\Big|_{\underline{\Omega}}\right)\right){}^{\sigma}\underset{\sim}{\boldsymbol{G}}^{i} \end{aligned} \tag{12.15} $$

Collecting coefficients of (the terms defined in the current configuration) $\boldsymbol{I}$, ${}^{\sigma}\underset{\sim}{I}^{j}\boldsymbol{I}$, ${}^{\sigma}\underset{\sim}{\boldsymbol{G}}^{i}$ and ${}^{\sigma}\underset{\sim}{I}^{j}({}^{\sigma}\underset{\sim}{\boldsymbol{G}}^{i})$; $i = 1, 2, \dots, N$; $j = 1, 2, \dots, M$ and defining

$$ \begin{aligned} \underset{\sim}{\sigma}^{0}\Big|_{\underline{\Omega}} &= {}^{\sigma}\underset{\sim}{\alpha}^{0}\Big|_{\underline{\Omega}} - \sum_{j=1}^{M} \frac{\partial\, {}^{\sigma}\underset{\sim}{\alpha}^{0}}{\partial\, {}^{\sigma}\underset{\sim}{I}^{j}}\Bigg|_{\underline{\Omega}} \left({}^{\sigma}\underset{\sim}{I}^{j}\Big|_{\underline{\Omega}}\right) \quad ; \quad {}^{\sigma}\underset{\sim}{a}_{j} = \frac{\partial\, {}^{\sigma}\underset{\sim}{\alpha}^{0}}{\partial\, {}^{\sigma}\underset{\sim}{I}^{j}}\Bigg|_{\underline{\Omega}} \\ {}^{\sigma}\underset{\sim}{b}_{i} &= {}^{\sigma}\underset{\sim}{\alpha}^{i}\Big|_{\underline{\Omega}} - \sum_{j=1}^{M} \frac{\partial\, {}^{\sigma}\underset{\sim}{\alpha}^{i}}{\partial\, {}^{\sigma}\underset{\sim}{I}^{j}}\Bigg|_{\underline{\Omega}} \left({}^{\sigma}\underset{\sim}{I}^{j}\Big|_{\underline{\Omega}}\right) \quad ; \quad {}^{\sigma}\underset{\sim}{c}_{ij} = \frac{\partial\, {}^{\sigma}\underset{\sim}{\alpha}^{i}}{\partial\, {}^{\sigma}\underset{\sim}{I}^{j}}\Bigg|_{\underline{\Omega}} \\ & i = 1, 2, \dots, N \quad ; \quad j = 1, 2 \dots, M \end{aligned} \tag{12.16} $$

equation (12.15) can be written as

$$ {}^{(0)}_{\;d}\bar{\boldsymbol{\sigma}} = \underset{\sim}{\sigma}^{0}\Big|_{\underline{\Omega}}\boldsymbol{I} + \sum_{j=1}^{M} {}^{\sigma}\underset{\sim}{a}_{j}({}^{\sigma}\underset{\sim}{I}^{j})\boldsymbol{I} + \sum_{i=1}^{N}\sum_{j=1}^{M} {}^{\sigma}\underset{\sim}{c}_{ij}({}^{\sigma}\underset{\sim}{I}^{j}){}^{\sigma}\underset{\sim}{\boldsymbol{G}}^{i} + \sum_{i=1}^{N} {}^{\sigma}\underset{\sim}{b}_{i}\,{}^{\sigma}\underset{\sim}{\boldsymbol{G}}^{i} \tag{12.17} $$

${}^{\sigma}\underset{\sim}{a}_{j}$, ${}^{\sigma}\underset{\sim}{b}_{i}$, ${}^{\sigma}\underset{\sim}{c}_{ij}$; $i = 1, 2, \dots, N$; $j = 1, 2, \dots, M$ are material coefficients defined in the known configuration $\underline{\Omega}$. $\underset{\sim}{\sigma}^{0}\Big|_{\underline{\Omega}}$ is the initial stress field in a known configuration

$\underline{\Omega}$. This constitutive theory for ${}^{(0)}{}_d\bar{\boldsymbol{\sigma}}$ requires $(M+MN+N)$ material coefficients. This constitutive theory is based on integrity (complete basis of the space of ${}^{(0)}{}_d\bar{\boldsymbol{\sigma}}$). Material coefficients can be functions of $\bar{\rho}|_{\underline{\Omega}}$, $\bar{\theta}|_{\underline{\Omega}}$ and ${}^{\sigma}\underset{\sim}{I}^j\big|_{\underline{\Omega}}$; $j=1,2,\dots,M$.

Remarks

(1) In the constitutive theory for the deviatoric Cauchy stress tensor ${}^{(0)}{}_d\bar{\boldsymbol{\sigma}}$, the dissipation mechanism is due to convected time derivatives ${}^{(i)}\boldsymbol{\gamma}$; $i=1,2,\dots,n$ of the strain tensor. Thus, we refer to this constitutive theory of up to order n as the *ordered rate constitutive theory of up to order n*.

(2) By choosing appropriate conjugate pairs for ${}^{(0)}{}_d\bar{\boldsymbol{\sigma}}$ and ${}^{(i)}\boldsymbol{\gamma}$; $i=1,2,\dots,n$ basis specific constitutive theories for deviatoric Cauchy stress tensor can be obtained. These are listed below.

$$
\begin{aligned}
&\text{Contravariant stress measure:} \quad {}_d\bar{\boldsymbol{\sigma}}^{(0)}, \boldsymbol{\gamma}_{(i)} \quad ; \quad i=1,2,\dots,n \\
&\text{Covariant stress measure:} \quad {}_d\bar{\boldsymbol{\sigma}}_{(0)}, \boldsymbol{\gamma}^{(i)} \quad ; \quad i=1,2,\dots,n \\
&\text{Jaumann rates:} \quad {}^{(0)}\bar{\boldsymbol{\sigma}}^J, {}^{(i)}\boldsymbol{\gamma}^J \quad ; \quad i=1,2,\dots,n
\end{aligned}
\tag{12.18}
$$

(3) Configuration $\underline{\Omega}$ is a suitable previously known configuration (generally corresponding to last value of time t for which deformation is known). Configuration $\underline{\Omega}$ can also be the reference configuration in which case the material coefficients will be constant, i.e., not dependent on varying $\bar{\rho}$, $\bar{\theta}$ and ${}^{\sigma}\underset{\sim}{I}^j$; $j=1,2,\dots,M$.

(4) The material coefficients as a function of unknown deformation in current configuration is not supported by CCM. This is rather obvious because if this were the case then ${}^{\sigma}\underset{\sim}{\alpha}^i$; $i=0,1,\dots,N$ in (12.12) would be material coefficients and there would be no need for Taylor series expansion of ${}^{\sigma}\underset{\sim}{\alpha}^i$ in ${}^{\sigma}\underset{\sim}{I}^j$; $j=1,2,\dots,M$ about a known configuration. In continuum theories, material coefficients are known properties of the matter; hence, cannot be treated as functions of unknown deformation.

Simplified constitutive theories for ${}^{(0)}{}_d\bar{\boldsymbol{\sigma}}$

Obviously different choices of n and specific choices of generators and invariants in (12.17) will yield different constitutive theories.

12.4.1 A constitutive theory of order one ($n=1$) for ${}^{(0)}{}_d\bar{\boldsymbol{\sigma}}$: Newtonian and generalized Newtonian fluids

In this case, we have

$$
{}^{(0)}{}_d\bar{\boldsymbol{\sigma}} = {}^{(0)}{}_d\bar{\boldsymbol{\sigma}}(\bar{\rho}, {}^{(1)}\boldsymbol{\gamma}, \bar{\theta}) = {}^{(0)}{}_d\bar{\boldsymbol{\sigma}}(\bar{\rho}, \bar{\boldsymbol{D}}, \bar{\theta}) \tag{12.19}
$$

Combined generators of the argument tensors $\bar{\rho}$, $\bar{\boldsymbol{D}}$ and $\bar{\theta}$ that are symmetric tensors of rank two are

$$
{}^{\sigma}\underset{\sim}{\boldsymbol{G}}^1 = \bar{\boldsymbol{D}} \quad , \quad {}^{\sigma}\underset{\sim}{\boldsymbol{G}}^2 = \bar{\boldsymbol{D}}^2 \quad ; \quad (N=2 \text{ in } (12.17)) \tag{12.20}
$$

Hence

$$
{}^{(0)}{}_d\bar{\boldsymbol{\sigma}} = {}^{\sigma}\underset{\sim}{\alpha}^0 \boldsymbol{I} + {}^{\sigma}\underset{\sim}{\alpha}^i \bar{\boldsymbol{D}} + {}^{\sigma}\underset{\sim}{\alpha}^2 \bar{\boldsymbol{D}}^2 \tag{12.21}
$$

The combined invariants of the argument tensors of ${}^{(0)}_{\;d}\bar{\boldsymbol{\sigma}}$ in (12.19) are given by

$$ {}^{\sigma}\underset{\sim}{I}^1 = \operatorname{tr}(\bar{\boldsymbol{D}}) \quad , \quad {}^{\sigma}\underset{\sim}{I}^2 = \operatorname{tr}(\bar{\boldsymbol{D}}^2) \quad , \quad {}^{\sigma}\underset{\sim}{I}^3 = \operatorname{tr}(\bar{\boldsymbol{D}}^3) \quad ; \quad (M = 3 \text{ in } (12.17)) \tag{12.22} $$

Using (12.20) and (12.22) in (12.17) the constitutive theory for ${}^{(0)}_{\;d}\bar{\boldsymbol{\sigma}}$ is completely defined for $n = 1$. This constitutive theory contain up to fifth degree terms of the components of tensor $\bar{\boldsymbol{D}}$ and requires $(3 + 3(2) + 2) = 11$ material coefficients. Explicit expression for ${}^{(0)}_{\;d}\bar{\boldsymbol{\sigma}}$ is given in the following

$$ \begin{aligned} {}^{(0)}_{\;d}\bar{\boldsymbol{\sigma}} = {} & \underset{\sim}{\sigma}^0\big|_{\underline{\Omega}} + {}^{\sigma}\underset{\sim}{a}_1({}^{\sigma}\underset{\sim}{I}^1)\boldsymbol{I} + {}^{\sigma}\underset{\sim}{a}_2({}^{\sigma}\underset{\sim}{I}^2)\boldsymbol{I} + {}^{\sigma}\underset{\sim}{a}_3({}^{\sigma}\underset{\sim}{I}^3)\boldsymbol{I} \\ & + {}^{\sigma}\underset{\sim}{c}_{11}({}^{\sigma}\underset{\sim}{I}^1)\bar{\boldsymbol{D}} + {}^{\sigma}\underset{\sim}{c}_{21}({}^{\sigma}\underset{\sim}{I}^1)\bar{\boldsymbol{D}}^2 + {}^{\sigma}\underset{\sim}{c}_{12}({}^{\sigma}\underset{\sim}{I}^2)\bar{\boldsymbol{D}} + {}^{\sigma}\underset{\sim}{c}_{22}({}^{\sigma}\underset{\sim}{I}^2)\bar{\boldsymbol{D}}^2 \\ & + {}^{\sigma}\underset{\sim}{c}_{13}({}^{\sigma}\underset{\sim}{I}^3)\bar{\boldsymbol{D}} + {}^{\sigma}\underset{\sim}{c}_{23}({}^{\sigma}\underset{\sim}{I}^3)\bar{\boldsymbol{D}}^2 + {}^{\sigma}\underset{\sim}{b}_1\bar{\boldsymbol{D}} + {}^{\sigma}\underset{\sim}{b}_2\bar{\boldsymbol{D}}^2 \end{aligned} \tag{12.23} $$

Remarks

(1) Constitutive theory (12.23) is the general form of the constitutive theory for Newtonian and generalized Newtonian fluid based on integrity. That is, this constitutive theory utilizes complete basis of space of ${}^{(0)}_{\;d}\bar{\boldsymbol{\sigma}}$.

(2) When the material coefficients are constant, (12.23) is referred to as the constitutive theory for Newtonian fluids.

(3) When the material coefficients are functions of $\bar{\rho}$, $\bar{\theta}$ and the invariants of the argument tensors, the constitutive theory (12.23) is referred to as constitutive theory for generalized Newtonian fluids.

(4) Since $\bar{\boldsymbol{D}}$ is basis independent, we can replace ${}^{(0)}_{\;d}\bar{\boldsymbol{\sigma}}$ by ${}_d\bar{\boldsymbol{\sigma}}$ in (12.23), basis independent measure of Cauchy stress.

12.4.2 Linear constitutive theory of order n for ${}^{(0)}_{\;d}\bar{\boldsymbol{\sigma}}$

From the general constitutive theory of order n based on integrity we can obtain a constitutive theory of order n that is linear in the components of ${}^{(i)}\boldsymbol{\gamma}$; $i = 1, 2, \ldots, n$. Redefining new material coefficients, we can write (using (12.17))

$$ {}^{(0)}_{\;d}\bar{\boldsymbol{\sigma}} = \underset{\sim}{\sigma}^0\big|_{\underline{\Omega}}\boldsymbol{I} + \sum_{i=1}^{n} a_i\,{}^{(i)}\boldsymbol{\gamma} + \sum_{i=1}^{n} b_i(\operatorname{tr}{}^{(i)}\boldsymbol{\gamma})\boldsymbol{I} \tag{12.24} $$

Using Voigt's notation (with the same arrangement of the components of ${}^{(0)}_{\;d}\bar{\boldsymbol{\sigma}}$ and ${}^{(i)}\boldsymbol{\gamma}$ as in Chapter 10, (10.16) we can write the following.

$$ \{{}^{(0)}_{\;d}\bar{\boldsymbol{\sigma}}\} = \underset{\sim}{\sigma}^0\big|_{\underline{\Omega}} \begin{Bmatrix} 1 \\ 1 \\ 1 \\ 0 \\ 0 \\ 0 \end{Bmatrix} + \sum_{i=1}^{n} [\underset{\sim}{a}_i]\{{}^{(i)}\boldsymbol{\gamma}\} \tag{12.25} $$

in which $[\underset{\sim}{a}_i]$ is given by

$$[\underset{\sim}{a}_i] = \begin{bmatrix} a_i + b_i & b_i & b_i & 0 & 0 & 0 \\ b_i & a_i + b_i & b_i & 0 & 0 & 0 \\ b_i & b_i & a_i + b_i & 0 & 0 & 0 \\ 0 & 0 & 0 & a_i & 0 & 0 \\ 0 & 0 & 0 & 0 & a_i & 0 \\ 0 & 0 & 0 & 0 & 0 & a_i \end{bmatrix} \tag{12.26}$$

In this constitutive theory, inclusion of the convected time derivative of each order of the strain tensor requires two additional material coefficients.

12.4.3 Linear constitutive theory of order one ($n = 1$): Newtonian and generalized Newtonian fluids

From the constitutive theory of order n given by (12.24) we can obtain a linear constitutive theory of order one by using $n = 1$ (neglecting $\underset{\sim}{\sigma}^0\big|_{\underline{\Omega}}\boldsymbol{I}$ term)

$$ {}^{(0)}{}_d\bar{\boldsymbol{\sigma}} = a_1 {}^{(1)}\boldsymbol{\gamma} + b_1(\operatorname{tr} {}^{(1)}\boldsymbol{\gamma})\boldsymbol{I} \tag{12.27}$$

redefining material coefficients

$$\begin{gathered} {}^{(0)}{}_d\bar{\boldsymbol{\sigma}} = a_1\bar{\boldsymbol{D}} + b_1(\operatorname{tr}\bar{\boldsymbol{D}})\boldsymbol{I} \\ a_1 = 2\eta\big|_{\underline{\Omega}} \quad \text{and} \quad b_1 = \kappa\big|_{\underline{\Omega}} \end{gathered} \tag{12.28}$$

and noting that deviatoric Cauchy stress is basis independent as $\bar{\boldsymbol{D}}$ is basis independent; hence, we can use ${}_d\bar{\boldsymbol{\sigma}}$ in place of ${}^{(0)}{}_d\bar{\boldsymbol{\sigma}}$, we can write (12.27) as

$$ {}_d\bar{\boldsymbol{\sigma}} = 2\eta\big|_{\underline{\Omega}}\bar{\boldsymbol{D}} + \kappa\big|_{\underline{\Omega}}(\operatorname{tr}\bar{\boldsymbol{D}})\boldsymbol{I} \tag{12.29}$$

$\eta\big|_{\underline{\Omega}}$ and $\kappa\big|_{\underline{\Omega}}$ are called first and second viscosities.

Remarks

(1) When $\eta\big|_{\underline{\Omega}}$ and $\kappa\big|_{\underline{\Omega}}$ are constant, (12.29) is a linear constitutive theory for compressible fluids referred to as *Newton's law of viscosity.* If the fluid is incompressible, then $\operatorname{tr}\bar{\boldsymbol{D}} = 0$, hence (12.29) reduces to the following constitutive theory.

$$ {}_d\bar{\boldsymbol{\sigma}} = 2\eta\big|_{\underline{\Omega}}\bar{\boldsymbol{D}} \tag{12.30}$$

(2) When $\eta\big|_{\underline{\Omega}}$ and $\kappa_{\underline{\Omega}}$ are functions of $\bar{\rho}\big|_{\underline{\Omega}}$, $\bar{\theta}\big|_{\underline{\Omega}}$ and invariants of $\bar{\boldsymbol{D}}$ in the known configuration $\underline{\Omega}$, then (12.29) is the linear constitutive theory for ${}_d\bar{\boldsymbol{\sigma}}$ for compressible generalized Newtonian fluids. Likewise, $\eta\big|_{\underline{\Omega}}$ in (12.30) is a function of $\bar{\theta}\big|_{\underline{\Omega}}$ and invariants of $\bar{\boldsymbol{D}}$ in $\underline{\Omega}$. Equation (12.30) is the linear constitutive theory for ${}_d\bar{\boldsymbol{\sigma}}$ for incompressible generalized Newtonian fluid.

12.4.4 Generalized Newtonian fluids: variable transport properties

For incompressible generalized Newtonian fluids $\eta\rfloor_\Omega$ and $\kappa\rfloor_\Omega$ can vary during the evolution. If we consider principal invariants of $\bar{\boldsymbol{D}}$, we can write

$$\begin{aligned}\eta\rfloor_\Omega &= \eta\rfloor_\Omega(\bar{\rho}\rfloor_\Omega, \bar{\theta}\rfloor_\Omega, (I_{\bar{D}})\rfloor_\Omega, (II_{\bar{D}})\rfloor_\Omega, (III_{\bar{D}})\rfloor_\Omega) \\ \kappa\rfloor_\Omega &= \kappa\rfloor_\Omega(\bar{\rho}\rfloor_\Omega, \bar{\theta}\rfloor_\Omega, (I_{\bar{D}})\rfloor_\Omega, (II_{\bar{D}})\rfloor_\Omega, (III_{\bar{D}})\rfloor_\Omega)\end{aligned} \tag{12.31}$$

In the case of incompressible, generalized Newtonian fluids, $\operatorname{tr}\bar{\boldsymbol{D}} = I_{\bar{D}} = i_{\bar{D}} = 0$ and $\bar{\rho}$ =constant, but η can vary during evolution as shown below

$$\eta\rfloor_\Omega = \eta\rfloor_\Omega(\bar{\theta}\rfloor_\Omega, (II_{\bar{D}})_\Omega, (III_{\bar{D}})_\Omega) \tag{12.32}$$

Equation (12.31) and (12.32) permit use of experimentally, empirically or analytically established relationships for $\eta\rfloor_\Omega$ and $\kappa\rfloor_\Omega$ as functions of their arguments as long as $\eta\rfloor_\Omega$ and $\kappa\rfloor_\Omega$ remain continuous and differentiable in their arguments. We remark that $\eta\rfloor_\Omega$ and $\kappa\rfloor_\Omega$ are only defined in a known configuration $\underline{\Omega}$. Instead of the principal invariant of $\bar{\boldsymbol{D}}$, we can also use the invariants $i_{\bar{D}}$, $ii_{\bar{D}}$ and $iii_{\bar{D}}$ as the two sets of invariants are related (Chapter 2).

12.4.4.1 Temperature dependent viscosity

Power law and Sutherland law [139] are commonly used temperature dependent models for viscosity η given in the following.

$$\eta\rfloor_\Omega = \eta^0 \left(\frac{\bar{\theta}\rfloor_\Omega}{\theta^0}\right)^{\underset{\sim}{\eta}} \quad ; \quad \text{Power Law} \tag{12.33}$$

$$\eta\rfloor_\Omega = \eta^0 \left(\frac{\bar{\theta}\rfloor_\Omega}{\theta^0}\right)^{3/2} \left(\frac{\theta^0 + s}{\bar{\theta}\rfloor_\Omega + s}\right) \quad ; \quad \text{Sutherland Law} \tag{12.34}$$

Parameters η^0, θ^0, $\underset{\sim}{\eta}$, s are known constants for a given fluid.

12.4.4.2 Power law model of viscosity

Power law and Carreau-Yasuda models [21] for shear thinning and shear thickening fluids are examples of the dependence of viscosity η on $II_{\bar{D}}$ or $ii_{\bar{D}}$.

The derivation of the power law model presented in the following is the same as what is used currently in polymer science and engineering. This derivation is important as the experimental data available in the published works for determining power law index use the final form of this power law model presented here. Let

$$\boldsymbol{\gamma} = \bar{\boldsymbol{L}} + \bar{\boldsymbol{L}}^T = 2\bar{\boldsymbol{D}} \tag{12.35}$$

$$ii_\gamma = \operatorname{tr}(\boldsymbol{\gamma}^2) = 4\operatorname{tr}(\bar{\boldsymbol{D}}^2) = 4ii_{\bar{D}} \tag{12.36}$$

We define a scalar γ as

$$\gamma = \sqrt{\frac{1}{2}ii_\gamma} = \sqrt{\frac{1}{2}4ii_{\bar{D}}} = \sqrt{2ii_{\bar{D}}} \tag{12.37}$$

η is expressed as a function of γ

$$\eta = m(\gamma)^{\widetilde{n}-1} = m(2ii_{\bar{D}})^{\frac{\widetilde{n}-1}{2}} \tag{12.38}$$

Equation 12.38 defines power law model of viscosity η. In (12.38), $\widetilde{n}$ is known as power law index. Taking log of (12.38)

$$\log \eta = \log m + (\widetilde{n} - 1) \log \gamma \tag{12.39}$$

$$\text{or} \quad \log \eta = \log m + \left(\frac{\widetilde{n} - 1}{2} \right) \log(2ii_{\bar{D}}) \tag{12.40}$$

In (12.38), m is known as zero shear rate viscosity. We conduct experiments with progressively increasing γ in (12.39) or progressively increasing $ii_{\bar{D}}$ in (12.40), generally using a viscometer and measure values of η for each γ or $ii_{\bar{D}}$. Based on (12.39), a graph of $\log n$ versus $\log \gamma$ is a straight line with intercept $\log m$ on $\log \eta$ axis and the slope of the straight line is $(\widetilde{n} - 1)$ from which we can determine $\widetilde{n}$. On the other hand we can also plot a graph of $\log \eta$ versus $\log(2ii_{\bar{D}})$. The intercept on $\log \eta$ axis in this case is also $\log m$. Setting the slope of this straight line equal to $\frac{\widetilde{n}-1}{2}$, we can determine the index $\widetilde{n}$. Both approach are identical. The index $\widetilde{n}$ is called power law index. Different generalized Newtonian power law fluids naturally will have different values of power law index $\widetilde{n}$.

When $\widetilde{n} = 1$, $\widetilde{n} - 1 = 0 = \frac{\widetilde{n}-1}{2}$, we have $\eta|_{\Omega} = m$, zero shear rate viscosity, a constant, hence we have a Newtonian fluid. For $\widetilde{n} < 1$, $\eta < m$ holds; hence, we have a shear thinning fluid. For such fluids, progressively increasing γ or $ii_{\bar{D}}$ for a fixed $\widetilde{n}$ (i.e., a chosen fluid) results in progressively reducing η. Progressively lower $\widetilde{n}$ values obviously correspond to progressively shear thinning fluids which will exhibit shear thinning behavior with increasing γ or $ii_{\bar{D}}$.

When $\widetilde{n} > 1$, $\eta > m$, hence we have shear thickening fluids. For such fluids progressively increasing γ or $ii_{\bar{D}}$ results in progressively increasing viscosity η for a fixed value of $\widetilde{n}$. Progressively increasing $\widetilde{n}$ values correspond to progressively shear thickening behavior with increasing γ or $ii_{\bar{D}}$.

Remarks

(1) Experiments for η versus $ii_{\bar{D}}$ or $II_{\bar{D}}$ can only be performed within a certain range of $ii_{\bar{D}}$ (range B-C in Figures 12.1(a) and 12.1(b)) In the range A-B, the shear rate is too low and in range C-D it is too high for the experiments to be valid or even the experiments to be performed. Power law model that is valid in range B-C, when extended to ranges A-B and C-D yields unrealistic values of viscosity. Nonetheless, the $ii_{\bar{D}}$ values in range A-B and C-D exists in almost all applications (lid driven cavity, asymmetric expansion, etc.). Thus, use of power law viscosity model is undoubtedly going to produce spurious results in applications that encounter $ii_{\bar{D}}$ values in the ranges A-B and C-D.

(2) Based on Remark (1), power law model of viscosity only remains valid for applications in which $ii_{\bar{D}}$ is always ensured to be in the range B-C for all spatial locations and for all values of time.

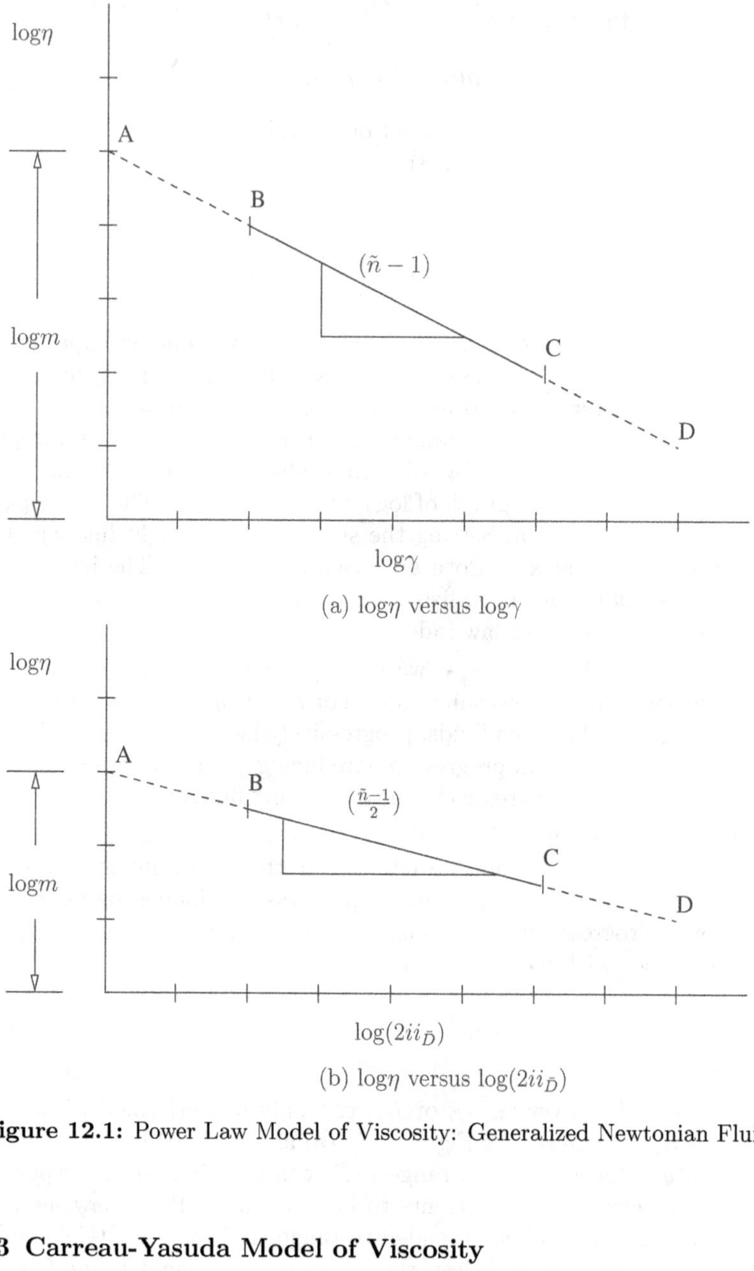

Figure 12.1: Power Law Model of Viscosity: Generalized Newtonian Fluid

12.4.4.3 Carreau-Yasuda Model of Viscosity

The Carreau-Yasuda empirical model of viscosity for generalized Newtonian fluids is designed to overcome the shortcomings of the power law model and is given by

$$\eta|_{\underline{\Omega}} = \eta^{\infty} + (\eta^0 - \eta^{\infty})\left(1 + (\lambda\gamma)^a\right)^{\frac{\tilde{n}-1}{a}} \tag{12.41}$$

$$\text{or} \quad \eta|_{\underline{\Omega}} = \eta^{\infty} + (\eta^0 - \eta^{\infty})\left(1 + (\lambda\sqrt{2ii_{\bar{D}}})^a\right)^{\frac{\tilde{n}-1}{a}} \tag{12.42}$$

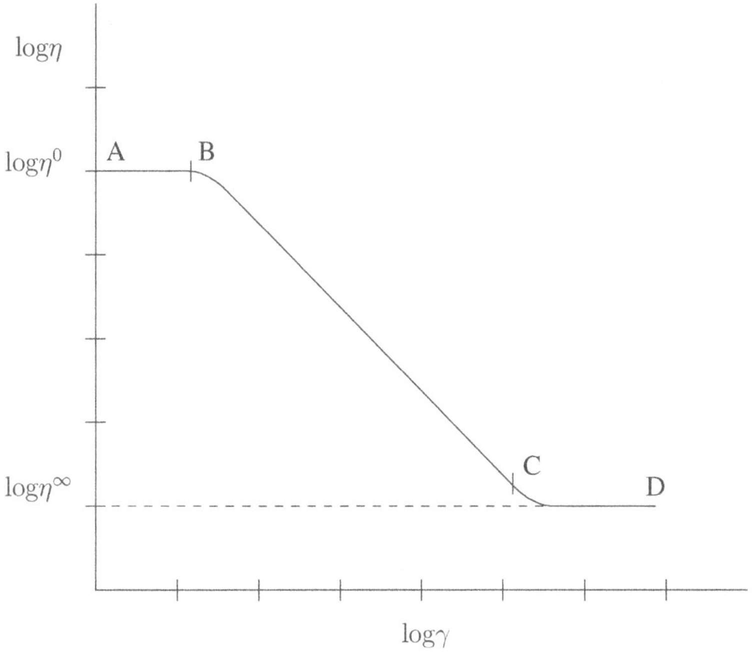

Figure 12.2: Carreua-Yasuda Model

In this model η^0, η^∞, a, λ and $\widetilde{n}$ are model constants [20, 21]. η^0 is zero shear rate viscosity, η^∞ is infinite shear rate viscosity, a is a constant used to generalize power law model, λ is a model coefficient and $\widetilde{n}$ is the power law index. To compare this model with power law we choose $a = 1$ for which (12.41) becomes

$$\eta\big|_{\Omega} = \eta^\infty + (\eta^0 - \eta^\infty)\,(1 + \lambda\gamma)^{\widetilde{n}-1} \tag{12.43}$$

when $\widetilde{n} = 1$, $\eta\big|_{\Omega} = \eta^0$, zero shear rate viscosity which is independent of γ or $ii_{\bar{D}}$, hence defines constant viscosity. When γ or $ii_{\bar{D}}$ is in the range A-B, $\lambda\gamma << 1$, hence $\lambda\gamma$ can be neglected compared to one and we have

$$\eta\big|_{\Omega} \simeq \eta^\infty + (\eta^0 - \eta^\infty)(1) = \eta^0 \tag{12.44}$$

When γ or ii_D is large, $\lambda\gamma >> 1$, 1 can be neglected compared to $\lambda\gamma$, hence (12.43) reduces to

$$\eta\big|_{\Omega} \simeq \eta^\infty + (\eta^0 - \eta^\infty)(\lambda\gamma)^{\widetilde{n}-1} \tag{12.45}$$

For shear thinning fluids $\widetilde{n} < 1$, (12.45) will yield $\eta\big|_{\Omega} \simeq \eta^\infty$, infinite shear rate viscosity. For shear thickening fluids $\widetilde{n} > 1$; hence, $\widetilde{n} - 1 > 0$. Thus, (12.45) will give $\eta\big|_{\Omega}$ higher than η^∞. Carreau-Yasuda model has been found to be in good agreement with experimental observation and eliminates the problems associated with power law model when γ or $ii_{\bar{D}}$ is in the ranges A-B and C-D.

12.5 Constitutive theory for heat vector $\bar{\boldsymbol{q}}$

The condition resulting from the entropy inequality requires

$$\bar{\boldsymbol{q}} \cdot \bar{\boldsymbol{g}} \leq 0 \tag{12.46}$$

The constitutive theory for $\bar{\boldsymbol{q}}$ must satisfy (12.46). This condition can be used to derive constitutive theory for $\bar{\boldsymbol{q}}$. In the second approach, we could consider the fact that $\bar{\boldsymbol{q}}$ and $\bar{\boldsymbol{g}}$ are a conjugate pair in (12.46); hence, we could consider

$$\bar{\boldsymbol{q}} = \bar{\boldsymbol{q}}(\bar{\rho}, \bar{\boldsymbol{g}}, \bar{\theta}) \tag{12.47}$$

and use representation theorem to derive constitutive theory for $\bar{\boldsymbol{q}}$. We consider both approaches in the following.

12.5.1 Constitutive theory for $\bar{\boldsymbol{q}}$ using entropy inequality

The derivation based on (12.46) is fundamental and can be found in any textbook on continuum mechanics [30, 31]. In this derivation we begin with (12.46) which implies that

$$\bar{\boldsymbol{q}} \cdot \bar{\boldsymbol{g}} = \bar{\beta} \leq 0 \tag{12.48}$$

using equality, we obtain

$$\frac{\partial \bar{\beta}}{\partial \bar{\boldsymbol{g}}} = \bar{\boldsymbol{q}} \tag{12.49}$$

$\bar{\beta}$ has maximum value at $\bar{\boldsymbol{g}} = 0$ (due to (12.48)), hence

$$\left. \frac{\partial \bar{\beta}}{\partial \bar{\boldsymbol{g}}} \right|_{\bar{\boldsymbol{g}}=0} = \bar{\boldsymbol{q}}|_{\bar{\boldsymbol{g}}=0} = 0 \tag{12.50}$$

That is, *heat flux vanishes in the absence of temperature gradient.* Thus, constitutive theory for $\bar{\boldsymbol{q}}$ must be a function of $\bar{\boldsymbol{g}}$. At this stage more than one possibility exists, the simplest of course is assuming that $\bar{\boldsymbol{q}}$ is proportional to $\bar{\boldsymbol{g}}$, i.e., $\bar{\boldsymbol{q}}$ is a linear function of $\bar{\boldsymbol{g}}$.

$$\bar{\boldsymbol{q}} = -k(\bar{\theta}) \bar{\boldsymbol{g}} \tag{12.51}$$

in (12.51) $k(\bar{\theta})$ is a temperature dependent constant of proportionality known as thermal conductivity of the volume of matter. Equation (12.51) with constant k or $k(\bar{\theta})$ is Fourier heat conduction law. Mathematically we can write a more general form of (12.51) given by

$$\bar{\boldsymbol{q}} = -\boldsymbol{k}(\bar{\theta}) \cdot \bar{\boldsymbol{g}} \quad \text{or} \quad \bar{q}_i = -k_{ij}(\bar{\theta}) \bar{g}_j \tag{12.52}$$

in which $\boldsymbol{k}(\bar{\theta})$ is a tensor of rank two. From which we define

$$\frac{\partial \bar{\boldsymbol{q}}}{\partial \bar{\boldsymbol{g}}} = -\boldsymbol{k}(\bar{\theta}) \quad \text{or} \quad \frac{\partial \bar{q}_i}{\partial \bar{g}_j} = -k_{ij}(\bar{\theta}) \tag{12.53}$$

Also from (12.49)

$$\frac{\partial^2 \bar{\beta}}{\partial \bar{\boldsymbol{g}}^2} = \frac{\partial \bar{\boldsymbol{q}}}{\partial \bar{\boldsymbol{g}}} = -\boldsymbol{k}(\bar{\theta}) \quad \text{or} \quad \frac{\partial^2 \bar{\beta}}{\partial \bar{g}_j \bar{g}_i} = \frac{\partial \bar{q}_i}{\partial \bar{g}_j} = -k_{ij}(\bar{\theta}) \tag{12.54}$$

From (12.54), we can conclude that $\boldsymbol{k}(\bar{\theta})$ is positive-semidefinite and all its eigenvalues are non-negative. $\boldsymbol{k}(\bar{\theta})$ or $[k(\bar{\theta})]$, the thermal conductivity matrix does not have to be symmetric but is generally assumed to be. Coefficients of $[k(\bar{\theta})]$ can be functions of temperature $\bar{\theta}$. While mathematically (12.52) is justified as a constitutive theory, we note that for isotropic homogeneous matter $\boldsymbol{q}$ cannot exhibit directional dependence. Thus, based on CCM for isotropic homogeneous matter, (12.52) and what follows, it is not valid. Hence, (12.52) must be reduced to

$$\bar{\boldsymbol{q}} = -k(\bar{\theta})\bar{\boldsymbol{g}} \quad \text{or} \quad \bar{q}_i = -k(\bar{\theta})\bar{g}_i \tag{12.55}$$

This constitutive theory (Fourier heat conduction law) requires only one material coefficient, thermal conductivity $k(\bar{\theta})$, that can be temperature dependent.

12.5.2 Constitutive theory for $\bar{\boldsymbol{q}}$ using representation theorem

In this derivation, we begin with (12.47). $\bar{\boldsymbol{q}}$ is a tensor of rank one. Its argument tensor $\bar{\rho}$ and $\bar{\theta}$ are tensors of rank zero and $\bar{\boldsymbol{q}}$ is a tensor of rank one. Thus, the combined generators of the argument tensors $\bar{\rho}$, $\bar{\theta}$ and $\bar{\boldsymbol{g}}$ that are tensors of rank one are only $\bar{\boldsymbol{g}}$. Thus, the tensor $\bar{\boldsymbol{g}}$ constitutes the basis of the space of $\bar{\boldsymbol{q}}$, hence we can represent $\bar{\boldsymbol{q}}$ as a linear combination of $\bar{\boldsymbol{q}}$ in the current configuration.

$$\bar{\boldsymbol{q}} = -{}^q\underset{\sim}{\alpha}\bar{\boldsymbol{g}} \tag{12.56}$$

${}^q\underset{\sim}{\alpha}$ can be a function of $\bar{\rho}$, $\bar{\theta}$ and combined invariants of $\bar{\rho}$, $\bar{\theta}$ and $\bar{\boldsymbol{q}}$. Let us define

$${}^q\underset{\sim}{I} = \bar{\boldsymbol{g}} \cdot \bar{\boldsymbol{g}} \tag{12.57}$$

${}^q\underset{\sim}{I}$ is the combined invariant of $\bar{\rho}$, $\bar{\theta}$, $\bar{\boldsymbol{g}}$. Thus,

$${}^q\underset{\sim}{\alpha} = {}^q\underset{\sim}{\alpha}(\bar{\rho}, \bar{\theta}, {}^q\underset{\sim}{I}) \tag{12.58}$$

The material coefficients in the constitutive theory (12.56) for $\bar{\boldsymbol{q}}$ is determined by expanding ${}^q\underset{\sim}{\alpha}(\bar{\rho}, \bar{\theta}, {}^q\underset{\sim}{I})$ in Taylor series in $\bar{\theta}$ and ${}^q\underset{\sim}{I}$ about a known configuration $\underline{\Omega}$ and retaining only up to linear terms in $\bar{\theta}$ and ${}^q\underset{\sim}{I}$ (for simplicity).

$${}^q\underset{\sim}{\alpha} = {}^q\underset{\sim}{\alpha}\Big|_{\underline{\Omega}} + \frac{\partial\, {}^q\underset{\sim}{\alpha}}{\partial\, {}^q\underset{\sim}{I}}\Bigg|_{\underline{\Omega}} \left({}^q\underset{\sim}{I} - {}^q\underset{\sim}{I}\Big|_{\underline{\Omega}}\right) + \frac{\partial\, {}^q\underset{\sim}{\alpha}}{\partial \bar{\theta}}\Bigg|_{\underline{\Omega}} (\bar{\theta} - \bar{\theta}|_{\underline{\Omega}}) \tag{12.59}$$

substituting from (12.59) into (12.56)

$$\bar{\boldsymbol{q}} = -\left({}^q\underset{\sim}{\alpha}|_{\underline{\Omega}} + \frac{\partial\, {}^q\underset{\sim}{\alpha}}{\partial\, {}^q\underset{\sim}{I}}\Bigg|_{\underline{\Omega}} \left({}^q\underset{\sim}{I} - {}^q\underset{\sim}{I}\Big|_{\underline{\Omega}}\right) + \frac{\partial\, {}^q\underset{\sim}{\alpha}}{\partial \bar{\theta}}\Bigg|_{\underline{\Omega}} (\bar{\theta} - \bar{\theta}|_{\underline{\Omega}})\right)\bar{\boldsymbol{g}} \tag{12.60}$$

From (12.60) we can write the following using ${}^q\underset{\sim}{I} = \bar{\boldsymbol{g}} \cdot \bar{\boldsymbol{g}}$ and regrouping terms.

$$\bar{\boldsymbol{q}} = -\,{}^q\underset{\sim}{\alpha}\Big|_{\underline{\Omega}}\bar{\boldsymbol{g}} - \frac{\partial\, {}^q\underset{\sim}{\alpha}}{\partial\, {}^q\underset{\sim}{I}}\Bigg|_{\underline{\Omega}} (\bar{\boldsymbol{g}} \cdot \bar{\boldsymbol{g}})\bar{\boldsymbol{g}} + \frac{\partial\, {}^q\underset{\sim}{\alpha}}{\partial\, {}^q\underset{\sim}{I}}\Bigg|_{\underline{\Omega}} (\bar{\boldsymbol{g}} \cdot \bar{\boldsymbol{g}})|_{\underline{\Omega}}\bar{\boldsymbol{g}} - \frac{\partial\, {}^q\underset{\sim}{\alpha}}{\partial \bar{\theta}}\Bigg|_{\underline{\Omega}} (\bar{\theta} - \bar{\theta}|_{\underline{\Omega}})\,\bar{\boldsymbol{g}} \tag{12.61}$$

or

$$\bar{\boldsymbol{q}} = -\left({}^q\underset{\sim}{\alpha}\big|_{\underline{\Omega}} - \left.\frac{\partial {}^q\underset{\sim}{\alpha}}{\partial {}^q\underset{\sim}{I}}\right|_{\underline{\Omega}} (\bar{\boldsymbol{g}} \cdot \bar{\boldsymbol{g}})\big|_{\underline{\Omega}} \right) \bar{\boldsymbol{g}} - \left.\frac{\partial {}^q\underset{\sim}{\alpha}}{\partial {}^q\underset{\sim}{I}}\right|_{\underline{\Omega}} (\bar{\boldsymbol{g}} \cdot \bar{\boldsymbol{g}})\bar{\boldsymbol{g}} - \left.\frac{\partial {}^q\underset{\sim}{\alpha}}{\partial \bar{\theta}}\right|_{\underline{\Omega}} (\bar{\theta} - \bar{\theta}\big|_{\underline{\Omega}})\bar{\boldsymbol{g}} \tag{12.62}$$

Let

$$\begin{aligned} k(\bar{\rho}\big|_{\underline{\Omega}}, \bar{\theta}_{\underline{\Omega}}, {}^q\underset{\sim}{I}_{\underline{\Omega}}) &= {}^q\underset{\sim}{\alpha}\big|_{\underline{\Omega}} - \left.\frac{\partial {}^q\underset{\sim}{\alpha}}{\partial {}^q\underset{\sim}{I}}\right|_{\underline{\Omega}} (\bar{\boldsymbol{g}} \cdot \bar{\boldsymbol{g}})\big|_{\underline{\Omega}} \\ k_1(\bar{\rho}\big|_{\underline{\Omega}}, \bar{\theta}\big|_{\underline{\Omega}}, {}^q\underset{\sim}{I}) &= \left.\frac{\partial {}^q\underset{\sim}{\alpha}}{\partial {}^q\underset{\sim}{I}}\right|_{\underline{\Omega}} \\ k_2(\bar{\rho}\big|_{\underline{\Omega}}, \bar{\theta}\big|_{\underline{\Omega}}, {}^q\underset{\sim}{I}\big|_{\underline{\Omega}}) &= \frac{\partial {}^q\underset{\sim}{\alpha}}{\partial \bar{\theta}} \end{aligned} \tag{12.63}$$

using (12.63), (12.62) can be written as

$$\bar{\boldsymbol{q}} = -k\bar{\boldsymbol{g}} - k_1(\bar{\boldsymbol{g}} \cdot \bar{\boldsymbol{g}})\bar{\boldsymbol{g}} - k_2(\bar{\theta} - \bar{\theta}\big|_{\underline{\Omega}})\bar{\boldsymbol{g}} \tag{12.64}$$

This is the simplest possible constitutive theory that can be derived using (12.47) and representation theorem that utilizes the complete basis of the space of $\bar{\boldsymbol{q}}$. The only assumption in this theory is the truncation of the Taylor series beyond linear terms in invariant ${}^q\underset{\sim}{I}$ and $\bar{\theta}$. We note that this constitutive theory is cubic in $\bar{\boldsymbol{g}}$. It contains linear and cubic terms in $\bar{\boldsymbol{g}}$ but does not contain the quadratic term in $\bar{\boldsymbol{g}}$. We note that the commonly used Fourier heat conduction law permits only temperature dependent thermal conductivity. Power law and Sutherland law are commonly used models for temperature dependent thermal conductivity.

$$k(\bar{\theta})\big|_{\underline{\Omega}} = k^0 \left(\frac{\bar{\theta}}{\theta^0}\right)^n_{\underline{\Omega}} \quad ; \quad \text{Power law} \tag{12.65}$$

$$k(\bar{\theta})\big|_{\underline{\Omega}} = k^0 \left(\frac{\bar{\theta}}{\theta^0}\right)^{3/2}_{\underline{\Omega}} \left(\frac{\bar{\theta} + s}{\bar{\theta} + s}\right)_{\underline{\Omega}} \quad ; \quad \text{Sutherland law} \tag{12.66}$$

$k(\bar{\theta})\big|_{\underline{\Omega}}$ is defined in a known configuration $\underline{\Omega}$.

12.6 General remarks

We present some observations and remarks that may be helpful in the use of mathematical models consisting of CBL in Eulerian description and the constitutive theories presented in this chapter.

(1) CBL used for thermoviscous fluent continua are in Eulerian description in which velocities $\bar{\boldsymbol{v}}$ are assumed to be observable quantities, thus, $\bar{\boldsymbol{v}} \neq \frac{D\bar{\boldsymbol{u}}}{Dt}$ (even though in the CBL in Eulerian description derived in Chapter 6, $\bar{\boldsymbol{u}}$ are observable quantities and $\bar{\boldsymbol{v}} = \frac{D\bar{\boldsymbol{u}}}{Dt}$ holds).

(2) The CBL and the constitutive theories for such materials do not contain $\bar{\boldsymbol{u}}$, $\left[\frac{\partial\{\bar{u}\}}{\partial\{\bar{x}\}}\right]$; hence, do not contain strains either. Thus, such materials do not have elasticity and stiffness, but naturally have mass ($\bar{\rho}$) and dissipation mechanism due to strain rate(s).

(3) Such materials cannot support sound waves due to lack of elasticity. Speed of propagation of a sound wave is infinity in an incompressible thermoviscous fluent continua. A compressible thermoviscous continua can support pressure waves. Hence, as a result, density and temperature waves also exist but such medium cannot support the stress waves due to lack of elasticity.

(4) Due to lack of elasticity in such continua

  (a) Only compression is possible.
  (b) No concept of vibrations, natural vibrations, natural modes etc. exist in such continua due to lack of elasticity or stiffness.
  (c) In unsteady flows where the fluid motion is referred to as vibration of the fluid, is in fact time dependent oscillating motion of the fluid volume.

(5) It is significant to point out that in solid continua (Lagrangian or Eulerian description) displacements $\boldsymbol{u}(\boldsymbol{x},t)$ or $\bar{\boldsymbol{u}}(\bar{\boldsymbol{x}},t)$ are observable quantities and the velocities are obtained using $\boldsymbol{v}(\boldsymbol{x},t) = \frac{D\boldsymbol{u}(\boldsymbol{x},t)}{Dt}$ or $\bar{\boldsymbol{v}}(\bar{\boldsymbol{x}},t) = \frac{D\bar{\boldsymbol{u}}(\boldsymbol{x},t)}{Dt}$, that is velocities are derived by differentiation of displacements; hence, they are not observable quantities. Whereas, the manner in which CBL in Eulerian description are used for thermoviscous fluent continua, $\bar{\boldsymbol{v}}$ are observable quantities and $\bar{\boldsymbol{v}} \neq \frac{D\bar{\boldsymbol{u}}}{Dt}$. Displacements $\bar{\boldsymbol{u}}$ are neither used in the CBL nor in the constitutive theories. Thus, $\bar{\boldsymbol{v}}$ in solid and fluent continua have totally different definitions.

These are the basic reasons due to which solid and fluid interactions cannot be studied as a single combined problem as the mathematical models for solids and fluids have different meaning of similar quantities (such as velocities $\bar{\boldsymbol{v}}$ as illustrated above).

12.7 Summary

Constitutive theories have been derived for Cauchy stress tensor and heat vector for compressible and incompressible thermoviscous fluent continua using entropy inequality in conjunction with representation theorem. Initial choices of constitutive tensors and their argument tensors are made using SLT and the principle of equipresence. Basis independent Cauchy stress tensor ${}^{(0)}\bar{\boldsymbol{\sigma}}$ that could be $\bar{\boldsymbol{\sigma}}^{(0)}$, $\bar{\boldsymbol{\sigma}}_{(0)}$ or ${}^{(0)}\bar{\boldsymbol{\sigma}}^J$ and basis independent convected time derivatives of strain tensors, ${}^{(i)}\boldsymbol{\gamma}$; $i = 1, 2, \dots, n$ that could be convected time derivatives of the Green's strain tensor $\boldsymbol{\varepsilon}_{[0]}$, $\boldsymbol{\gamma}_{(i)}$; $i = 1, 2, \dots, n$ or the convected time derivatives of the Almansi strain tensor $\bar{\boldsymbol{\varepsilon}}^{[0]}$, $\boldsymbol{\gamma}^{(0)}$; $i = 1, 2, \dots, n$ or Jaumann rates ${}^{(i)}\boldsymbol{\gamma}^J$; $i = 1, 2, \dots, n$ are used in the derivation of the constitutive theory for the Cauchy stress tensor. Basis independent Cauchy stress tensor ${}^{(0)}\bar{\boldsymbol{\sigma}}$ is decomposed into equilibrium stress tensor ${}^{(0)}_{\ e}\bar{\boldsymbol{\sigma}}$ and deviatoric stress tensor ${}^{(0)}_{\ d}\bar{\boldsymbol{\sigma}}$. Constitutive theory for ${}^{(0)}_{\ e}\bar{\boldsymbol{\sigma}}$ is derived using $\bar{\Phi}$ and incompressibility condition for compressible and incompressible thermoviscous fluids. Ordered rate constitutive theory for ${}^{(0)}_{\ d}\bar{\boldsymbol{\sigma}}$ of order n with argument tensors $\bar{\rho}$, $\bar{\theta}$ and ${}^{(i)}\boldsymbol{\gamma}$; $i = 1, 2, \dots, n$ is derived using representation theorem. Bases dependent constitutive theories for deviatoric Cauchy stress tensor are easily obtained using the basis independent constitutive theory. Material coefficients are derived using Taylor series expansion of the coefficients used in linear combination of combine generators. Simplified forms of the constitutive theories for ${}^{(0)}_{\ d}\bar{\boldsymbol{\sigma}}$: (i) Constitutive theory of order one ($n = 1$) (ii) Linear constitutive theory of order n (iii) Linear constitutive theory of order one ($n = 1$) (iv) Constitutive theories for Newtonian and generalized Newtonian fluids, power law and Carreau-Yasuda models of viscosity have been presented. Models of viscosity dependent on $\bar{\theta}$ have also been presented.

For ordered rate constitutive theories of order greater than one, i.e., when $\boldsymbol{\gamma}^{(2)}$, $\boldsymbol{\gamma}^{(3)}, \dots, \boldsymbol{\gamma}^{(n)}$ or $\boldsymbol{\gamma}^{(2)}, \boldsymbol{\gamma}_{(3)}, \dots, \boldsymbol{\gamma}^{(n)}$ or ${}^{(2)}\boldsymbol{\gamma}^J$, ${}^{(3)}\boldsymbol{\gamma}^J, \dots, {}^{(n)}\boldsymbol{\gamma}^J$ are argument tensors of the deviatoric Cauchy stress tensors, the stress measures are not the same, i.e for this case ${}^{(0)}_{\;d}\bar{\boldsymbol{\sigma}} \neq {}_d\bar{\boldsymbol{\sigma}}_{(0)} \neq {}_d{}^{(0)}\bar{\boldsymbol{\sigma}}^J$ even though they are all defined in x-frame with the same dyads. For ordered rate constitutive theory of order one ($n = 1$), we have ${}^{(1)}\boldsymbol{\gamma} = \boldsymbol{\gamma}^{(1)} = \boldsymbol{\gamma}_{(1)} = \bar{\boldsymbol{D}}$. Hence, in this case, ${}^{(0)}_{\;d}\bar{\boldsymbol{\sigma}} = {}_d\bar{\boldsymbol{\sigma}}_{(0)} = {}^{(0)}_{\;d}\bar{\boldsymbol{\sigma}}^J = {}_d\bar{\boldsymbol{\sigma}}$, ${}_d\bar{\boldsymbol{\sigma}}$ being basis independent deviatoric Cauchy stress tensor. The constitutive theories for Newtonian and generalized Newtonian fluids are linear constitutive theories of order ($n = 1$). Hence, the deviatoric Cauchy stress tensor in these theories is basis independent.

Two derivations are presented for the constitutive theory for heat vector $\bar{\boldsymbol{q}}$. The derivation in the absence of the use of representation theorem is clearly a subset of the constitutive theory for $\bar{\boldsymbol{q}}$ derived using representation theorem. The constitutive theory based on representation theorem uses complete basis and invariants. This constitutive theory for $\bar{\boldsymbol{q}}$ is cubic in $\bar{\boldsymbol{q}}$ (a nonlinear constitutive theory), pointing out the deficiency in the currently used linear constitutive theory for $\bar{\boldsymbol{q}}$. Material coefficients are derived and their dependence on $\bar{\rho}$, $\bar{\theta}$ and invariant ${}^q\underset{\sim}{I}$ is established.

The constitutive theories for the deviatoric stress tensor for compressible thermoviscous fluids can be easily reduced for incompressible thermoviscous fluids by using the incompressibility condition $\bar{\boldsymbol{\nabla}} \cdot \bar{\boldsymbol{v}} = 0$. We note that the constitutive theories presented for heat vector $\bar{\boldsymbol{q}}$ hold for both compressible and incompressible thermoviscous fluids.

Problems

12.1 Derive a linear constitutive theory for compressible thermofluids from first principles using the theory of generators and invariants in which the deviatoric Cauchy stress tensor is a linear function of the symmetric part of the velocity gradient tensor, temperature gradient and temperature $\bar{\theta}$. Derive material coefficients and establish their dependence on the arguments and/or their invariants. Also derive a consistent constitutive theory for the heat vector. Specialize the final results for the incompressible case. Are the constitutive theories basis dependent? Discuss.

12.2 Consider a constitutive theory for compressible thermofluids from first principles using the theory of generators and invariants in which the symmetric part of the velocity gradient tensor, temperature gradient and temperature are arguments of the dependent variables in the constitutive theory.

- **(a)** Derive a constitutive theory for the stress tensor that is quadratic in the symmetric part of the velocity gradient tensor, linear in temperature gradient and temperature.
- **(b)** Specialize the theory derived in (a) for an incompressible thermofluid.

(c) Derive a constitutive theory for the stress tensor that is linear in the symmetric part of the velocity gradient tensor and temperature $\bar{\theta}$ but quadratic in temperature gradient.
(d) Specialize the theory in (c) for incompressible thermofluids.
(e) Derive a constitutive theory for the heat vector $\bar{\boldsymbol{q}}$ that is only dependent on temperature gradient and temperature such that

 (i) $\bar{\boldsymbol{q}}$ is a linear function of the temperature gradient $\bar{\boldsymbol{g}}$ and temperature $\bar{\theta}$;
 (ii) $\bar{\boldsymbol{q}}$ is quadratic in temperature gradient $\bar{\boldsymbol{g}}$ but linear in temperature $\bar{\theta}$;
 (iii) $\bar{\boldsymbol{q}}$ is cubic in temperature gradient $\bar{\boldsymbol{g}}$ but linear in temperature $\bar{\theta}$.

In each case, derive material coefficients and establish their dependence on the arguments and/or their invariants. Are any of these theories basis dependent? Discuss.

12.3 Consider the flow of a compressible thermofluid in $\mathbb{R}^2$, i.e., x_1x_2 plane of the x-frame.

(a) Present explicit forms of the constitutive equations for the deviatoric Cauchy stress tensor that is

 (i) linear in $[\bar{D}]$ and $\bar{\theta}$;
 (ii) quadratic in $[\bar{D}]$ and linear in $\bar{\theta}$.

 Express these constitutive theories in matrix and vector form using Voigt's notation. Discuss variable and constant material coefficients.
(b) Present explicit forms of the constitutive equations for the heat vector $\bar{\boldsymbol{q}}$ that is

 (i) linear in $\bar{\boldsymbol{g}}$ and $\bar{\theta}$;
 (ii) quadratic in $\bar{\boldsymbol{g}}$ but linear in $\bar{\theta}$;
 (iii) cubic in $\bar{\boldsymbol{g}}$ but linear in $\bar{\theta}$.

 Express these constitutive theories in matrix and vector form using Voigt's notation if possible. Discuss variable and constant material coefficients.

Discuss the basis dependency (or lack of it) of these constitutive theories.

12.4 Consider constitutive theories for the deviatoric stress tensor and heat vector in contravariant and covariant bases and Jaumann rates.

(a) Derive constitutive theories for the deviatoric Cauchy stress tensor in these bases and using Jaumann rates that are linear in the first and second convected time derivative of the conjugate strain tensors, linear in temperature gradient and temperature $\bar{\theta}$.
(b) Derive a constitutive theory for the heat vector $\bar{\boldsymbol{q}}$ that is consistent with (a) in terms of argument tensors.

Derive material coefficients in each case. Discuss constant and variable material coefficients.

13
CONSTITUTIVE THEORIES FOR THERMOVISCOELASTIC FLUIDS

13.1 Introduction

The thermoviscoelastic fluids or polymeric fluids are viscous as well as elastic. Composition of such fluids is generally due to synthesis of a solvent and a polymer. A solvent is generally a dilute solution primarily consisting of short molecules, hence can be viewed as Newtonian fluid. The polymer, on the other hand, consists of long chain molecules, hence has much higher viscosity than the solvent. A thermoviscoelastic fluid may be solvent dominated with some smaller volume fraction of polymer. Such thermoviscoelastic fluids are called *dilute polymeric fluids.* On the other hand, if the composition of the thermoviscoelastic fluid is heavily dominated by the long chain molecules of polymer, then the fluid is generally referred to as *dense polymeric fluid.*

When a polymeric fluid is in relaxed state (unstressed or undisturbed), the long chain molecules are in relaxed or unstressed state. Upon application of disturbance to a thermoviscoelastic fluid, the short chain molecules can be assumed not to stretch but can experience motion relative to other short chain molecules as well as with respect to long chain molecules of the polymer in its vicinity. This motion of course requires overcoming the viscous drag forces between the neighboring molecules. Experimental studies of polymeric fluid motion under microscope shows complex motion of polymer molecules generally referred to as Brownian motion [23, 24]. A detailed study of this motion reveals that polymer molecules in relaxed state (generally in somewhat coiled form) are separated by solvent but are generally congregated in distinct colonies and these colonies are connected to the neighboring colonies. Upon application of the disturbance, the polymer molecules in relaxed states stretch (uncoil). In doing so, they need to overcome the viscous resistance offered by the neighboring solvent and polymer molecules, thus, creating a spring like action in the direction of the application of disturbance. The polymer molecules may get dislodged from the existing colonies and may reattach to the new colonies, and the existing connections between the colonies may break and/or new colonies and new connections between them may form. This physics of polymer motion is referred to as *mobility* is primarily responsible for elasticity in the thermoviscoelastic fluids. In other words, polymer molecules and their motion in the presence of resisting forces due to viscous drag is similar to deformation of a one dimensional spring. Thus, in polymeric fluids, the dominant elasticity is only in the direction of fluid motion. The elastic effects in the transverse directions to the fluid motion are weak as stretching of the polymer molecules in these directions is not significant. The elastic effects in the direction transverse to the direction of application of the load in solids is due to Poisson's effect. This physics does not exist in thermoviscoelastic fluids.

DOI: 10.1201/9781003105336-13

We note that since mobility physics is primarily due to long chain molecules, mobility is not significant in dilute polymeric fluids such as Maxwell fluid, Oldroyd-B fluid. Hence, in deriving constitutive theories for such fluids, mobility physics can be neglected. Polymeric fluids like Giesekus fluids that have higher concentration of polymer molecules, mobility physics is significant, hence must be considered in deriving constitutive theories for such polymeric fluids. Thus, we note that consideration in deriving constitutive theories for dilute polymeric fluids and dense polymeric fluids may be quite different. Thermoviscoelastic fluids are of significant industrial importance. Flow of food slurries in food processing industries, and flow of molten plastic and metals are some examples.

Upon cessation of disturbance to a polymeric fluid, the stretched (hence stressed) long chain polymer molecules try to return to their initial unstressed (hence relaxed) state. In doing so they must overcome the viscous drag forces exerted by the neighboring short chain and long chain molecules. Thus, the process of long chain molecules to returning to unstressed state does not happen instantaneously upon cessation of disturbance, but requires a finite amount of time. This time required for the polymer molecules to reach the relaxed or unstressed state depends upon a characteristic material property of the polymeric fluid called relaxation time. The process of polymer molecules returning to relaxed or unstressed state is referred to as relaxation or rheology. Thus, thermoviscoelastic fluids are called rheological fluids as they exhibit rheology. We must note that relaxation time is not the time it takes for a fluid to relax, but the actual time it takes for a fluid to relax is a function of relaxation time. Different polymeric fluids naturally have different relation times, a property of the polymeric fluid that must be determined experimentally.

From the point of view of CCM we can only undertake derivation of the constitutive theories for polymeric fluids if we assume that the polymeric fluid (though consisting of short chain and long chain molecules) is homogeneous and isotropic. Even though the solvent and the polymer have their own viscosities and other material properties, the use of principles of CCM requires that we treat the final synthesized polymeric fluid as isotropic and homogeneous. Hence, in deriving constitutive theories for polymeric fluids, we consider a single viscosity as well as other properties to be a single property for the overall polymeric fluid.

13.2 Preliminary considerations

The constitutive theories in the Eulerian description necessitate their derivations in the current configuration in which co- and/or contra-variant bases must be considered with consistent measures of stress and strain as well as their convected time derivatives in the bases in which they are defined. Let $\boldsymbol{\gamma}_{(i)}$, $\boldsymbol{\gamma}^{(i)}$ and ${}^{(i)}\boldsymbol{\gamma}^J$; $i = 1, 2, \ldots, n$ be the convected time derivatives of Green's strain tensor, Almansi strain tensor and Jaumann rate up to orders n and $\bar{\boldsymbol{\sigma}}^{(k)}$, $\bar{\boldsymbol{\sigma}}_{(k)}$, ${}^{(k)}\bar{\boldsymbol{\sigma}}^J$; $k = 0, 1, \ldots, m$ be the convected time derivatives of the contravariant Cauchy stress tensor, covariant Cauchy stress tensor and Jaumann stress rates up to orders m. Similar to Chapter 11, here also we introduce basis independent convected time derivatives of the strain tensor ${}^{(i)}\boldsymbol{\gamma}$; $i = 1, 2, \ldots, n$ and basis independent convected time derivatives of the Cauchy stress tensors, ${}^{(k)}\bar{\boldsymbol{\sigma}}$; $k = 0, 1, \ldots, m$. Clearly for $j = 0$ we have ${}^{(0)}\bar{\boldsymbol{\sigma}}$, convected time derivative of order zero, i.e., stress tensor itself. Constitutive theory for the Cauchy stress tensors is derived using the basis independent convected time derivatives ${}^{(i)}\boldsymbol{\gamma}$; $i = 1,2\dots,n$ and ${}^{(k)}\bar{\boldsymbol{\sigma}}$; $k = 0,1,\dots,m$. Specific forms of the basis dependent constitutive theories can be obtained by choosing:

$$\begin{aligned} &{}^{(k)}\bar{\boldsymbol{\sigma}}, {}^{(i)}\boldsymbol{\gamma} \quad \text{as} \quad \bar{\boldsymbol{\sigma}}^{(k)}, \boldsymbol{\gamma}_{(i)} \\ &{}^{(k)}\bar{\boldsymbol{\sigma}}, {}^{(i)}\boldsymbol{\gamma} \quad \text{as} \quad \bar{\boldsymbol{\sigma}}_{(k)}, \boldsymbol{\gamma}^{(i)} \quad ; \quad i = 0,1,\dots,n \quad k = 0,1,\dots,m \\ &{}^{(k)}\bar{\boldsymbol{\sigma}}, {}^{(i)}\boldsymbol{\gamma} \quad \text{as} \quad {}^{(k)}\bar{\boldsymbol{\sigma}}^J, {}^{(i)}\boldsymbol{\gamma}^J \end{aligned} \tag{13.1}$$

We recall that

$${}^{(1)}\boldsymbol{\gamma} = \boldsymbol{\gamma}_{(1)} = \boldsymbol{\gamma}^{(1)} = \boldsymbol{\gamma}_{(1)} = {}^{(1)}\boldsymbol{\gamma}^J = \bar{\boldsymbol{D}} \tag{13.2}$$

in which $\bar{\boldsymbol{D}}$ is the symmetric part of the velocity gradient tensor. Rationale for considering convected time derivatives of strain and stress tensors up to orders n and m is similar to Chapter 11, but is explained in the following sections as well.

It is shown that the constitutive theories of orders (m,n) for Cauchy stress tensor derived here can be reduced to Maxwell model, Oldroyd-B model, Giesekus model after many simplifications.

13.3 Considerations in the constitutive theories

Consider entropy inequality in Eulerian description (Chapter 6)

$$\bar{\rho}\left(\frac{D\bar{\Phi}}{Dt} + \bar{\eta}\frac{D\bar{\theta}}{Dt}\right) - {}^{(0)}\bar{\boldsymbol{\sigma}} : \bar{\boldsymbol{D}} \leq 0 + \frac{\bar{\boldsymbol{q}} \cdot \bar{\boldsymbol{g}}}{\bar{\theta}} \tag{13.3}$$

Conjugate pairs in (13.3) suggest that

$${}^{(0)}\bar{\boldsymbol{\sigma}} = {}^{(0)}\bar{\boldsymbol{\sigma}}(\bar{\rho}, \bar{\boldsymbol{D}}, \bar{\theta}) \tag{13.4}$$

$$\bar{\boldsymbol{q}} = \bar{\boldsymbol{q}}(\bar{\rho}, \bar{\boldsymbol{g}}, \bar{\theta}) \tag{13.5}$$

First, we additively decompose Cauchy stress tensor ${}^{(0)}\bar{\boldsymbol{\sigma}}$ into equilibrium and deviatoric Cauchy stress tensors, ${}^{(0)}_{\ e}\bar{\boldsymbol{\sigma}}$ and ${}^{(0)}_{\ d}\bar{\boldsymbol{\sigma}}$ to facilitate derivation of constitutive theories for volumetric deformation using ${}^{(0)}_{\ e}\bar{\boldsymbol{\sigma}}$ and distortional deformation, dissipation and memory using constitutive theory for ${}^{(0)}_{\ d}\bar{\boldsymbol{\sigma}}$. Thus, now we have

$${}^{(0)}_{\ e}\bar{\boldsymbol{\sigma}} = {}^{(0)}_{\ e}\bar{\boldsymbol{\sigma}}(\bar{\rho}, \bar{\theta}) \tag{13.6}$$

$${}^{(0)}_{\ d}\bar{\boldsymbol{\sigma}} = {}^{(0)}_{\ d}\bar{\boldsymbol{\sigma}}(\bar{\rho}, \bar{\boldsymbol{D}}, \bar{\theta}) \tag{13.7}$$

$$\text{and} \quad {}^{(0)}_{\ d}\bar{\boldsymbol{\sigma}} = {}^{(0)}_{\ d}\bar{\boldsymbol{\sigma}}(\bar{\rho}, 0, \bar{\theta}) = 0 \tag{13.8}$$

Equations (13.6) and (13.7) imply that ${}^{(0)}_{\ e}\bar{\boldsymbol{\sigma}}$ and ${}^{(0)}_{\ d}\bar{\boldsymbol{\sigma}}$ are mutually exclusive. We know from polymer science [21] that in order for a fluid to have memory mechanism or rheology, the constitutive theory for the Cauchy stress tensor must be at least a first order differential equation in time. Only then is the existence of memory modulus is possible; thereby, establishing relaxation behavior of the thermoviscoelastic fluid. Thus, instead of ${}^{(0)}\bar{\boldsymbol{\sigma}}$, we must at least consider ${}^{(1)}_{\ d}\bar{\boldsymbol{\sigma}}$ as constitutive tensor with ${}^{(0)}_{\ d}\bar{\boldsymbol{\sigma}}$ as its argument tensor. We can modify (13.7).

$${}^{(1)}_{\ d}\bar{\boldsymbol{\sigma}} = {}^{(1)}_{\ d}\bar{\boldsymbol{\sigma}}(\bar{\rho}, {}^{(0)}_{\ d}\bar{\boldsymbol{\sigma}}, {}^{(i)}\boldsymbol{\gamma}, \bar{\theta}) \quad ; \quad i = 1,2,\dots,n \tag{13.9}$$

in which ${}^{(1)}_{\ d}\bar{\boldsymbol{\sigma}}$ is the first convected time derivative of the deviatoric Cauchy stress tensor ${}^{(0)}_{\ d}\bar{\boldsymbol{\sigma}}$. In (13.8), we have replaced $\bar{\boldsymbol{D}}$ with ${}^{(i)}\boldsymbol{\gamma}$; $i = 1,2,\dots,n$, convected time derivatives of the strain tensor. We note that $\boldsymbol{\gamma} = \bar{\boldsymbol{D}}$. Using generalization of (13.9) to incorporate convected time derivatives of ${}^{(0)}_{d}\bar{\boldsymbol{\sigma}}$ up to order m leads to the following.

$$\begin{gathered} {}^{(m)}_{d}\bar{\boldsymbol{\sigma}} = {}^{(m)}_{d}\bar{\boldsymbol{\sigma}}(\bar{\rho}, {}^{(k)}_{d}\bar{\boldsymbol{\sigma}}, {}^{(i)}\boldsymbol{\gamma}, \bar{\theta}) \\ i = 1, 2, \dots, n \quad ; \quad k = 0, 1, \dots, m-1 \end{gathered} \tag{13.10}$$

$$\text{and} \quad \bar{\boldsymbol{q}} = \bar{\boldsymbol{q}}(\bar{\rho}, \bar{\boldsymbol{g}}, \bar{\theta}) \tag{13.11}$$

13.4 Constitutive theory for equilibrium stress ${}^{(0)}_{e}\bar{\boldsymbol{\sigma}}$

Constitutive theory for ${}^{(0)}_{e}\bar{\boldsymbol{\sigma}}$ is derived using the entropy inequality in Eulerian description. Following Chapter 8 we have the following constitutive theories for ${}^{(0)}_{e}\bar{\boldsymbol{\sigma}}$.

Compressible matter

$${}^{(0)}_{e}\bar{\boldsymbol{\sigma}} = \bar{p}(\bar{\rho}, \bar{\theta})\boldsymbol{\delta} \tag{13.12}$$

$$\bar{p} = -\bar{\rho}^2 \frac{\partial \bar{\Phi}(\bar{\rho}, \bar{\theta})}{\partial \bar{\rho}} \tag{13.13}$$

$\bar{p}(\bar{\rho}, \bar{\theta})$ is thermodynamic pressure or equation of state.

Incompressible matter

$${}^{(0)}_{e}\bar{\boldsymbol{\sigma}} = \bar{p}(\bar{\theta})\boldsymbol{\delta} \tag{13.14}$$

We note that ${}^{(0)}_{e}\bar{\boldsymbol{\sigma}}$ is basis independent. Hence, we can use ${}^{(0)}_{e}\bar{\boldsymbol{\sigma}} = {}_{e}\bar{\boldsymbol{\sigma}}$, but we continue with the notation ${}^{(0)}_{e}\bar{\boldsymbol{\sigma}}$. The reduced form of entropy inequality is given by (Chapter 8).

$$-{}^{(0)}_{d}\bar{\boldsymbol{\sigma}} : \bar{\boldsymbol{D}} + \frac{\bar{\boldsymbol{q}} \cdot \bar{\boldsymbol{g}}}{\bar{\theta}} \leq 0 \tag{13.15}$$

The entropy inequality (13.15) is satisfied if ${}^{(0)}_{d}\bar{\boldsymbol{\sigma}} : \bar{\boldsymbol{D}} > 0$ and $\frac{\bar{\boldsymbol{q}} \cdot \bar{\boldsymbol{g}}}{\bar{\theta}} \leq 0$. These serve as restrictions on the constitutive theories for ${}^{(0)}_{d}\bar{\boldsymbol{\sigma}}$ and $\bar{\boldsymbol{q}}$.

13.5 Constitutive theory for deviatoric stress ${}^{(0)}_{d}\bar{\boldsymbol{\sigma}}$

Constitutive theory for deviatoric Cauchy stress tensor is derived using (13.10) and the representation theorem. Let ${}^{\sigma}\underset{\sim}{\boldsymbol{G}}^i$; $i = 1, 2, \dots, N$ be the combined generators of the argument tensors of ${}^{(m)}_{d}\bar{\boldsymbol{\sigma}}$ that are symmetric tensors of rank two. Then, $\boldsymbol{I}$, ${}^{\sigma}\underset{\sim}{\boldsymbol{G}}^i$; $i = 1, 2, \dots, N$ constitute the basis of the space of ${}^{(m)}_{d}\bar{\boldsymbol{\sigma}}$. Hence, we can express ${}^{(m)}_{d}\bar{\boldsymbol{\sigma}}$ as a linear combination of the generators $\boldsymbol{I}$ and ${}^{\sigma}\underset{\sim}{\boldsymbol{G}}^i$; $i = 1, 2, \dots, N$ in the current configuration.

$${}^{(m)}_{d}\bar{\boldsymbol{\sigma}} = {}^{\sigma}\underset{\sim}{\alpha}^0 \boldsymbol{I} + \sum_{i=1}^{N} {}^{\sigma}\underset{\sim}{\alpha}^i ({}^{\sigma}\underset{\sim}{\boldsymbol{G}}^i) \tag{13.16}$$

in which the coefficients ${}^{\sigma}\underset{\sim}{\alpha}^i$; $i = 0, 1, \dots, N$ in the linear combination (13.16) are functions of $\bar{\rho}$, $\bar{\theta}$ and the combined invariants ${}^{\sigma}\underset{\sim}{I}^j$; $j = 1, 2, \dots, M$ of the argument tensors of ${}^{(m)}_{\ d}\bar{\boldsymbol{\sigma}}$ in (13.47) in the current configuration, i.e.

$$ {}^{\sigma}\underset{\sim}{\alpha}^i = {}^{\sigma}\underset{\sim}{\alpha}^i(\bar{\rho}, \bar{\theta}, {}^{\sigma}\underset{\sim}{I}^j) \quad ; \quad ; i = 0, 1, \dots, N \quad ; \quad j = 1, 2, \dots, M \tag{13.17} $$

Material coefficients

To determine material coefficients in the constitutive theory (13.16), we expand ${}^{\sigma}\underset{\sim}{\alpha}^i$; $i = 0, 1, \dots, N$ in Taylor series in ${}^{\sigma}\underset{\sim}{I}^j$; $j = 1, 2, \dots, M$ about a known configuration $\underline{\Omega}$ and retaining only up to linear terms in ${}^{\sigma}\underset{\sim}{I}^j$; $j = 1, 2, \dots, M$ (for simplicity). We do not consider Taylor series expansion of ${}^{\sigma}\underset{\sim}{\alpha}^i$; $i = 0, 1, \dots, N$ in $\bar{\rho}$ and $\bar{\theta}$ as their influence on the Cauchy stress tensor has already been accounted for in deriving the constitutive theory for ${}^{(0)}_{\ e}\bar{\boldsymbol{\sigma}}$.

$$ {}^{\sigma}\underset{\sim}{\alpha}^i = {}^{\sigma}\underset{\sim}{\alpha}^i\Big|_{\underline{\Omega}} + \sum_{j=1}^{M} \frac{\partial\, {}^{\sigma}\underset{\sim}{\alpha}^i}{\partial\, {}^{\sigma}\underset{\sim}{I}^j}\Bigg|_{\underline{\Omega}} \left({}^{\sigma}\underset{\sim}{I}^j - {}^{\sigma}\underset{\sim}{I}^j\Big|_{\underline{\Omega}} \right) \quad ; \quad i = 0, 1, \dots, N \tag{13.18} $$

Substituting ${}^{\sigma}\underset{\sim}{\alpha}^i$; $i = 0, 1, \dots, N$ in (13.16)

$$ \begin{aligned} {}^{(m)}_{\ d}\bar{\boldsymbol{\sigma}} = & \left({}^{\sigma}\underset{\sim}{\alpha}^0\Big|_{\underline{\Omega}} + \sum_{j=1}^{M} \frac{\partial\, {}^{\sigma}\underset{\sim}{\alpha}^0}{\partial\, {}^{\sigma}\underset{\sim}{I}^j}\Bigg|_{\underline{\Omega}} \left({}^{\sigma}\underset{\sim}{I}^j - {}^{\sigma}\underset{\sim}{I}^j\Big|_{\underline{\Omega}} \right) \right) \boldsymbol{I} \\ & + \sum_{i=1}^{N} \left({}^{\sigma}\underset{\sim}{\alpha}^i\Big|_{\underline{\Omega}} + \sum_{j=1}^{M} \frac{\partial\, {}^{\sigma}\underset{\sim}{\alpha}^i}{\partial\, {}^{\sigma}\underset{\sim}{I}^j}\Bigg|_{\underline{\Omega}} \left({}^{\sigma}\underset{\sim}{I}^j - {}^{\sigma}\underset{\sim}{I}^j\Big|_{\underline{\Omega}} \right) \right) {}^{\sigma}\underset{\sim}{\boldsymbol{G}}^i \end{aligned} \tag{13.19} $$

Collecting coefficients of (the terms defined in the current configuration) $\boldsymbol{I}$, ${}^{\sigma}\underset{\sim}{I}^j\boldsymbol{I}$, ${}^{\sigma}\underset{\sim}{\boldsymbol{G}}^i$ and ${}^{\sigma}\underset{\sim}{I}^j({}^{\sigma}\underset{\sim}{\boldsymbol{G}}^i)$; $i = 1, 2, \dots, N$; $j = 1, 2, \dots, M$ and defining

$$ \begin{aligned} & \underset{\sim}{\sigma}^0\Big|_{\underline{\Omega}} = {}^{\sigma}\underset{\sim}{\alpha}^0\Big|_{\underline{\Omega}} - \sum_{j=1}^{M} \frac{\partial\, {}^{\sigma}\underset{\sim}{\alpha}^0}{\partial\, {}^{\sigma}\underset{\sim}{I}^j}\Bigg|_{\underline{\Omega}} \left({}^{\sigma}\underset{\sim}{I}^j\Big|_{\underline{\Omega}} \right) \quad ; \quad {}^{\sigma}\underset{\sim}{a}_j = \frac{\partial\, {}^{\sigma}\underset{\sim}{\alpha}^0}{\partial\, {}^{\sigma}\underset{\sim}{I}^j}\Bigg|_{\underline{\Omega}} \\ & {}^{\sigma}\underset{\sim}{b}_i = {}^{\sigma}\underset{\sim}{\alpha}^i\Big|_{\underline{\Omega}} - \sum_{j=1}^{M} \frac{\partial\, {}^{\sigma}\underset{\sim}{\alpha}^i}{\partial\, {}^{\sigma}\underset{\sim}{I}^j}\Bigg|_{\underline{\Omega}} \left({}^{\sigma}\underset{\sim}{I}^j\Big|_{\underline{\Omega}} \right) \quad ; \quad {}^{\sigma}\underset{\sim}{c}_{ij} = \frac{\partial\, {}^{\sigma}\underset{\sim}{\alpha}^i}{\partial\, {}^{\sigma}\underset{\sim}{I}^j}\Bigg|_{\underline{\Omega}} \\ & i = 1, 2 \dots, N \quad ; \quad j = 1, 2 \dots, M \end{aligned} \tag{13.20} $$

equation (13.16) can be written as

$$ {}^{(m)}_{\ d}\bar{\boldsymbol{\sigma}} = \underset{\sim}{\sigma}\Big|_{\underline{\Omega}} \boldsymbol{I} + \sum_{j=1}^{M} {}^{\sigma}\underset{\sim}{a}_j ({}^{\sigma}\underset{\sim}{I}^j) \boldsymbol{I} + \sum_{i=1}^{N}\sum_{j=1}^{M} {}^{\sigma}\underset{\sim}{c}_{ij} ({}^{\sigma}\underset{\sim}{I}^j) {}^{\sigma}\underset{\sim}{\boldsymbol{G}}^i + \sum_{i=1}^{N} {}^{\sigma}\underset{\sim}{b}_i\, {}^{\sigma}\underset{\sim}{\boldsymbol{G}}^i \tag{13.21} $$

${}^{\sigma}\underset{\sim}{a}_j$, ${}^{\sigma}\underset{\sim}{b}_i$, ${}^{\sigma}\underset{\sim}{c}_{ij}$; $i = 1, 2, \dots, N$; $j = 1, 2, \dots, M$ are material coefficients defined in a known configuration $\underline{\Omega}$. This constitutive theory for ${}^{(m)}_{\ d}\bar{\boldsymbol{\sigma}}$ requires $(M + MN + N)$ material coefficients. This constitutive theory is based on integrity (complete basis of the space of ${}^{(m)}_{\ d}\bar{\boldsymbol{\sigma}}$). Material coefficients can be functions of $\bar{\rho}|_{\underline{\Omega}}$, $\bar{\theta}|_{\underline{\Omega}}$ and ${}^{\sigma}\underset{\sim}{I}^j\Big|_{\underline{\Omega}}$; $j = 1, 2, \dots, M$.

Remarks

(1) The constitutive theory derived here is an ordered rate constitutive theory of orders m and n of the convected time derivatives of the Cauchy stress tensor and the Green's or Almansi strain tensor respectively for both compressible and incompressible thermoviscoelastic fluent continua.

(2) The dissipation mechanism is due to convected time derivatives ${}^{(i)}\boldsymbol{\gamma}$; $i = 1, 2, \dots, n$ of the strain tensor. The memory mechanism is due to both, the convected time derivatives of the Cauchy stress tensor as well as the convected time derivatives of the Green's or Almansi strain tensor.

(3) The derivation of the constitutive theory has been presented using basis independent measures of stress (and its rates) as well as basis independent strain rates. By choosing appropriate conjugate pairs for ${}^{(k)}_{\;\;d}\bar{\boldsymbol{\sigma}}$ and ${}^{(j)}\boldsymbol{\gamma}$; $k = 0, 1, \dots, m$ and $j = 1, 2, \dots, n$, basis specific constitutive theories for the deviatoric Cauchy stress tensor can be obtained.

$$
\begin{aligned}
\bar{\boldsymbol{\sigma}}^{(k)}, \boldsymbol{\gamma}_{(i)} \quad &; \quad \text{Contravariant Cauchy stress measure} \\
\bar{\boldsymbol{\sigma}}_{(k)}, \boldsymbol{\gamma}^{(i)} \quad &; \quad \text{Covariant Cauchy stress measure} \\
{}^{(k)}\bar{\boldsymbol{\sigma}}^{J}, {}^{(i)}\boldsymbol{\gamma}^{J} \quad &; \quad \text{Jaumann rates} \\
k = 0, 1, \dots, m \quad &; \quad i = 1, 2, \dots, n
\end{aligned}
\tag{13.22}
$$

(4) Known configuration $\underline{\Omega}$ is a suitable previously known configuration (generally last value of time for which deformation is known). Configuration $\underline{\Omega}$ can be reference configuration in which case the material coefficients will be constant, i.e., not dependent on varying $\bar{\rho}$, $\bar{\theta}$ and ${}^{\sigma}\underset{\sim}{I}^{j}$; $j = 1, 2, \dots, M$.

(5) The material coefficients as functions of unknown deformation in current configuration is not supported by CCM. This is rather obvious because, if this is the case, then ${}^{\sigma}\underset{\sim}{\alpha}^{i}$; $i = 0, 1, \dots, N$ would be material coefficients, and there would be no need for Taylor series expansion of ${}^{\sigma}\underset{\sim}{\alpha}^{i}$ in $\bar{\rho}$, $\bar{\theta}$ and ${}^{\sigma}\underset{\sim}{I}^{j}$; $j = 1, 2, \dots, M$ about a known configuration. In continuum theories, material coefficients are known properties; hence, they cannot be treated as functions of unknown deformation.

(6) The constitutive theory (13.21), though based on integrity, requires too many material coefficients which are quite difficult to determine experimentally or otherwise. Thus, in order to use the constitutive theory (13.21) in applications, it is necessary to extract simpler constitutive theories from it. Maxwell and Oldroyd-B constitutive models for dilute polymeric incompressible fluids, Giesekus constitutive model for incompressible, dense polymeric fluids, etc. used currently in polymer science are shown to be much more simplified subsets of the constitutive theory (13.21) that is based on integrity.

(7) In the following section we show that Maxwell and Giesekus constitutive models are a simplified subset of the ordered rate constitutive theory of $m = 1$ and $n = 1$ that is based on integrity. Where as Oldroyd-B constitutive model is a simplified subset of the constitutive theory of orders $m = 1$ and $n = 2$.

(8) Finally, we show that all three constitutive models used currently: Maxwell, Oldroyd-B and Giesekus can be described by a single simplified constitutive

theory of orders $m = 1$ and $n = 2$. Appropriate choice of material coefficients yield the specific desired constitutive theory.

13.5.1 Simplified constitutive theories for ${}^{(m)}_{\ d}\bar{\boldsymbol{\sigma}}$: $m = 1$, $n = 1$

Different choices of m and n in the general ordered rate constitutive theory of orders m and n for ${}^{(m)}_{\ d}\bar{\boldsymbol{\sigma}}$ and the specific choices of generators and invariants yield the desired constitutive theories. In this section, we derive: (1) constitutive theory for $m = 1$ and $n = 1$ based on integrity first and then show that Maxwell and Giesekus constitutive models can be extracted from this with appropriate assumptions, (2) ordered rate constitutive theory for $m = 1$ and $n = 2$ based on integrity from which the Oldroyd-B constitutive model is extracted. When $m = 1$ and $n = 1$, we have

$$ {}^{(1)}_{\ d}\bar{\boldsymbol{\sigma}} = {}^{(1)}_{\ d}\bar{\boldsymbol{\sigma}}(\bar{\rho}, {}^{(0)}_{\ d}\bar{\boldsymbol{\sigma}}, {}^{(1)}\boldsymbol{\gamma}, \bar{\theta}) \tag{13.23} $$

Let the combined generators of the argument tensors $\bar{\rho}$, ${}^{(0)}_{\ d}\bar{\boldsymbol{\sigma}}$, ${}^{(1)}\boldsymbol{\gamma}$, $\bar{\theta}$ (tensors of ranks zero, two, two and zero respectively) that are symmetric tensors of rank two be ${}^{\sigma}\underset{\sim}{\boldsymbol{G}}^i$; $i = 1, 2, \dots, N$. These are listed in Table 13.1. In this case, $N = 7$. The combined invariants of the argument tensors in (13.23) are listed in Table 13.2. We have $M = 10$. Entries in Tables 13.1 and 13.2 can be generated using tables in Appendix A.

Remarks

(1) We note that the invariants in Table 13.2 under (2) marked (a) need not be included due to the fact that

$$ \mathrm{tr}([{}^{(0)}_{\ d}\bar{\sigma}][{}^{(1)}\gamma] + [{}^{(1)}\gamma][{}^{(0)}_{\ d}\bar{\sigma}]) + \mathrm{tr}([{}^{(0)}_{\ d}\bar{\sigma}][{}^{(1)}\gamma] - [{}^{(1)}\gamma][{}^{(0)}_{\ d}\bar{\sigma}]) = 2\,\mathrm{tr}([{}^{(0)}_{\ d}\bar{\sigma}][{}^{(1)}\gamma]) $$

which is the same as ${}^{\sigma}\underset{\sim}{I}^7$ (except for the factor 2, which is of no consequence).

(2) In many published works (a) are also included in the list of invariants in addition to ${}^{\sigma}\underset{\sim}{I}^7$, which is redundant.

Using the generators in Table 13.1 we can express ${}^{(1)}_{\ d}\bar{\boldsymbol{\sigma}}$ as a linear combination of $\boldsymbol{I}$ and the combined generators ${}^{\sigma}\underset{\sim}{\boldsymbol{G}}^i$; $i = 1, 2, \dots, 7$ $(N = 7)$.

$$ {}^{(1)}_{\ d}\bar{\boldsymbol{\sigma}} = {}^{\sigma}\underset{\sim}{\alpha}^0 \boldsymbol{I} + \sum_{i=1}^{7} {}^{\sigma}\underset{\sim}{\alpha}^i ({}^{\sigma}\underset{\sim}{\boldsymbol{G}}^i) \tag{13.24} $$

in which the coefficients ${}^{\sigma}\underset{\sim}{\alpha}^i$; $i = 0, 1, \dots, 7$ are functions of the combined invariants ${}^{\sigma}\underset{\sim}{I}^j$; $j = 1, 2, \dots, 10$ $(M = 10)$, density $\bar{\rho}$ and temperature $\bar{\theta}$. The material coefficients from (13.24) are determined using the same procedure as described for the ordered rate theory of orders m and n in Section 13.5, and we obtain (13.21) in which $N = 7$ and $M = 10$. Explicit expressions for the generators and the invariants are given in Tables 13.1 and 13.2. This constitutive theory requires $M + MN + N = 10 + 10 \times 7 + 7 = 87$ material coefficients. This constitutive theory is based on integrity, i.e., complete basis of the space of ${}^{(1)}_{\ d}\bar{\boldsymbol{\sigma}}$.

Table 13.1: Combined generators for ${}^{(1)}_{\,d}\bar{\boldsymbol{\sigma}}$: $m = 1$, $n = 1$; first order rate theory

	Arguments	**Generators**
(1)	none	$[I]$
(2)	one at a time	(including (1))
	$[{}^{(0)}_{\,d}\bar{\sigma}]$	$[{}^{\sigma}G^1] = [{}^{(0)}_{\,d}\bar{\sigma}]$; $[{}^{\sigma}G^2] = [{}^{(0)}_{\,d}\bar{\sigma}]^2$
	$[{}^{(1)}\gamma]$	$[{}^{\sigma}G^3] = [{}^{(1)}\gamma]$; $[{}^{\sigma}G^4] = [{}^{(1)}\gamma]^2$
(3)	two at a time	(including (1) and (2))
	$[{}^{(0)}_{\,d}\bar{\sigma}]$, $[{}^{(1)}\gamma]$	$[{}^{\sigma}G^5] = [{}^{(0)}_{\,d}\bar{\sigma}][{}^{(1)}\gamma] + [{}^{(1)}\gamma][{}^{(0)}_{\,d}\bar{\sigma}]$
		$[{}^{\sigma}G^6] = [{}^{(0)}_{\,d}\bar{\sigma}]^2[{}^{(1)}\gamma] + [{}^{(1)}\gamma][{}^{(0)}_{\,d}\bar{\sigma}]^2$
		$[{}^{\sigma}G^7] = [{}^{(0)}_{\,d}\bar{\sigma}][{}^{(1)}\gamma]^2 + [{}^{(1)}\gamma]^2[{}^{(0)}_{\,d}\bar{\sigma}]$

Table 13.2: Combined invariants for ${}^{(1)}_{\,d}\bar{\boldsymbol{\sigma}}$: $m = 1$, $n = 1$; first order rate theory

	Arguments	**Invariants**
(1)	one at a time	
	$[{}^{(0)}_{\,d}\bar{\sigma}]$	${}^{\sigma}\mathcal{I}^1 = \text{tr}([{}^{(0)}_{\,d}\bar{\sigma}])$; ${}^{\sigma}\mathcal{I}^2 = \text{tr}([{}^{(0)}_{\,d}\bar{\sigma}]^2)$
		${}^{\sigma}\mathcal{I}^3 = \text{tr}([{}^{(0)}_{\,d}\bar{\sigma}]^3)$
	$[{}^{(1)}\gamma]$	${}^{\sigma}\mathcal{I}^4 = \text{tr}([{}^{(1)}\gamma])$; ${}^{\sigma}\mathcal{I}^5 = \text{tr}([{}^{(1)}\gamma]^2)$
		${}^{\sigma}\mathcal{I}^6 = \text{tr}([{}^{(1)}\gamma]^3)$
(2)	two at a time	(including (1))
	$[{}^{(0)}_{\,d}\bar{\sigma}]$, $[{}^{(1)}\gamma]$	${}^{\sigma}\mathcal{I}^7 = \text{tr}([{}^{(0)}_{\,d}\bar{\sigma}][{}^{(1)}\gamma])$; ${}^{\sigma}\mathcal{I}^8 = \text{tr}([{}^{(0)}_{\,d}\bar{\sigma}]^2[{}^{(1)}\gamma])$
		${}^{\sigma}\mathcal{I}^9 = \text{tr}([{}^{(0)}_{\,d}\bar{\sigma}][{}^{(1)}\gamma]^2)$; ${}^{\sigma}\mathcal{I}^{10} = \text{tr}([{}^{(0)}_{\,d}\bar{\sigma}]^2[{}^{(1)}\gamma]^2)$
		(a) ${}^{\sigma}\mathcal{I} = \text{tr}([{}^{(0)}_{\,d}\bar{\sigma}][{}^{(1)}\gamma] + [{}^{(1)}\gamma][{}^{(0)}_{\,d}\bar{\sigma}])$
		${}^{\sigma}\mathcal{I} = \text{tr}([{}^{(0)}_{\,d}\bar{\sigma}][{}^{(1)}\gamma] - [{}^{(1)}\gamma][{}^{(0)}_{\,d}\bar{\sigma}])$

13.5.2 Maxwell constitutive model

The Maxwell constitutive model is a linear viscoelastic model generally suitable for dilute polymeric fluids. It can be derived from the ordered rate constitutive model for $m = 1$ and $n = 1$ (presented in Section 13.5.1) and by suitable choices of the combined generators and the invariants listed in Tables 13.1 and 13.2. Since *Maxwell constitutive model is a linear viscoelastic model*, ${}^{(1)}_{\;d}\bar{\boldsymbol{\sigma}}$ can only be a function of generators ${}^{\sigma}\underset{\sim}{\boldsymbol{G}}^1 = {}^{(0)}_{\;d}\bar{\boldsymbol{\sigma}}$ and ${}^{\sigma}\underset{\sim}{\boldsymbol{G}}^3 = {}^{(1)}\boldsymbol{\gamma}$ (Table 13.1). Thus, we can begin with

$$
{}^{(1)}_{\;d}\bar{\boldsymbol{\sigma}} = {}^{\sigma}\underset{\sim}{\alpha}^0 \boldsymbol{I} + {}^{\sigma}\underset{\sim}{\alpha}^1({}^{(0)}_{\;d}\bar{\boldsymbol{\sigma}}) + {}^{\sigma}\underset{\sim}{\alpha}^2({}^{(1)}\boldsymbol{\gamma}) \tag{13.25}
$$

In deriving material coefficients, we must observe the restriction that is a linear viscoelastic model; hence

(a) The only invariants that are admissible in this theory are due to ${}^{(0)}_{\;d}\bar{\boldsymbol{\sigma}}$ and ${}^{(1)}\boldsymbol{\gamma}$, but linear functions of ${}^{(0)}_{\;d}\bar{\boldsymbol{\sigma}}$ and ${}^{(1)}\boldsymbol{\gamma}$. Thus, $\mathrm{tr}({}^{(0)}_{\;d}\bar{\boldsymbol{\sigma}})$ and $\mathrm{tr}({}^{(1)}\boldsymbol{\gamma})$ (${}^{\sigma}\underset{\sim}{I}^1$ and ${}^{\sigma}\underset{\sim}{I}^4$ in Table 13.2) are the only admissible invariants.

(b) The products of invariants and generators are neglected in this constitutive theory, even though $\mathrm{tr}({}^{(0)}_{\;d}\bar{\boldsymbol{\sigma}}){}^{(1)}\boldsymbol{\gamma}$ and $\mathrm{tr}({}^{(1)}\boldsymbol{\gamma}){}^{(0)}_{\;d}\bar{\boldsymbol{\sigma}}$ are linear in both ${}^{(0)}_{\;d}\bar{\boldsymbol{\sigma}}$ and ${}^{(1)}\boldsymbol{\gamma}$.

Thus, the Maxwell constitutive model consists of (redefining material coefficients in (13.21) for $n = 1$ and $m = 1$) the following

$$
{}^{(1)}_{\;d}\bar{\boldsymbol{\sigma}} = a_0\boldsymbol{I} + a_1{}^{(0)}_{\;d}\bar{\boldsymbol{\sigma}} + a_2{}^{(1)}\boldsymbol{\gamma} + b_1\,\mathrm{tr}({}^{(1)}\boldsymbol{\gamma})\boldsymbol{I} + b_2\,\mathrm{tr}({}^{(0)}_{\;d}\bar{\boldsymbol{\sigma}})\boldsymbol{I} \tag{13.26}
$$

We redefine material coefficients coefficients to conform to the commonly used notations in polymer science. Divide (13.26) by $-a_1$ ($a_1 \neq 0$) and rearrange terms

$$
\begin{aligned}
{}^{(0)}_{\;d}\bar{\boldsymbol{\sigma}} + \left(-\frac{1}{a_1}\right){}^{(1)}_{\;d}\bar{\boldsymbol{\sigma}} &= \left(-\frac{a_0}{a_1}\right)\boldsymbol{I} + \left(-\frac{a_2}{a_1}\right){}^{(1)}\boldsymbol{\gamma} \\
&\quad + \left(-\frac{b_1}{a_1}\right)\mathrm{tr}({}^{(1)}\boldsymbol{\gamma})\boldsymbol{I} + \left(-\frac{b_2}{a_1}\right)\mathrm{tr}({}^{(0)}_{\;d}\bar{\boldsymbol{\sigma}})\boldsymbol{I}
\end{aligned} \tag{13.27}
$$

We define the following material coefficients. Let

$$
\begin{aligned}
&\left(-\frac{1}{a_1}\right) = \lambda \quad \text{or} \quad \lambda\big|_{\underline{\Omega}} \;\; ; \quad \left(-\frac{a_0}{a_1}\right) = \underset{\sim}{\sigma}^0 \quad \text{or} \quad \underset{\sim}{\sigma}^0\big|_{\underline{\Omega}} \\
&\left(-\frac{a_2}{a_1}\right) = 2\eta \quad \text{or} \quad 2\eta\big|_{\underline{\Omega}} \;\; ; \quad \left(-\frac{b_1}{a_1}\right) = \kappa \quad \text{or} \quad \kappa\big|_{\underline{\Omega}} \\
&\left(-\frac{b_2}{a_1}\right) = \underset{\sim}{\beta} \quad \text{or} \quad \underset{\sim}{\beta}\big|_{\underline{\Omega}}
\end{aligned} \tag{13.28}
$$

Using (13.28) in (13.27), we obtain

$$
{}^{(0)}_{\;d}\bar{\boldsymbol{\sigma}} + \lambda({}^{(1)}_{\;d}\bar{\boldsymbol{\sigma}}) = \underset{\sim}{\sigma}^0\big|_{\underline{\Omega}}\boldsymbol{I} + 2\eta({}^{(1)}\boldsymbol{\gamma}) + \kappa\,\mathrm{tr}({}^{(1)}\boldsymbol{\gamma})\boldsymbol{I} + \underset{\sim}{\beta}\,\mathrm{tr}({}^{(0)}_{\;d}\bar{\boldsymbol{\sigma}})\boldsymbol{I} \tag{13.29}
$$

The constitutive model (13.29) is further simplified by neglecting $\underset{\sim}{\sigma}^0\big|_{\underline{\Omega}}\boldsymbol{I}$, $\underset{\sim}{\beta}\,\mathrm{tr}({}^{(0)}_{\;d}\bar{\boldsymbol{\sigma}})$ terms to obtain

$$
{}^{(0)}_{\;d}\bar{\boldsymbol{\sigma}} + \lambda({}^{(1)}_{\;d}\bar{\boldsymbol{\sigma}}) = 2\eta({}^{(1)}\boldsymbol{\gamma}) + \kappa\,\mathrm{tr}({}^{(1)}\boldsymbol{\gamma})\boldsymbol{I} \tag{13.30}
$$

This constitutive theory is valid for compressible Maxwell polymeric fluids. If the fluid is incompressible, then $\mathrm{tr}({}^{(1)}\boldsymbol{\gamma}) = \mathrm{tr}(\bar{\boldsymbol{D}}) = 0$; hence, (13.30) reduces to

$$ {}^{(0)}_{\ d}\bar{\boldsymbol{\sigma}} + \lambda({}^{(1)}_{\ d}\bar{\boldsymbol{\sigma}}) = 2\eta({}^{(1)}\boldsymbol{\gamma}) \tag{13.31} $$

Where λ is called relaxation time, η is the viscosity of the polymeric fluid, κ is the second viscosity if the fluid is compressible. First and second viscosities η and κ can be functions of any desired invariants in Table 13.2 as well as $\bar{\rho}$ and $\bar{\theta}$ in a known configuration $\underline{\Omega}$.

13.5.2.1 Maxwell constitutive model: Memory modulus

If we consider a simplified form of (13.31) for incompressible fluid then we can derive the expression for relaxation or memory modulus. Consider (13.31) for incompressible fluids in which the material derivative is replaced with the time derivative.

$$ {}^{(0)}_{\ d}\bar{\boldsymbol{\sigma}} + \lambda \frac{\partial\, {}^{(0)}_{\ d}\bar{\boldsymbol{\sigma}}}{\partial t} = 2\eta({}^{(1)}\boldsymbol{\gamma}) \tag{13.32} $$

Following Section 11.4 in Chapter 11, we can derive

$$ {}^{(0)}_{\ d}\bar{\boldsymbol{\sigma}} = \int_{-\infty}^{t} \boldsymbol{Q}(t') e^{-\frac{(t-t')}{\lambda}}\, dt' \tag{13.33} $$

Where

$$ \boldsymbol{Q}(t) = \frac{1}{\lambda} 2\eta({}^{(1)}\boldsymbol{\gamma}) \tag{13.34} $$

Hence

$$ {}^{(0)}_{\ d}\bar{\boldsymbol{\sigma}} = \int_{-\infty}^{t} \left(\frac{2\eta}{\lambda} e^{-\frac{(t-t')}{\lambda}} \right) {}^{(1)}\boldsymbol{\gamma}\, dt' \tag{13.35} $$

$$ \text{or} \quad {}^{(0)}_{\ d}\bar{\boldsymbol{\sigma}} = \int_{-\infty}^{t} R_m\, {}^{(1)}\boldsymbol{\gamma}\, dt' \tag{13.36} $$

In which

$$ R_m = \frac{2\eta}{\lambda} e^{-\frac{(t-t')}{\lambda}} \tag{13.37} $$

Where R_m is called the *relaxation modulus*. We can also define the entire integrand in (13.33) as relaxation modulus.

13.5.3 Giesekus Constitutive Model

Giesekus constitutive model is generally suitable for dense polymeric fluids such as polymer melts. This is a non-linear viscoelastic constitutive model. The model uses generators ${}^{\sigma}\underset{\sim}{\boldsymbol{G}}^1 = {}^{(0)}_{\ d}\bar{\boldsymbol{\sigma}}$, ${}^{\sigma}\underset{\sim}{\boldsymbol{G}}^3 = {}^{(1)}\boldsymbol{\gamma}$ and ${}^{\sigma}\underset{\sim}{\boldsymbol{G}}^2 = ({}^{(0)}_{\ d}\bar{\boldsymbol{\sigma}})^2$ (see Table 13.1). Using these generators and tensor $\boldsymbol{I}$ we can write

$$ {}^{(1)}_{\ d}\bar{\boldsymbol{\sigma}} = {}^{\sigma}\underset{\sim}{\alpha}^0 \boldsymbol{I} + {}^{\sigma}\underset{\sim}{\alpha}^1({}^{(0)}_{\ d}\bar{\boldsymbol{\sigma}}) + {}^{\sigma}\underset{\sim}{\alpha}^2({}^{(1)}\boldsymbol{\gamma}) + {}^{\sigma}\underset{\sim}{\alpha}^3({}^{(0)}_{\ d}\bar{\boldsymbol{\sigma}})^2 \tag{13.38} $$

This constitutive model contains $({}^{(0)}_{\ d}\bar{\boldsymbol{\sigma}})^2$ as additional generator (non-linear) compared to the Maxwell model. In deriving material coefficients, we retain linear terms in ${}^{(0)}_{\ d}\bar{\boldsymbol{\sigma}}$ and ${}^{(1)}\boldsymbol{\gamma}$ but exclude the product of linear terms (as in Maxwell model). The Giesekus model consists of (redefining material coefficients in (13.21) for $n = 1$ and $m = 1$)

$$ {}^{(1)}_{\ d}\bar{\boldsymbol{\sigma}} = a_0\boldsymbol{I} + a_1\,{}^{(0)}_{\ d}\bar{\boldsymbol{\sigma}} + a_2\,{}^{(1)}\boldsymbol{\gamma} + c_1({}^{(0)}_{\ d}\bar{\boldsymbol{\sigma}})^2 + b_2\,\mathrm{tr}({}^{(0)}_{\ d}\bar{\boldsymbol{\sigma}})\boldsymbol{I} + b_1\,\mathrm{tr}({}^{(1)}\boldsymbol{\gamma})\boldsymbol{I} \tag{13.39} $$

We redefine material coefficients to conform to commonly used notations in polymer science. Divide (13.39) by $-a_1$ ($a_1 \neq 0$) and rearrange terms.

$$ \begin{aligned} {}^{(0)}_{\ d}\bar{\boldsymbol{\sigma}} + \left(-\frac{1}{a_1}\right){}^{(1)}_{\ d}\bar{\boldsymbol{\sigma}} = \left(-\frac{a_0}{a_1}\right)\boldsymbol{I} + \left(-\frac{a_2}{a_1}\right){}^{(1)}\boldsymbol{\gamma} + \left(-\frac{c_1}{a_1}\right)({}^{(0)}_{\ d}\bar{\boldsymbol{\sigma}})^2 \\ + \left(-\frac{b_2}{a_1}\right)\mathrm{tr}({}^{(0)}_{\ d}\bar{\boldsymbol{\sigma}})\boldsymbol{I} + \left(-\frac{b_1}{a_1}\right)\mathrm{tr}({}^{(1)}\boldsymbol{\gamma})\boldsymbol{I} \end{aligned} \tag{13.40} $$

We define the following material coefficients

$$ \begin{aligned} \left(-\frac{1}{a_1}\right) &= \lambda \quad \text{or} \quad \lambda\big|_{\underline{\Omega}} \quad ; \quad \left(-\frac{a_0}{a_1}\right) = \underset{\sim}{\sigma}^0 \quad \text{or} \quad \underset{\sim}{\sigma}^0\big|_{\underline{\Omega}} \\ \left(-\frac{a_2}{a_1}\right) &= 2\eta \quad \text{or} \quad 2\eta\big|_{\underline{\Omega}} \quad ; \quad \left(-\frac{b_2}{a_1}\right) = \underset{\sim}{\beta} \quad \text{or} \quad \underset{\sim}{\beta}\big|_{\underline{\Omega}} \\ \left(-\frac{b_1}{a_1}\right) &= \kappa \quad \text{or} \quad \kappa\big|_{\underline{\Omega}} \end{aligned} $$

Substituting (13.40) in (13.39)

$$ {}^{(0)}_{\ d}\bar{\boldsymbol{\sigma}} + \lambda({}^{(1)}_{\ d}\bar{\boldsymbol{\sigma}}) = \underset{\sim}{\sigma}^0\boldsymbol{I} + 2\eta({}^{(1)}\boldsymbol{\gamma}) + \left(-\frac{c_1}{a_1}\right)({}^{(0)}_{\ d}\bar{\boldsymbol{\sigma}})^2 + \underset{\sim}{\beta}\,\mathrm{tr}({}^{(0)}_{\ d}\bar{\boldsymbol{\sigma}})\boldsymbol{I} + \kappa\,\mathrm{tr}({}^{(1)}\boldsymbol{\gamma})\boldsymbol{I} \tag{13.41} $$

Where λ is the relaxation time, η is the viscosity of the polymeric, κ is the second viscosity (for compressible case) and $\underset{\sim}{\beta}$ is similar to κ. We note that each term in (13.41) has dimensions of stress, thus, the coefficient of the $({}^{(0)}_{\ d}\bar{\boldsymbol{\sigma}})^2$ term must have dimensions of "1/stress", which is the same as "time/dimensions of viscosity." Thus, the choice of λ/η for $\left(-\frac{c_1}{a_1}\right)$ is dimensionally admissible. We choose

$$ \left(-\frac{c_1}{a_1}\right) = \frac{\lambda}{\eta}\alpha \tag{13.42} $$

Here, α is called the mobility factor, a measure of the formation and break up of the polymer molecule colonies and their mobility. Thus, we can write (13.41) as

$$ {}^{(0)}_{\ d}\bar{\boldsymbol{\sigma}} + \lambda({}^{(1)}_{\ d}\bar{\boldsymbol{\sigma}}) = \underset{\sim}{\sigma}^0\boldsymbol{I} + 2\eta({}^{(1)}\boldsymbol{\gamma}) + \frac{\lambda}{\eta}\alpha({}^{(0)}_{\ d}\bar{\boldsymbol{\sigma}})^2 + \underset{\sim}{\beta}\,\mathrm{tr}({}^{(0)}_{\ d}\bar{\boldsymbol{\sigma}})\boldsymbol{I} + \kappa\,\mathrm{tr}({}^{(1)}\boldsymbol{\gamma})\boldsymbol{I} \tag{13.43} $$

This is Giesekus constitutive model for dense polymeric compressible fluids. In polymer science works $\underset{\sim}{\sigma}^0\boldsymbol{I}$, $\underset{\sim}{\beta}\,\mathrm{tr}({}^{(0)}_{\ d}\bar{\boldsymbol{\sigma}})$ terms are neglected. Thus, for compressible thermoviscoelastic fluid, (13.43) reduces to the following.

$$ {}^{(0)}_{\ d}\bar{\boldsymbol{\sigma}} + \lambda({}^{(1)}_{\ d}\bar{\boldsymbol{\sigma}}) = 2\eta({}^{(1)}\boldsymbol{\gamma}) + \frac{\lambda}{\eta}\alpha({}^{(0)}_{\ d}\bar{\boldsymbol{\sigma}})^2 + \kappa\,\mathrm{tr}({}^{(1)}\boldsymbol{\gamma})\boldsymbol{I} \tag{13.44} $$

If the polymeric fluid is treated as incompressible, then $\text{tr}({}^{(1)}\boldsymbol{\gamma}) = \text{tr}(\bar{\boldsymbol{D}}) = 0$, hence (13.44) reduces to

$$ {}^{(0)}_{\ d}\bar{\boldsymbol{\sigma}} + \lambda({}^{(1)}_{\ d}\bar{\boldsymbol{\sigma}}) = 2\eta({}^{(1)}\boldsymbol{\gamma}) + \frac{\lambda}{\eta}\alpha({}^{(0)}_{\ d}\bar{\boldsymbol{\sigma}})^2 \tag{13.45}$$

Remarks

(1) By choosing ${}^{(0)}_{\ d}\bar{\boldsymbol{\sigma}}$ and ${}^{(1)}_{\ d}\bar{\boldsymbol{\sigma}}$ as ${}_d\bar{\boldsymbol{\sigma}}^{(0)}$, ${}_d\bar{\boldsymbol{\sigma}}^{(1)}$ or ${}_d\bar{\boldsymbol{\sigma}}_{(0)}$, ${}_d\bar{\boldsymbol{\sigma}}_{(1)}$, (13.45) will result in what are commonly referred to as upper convected and lower convected constitutive models.

(2) Material coefficients λ, η, κ can be functions of the combined invariants of ${}^{(0)}_{\ d}\bar{\boldsymbol{\sigma}}$ and ${}^{(1)}\boldsymbol{\gamma}$, but in a known configuration $\underline{\Omega}$.

(3) We remark that choice of $\left(-\frac{c_1}{a_1}\right) = \frac{\lambda}{\eta}\alpha$, makes this material coefficient dependent on other two, λ and η. This is not permissible in continuum mechanics. All material coefficients must be independent material properties, but (13.42) is used almost universally in polymer science published works.

(4) In the derivation of the constitutive theory presented here, the polymeric fluid has been treated as isotropic and homogeneous fluent continua. Hence η, κ, $\underset{\sim}{\beta}$ are unique properties of the synthesized fluid (using solvent and polymer). Even though η, κ, $\underset{\sim}{\beta}$ of the polymeric fluid most certainly depend on the properties of the solvent and the polymer, determination of η, κ, $\underset{\sim}{\beta}$ of the polymeric fluid using the corresponding properties of the solvent and the polymer is not trivial [78] and may not even be possible. One must determine η, κ, $\underset{\sim}{\beta}$ for the polymeric fluid (treating it homogeneous and isotropic) experimentally.

13.5.4 Discussion on the Giesekus constitutive model derived here and the constitutive model used currently

We note that the entropy inequality requires decomposition of the Cauchy stress tensor (in contra- or co-variant basis or Jaumann stress) into equilibrium stress and deviatoric stress. The constitutive theory for the equilibrium stress using entropy inequality results in thermodynamic pressure $\bar{p}(\bar{\rho},\bar{\theta})$ for compressible thermoviscoelastic fluids and mechanical pressure $\bar{p}(\bar{\theta})$ for the incompressible case. Since the entropy inequality only requires the conversion of mechanical energy due to the deviatoric Cauchy stress to be positive but provides no mechanism for establishing the constitutive theory for it, representation theorem is used for deriving the constitutive theory for it. The use of the deviatoric Cauchy stress tensor in the Giesekus constitutive model derived here is necessitated due to the conjugate pair ${}^{(0)}_{\ d}\bar{\boldsymbol{\sigma}} : \bar{\boldsymbol{D}}$ in the entropy inequality. In the currently used Giesekus constitutive model for the stress tensor, the deviatoric Cauchy stress is further decomposed into deviatoric solvent stress and deviatoric polymer stress.

$$ {}^{(0)}_{\ d}\bar{\boldsymbol{\sigma}} = ({}^{(0)}_{\ d}\bar{\boldsymbol{\sigma}})_s + ({}^{(0)}_{\ d}\bar{\boldsymbol{\sigma}})_p \tag{13.46}$$

in which s and p stand for solvent and polymer. The currently used Giesekus constitutive model contains exactly the same form as presented here but uses $({}^{(0)}_{\ d}\bar{\boldsymbol{\sigma}})_p$ instead of ${}^{(0)}_{\ d}\bar{\boldsymbol{\sigma}}$ and is derived using Brownian motion of polymer molecules and kinetic theory [21, 35]. For the solvent stress $({}^{(0)}_{\;d}\bar{\boldsymbol{\sigma}})_s$, Newton's law of viscosity is assumed as a constitutive theory. We note the following:

(1) Based on Remark (3), decomposition in (13.46) is not possible as it implies heterogeneity which cannot be supported by CCM.
(2) If we use the decomposition shown above and substitute it in the conditions resulting from the entropy inequality we still have the same restriction that the conversion of mechanical energy due to both solvent and polymer deviatoric Cauchy stress tensors be positive, but we have no mechanism for deriving constitutive theories for either one of them.
(3) If we derive the Giesekus constitutive model based on representation theorem and $({}^{(0)}_{\;d}\bar{\boldsymbol{\sigma}})_p$ as a constitutive variable and if we assume Newton's law of viscosity for $({}^{(0)}_{\;d}\bar{\boldsymbol{\sigma}})_s$, then of course we would obtain exactly the same Giesekus constitutive model as used currently. The question is: "Is this permissible within the framework of the axioms of the constitutive theory and principles of continuum mechanics?".
(4) Based on the axioms of the constitutive theory and the conjugate pairs in the entropy inequality, ${}^{(0)}_{\;d}\bar{\boldsymbol{\sigma}}$ is a variable in the rate constitutive theory for thermoviscoelastic fluids, and hence, must be used as a constitutive variable in the derivation of the rate theory.
(5) If we follow (3), i.e., if we use $({}^{(0)}_{\;d}\bar{\boldsymbol{\sigma}})_p$ as a dependent variable in the rate theory for the Giesekus constitutive model, then the constitutive theory for $({}^{(0)}_{\;d}\bar{\boldsymbol{\sigma}})_s$ must be derivable as well from the entropy inequality (and not assuming Newton's law of viscosity for it). This is obviously not possible.
(6) Thus, based on the work presented here, we conclude that the use of the deviatoric Cauchy stress tensor as a constitutive tensor is necessary in the derivation of the Giesekus constitutive model. This is consistent with the conditions resulting from the entropy inequality and the axioms of the constitutive theory based on continuum mechanics. Furthermore, there is no justification for the decomposition (13.46) for isotropic, homogeneous matter. The use of Newton's law of viscosity for deviatoric solvent stress may be a good engineering assumption but is not supported by CCM.
(7) It is rather obvious that the use of the Giesekus constitutive model presented in this chapter and that used currently in conjunction with the conservation laws for deforming thermoviscoelastic fluids will undoubtedly produce different behaviors.

13.5.5 Simplified constitutive theory for deviatoric Cauchy stress tensor ${}_{d}\bar{\boldsymbol{\sigma}}^{(0)}$: $m = 1$, $n = 2$

The motivation for presenting this ordered rate constitutive theory is due to the fact that Oldroyd-B model is a simplified subset of this constitutive model. For $m = 1$ and $n = 2$, we have

$$ {}^{(1)}_{\;d}\bar{\boldsymbol{\sigma}} = {}^{(1)}_{\;d}\bar{\boldsymbol{\sigma}}(\bar{\rho}, {}^{(0)}_{\;d}\bar{\boldsymbol{\sigma}}, {}^{(1)}\boldsymbol{\gamma}, {}^{(2)}\boldsymbol{\gamma}, \bar{\theta}) \tag{13.47} $$

We consider representation theorem in deriving the constitutive theory for ${}^{(1)}_{\;d}\bar{\boldsymbol{\sigma}}$. Following Appendix A, we have 15 combined generators ($N = 15$) and 22 combined invariants ($M = 22$). Their count and combination of argument tensors is shown in Tables 13.3 and 13.4.

Table 13.3: Number of generators ($N = 15$)

One at a time	${}^{(0)}_{\;d}\bar{\boldsymbol{\sigma}}$	${}^{(1)}\boldsymbol{\gamma}$	${}^{(2)}\boldsymbol{\gamma}$
	2	2	2
Two at a time	${}^{(0)}_{\;d}\bar{\boldsymbol{\sigma}}, {}^{(1)}\boldsymbol{\gamma}$	${}^{(1)}\boldsymbol{\gamma}, {}^{(2)}\boldsymbol{\gamma}$	${}^{(2)}\boldsymbol{\gamma}, {}^{(0)}_{\;d}\bar{\boldsymbol{\sigma}}$
	3	3	3
Three at a time	${}^{(0)}_{\;d}\bar{\boldsymbol{\sigma}}, {}^{(1)}\boldsymbol{\gamma}, {}^{(2)}\boldsymbol{\gamma}$		
	0		

Table 13.4: Number of invariants ($M = 22$)

One at a time	${}^{(0)}_{\;d}\bar{\boldsymbol{\sigma}}$	${}^{(1)}\boldsymbol{\gamma}$	${}^{(2)}\boldsymbol{\gamma}$
	3	3	3
Two at a time	${}^{(0)}_{\;d}\bar{\boldsymbol{\sigma}}, {}^{(1)}\boldsymbol{\gamma}$	${}^{(1)}\boldsymbol{\gamma}, {}^{(2)}\boldsymbol{\gamma}$	${}^{(2)}\boldsymbol{\gamma}, {}^{(0)}_{\;d}\bar{\boldsymbol{\sigma}}$
	4	4	4
Three at a time	${}^{(0)}_{\;d}\bar{\boldsymbol{\sigma}}, {}^{(1)}\boldsymbol{\gamma}, {}^{(2)}\boldsymbol{\gamma}$		
	1		

Using $N = 15$ and $M = 22$ in (13.21) we have a constitutive theory for ${}^{(1)}_{\;d}\bar{\boldsymbol{\sigma}}$ based on integrity that requires $(15 + 22 \times 15 + 22)$ material coefficients. Oldroyd-B constitutive model is a simplified subset of the constitutive theory for $m = 1$ and $n = 2$ based on integrity.

13.5.6 Oldroyd-B constitutive model ($m = 1$, $n = 2$)

The Oldroyd-B constitutive model is referred to as quasi-linear constitutive model believed to be suitable for dilute polymeric fluids. This constitutive model uses generators ${}^{(0)}_{\;d}\bar{\boldsymbol{\sigma}}$, ${}^{(1)}\boldsymbol{\gamma}$ and ${}^{(2)}\boldsymbol{\gamma}$, thus, we can write the following in the current configuration.

$$ {}^{(1)}_{\;d}\bar{\boldsymbol{\sigma}} = {}^{\sigma}\underset{\sim}{\alpha}^{0}\boldsymbol{I} + {}^{\sigma}\underset{\sim}{\alpha}^{1}({}^{(0)}_{\;d}\bar{\boldsymbol{\sigma}}) + {}^{\sigma}\underset{\sim}{\alpha}^{2}({}^{(1)}\boldsymbol{\gamma}) + {}^{\sigma}\underset{\sim}{\alpha}^{3}({}^{(2)}\boldsymbol{\gamma}) \tag{13.48} $$

Based on the currently used Oldroyd-B model, we only consider invariants $\operatorname{tr}({}^{(0)}_{\ d}\bar{\boldsymbol{\sigma}})$ and $\operatorname{tr}({}^{(1)}\boldsymbol{\gamma})$ and neglect all product terms between the generators and the invariants, giving us the following constitutive theory

$$ {}^{(1)}_{\ d}\bar{\boldsymbol{\sigma}} = a_0\boldsymbol{I} + a_1({}^{(0)}_{\ d}\bar{\boldsymbol{\sigma}}) + a_2({}^{(1)}\boldsymbol{\gamma}) + d_1({}^{(2)}\boldsymbol{\gamma}) + b_2\operatorname{tr}({}^{(0)}_{\ d}\bar{\boldsymbol{\sigma}})\boldsymbol{I} + b_1\operatorname{tr}({}^{(1)}\boldsymbol{\gamma})\boldsymbol{I} \tag{13.49} $$

We define material coefficients to conform to commonly used notations in polymer science. Dividing (13.49) by $-a_1$ $(a_1 \neq 0)$ and rearranging terms.

$$ \begin{aligned} {}^{(0)}_{\ d}\bar{\boldsymbol{\sigma}} + \left(-\frac{1}{a_1}\right){}^{(1)}_{\ d}\bar{\boldsymbol{\sigma}} &= \left(-\frac{a_0}{a_1}\right)\boldsymbol{I} + \left(-\frac{a_2}{a_1}\right){}^{(1)}\boldsymbol{\gamma} + \left(-\frac{d_1}{a_1}\right){}^{(2)}\boldsymbol{\gamma} \\ &\quad + \left(-\frac{b_2}{a_1}\right)\operatorname{tr}({}^{(0)}_{\ d}\bar{\boldsymbol{\sigma}})\boldsymbol{I} + \left(-\frac{b_1}{a_1}\right)\operatorname{tr}({}^{(1)}\boldsymbol{\gamma})\boldsymbol{I} \end{aligned} \tag{13.50} $$

We define the following material coefficients

$$ \begin{aligned} &\left(-\frac{1}{a_1}\right) = \lambda_1 \quad \text{or} \quad \lambda_1\big|_{\underline{\Omega}} \quad ; \quad \left(-\frac{a_0}{a_1}\right) = \underset{\sim}{\sigma}^0 \quad \text{or} \quad \underset{\sim}{\sigma}^0\big|_{\underline{\Omega}} \\ &\left(-\frac{a_2}{a_1}\right) = 2\eta \quad \text{or} \quad 2\eta\big|_{\underline{\Omega}} \quad ; \quad \left(-\frac{b_2}{a_1}\right) = \underset{\sim}{\beta} \quad \text{or} \quad \underset{\sim}{\beta}\big|_{\underline{\Omega}} \\ &\left(-\frac{b_1}{a_1}\right) = \kappa \quad \text{or} \quad \kappa\big|_{\underline{\Omega}} \end{aligned} \tag{13.51} $$

Using (13.51) in (13.50) we obtain

$$ \begin{aligned} {}^{(0)}_{\ d}\bar{\boldsymbol{\sigma}} + \lambda_1({}^{(1)}_{\ d}\bar{\boldsymbol{\sigma}}) &= \underset{\sim}{\sigma}^0\big|_{\underline{\Omega}}\boldsymbol{I} + 2\eta({}^{(1)}\boldsymbol{\gamma}) + \left(-\frac{d_1}{a_1}\right){}^{(2)}\boldsymbol{\gamma} \\ &\quad + \underset{\sim}{\beta}\operatorname{tr}({}^{(0)}_{\ d}\bar{\boldsymbol{\sigma}})\boldsymbol{I} + \kappa\operatorname{tr}({}^{(1)}\boldsymbol{\gamma})\boldsymbol{I} \end{aligned} \tag{13.52} $$

We note that ${}^{(1)}\boldsymbol{\gamma}$ and ${}^{(2)}\boldsymbol{\gamma}$ are first and second convected time derivatives of the strain tensor (in a chosen basis). Thus, dimensionally (or in terms of units), if we multiply ${}^{(2)}\boldsymbol{\gamma}$ by time we obtain units of ${}^{(1)}\boldsymbol{\gamma}$. Since ${}^{(1)}\boldsymbol{\gamma}$ is already multiplied with 2η, ${}^{(2)}\boldsymbol{\gamma}$ can multiplied with $2\eta\lambda_2$ where λ_2 is a time constant. Therefore, we can choose

$$ \left(-\frac{d_1}{a_1}\right) = 2\eta\lambda_2 \tag{13.53} $$

$$ \begin{aligned} {}^{(0)}_{\ d}\bar{\boldsymbol{\sigma}} + \lambda_1({}^{(1)}_{\ d}\bar{\boldsymbol{\sigma}}) &= \underset{\sim}{\sigma}^0\big|_{\underline{\Omega}}\boldsymbol{I} + 2\eta({}^{(1)}\boldsymbol{\gamma}) + 2\eta\lambda_2({}^{(2)}\boldsymbol{\gamma}) \\ &\quad + \underset{\sim}{\beta}\operatorname{tr}({}^{(0)}_{\ d}\bar{\boldsymbol{\sigma}})\boldsymbol{I} + \kappa\operatorname{tr}({}^{(1)}\boldsymbol{\gamma})\boldsymbol{I} \end{aligned} \tag{13.54} $$

In currently used Oldroyd-B model, $\underset{\sim}{\sigma}^0\boldsymbol{I}$ and $\underset{\sim}{\beta}\operatorname{tr}({}^{(0)}_{\ d}\bar{\boldsymbol{\sigma}})$ terms are neglected. Thus, we now have

$$ {}^{(0)}_{\ d}\bar{\boldsymbol{\sigma}} + \lambda_1({}^{(1)}_{\ d}\bar{\boldsymbol{\sigma}}) = 2\eta({}^{(1)}\boldsymbol{\gamma}) + 2\eta\lambda_2({}^{(2)}\boldsymbol{\gamma}) + \kappa\operatorname{tr}({}^{(1)}\boldsymbol{\gamma})\boldsymbol{I} \tag{13.55} $$

Equation (13.55) is the Oldroyd-B constitutive model for compressible dilute polymeric fluids. If the fluid is incompressible, $\operatorname{tr}({}^{(1)}\boldsymbol{\gamma}) = \operatorname{tr}\bar{\boldsymbol{D}} = 0$, and equation (13.55) reduces to

$$ {}^{(0)}_{\;d}\bar{\boldsymbol{\sigma}} + \lambda_1({}^{(1)}_{\;d}\bar{\boldsymbol{\sigma}}) = 2\eta({}^{(1)}\boldsymbol{\gamma}) + 2\eta\lambda_2{}^{(2)}\boldsymbol{\gamma} \tag{13.56} $$

Here, λ_1 is relaxation time, η is viscosity of the polymeric fluid, λ_2 is retardation time.

13.5.7 A single constitutive theory for dilute and dense polymeric fluids ($m = 1$, $n = 2$)

We have seen that Maxwell, Giesekus and Oldroyd-B constitutive models are a subset of the ordered rate constitutive theory of orders $m = 1$ and $n = 2$ that is based on integrity. Thus, we begin with

$$ {}^{(1)}_{\;d}\bar{\boldsymbol{\sigma}} = {}^{(1)}_{\;d}\bar{\boldsymbol{\sigma}}(\bar{\rho}, {}^{(0)}_{\;d}\bar{\boldsymbol{\sigma}}, {}^{(1)}\boldsymbol{\gamma}, {}^{(2)}\boldsymbol{\gamma}, \bar{\theta}) \tag{13.57} $$

By choosing ${}^{(0)}_{\;d}\bar{\boldsymbol{\sigma}}$, ${}^{(1)}\boldsymbol{\gamma}$, ${}^{(2)}\boldsymbol{\gamma}$ and $({}^{(0)}_{\;d}\bar{\boldsymbol{\sigma}})^2$ as combined generators and $\mathrm{tr}({}^{(1)}\boldsymbol{\gamma})$ as the only invariant and neglecting all product terms of generators and invariants we can derive the following constitutive theory

$$ {}^{(1)}_{\;d}\bar{\boldsymbol{\sigma}} = a_0\boldsymbol{I} + a_1({}^{(0)}_{\;d}\bar{\boldsymbol{\sigma}}) + a_2{}^{(1)}\boldsymbol{\gamma} + d_1({}^{(2)}\boldsymbol{\gamma}) + c_1({}^{(0)}_{\;d}\bar{\boldsymbol{\sigma}})^2 + b_1\,\mathrm{tr}({}^{(1)}\boldsymbol{\gamma})\boldsymbol{I} \tag{13.58} $$

dividing by $-a_1$ ($a_1 \neq 0$) and defining the new material coefficients

$$ \begin{aligned} &\left(-\frac{1}{a_1}\right) = \lambda \;\text{ or }\; \lambda\big|_{\underline{\Omega}} \;\; ; \;\; \left(-\frac{a_2}{a_1}\right) = 2\eta \;\text{ or }\; 2\eta\big|_{\underline{\Omega}} \\ &\left(-\frac{d_1}{a_1}\right) = 2\eta\lambda_2 \;\text{ or }\; 2\eta\big|_{\underline{\Omega}}\lambda_2\big|_{\underline{\Omega}} \;\; ; \;\; \left(-\frac{c_1}{a_1}\right) = \frac{\lambda}{\eta}\alpha \;\text{ or }\; \frac{\lambda\big|_{\underline{\Omega}}}{\eta\big|_{\underline{\Omega}}}\alpha \\ &\left(-\frac{b_1}{a_1}\right) = \kappa \;\text{ or }\; \kappa_{\underline{\Omega}} \;\; ; \;\; \left(-\frac{a_0}{a_1}\right) = \underset{\sim}{\sigma}^0 \;\text{ or }\; \underset{\sim}{\sigma}^0\big|_{\underline{\Omega}} \end{aligned} \tag{13.59} $$

using (13.59) in (13.58) we obtain (setting $\underset{\sim}{\sigma}^0\big|_{\underline{\Omega}}\boldsymbol{I}$ term to zero)

$$ {}^{(0)}_{\;d}\bar{\boldsymbol{\sigma}} + \lambda({}^{(1)}_{\;d}\bar{\boldsymbol{\sigma}}) = 2\eta({}^{(1)}\boldsymbol{\gamma}) + 2\eta\lambda_2({}^{(2)}\boldsymbol{\gamma}) + \frac{\lambda}{\eta}\alpha({}^{(0)}_{\;d}\bar{\boldsymbol{\sigma}})^2 + \kappa\,\mathrm{tr}({}^{(1)}\boldsymbol{\gamma}) \tag{13.60} $$

For compressible thermoviscoelastic polymeric fluid:

Maxwell Model

$$ \lambda_2 = 0 \quad , \quad \alpha = 0 \tag{13.61} $$

Giesekus Model

$$ \lambda_2 = 0 \tag{13.62} $$

Oldroyd-B Model

$$ \alpha = 0 \quad \text{and} \quad \lambda = \lambda_1 \tag{13.63} $$

For incompressible thermoviscoelastic polymeric fluid: (13.61)–(13.63) hold, but additionally $\mathrm{tr}({}^{(1)}\boldsymbol{\gamma}) = 0$ in (13.60).

13.6 Constitutive theory for heat vector $\bar{\boldsymbol{q}}$

The details of the constitutive theory for $\boldsymbol{q}$ and the resulting constitutive theories are exactly the same as those presented in Section 12.5 of Chapter 12, hence not repeated here for the sake of brevity.

13.7 Numerical studies using Giesekus constitutive model

In this section, we consider fully developed flow of an incompressible Giesekus fluid between parallel plates as a model problem [113].

We use contravariant Cauchy stress tensor and Almansi strain tensors as conjugate measures of the stress and strain tensors in Eulerian description. If we decompose the contravariant Cauchy stress tensor in equilibrium stress and deviatoric contravariant Cauchy stress tensor, then the equilibrium stress is mechanical pressure p and the deviatoric contravariant Cauchy stress tensor becomes a dependent variable in the constitutive theory. This yields the upper convected Giesekus (UCG1) constitutive model. Numerical results are presented using the upper convected Giesekus constitutive model derived in this chapter (UCG1) as well as using the currently used constitutive model in deviatoric polymer stress (UCG2).

Since the description is understood to be Eulerian, we drop the overbar $(\bar{\ })$ on all quantities for simplicity of notation and replace it with hat $(\hat{\ })$ to emphasize that these quantities have dimensions. Quantities without hat $(\hat{\ })$ are dimensionless. To conform to commonly used engineering notations we replace x_i ; $i = 1, 2, 3$ by x, y, z and $\bar{v}_i$; $i = 1, 2, 3$ by u, v, w and ${}^{(0)}_{\ d}\bar{\sigma}_{ij}$; $i, j = 1, 2, 3$ by τ_{ij} ; $i, j = 1, 2, 3$ in the mathematical model.

We consider an incompressible Giesekus fluid PIB/C14 [75] with the following material coefficients (assumed constant):

$$\hat{\rho} = 800 \text{ kg/m}^3 \quad ; \quad \hat{\eta}_s = 0.002 \text{ Pa s} \quad ; \quad \hat{\eta}_p = 1.424 \text{ Pa s}$$
$$\hat{\eta} = 1.426 \text{ Pa s} \quad ; \quad \hat{\lambda} = 0.06 \text{ s} \quad ; \quad \alpha = 0.15$$

In which $\hat{\rho}$, $\hat{\eta}_s$, $\hat{\eta}_p$, $\hat{\eta}$, $\hat{\lambda}$ and α are density, solvent viscosity, polymer viscosity, total viscosity, relaxation time and mobility factor.

For a fixed configuration and a given fluid we can study the influence of the constitutive models on the flow physics in at least two ways:

(i) For a fixed flow rate, the differences in the constitutive equation in the two models will produce different $\partial p/\partial x$ and other dependent variables in the two cases. As the flow rate increases, the differences in $\partial p/\partial x$ and the other dependent variables in the two cases are expected to increase as well.

(ii) In the second approach, we could choose a value of $\partial p/\partial x$ that is the same in both cases and compute results. Both models are bound to produce different velocity fields, and hence, different flow rates. For very low values of $\partial p/\partial x$ we expect the velocity field in the two cases to be not drastically different from each other, but as $\partial p/\partial x$ increases, the differences are expected to be significant.

Obviously, (ii) is easier as it merely requires specification of $\partial p/\partial x$ as input and the rest of the details of the flow are computed. We use this approach to study the influence of the two constitutive models (UCG1 and UCG2) on the flow physics of fully developed flow between parallel plates (model problem 1) and fully developed flow between parallel plates using a two-dimensional formulation (model problem 2). It is obvious that both model problems will be in agreement when the same constitutive model is used.

13.7.1 Model Problem 1: fully developed flow between parallel plates

Figure 13.1 shows a schematic using dimensionless quantities. The plates are separated by a distance $2H$. The origin of the xy-coordinate is located at the center of the plates and the positive x-direction is the direction of the flow. The flow is pressure driven, i.e., $\partial p/\partial x$ (negative) is specified. The mathematical model describing the flow physics (for incompressible case with isothermal flow assumption) consists of x- and y-momentum equations and the constitutive equations. The continuity equation in this case is satisfied identically.

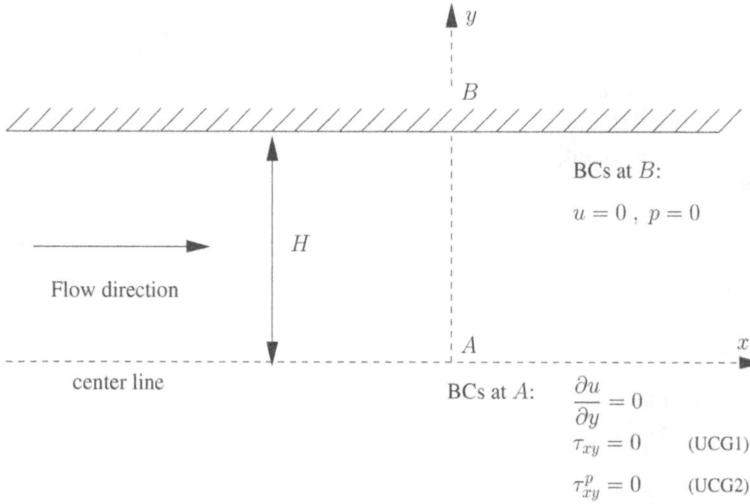

Figure 13.1: Schematic of 1D fully developed flow between parallel plates (half domain)

We begin with all quantities with their usual dimensions (units) in the development of the mathematical model and then non-dimensionalize them using the following. The quantities with the subscript zero are the reference quantities.

$$x=\frac{\hat{x}}{L_0}\;,\; y=\frac{\hat{y}}{L_0}\;,\; \eta=\frac{\hat{\eta}}{\eta_0}\;,\; \eta_s=\frac{\hat{\eta}_s}{\eta_0}\;,\; \eta_p=\frac{\hat{\eta}_p}{\eta_0}\;,\; \rho=\frac{\hat{\rho}}{\rho_0}\;,\; u=\frac{\hat{u}}{u_0}$$

$$v=\frac{\hat{v}}{u_0}\;,\; p=\frac{\hat{p}}{p_0}\;,\; \boldsymbol{\tau}=\frac{\hat{\boldsymbol{\tau}}}{\tau_0}\;,\; p_0=\tau_0=\begin{cases}\rho_0 u_0^2\,;\text{Ch. kinetic energy}\\ \quad\text{or}\\ \dfrac{\eta_0 u_0}{L_0}\,;\text{Ch. viscous stress}\end{cases} \tag{13.64}$$

In which $\hat{u}$, $\hat{v}$ are velocities in the x- and y-direction, $\hat{p}$ is mechanical pressure and $\hat{\boldsymbol{\tau}}$ is deviatoric stress tensor, all in the current configuration. We choose the larger of the two for p_0 (and τ_0). This results in the dimensionless form of the mathematical model given in the following.

Momentum equations

In the absence of body forces

$$\left(\frac{p_0}{\rho_0 u_0^2}\right)\frac{\partial p}{\partial x} - \left(\frac{\tau_0}{\rho_0 u_0^2}\right)\frac{\partial \tau_{xy}}{\partial y} = 0 \tag{13.65}$$

$$\left(\frac{p_0}{\rho_0 u_0^2}\right)\frac{\partial p}{\partial y} - \left(\frac{\tau_0}{\rho_0 u_0^2}\right)\frac{\partial \tau_{yy}}{\partial y} = 0 \tag{13.66}$$

Giesekus constitutive model

We consider the upper convected Giesekus constitutive model derived in this chapter (UCG1) and the upper convected Giesekus constitutive model used currently (UCG2).

UCG1

In this model, the first convected time derivative of $\boldsymbol{\tau}$, the deviatoric contravariant Cauchy stress tensor, is a dependent variable in the constitutive theory. Dimensionless form of the constitutive model is given by

$$\begin{aligned}
&\tau_{xx} - 2De\tau_{xy}\frac{\partial u}{\partial y} - \alpha\frac{De}{\eta}\left(\frac{L_0\tau_0}{u_0\eta_0}\right)\left((\tau_{xx})^2 + (\tau_{xy})^2\right) = 0\\
&\tau_{yy} - \alpha\frac{De}{\eta}\left(\frac{L_0\tau_0}{u_0\eta_0}\right)\left((\tau_{yy})^2 + (\tau_{xy})^2\right) = 0\\
&\tau_{xy} - De\tau_{yy}\frac{\partial u}{\partial y} - \alpha\frac{De}{\eta}\left(\frac{L_0\tau_0}{u_0\eta_0}\right)\tau_{xy}(\tau_{xx} + \tau_{yy}) = \eta\left(\frac{u_0\eta_0}{L_0\tau_0}\right)\frac{\partial u}{\partial y}
\end{aligned} \tag{13.67}$$

Equations (13.65)–(13.67) constitute the complete mathematical model in dependent variables u, p, τ_{xx}, τ_{yy} and τ_{xy} for fully developed flow between parallel plates when using UCG1.

UCG2

This constitutive model is used currently [20]. In this model $\boldsymbol{\tau}$ is decomposed into solvent and polymer stresses.

$$\boldsymbol{\tau} = \boldsymbol{\tau}^s + \boldsymbol{\tau}^p \tag{13.68}$$

Newton's law of viscosity is assumed as a constitutive model for $\boldsymbol{\tau}^s$. Stresses τ_{xx}^s and τ_{yy}^s are zero for this model problem and we only have τ_{xy}^s in the constitutive model for solvent stress.

$$\tau_{xy}^s = \left(\frac{u_0\eta_0}{L_0\tau_0}\right)\eta_s\frac{\partial u}{\partial y} \tag{13.69}$$

and hence, from (13.68)

$$\tau_{xx} = \tau_{xx}^p \quad ; \quad \tau_{yy} = \tau_{yy}^p \quad ; \quad \tau_{xy} = \tau_{xy}^p + \left(\frac{u_0\eta_0}{L_0\tau_0}\right)\eta_s\frac{\partial u}{\partial y} \tag{13.70}$$

For polymer stress $\boldsymbol{\tau}^p$, the dimensionless form of the constitutive equations are given by (obtained by replacing $\boldsymbol{\tau}$ with $\boldsymbol{\tau}^p$ and η by η_p in (13.67)) the following [20]:

$$
\begin{aligned}
&\tau^p_{xx} - 2De\tau^p_{xy}\frac{\partial u}{\partial y} - \alpha\frac{De}{\eta_p}\Big(\frac{L_0\tau_0}{u_0\eta_0}\Big)\big((\tau^p_{xx})^2 + (\tau^p_{xy})^2\big) = 0\\
&\tau^p_{yy} - \alpha\frac{De}{\eta_p}\Big(\frac{L_0\tau_0}{u_0\eta_0}\Big)\big((\tau^p_{yy})^2 + (\tau^p_{xy})^2\big) = 0\\
&\tau^p_{xy} - De\tau^p_{yy}\frac{\partial u}{\partial y} - \alpha\frac{De}{\eta_p}\Big(\frac{L_0\tau_0}{u_0\eta_0}\Big)\tau^p_{xy}\big(\tau^p_{xx} + \tau^p_{yy}\big) = \eta_p\Big(\frac{u_0\eta_0}{L_0\tau_0}\Big)\frac{\partial u}{\partial y}
\end{aligned}
\tag{13.71}
$$

Using (13.70) in the momentum equations (13.65) and (13.66), we can express the momentum equations in terms of τ^p_{yy}, τ^p_{xy} and velocity gradients.

$$
\Big(\frac{p_0}{\rho_0 u_0^2}\Big)\frac{\partial p}{\partial x} - \Big(\frac{\tau_0}{\rho_0 u_0^2}\Big)\frac{\partial \tau^p_{xy}}{\partial y} - \Big(\frac{\eta_0}{L_0\rho_0 u_0}\Big)\eta_s\frac{\partial^2 u}{\partial y^2} = 0 \tag{13.72}
$$

$$
\Big(\frac{p_0}{\rho_0 u_0^2}\Big)\frac{\partial p}{\partial y} - \Big(\frac{\tau_0}{\rho_0 u_0^2}\Big)\frac{\partial \tau^p_{yy}}{\partial y} = 0 \tag{13.73}
$$

Equations (13.71)–(13.73) constitute the complete mathematical model in dependent variables u, p, τ^p_{xx}, τ^p_{yy} and τ^p_{xy} for fully developed flow between parallel plates when using UCG2.

Solutions of the BVPs

In this section, we consider solutions of the BVPs described by (13.65)–(13.67) for UCG1 and (13.71)–(13.73) for UCG2. Since $\partial p/\partial x$ is constant (specified), from (13.65) we can determine τ_{xy} by integrating with respect to y and using the boundary condition $\tau_{xy} = 0$ at $y = 0$ (due to symmetry)

$$
\tau_{xy} = \Big(\frac{\partial p}{\partial x}\Big)y \tag{13.74}
$$

A theoretical solution for the remaining dependent variables is not readily possible due to the complexity of the constitutive equations in both boundary value problems (UCG1 and UCG2). Hence, we consider their numerical solutions using finite element processes based on the residual functional (least squares finite element method). The local approximations are considered in higher order spaces $H^{k,p}(\bar{\Omega}^e_x)$ in which $\bar{\Omega}^e_x$ is the spatial domain of a typical element e of the discretization. The resulting non-linear algebraic equations from the least squares process are solved using Newton's linear method. The computational processes in this approach are unconditionally stable and permit higher order global differentiability local approximations. See references [16,17,140] for details of local approximations and the least squares process for non-linear PDEs and higher order spaces. In the computations of the numerical solutions we choose

$$
\hat{H}=L_0=3.175 \text{ mm} \; ; \; \rho_0=\hat{\rho}=800 \text{ kg/m}^3 \; ; \; \eta_0=\hat{\eta}=1.426 \text{ Pa s} \; ; \; u_0=0.5 \text{ m/s}
$$

which gives

$$
H = 1 \quad ; \quad p_0 = \tau_0 = \rho_0 u_0^2 = 200 \text{ Pa} \quad ; \quad Re = \frac{\rho_0 L_0 u_0}{\eta_0} = 0.8906
$$

$$
De = \frac{\hat{\lambda} u_0}{L_0} = 9.45 \quad \text{or} \quad De = \frac{\hat{\lambda} u_{max}}{L_0} = 18.89764
$$

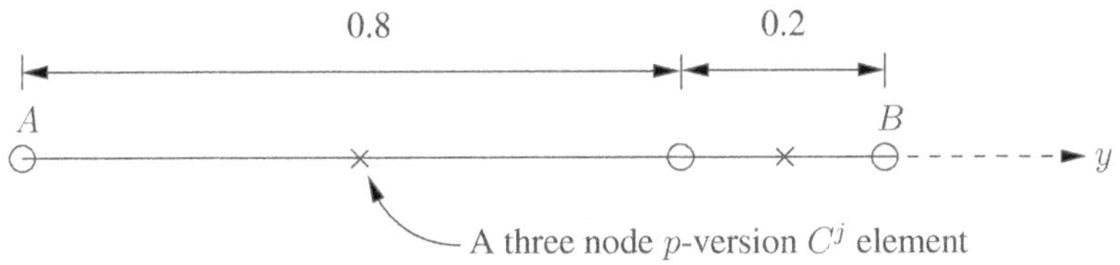

Figure 13.2: Graded mesh discretization using two 3-node p-version elements

A good discretization of the spatial domain $0 \leq y \leq 1$ is important in ensuring satisfactory convergence of Newton's linear method for the system of non-linear algebraic equations and good accuracy of computed solutions. With progressively increasing $\partial p/\partial x$, we expect development of a constant velocity core at the center of the flow. This suggests the use of a highly biased finer discretization toward the walls. A two element graded mesh with element length of 0.2 and 0.8 starting from the wall (see Figure 13.2) is used with local approximations that are p-version (3-node elements) in higher order spaces.

Initial p-convergence studies with this discretization suggest $p = 9$ with $k = 2$, local approximations of class $C^1(\bar{\Omega}_x^e)$, to be sufficient for good accuracy of results. For this choice of mesh, p-level ($p = 9$) and order of the space ($k = 2$), the residual or least squares functional values remain $O(10^{-8})$–$O(10^{-20})$ indicating that the PDEs are satisfied very accurately (in the pointwise sense for UCG1 as the integrals are Riemann, and not strictly in the pointwise sense for UCG2 since the integrals are Lebesgue) when local approximations for u, p, τ_{xx}, τ_{yy} and τ_{xy} are of class $C^1(\bar{\Omega}_x^e)$. Newton's linear method is used for solving the non-linear algebraic equations converges in less than 10 iterations for all numerical studies presented here. In the numerical studies, we begin with $\partial p/\partial x = -0.1$ for which a converged solution is obtained and then progressively increase it up to $\partial p/\partial x = -0.275$ using a continuation procedure in which converged solutions at lower $\partial p/\partial x$ are used as initial (or starting) solution in the Newton's linear method.

Figure 13.3 shows graphs of velocity u versus y for different values of $\partial p/\partial x$ for both UCG1 and UCG2. Graphs of velocity gradient $\partial u/\partial y$ versus y for different values of $\partial p/\partial x$ are shown in Figure 13.4.

For $\partial p/\partial x$ values up to -0.2, the results from both UCG1 and UCG2 are in good agreement (Figures 13.3 and 13.4). Beyond $\partial p/\partial x$ values of -0.2, the results from the two BVPs begin to deviate. Higher values of $\partial p/\partial x$ result in larger deviations between the two models. At $\partial p/\partial x = -0.275$, u_{max} at $y = 0$ from UCG1 is more than twice u_{max} at $y = 0$ from UCG2.

This of course implies drastically different flow rates resulting from the two models for the same pressure gradient. Figures 13.5–13.7 show plots of τ_{xx}, τ_{yy} and τ_{xy} versus y for both UCG1 and UCG2. For $\partial p/\partial x$ values beyond -0.2, we observe progressively increasing deviations between solutions obtained from the two BVPs for τ_{xx} and τ_{yy}. For $\partial p/\partial x = -0.275$, τ_{xx} and τ_{yy} from UCG1 are roughly more than twice those from UCG2. Computed τ_{xy} from the numerical solutions of both BVPs are in perfect agreement with the theoretical solution (13.74) for all values of $\partial p/\partial x$ as τ_{xy} only depends on $\partial p/\partial x$, which is the same in both models.

The residual (I) values of $O(10^{-8})$ or lower and the use of $C^1(\bar{\Omega}_x^e)$ ensure that the computed solutions satisfy the GDEs accurately.

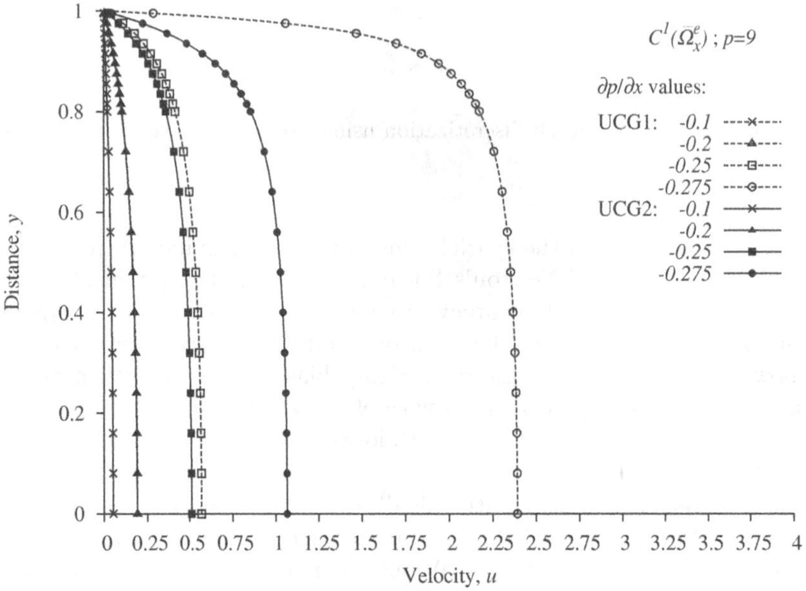

Figure 13.3: Velocity u versus distance y

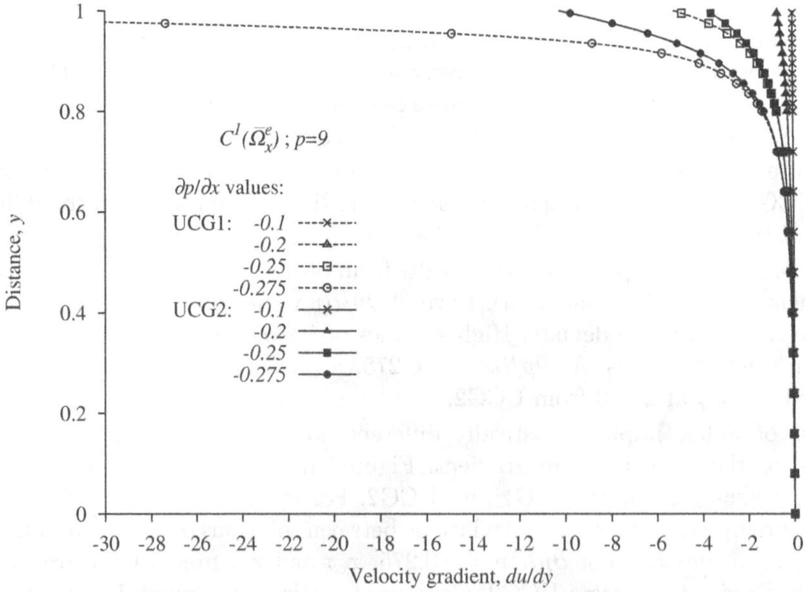

Figure 13.4: Velocity gradient du/dy versus distance y

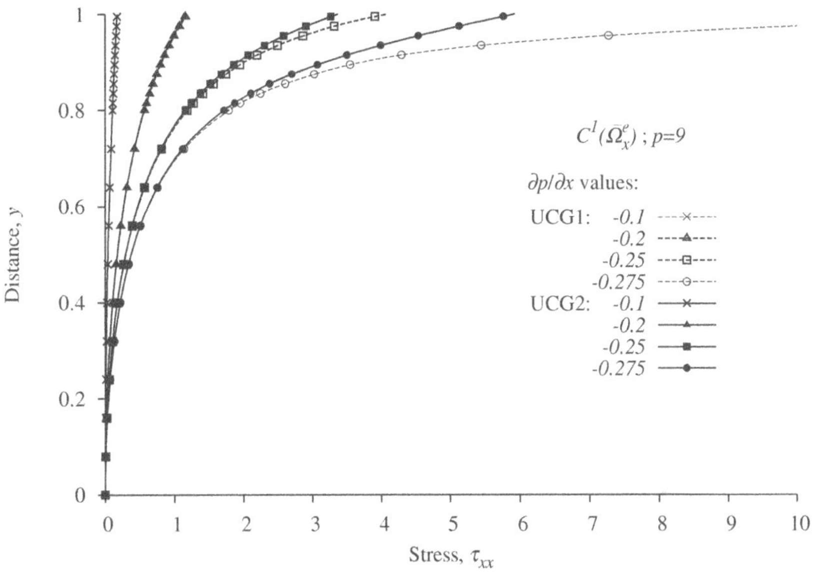

Figure 13.5: Stress component τ_{xx} versus distance y

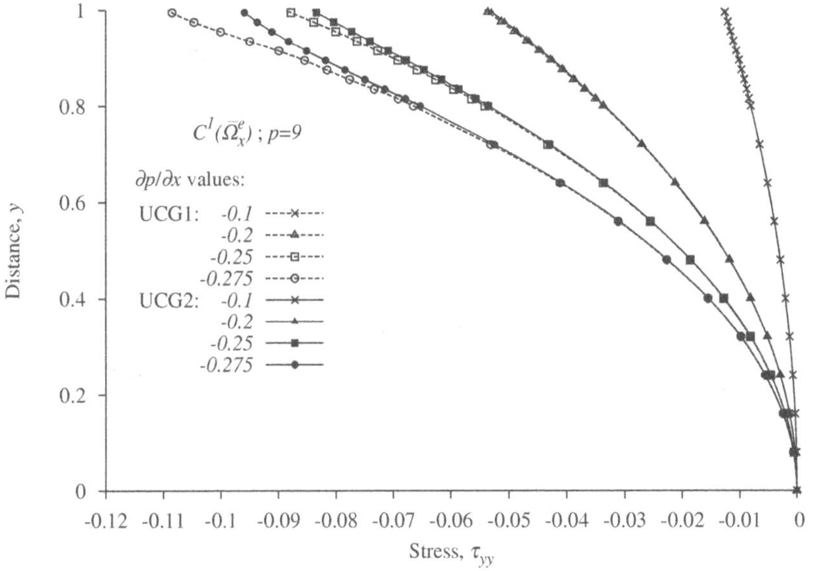

Figure 13.6: Stress component τ_{yy} versus distance y

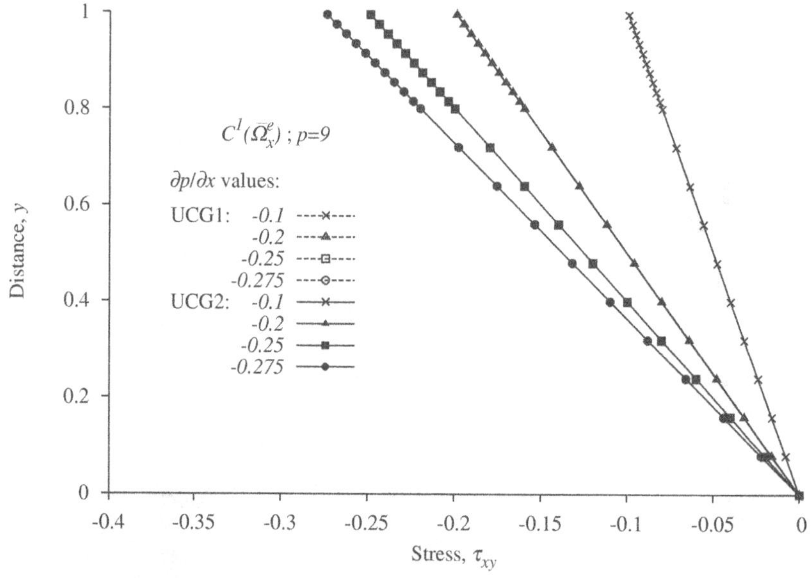

Figure 13.7: Stress component τ_{xy} versus distance y

13.7.2 Model Problem 2: fully developed flow between parallel plates using 2D formulation

In this numerical study we consider the same model problem as considered for Model Problem 1, i.e., fully developed flow between parallel plates but we use a 2D mathematical model. The purpose of this study is to show performance of the full mathematical model and to demonstrate that for fully developed flow between parallel plates. This full model produces precisely the same results as the degenerate model used in Section 13.7.1. Figure 13.8 shows a schematic using dimensionless quantities in which $ABCD$ is the computational domain. Origin of the coordinate system x, y is located at A. Positive x-direction is the direction of the flow.

In this case, the mathematical model describing the flow physics (for the incompressible case with isothermal flow assumption) consists of the continuity equation, x- and y-momentum equations and the constitutive equations. We begin with all quantities with their usual dimensions (units) in the development of the mathematical model and then non-dimensionalize them using (13.64) in which $\hat{u}$, $\hat{v}$ are velocities in the x- and y-direction, $\hat{p}$ is mechanical pressure and $\hat{\boldsymbol{\tau}}$ is deviatoric stress tensor, all in the current configuration. We choose the larger of the two for p_0 (and τ_0). This results in the following dimensionless form of the mathematical model:

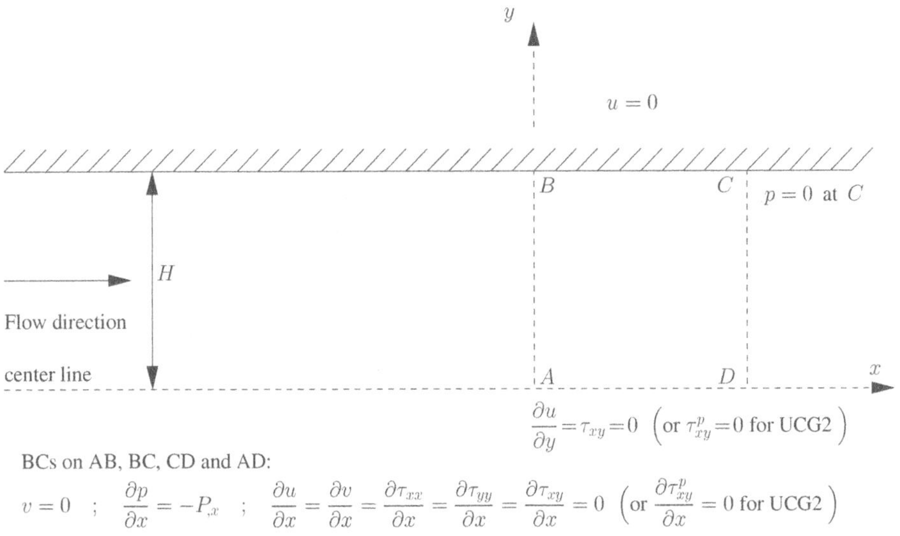

Figure 13.8: Schematic of 2D fully developed flow between parallel plates (half domain)

Continuity equation

$$\rho\Big(\frac{\partial u}{\partial x} + \frac{\partial v}{\partial y}\Big) = 0 \tag{13.75}$$

Momentum equations

In the absence of body forces

$$\begin{aligned}
&\rho\Big(u\frac{\partial u}{\partial x} + v\frac{\partial u}{\partial y}\Big) + \Big(\frac{p_0}{\rho_0 u_0^2}\Big)\frac{\partial p}{\partial x} - \Big(\frac{\tau_0}{\rho_0 u_0^2}\Big)\Big(\frac{\partial \tau_{xx}}{\partial x} + \frac{\partial \tau_{xy}}{\partial y}\Big) = 0\\
&\rho\Big(u\frac{\partial v}{\partial x} + v\frac{\partial v}{\partial y}\Big) + \Big(\frac{p_0}{\rho_0 u_0^2}\Big)\frac{\partial p}{\partial y} - \Big(\frac{\tau_0}{\rho_0 u_0^2}\Big)\Big(\frac{\partial \tau_{xy}}{\partial x} + \frac{\partial \tau_{yy}}{\partial y}\Big) = 0
\end{aligned} \tag{13.76}$$

Giesekus constitutive model

We consider the upper convected Giesekus constitutive model derived in this chapter (UCG1) and the upper convected Giesekus constitutive model used currently (UCG2).

UCG1

In this model, the first convected time derivative of $\boldsymbol{\tau}$, the deviatoric contravariant Cauchy stress tensor, is a dependent variable in the constitutive theory. Dimensionless form of the constitutive model is given by

$$
\begin{aligned}
&\tau_{xx}+De\Big(u\frac{\partial \tau_{xx}}{\partial x}+v\frac{\partial \tau_{xx}}{\partial y}-2\tau_{xy}\frac{\partial u}{\partial y}-2\tau_{xx}\frac{\partial u}{\partial x}\Big)\\
&\qquad-\alpha\frac{De}{\eta}\Big(\frac{L_0\tau_0}{u_0\eta_0}\Big)\big((\tau_{xx})^2+(\tau_{xy})^2\big)=2\eta\Big(\frac{u_0\eta_0}{L_0\tau_0}\Big)\frac{\partial u}{\partial x}\\
&\tau_{yy}+De\Big(u\frac{\partial \tau_{yy}}{\partial x}+v\frac{\partial \tau_{yy}}{\partial y}-2\tau_{xy}\frac{\partial v}{\partial x}-2\tau_{yy}\frac{\partial v}{\partial y}\Big)\\
&\qquad-\alpha\frac{De}{\eta}\Big(\frac{L_0\tau_0}{u_0\eta_0}\Big)\big((\tau_{yy})^2+(\tau_{xy})^2\big)=2\eta\Big(\frac{u_0\eta_0}{L_0\tau_0}\Big)\frac{\partial v}{\partial y}\\
&\tau_{xy}+De\Big(u\frac{\partial \tau_{xy}}{\partial x}+v\frac{\partial \tau_{xy}}{\partial y}-\tau_{xy}\big(\frac{\partial u}{\partial x}+\frac{\partial v}{\partial y}\big)-\tau_{xx}\frac{\partial v}{\partial x}-\tau_{yy}\frac{\partial u}{\partial y}\Big)\\
&\qquad-\alpha\frac{De}{\eta}\Big(\frac{L_0\tau_0}{u_0\eta_0}\Big)\tau_{xy}(\tau_{xx}+\tau_{yy})=\eta\Big(\frac{u_0\eta_0}{L_0\tau_0}\Big)\Big(\frac{\partial u}{\partial y}+\frac{\partial v}{\partial x}\Big)
\end{aligned}
\tag{13.77}
$$

Equations (13.75)–(13.77) constitute the complete mathematical model in dependent variables u, v, p, τ_{xx}, τ_{yy} and τ_{xy} for two-dimensional steady flow using the constitutive model UCG1.

UCG2

This constitutive model is used currently [20]. As in Model Problem 1, here also, $\boldsymbol{\tau}$ is decomposed into solvent and polymer stresses.

$$
\boldsymbol{\tau}=\boldsymbol{\tau}^s+\boldsymbol{\tau}^p \tag{13.78}
$$

and Newton's law of viscosity is assumed as a constitutive theory for $\boldsymbol{\tau}^s$.

$$
\begin{aligned}
&\tau^s_{xx}=2\Big(\frac{u_0\eta_0}{L_0\tau_0}\Big)\eta_s\frac{\partial u}{\partial x}\quad;\quad \tau^s_{yy}=2\Big(\frac{u_0\eta_0}{L_0\tau_0}\Big)\eta_s\frac{\partial v}{\partial y}\\
&\tau^s_{xy}=\Big(\frac{u_0\eta_0}{L_0\tau_0}\Big)\eta_s\Big(\frac{\partial u}{\partial y}+\frac{\partial v}{\partial x}\Big)
\end{aligned}
\tag{13.79}
$$

Hence, from (13.78),

$$
\begin{aligned}
&\tau_{xx}=\tau^p_{xx}+2\Big(\frac{u_0\eta_0}{L_0\tau_0}\Big)\eta_s\frac{\partial u}{\partial x}\quad;\quad \tau_{yy}=\tau^p_{yy}+2\Big(\frac{u_0\eta_0}{L_0\tau_0}\Big)\eta_s\frac{\partial v}{\partial y}\\
&\tau_{xy}=\tau^p_{xy}+\Big(\frac{u_0\eta_0}{L_0\tau_0}\Big)\eta_s\Big(\frac{\partial u}{\partial y}+\frac{\partial v}{\partial x}\Big)
\end{aligned}
\tag{13.80}
$$

For polymer stress $\boldsymbol{\tau}^p$, the dimensionless form of the constitutive equations are given by (obtained by replacing $\boldsymbol{\tau}$ with $\boldsymbol{\tau}^p$ and η by η_p in (13.77)) the following [20]:

$$
\begin{aligned}
&\tau^p_{yy}+De\Big(u\frac{\partial \tau^p_{yy}}{\partial x}+v\frac{\partial \tau^p_{yy}}{\partial y}-2\tau^p_{xy}\frac{\partial v}{\partial x}-2\tau^p_{yy}\frac{\partial v}{\partial y}\Big)\\
&\qquad-\alpha\frac{De}{\eta_p}\Big(\frac{L_0\tau_0}{u_0\eta_0}\Big)\big((\tau^p_{yy})^2+(\tau^p_{xy})^2\big)=2\eta_p\Big(\frac{u_0\eta_0}{L_0\tau_0}\Big)\frac{\partial v}{\partial y}
\end{aligned}
\tag{13.81}
$$

$$
\begin{aligned}
&\tau^p_{xy}+De\Big(u\frac{\partial \tau^p_{xy}}{\partial x}+v\frac{\partial \tau^p_{xy}}{\partial y}-\tau^p_{xy}\big(\frac{\partial u}{\partial x}+\frac{\partial v}{\partial y}\big)-\tau^p_{xx}\frac{\partial v}{\partial x}-\tau^p_{yy}\frac{\partial u}{\partial y}\Big)\\
&\qquad-\alpha\frac{De}{\eta_p}\Big(\frac{L_0\tau_0}{u_0\eta_0}\Big)\tau^p_{xy}(\tau^p_{xx}+\tau^p_{yy})=\eta_p\Big(\frac{u_0\eta_0}{L_0\tau_0}\Big)\Big(\frac{\partial u}{\partial y}+\frac{\partial v}{\partial x}\Big)
\end{aligned}
\tag{13.82}
$$

Using (13.80) in the momentum equations (13.76), we can express the momentum equations in terms of τ^p_{xx}, τ^p_{yy}, τ^p_{xy} and velocity gradients.

$$
\begin{aligned}
&\rho\Big(u\frac{\partial u}{\partial x}+v\frac{\partial u}{\partial y}\Big)+\Big(\frac{p_0}{\rho_0 u_0^2}\Big)\frac{\partial p}{\partial x}-\Big(\frac{\tau_0}{\rho_0 u_0^2}\Big)\Big(\frac{\partial \tau^p_{xx}}{\partial x}+\frac{\partial \tau^p_{xy}}{\partial y}\Big)\\
&\qquad-\Big(\frac{\eta_0}{L_0\rho_0 u_0}\Big)\eta_s\Big(2\frac{\partial^2 u}{\partial x^2}+\frac{\partial^2 u}{\partial y^2}+\frac{\partial^2 v}{\partial y\partial x}\Big)=0\\
&\rho\Big(u\frac{\partial v}{\partial x}+v\frac{\partial v}{\partial y}\Big)+\Big(\frac{p_0}{\rho_0 u_0^2}\Big)\frac{\partial p}{\partial y}-\Big(\frac{\tau_0}{\rho_0 u_0^2}\Big)\Big(\frac{\partial \tau^p_{xy}}{\partial x}+\frac{\partial \tau^p_{yy}}{\partial y}\Big)\\
&\qquad-\Big(\frac{\eta_0}{L_0\rho_0 u_0}\Big)\eta_s\Big(2\frac{\partial^2 v}{\partial y^2}+\frac{\partial^2 v}{\partial x^2}+\frac{\partial^2 u}{\partial y\partial x}\Big)=0
\end{aligned}
\tag{13.83}
$$

Equations (13.75), (13.81)–(13.83) constitute the complete mathematical model in dependent variables u, v, p, τ^p_{xx}, τ^p_{yy} and τ^p_{xy} for two-dimensional steady flow using the constitutive model UCG2, used currently for incompressible Giesekus fluids.

Solutions of the BVPs

In this section, we consider solutions of the BVPs described by (13.75)–(13.77) for UCG1 and (13.75), (13.81), (13.83) for UCG2. A theoretical solution for dependent variables is not possible due to the complexity of the constitutive equations in both boundary value problems. Hence, we consider their numerical solutions using finite element processes based on the residual functional (least squares finite element method) in which the resulting non-linear algebraic equations from the least squares process are solved using Newton's linear method. The computational processes in this approach are unconditionally stable and permit higher order global differentiability local approximations. Details of the local approximations and the least squares finite element processes for non-linear PDEs and higher order spaces can be found in references [16, 17, 104–106, 140]. The local approximations are considered in higher order spaces $H^{k,p}(\bar{\Omega}^e_{xy})$ in which $\bar{\Omega}^e_{xy}$ is the spatial domain of a typical element e of the discretization.

In the computations of the numerical solutions we choose

$$\hat{H}=L_0=3.175 \text{ mm} \ ; \ \rho_0=\hat{\rho}=800 \text{ kg/m}^3 \ ; \ \eta_0=\hat{\eta}=1.426 \text{ Pa s} \ ; \ u_0=0.5 \text{ m/s}$$

where $H = 1$, $P_0 = 200$ Pa, $Re = 0.8906$ and $De = 9.45$, the same as Model Problem 1.

In this case, the rectangular domain $ABCD$ is discretized using two 9-node p-version elements of lengths 0.2 and 0.8 (Figure 13.9) in the y-direction. Length AD is chosen as 1.0 (arbitrary). The local approximations are considered to be of equal degree for all variables. We consider $p = (p_1, p_2) = (9, 9)$ with $k = (k_1, k_2) = (2, 2)$, i.e., local approximations of class $C^{1,1}(\bar{\Omega}^e_{xy})$. For this choice of mesh, p-level and order of space, the residual functional values are of orders of $O(10^{-8})$–$O(10^{-16})$ indicating that the PDEs are satisfied very accurately (in the pointwise sense for UCG1 as the integrals are Riemann, and not strictly in the pointwise sense for UCG2 since the integrals are Lebesgue) when the local approximations for u, v, p, τ_{xx}, τ_{yy} and τ_{xy} are of class $C^{1,1}(\bar{\Omega}^e_{xy})$.

In the numerical studies, we begin with $\partial p/\partial x = -0.1$ for which a converged solution is obtained and then progressively increase it up to $\partial p/\partial x = -0.275$ using

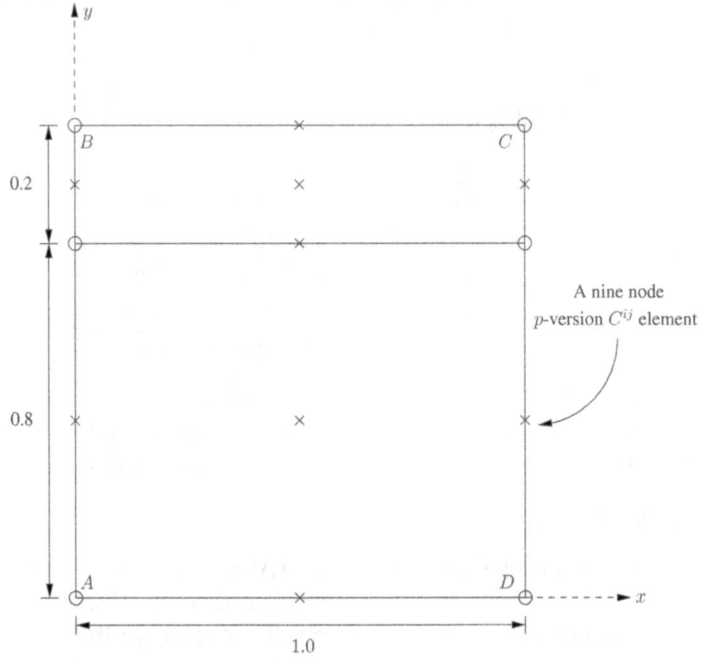

Figure 13.9: Graded mesh discretization using two 9-node p-version elements

a continuation procedure in which converged solutions at lower $\partial p/\partial x$ are used as initial (or starting) solution in Newton's linear method. For all values of $\partial p/\partial x$, the computed numerical solutions confirm that u, $\partial u/\partial x$, τ_{xx}, τ_{yy} and τ_{xy} versus y are invariant of spatial location x along AD and are in perfect agreement with those obtained in Model Problem 1 using fully developed flow 1D numerical studies, hence are not repeated for the sake of brevity.

13.8 General remarks

In this section, we make some remarks that may be helpful in the use of mathematical models consisting of CBL in Eulerian description in conjunction with the constitutive theories derived in this chapter.

(1) As in Chapter 11 and 12, here also, CBL are based on Eulerian description in which velocities $\bar{\boldsymbol{v}}$ are observable quantities; hence, Remark (1) in Section 12.6 holds here as well.

(2) CBL and the constitutive theories for such materials do not contain $\bar{\boldsymbol{u}}$, $\left[\frac{\partial\{\bar{u}\}}{\partial\{x\}}\right]$; hence, they do not contain strains either, but such materials have elasticity due to motion of long chain molecules. As a consequence, we have stiffness, mass (ρ), dissipation and rheology (relaxation phenomena) due to memory modulus which is a function of the relaxation time, a property of the material.

(3) Upon cessation of disturbance, such material takes a finite amount of time to come to stress free relaxed state. This time is a function of the relaxation time.

The memory modulus, a function of relaxation time, provides quantitative measure of time it takes for fluid to relax.

(4) Due to elasticity and mass (ρ), such fluent continua support propagation of stress waves. Due to dissipation, stress waves naturally attenuate during evolution (amplitude decay and base elongation).

(5) There is no concept of natural vibrations, resonance, modes of vibration and dynamics in general as in case of solid continua. This is obviously due to the fact that such continua lacks stiffness due to strains that is necessary in conjunction with mass for classical dynamic motion to exist.

(6) Even though such fluent continua have elasticity and support wave propagation, but still, wave propagation in fluid-solid interaction using this fluid model remains prohibitive due to $\bar{\boldsymbol{u}}$ and $\bar{\boldsymbol{v}}$ being observable quantities in solid continua and this fluent continua and that $\bar{\boldsymbol{v}} \neq \frac{D\bar{\boldsymbol{u}}}{Dt}$ in case of fluent continua.

13.9 Summary

Ordered rate constitutive theories of orders m and n in stress and strain rates have been derived for compressible and incompressible thermoviscoelastic (polymeric) fluent continua using entropy inequality in conjunction with representation theorem. Initial choice of constitutive tensors and their argument tensors are made using SLT and the principle of equipresence. Basis independent convected time derivatives ${}^{(k)}_{\;d}\bar{\boldsymbol{\sigma}}$; $k = 0, 1, 2 \dots, m$ of the deviatoric Cauchy stress tensor and the basis independent convected time derivatives ${}^{(j)}\boldsymbol{\gamma}$; $j = 1, 2, \dots, n$ of the strain tensor are used in the derivation of the constitutive theory for the deviatoric Cauchy stress tensor. The constitutive theories for the equilibrium Cauchy stress tensor ${}^{(0)}_{\;e}\bar{\boldsymbol{\sigma}}$ are derived using $\bar{\Phi}$ and the incompressibility condition for compressible and incompressible polymeric fluids. It is shown that the basis independent constitutive theories can be made basis specific by choosing appropriate conjugate pairs ${}^{(k)}_{\;d}\bar{\boldsymbol{\sigma}}$, ${}^{(j)}\boldsymbol{\gamma}$; $k = 0, 1, \dots, m$; $j = 1, 2, \dots, n$.

Simplified forms of the constitutive theories for deviatoric Cauchy stress tensors of orders $m = 1$, $n = 1$ and $m = 1$, $n = 2$ are also derived and are shown to be subsets of the general rate constitutive theory of orders m and n. We make the following specific remarks or observations based on the work presented in this chapter.

(1) The general ordered rate theory of orders m and n for deviatoric Cauchy stress tensor is based on integrity (complete basis of the space of the deviatoric Cauchy stress tensor).

(2) Maxwell and Giesekus constitutive models used currently are derived and are shown to be a small subset of the general theory of orders $m = 1$ and $n = 1$.

(3) The Oldroyd-B constitutive model used currently is derived and is shown to be a small subset of the general constitutive theory of orders $m = 1$, $n = 2$.

(4) Since the constitutive theory for deviatoric stress tensor of orders $m = 1$, $n = 1$ is a subset of the constitutive theory of orders $m = 1$ and $n = 2$, it is shown that only one constitutive theory, a subset of $m = 1$ and $n = 2$ is needed for dilute as well as dense polymers. Maxwell, Oldroyd-B and Giesekus constitutive models are also contained in this single constitutive theory.

(5) It is important to note that the Giesekus constitutive model derived here uses deviatoric Cauchy stress tensor and its convected time derivatives up to orders m. In polymer science, deviatoric Cauchy stress is decomposed into deviatoric Cauchy stress tensor for solvent and polymer $(({}^{(0)}_{\ d}\bar{\boldsymbol{\sigma}})_s,\ ({}^{(0)}_{\ d}\bar{\boldsymbol{\sigma}})_p)$. The constitutive theory for $({}^{(0)}_{\ d}\bar{\boldsymbol{\sigma}})_s$ is assumed to be Newton's law of viscosity and the constitutive theory for $({}^{(0)}_{\ d}\bar{\boldsymbol{\sigma}})_p$ has the same form as the constitutive theory derived in this chapter. This decomposition is not valid if the polymeric fluid is isotropic and homogeneous, and if not then CCM cannot be used to derive constitutive theories.

(6) In polymer science, it is argued [76, 77] that decomposition of the deviatoric Cauchy stress in terms of viscous (both solvent and polymer) and elastic components, and then, expressing viscous stress using Newton's law of viscosity; thus, obtaining constitutive equations in terms of deviatoric elastic stress is meritorious (computationally). This decomposition is not valid if the polymeric fluid is isotropic and homogeneous.

(7) Material coefficients are derived in all cases and are shown to be dependent on $\bar{\rho}$, $\bar{\theta}$ and the invariants of the argument tensors but in a known configuration $\underline{\Omega}$.

(8) Constitutive theory for heat vector $\boldsymbol{q}$ remains the same as in Chapter 12.

(9) Numerical studies are presented for fully developed flow between parallel plates, and fully developed flow between parallel plates using a two-dimensional formulation for a dense polymeric liquid (PIB/C14) using the Giesekus constitutive model derived in this chapter as well as the currently used Giesekus constitutive model. We use contravariant Cauchy stress tensor and Almansi strain tensors as conjugate measures of the stress and strain tensors in Eulerian description. This yields upper convected Giesekus constitutive models. Numerical results are presented using the upper convected Giesekus constitutive model derived in this chapter (UCG1) as well as using the currently used constitutive model in deviatoric polymer stress (UCG2).

(10) We choose a value of $\partial p/\partial x$ that is the same in both constitutive models and compute the results. For very low values of $\partial p/\partial x$, velocity fields in the two cases are not drastically different, but as $\partial p/\partial x$ increases, both models produced significantly different velocity fields, and hence, different flow rates. Computed results from fully developed flow between parallel plates using studies in $\mathbb{R}^1$ and fully developed flow between parallel plates using the two-dimensional formulation, i.e., $\mathbb{R}^2$, are in perfect agreement when the same constitutive model is used.

Problems

13.1 Consider a compressible thermoviscoelastic fluid. Derive a first order linear rate constitutive theory for deviatoric Cauchy stress tensor in contravariant and covariant bases and Jaumann rates from first principles using the theory of generators and invariants in which the argument tensors are first and second convected time derivatives of the conjugate strain tensor, deviatoric Cauchy stress tensor and temperature $\bar{\theta}$. Derive material coefficients. Give explicit forms of the equations in $\mathbb{R}^1$ and $\mathbb{R}^2$. Also, express these in matrix and vector form using Voigt's notation if possible. Specialize the final equations for the incompressible case.

13.2 Consider the argument tensors in 13.1 and temperature gradient as an additional argument to derive a linear constitutive theory for the heat vector in contravariant and covariant bases as well as using Jaumann rates. Derive material coefficients. Give explicit forms of the equations in $\mathbb{R}^1$ and $\mathbb{R}^2$. Also, express these in the matrix and vector form using Voigt's notation if possible. Specialize the final equations for the incompressible case.

13.3 Consider a compressible thermoviscoelastic fluid and ordered linear rate constitutive theories in contravariant and covariant bases and using Jaumann rates from first principles using the theory of generators and invariants.

(a) Derive a constitutive theory for the deviatoric Cauchy stress tensor in which the convected time derivative of order m of the deviatoric Cauchy stress tensor is a linear function of the Cauchy stress tensor and its convected time derivatives up to orders $(m-1)$, convected time derivatives of the conjugate strain tensor up to orders n, temperature gradient and temperature $\bar{\theta}$.

(b) Also derive a constitutive theory for the heat vector $\bar{\boldsymbol{q}}$ that is consistent with the choice of arguments in (a).

Derive material coefficients in both cases. Discuss constant and variable material coefficients.

13.4 Derive expanded forms of the constitutive equations for compressible Maxwell fluid in contravariant and covariant bases and using Jaumann rates for $\mathbb{R}^1$, $\mathbb{R}^2$ and $\mathbb{R}^3$. Simplify these for the incompressible case.

13.5 Derive expanded forms of the constitutive equations for compressible Oldroyd-B fluid in contravariant and covariant bases and using Jaumann rates for $\mathbb{R}^1$, $\mathbb{R}^2$ and $\mathbb{R}^3$. Simplify these for the incompressible case.

13.6 Derive expanded forms of the constitutive equations for compressible Giesekus fluid in contravariant and covariant bases and in Jaumann rates for $\mathbb{R}^1$, $\mathbb{R}^2$ and $\mathbb{R}^3$. Simplify these for the incompressible case.

14
CONSTITUTIVE THEORIES FOR THERMO HYPO-ELASTIC SOLIDS

14.1 Introduction

Thermo hypo-elastic solids are a special class of materials in which the stress rates are a function of strain rates. In such materials one assumes strains are not significant in the kinematics of deformation, hence can be neglected. Thus, there is no need to follow material particles either. This prompts us to consider CBL in Eulerian description in which we assume velocities $\bar{\boldsymbol{v}}(\bar{\boldsymbol{x}},t)$ to be observable quantities; hence, the displacements (and thus strains) need not be considered at all. Additive decomposition of the deformation (of volume $\bar{V}(t)$) into volumetric and distortional deformations and Cauchy stress tensor into equilibrium and deviatoric stress tensors facilitates the derivation of the constitutive theories. As usual, the constitutive theory for the equilibrium stress tensor addresses volumetric deformation while the distortional deformation physics is addressed by the constitutive theory for deviatoric stress tensor. We always begin with the entropy inequality in Eulerian description followed by the constitutive theories for equilibrium Cauchy stress tensor and the reduced form of entropy inequality. The reduced form of the entropy inequality in conjunction with representation theorem is used to derive the constitutive theory for deviatoric Cauchy stress tensor based on integrity from which many simplified constitutive theories are possible (some are presented). Constitutive theories for heat vector are also considered.

14.2 Preliminary considerations

We consider CBL in Eulerian description in which velocities $\bar{\boldsymbol{v}}(\bar{\boldsymbol{x}},t)$ are observable quantities. Choice of appropriate stress measure(s) and convected time derivatives of the strain tensors is crucial as well. In the CBL and the derivation of the constitutive theories we choose basis independent Cauchy stress tensor notation ${}^{(0)}\bar{\boldsymbol{\sigma}}$ that could be $\bar{\boldsymbol{\sigma}}^{(0)}$, $\bar{\boldsymbol{\sigma}}_{(0)}$ or ${}^{(0)}\bar{\boldsymbol{\sigma}}^J$. We also consider the basis independent notation for convected time derivatives of the strain tensor(s) ${}^{(i)}\boldsymbol{\gamma}$; $i = 1,2,\dots,n$ that could be covariant convected time derivatives $\boldsymbol{\gamma}_{(i)}$; $i = 1,2,\dots,n$ of the Green's strain tensors $\boldsymbol{\varepsilon}_{[0]}$ or could be contravariant convected time derivatives $\boldsymbol{\gamma}^{(i)}$; $i = 1,2,\dots,n$ of the Almansi strain tensor $\bar{\boldsymbol{\varepsilon}}^{[0]}$ or Jaumann rates ${}^{(i)}\boldsymbol{\gamma}^J$; $i = 1,2,\dots,n$ which are averages of $\boldsymbol{\gamma}_{(i)}$ and $\boldsymbol{\gamma}^{(i)}$; $i = 1,2,\dots,n$ rates. We note that (see Chapter 5).

$$\boldsymbol{\gamma}^{(1)} = \boldsymbol{\gamma}_{(1)} = \bar{\boldsymbol{D}} = {}^{(1)}\boldsymbol{\gamma}^J \tag{14.1}$$

Where $\bar{\boldsymbol{D}}$ is the symmetric parts of the velocity gradient tensor. Thus, ${}^{(0)}\bar{\boldsymbol{\sigma}}$, $\bar{\boldsymbol{D}}$ or $\bar{\boldsymbol{\sigma}}^{(0)}$, $\bar{\boldsymbol{D}}$ or $\bar{\boldsymbol{\sigma}}_{(0)}$, $\bar{\boldsymbol{D}}$ or ${}^{(0)}\bar{\boldsymbol{\sigma}}^J$, $\bar{\boldsymbol{D}}$ are rate of work conjugate pairs. Thus, $\bar{\boldsymbol{D}}$ or ${}^{(1)}\boldsymbol{\gamma}$ can be an argument tensor of ${}^{(0)}\bar{\boldsymbol{\sigma}}$. We generalize the choice of ${}^{(1)}\boldsymbol{\gamma}$ as an argument tensor by replacing it with the convected time derivatives ${}^{(i)}\boldsymbol{\gamma}$; $i = 2,3,\dots,n$.

DOI: 10.1201/9781003105336-14

Thus, we must consider

$$ {}^{(0)}\bar{\boldsymbol{\sigma}} \quad \text{and} \quad {}^{(i)}\boldsymbol{\gamma} \quad ; \quad i = 1, 2, \dots, n \tag{14.2}$$

as basis independent rate of work conjugate pairs in the derivation of the constitutive theories. Basis dependent constitutive theories can be obtained by choosing

$$\begin{aligned} &{}^{(0)}\bar{\boldsymbol{\sigma}}, {}^{(i)}\boldsymbol{\gamma} \quad \text{as} \quad \bar{\boldsymbol{\sigma}}^{(0)}, \boldsymbol{\gamma}_{(i)} \\ &{}^{(0)}\bar{\boldsymbol{\sigma}}, {}^{(i)}\boldsymbol{\gamma} \quad \text{as} \quad \bar{\boldsymbol{\sigma}}_{(0)}, \boldsymbol{\gamma}^{(i)} \\ &{}^{(0)}\bar{\boldsymbol{\sigma}}, {}^{(i)}\boldsymbol{\gamma} \quad \text{as} \quad {}^{(0)}\bar{\boldsymbol{\sigma}}^{J}, {}^{(i)}\boldsymbol{\gamma}^{J} \\ &i = 1, 2, \dots, n \end{aligned} \tag{14.3}$$

We perform additive stress decomposition of Cauchy stress tensor ${}^{(0)}\bar{\boldsymbol{\sigma}}$ into equilibrium stress tensor ${}^{(0)}_{e}\bar{\boldsymbol{\sigma}}$ and deviatoric stress tensor ${}_{d}\bar{\boldsymbol{\sigma}}^{(0)}$. Volumetric deformation physics is described by the constitutive theory for ${}^{(0)}_{e}\bar{\boldsymbol{\sigma}}$. Distortional deformation physics and dissipation mechanism are due to the constitutive theory for ${}^{(0)}_{d}\bar{\boldsymbol{\sigma}}$. Furthermore, volumetric deformation physics in Eulerian description is invariant of the type of deformation and the constitution of the matter (Chapter 8).

14.3 Constitutive theory for equilibrium Cauchy stress ${}^{(0)}_{e}\bar{\boldsymbol{\sigma}}$

The derivation of the constitutive theory for ${}^{(0)}_{e}\bar{\boldsymbol{\sigma}}$ has been presented in Chapter 8, see Section 8.5.4. We can write

$$ {}^{(0)}_{e}\bar{\boldsymbol{\sigma}} = \bar{p}(\bar{\rho}, \bar{\theta})\boldsymbol{\delta} \quad ; \quad \text{compressible} \tag{14.4}$$

$$ {}^{(0)}_{e}\bar{\boldsymbol{\sigma}} = \bar{p}(\bar{\theta})\boldsymbol{\delta} \quad ; \quad \text{incompressible} \tag{14.5}$$

$$ \text{in which} \quad \bar{p}(\bar{\rho}, \bar{\theta}) = -\bar{\rho}^2 \frac{\partial \bar{\Phi}(\bar{\rho}, \bar{\theta})}{\partial \bar{\rho}} \tag{14.6}$$

Here, $\bar{p}(\bar{\rho}, \bar{\theta})$ is thermodynamic pressure or equation of state for the deforming matter. We note that the equilibrium Cauchy stress tensor constitutive theory in (14.4) and (14.5) is basis independent as it is pressure field. We continue using the notation in (14.4) and (14.5) for more clarity when the equilibrium Cauchy stress tensor is added to the deviatoric Cauchy stress tensor. The reduced form of the entropy inequality is given by (equation (8.76))

$$ -{}_{d}\bar{\boldsymbol{\sigma}}^{(0)} : \bar{\boldsymbol{D}} + \frac{\bar{\boldsymbol{q}} \cdot \bar{\boldsymbol{g}}}{\bar{\theta}} \leq 0 \tag{14.7}$$

The entropy inequality (14.7) is satisfied if ${}_{d}\bar{\boldsymbol{\sigma}}^{(0)} : \bar{\boldsymbol{D}} > 0$ and $\frac{\bar{\boldsymbol{q}} \cdot \bar{\boldsymbol{g}}}{\bar{\theta}} \leq 0$. These serve as restrictions on the constitutive theories for ${}_{d}\bar{\boldsymbol{\sigma}}^{(0)}$ and $\bar{\boldsymbol{q}}$. In Eulerian description, density $\bar{\rho}$ is a dependent variable in CBL. Thus, we can include $\bar{\rho}$ as an argument tensor of the constitutive tensors. Due to non-isothermal physics, $\bar{\theta}$ as argument tensor of the constitutive tensors is a natural choice. Thus, strictly based on the conjugate pairs in (14.7) we can write

$$ {}^{(0)}_{d}\bar{\boldsymbol{\sigma}} = {}^{(0)}_{d}\bar{\boldsymbol{\sigma}}(\bar{\rho}, \bar{\boldsymbol{D}}, \bar{\theta}) \tag{14.8}$$

$$ \text{and} \quad \bar{\boldsymbol{q}} = \bar{\boldsymbol{q}}(\bar{\rho}, \bar{\boldsymbol{g}}, \bar{\theta}) \tag{14.9}$$

14.4 Constitutive theory for deviatoric Cauchy stress ${}^{(0)}_{\,d}\bar{\boldsymbol{\sigma}}$

We recall that

$$\bar{\boldsymbol{D}} = \boldsymbol{\gamma}^{(i)} = \boldsymbol{\gamma}_{(i)} = {}^{(i)}\boldsymbol{\gamma} \tag{14.10}$$

We replace $\bar{\boldsymbol{D}}$ or ${}^{(1)}\boldsymbol{\gamma}$ in (14.8) by ${}^{(i)}\boldsymbol{\gamma}$; $i = 1, 2, \dots, n$ (generalizing to include convected time derivatives of the strain tensor up to order n).

$${}^{(0)}_{\,d}\bar{\boldsymbol{\sigma}} = {}^{(0)}_{\,d}\bar{\boldsymbol{\sigma}}(\bar{\rho}, {}^{(i)}\boldsymbol{\gamma}, \bar{\theta}) \quad ; \quad i = 1, 2, \dots, n \tag{14.11}$$

For thermo hypo-elastic solids, the first convected time derivatives of the Cauchy stress tensor (basis dependent) must be a function of at least the first convected time derivative of the conjugate strain tensor. Considering convected time derivatives of the strain tensor up to order n, (14.11) can be modified as

$${}^{(1)}_{\,d}\bar{\boldsymbol{\sigma}} = {}^{(1)}_{\,d}\bar{\boldsymbol{\sigma}}(\bar{\rho}, {}^{(i)}\boldsymbol{\gamma}, \bar{\theta}) \quad ; \quad i = 1, 2, \dots, n \tag{14.12}$$

$$\bar{\boldsymbol{q}} = \bar{\boldsymbol{q}}(\bar{\rho}, \bar{\boldsymbol{g}}, \bar{\theta}) \tag{14.13}$$

We consider (14.12) and representation theorem to derive constitutive theory for ${}^{(1)}\bar{\boldsymbol{\sigma}}$. Let ${}^{\sigma}\underset{\sim}{\boldsymbol{G}}^i$; $i = 1, 2, \dots, N$ be the combined generators of the argument tensors of ${}^{(1)}_{\,d}\bar{\boldsymbol{\sigma}}$ in (14.12) that are symmetric tensors of rank two and let ${}^{\sigma}\underset{\sim}{I}^j$; $j = 1, 2, \dots, M$ be the combined invariants of the same argument tensors of ${}^{(1)}_{\,d}\bar{\boldsymbol{\sigma}}$ in (14.12). Then, $\boldsymbol{I}$, ${}^{\sigma}\underset{\sim}{\boldsymbol{G}}^i$; $i = 1, 2, \dots, N$ constitute the basis of the space of ${}^{(1)}_{\,d}\bar{\boldsymbol{\sigma}}$. Hence, we can express ${}^{(1)}_{\,d}\bar{\boldsymbol{\sigma}}$ as a linear combination of the generators $\boldsymbol{I}$, ${}^{\sigma}\underset{\sim}{\boldsymbol{G}}^i$; $i = 1, 2, \dots, N$ in the current configuration.

$${}^{(1)}_{\,d}\bar{\boldsymbol{\sigma}} = {}^{\sigma}\underset{\sim}{\alpha}^0 \boldsymbol{I} + \sum_{i=1}^{N} {}^{\sigma}\underset{\sim}{\alpha}^i ({}^{\sigma}\underset{\sim}{\boldsymbol{G}}^i) \tag{14.14}$$

In which the coefficients ${}^{\sigma}\underset{\sim}{\alpha}^i$; $i = 0, 1, \dots, N$ in the linear combination (14.14) are function of $\bar{\rho}$, $\bar{\theta}$ and combined invariants of the argument tensors of ${}^{(0)}_{\,d}\bar{\boldsymbol{\sigma}}$ in (14.11), ${}^{\sigma}\underset{\sim}{I}^j$; $j = 1, 2, \dots, M$ in the current configuration. That is

$${}^{\sigma}\underset{\sim}{\alpha}^i = {}^{\sigma}\underset{\sim}{\alpha}^i(\bar{\rho}, \bar{\theta}, {}^{\sigma}\underset{\sim}{I}^j) \quad ; \quad j = 1, 2, \dots, M \quad ; \quad i = 0, 1, \dots, N \tag{14.15}$$

Material coefficients

To determine material coefficients in the constitutive theory (14.14), we expand ${}^{\sigma}\underset{\sim}{\alpha}^i$; $i = 0, 1, \dots, N$ in ${}^{\sigma}\underset{\sim}{I}^j$; $j = 1, 2, \dots, M$ about a known configuration $\underline{\Omega}$ and retain only up to linear terms in ${}^{\sigma}\underset{\sim}{I}^j$; $j = 1, 2, \dots, M$ (for simplicity). We do not consider Taylor series expansion of ${}^{\sigma}\underset{\sim}{\alpha}^i$; $i = 0, 1, \dots, N$ in $\bar{\rho}$ and $\bar{\theta}$ as their influence on the Cauchy stress tensor has already been accounted for in the constitutive theory for ${}^{(0)}_{\,e}\bar{\boldsymbol{\sigma}}$.

$${}^{\sigma}\underset{\sim}{\alpha}^i = {}^{\sigma}\underset{\sim}{\alpha}^i\Big|_{\underline{\Omega}} + \sum_{j=1}^{M} \frac{\partial\, {}^{\sigma}\underset{\sim}{\alpha}^i}{\partial\, {}^{\sigma}\underset{\sim}{I}^j}\Bigg|_{\underline{\Omega}} \left({}^{\sigma}\underset{\sim}{I}^j - {}^{\sigma}\underset{\sim}{I}^j\Big|_{\underline{\Omega}} \right) \quad ; \quad i = 0, 1, \dots, N \tag{14.16}$$

Substituting ${}^{\sigma}\underset{\sim}{\alpha}^{i}$ from (14.16) in (14.14)

$$
\begin{aligned}
{}^{(1)}_{\ d}\bar{\boldsymbol{\sigma}} = & \left({}^{\sigma}\underset{\sim}{\alpha}^{0}\Big|_{\underline{\Omega}} + \sum_{j=1}^{M} \frac{\partial {}^{\sigma}\underset{\sim}{\alpha}^{0}}{\partial {}^{\sigma}\underset{\sim}{I}^{j}}\bigg|_{\underline{\Omega}} \left({}^{\sigma}\underset{\sim}{I}^{j} - {}^{\sigma}\underset{\sim}{I}^{j}\Big|_{\underline{\Omega}} \right) \right) \boldsymbol{I} \\
& + \sum_{i=1}^{N} \left({}^{\sigma}\underset{\sim}{\alpha}^{i}\Big|_{\underline{\Omega}} + \sum_{j=1}^{M} \frac{\partial {}^{\sigma}\underset{\sim}{\alpha}^{i}}{\partial {}^{\sigma}\underset{\sim}{I}^{j}}\bigg|_{\underline{\Omega}} \left({}^{\sigma}\underset{\sim}{I}^{j} - {}^{\sigma}\underset{\sim}{I}^{j}\Big|_{\underline{\Omega}} \right) \right) {}^{\sigma}\boldsymbol{G}^{i}
\end{aligned}
\tag{14.17}
$$

Collecting coefficients of (the terms defined in the current configuration) $\boldsymbol{I}$, ${}^{\sigma}\underset{\sim}{I}^{j}\boldsymbol{I}$, ${}^{\sigma}\boldsymbol{G}^{i}$ and ${}^{\sigma}\underset{\sim}{I}^{j}({}^{\sigma}\boldsymbol{G}^{i})$; $i = 1, 2, \dots, N$; $j = 1, 2, \dots, M$ and defining

$$
\begin{aligned}
& \underset{\sim}{\sigma}^{0}\Big|_{\underline{\Omega}} = {}^{\sigma}\underset{\sim}{\alpha}^{0}\Big|_{\underline{\Omega}} - \sum_{j=1}^{M} \frac{\partial {}^{\sigma}\underset{\sim}{\alpha}^{0}}{\partial {}^{\sigma}\underset{\sim}{I}^{j}}\bigg|_{\underline{\Omega}} \left({}^{\sigma}\underset{\sim}{I}^{j}\Big|_{\underline{\Omega}} \right) \quad ; \quad {}^{\sigma}\underset{\sim}{a}_{j} = \frac{\partial {}^{\sigma}\underset{\sim}{\alpha}^{0}}{\partial {}^{\sigma}\underset{\sim}{I}^{j}}\bigg|_{\underline{\Omega}} \\
& {}^{\sigma}\underset{\sim}{b}_{i} = {}^{\sigma}\underset{\sim}{\alpha}^{i}\Big|_{\underline{\Omega}} - \sum_{j=1}^{M} \frac{\partial {}^{\sigma}\underset{\sim}{\alpha}^{i}}{\partial {}^{\sigma}\underset{\sim}{I}^{j}}\bigg|_{\underline{\Omega}} \left({}^{\sigma}\underset{\sim}{I}^{j}\Big|_{\underline{\Omega}} \right) \quad ; \quad {}^{\sigma}\underset{\sim}{c}_{ij} = \frac{\partial {}^{\sigma}\underset{\sim}{\alpha}^{i}}{\partial {}^{\sigma}\underset{\sim}{I}^{j}}\bigg|_{\underline{\Omega}} \\
& i = 1, 2 \dots, N \quad ; \quad j = 1, 2 \dots, M
\end{aligned}
\tag{14.18}
$$

Equation (14.17) can be written as

$$
{}^{(1)}_{\ d}\bar{\boldsymbol{\sigma}} = \underset{\sim}{\sigma}^{0}\Big|_{\underline{\Omega}} \boldsymbol{I} + \sum_{j=1}^{M} {}^{\sigma}\underset{\sim}{a}_{j} ({}^{\sigma}\underset{\sim}{I}^{j}) \boldsymbol{I} + \sum_{i=1}^{N} \sum_{j=1}^{M} {}^{\sigma}\underset{\sim}{c}_{ij} ({}^{\sigma}\underset{\sim}{I}^{j}) {}^{\sigma}\boldsymbol{G}^{i} + \sum_{i=1}^{N} {}^{\sigma}\underset{\sim}{b}_{i} {}^{\sigma}\boldsymbol{G}^{i} \tag{14.19}
$$

Where ${}^{\sigma}\underset{\sim}{a}_{j}$, ${}^{\sigma}\underset{\sim}{b}_{i}$, ${}^{\sigma}\underset{\sim}{c}_{ij}$; $i = 1, 2, \dots, N$; $j = 1, 2, \dots, M$ are material coefficients defined in a known configuration $\underline{\Omega}$. $\underset{\sim}{\sigma}^{0}$ is the initial stress field in the known configuration $\underline{\Omega}$. This constitutive theory for ${}^{(1)}_{\ d}\bar{\boldsymbol{\sigma}}$ requires $(M + MN + N)$ material coefficients. This constitutive theory is based on integrity (complete basis of the space of ${}^{(1)}_{\ d}\bar{\boldsymbol{\sigma}}$). Material coefficients can be functions of $\bar{\rho}|_{\underline{\Omega}}$, $\bar{\theta}|_{\underline{\Omega}}$ and ${}^{\sigma}\underset{\sim}{I}^{j}\Big|_{\underline{\Omega}}$; $j = 1, 2, \dots, M$.

Remarks

(1) In the constitutive theory for the deviatoric Cauchy stress tensor ${}^{(1)}_{\ d}\bar{\boldsymbol{\sigma}}$, the dissipation mechanism is due to convected time derivatives ${}^{(i)}\boldsymbol{\gamma}$; $i = 1, 2, \dots, n$ of the strain tensor. Thus, we refer to this constitutive theory of up to order n as an *ordered rate constitutive theory of up to order* n.

(2) By choosing appropriate conjugate pairs for ${}^{(1)}_{\ d}\bar{\boldsymbol{\sigma}}$ and ${}^{(i)}\boldsymbol{\gamma}$; $i = 1, 2, \dots, n$ basis specific constitutive theories for deviatoric Cauchy stress tensor can be obtained. These are listed below.

$$
\begin{aligned}
& \text{Contravariant stress measure:} \quad {}_{d}\bar{\boldsymbol{\sigma}}^{(0)}, {}_{d}\bar{\boldsymbol{\sigma}}^{(1)}, \boldsymbol{\gamma}_{(i)} \quad ; \quad i = 1, 2, \dots, n \\
& \text{Covariant stress measure:} \quad {}_{d}\bar{\boldsymbol{\sigma}}_{(0)}, {}_{d}\bar{\boldsymbol{\sigma}}_{(1)}, \boldsymbol{\gamma}^{(i)} \quad ; \quad i = 1, 2, \dots, n \\
& \text{Jaumann rates:} \quad {}^{(0)}\bar{\boldsymbol{\sigma}}^{J}, {}^{(1)}\bar{\boldsymbol{\sigma}}^{J}, {}^{(i)}\boldsymbol{\gamma}^{J} \quad ; \quad i = 1, 2, \dots, n
\end{aligned}
\tag{14.20}
$$

(3) Configuration $\underline{\Omega}$ is a suitable previously known configuration (generally corresponding to last value of time t for which deformation is known). Configuration $\underline{\Omega}$ can also be the reference configuration in which case the material

coefficients will be constant, i.e., not dependent on varying $\bar{\rho}$, $\bar{\theta}$ and ${}^{\sigma}\underset{\sim}{I}^{j}$; $j = 1, 2, \dots, M$.

(4) The material coefficients as a function of unknown deformation in current configuration is not supported by CCM. This is rather obvious because, if this were the case, then ${}^{\sigma}\underset{\sim}{\alpha}^{i}$; $i = 0, 1, \dots, N$ in (14.14) would be material coefficients and there would be no need for Taylor series expansion of ${}^{\sigma}\underset{\sim}{\alpha}^{i}$ in ${}^{\sigma}\underset{\sim}{I}^{j}$; $j = 1, 2, \dots, M$ about a known configuration. In continuum theories, material coefficients are known properties of the matter, hence cannot be treated as functions of unknown deformation.

Simplified constitutive theories for ${}^{(1)}_{d}\bar{\boldsymbol{\sigma}}$

Different choices of n, generators and invariants in (14.19) can be used to obtain desired constitutive theories for ${}^{(1)}_{d}\bar{\boldsymbol{\sigma}}$.

14.4.1 Linear constitutive theory of order n for ${}^{(1)}_{d}\bar{\boldsymbol{\sigma}}$

From the general constitutive theory (14.19) of order n based on integrity we can obtain a constitutive theory of order n that is linear in the components of ${}^{(i)}\boldsymbol{\gamma}$; $i = 1, 2, \dots, n$ by retaining generators ${}^{(i)}\boldsymbol{\gamma}$; $i = 1, 2, \dots, n$ and invariants $\text{tr}({}^{(i)}\boldsymbol{\gamma})$; $i = 1, 2, \dots, n$. Redefining new material coefficients, we can write (using (14.19))

$$
{}^{(1)}_{d}\bar{\boldsymbol{\sigma}} = \underset{\sim}{\sigma}^{0}\Big|_{\underset{\sim}{\Omega}} \boldsymbol{I} + \sum_{i=1}^{n} a_i \, {}^{(i)}\boldsymbol{\gamma} + \sum_{i=1}^{n} b_i (\text{tr}\, {}^{(i)}\boldsymbol{\gamma}) \boldsymbol{I} \tag{14.21}
$$

Using Voigt's notation (with the same arrangement of the components of ${}^{(0)}_{d}\bar{\boldsymbol{\sigma}}$ and ${}^{(i)}\boldsymbol{\gamma}$ as in Chapter 10, equation (10.16) we can write the following.

$$
\{{}^{(1)}_{d}\bar{\sigma}\} = \underset{\sim}{\sigma}^{0}\Big|_{\underset{\sim}{\Omega}} \begin{Bmatrix} 1 \\ 1 \\ 1 \\ 0 \\ 0 \\ 0 \end{Bmatrix} + \sum_{i=1}^{n} [\underset{\sim}{a}_i]\{{}^{(i)}\gamma\} \tag{14.22}
$$

In which $[\underset{\sim}{a}_i]$ is given by

$$
[\underset{\sim}{a}_i] = \begin{bmatrix} a_i + b_i & b_i & b_i & 0 & 0 & 0 \\ b_i & a_i + b_i & b_i & 0 & 0 & 0 \\ b_i & b_i & a_i + b_i & 0 & 0 & 0 \\ 0 & 0 & 0 & a_i & 0 & 0 \\ 0 & 0 & 0 & 0 & a_i & 0 \\ 0 & 0 & 0 & 0 & 0 & a_i \end{bmatrix} \tag{14.23}
$$

In this constitutive theory inclusion of the convected time derivative of each order of the strain tensor requires two additional material coefficients.

14.4.2 Linear constitutive theory of order one ($n = 1$)

From the constitutive theory of order n given by (14.21) we can obtain a linear constitutive theory of order one by using $n = 1$ (neglecting $\underset{\sim}{\sigma}^0\big|_{\Omega}\boldsymbol{I}$ term)

$$ {}^{(1)}_{\ d}\bar{\boldsymbol{\sigma}} = a_1 {}^{(1)}\boldsymbol{\gamma} + b_1(\operatorname{tr}{}^{(1)}\boldsymbol{\gamma})\boldsymbol{I} \tag{14.24} $$

Redefining material coefficients

$$ \begin{gathered} {}^{(1)}\bar{\boldsymbol{\sigma}} = a_1\bar{\boldsymbol{D}} + b_1(\operatorname{tr}\bar{\boldsymbol{D}})\boldsymbol{I} \\ \text{defining} \quad a_1 = 2\eta\big|_{\Omega} \quad \text{and} \quad b_1 = \kappa\big|_{\Omega} \end{gathered} \tag{14.25} $$

$$ {}^{(1)}_{\ d}\bar{\boldsymbol{\sigma}} = 2\eta\big|_{\Omega}\bar{\boldsymbol{D}} + \kappa\big|_{\Omega}(\operatorname{tr}\bar{\boldsymbol{D}})\boldsymbol{I} \tag{14.26} $$

Here, $\eta\big|_{\Omega}$ and $\kappa\big|_{\Omega}$ are similar to first and second viscosities. When $\eta\big|_{\Omega}$ and $\kappa\big|_{\Omega}$ are constant, (14.26) is a linear constitutive theory in $\bar{\boldsymbol{D}}$ for compressible matter. If the matter is incompressible, then $\operatorname{tr}\bar{\boldsymbol{D}} = 0$; hence, (14.26) reduces to the following constitutive theory.

$$ {}^{(1)}_{\ d}\bar{\boldsymbol{\sigma}} = 2\eta\big|_{\Omega}\bar{\boldsymbol{D}} \tag{14.27} $$

Remarks

(1) In this constitutive theory for hypo-elastic solid continua, we consider first convected time derivative of the Cauchy stress as a function of the first convected time derivative of the conjugate strain tensor.

(2) Since strains are not part of the constitutive theory for the stress tensor, the CBL in Eulerian description with velocities $\bar{\boldsymbol{v}}(\bar{\boldsymbol{x}}, t)$ as observable quantities are ideally suited.

(3) The constitutive theory for ${}^{(0)}_{\ e}\bar{\boldsymbol{\sigma}}$ defining volumetric deformation physics is borrowed from Chapter 8, as this physics is independent of the type of continua, type of deformation and the material constitution.

(4) The constitutive theories derived here hold for compressible thermoviscoelastic solid continua without memory. These constitutive theories also hold for incompressible hypo-elastic solid continua with the restriction $\bar{\boldsymbol{\nabla}} \cdot \bar{\boldsymbol{v}} = 0$ (divergence free velocity field).

(5) We keep in mind that solid continua without strains is rather a hypothetical solid material; hence, care must be taken in using this constitutive theory.

14.5 Constitutive theory for heat vector $\bar{\boldsymbol{q}}$

The constitutive theories for the heat vector $\bar{\boldsymbol{q}}$ derived in Chapter 12 in Eulerian description hold here as well, hence is not repeated.

14.6 General remarks

In the following we make some general remarks regarding the use of the mathematical model consisting of CBL in Eulerian description and the constitutive theory derived in this chapter.

(1) The mathematical model and the constitutive theories in Eulerian description use velocities $\bar{\boldsymbol{v}}(\bar{\boldsymbol{x}},t)$. Displacements $\bar{\boldsymbol{u}}$ and their gradients $\left[\frac{\partial\{\bar{u}\}}{\partial\{\bar{x}\}}\right]$ are neither considered nor used in the final equations constituting the mathematical model.

(2) Such materials have have no elasticity, hence do not have stiffness but naturally possess mass (ρ) and a dissipation mechanism due to strain rate dependent constitutive theory. Thus, strictly speaking such material cannot be classified in the category of any type of 'solid' as solid continua must have elasticity and stiffness.

(3) This misconception of calling such materials 'solids' goes back to the original thinking that for hyper-elastic material, the constitutive theories are derivable from strain energy density functionals, and those materials for which this is not possible are hypo-elastic solids. We keep in mind that only for isothermal elastic solids with further restriction of zero equilibrium stress, the constitutive theory derivation is possible using strain energy density functional. Thus, based on this definition all other materials (solids and fluids alike) are hypo-elastic solids. This assertion and conclusion is obviously not true. In this chapter, we have retained 'hypo-elastic solids' designation for such materials to be in conformity with what is commonly used in published works.

(4) Based on Remark (2), such materials cannot support stress waves and their propagation, and when the matter is incompressible, the theoretical speed of pressure wave is infinity. In case of compressible matter, compression waves due to pressure difference form pressure waves (referred to as shock) and corresponding density and temperature difference fronts that propagate during evolution.

14.7 Summary

The ordered rate constitutive theories have been derived for compressible and incompressible thermohypoelastic solid continua using convected time derivatives of the strain tensor up to order n and the first convected time derivative of the Cauchy stress tensor as constitutive variable. The basis independent measure of stress, its convected time derivative (${}^{(0)}\bar{\boldsymbol{\sigma}}$, ${}^{(1)}\bar{\boldsymbol{\sigma}}$) and basis independent convected time derivative (${}^{(i)}\boldsymbol{\gamma}$; $i = 1, 2, \dots, n$) of the conjugate strain tensor are used in the derivation of the constitutive theories.

Decomposition of Cauchy stress tensor ${}^{(0)}\bar{\boldsymbol{\sigma}}$ into equilibrium Cauchy stress ${}^{(0)}_{\;e}\bar{\boldsymbol{\sigma}}$ and deviatoric Cauchy stress tensor ${}^{(0)}_{\;d}\bar{\boldsymbol{\sigma}}$ facilitates constitutive theories for volumetric and distortional physics. Since the constitutive theory for ${}^{(0)}_{\;e}\bar{\boldsymbol{\sigma}}$ addressing volumetric deformation physics is invariant of the type of application and material constitution, its derivation in Chapter 8 holds here as well. The constitutive theory for the deviatoric Cauchy stress tensor is derived using the reduced form of the entropy inequality, ${}^{(1)}\bar{\boldsymbol{\sigma}}$ as a constitutive variable and the representation theorem. Material coefficients are derived. The constitutive theory based on integrity resulting from the representation theorem is used to present simplified linear constitutive theories for ${}^{(1)}\bar{\boldsymbol{\sigma}}$ with fewer material coefficients.

15 THERMODYNAMIC RELATIONS AND COMPLETE MATHEMATICAL MODELS

15.1 Introduction

From the development of the mathematical models based on CBL in Eulerian and Lagrangian descriptions in Chapters 6 and 7 and the derivations of the constitutive theories in Chapters 9–14, we note that we need additional information regarding thermodynamic pressure $p(\rho,\theta)$ or $\bar{p}(\bar{\rho},\bar{\theta})$, specific internal energy e or $\bar{e}$ and more specific details of the material coefficients such as specific heats, thermal conductivity, viscosity, modulus of elasticity, Poisson's ratio, etc. We consider some of the details in this chapter. Without this information, the CBL including the constitutive theories generally cannot be used in applications.

We have seen in Chapter 6 that CBL in Eulerian description can be derived in integral as well as in differential forms. Differential forms are only possible when the integrand in the integral forms satisfy localization theorem (Chapter 6). From the CBL in Chapter 6, we note that the integral and the differential forms are always not the same. The differential form of the CBL may be needed for further simplification of integral forms. In case of Lagrangian description, the integral form is only possible for CM.

Differential forms of the conservation and balance laws are presented in Eulerian as well as Lagrangian descriptions. The integral forms of the CBL in the Eulerian description can be obtained from Chapter 6 (not repeated here). We remark that the differential forms of the CBL require that localization theorem must hold in the derivation of CBL. This can be ensured if the deforming volume contains homogeneous and isotropic matter. If the deforming volume contains inhomogeneous and non-isotropic matter then we must ensure that the integrand in the integral form of CBL is continuous everywhere in the deforming volume, otherwise the differential forms of the CBL are not valid.

15.2 Thermodynamic pressure: equation of state

For compressible matter, when the additive decomposition of the Cauchy stress tensor ($\boldsymbol{\sigma}^{(0)}$ or $\bar{\boldsymbol{\sigma}}^{(0)}$) is used in the derivations of the constitutive theories, the constitutive theory for the equilibrium Cauchy stress tensor describing volumetric deformation results in thermodynamic pressure p or $\bar{p}$ for compressible matter that depends on density and temperature in the current configuration, i.e., $p = p(\rho,\theta)$ or $\bar{p} = \bar{p}(\bar{\rho},\bar{\theta})$. The dependence of thermodynamic pressure on density and temperature is generally called the *equation of state* and can be established analytically or experimentally using methods such as *kinetic theory* or *empiricism*. Continuum mechanics places no restrictions whatsoever on the equation of state except that the thermodynamic pressure must be continuous and differentiable in its arguments.

DOI: 10.1201/9781003105336-15

For gases, the equations of state are well established for the most part. We present these in the following without bias to Lagrangian or Eulerian description (hence overbar is removed from all quantities).

15.2.1 Perfect or ideal gas law

For all common gases, the relationship between p, ρ and θ can be described with reasonable accuracy in some finite range of p, ρ and θ by

$$p(\rho, \theta) = \rho \, R \, \theta \tag{15.1}$$

where R is the gas constant defined by the ratio of *Boltzmann's constant* K and the mass m of a single molecule. Equation (15.1) describes the perfect or ideal gas law.

$$R = \frac{K}{m} \tag{15.2}$$

15.2.2 Real gas models

In high temperature regions, vibrational and electronic excitations, molecular dissociation and ionization become important in the description of the behavior of gases. It has been shown that on an equilibrium basis, many of these effects can be accounted for with reasonable accuracy by introducing several free parameters in the equations of state that are determined experimentally or analytically. These types of mathematical descriptions describing the thermodynamic pressure as a function of density and temperature are generally referred to as *real gas models*. Equations of state such as *van der Waals*, *Redlich-Kwong*, *Beattie-Bridgman*, *Benedict-Webb-Rubin* etc. are some examples of real gas models. These are described in the following.

15.2.2.1 Van der Waals equation of state

$$p(\rho, \theta) = \frac{\rho R \theta}{1 - b\rho} - a\rho^2 \tag{15.3}$$

a, b are constants of the model.

15.2.2.2 Redlich-Kwong equation of state

$$p(\rho, \theta) = \frac{\rho R \theta}{1 - b\rho} - \frac{a\rho^2}{1 + b\rho}\theta^{-1/2} \tag{15.4}$$

a, b are model constants (different than those in (15.3)).

15.2.2.3 Beattie-Bridgman equation of state

$$p(\rho, \theta) = \rho R \theta \Big(1 - \frac{C\rho}{\theta^3}\Big)\Big(\frac{1}{\rho} + B_0(1 - b\rho)\Big) - \rho^2 A_0 (1 - a\rho) \tag{15.5}$$

C, B_0, b, A_0 and a are constants of the model.

15.2.2.4 Benedict-Webb-Rubin equation of state

$$\begin{aligned} p(\rho, \theta) &= \rho R \theta + \rho^2 \big(B_0 R \theta - A_0 - \frac{c_0}{\theta^2}\big) + \rho^3 (b R \theta - a) \\ &+ a\alpha\rho^6 + \frac{\rho c (1 + \gamma\rho^2)}{\theta^2} e^{-\nu\rho^2} \end{aligned} \tag{15.6}$$

B_0, A_0, c_0, b, a, α, c, γ and ν are constants of the model.

For more details on these models and others, see reference [72].

Remarks

(1) In the case of ideal gases with the ideal gas law, for a given temperature θ, pressure p versus ρ (or $v = 1/\rho$; specific volume) can be described by a straight line.

(2) In the case of real gas models, the relationship between p, ρ and θ is nonlinear.

(3) For a given temperature θ, one could study the behavior of p versus ρ (or $v = 1/\rho$). Figures 15.1(a) and 15.1(b) show typical behaviors of p versus ρ and p versus v at a fixed temperature θ. We make the following observations.

   (a) In the range AB, an increase in ρ results in increase in p.
   (b) In range BC, decreasing pressure results with increasing density ρ, thus B is an inflection or bifurcation point.
   (c) The real gas models are believed to produce anomalous behaviors in gases.
   (d) In Figure 15.1(a), the linear behavior of p versus ρ in the vicinity of the origin is in fact the ideal gas law.

(4) Derivation of the constitutive theories using the entropy inequality necessitates decomposition of the Cauchy stress tensor into equilibrium and deviatoric stress tensors. The equilibrium stress tensor addressing volumetric deformation is further established as thermodynamic pressure or a pressure field. Since the equation of state establishes $p = p(\rho, \theta)$, at this stage we have two choices:

   (a) Retain p as a dependent variable in the balance laws and introduce $p = p(\rho, \theta)$ as an additional equation in the mathematical model;
   (b) Eliminate p from the mathematical model using

$$\begin{aligned} p &= p(\rho, \theta) \\ \frac{\partial p}{\partial x_i} &= \frac{\partial p}{\partial \rho}\frac{\partial \rho}{\partial x_i} + \frac{\partial p}{\partial \theta}\frac{\partial \theta}{\partial x_i} \end{aligned} \tag{15.7}$$

Thus, *in the case of compressible fluids, we can maintain or eliminate p as a dependent variable from the mathematical model.*

15.2.3 Compressible solids

We have seen in Chapter 8 that entropy inequality in Eulerian description, CM in Eulerian description and Cauchy stress tensor ${}^{(0)}\bar{\boldsymbol{\sigma}}$ (basis independent notation) are the most fundamental relations and measures in which the physics of compressibility, incompressibility and non-isothermal physics are clearly observed. Necessity of the additive Cauchy stress decomposition

$${}^{(0)}\bar{\boldsymbol{\sigma}} = {}^{(0)}_{e}\bar{\boldsymbol{\sigma}} + {}^{(0)}_{d}\bar{\boldsymbol{\sigma}} \tag{15.8}$$

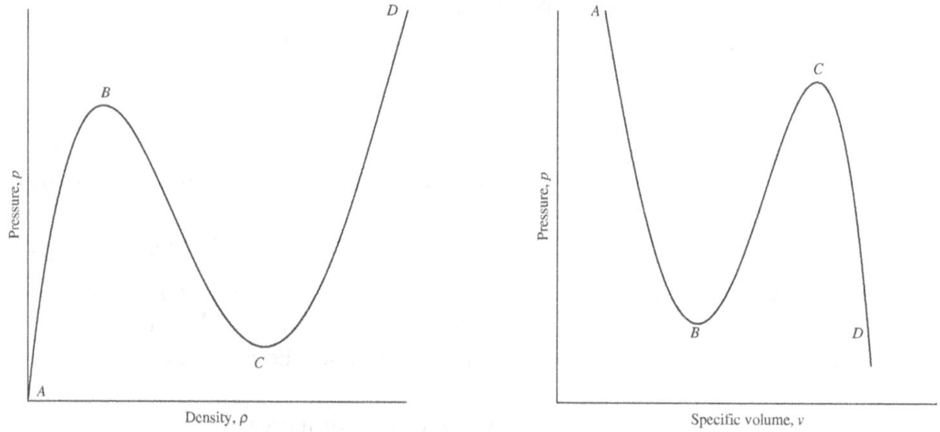

(a) p versus ρ (fixed θ) (b) p versus v (fixed θ)

Figure 15.1: Behavior of p versus ρ and p versus v at a fixed θ for real gases

arises to separate volumetric and distortional deformation physics. Compressibility is important in both fluid and solid continua, but in solid continua, compressibility is significant and important at higher pressures. Thus

$$ {}_{e}\boldsymbol{\sigma}^{(0)} = p(\rho, \theta)\boldsymbol{\delta} \tag{15.9}$$

holds for solid continua in which $p(\rho, \theta)$ is the equation of state for thermodynamic pressure $p(\rho, \theta)$ in Lagrangian description. Equations of state such as Mie-Grüneisen, the alternate form of the Mie-Grüneisen equation of state, equation of state for plastic, rubbers, glasses and polymers are some examples.

15.2.3.1 Mie-Grüneisen equation of state

$$p(\rho, \theta) = \frac{\rho_0 C_0^2 (1 - \rho/\rho_0)}{\left(1 - s(1 - \rho/\rho_0)\right)^2} \left(1 - \frac{\Gamma_0}{2}(1 - \rho/\rho_0)\right) + \Gamma_0 \rho_0 c_v (\theta - \theta_0) \tag{15.10}$$

15.2.3.2 Alternate Mie-Grüneisen equation of state

$$p(\rho, \theta) = \frac{\rho_0 C_0^2 (\rho/\rho_0 - 1)}{\left(\rho/\rho_0 - s(\rho/\rho_0 - 1)\right)^2} \left(\rho/\rho_0 - \frac{\Gamma_0}{2}(\rho/\rho_0 - 1)\right) + \Gamma_0 \rho_0 c_v (\theta - \theta_0) \tag{15.11}$$

In both cases, C_0, Γ_0, s and c_v are the bulk speed of sound, Grüneisen's gamma, Hugoniot slope coefficient and specific heat at constant volume. These are constants of the models.

For more details on these models and their validity for specific types of physics, see reference [25].

15.2.3.3 Equation of state for plastics, rubbers, glasses and polymers

In reference [51], the authors proposed the following equation of state for the thermodynamic pressure for plastics, rubbers, glasses and polymers

$$\ln(\underset{\sim}{v}) = -a_1(\ln(1 + a_2 a_3)) - a_4 \tag{15.12}$$

Where $\underset{\sim}{v}$ is dimensionless specific volume, ratio of specific volume v at pressure p (in kbar) to the specific volume (v_0) at reference temperature and pressure.

$$\underset{\sim}{v} = \frac{v}{v_0} = \frac{1/\rho}{1/\rho_0} = \frac{\rho_0}{\rho} = \frac{1}{\underset{\sim}{\rho}} \tag{15.13}$$

Where ρ is density at pressure p and $\underset{\sim}{\rho}$ is dimensionless density.

$$\begin{aligned} a_1 &= \frac{1}{B_\theta(0,\theta_0)Z + B'_\theta(0,\theta_0)} \quad ; \quad a_2 = \frac{B'_\theta(0,\theta_0)}{B(0,\theta_0)Z} \\ a_3 &= \frac{Zp}{B_\theta(0,\theta)Z + B'_\theta(0,\theta_0)} \quad ; \quad a_4 = \frac{Zp}{B_\theta(0,\theta_0)Z + B_\theta(0,\theta)} \end{aligned} \tag{15.14}$$

Here, θ_0 is reference temperature at which calculations are done (25°C in reference [51]). $B_\theta(p,\theta_0)$ is bulk modulus, $B'_\theta(p,\theta_0)$ is pressure derivative of bulk modulus, both at $p = 0$. Z is a pressure dependent parameter. Values of $B_\theta(0,\theta_0)$, $B'_\theta(0,\theta_0)$ and Z are listed in reference [51] for seventeen different materials. Compression tests were performed for the pressure p in the range of 0-4500 kbar. For p in the range 0-4500 kbar, we calculate $\underset{\sim}{v}$ using (15.12) for hard rubber [51] for which

$$B_\theta(0,\theta_0) = 56.471 \text{ kbar} \quad ; \quad B'_\theta(0,\theta_0) = 7.843 \quad ; \quad Z = 6.723 \times 10^{-3}$$

are used (obtained from Table 1 in reference [51]). Figure 15.2 shows a plot of $\underset{\sim}{v}$ versus pressure p.

Alternate form of equation of state

Equation of state (15.12) is an implicit expression for pressure as a function of density. When using the equation of state in CBL, it is preferable to have an explicit expression for $p = p(\rho)$. This can be done using a polynomial fit to the $\underset{\sim}{v}$ versus p data obtained using equation of state (15.12). Let us consider $p(\underset{\sim}{v})$ as a fifth degree polynomial in $\underset{\sim}{v}$.

$$p(\underset{\sim}{v}) = c_0 + c_1\underset{\sim}{v} + c_2\underset{\sim}{v}^2 + c_3\underset{\sim}{v}^3 + c_4\underset{\sim}{v}^4 + c_5\underset{\sim}{v}^5 \tag{15.15}$$

we consider following five $(\underset{\sim}{v}, p)$ data points obtained using (15.12).

$\underset{\sim}{v}$	0.10000E+01	0.78242E+00	0.58437E+00	0.20181E+00	0.17365E-01
p	0.00000E+00	0.90909E+02	0.27273E+03	0.15000E+04	0.45000E+04

(15.16)

and additionally we use $\left.\frac{\partial p}{\partial \underset{\sim}{v}}\right|_{\underset{\sim}{v}=1} = 0$ as the sixth condition to calculate coefficients c_i ; $i = 0, 1 \dots, 5$. The calculated coefficients are given in the following.

$$\begin{aligned} c_0 &= 0.49745E+04 \quad ; \quad c_1 = -0.28567E+05 \quad ; \quad c_2 = 0.73300E+05 \\ c_3 &= -0.96498E+05 \quad ; \quad c_4 = 0.62490E+05 \quad ; \quad c_5 = -0.15700E+05 \end{aligned} \tag{15.17}$$

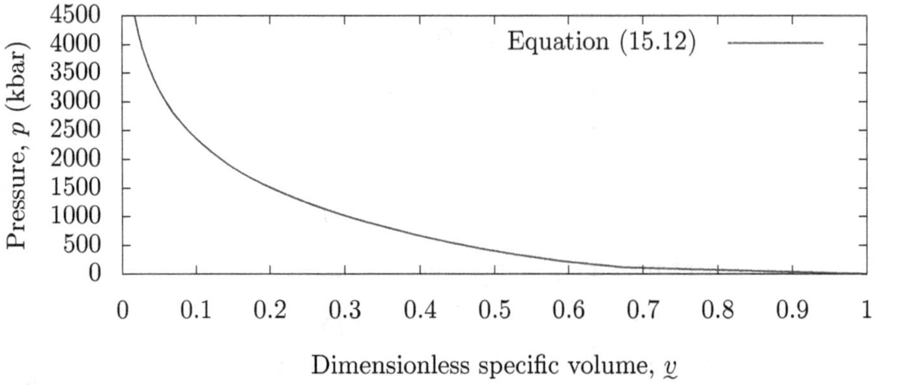

Figure 15.2: Pressure p (kbar) versus dimensionless specific volume $\underset{\sim}{v}$ using equation (15.12) of reference [51] (compression test)

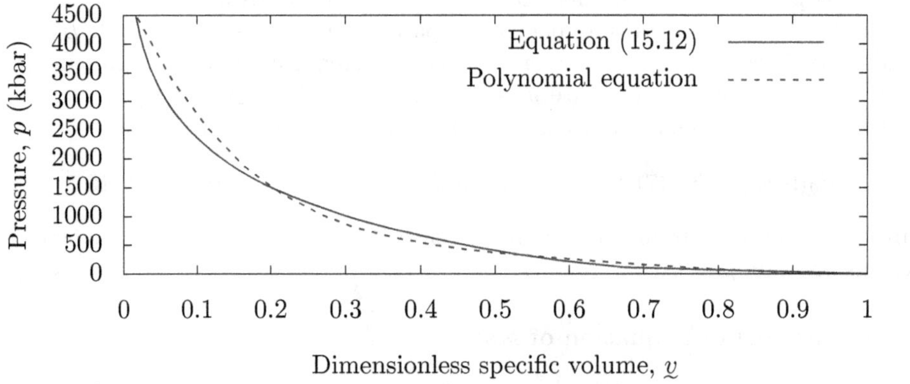

Figure 15.3: Pressure p (kbar) versus dimensionless specific volume $\underset{\sim}{v}$ using equations (15.12) and polynomial expression (15.15)

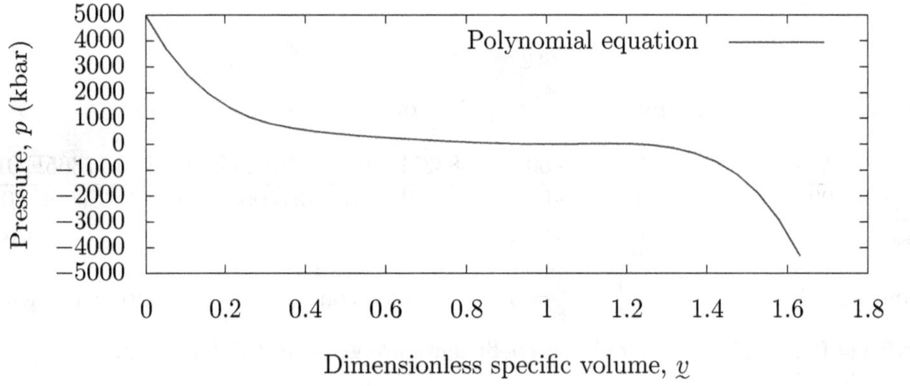

Figure 15.4: Pressure p (kbar) versus $\underset{\sim}{v}$ using polynomial expression (15.17) tension and compression

Finally, our equation of state becomes (15.15) with c_i ; $i = 0, 1 \ldots, 5$ defined in (15.17). We keep in mind that p in (15.12) is in kbar and $\underset{\sim}{v}$ is dimensionless specific volume. Figure 15.3 shows $\underset{\sim}{v}$ calculated using (15.12) and (15.15); the agreement is good.

Figure 15.4 shows plots of p versus $\underset{\sim}{v}$ in tension and compression. We note p versus $\underset{\sim}{v}$ behavior is different in tension and compression as expected. Equation (15.12) proposed in reference [51] is only valid for compression test data, p versus $\underset{\sim}{v}$ in tension reported in Figure 15.4 is obtained using equation (15.15).

15.2.3.4 Equation of state for natural and synthetic rubber like materials

In reference [12], the authors present experimental investigation for rubber like materials to derive equation of state. Their work shows resemblance of their equation of state to the van der Waals model of equation of state. They proposed the following

$$p(L, \theta) = A(L) + B(L)\theta$$

In which L is the percentage elongation; hence, it is related to density. They presented graphs for $A(L)$ and $B(L)$ (see reference [12] for details).

15.3 Internal energy

15.3.1 Compressible matter

In general, the specific internal energy $\bar{e}$ or e is a function of $\bar{p}$, $\bar{\rho}$ and $\bar{\theta}$ or p, ρ and θ, i.e.

$$\begin{aligned} \bar{e} &= \bar{e}(\bar{p}\,,\ \bar{\rho}\,,\ \bar{\theta}) \\ e &= e(p\,,\ \rho\,,\ \theta) \end{aligned} \tag{15.18}$$

Since $\bar{p} = \bar{p}(\bar{\rho}, \bar{\theta})$ or $p = p(\rho, \theta)$, the equation of state, $\bar{p}$ or p can be eliminated from (15.18) and we can write

$$\begin{aligned} \bar{e} &= \bar{e}(\bar{\rho}\,,\ \bar{\theta}) \\ e &= e(\rho\,,\ \theta) \end{aligned} \tag{15.19}$$

For example, in the case of compressible gases with variable specific heat we consider (using Eulerian description and considering $\bar{p} = \bar{p}(\bar{\rho}, \bar{\theta})$) [72].

$$\bar{e}(\bar{\rho}, \bar{\theta}) = \int_{\bar{\theta}_0}^{\bar{\theta}} \bar{c}_v \, d\bar{\theta} - \bar{\theta} \int_{\bar{\rho}_0}^{\bar{\rho}} \frac{1}{\bar{\rho}^2} \left(\left(\frac{\partial \bar{p}}{\partial \bar{\theta}} \right) \Big|_{\bar{\rho}} - \bar{p} \right) d\bar{\rho} \tag{15.20}$$

With specific heat $\bar{c}_v = \bar{c}_v(\bar{\theta})$, we can write (15.20) as

$$\bar{e}(\bar{\rho}, \bar{\theta}) = \bar{c}_v^* - \int_{\bar{\rho}_0}^{\bar{\rho}} \frac{1}{\bar{\rho}^2} \left(\left(\bar{\theta} \frac{\partial \bar{p}}{\partial \bar{\theta}} \right) \Big|_{\bar{\rho}} - \bar{p} \right) d\bar{\rho} \tag{15.21}$$

Where $\bar{c}_v^*$ is defined as

$$\bar{c}_v^* = \sum_{j=1}^{l} \bar{c}_j \bar{\theta}^j \tag{15.22}$$

Remarks

(1) Equation (15.20) is an example to illustrate how thermodynamic pressure affects $\bar{e}(\bar{\rho},\bar{\theta})$. Other suitable relations can also be used instead.

(2) Using the desired equation of state $\bar{p} = \bar{p}(\bar{\rho},\bar{\theta})$ and (15.21) we can obtain an explicit expression for $\bar{e} = \bar{e}(\bar{\rho},\bar{\theta})$ that corresponds to the chosen equation of state.

(3) This expression for $\bar{e} = \bar{e}(\bar{\rho},\bar{\theta})$ can be substituted in the energy equation to expand $\frac{D\bar{e}}{Dt}$.

15.3.2 Incompressible matter

For incompressible matter, there is no change in volume; hence, the density is constant. Therefore,

$$\begin{aligned} \bar{e} &= \bar{e}(\bar{\theta}) \\ e &= e(\theta) \end{aligned} \tag{15.23}$$

and the following expression (using Eulerian description at constant volume)

$$\bar{e}(\bar{\theta}) = \bar{c}_v \, (\bar{\theta} - \bar{\theta}_0) \tag{15.24}$$

is found to be adequate for specific internal energy. In (15.24) the specific heat can be a function of temperature, i.e., $\bar{c}_v(\bar{\theta})$ is valid.

15.4 Differential form of complete mathematical models in Lagrangian description

In the following we present differential form of complete mathematical models consisting of CBL and the constitutive theories in Lagrangian description for different types of deformation physics and material constitution. These mathematical models are ideally suited for solid continua. We present details of CBL and the constitutive theories. Argument tensors of the constitutive variables are given. The complete constitutive theory for deviatoric stress tensor based on integrity using representation theorem is presented first in terms of combined generators ${}^{\sigma}\underset{\sim}{\boldsymbol{G}}^i$; $i = 1, 2, \dots, N$ and combined invariants ${}^{\sigma}\underset{\sim}{I}^j$; $j = 1, 2, \dots, M$. Simplified forms of these constitutive theories are also given. Constitutive theories for equilibrium stress tensor are summarized also. Lastly, the constitutive theory for heat vector $\boldsymbol{q}$ is given. In all cases the material coefficients are functions of the invariants and θ (non-isothermal) in a known configuration $\underset{\sim}{\Omega}$.

15.4.1 CBL: Finite deformation, finite strain

Using $\boldsymbol{\sigma}^{[0]}$ and $\dot{\boldsymbol{\varepsilon}}_{[0]}$ as the rate of work conjugate pair, we have the following CBL (Chapter 7).

$$\rho_0(\boldsymbol{x}) = |\boldsymbol{J}|\rho(\boldsymbol{x}, t) \quad \text{(CM)} \tag{15.25}$$

$$\rho_0 \frac{\partial \boldsymbol{v}}{\partial t} - \rho_0 \boldsymbol{F}^b - \boldsymbol{\nabla} \cdot ([J] \cdot [\sigma^{[0]}]^T) = 0 \quad \text{(BLM)} \tag{15.26}$$

$$\boldsymbol{\sigma}^{(0)} = (\boldsymbol{\sigma}^{(0)})^T \implies \boldsymbol{\sigma}^{[0]} = (\boldsymbol{\sigma}^{[0]})^T \quad \text{(BAM)} \tag{15.27}$$

$$\rho_0 \frac{\partial e}{\partial t} + \boldsymbol{\nabla} \cdot \boldsymbol{q} - \boldsymbol{\sigma}^{[0]} : \dot{\boldsymbol{\varepsilon}}_{[0]} = 0 \quad \text{(FLT)} \tag{15.28}$$

$$-{}_d\boldsymbol{\sigma}^{[0]} : \dot{\boldsymbol{\varepsilon}}_{[0]} + \frac{\boldsymbol{q} \cdot \boldsymbol{g}}{\theta} \leq 0 \quad \text{(SLT)} \tag{15.29}$$

$$\boldsymbol{\sigma}^{[0]} = {}_e\boldsymbol{\sigma}^{[0]} + {}_d\boldsymbol{\sigma}^{[0]} \tag{15.30}$$

15.4.2 Constitutive theory for ${}_e\boldsymbol{\sigma}^{[0]}$: finite deformation, finite strain

$$ {}_e\boldsymbol{\sigma}^{[0]} = |\boldsymbol{J}|(\boldsymbol{J})^{-1} \cdot {}_e\boldsymbol{\sigma}^{(0)} \cdot (\boldsymbol{J}^T)^{-1} \tag{15.31}$$

where

$$\left.\begin{aligned}
{}_e\boldsymbol{\sigma}^{(0)} &= p(\rho,\theta)\boldsymbol{\delta} \quad ; && \text{compressible, non-isothermal} \\
{}_e\boldsymbol{\sigma}^{(0)} &= p(\rho)\boldsymbol{\delta} \quad ; && \text{compressible, isothermal} \\
{}_e\boldsymbol{\sigma}^{(0)} &= p(\theta)\boldsymbol{\delta} \quad ; && \text{incompressible, non-isothermal} \\
{}_e\boldsymbol{\sigma}^{(0)} &= p\boldsymbol{\delta} \quad ; && \text{incompressible, isothermal}
\end{aligned}\right. \tag{15.32}$$

15.4.3 Constitutive theory for ${}_d\boldsymbol{\sigma}^{[0]}$: finite deformation, finite strain

15.4.3.1 Thermoelastic solid continua

$$ {}_d\boldsymbol{\sigma}^{[0]} = {}_d\boldsymbol{\sigma}^{[0]}(\boldsymbol{\varepsilon}_{[0]}, \theta) \tag{15.33}$$

Based on integrity

$$\begin{gathered}
{}_d\boldsymbol{\sigma}^{[0]} = \underset{\sim}{\sigma}^0\big|_{\underline{\Omega}} \boldsymbol{I} + \sum_{j=1}^{M} {}^{\sigma}\underset{\sim}{a}_j ({}^{\sigma}\underset{\sim}{I}^j)\boldsymbol{I} + \sum_{j=1}^{M}\sum_{i=1}^{N} {}^{\sigma}\underset{\sim}{c}_{ij} ({}^{\sigma}\underset{\sim}{I}^j) {}^{\sigma}\underset{\sim}{\boldsymbol{G}}^i + \sum_{i=1}^{N} {}^{\sigma}\underset{\sim}{b}_i {}^{\sigma}\underset{\sim}{\boldsymbol{G}}^i \\
N = 2 \quad ; \quad {}^{\sigma}\underset{\sim}{\boldsymbol{G}}^1 = \boldsymbol{\varepsilon}_{[0]} \quad ; \quad {}^{\sigma}\underset{\sim}{\boldsymbol{G}}^2 = (\boldsymbol{\varepsilon}_{[0]})^2 \\
M = 3 \quad ; \quad {}^{\sigma}\underset{\sim}{I}^1 = I_{\varepsilon_{[0]}} \quad ; \quad {}^{\sigma}\underset{\sim}{I}^2 = II_{\varepsilon_{[0]}} \quad ; \quad {}^{\sigma}\underset{\sim}{I}^3 = III_{\varepsilon_{[0]}}
\end{gathered} \tag{15.34}$$

Linear constitutive theory

$$ {}_d\boldsymbol{\sigma}^{[0]} = \underset{\sim}{\sigma}^0\big|_{\underline{\Omega}} \boldsymbol{I} + 2\mu\boldsymbol{\varepsilon}_{[0]} + \lambda(\operatorname{tr}\boldsymbol{\varepsilon}_{[0]})\boldsymbol{I} \tag{15.35}$$

15.4.3.2 Thermoviscoelastic solid continua without memory

$$ {}_d\boldsymbol{\sigma}^{[0]} = {}_d\boldsymbol{\sigma}^{[0]}(\boldsymbol{\varepsilon}_{[0]}, \boldsymbol{\varepsilon}_{[i]}, \theta) \quad ; \quad i = 1, 2, \ldots, n \tag{15.36}$$

Based on integrity

$$ {}_d\boldsymbol{\sigma}^{[0]} = \underset{\sim}{\sigma}^0\big|_{\underline{\Omega}} \boldsymbol{I} + \sum_{j=1}^{M} {}^{\sigma}\underset{\sim}{a}_j ({}^{\sigma}\underset{\sim}{I}^j)\boldsymbol{I} + \sum_{j=1}^{M}\sum_{i=1}^{N} {}^{\sigma}\underset{\sim}{c}_{ij} ({}^{\sigma}\underset{\sim}{I}^j) {}^{\sigma}\underset{\sim}{\boldsymbol{G}}^i + \sum_{i=1}^{N} {}^{\sigma}\underset{\sim}{b}_i {}^{\sigma}\underset{\sim}{\boldsymbol{G}}^i \tag{15.37}$$

Linear constitutive theory of order n

$$_d\boldsymbol{\sigma}^{[0]} = \underset{\sim}{\sigma}^0\big|_{\underline{\Omega}}\boldsymbol{I} + a_1\boldsymbol{\varepsilon}_{[0]} + a_2(\operatorname{tr}\boldsymbol{\varepsilon}_{[0]})\boldsymbol{I} + \sum_{i=1}^{n} b_i^1\boldsymbol{\varepsilon}_{[i]} + \sum_{i=1}^{n} b_i^2(\operatorname{tr}\boldsymbol{\varepsilon}_{[i]})\boldsymbol{I} \tag{15.38}$$

Linear constitutive theory of order $n = 1$

$$_d\boldsymbol{\sigma}^{[0]} = \underset{\sim}{\sigma}^0\big|_{\underline{\Omega}}\boldsymbol{I} + a_1\boldsymbol{\varepsilon}_{[0]} + a_2(\operatorname{tr}\boldsymbol{\varepsilon}_{[0]})\boldsymbol{I} + b_1^1\boldsymbol{\varepsilon}_{[1]} + b_1^2(\operatorname{tr}\boldsymbol{\varepsilon}_{[1]})\boldsymbol{I} \tag{15.39}$$

$$\text{or} \quad _d\boldsymbol{\sigma}^{[0]} = \underset{\sim}{\sigma}^0\big|_{\underline{\Omega}}\boldsymbol{I} + 2\underset{\sim}{\mu}\boldsymbol{\varepsilon}_{[0]} + \underset{\sim}{\lambda}(\operatorname{tr}\boldsymbol{\varepsilon}_{[0]})\boldsymbol{I} + 2\underset{\sim}{\eta}\boldsymbol{\varepsilon}_{[1]} + \underset{\sim}{\kappa}(\operatorname{tr}\boldsymbol{\varepsilon}_{[1]})\boldsymbol{I} \tag{15.40}$$

15.4.3.3 Thermoviscoelastic solid continua with memory

$$\begin{gathered} _d\boldsymbol{\sigma}^{[m]} = {_d\boldsymbol{\sigma}^{[m]}}(\boldsymbol{\varepsilon}_{[0]}, \boldsymbol{\varepsilon}_{[i]}, {_d\boldsymbol{\sigma}^{[j]}}, \theta) \\ i = 1, 2, \dots, n \quad ; \quad j = 1, 2, \dots, m-1 \end{gathered} \tag{15.41}$$

Based on integrity

$$_d\boldsymbol{\sigma}^{[m]} = \underset{\sim}{\sigma}^0\big|_{\underline{\Omega}}\boldsymbol{I} + \sum_{j=1}^{M} {^\sigma}\underset{\sim}{a}_j({^\sigma}\underset{\sim}{I}^j)\boldsymbol{I} + \sum_{j=1}^{M}\sum_{i=1}^{N} {^\sigma}\underset{\sim}{c}_{ij}({^\sigma}\underset{\sim}{I}^j){^\sigma}\underset{\sim}{\boldsymbol{G}}^i + \sum_{i=1}^{N} {^\sigma}\underset{\sim}{b}_i{^\sigma}\underset{\sim}{\boldsymbol{G}}^i \tag{15.42}$$

Linear constitutive theory of orders m and n

$$\begin{aligned} _d\boldsymbol{\sigma}^{[m]} = {} & \underset{\sim}{\sigma}^0\big|_{\underline{\Omega}}\boldsymbol{I} + a_1\boldsymbol{\varepsilon}_{[0]} + a_2(\operatorname{tr}\boldsymbol{\varepsilon}_{[0]})\boldsymbol{I} + \sum_{i=1}^{n} b_i^1\boldsymbol{\varepsilon}_{[i]} + \sum_{i=1}^{n} b_i^2(\operatorname{tr}\boldsymbol{\varepsilon}_{[i]})\boldsymbol{I} \\ & + \sum_{j=0}^{m-1} c_j^1({_d\boldsymbol{\sigma}^{[j]}}) + \sum_{j=0}^{m-1} c_j^2(\operatorname{tr}({_d\boldsymbol{\sigma}^{[j]}}))\boldsymbol{I} \end{aligned} \tag{15.43}$$

Linear constitutive theory of orders $m = 1$, $n = 1$

$$_d\boldsymbol{\sigma}^{[0]} + \lambda({_d\boldsymbol{\sigma}^{[1]}}) = 2\underset{\sim}{\mu}\boldsymbol{\varepsilon}_{[0]} + \underset{\sim}{\lambda}(\operatorname{tr}\boldsymbol{\varepsilon}_{[0]})\boldsymbol{I} + 2\underset{\sim}{\eta}\boldsymbol{\varepsilon}_{[1]} + \underset{\sim}{\kappa}(\operatorname{tr}\boldsymbol{\varepsilon}_{[1]})\boldsymbol{I} + \underset{\sim}{\beta}(\operatorname{tr}{_d\boldsymbol{\sigma}^{[0]}})\boldsymbol{I} \tag{15.44}$$

In case of isothermal deformation physics, dependence of all quantities in Sections 15.4.1 - 15.4.3 on θ is removed.

15.4.4 CBL: Small deformation, small strain

$$\rho_0(\boldsymbol{x}) = |\boldsymbol{J}|\rho(\boldsymbol{x}, t) \quad \text{(CM)} \tag{15.45}$$

$$\rho_0\frac{\partial \boldsymbol{v}}{\partial t} - \rho_0\boldsymbol{F}^b - \boldsymbol{\nabla}\cdot\boldsymbol{\sigma} = 0 \quad \text{(BLM)} \tag{15.46}$$

$$\boldsymbol{\sigma} = \boldsymbol{\sigma}^T \quad \text{(BAM)} \tag{15.47}$$

$$\rho_0\frac{\partial e}{\partial t} + \boldsymbol{\nabla}\cdot\boldsymbol{q} - \boldsymbol{\sigma} : \dot{\boldsymbol{\varepsilon}} = 0 \quad \text{(FLT)} \tag{15.48}$$

$$-\boldsymbol{\sigma} : \dot{\boldsymbol{\varepsilon}} + \frac{\boldsymbol{q}\cdot\boldsymbol{g}}{\theta} \leq 0 \quad \text{(SLT)} \tag{15.49}$$

$$\boldsymbol{\sigma} = {_e\boldsymbol{\sigma}} + {_d\boldsymbol{\sigma}} \tag{15.50}$$

15.4.5 Constitutive theory for ${}_e\boldsymbol{\sigma}$: small deformation, small strain

$$\begin{aligned}
{}_e\boldsymbol{\sigma} &= p(\rho,\theta)\boldsymbol{\delta} \quad ; \quad \text{compressible, non-isothermal} \\
{}_e\boldsymbol{\sigma} &= p(\rho)\boldsymbol{\delta} \quad ; \quad \text{compressible, isothermal} \\
{}_e\boldsymbol{\sigma} &= p(\theta)\boldsymbol{\delta} \quad ; \quad \text{incompressible, non-isothermal} \\
{}_e\boldsymbol{\sigma} &= p\boldsymbol{\delta} \quad ; \quad \text{incompressible, isothermal}
\end{aligned} \tag{15.51}$$

15.4.6 Constitutive theory for ${}_d\boldsymbol{\sigma}$: small deformation, small strain

15.4.6.1 Thermoelastic solid continua

$$ {}_d\boldsymbol{\sigma} = {}_d\boldsymbol{\sigma}(\boldsymbol{\varepsilon},\theta) \tag{15.52}$$

Based on integrity

$$\begin{gathered}
{}_d\boldsymbol{\sigma} = \underset{\sim}{\sigma}^0\big|_{\underset{\sim}{\Omega}}\boldsymbol{I} + \sum_{j=1}^{M} {}^{\sigma}\underset{\sim}{a}_j({}^{\sigma}\underset{\sim}{I}^j)\boldsymbol{I} + \sum_{j=1}^{M}\sum_{i=1}^{N} {}^{\sigma}\underset{\sim}{c}_{ij}({}^{\sigma}\underset{\sim}{I}^j)\,{}^{\sigma}\underset{\sim}{\boldsymbol{G}}^i + \sum_{i=1}^{N} {}^{\sigma}\underset{\sim}{b}_i\,{}^{\sigma}\underset{\sim}{\boldsymbol{G}}^i \\
N = 2 \quad ; \quad {}^{\sigma}\underset{\sim}{\boldsymbol{G}}^1 = \boldsymbol{\varepsilon} \quad ; \quad {}^{\sigma}\underset{\sim}{\boldsymbol{G}}^2 = (\boldsymbol{\varepsilon})^2 \\
M = 3 \quad ; \quad {}^{\sigma}\underset{\sim}{I}^1 = I_{\varepsilon} \quad ; \quad {}^{\sigma}\underset{\sim}{I}^2 = II_{\varepsilon} \quad ; \quad {}^{\sigma}\underset{\sim}{I}^3 = III_{\varepsilon}
\end{gathered} \tag{15.53}$$

Linear constitutive theory

$$ {}_d\boldsymbol{\sigma} = \underset{\sim}{\sigma}^0\big|_{\underset{\sim}{\Omega}}\boldsymbol{I} + 2\mu\boldsymbol{\varepsilon} + \lambda(\operatorname{tr}\boldsymbol{\varepsilon})\boldsymbol{I} \tag{15.54}$$

15.4.6.2 Thermoviscoelastic solid continua without memory

$$ {}_d\boldsymbol{\sigma} = {}_d\boldsymbol{\sigma}(\boldsymbol{\varepsilon},\boldsymbol{\varepsilon}_{(i)},\theta) \quad ; \quad i = 1,2,\dots,n \tag{15.55}$$

Based on integrity

$$ {}_d\boldsymbol{\sigma} = \underset{\sim}{\sigma}^0\big|_{\underset{\sim}{\Omega}}\boldsymbol{I} + \sum_{j=1}^{M} {}^{\sigma}\underset{\sim}{a}_j({}^{\sigma}\underset{\sim}{I}^j)\boldsymbol{I} + \sum_{j=1}^{M}\sum_{i=1}^{N} {}^{\sigma}\underset{\sim}{c}_{ij}({}^{\sigma}\underset{\sim}{I}^j)\,{}^{\sigma}\underset{\sim}{\boldsymbol{G}}^i + \sum_{i=1}^{N} {}^{\sigma}\underset{\sim}{b}_i\,{}^{\sigma}\underset{\sim}{\boldsymbol{G}}^i \tag{15.56}$$

Linear constitutive theory of order n

$$ {}_d\boldsymbol{\sigma} = \underset{\sim}{\sigma}^0\big|_{\underset{\sim}{\Omega}}\boldsymbol{I} + a_1\boldsymbol{\varepsilon} + a_2(\operatorname{tr}\boldsymbol{\varepsilon})\boldsymbol{I} + \sum_{i=1}^{n} b_i^1\boldsymbol{\varepsilon}_{(i)} + \sum_{i=1}^{n} b_i^2(\operatorname{tr}\boldsymbol{\varepsilon}_{(i)})\boldsymbol{I} \tag{15.57}$$

Linear constitutive theory of order $n = 1$

$$ {}_d\boldsymbol{\sigma} = \underset{\sim}{\sigma}^0\big|_{\underset{\sim}{\Omega}}\boldsymbol{I} + a_1\boldsymbol{\varepsilon} + a_2(\operatorname{tr}\boldsymbol{\varepsilon})\boldsymbol{I} + b_1^1\boldsymbol{\varepsilon}_{(1)} + b_1^2(\operatorname{tr}\boldsymbol{\varepsilon}_{(1)})\boldsymbol{I} \tag{15.58}$$

$$ \text{or} \quad {}_d\boldsymbol{\sigma} = \underset{\sim}{\sigma}^0\big|_{\underset{\sim}{\Omega}}\boldsymbol{I} + 2\underset{\sim}{\mu}\boldsymbol{\varepsilon} + \underset{\sim}{\lambda}(\operatorname{tr}\boldsymbol{\varepsilon})\boldsymbol{I} + \underset{\sim}{\eta}\boldsymbol{\varepsilon}_{(1)} + \underset{\sim}{\kappa}(\operatorname{tr}\boldsymbol{\varepsilon}_{(1)})\boldsymbol{I} \tag{15.59}$$

15.4.6.3 Thermoviscoelastic solid continua with memory

$$ {}_d\boldsymbol{\sigma}^{(m)} = {}_d\boldsymbol{\sigma}^{(m)}(\boldsymbol{\varepsilon},\boldsymbol{\varepsilon}_{(i)},{}_d\boldsymbol{\sigma}^{(j)},\theta) \quad ; \quad i = 1,2,\dots,n \quad ; \quad j = 0,1,\dots,m-1 \tag{15.60}$$

Based on integrity

$$
{}_d\boldsymbol{\sigma}^{(m)} = \underset{\sim}{\sigma}^0\Big|_{\underline{\Omega}}\boldsymbol{I} + \sum_{j=1}^{M} {}^{\sigma}\underset{\sim}{a}_j({}^{\sigma}\underset{\sim}{I}^j)\boldsymbol{I} + \sum_{j=1}^{M}\sum_{i=1}^{N} {}^{\sigma}\underset{\sim}{c}_{ij}({}^{\sigma}\underset{\sim}{I}^j){}^{\sigma}\underset{\sim}{\boldsymbol{G}}^i + \sum_{i=1}^{N} {}^{\sigma}\underset{\sim}{b}_i\,{}^{\sigma}\underset{\sim}{\boldsymbol{G}}^i \tag{15.61}
$$

Linear constitutive theory of orders m and n

$$
\begin{aligned}
{}_d\boldsymbol{\sigma}^{(m)} = {} & \underset{\sim}{\sigma}^0\Big|_{\underline{\Omega}}\boldsymbol{I} + a_1\boldsymbol{\varepsilon} + a_2(\operatorname{tr}\boldsymbol{\varepsilon})\boldsymbol{I} + \sum_{i=1}^{n} b_i^1\boldsymbol{\varepsilon}_{(i)} + \sum_{i=1}^{n} b_i^2(\operatorname{tr}\boldsymbol{\varepsilon}_{(i)})\boldsymbol{I} \\
& + \sum_{j=0}^{m-1} c_j^1({}_d\boldsymbol{\sigma}^{(j)}) + \sum_{j=0}^{m-1} c_j^2(\operatorname{tr}({}_d\boldsymbol{\sigma}^{(j)}))\boldsymbol{I}
\end{aligned} \tag{15.62}
$$

Linear constitutive theory of orders $m = 1$, $n = 1$

$$
{}_d\boldsymbol{\sigma}^{(0)} + \lambda({}_d\boldsymbol{\sigma}^{(1)}) = 2\underset{\sim}{\mu}\boldsymbol{\varepsilon} + \underset{\sim}{\lambda}(\operatorname{tr}\boldsymbol{\varepsilon})\boldsymbol{I} + \underset{\sim}{\eta}\boldsymbol{\varepsilon}_{(1)} + \underset{\sim}{\kappa}(\operatorname{tr}\boldsymbol{\varepsilon}_{(1)})\boldsymbol{I} + \underset{\sim}{\beta}(\operatorname{tr}({}_d\boldsymbol{\sigma}^{(0)}))\boldsymbol{I} \tag{15.63}
$$

In the case of isothermal deformation physics details in Sections 15.4.1 - 15.4.3 remain valid except that dependence of all quantities in Sections 15.4.1 - 15.4.3 on θ is removed.

15.4.7 Constitutive theory for heat vector $\boldsymbol{q}$

Strictly based on entropy inequality

$$
\frac{\boldsymbol{q}\cdot\boldsymbol{g}}{\theta} \leq 0 \tag{15.64}
$$

$$
\boldsymbol{q} = -k(\theta)\boldsymbol{g} \tag{15.65}
$$

Based on representation theorem

$$
\boldsymbol{q} = \boldsymbol{q}(\boldsymbol{g},\theta) \tag{15.66}
$$

$$
\boldsymbol{q} = -k\boldsymbol{g} - k_1(\boldsymbol{g}\cdot\boldsymbol{g})\boldsymbol{g} - k_2(\theta - \theta|_{\underline{\Omega}})\boldsymbol{g} \tag{15.67}
$$

Material coefficients k, k_1, k_2 can be functions of $(\boldsymbol{g}\cdot\boldsymbol{g})|_{\underline{\Omega}}$ and $\theta|_{\underline{\Omega}}$

15.5 Differential form of complete mathematical model in Eulerian description

In the following, we present conservation and balance laws and constitutive theories in Eulerian description in which velocities $\bar{\boldsymbol{v}}(\bar{\boldsymbol{x}},t)$ are observable quantities. These mathematical models are suitable for fluent continua in which the complex motion (displacements) of the material points is difficult to monitor. The constitutive theories in fluent continua do not consider strain. Thus, displacement gradients are not needed either. Derivation of the constitutive theories follow the same procedure as in the case of solid continua (Section 15.4) and the same concepts related to volumetric and distortion deformation physics as used in Section 15.4 hold here as well. The CBL and the constitutive theories for the deviatoric Cauchy stress tensor are presented using basis independent notation for the Cauchy stress measure and its convected time derivatives, i.e., ${}^{(0)}\bar{\boldsymbol{\sigma}}$; $j = 0, 1, \ldots, m$ and ${}^{(i)}\boldsymbol{\gamma}$; $i = 1, 2, \ldots, n$, the convected time derivatives of the conjugate strain tensor using the reduced form of the entropy inequality and the representation theorem. The constitutive theories derived based on integrity are simplified to present linear theories. Constitutive theories for heat conduction vector $\bar{\boldsymbol{q}}$ are also given.

15.5.1 CBL: Conservation and balance laws

Using ${}^{(0)}\bar{\boldsymbol{\sigma}}$ (basis independent notation) and $\bar{\boldsymbol{D}}$ as rate of work conjugate pair and noting that $\bar{\boldsymbol{D}} = {}^{(1)}\boldsymbol{\gamma} = \boldsymbol{\gamma}^{(1)} = \boldsymbol{\gamma}_{(1)} = {}^{(1)}\boldsymbol{\gamma}^J$ we can write the following for CBL in Eulerian description (Chapter 6)

$$\dot{\bar{\rho}} + \bar{\rho}(\bar{\boldsymbol{\nabla}} \cdot \bar{\boldsymbol{v}}) = 0 \quad \text{(CM)} \tag{15.68}$$

$$\bar{\rho}\frac{D\bar{\boldsymbol{v}}}{Dt} - \bar{\rho}\bar{\boldsymbol{F}}^b - \bar{\boldsymbol{\nabla}} \cdot ({}^{(0)}\bar{\boldsymbol{\sigma}}) = 0 \quad \text{(BLM)} \tag{15.69}$$

$${}^{(0)}\bar{\boldsymbol{\sigma}} = {}^{(0)}\bar{\boldsymbol{\sigma}}^T \quad \text{(BAM)} \tag{15.70}$$

$$\bar{\rho}\frac{D\bar{e}}{Dt} + \bar{\boldsymbol{\nabla}} \cdot \bar{\boldsymbol{q}} - {}^{(0)}\bar{\boldsymbol{\sigma}} : \bar{\boldsymbol{D}} = 0 \quad \text{(FLT)} \tag{15.71}$$

$$-{}^{(0)}_{\ d}\bar{\boldsymbol{\sigma}} : \bar{\boldsymbol{D}} + \frac{\bar{\boldsymbol{q}} \cdot \bar{\boldsymbol{g}}}{\bar{\theta}} \leq 0 \quad \text{(SLT)} \tag{15.72}$$

$${}^{(0)}\bar{\boldsymbol{\sigma}} = {}^{(0)}_{\ e}\bar{\boldsymbol{\sigma}} + {}^{(0)}_{\ d}\bar{\boldsymbol{\sigma}} \tag{15.73}$$

15.5.2 Constitutive theory for ${}^{(0)}_{\ e}\bar{\boldsymbol{\sigma}}$

Following the derivations in Chapter 8, we have

$${}^{(0)}_{\ e}\bar{\boldsymbol{\sigma}} = \bar{p}(\bar{\rho},\bar{\theta})\boldsymbol{\delta} \qquad \text{compressible, non-isothermal} \tag{15.74}$$

$${}^{(0)}_{\ e}\bar{\boldsymbol{\sigma}} = \bar{p}(\bar{\rho})\boldsymbol{\delta} \qquad \text{compressible, isothermal} \tag{15.75}$$

$${}^{(0)}_{\ e}\bar{\boldsymbol{\sigma}} = \bar{p}(\bar{\theta})\boldsymbol{\delta} \qquad \text{incompressible, non-isothermal} \tag{15.76}$$

$${}^{(0)}_{\ e}\bar{\boldsymbol{\sigma}} = \bar{p}\boldsymbol{\delta} \qquad \text{incompressible, isothermal} \tag{15.77}$$

Fluent continua always have a dissipation mechanism due to viscosity, thus, the isothermal case is rather hypothetical.

15.5.3 Constitutive theory for ${}^{(0)}_{\ d}\bar{\boldsymbol{\sigma}}$

15.5.3.1 Thermoviscous fluent continua

Such fluids have a dissipation mechanism but no memory.

$${}^{(0)}_{\ d}\bar{\boldsymbol{\sigma}} = {}^{(0)}_{\ d}\bar{\boldsymbol{\sigma}}(\bar{\rho}, {}^{(i)}\boldsymbol{\gamma}, \bar{\theta}) \quad ; \quad i = 1, 2, \dots, n \tag{15.78}$$

Based on integrity (of order n)

$${}^{(0)}_{\ d}\bar{\boldsymbol{\sigma}} = \underset{\sim}{\sigma}^0\Big|_{\underset{\sim}{\Omega}} \boldsymbol{I} + \sum_{j=1}^{M} {}^{\sigma}\underset{\sim}{a}_j({}^{\sigma}\underset{\sim}{I}^j)\boldsymbol{I} + \sum_{j=1}^{M}\sum_{i=1}^{N} {}^{\sigma}\underset{\sim}{c}_{ij}({}^{\sigma}\underset{\sim}{I}^j){}^{\sigma}\underset{\sim}{\boldsymbol{G}}^i + \sum_{i=1}^{N} {}^{\sigma}\underset{\sim}{b}_i{}^{\sigma}\underset{\sim}{\boldsymbol{G}}^i \tag{15.79}$$

Linear constitutive theory of order n

$${}^{(0)}_{\ d}\bar{\boldsymbol{\sigma}} = \underset{\sim}{\sigma}^0\Big|_{\underset{\sim}{\Omega}} \boldsymbol{I} + \sum_{i=1}^{n} a_i{}^{(i)}\boldsymbol{\gamma} + \sum_{i=1}^{n} b_i(\text{tr}({}^{(i)}\boldsymbol{\gamma}))\boldsymbol{I} \tag{15.80}$$

Linear constitutive theory of order $n = 1$

$$ {}^{(0)}_{\ d}\bar{\boldsymbol{\sigma}} = \underset{\sim}{\sigma}^0\big|_{\underline{\Omega}}\boldsymbol{I} + 2\eta\bar{\boldsymbol{D}} + \kappa(\operatorname{tr}\bar{\boldsymbol{D}})\boldsymbol{I} \tag{15.81} $$

Where η and κ are usual first and second viscosities that can be functions of the invariants of $\bar{\boldsymbol{D}}$, $\bar{\rho}$ and $\bar{\theta}$ in a known configuration $\underline{\Omega}$.

15.5.3.2 Thermoviscoelastic fluent continua with memory

We consider rate constitutive theory of up to order m in deviatoric Cauchy stress tensor and of up to n of the convected time derivative of the conjugate strain tensor. Choosing basis independent notation.

$$ \begin{gathered} {}^{(m)}_{\ d}\bar{\boldsymbol{\sigma}} = {}^{(m)}_{\ d}\bar{\boldsymbol{\sigma}}(\bar{\rho}, {}^{(i)}\boldsymbol{\gamma}, {}^{(j)}_{\ d}\bar{\boldsymbol{\sigma}}, \theta) \\ i = 1, 2, \ldots, n \quad ; \quad j = 0, 1, \ldots, m-1 \end{gathered} \tag{15.82} $$

Based on integrity

$$ {}^{(m)}_{\ d}\bar{\boldsymbol{\sigma}} = \underset{\sim}{\sigma}^0\big|_{\underline{\Omega}}\boldsymbol{I} + \sum_{j=1}^{M} {}^{\sigma}\underset{\sim}{a}_j({}^{\sigma}\underset{\sim}{I}^j)\boldsymbol{I} + \sum_{j=1}^{M}\sum_{i=1}^{N} {}^{\sigma}\underset{\sim}{c}_{ij}({}^{\sigma}\underset{\sim}{I}^j)\,{}^{\sigma}\underset{\sim}{\boldsymbol{G}}^i + \sum_{i=1}^{N} {}^{\sigma}\underset{\sim}{b}_i\,{}^{\sigma}\underset{\sim}{\boldsymbol{G}}^i \tag{15.83} $$

Linear constitutive theory of orders m and n

$$ \begin{aligned} {}^{(m)}_{\ d}\bar{\boldsymbol{\sigma}} = {} & \underset{\sim}{\sigma}^0\big|_{\underline{\Omega}}\boldsymbol{I} + \sum_{i=1}^{n} a_i^1({}^{(i)}\boldsymbol{\gamma}) + \sum_{i=1}^{n} a_i^2(\operatorname{tr}({}^{(i)}\boldsymbol{\gamma}))\boldsymbol{I} \\ & + \sum_{j=0}^{m-1} b_j^1({}^{(j)}_{\ d}\bar{\boldsymbol{\sigma}}) + \sum_{j=0}^{m-1} b_j^2(\operatorname{tr}({}^{(j)}_{\ d}\bar{\boldsymbol{\sigma}}))\boldsymbol{I} \end{aligned} \tag{15.84} $$

Linear constitutive theory of orders $m = 1$, $n = 1$

$$ {}^{(0)}_{\ d}\bar{\boldsymbol{\sigma}} + \lambda({}^{(1)}_{\ d}\bar{\boldsymbol{\sigma}}) = \underset{\sim}{\sigma}^0\big|_{\underline{\Omega}}\boldsymbol{I} + 2\eta({}^{(1)}\boldsymbol{\gamma}) + \kappa(\operatorname{tr}({}^{(1)}\boldsymbol{\gamma}))\boldsymbol{I} + \underset{\sim}{\beta}(\operatorname{tr}({}^{(0)}_{\ d}\bar{\boldsymbol{\sigma}}))\boldsymbol{I} \tag{15.85} $$

15.5.3.3 Maxwell model: linear model ($m = 1$, $n = 1$)

This constitutive model is primarily used for dilute polymeric fluids

$$ {}^{(0)}_{\ d}\bar{\boldsymbol{\sigma}} + \lambda({}^{(1)}_{\ d}\bar{\boldsymbol{\sigma}}) = 2\eta({}^{(1)}\boldsymbol{\gamma}) + \kappa(\operatorname{tr}({}^{(1)}\boldsymbol{\gamma}))\boldsymbol{I} \tag{15.86} $$

15.5.3.4 Oldroyd-B model: quasi-linear model ($m = 1$, $n = 2$)

This constitutive model is also used for dilute polymeric fluids.

$$ {}^{(0)}_{\ d}\bar{\boldsymbol{\sigma}} + \lambda({}^{(1)}_{\ d}\bar{\boldsymbol{\sigma}}) = 2\eta({}^{(1)}\boldsymbol{\gamma}) + \kappa(\operatorname{tr}({}^{(1)}\boldsymbol{\gamma}))\boldsymbol{I} + 2\eta\lambda_2({}^{(2)}\boldsymbol{\gamma}) \tag{15.87} $$

15.5.3.5 Giesekus model: non-linear model ($m = 1$, $n = 2$)

This constitutive model is primarily used for dense polymeric fluids.

$$ {}^{(0)}_{\ d}\bar{\boldsymbol{\sigma}} + \lambda({}^{(1)}_{\ d}\bar{\boldsymbol{\sigma}}) = 2\eta\bar{\boldsymbol{D}} + \frac{\lambda}{\eta}\alpha({}^{(0)}_{\ d}\bar{\boldsymbol{\sigma}})^2 + \kappa(\operatorname{tr}({}^{(1)}\boldsymbol{\gamma}))\boldsymbol{I} \tag{15.88} $$

15.5.4 Constitutive theory for heat vector $\bar{\boldsymbol{q}}$

Strictly based on entropy inequality

$$\frac{\bar{\boldsymbol{q}} \cdot \bar{\boldsymbol{g}}}{\bar{\theta}} \leq 0 \tag{15.89}$$

Based on representation theorem

$$\bar{\boldsymbol{q}} = \bar{\boldsymbol{q}}(\bar{\boldsymbol{g}}, \bar{\theta}) \tag{15.90}$$

$$\bar{\boldsymbol{q}} = -k\bar{\boldsymbol{g}} - k_1(\bar{\boldsymbol{g}} \cdot \bar{\boldsymbol{g}})\bar{\boldsymbol{g}} - k_2(\bar{\theta} - \bar{\theta}\lfloor_{\Omega})\bar{\boldsymbol{g}} \tag{15.91}$$

Material coefficients k, k_1, k_2 can be functions of $(\bar{\boldsymbol{g}} \cdot \bar{\boldsymbol{g}})\lfloor_{\Omega}$ and $\bar{\theta}\lfloor_{\Omega}$

15.6 Summary

In this chapter, a summary of the mathematical models resulting from conservation and balance laws, constitutive theories and thermodynamic relations is presented in Lagrangian and Eulerian descriptions. In all cases, the mathematical models have closure, i.e., there are as many equations as the number of dependent variables. In all cases, the constitutive theories based on integrity as well as simplified constitutive theories are presented.

Problems

15.1 Consider a compressible, thermoelastic solid with finite deformation. Derive explicit forms of the conservation and balance laws and the constitutive equations in $\mathbb{R}^1$, $\mathbb{R}^2$, $\mathbb{R}^3$ in x-frame. Assume the second Piola-Kirchhoff stress to be a linear function of the Green's strain tensor and Fourier heat conduction law for the heat vector. Simplify the mathematical models for the incompressible case and for small deformation. Derive mathematical models in which stresses and heat vector are eliminated by substituting them in the momentum equations, energy equation and entropy inequality.

15.2 Consider a compressible thermoviscoelastic solid without memory but with finite deformation. Derive explicit forms of the conservation and balance laws and the constitutive equations in $\mathbb{R}^1$, $\mathbb{R}^2$, $\mathbb{R}^3$ in x-frame. Assume the deviatoric second Piola-Kirchhoff stress tensor to be a linear function of the Green's strain and Green's strain rate and the Fourier heat conduction law for the heat vector. Simplify the mathematical models for the incompressible case and for small deformation. Derive mathematical models in which stresses and heat vector are eliminated by substituting them in the momentum equations, energy equation and entropy inequality.

15.3 Consider a compressible thermoviscoelastic solid with memory and finite deformation. Derive explicit forms of the linear constitutive equations assuming the deviatoric second Piola-Kirchhoff stress, Green's strain tensor and its rate to be the arguments of the time rate of the deviatoric second Piola-Kirchhoff stress. Assume the Fourier heat conduction law for the heat vector. Use these constitutive equations with the conservation and balance laws from 15.2 to obtain complete mathematical models in $\mathbb{R}^1$, $\mathbb{R}^2$, $\mathbb{R}^3$ in x-frame. Simplify the mathematical model for the incompressible case and for small deformation.

15.4 Consider developing flow of an incompressible Newtonian fluid between parallel plates (x_1x_2-plane). Derive explicit forms of the conservation and balance laws and the constitutive equations. Use Newton's law of viscosity and the Fourier heat conduction law for the constitutive theories for the deviatoric Cauchy stress tensor and the heat vector. Use x_1 as the direction of the flow. Obtain the compact form of the mathematical model by eliminating stresses and heat vector as dependent variables.

15.5 Consider fully developed flow of an incompressible Newtonian fluid between parallel plates (x_1x_2-plane). Consider x_1-direction as the direction of the flow. Derive explicit forms of the conservation and balance laws and the constitutive equations. Use Newton's law of viscosity and the Fourier heat conduction law for the constitutive theories for the deviatoric Cauchy stress tensor and heat vector. Obtain the compact form of the mathematical model by eliminating stresses and heat vector as dependent variables.

15.6 Consider purely one-dimensional flow (x_1-direction) of a compressible Newtonian fluid. Derive explicit forms of the conservation and balance laws and the constitutive equations. Use Newton's law of viscosity and the Fourier heat conduction law for constitutive theories for the deviatoric Cauchy stress tensor and heat vector. Obtain explicit forms of the mathematical model by eliminating the deviatoric stress and heat vector as dependent variables.

15.7 Present the complete mathematical models for compressible Maxwell fluid, Oldroyd-B fluid and Giesekus fluid in contravariant and covariant bases in

(a) $\mathbb{R}^1$; x_1-direction;
(b) $\mathbb{R}^2$; x_1x_2-plane;
(c) $\mathbb{R}^3$; x-frame.

15.8 Determine explicit expression for the specific internal energy as a function of density and temperature using

(a) van der Waals equation of state;
(b) Redlich-Kwong equation of state.

16
ENERGY METHODS, PRINCIPLE OF VIRTUAL WORK AND CALCULUS OF VARIATIONS

16.1 Introduction

Energy methods and the principle of virtual work are commonly used in structural mechanics (beams, plates, shells, etc.) for obtaining solutions of boundary value problems (BVPs) and the initial value problems (IVPs) for homogeneous, isotropic and non-homogeneous, non-isotropic materials. In these methods, one can write the statement of energy or virtual work for a volume V with boundary surface ∂V directly without using the differential form of the mathematical models. Energy methods are also used for deriving the mathematical models for BVPs and IVPs for homogeneous, isotropic as well as non-homogeneous, non-isotropic matter.

We have seen in the preceding chapters that CBL of CCM and the constitutive theories provide a thermodynamically consistent framework for deriving the mathematical description (models) of deforming continua. We ask, what is the motivation and need for energy methods and the principle of virtual work for continuous matter? Is there any relationship between the energy methods, the principle of virtual work and the mathematical models based on CBL of CCM for continua? Are there applications for which the assumptions employed in CCM for continuous matter prohibit derivation of the mathematical models for the physics under consideration; hence, leaving one with no choice except to use energy methods and principle of virtual work?

Answers to all these equations require that we consider and examine the mathematical foundation of energy methods and principle of virtual work to determine: (1) if there is any possible link between these methods and the mathematical models derived using CBL of CCM and (2) determine based on calculus of variations when the use of these methods is valid and justified mathematically. In the following we consider BVPs and IVPs in two separate sections to address these questions.

16.2 Boundary value problems (BVPs)

We begin by assuming that the physics under consideration for deforming continuous matter is such that a mathematical model in the differential form is possible using CBL and the constitutive theories based on CCM. These mathematical models would be a system of partial differential equations (PDEs) in dependent variables and space coordinates x_i (we consider Lagrangian description without the loss of generality). We can symbolically represent the PDEs as

$$[A]\{\phi\} = \{f\} \quad \forall x \in \Omega_x \tag{16.1}$$

DOI: 10.1201/9781003105336-16

In which $[A]$ is the differential operator matrix and $\{\phi\}$ is the vector of dependent variables. For the sake of simplicity in considering mathematical details, it suffices to consider a scalar form of (16.1), i.e.

$$A\phi = f \quad \forall x \in \Omega_x \tag{16.2}$$

In which, A is the differential operator.

16.2.1 Mathematical classification of differential operators

All differential operators appearing in totality of all BVPs can be mathematically classified in three categories [121]: (1) self-adjoint, linear and symmetric, for such operators their adjoint $A^* = A$ (2) non-self adjoint, linear but not symmetric, i.e., $A^* \neq A$ and (3) non-linear, hence cannot be symmetric. This mathematical classification of differential operators is absolutely essential in using calculus of variations which helps us better understand energy methods and the principle of virtual work. We note that solutions of differential systems are naturally integral systems. This is the incentive for us to investigate calculus of variations for obtaining solutions of PDEs describing BVPs.

16.2.2 Calculus of variations

As is well known, calculus of variations [15, 33, 62, 85, 91] is a branch of applied mathematics devoted to the study of extremums of functionals. If I is a functional (for example definite integrals) that is continuous in its arguments (that can be functions and/or their derivatives), then the first variation of I, i.e., δI, is unique and $\delta I = 0$ is a necessary condition for an extremum of I and $\delta^2 I > 0$, $= 0$, < 0 corresponding to minimum, saddle point and maximum of I is the extremum principle or sufficient condition. When a unique extremum principle exists, then a solution obtained from $\delta I = 0$ gives a unique extremum of I and is also a unique solution of Euler's equation, a differential equation that is derived from $\delta I = 0$. Thus, if we construct I using BVP $A\phi - f = 0$, such that Euler's equation for $\delta I = 0$ is $A\phi - f = 0$, and if we have a unique extremum principle, then a ϕ from $\delta I = 0$ will yield unique extremum of I as well as a unique solution of the BVP. This approach forms the basis for classical methods of approximation as well as finite element method for BVPs. We note that I is a functional, a definite integral; hence, $\delta I = 0$ is also an integral form that is used in determining ϕ. When I exists, δI is a unique and $\delta^2 I$ is a unique extremum principle, then the integral form $\delta I = 0$ used for determining ϕ is called a variationally consistent (VC) integral form that yields a unique solution ϕ. If $\delta^2 I$ does not yield unique extremum principle, then ϕ obtained from $\delta I = 0$ is not unique or may not be even possible to obtain. In this case, the integral form $\delta I = 0$ is called a variationally inconsistent integral form (VIC) (see reference [121] for more details).

16.2.2.1 Fundamental lemma

The integral form corresponding to BVP $A\phi - f = 0$ in Ω_x can also be constructed directly using the fundamental lemma of the calculus of variations [121].

For the BVP $A\phi - f = 0$ in Ω_x, we can write

$$(A\phi - f, v)_{\bar{\Omega}_x} = \int_{\bar{\Omega}_x} (A\phi - f) v d\Omega = 0 \tag{16.3}$$

In which v is called the test function. The test function $v = 0$ where ϕ is given or specified. We note that (16.3) is same as $\delta I = 0$, i.e., necessary condition. Different choices of v lead to different methods. When $v = \delta\phi$ and some differentiation is transferred from ϕ to v using integration by parts, we have Galerkin method with weak form (GM/WF). When $v = \psi \neq \delta\phi$ but $v = 0$ wherever ϕ is given, we have Petrov-Galerkin method or weighted residual method (WRM). One could show that GM, PGM and WRM yields VIC integral form [121]. Only when using GM/WF it is possible to construct a VC integral form for a self-adjoint differential operator. In GM/WF when A is linear, we have

$$(A\phi - f, v) = B(\phi, v) - l(v) = 0 \tag{16.4}$$

In which $B(\cdot,\cdot)$ is bilinear and $l(v)$ is linear. When A is self-adjoint, i.e., linear and its adjoint $A^* = A$, then it is possible to transfer half of the differentiation from ϕ to v in such a way that $B(\phi, v)$ is bilinear and symmetric. Consider (16.4) in which $B(\cdot,\cdot)$ is bilinear and symmetric and $l(v)$ is linear.

16.2.2.2 Quadratic functional and energy methods

When $B(\phi, v)$ is bilinear and symmetric and $l(v)$ is linear, we can construct a functional $I(\phi)$

$$I(\phi) = \frac{1}{2}B(\phi, \phi) - l(\phi) \tag{16.5}$$

Clearly

$$\delta I(\phi) = B(\phi, v) - l(v) = 0 \quad ; \quad \text{the weak form} \tag{16.6}$$

$$\delta^2 I(\phi) = B(v, v) > 0 \quad ; \quad \text{unique extremum principle} \tag{16.7}$$

Thus, weak form (16.4) is VC. The functional $I(\phi)$ is called quadratic functional. In case of linear elasticity, $\frac{1}{2}B(\phi, \phi)$ represents strain energy and $l(\phi)$ represents potential energy of loads; hence, $I(\phi)$ is the total potential energy. Equation (16.7) implies that a ϕ obtained from (16.6) minimizes $I(\phi)$, i.e., minimizes total potential energy. Thus, in this case, we have principle of minimum total potential energy.

Remarks

(1) When the differential operator A in $A\phi - f = 0$ is self-adjoint, then it is possible to construct total potential energy functional $I(\phi)$ using (16.6). Euler's equation from the $\delta I = 0$ is naturally $A\phi - f = 0$, the BVP.

(2) The integrand from (16.5) can be used in classical methods of approximation or finite element method of approximation for obtaining approximate solutions.

(3) Here, we note the connection between energy methods and differential form of the mathematical model. For the energy methods to be valid, the differential operator in the mathematical description of the physical process must be

self-adjoint. In isotropic, homogeneous matter with small deformation, small strain, incompressible, isothermal, linear elasticity (without dissipation) the differential operators are self-adjoint. In all other deformation physics of continuous matter, the differential operators are either non-self adjoint or non-linear, hence preclude the use of energy methods.

(4) If one could write a correct statement of $I(\phi)$, the total potential energy functional directly without using the differential form of the mathematical model, then of course we could use $\delta I = 0$ to derive Euler's equation which would be the mathematical model corresponding to functional $I(\phi)$. We note that this may be possible only for simple and trivial physics.

(5) From the SLT, we know that for small deformation, small strain, tr$(\boldsymbol{\sigma} : \dot{\boldsymbol{\varepsilon}})$ is the rate of work per unit volume for isotropic, homogeneous matter (as the constitutive theories for non-isotropic are not possible using rate of work conjugate pair $\boldsymbol{\sigma} : \dot{\boldsymbol{\varepsilon}}$). It is important to note that we only come to this conclusion through the derivation of entropy inequality. Thus, if we consider non-homogeneous, non-isotropic matter with an orthotropic constitutive theory for $\boldsymbol{\sigma}$ (for example) derived using some non-thermodynamic approach, we cannot be sure if $\boldsymbol{\sigma} : \dot{\boldsymbol{\varepsilon}}$ is rate of work. Hence, in this case, construction of $B(\phi, \phi)$ in $I(\phi)$ using $\boldsymbol{\sigma} : \boldsymbol{\varepsilon}$ is questionable.

(6) For an energy method to be valid, if we construct $I(\phi)$ directly without using differential mathematical model, then we must show the Euler's equation derived from $\delta I(\phi) = 0$ is indeed a differential form that is same as what could have been derived using CBL of CCM.

Concluding remark

In view of Remarks (1)–(6), it is rather obvious that use of energy methods without a check on the resulting Euler's equations using CBL is quite risky and is highly prone to misjudgments and errors. This clearly suggests a lack of usefulness of the energy methods except in the case of very simple and trivial deformation physics.

16.2.3 Principle of Virtual work: BVPs

It is a rather trivial exercise to show that the principle of virtual work is in fact the same as the integral statement resulting from the fundamental lemma of the calculus of variations [15,33,62,85,91] when the differential form of the mathematical model is known.

We use balance of linear momenta as an example to illustrate this. Consider BLM in Lagrangian description (small deformation, small strain)

$$\rho_0 F_i^b + \frac{\partial \sigma_{ji}}{\partial x_j} = 0 \quad \forall x \in \Omega_x \tag{16.8}$$

Let w_i be test functions such that $w_i = \delta u_i$ (Galerkin method, Galerkin method with weak form). We can construct scalar product of (16.8) with test function w_i over domain Ω_x and set it to zero (based on fundamental lemma), using $w_i = \delta u_i$

$$\left(\rho_0 F_i^b + \frac{\partial \sigma_{ji}}{\partial x_j}, \delta u_i \right)_{\bar{\Omega}_x} = 0 \quad ; \quad i = 1, 2, 3 \quad \text{(no sum over } i\text{)} \tag{16.9}$$

More explicitly

$$\int_V \left(\rho_0 F_i^b \delta u_i + \frac{\partial \sigma_{ji}}{\partial x_j} \delta u_i \right) = 0 \quad ; \quad i = 1, 2, 3 \quad \text{(no sum over } i\text{)} \tag{16.10}$$

The first term, $\rho_0 F_i^b \delta u_i$, in (16.10) is the virtual work per unit volume of body forces due to virtual displacements δu_i and the second term in the integrand (16.10) is the virtual work of stress per unit volume due to virtual displacements δu_i. Thus, the virtual work statement (16.10) is generally not possible without the knowledge of the differential forms of the mathematical model. In mechanics of beams, plates, and shells, one writes the virtual work statement like (16.10) without the knowledge of the differential form of the mathematical model (16.8) that can be derived using the CBL of CCM. This may be possible for trivial physics but in general, relying on intuition to write a correct statement like (16.10) is not possible and may be quite risky.

16.2.4 Final remarks (BVPs)

Thus, we can see that in the case of BVPs, energy methods and the principle of virtual work both are descendant from the calculus of variations. They require mathematical classification of the differential operators and the mathematical model from CBL in order to ensure that $B(\cdot, \cdot)$ and $l(\cdot)$ in energy methods and the integral statement of the principle of virtual work are correctly incorporating the deformation physics. In the author's opinion, energy methods and principle of virtual work lack rigor (in the absence of differential models) and serve no useful purpose in deriving mathematical models or in obtaining the solution of the BVPs. Both of these methods cannot be used correctly without calculus of variations (fundamental lemma) and differential operator classification. Their use can be detrimental without precise understanding of the differential models based on CBL of CCM and the mathematical nature of the differential operators.

16.3 IVPs, energy methods and principle of virtual work

Consider a simple abstract initial value problem in $\mathbb{R}^1$

$$A\phi(x_1, t) - f(x_1, t) = 0 \quad \forall x_1, t \in \Omega_{x_1,t} = \Omega_{x_1} \times \Omega_t = (0, L) \times (0, \tau) \tag{16.11}$$

in which A is a space-time differential operator and $\Omega_{x_1,t}$ is the space-time domain. The space-time differential operator are either non-self adjoint or non-linear [122]. This mathematical classification has two important consequences.

(1) Given the differential form of the IVP, it is not possible to show existence of the energy functional (I) using fundamental lemma of the calculus of variations [122].

(2) If one constructs the energy functional I (for some IVP directly using physical arguments), then we can show that the Euler's equation resulting from the $\delta I = 0$ is not the differential form of the IVP, implying that the energy functional constructed does not correspond to the IVP.

Both Remarks (1) and (2) suggest that for IVPs, the concept of energy functional is not valid, hence can neither be used for deriving mathematical models of the IVPs, nor for obtaining their solutions. We give some examples in the following.

16.3.1 Energy functional from the differential form of IVP

Consider simplified form of 1D BLM in the dimensionless form in which all coefficients are unity (1D wave equation).

$$\begin{aligned} &\frac{\partial^2 u_1}{\partial t^2} - \frac{\partial^2 u_1}{\partial x_1^2} - F_1^b = 0 \\ &\forall (x_1, t) \in \Omega_{x_1,t} = \Omega_{x_1} \times \Omega_t = (0, L) \times (0, \tau) \end{aligned} \tag{16.12}$$

$$\text{or} \quad Au_1 - f = 0$$

$$\begin{aligned} \text{BCs:} \quad & u_1(0,t) = 0 \quad \forall t \in [0, \tau] \\ & u_1(L,t) = u_L \quad \forall t \in [0, \tau] \end{aligned} \tag{16.13}$$

$$\begin{aligned} \text{ICs:} \quad & u_1(x_1, 0) = 0 \quad \forall x_1 \in [0, L] \\ & \frac{\partial u_1}{\partial t}(x_1, 0) = 0 \quad \forall x_1 \in [0, L] \end{aligned} \tag{16.14}$$

Figure 16.1 shows the space-time domain $\bar{\Omega}_{x_1,t}$. The adjoint $A^* \neq A$ due to the open boundary at $t = \tau$ [122]. In order to investigate construction of energy functional corresponding to (16.12)–(16.14), we use fundamental lemma of the calculus of variations to construct space-time integral form using (16.12).

$$(Au_1 - f, v)_{\bar{\Omega}_{x_1,t}} = 0 \tag{16.15}$$

Here, $v(x_1, t)$ is the test function. The test function v is zero where u_1 is given or defined. We can use space-time Galerkin method, Petrov-Galerkin method, weighted residual method or Galerkin method with weak form (STGM, STPGM, STWRM, STGM/WF) to obtain the following

$$(Au_1 - f, v)_{\bar{\Omega}_{x_1,t}} = B(u_1, v) - l(v) = 0 \tag{16.16}$$

In which $B(u_1, v)$ is bilinear (when A is linear) but is not symmetric, i.e., $B(u_1, v) \neq B(v, u_1)$ and $l(v)$ is linear. This holds regardless of the method used (as long as A is linear, which is the case here). In STGM and STGM/WF, $v = \delta u_1$. Hence, $v = 0$ is automatically satisfied wherever u_1 is specified or given. In STPGM and STWRM, v is chosen such that $v = 0$ wherever u_1 is given or known. We consider STGM/WF (meritorious as the operator A has even order derivatives in space x_1 and time t [122]) to obtain after transferring half of the differentiation from ϕ to v.

$$(Au_1 - f, v)_{\bar{\Omega}_{x_1,t}} = B(u_1, v) - l(v) = 0 \tag{16.17}$$

in which

$$B(u_1, v) = \int\limits_{\bar{\Omega}_{x_1,t}} \left(-\frac{\partial u_1}{\partial t}\frac{\partial v}{\partial t} + \frac{\partial u_1}{\partial x_1}\frac{\partial v}{\partial x_1} \right) d\Omega + \left(v(x_1, \tau), \left.\frac{\partial u_1}{\partial t}\right|_{\tau} \right)_{\Gamma_4} \tag{16.18}$$

$$l(v) = (F_1^b, v)_{\bar{\Omega}_{x_1,t}} + < Au_1, v >_{\Gamma} \tag{16.19}$$

$$< Au_1, v >_\Gamma = \int_{\Gamma_1} \left.\frac{\partial u_1}{\partial x_1}\right|_{x_1=0} v(0,t)(-1)d\Gamma + \int_{\Gamma_3} \left.\frac{\partial u_1}{\partial x_1}\right|_{x_1=L} v(L,t)(1)d\Gamma \tag{16.20}$$

We clearly see that $B(\cdot,\cdot)$ is bilinear but not symmetric, i.e., $B(u_1, v) \neq B(v, u_1)$. Hence, the quadratic functional (or energy functional) is not possible.

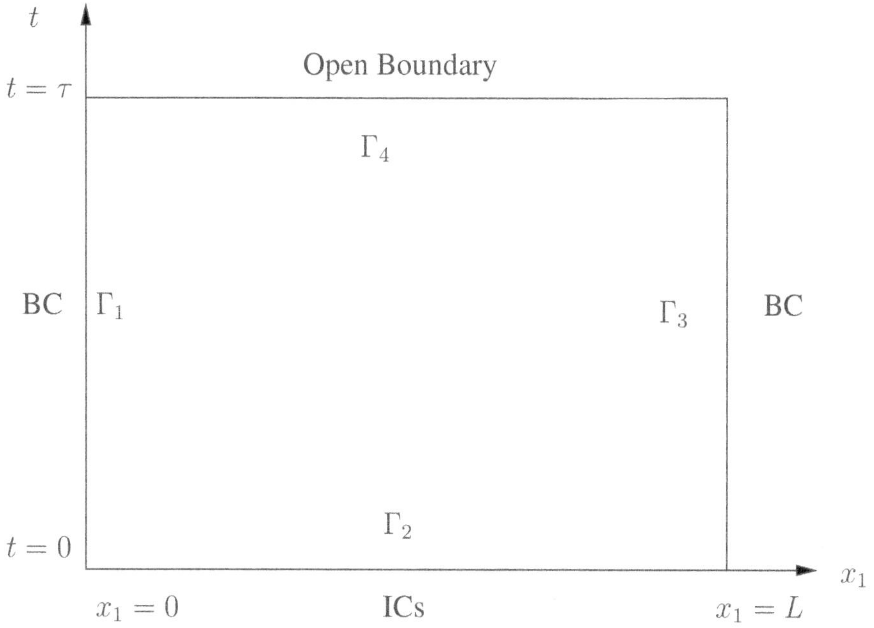

Figure 16.1: Space time domain $\bar{\Omega}_{x_1,t} \ : \ \Gamma = \bigcup_i^4 \Gamma_i$

$$I(u_1) \neq \frac{1}{2} B(u_1, u_1) - l(u_1) \tag{16.21}$$

$$\text{and naturally} \quad \delta I(u_1) \neq B(u_1, v) - l(v) \tag{16.22}$$

From (16.18), we note that it is the open boundary Γ_4 that destroys the symmetry of $B(\cdot,\cdot)$. Since the open boundary is present in all IVPs, it is not possible to construct energy functional in case of IVPs when the differential form of the mathematical of the IVP is known.

16.3.2 Differential form of IVP from energy functional

We consider the IVP corresponding to the BVP of Section 16.2.3 in $\mathbb{R}^1$, but we construct energy functional using kinetic energy, strain energy and the potential energy of loads directly without using the differential form of the mathematical model.

$$I = \int_{\bar{\Omega}_{x_1,t}} \left(\frac{1}{2} \left(\frac{\partial u_1}{\partial t} \right)^2 - \sigma_{11}\varepsilon_{11} + F_1^b u_1 \right) d\Omega \tag{16.23}$$

$$\varepsilon_{11} = \frac{\partial u_1}{\partial x_1} \tag{16.24}$$

Substituting ε_{11} in (16.23)

$$I = \int_{\bar{\Omega}_{x_1,t}} \left(\frac{1}{2} \left(\frac{\partial u_1}{\partial t} \right)^2 - \sigma_{11}\frac{\partial u_1}{\partial x_1} + F_1^b u_1 \right) d\Omega \tag{16.25}$$

$$\therefore \quad \delta I = \int_{\bar{\Omega}_{x_1,t}} \left(\frac{\partial u_1}{\partial t}\frac{\partial(\delta u_1)}{\partial t} - \sigma_{11}\frac{\partial \delta u_1}{\partial x_1} + F_1^b \delta u_1 \right) d\Omega \tag{16.26}$$

We transfer one order of differentiation from δu_1 to $\frac{\partial u_1}{\partial t}$ with respect to t and from δu_1 to σ_{11} with respect to x_1 using integration by parts

$$\begin{aligned} \delta I = & \int_{\bar{\Omega}_{x_1,t}} \left(-\frac{\partial^2 u_1}{\partial t^2}\delta u_1 + \frac{\partial \sigma_{11}}{\partial x_1}\delta u_1 + F_1^b \delta u_1 \right) d\Omega \\ & + \oint_{\Gamma} \frac{\partial u_1}{\partial t}\delta u_1 n_t d\Gamma - \oint_{\Gamma} \sigma_{11} n_{x_1} \delta u_1 d\Gamma \end{aligned} \tag{16.27}$$

We can show that

$$\oint_{\Gamma} \frac{\partial u_1}{\partial t} n_t \delta u_1 = \left(\delta u_1|_{\tau}, \left.\frac{\partial u_1}{\partial t}\right|_{\tau} \right)_{\Gamma_4} \tag{16.28}$$

$$\text{and} \quad \oint_{\Gamma} \sigma_{11} n_{x_1} \delta u_1 d\Gamma = 0 \tag{16.29}$$

Using (16.28) and (16.29) in (16.27)

$$\delta I = \int_{\bar{\Omega}_{x_1,t}} \left(-\frac{\partial^2 u_1}{\partial t^2} + \frac{\partial \sigma_{11}}{\partial x_1} + F_1^b \right) \delta u_1 d\Omega + \left(\delta u_1|_{\tau}, \left.\frac{\partial u_1}{\partial t}\right|_{\tau} \right)_{\Gamma_4} \tag{16.30}$$

In the absence of the scalar product term over Γ_4, the Euler's equation [122] resulting from (16.30) would be

$$\frac{\partial^2 u_1}{\partial t^2} - \frac{\partial \sigma_{11}}{\partial x_1} - F_1^b = 0 \tag{16.31}$$

$$\text{or} \quad \frac{\partial^2 u_1}{\partial t^2} - \frac{\partial^2 u_1}{\partial x_1^2} - F_1^b = 0 \tag{16.32}$$

which is the same as IVP (16.12), however because of the open boundary Γ_4, the term $(\cdot,\cdot)_{\Gamma_4}$ in (16.30) is not zero.

Remarks

(1) We clearly see that from Section 16.3.2 and 16.3.1 that there is no correspondence between the differential form of the IVP and the energy functional as well as the energy functional from the differential form of the IVP.

(2) Extension of material in Section 16.3.1 and 16.3.2 to $\mathbb{R}^3$ is straightforward. Thus, we conclude that energy functional consisting of kinetic energy, strain energy and the potential energy of loads cannot be used to derive the differential form of the BLM in $\mathbb{R}^3$ without neglecting the influence of open boundary. This illustration confirms that Hamilton's principle as used currently is mathematically not valid as it involves approximation by neglecting the integral over the open boundary Γ_4.

(3) In conclusion, energy methods have no usefulness in IVPs primarily due to the fact that the space-time operators are not self-adjoint.

16.3.3 Principle of virtual work in IVPs

As in case of BVPs, here also it is straightforward to show that the principle of virtual work is in fact application of space-time fundamental lemma of the calculus of variations [122] when the differential form of the mathematical model (IVP) is known.

We consider differential form of BLM as an example (small deformation, small strain).

$$\rho_0 \frac{\partial^2 u_i}{\partial t^2} - \rho_0 F_i^b - \frac{\partial \sigma_{ji}}{\partial x_j} = 0 \quad \forall x_i \in \Omega_{x_i,t} = \Omega_{x_i} \times \Omega_t \tag{16.33}$$

Let w_i be the test functions such that $w_i = \delta u_i$ (STGM or STGM/WF). We construct scalar product of (16.33) with test functions w_i over the space-time domain $\bar{\Omega}_{x_i,t}$ and set it to zero (based on fundamental lemma). Using $w_i = \delta u_i$

$$\left(\rho_0 \frac{\partial^2 u_i}{\partial t^2} - \rho_0 F_i^b - \frac{\partial \sigma_{ji}}{\partial x_j}, \delta u_i \right)_{\bar{\Omega}_{x_i,t}} = 0 \quad ; \quad i = 1, 2, 3 \text{ (no sum over } i) \tag{16.34}$$

or

$$\int_V \left(\rho_0 \frac{\partial^2 u_i}{\partial t^2} \delta u_i - \rho_0 F_i^b \delta u_i - \frac{\partial \sigma_{ji}}{\partial x_j} \delta u_i \right) dV = 0 \quad ; \quad i = 1, 2, 3 \text{ (no sum over } i) \tag{16.35}$$

We note that $\rho_0 \frac{\partial^2 u_i}{\partial t^2} \delta u_i$ is the virtual work per unit volume of the inertial forces in x_i directions due to virtual displacements δu_i, $\rho_0 F_i^b \delta u_i$ is the virtual work per unit volume of body forces in x_i directions due to virtual displacements δu_i and the last term in the integrand in (16.35) is the virtual work per unit volume of stresses due to virtual displacements δu_i. This virtual work statement requires the knowledge of the differential form of the mathematical model. In dealing with IVPs in mechanics of beams, plates, shells, one writes the virtual work statement like (16.35) without the knowledge of the differential form of the mathematical model. This approach is mathematically not rigorous, is subjective due to one's understanding of the physics and is likely to result in erroneous statement for applications that are not simple as there is no theory that supports this approach.

16.4 Summary

Energy methods and the principle of virtual work have been investigated for continuous matter for BVPs and IVPs using mathematical classification of the differential operators and the calculus of variations. We have shown that the use of energy methods is only possible for BVPs when the differential operators are self-adjoint. Use of principle of virtual work for BVPs requires knowledge of the differential form of the mathematical model for the BVP. It is shown that principle of virtual work is statement of fundamental lemma which requires knowledge of the differential form of the mathematical model for the BVP. When the differential operator is self-adjoint, the use of fundamental lemma and GM/WF in which $B(\cdot,\cdot)$ is bilinear and symmetric, automatically provides correct means of constructing the quadratic functional (a statement of total potential energy associated with the BVP). In this case, Euler's equation from $\delta I = 0$ is always the BVP. Thus, in our view energy methods and principle of virtual work have no usefulness over and beyond differential form of the BVP description and the mathematical classification of differential operators in conjunction with fundamental lemma of the calculus of variations.

In case of IVPs, the space-time differential operators are not self-adjoint; hence, energy methods in any form have no mathematical justification, thus, are ruled out. Regarding principle of virtual work, all we need is use of fundamental lemma in space and time in conjunction with differential form of the mathematical model of the IVPs as the integral statement from the fundamental lemma using differential form of the IVP statement is a statement of virtual work. Naturally, this cannot be constructed correctly without the knowledge of the differential form of the mathematical model of the IVP.

Since the energy functional for IVPs is not possible, Hamilton's principle in deriving mathematical models is only approximate as it neglects the integral(s) over the open boundary of the space-time domain. We also point out that in non-structural applications, energy methods are not possible and play no role at all. It is straightforward to derive mathematical models using CBL of CCM for solid and fluent continua, then use fundamental lemma regardless of whether we have a BVP or an IVP, this is in fact the principle of virtual work. This approach avoids all issues and uncertainties in writing the statement of the principle of virtual work.

Appendix A: Combined generators and invariants

Table A.1: Complete set of irreducible invariants of symmetric tensor of rank two $[s]$, vector (tensor of rank one) $\{v\}$ and skew-symmetric tensor $[w]$

(1) Invariants depending on one variable

Variables	**Invariants**
$[s]$	$\text{tr}([s])\,,\quad \text{tr}([s]^2)\,,\quad \text{tr}([s]^3)$
$\{v\}$	$\{v\}^T\{v\}$
$[w]$	$\text{tr}([w]^2)$

(2) Invariants depending on two variables when (1) holds

Variables	**Invariants**
$[s_1]\,,\quad [s_2]$	$\text{tr}([s_1][s_2])\,,\quad \text{tr}([s_1]^2[s_2])\,,$ $\text{tr}([s_1][s_2]^2)\,,\quad \text{tr}([s_1]^2[s_2]^2)\,,$ $\text{tr}([s_1][s_2]+[s_2][s_1])\,,\quad \text{tr}([s_1][s_2]-[s_2][s_1])$
$[s]\,,\quad \{v\}$	$\{v\}^T\{[s]\{v\}\}\,,\quad \{v\}^T\{[s]^2\{v\}\}$
$[s]\,,\quad [w]$	$\text{tr}([s][w]^2)\,,\quad \text{tr}([s]^2[w]^2)\,,$ $\text{tr}([s]^2[w]^2[s][w])$
$\{v_1\}\,,\quad \{v_2\}$	$\{v_1\}^T\{v_2\}$
$\{v\}\,,\quad [w]$	$\{v\}^T\{[w]^2\{v\}\}$
$[w_1]\,,\quad [w_2]$	$\text{tr}([w_1][w_2])$

DOI: 10.1201/9781003105336-A

Table A.1: (Contd.) Complete set of irreducible invariants of symmetric tensor of rank two $[s]$, vector (tensor of rank one) $\{v\}$ and skew symmetric tensor $[w]$

(3) Invariants depending on three variables when (1) and (2) hold

Variables	**Invariants**
$[s_1]\,,\ [s_2]\,,\ [s_3]$	$\text{tr}([s_1][s_2][s_3])$
$[s_1]\,,\ [s_2]\,,\ \{v\}$	$\{v\}^T\{[s_1][s_2]\{v\}\}$
$[s]\,,\ \{v_1\}\,,\ \{v_2\}$	$\{v_1\}^T\{[s]\{v_2\}\}\,,$ $\{v_1\}^T\{[s]^2\{v_2\}\}$
$[s]\,,\ [w_1]\,,\ [w_2]$	$\text{tr}([s][w_1][w_2])\,,$ $\text{tr}([s][w_1][w_2]^2)\,,$ $\text{tr}([s][w_1]^2[w_2])$
$[s_1]\,,\ [s_2]\,,\ [w]$	$\text{tr}([s_1][s_2][w])\,,$ $\text{tr}([s_1]^2[s_2][w])\,,$ $\text{tr}([s_1][w]^2[s_2][w])\,,$ $\text{tr}([s_1][s_2]^2[w])$
$[w_1]\,,\ [w_2]\,,\ [w_3]$	$\text{tr}([w_1][w_2][w_3])$
$\{v_1\}\,,\ \{v_2\}\,,\ [w]$	$\{v_1\}^T\{[w]\{v_2\}\}\,,$ $\{v_1\}^T\{[w]^2\{v_2\}\}$
$\{v\}\,,\ [w_1]\,,\ [w_2]$	$\{v\}^T\{[w_1][w_2]\{v\}\}\,,$ $\{v\}^T\{[w_1]^2[w_2]\{v\}\}\,,$ $\{v\}^T\{[w_1][w_2]^2\{v\}\}$
$[s]\,,\ \{v\}\,,\ [w]$	$\{v\}^T\{[s][w]\{v\}\}\,,$ $\{v\}^T\{[s]^2[w]\{v\}\}\,,$ $\{v\}^T\{[w][s][w]^2\{v\}\}$

Table A.1: (Contd.) Complete set of irreducible invariants of symmetric tensor of rank two $[s]$, vector (tensor of rank one) $\{v\}$ and skew symmetric tensor $[w]$

(4) Invariants depending on four variables when (1) to (3) hold

Variables	Invariants
$[s_1]\ ,\ [s_2]\ ,\ \{v_1\}\ ,\ \{v_2\}$	$\{v_1\}^T\{\,[[s_1][s_2] - [s_2][s_1]]\,\{v_2\}\}$
$[s]\ ,\ \{v_1\}\ ,\ \{v_2\}\ ,\ [w]$	$\{v_1\}^T\{\,[[s][w] - [w][s]]\,\{v_2\}\}$
$\{v_1\}\ ,\ \{v_2\}\ ,\ [w_1]\ ,\ [w_2]$	$\{v_1\}^T\{\,[[w_1][w_2] - [w_2][w_1]]\,\{v_2\}\}$

Table A.2: Generators of rank one (vector valued isotropic functions)

(1) Generators depending on one variable

Variables	Generators
$\{v\}$	$\{v\}$

(2) Generators depending on two variables when (1) holds

Variables	Generators
$[s]\,,\quad \{v\}$	$[s]\{v\}\,,$ $[s]^2\{v\}$
$[w]\,,\quad \{v\}$	$[w]\{v\}\,,$ $[w]^2\{v\}$

(3) Generators depending on three variables when (1) and (2) hold

Variables	Generators
$[s_1]\,,\quad [s_2]\,,\quad \{v\}$	$\big[[s_1][s_2]+[s_2][s_1]\big]\{v\}\,,$ $\big[[s_1][s_2]-[s_2][s_1]\big]\{v\}$
$[w_1]\,,\quad [w_2]\,,\quad \{v\}$	$\big[[w_1][w_2]-[w_2][w_1]\big]\{v\}$
$[s]\,,\quad \{v\}\,,\quad [w]$	$\big[[s][w]-[w][s]\big]\{v\}$

Table A.3: Generators for symmetric tensor-valued isotropic functions

(1) Generators depending on no variables: $[I]$

(2) Generators depending on one variable

Variables	Generators
$[s]$	$[s]\,,\quad [s]^2$
$\{v\}$	$\{v\}\{v\}^T$
$[w]$	$[w]^2$

(3) Generators depending on two variables when (1) holds

Variables	Generators
$[s_1]\,,\quad [s_2]$	$[s_1][s_2]+[s_2][s_1]\,,\quad [s_1]^2[s_2]+[s_1][s_2]^2\,,$ $[s_1][s_2]^2+[s_2]^2[s_1]$
$[s]\,,\quad \{v\}$	$\{v\}\big\{[s]\{v\}\big\}^T+\{[s]\{v\}\}\{v\}^T\,,$ $\{v\}\big\{[s]^2\{v\}\big\}^T+\big\{[s]^2\{v\}\big\}\{v\}^T$
$[s]\,,\quad [w]$	$[s][w]-[w][s]\,,\quad [w][s][w]\,,$ $[s]^2[w]-[w][s]^2\,,\quad [w][s][w]^2-[w]^2[s][w]$
$\{v_1\}\,,\ \{v_2\}$	$\{v_1\}\{v_2\}^T+\{v_2\}\{v_1\}^T$
$\{v\}\,,\quad [w]$	$\big\{[w]\{v\}\big\}\big\{[w]\{v\}\big\}^T\,,$ $\{v\}\big\{[w]\{v\}\big\}^T+\big\{[w]\{v\}\big\}\{v\}^T\,,$ $\big\{[w]\{v\}\big\}\big\{[w]^2\{v\}\big\}^T+\big\{[w]^2\{v\}\big\}\big\{[w]\{v\}\big\}^T$
$[w_1]\,,\ [w_2]$	$[w_1][w_2]+[w_2][w_1]\,,\quad [w_1][w_2]^2-[w_2]^2[w_1]\,,$ $[w_1]^2[w_2]-[w_2][w_1]^2$

Table A.3: (Contd.) Generators for symmetric tensor-valued isotropic functions

(4) Generators depending on three variables when (1) and (2) hold

Variables	Generators
$[s], \{v_1\}, \{v_2\}$	$[s]\left[\{v_1\}\{v_2\}^T - \{v_2\}\{v_1\}^T\right] -$ $\left[\{v_1\}\{v_2\}^T - \{v_2\}\{v_1\}^T\right][s]$
$[w], \{v_1\}, \{v_2\}$	$[w]\left[\{v_1\}\{v_2\}^T - \{v_2\}\{v_1\}^T\right] -$ $\left[\{v_1\}\{v_2\}^T - \{v_2\}\{v_1\}^T\right][w]$

Table A.4: Generators for skew symmetric tensor-valued isotropic functions

(1) Generators depending on one variable

Variables	**Generators**
$[w]$	$[w]$

(2) Generators depending on two variables when (1) holds

Variables	**Generators**
$[s_1]\,,\quad [s_2]$	$[s_1][s_2]-[s_2][s_1]\,,\quad [s_1]^2[s_2]-[s_2][s_1]^2\,,$ $[s_1][s_2]^2-[s_2]^2[s_1]\,,\quad [s_1][s_2][s_1]^2-[s_1]^2[s_2][s_1]\,,$ $[s_2][s_1][s_2]^2-[s_2]^2[s_1][s_2]$
$[s]\,,\quad \{v\}$	$\{v\}\{[s]\{v\}\}^T-\{[s]\{v\}\}\{v\}^T\,,$ $\{v\}\{[s]^2\{v\}\}^T-\{[s]^2\{v\}\}\{v\}^T\,,$ $\{[s]\{v\}\}\{[s]^2\{v\}\}^T-\{[s]^2\{v\}\}\{[s]\{v\}\}^T$
$[s]\,,\quad [w]$	$[s][w]+[w][s]\,,\quad [s][w]^2-[w]^2[s]$
$[w]\,,\quad \{v\}$	$\{v\}\{[w]\{v\}\}^T-\{[w]\{v\}\}\{v\}^T\,,$ $\{v\}\{[w]^2\{v\}\}^T-\{[w]^2\{v\}\}\{v\}^T$
$\{v_1\}\,,\ \{v_2\}$	$\{v_1\}\{v_2\}^T-\{v_2\}\{v_1\}^T$
$[w_1]\,,\ [w_2]$	$[w_1][w_2]-[w_2][w_1]$

Table A.4: (Contd.) Generators for skew symmetric tensor-valued isotropic functions

(3) Generators depending on three variables when (1) and (2) hold

Variables	**Generators**
$[s_1], [s_2], [s_3]$	$[s_1][s_2][s_3] + [s_2][s_3][s_1] + [s_3][s_1][s_2] -$ $[s_3][s_2][s_1] - [s_1][s_3][s_2] - [s_2][s_1][s_3]$
$[s_1], [s_2], \{v\}$	$\{[s_1]\{v\}\}\{[s_2]\{v\}\}^T -$ $\{[s_2]\{v\}\}\{[s_1]\{v\}\}^T +$ $\{v\}\{[[s_1][s_2] - [s_2][s_1]]\{v\}\}^T -$ $\{[[s_1][s_2] - [s_2][s_1]]\{v\}\}\{v\}^T$
$[s], \{v_1\}, \{v_2\}$	$[s][\{v_1\}\{v_2\}^T - \{v_2\}\{v_1\}^T] +$ $[\{v_1\}\{v_2\}^T - \{v_2\}\{v_1\}^T][s]$
$[w], \{v_1\}, \{v_2\}$	$[w][\{v_1\}\{v_2\}^T - \{v_2\}\{v_1\}^T] -$ $[\{v_1\}\{v_2\}^T - \{v_2\}\{v_1\}^T][w]$

Appendix B: Transformations and operations in Cartesian, cylindrical and spherical coordinate systems

The material presented in this appendix is helpful in converting quantities defined in the Cartesian frame x_1, x_2, x_3 to cylindrical r, θ, z- and spherical r, θ, ϕ-frames.

Figures B.1 and B.2 show cylindrical and spherical frames in reference to the Cartesian frame. If we consider a point P located at a distance $\|\boldsymbol{R}\|$ from the origin, then we can express its coordinates in the three frames using the basis in them.

Let $\boldsymbol{e}_{x_1}, \boldsymbol{e}_{x_2}, \boldsymbol{e}_{x_3}$ be unit vectors (basis) in x_1, x_2, x_3-frame, $\boldsymbol{e}_r, \boldsymbol{e}_\theta, \boldsymbol{e}_z$ be unit vectors (basis) in r, θ, z-frame and $\boldsymbol{e}_r, \boldsymbol{e}_\theta, \boldsymbol{e}_\phi$ be unit vectors (basis) in the r, θ, ϕ-frame. Then a vector $\boldsymbol{R}$ defining the position of point P can be represented in these coordinate systems using respective bases with their respective Eulerian norms $\|\boldsymbol{R}\|$.

$$\text{Cartersian: } \boldsymbol{R} = x_1 \boldsymbol{e}_{x_1} + x_2 \boldsymbol{e}_{x_2} + x_3 \boldsymbol{e}_{x_3} \; ; \; \|\boldsymbol{R}\| = \sqrt{x_1^2 + x_2^2 + x_3^2} = R \tag{B.1}$$

$$\text{cylindrical: } \boldsymbol{R} = r\boldsymbol{e}_r + \boldsymbol{e}_\theta + z\boldsymbol{e}_z \; ; \; \|\boldsymbol{R}\| = \sqrt{r^2 + z^2} = R \tag{B.2}$$

$$\text{spherical: } \boldsymbol{R} = r\boldsymbol{e}_r + \boldsymbol{e}_\theta + \boldsymbol{e}_\phi \; ; \; \|\boldsymbol{R}\| = r = R \tag{B.3}$$

B.1 Cartesian frame x_1, x_2, x_3 and cylindrical frame r, θ, z

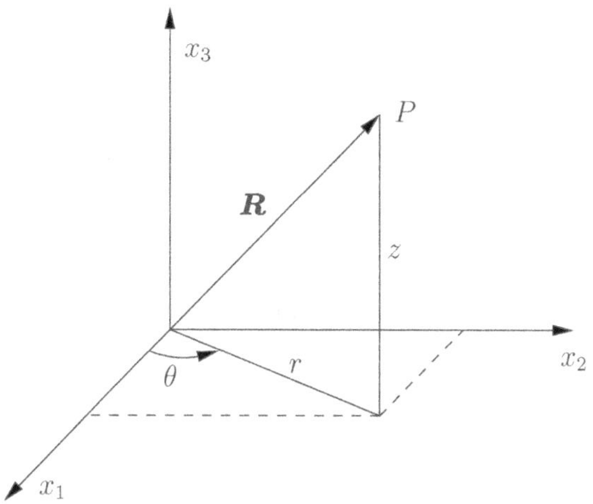

Figure B.1: x_1, x_2, x_3-frame and cylindrical r, θ, z-frame

DOI: 10.1201/9781003105336-B

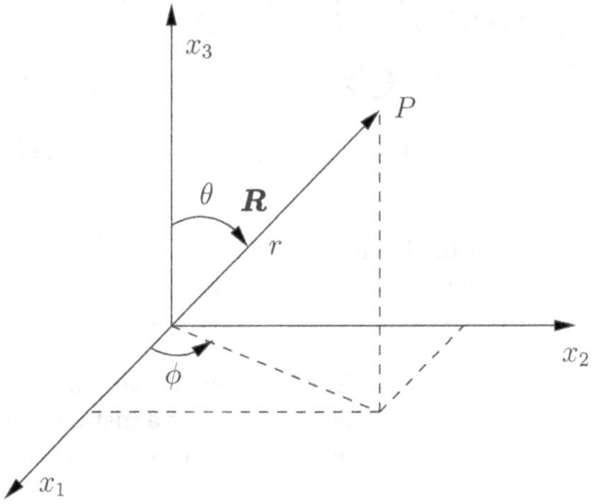

Figure B.2: x_1, x_2, x_3-frame and spherical r, θ, ϕ-frame

B.1.1 Relationship between coordinates of a point in x_1, x_2, x_3 and r, θ, z-frames

From Figure B.1 we can write

$$
\begin{aligned}
x_1 &= r\ \cos\theta = x_1(r,\theta) \\
x_2 &= r\ \sin\theta = x_2(r,\theta) \\
x_3 &= z = x_3(z)
\end{aligned}
\tag{B.4}
$$

and the inverse

$$
\begin{aligned}
r &= \sqrt{x_1^2 + x_2^2} = r(x_1, x_2) \\
\theta &= \arctan\left(\frac{x_2}{x_1}\right) = \theta(x_1, x_2) \\
z &= x_3 = z(x_3)
\end{aligned}
\tag{B.5}
$$

B.1.2 Converting derivatives of a scalar with respect to x_1, x_2, x_3 into its derivatives with respect to r, θ, z

We use the chain rule of differentiation to accomplish this. Consider a scalar Ψ to be a function of x_1, x_2, x_3 and also to be a function of r, θ, z, i.e., consider $\Psi(x_1, x_2, x_3)$ and $\Psi(r, \theta, z)$. We wish to obtain a relationship between $\frac{\partial\Psi}{\partial x_1}$, $\frac{\partial\Psi}{\partial x_2}$, $\frac{\partial\Psi}{\partial x_3}$ and $\frac{\partial\Psi}{\partial r}$, $\frac{\partial\Psi}{\partial\theta}$, $\frac{\partial\Psi}{\partial z}$.

Consider $\frac{\partial\Psi}{\partial x_1}$ and use $\Psi(r, \theta, z)$ to obtain

$$
\frac{\partial\Psi}{\partial x_1} = \frac{\partial\Psi}{\partial r}\frac{\partial r}{\partial x_1} + \frac{\partial\Psi}{\partial\theta}\frac{\partial\theta}{\partial x_1} + \frac{\partial\Psi}{\partial z}\frac{\partial z}{\partial x_1}
\tag{B.6}
$$

and using r, θ and z from (B.5) we obtain

$$\begin{aligned}\frac{\partial r}{\partial x_1} &= \frac{\partial}{\partial x_1}\Big(\sqrt{x_1^2+x_2^2}\,\Big) = \frac{x_1}{\sqrt{x_1^2+x_2^2}} = \frac{r\ \cos\theta}{r} = \cos\theta \\ \frac{\partial \theta}{\partial x_1} &= \frac{\partial}{\partial x_1}\Big(\arctan\big(\frac{x_2}{x_1}\big)\Big) = -\frac{x_2}{x_1^2+x_2^2} = -\frac{r\ \sin\theta}{r^2} = -\frac{\sin\theta}{r} \\ \frac{\partial z}{\partial x_1} &= \frac{\partial x_3}{\partial x_1} = 0\end{aligned} \tag{B.7}$$

Substituting from (B.7) into (B.6)

$$\frac{\partial \Psi}{\partial x_1} = (\cos\theta)\frac{\partial \Psi}{\partial r} + \Big(-\frac{\sin\theta}{r}\Big)\frac{\partial \Psi}{\partial \theta} + (0)\frac{\partial \Psi}{\partial z} \tag{B.8}$$

Hence, we have

$$\frac{\partial}{\partial x_1} = (\cos\theta)\frac{\partial}{\partial r} + \Big(-\frac{\sin\theta}{r}\Big)\frac{\partial}{\partial \theta} + (0)\frac{\partial}{\partial z} \tag{B.9}$$

Equation (B.9) defines $\frac{\partial}{\partial x_1}$ in terms of $\frac{\partial}{\partial r}$, $\frac{\partial}{\partial \theta}$ and $\frac{\partial}{\partial z}$. Similarly consider $\frac{\partial \Psi}{\partial x_2}$ and use $\Psi(r,\theta,z)$ to obtain

$$\frac{\partial \Psi}{\partial x_2} = \frac{\partial \Psi}{\partial r}\frac{\partial r}{\partial x_2} + \frac{\partial \Psi}{\partial \theta}\frac{\partial \theta}{\partial x_2} + \frac{\partial \Psi}{\partial z}\frac{\partial z}{\partial x_2} \tag{B.10}$$

and using r, θ and z from (B.5) we obtain

$$\begin{aligned}\frac{\partial r}{\partial x_2} &= \frac{\partial}{\partial x_2}\Big(\sqrt{x_1^2+x_2^2}\,\Big) = \frac{x_2}{\sqrt{x_1^2+x_2^2}} = \frac{r\ \sin\theta}{r} = \sin\theta \\ \frac{\partial \theta}{\partial x_2} &= \frac{\partial}{\partial x_2}\Big(\arctan\big(\frac{x_2}{x_1}\big)\Big) = \frac{x_1}{x_1^2+x_2^2} = \frac{r\ \cos\theta}{r^2} = \frac{\cos\theta}{r} \\ \frac{\partial z}{\partial x_2} &= \frac{\partial x_3}{\partial x_2} = 0\end{aligned} \tag{B.11}$$

Substituting from (B.11) into (B.10)

$$\frac{\partial \Psi}{\partial x_2} = (\sin\theta)\frac{\partial \Psi}{\partial r} + \big(\frac{\cos\theta}{r}\big)\frac{\partial \Psi}{\partial \theta} + (0)\frac{\partial \Psi}{\partial z} \tag{B.12}$$

Hence, we have

$$\frac{\partial}{\partial x_2} = (\sin\theta)\frac{\partial}{\partial r} + \Big(\frac{\cos\theta}{r}\Big)\frac{\partial}{\partial \theta} + (0)\frac{\partial}{\partial z} \tag{B.13}$$

Equation (B.13) defines $\frac{\partial}{\partial x_2}$ in terms of $\frac{\partial}{\partial r}$, $\frac{\partial}{\partial \theta}$ and $\frac{\partial}{\partial z}$. Likewise consider $\frac{\partial \Psi}{\partial x_3}$ and use $\Psi(r,\theta,z)$ to obtain

$$\frac{\partial \Psi}{\partial x_3} = \frac{\partial \Psi}{\partial r}\frac{\partial r}{\partial x_3} + \frac{\partial \Psi}{\partial \theta}\frac{\partial \theta}{\partial x_3} + \frac{\partial \Psi}{\partial z}\frac{\partial z}{\partial x_3} \tag{B.14}$$

and using r, θ and z from (B.5) we obtain

$$\frac{\partial r}{\partial x_3} = \frac{\partial}{\partial x_3}\Big(\sqrt{x_1^2+x_2^2}\,\Big) = 0 \tag{B.15}$$

$$\frac{\partial \theta}{\partial x_3} = \frac{\partial}{\partial x_3}\Big(\arctan(\frac{x_2}{x_1})\Big) = 0$$
$$\frac{\partial z}{\partial x_3} = \frac{\partial x_3}{\partial x_3} = 1$$

Substituting from (B.15) into (B.14)

$$\frac{\partial \Psi}{\partial x_3} = (0)\frac{\partial \Psi}{\partial r} + (0)\frac{\partial \Psi}{\partial \theta} + (1)\frac{\partial \Psi}{\partial z} \tag{B.16}$$

Hence, we have

$$\frac{\partial}{\partial x_3} = (0)\frac{\partial}{\partial r} + (0)\frac{\partial}{\partial \theta} + (1)\frac{\partial}{\partial z} \tag{B.17}$$

Equation (B.17) defines $\frac{\partial}{\partial x_3}$ in terms of $\frac{\partial}{\partial r}$, $\frac{\partial}{\partial \theta}$ and $\frac{\partial}{\partial z}$. Collectively (B.8), (B.12) and (B.16) can be written in a more convenient vector and matrix form.

$$\begin{Bmatrix} \dfrac{\partial \Psi}{\partial x_1} \\ \dfrac{\partial \Psi}{\partial x_2} \\ \dfrac{\partial \Psi}{\partial x_3} \end{Bmatrix} = \begin{bmatrix} \cos\theta & -\dfrac{\sin\theta}{r} & 0 \\ \sin\theta & \dfrac{\cos\theta}{r} & 0 \\ 0 & 0 & 1 \end{bmatrix} \begin{Bmatrix} \dfrac{\partial \Psi}{\partial r} \\ \dfrac{\partial \Psi}{\partial \theta} \\ \dfrac{\partial \Psi}{\partial z} \end{Bmatrix} \tag{B.18}$$

Using (B.9), (B.13) and (B.17) or alternately using (B.18) the operator from (B.18) can be written as

$$\begin{Bmatrix} \dfrac{\partial}{\partial x_1} \\ \dfrac{\partial}{\partial x_2} \\ \dfrac{\partial}{\partial x_3} \end{Bmatrix} = \begin{bmatrix} \cos\theta & -\dfrac{\sin\theta}{r} & 0 \\ \sin\theta & \dfrac{\cos\theta}{r} & 0 \\ 0 & 0 & 1 \end{bmatrix} \begin{Bmatrix} \dfrac{\partial}{\partial r} \\ \dfrac{\partial}{\partial \theta} \\ \dfrac{\partial}{\partial z} \end{Bmatrix} \tag{B.19}$$

B.1.3 Relationship between bases in x_1, x_2, x_3- and r, θ, z-frames

The simplest way to derive relationships between $\boldsymbol{e}_r$, $\boldsymbol{e}_\theta$, $\boldsymbol{e}_z$ and $\boldsymbol{e}_{x_1}$, $\boldsymbol{e}_{x_2}$, $\boldsymbol{e}_{x_3}$ is to refer to Figure B.3.

It is obvious that a unit vector $\boldsymbol{e}_r$ has projections $\cos\theta$, $\sin\theta$ and 0 along $o - x_1$, $o - x_2$ and $o - x_3$ axes. Hence

$$\begin{aligned} \boldsymbol{e}_r &= (\cos\theta)\, \boldsymbol{e}_{x_1} + (\sin\theta)\, \boldsymbol{e}_{x_2} + (0)\, \boldsymbol{e}_{x_3} \\ \text{and} \quad \boldsymbol{e}_z &= (0)\, \boldsymbol{e}_{x_1} + (0)\, \boldsymbol{e}_{x_2} + (1)\, \boldsymbol{e}_{x_3} \\ \text{hence,} \quad \boldsymbol{e}_\theta &= \boldsymbol{e}_z \times \boldsymbol{e}_r = (-\sin\theta)\, \boldsymbol{e}_{x_1} + (\cos\theta)\, \boldsymbol{e}_{x_2} + (0)\, \boldsymbol{e}_{x_3} \end{aligned} \tag{B.20}$$

Equation (B.20) can be written in the matrix and vector form

$$\begin{Bmatrix} \boldsymbol{e}_r \\ \boldsymbol{e}_\theta \\ \boldsymbol{e}_z \end{Bmatrix} = \begin{bmatrix} \cos\theta & \sin\theta & 0 \\ -\sin\theta & \cos\theta & 0 \\ 0 & 0 & 1 \end{bmatrix} \begin{Bmatrix} \boldsymbol{e}_{x_1} \\ \boldsymbol{e}_{x_2} \\ \boldsymbol{e}_{x_3} \end{Bmatrix} \tag{B.21}$$

Equation (B.21) expresses r, θ, z-frame basis in terms of x_1, x_2, x_3-frame basis. We can also obtain the inverse of (B.21).

$$\begin{Bmatrix} \boldsymbol{e}_{x_1} \\ \boldsymbol{e}_{x_2} \\ \boldsymbol{e}_{x_3} \end{Bmatrix} = \begin{bmatrix} \cos\theta & -\sin\theta & 0 \\ \sin\theta & \cos\theta & 0 \\ 0 & 0 & 1 \end{bmatrix} \begin{Bmatrix} \boldsymbol{e}_r \\ \boldsymbol{e}_\theta \\ \boldsymbol{e}_z \end{Bmatrix} \tag{B.22}$$

Equation (B.22) expresses x_1, x_2, x_3-frame basis in terms of r, θ, z-frame basis.

Relations (B.21) and (B.22) can also be derived in an alternate way given in the following. Let $\boldsymbol{e}_r$ be the base vector along r-direction. If l_1, l_2, l_3 *are the direction cosines of the unit vectors* $\boldsymbol{e}_{x_1}$, $\boldsymbol{e}_{x_2}$, $\boldsymbol{e}_{x_3}$ *in the* r*-direction*, then we can write

$$\boldsymbol{e}_r = l_1\, \boldsymbol{e}_{x_1} + l_2\, \boldsymbol{e}_{x_2} + l_3\, \boldsymbol{e}_{x_3} \tag{B.23}$$

where l_1, l_2 and l_3 can be obtained using x_1, x_2 and x_3 in (B.4).

$$l_1 = \frac{\partial x_1}{\partial r} = \cos\theta \quad ; \quad l_2 = \frac{\partial x_2}{\partial r} = \sin\theta \quad ; \quad l_3 = \frac{\partial x_3}{\partial r} = 0 \tag{B.24}$$

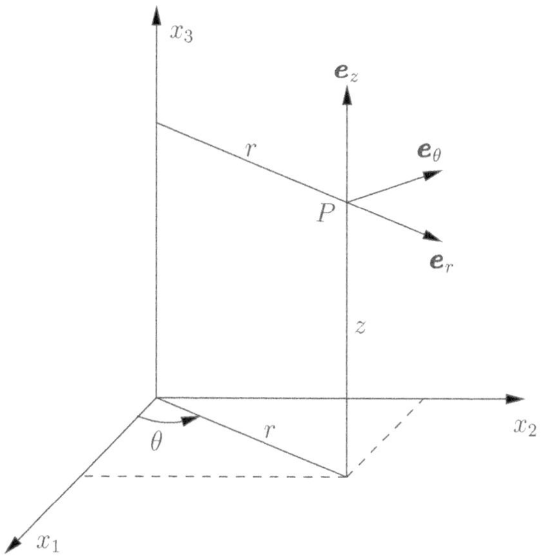

Figure B.3: Basis in cylindrical r, θ, z-frame

Hence

$$\boldsymbol{e}_r = (\cos\theta)\,\boldsymbol{e}_{x_1} + (\sin\theta)\,\boldsymbol{e}_{x_2} + (0)\,\boldsymbol{e}_{x3} \tag{B.25}$$

Let $\boldsymbol{e}_\theta$ be the base vector in the direction of increasing θ. Similarly, if m_1, m_2, m_3 *are the direction cosines of the unit vectors* $\boldsymbol{e}_{x_1}$, $\boldsymbol{e}_{x_2}$, $\boldsymbol{e}_{x_3}$ *in the* θ*-direction*, then we can write

$$\boldsymbol{e}_\theta = m_1\,\boldsymbol{e}_{x_1} + m_2\,\boldsymbol{e}_{x_2} + m_3\,\boldsymbol{e}_{x_3} \tag{B.26}$$

where m_1, m_2 and m_3 can be obtained using x_1, x_2 and x_3 in (B.4).

$$m_1 = \frac{\partial x_1}{\partial\theta} = -r\,\sin\theta \quad ; \quad m_2 = \frac{\partial x_2}{\partial\theta} = r\cos\theta \quad ; \quad m_3 = \frac{\partial x_3}{\partial\theta} = 0 \tag{B.27}$$

Hence we can write (using $r = 1$ as we need to consider unit length for basis)

$$\boldsymbol{e}_\theta = (-\,\sin\theta)\,\boldsymbol{e}_{x_1} + (\,\cos\theta)\,\boldsymbol{e}_{x_2} + (0)\,\boldsymbol{e}_{x_3} \tag{B.28}$$

Lastly, let $\boldsymbol{e}_z$ be the base vector along the z-direction. If s_1, s_2, s_3 *are the direction cosines of the unit vectors* $\boldsymbol{e}_{x_1}$, $\boldsymbol{e}_{x_2}$, $\boldsymbol{e}_{x_3}$ *in the* z*-direction*, then we can write

$$\boldsymbol{e}_z = s_1\,\boldsymbol{e}_{x_1} + s_2\,\boldsymbol{e}_{x_2} + s_3\,\boldsymbol{e}_{x_3} \tag{B.29}$$

where s_1, s_2 and s_3 can be obtained using x_1, x_2 and x_3 in (B.4).

$$s_1 = \frac{\partial x_1}{\partial z} = 0 \quad ; \quad s_2 = \frac{\partial x_2}{\partial z} = 0 \quad ; \quad s_3 = \frac{\partial x_3}{\partial z} = 1 \tag{B.30}$$

Then we can write

$$\boldsymbol{e}_z = (0)\,\boldsymbol{e}_{x_1} + (0)\,\boldsymbol{e}_{x_2} + (1)\,\boldsymbol{e}_{x_3} \tag{B.31}$$

We note that $\boldsymbol{e}_\theta = \boldsymbol{e}_z \times \boldsymbol{e}_r$, hence $r = 1$ in (B.27) is necessary in this derivation so that we have unit vector length.

Equations (B.25), (B.28) and (B.31) can be written in matrix and vector form to give (B.21), whose inverse is obviously (B.22).

B.2 Cartesian frame x_1, x_2, x_3 and spherical frame r, θ, ϕ

B.2.1 Relationship between coordinates of a point in x_1, x_2, x_3 and r, θ, ϕ-frames

From Figure B.2 we can write

$$\begin{aligned} x_1 &= r\sin\theta\cos\phi = x_1(r,\theta,\phi) \\ x_2 &= r\sin\theta\sin\phi = x_2(r,\theta,\phi) \\ x_3 &= r\cos\theta = x_3(r,\theta) \end{aligned} \tag{B.32}$$

and its inverse

$$\begin{aligned} r &= \sqrt{x_1^2 + x_2^2 + x_3^2} = r(x_1,x_2,x_3) \\ \theta &= \arctan\Big(\frac{\sqrt{x_1^2 + x_2^2}}{x_3}\Big) = \theta(x_1,x_2,x_3) \\ \phi &= \arctan\big(\frac{x_2}{x_1}\big) = \phi(x_1,x_2) \end{aligned} \tag{B.33}$$

B.2.2 Converting derivatives of a scalar with respect to x_1, x_2, x_3 into derivatives with respect to r, θ, ϕ

Consider the scalar function $\psi = \psi(x_1, x_2, x_3)$ and also $\psi = \psi(r, \theta, \phi)$. Then, in matrix and vector notation we can write the following using the chain rule of differentiation:

$$\begin{Bmatrix} \dfrac{\partial \psi}{\partial x_1} \\[2ex] \dfrac{\partial \psi}{\partial x_2} \\[2ex] \dfrac{\partial \psi}{\partial x_3} \end{Bmatrix} = \begin{bmatrix} \dfrac{\partial r}{\partial x_1} & \dfrac{\partial \theta}{\partial x_1} & \dfrac{\partial \phi}{\partial x_1} \\[2ex] \dfrac{\partial r}{\partial x_2} & \dfrac{\partial \theta}{\partial x_2} & \dfrac{\partial \phi}{\partial x_2} \\[2ex] \dfrac{\partial r}{\partial x_3} & \dfrac{\partial \theta}{\partial x_3} & \dfrac{\partial \phi}{\partial x_3} \end{bmatrix} \begin{Bmatrix} \dfrac{\partial \psi}{\partial r} \\[2ex] \dfrac{\partial \psi}{\partial \theta} \\[2ex] \dfrac{\partial \psi}{\partial \phi} \end{Bmatrix} \tag{B.34}$$

Using (B.33) we can obtain the following

$$\begin{aligned} \frac{\partial r}{\partial x_1} &= \frac{\partial}{\partial x_1}\left(\sqrt{x_1^2 + x_2^2 + x_3^2}\right) = \frac{x_1}{\sqrt{x_1^2 + x_2^2 + x_3^2}} \\ &= \frac{r \ \sin\theta \sin\phi}{r} = \sin\theta \sin\phi \\ \frac{\partial \theta}{\partial x_1} &= \frac{\partial}{\partial x_1}\left(\arctan\Big(\frac{\sqrt{x_1^2 + x_2^2}}{x_3}\Big)\right) = \frac{x_1 x_3}{(x_1^2 + x_2^2 + x_3^2)\sqrt{x_1^2 + x_2^2}} \\ &= \frac{(r \sin\theta \cos\phi)(r \cos\theta)}{r^2 \ r \sin\theta} = \frac{\cos\theta \cos\phi}{r} \\ \frac{\partial \phi}{\partial x_1} &= \frac{\partial}{\partial x_1}\left(\arctan\Big(\frac{x_2}{x_1}\Big)\right) = -\frac{x_2}{x_1^2 + x_2^2} \\ &= -\frac{r \sin\theta \sin\phi}{(r \sin\theta)^2} = -\frac{\sin\phi}{r \sin\theta} \end{aligned} \tag{B.35}$$

Additionally

$$\begin{aligned} \frac{\partial r}{\partial x_2} &= \frac{\partial}{\partial x_2}\left(\sqrt{x_1^2 + x_2^2 + x_3^2}\right) = \frac{x_2}{\sqrt{x_1^2 + x_2^2 + x_3^2}} \\ &= \frac{r \ \sin\theta \sin\phi}{r} = \sin\theta \sin\phi \\ \frac{\partial \theta}{\partial x_2} &= \frac{\partial}{\partial x_2}\left(\arctan\Big(\frac{\sqrt{x_1^2 + x_2^2}}{x_3}\Big)\right) = \frac{x_2 x_3}{(x_1^2 + x_2^2 + x_3^2)\sqrt{x_1^2 + x_2^2}} \\ &= \frac{(r \sin\theta \sin\phi)(r \cos\theta)}{r^2 \ r \sin\theta} = \frac{\cos\theta \sin\phi}{r} \\ \frac{\partial \phi}{\partial x_2} &= \frac{\partial}{\partial x_2}\left(\arctan\Big(\frac{x_2}{x_1}\Big)\right) = \frac{x_1}{x_1^2 + x_2^2} \\ &= \frac{r \ \sin\theta \cos\phi}{(r \ \sin\theta)^2} = \frac{\cos\phi}{r \ \sin\theta} \end{aligned} \tag{B.36}$$

and

$$
\begin{aligned}
\frac{\partial r}{\partial x_3} &= \frac{\partial}{\partial x_3}\left(\sqrt{x_1^2+x_2^2+x_3^2}\,\right) = \frac{x_3}{\sqrt{x_1^2+x_2^2+x_3^2}} \\
&= \frac{r\,\cos\theta}{r} = \cos\theta \\
\frac{\partial \theta}{\partial x_3} &= \frac{\partial}{\partial x_3}\left(\arctan\left(\frac{\sqrt{x_1^2+x_2^2}}{x_3}\right)\right) = -\frac{\sqrt{x_1^2+x_2^2}}{x_1^2+x_2^2+x_3^2} \\
&= -\frac{r\,\sin\theta}{r^2} = -\frac{\sin\theta}{r} \\
\frac{\partial \phi}{\partial x_3} &= \frac{\partial}{\partial x_3}\left(\arctan\left(\frac{x_2}{x_1}\right)\right) = 0
\end{aligned}
\tag{B.37}
$$

Equations (B.35)–(B.37) define the coefficients of the matrix in (B.34). Hence (B.34) can now be written as

$$
\begin{Bmatrix} \dfrac{\partial\psi}{\partial x_1} \\[2ex] \dfrac{\partial\psi}{\partial x_2} \\[2ex] \dfrac{\partial\psi}{\partial x_3} \end{Bmatrix} = \begin{bmatrix} \sin\theta\cos\phi & \dfrac{\cos\theta\cos\phi}{r} & -\dfrac{\sin\phi}{r\,\sin\theta} \\[2ex] \sin\theta\sin\phi & \dfrac{\cos\theta\sin\phi}{r} & \dfrac{\cos\phi}{r\sin\theta} \\[2ex] \cos\theta & -\dfrac{\sin\theta}{r} & 0 \end{bmatrix} \begin{Bmatrix} \dfrac{\partial\psi}{\partial r} \\[2ex] \dfrac{\partial\psi}{\partial \theta} \\[2ex] \dfrac{\partial\psi}{\partial \phi} \end{Bmatrix}
\tag{B.38}
$$

B.2.3 Relationship between bases, i.e., unit vectors in x_1, x_2, x_3- and r, θ, ϕ-frames

Following the procedure similar to Section B.1.3 we can derive the following relations:

$$
\begin{Bmatrix} \boldsymbol{e}_r \\ \boldsymbol{e}_\theta \\ \boldsymbol{e}_\phi \end{Bmatrix} = \begin{bmatrix} \sin\theta\cos\phi & \sin\theta\sin\phi & \cos\theta \\ \cos\theta\cos\phi & \cos\theta\cos\phi & -\sin\theta \\ -\sin\phi & \cos\phi & 0 \end{bmatrix} \begin{Bmatrix} \boldsymbol{e}_{x_1} \\ \boldsymbol{e}_{x_2} \\ \boldsymbol{e}_{x_3} \end{Bmatrix}
\tag{B.39}
$$

Equations (B.39) express x_1, x_2, x_3-frame basis in terms of r, θ, ϕ-frame basis. We can also obtain the inverse of (B.39).

$$
\begin{Bmatrix} \boldsymbol{e}_{x_1} \\ \boldsymbol{e}_{x_2} \\ \boldsymbol{e}_{x_3} \end{Bmatrix} = \begin{bmatrix} \sin\theta\cos\phi & \cos\theta\cos\phi & -\sin\phi \\ \sin\theta\sin\phi & \cos\theta\sin\phi & \cos\phi \\ \cos\theta & -\sin\theta & 0 \end{bmatrix} \begin{Bmatrix} \boldsymbol{e}_r \\ \boldsymbol{e}_\theta \\ \boldsymbol{e}_\phi \end{Bmatrix}
\tag{B.40}
$$

Equations (B.40) express r, θ, ϕ-frame basis in terms of x_1, x_2, x_3-frame basis.

B.3 Differential operations in r, θ, z- and r, θ, ϕ-frames

B.3.1 r, θ, z-frame

(a) *Derivatives of the base vectors in r, θ, z-frame*

These can be obtained using (B.21).

$$
\begin{aligned}
&\frac{\partial \boldsymbol{e}_r}{\partial r} = 0 \quad ; \quad \frac{\partial \boldsymbol{e}_r}{\partial \theta} = -\sin\theta \boldsymbol{e}_{x_1} + \cos\theta \boldsymbol{e}_{x_2} = \boldsymbol{e}_\theta \quad ; \quad \frac{\partial \boldsymbol{e}_r}{\partial z} = 0 \\
&\frac{\partial \boldsymbol{e}_\theta}{\partial r} = 0 \quad ; \quad \frac{\partial \boldsymbol{e}_\theta}{\partial \theta} = -\cos\theta \boldsymbol{e}_{x_1} - \sin\theta \boldsymbol{e}_{x_2} = -\boldsymbol{e}_r \quad ; \quad \frac{\partial \boldsymbol{e}_\theta}{\partial z} = 0 \\
&\frac{\partial \boldsymbol{e}_z}{\partial r} = 0 \quad ; \quad \frac{\partial \boldsymbol{e}_z}{\partial \theta} = 0 \quad ; \quad \frac{\partial \boldsymbol{e}_z}{\partial z} = 0
\end{aligned}
\tag{B.41}
$$

(b) *Differential operator $\boldsymbol{\nabla}$ and gradient of a scalar field $\boldsymbol{\nabla}\psi$ in r, θ, z-frame*

In x_1, x_2, x_3-frame we can write for a scalar ψ

$$
\boldsymbol{\nabla}\psi = \frac{\partial \psi}{\partial x_1}\boldsymbol{e}_{x_1} + \frac{\partial \psi}{\partial x_2}\boldsymbol{e}_{x_2} + \frac{\partial \psi}{\partial x_3}\boldsymbol{e}_{x_3} \tag{B.42}
$$

Substituting from (B.22) for $\boldsymbol{e}_{x_1}, \boldsymbol{e}_{x_2}, \boldsymbol{e}_{x_3}$ in terms of $\boldsymbol{e}_r, \boldsymbol{e}_\theta, \boldsymbol{e}_z$ and substituting from (B.18) for $\frac{\partial \psi}{\partial x_1}, \frac{\partial \psi}{\partial x_2}, \frac{\partial \psi}{\partial x_3}$ in terms of $\frac{\partial \psi}{\partial r}, \frac{\partial \psi}{\partial \theta}, \frac{\partial \psi}{\partial z}$ yields

$$
\begin{aligned}
\boldsymbol{\nabla}\psi =& \left(\cos\theta\frac{\partial \psi}{\partial r} - \frac{\sin\theta}{r}\frac{\partial \psi}{\partial \theta}\right)(\cos\theta \boldsymbol{e}_r - \sin\theta \boldsymbol{e}_\theta) \\
&+ \left(\sin\theta\frac{\partial \psi}{\partial r} + \frac{\cos\theta}{r}\frac{\partial \psi}{\partial \theta}\right)(\sin\theta \boldsymbol{e}_r + \cos\theta \boldsymbol{e}_\theta) + \frac{\partial \psi}{\partial z}\boldsymbol{e}_z
\end{aligned}
\tag{B.43}
$$

and upon simplifying the right side we obtain $\boldsymbol{\nabla}\psi$ in r, θ, z-frame

$$
\boldsymbol{\nabla}\psi = \frac{\partial \psi}{\partial r}\boldsymbol{e}_r + \frac{1}{r}\frac{\partial \psi}{\partial \theta}\boldsymbol{e}_\theta + \frac{\partial \psi}{\partial z}\boldsymbol{e}_z \tag{B.44}
$$

Hence the $\boldsymbol{\nabla}$ operator in $\boldsymbol{\nabla}\psi$ in r, θ, z-frame is given by

$$
\boldsymbol{\nabla} = \boldsymbol{e}_r\frac{\partial}{\partial r} + \boldsymbol{e}_\theta\frac{1}{r}\frac{\partial}{\partial \theta} + \boldsymbol{e}_z\frac{\partial}{\partial z} \tag{B.45}
$$

(c) *Divergence of a tensor of rank one $\boldsymbol{\nabla}\cdot\boldsymbol{v}$ in r, θ, z-frame*

Using (B.45) for $\boldsymbol{\nabla}$ in r, θ, z-frame and $\boldsymbol{v} = v_r\boldsymbol{e}_r + v_\theta\boldsymbol{e}_\theta + v_z\boldsymbol{e}_z$ we can write

$$
\boldsymbol{\nabla}\cdot\boldsymbol{v} = \left(\boldsymbol{e}_r\frac{\partial}{\partial r} + \boldsymbol{e}_\theta\frac{1}{r}\frac{\partial}{\partial \theta} + \boldsymbol{e}_z\frac{\partial}{\partial z}\right)\cdot(v_r\boldsymbol{e}_r + v_\theta\boldsymbol{e}_\theta + v_z\boldsymbol{e}_z) \tag{B.46}
$$

Expanding the right side by using derivatives of the base vectors in r, θ, z-frame given in Equation (B.41) gives

$$
\begin{aligned}
\boldsymbol{\nabla}\cdot\boldsymbol{v} = \boldsymbol{e}_r\cdot\boldsymbol{e}_r\frac{\partial v_r}{\partial r} + \frac{1}{r}\boldsymbol{e}_\theta\cdot\left(\boldsymbol{e}_r\frac{\partial v_r}{\partial\theta} + \boldsymbol{e}_\theta v_r\right) + \boldsymbol{e}_z\cdot\boldsymbol{e}_r\frac{\partial v_r}{\partial z} \\
+ \boldsymbol{e}_r\cdot\boldsymbol{e}_\theta\frac{\partial v_\theta}{\partial r} + \frac{1}{r}\boldsymbol{e}_\theta\cdot\left(\boldsymbol{e}_\theta\frac{\partial v_\theta}{\partial\theta} - \boldsymbol{e}_r v_\theta\right) + \boldsymbol{e}_z\cdot\boldsymbol{e}_\theta\frac{\partial v_\theta}{\partial z} \\
+ \boldsymbol{e}_r\cdot\boldsymbol{e}_z\frac{\partial v_z}{\partial r} + \boldsymbol{e}_\theta\cdot\boldsymbol{e}_z\frac{1}{r}\frac{\partial v_z}{\partial\theta} + \boldsymbol{e}_z\cdot\boldsymbol{e}_z\frac{\partial v_z}{\partial z}
\end{aligned}
\tag{B.47}
$$

Since $\boldsymbol{e}_r\cdot\boldsymbol{e}_r = 1$, $\boldsymbol{e}_\theta\cdot\boldsymbol{e}_\theta = 1$, $\boldsymbol{e}_r\cdot\boldsymbol{e}_\theta = 0$, $\boldsymbol{e}_r\cdot\boldsymbol{e}_z = 0$ etc. then

$$
\boldsymbol{\nabla}\cdot\boldsymbol{v} = \frac{v_r}{r} + \frac{\partial v_r}{\partial r} + \frac{1}{r}\frac{\partial v_\theta}{\partial\theta} + \frac{\partial v_z}{\partial z}
\tag{B.48}
$$

(d) *Gradient of a tensor of rank one or dyadic product* $\boldsymbol{\nabla v}$ *in* r,θ,z*-frame*

Using (B.45) for $\boldsymbol{\nabla}$ in r,θ,z-frame and $\boldsymbol{v} = v_r\boldsymbol{e}_r + v_\theta\boldsymbol{e}_\theta + v_z\boldsymbol{e}_z$ we can write

$$
\boldsymbol{\nabla v} = \left(\boldsymbol{e}_r\frac{\partial}{\partial r} + \boldsymbol{e}_\theta\frac{1}{r}\frac{\partial}{\partial\theta} + \boldsymbol{e}_z\frac{\partial}{\partial z}\right)\left(v_r\boldsymbol{e}_r + v_\theta\boldsymbol{e}_\theta + v_z\boldsymbol{e}_z\right)
\tag{B.49}
$$

We expand the right side by using derivatives of the base vectors in r,θ,z-frame given in Equation (B.41).

$$
\begin{aligned}
\boldsymbol{\nabla v} = \boldsymbol{e}_r\boldsymbol{e}_r\frac{\partial v_r}{\partial r} + \frac{1}{r}\boldsymbol{e}_\theta\left(\boldsymbol{e}_r\frac{\partial v_r}{\partial\theta} + \boldsymbol{e}_\theta v_r\right) + \boldsymbol{e}_z\boldsymbol{e}_r\frac{\partial v_r}{\partial z} \\
+ \boldsymbol{e}_r\boldsymbol{e}_\theta\frac{\partial v_\theta}{\partial r} + \frac{1}{r}\boldsymbol{e}_\theta\left(\boldsymbol{e}_\theta\frac{\partial v_\theta}{\partial\theta} - \boldsymbol{e}_r v_\theta\right) + \boldsymbol{e}_z\boldsymbol{e}_\theta\frac{\partial v_\theta}{\partial z} \\
+ \boldsymbol{e}_r\boldsymbol{e}_z\frac{\partial v_z}{\partial r} + \boldsymbol{e}_\theta\boldsymbol{e}_z\frac{1}{r}\frac{\partial v_z}{\partial\theta} + \boldsymbol{e}_z\boldsymbol{e}_z\frac{\partial v_z}{\partial z}
\end{aligned}
\tag{B.50}
$$

which can be written as

$$
\begin{aligned}
\boldsymbol{\nabla v} = \boldsymbol{e}_r\boldsymbol{e}_r\frac{\partial v_r}{\partial r} + \boldsymbol{e}_r\boldsymbol{e}_\theta\frac{\partial v_\theta}{\partial r} + \boldsymbol{e}_r\boldsymbol{e}_z\frac{\partial v_z}{\partial r} \\
+ \boldsymbol{e}_\theta\boldsymbol{e}_r\left(\frac{1}{r}\frac{\partial v_r}{\partial\theta} - \frac{v_\theta}{r}\right) + \boldsymbol{e}_\theta\boldsymbol{e}_\theta\left(\frac{1}{r}\frac{\partial v_\theta}{\partial\theta} + \frac{v_r}{r}\right) + \boldsymbol{e}_\theta\boldsymbol{e}_z\frac{1}{r}\frac{\partial v_z}{\partial\theta} \\
+ \boldsymbol{e}_z\boldsymbol{e}_r\frac{\partial v_r}{\partial z} + \boldsymbol{e}_z\boldsymbol{e}_\theta\frac{\partial v_\theta}{\partial z} + \boldsymbol{e}_z\boldsymbol{e}_z\frac{\partial v_z}{\partial z}
\end{aligned}
\tag{B.51}
$$

We note that each term in (B.51) has dyads associated with it. Also $\mathrm{tr}(\boldsymbol{\nabla v}) = \boldsymbol{\nabla}\cdot\boldsymbol{v}$.

B.3.2 r,θ,ϕ-frame

(a) *Derivatives of the base vectors in* r,θ,ϕ*-frame*

These can be obtained using (B.39).

$$
\frac{\partial\boldsymbol{e}_r}{\partial r} = 0 \quad ; \quad \frac{\partial\boldsymbol{e}_r}{\partial\theta} = \cos\theta\cos\phi\boldsymbol{e}_{x_1} + \cos\theta\sin\phi\boldsymbol{e}_{x_2} - \sin\theta\boldsymbol{e}_{x_3} = \boldsymbol{e}_\theta
$$

$$
\begin{aligned}
&\quad ; \quad \frac{\partial \boldsymbol{e}_r}{\partial \phi} = -\sin\theta\sin\phi\boldsymbol{e}_{x_1} + \sin\theta\cos\phi\boldsymbol{e}_{x_2} \\
&\qquad\qquad = \sin\phi\boldsymbol{e}_\phi \\
&\frac{\partial \boldsymbol{e}_\theta}{\partial r} = 0 \quad ; \quad \frac{\partial \boldsymbol{e}_\theta}{\partial \theta} = -\sin\theta\cos\phi\boldsymbol{e}_{x_1} - \sin\theta\sin\phi\boldsymbol{e}_{x_2} - \cos\theta\boldsymbol{e}_{x_3} \\
&\qquad\qquad = -\boldsymbol{e}_r \\
&\quad ; \quad \frac{\partial \boldsymbol{e}_\theta}{\partial \phi} = -\cos\theta\sin\phi\boldsymbol{e}_{x_1} + \cos\theta\cos\phi\boldsymbol{e}_{x_2} \\
&\qquad\qquad = \cos\theta\boldsymbol{e}_\phi \\
&\frac{\partial \boldsymbol{e}_\phi}{\partial r} = 0 \quad ; \quad \frac{\partial \boldsymbol{e}_\phi}{\partial \theta} = 0 \\
&\quad ; \quad \frac{\partial \boldsymbol{e}_\phi}{\partial \phi} = -\cos\phi\boldsymbol{e}_{x_1} - \sin\phi\boldsymbol{e}_{x_2} \\
&\qquad\qquad = -\cos\phi\big(\sin\theta\cos\phi\boldsymbol{e}_r + \cos\theta\cos\phi\boldsymbol{e}_\theta - \sin\phi\boldsymbol{e}_\phi\big) \\
&\qquad\qquad\quad - \sin\phi\big(\sin\theta\sin\phi\boldsymbol{e}_r + \cos\theta\sin\phi\boldsymbol{e}_\theta + \cos\phi\boldsymbol{e}_\phi\big) \\
&\qquad\qquad = -\sin\theta\boldsymbol{e}_r - \cos\theta\boldsymbol{e}_\theta
\end{aligned}
\tag{B.52}
$$

(b) *Differential operator* $\boldsymbol{\nabla}$ *and gradient of a scalar field* $\boldsymbol{\nabla}\psi$ *in* r, θ, ϕ*-frame*

As in the case of r, θ, z-frame, here also we can write for a scalar ψ

$$
\boldsymbol{\nabla}\psi = \frac{\partial \psi}{\partial x_1}\boldsymbol{e}_{x_1} + \frac{\partial \psi}{\partial x_2}\boldsymbol{e}_{x_2} + \frac{\partial \psi}{\partial x_3}\boldsymbol{e}_{x_3} \tag{B.53}
$$

Substituting from (B.40) for $\boldsymbol{e}_{x_1}, \boldsymbol{e}_{x_2}, \boldsymbol{e}_{x_3}$ in terms of $\boldsymbol{e}_r, \boldsymbol{e}_\theta, \boldsymbol{e}_\phi$ and substituting from (B.38) for $\frac{\partial \psi}{\partial x_1}, \frac{\partial \psi}{\partial x_2}, \frac{\partial \psi}{\partial x_3}$ in terms of $\frac{\partial \psi}{\partial r}, \frac{\partial \psi}{\partial \theta}, \frac{\partial \psi}{\partial \phi}$ yields

$$
\begin{aligned}
\boldsymbol{\nabla}\psi = &\big(\sin\theta\cos\phi\boldsymbol{e}_r + \cos\theta\cos\phi\boldsymbol{e}_\theta - \sin\phi\boldsymbol{e}_\phi\big)\big(\sin\theta\cos\phi\frac{\partial \psi}{\partial r} \\
&+ \frac{\cos\theta\cos\phi}{r}\frac{\partial \psi}{\partial \theta} - \frac{\sin\phi}{r\sin\theta}\frac{\partial \psi}{\partial \phi}\big) \\
&+ \big(\sin\theta\sin\phi\boldsymbol{e}_r + \cos\theta\sin\phi\boldsymbol{e}_\theta + \cos\phi\boldsymbol{e}_\phi\big)\big(\sin\theta\sin\phi\frac{\partial \psi}{\partial r} \\
&+ \frac{\cos\theta\cos\phi}{r}\frac{\partial \psi}{\partial \theta} + \frac{\cos\phi}{r\sin\theta}\frac{\partial \psi}{\partial \phi}\big) \\
&+ \big(\cos\theta\boldsymbol{e}_r - \sin\theta\boldsymbol{e}_\theta\big)\big(\cos\theta\frac{\partial \psi}{\partial r} - \frac{\sin\theta}{r}\frac{\partial \psi}{\partial \theta}\big)
\end{aligned}
\tag{B.54}
$$

and upon simplifying the right side we obtain $\boldsymbol{\nabla}\psi$ in r, θ, ϕ-frame.

$$
\boldsymbol{\nabla}\psi = \frac{\partial \psi}{\partial r}\boldsymbol{e}_r + \frac{1}{r}\frac{\partial \psi}{\partial \theta}\boldsymbol{e}_\theta + \frac{1}{r\sin\theta}\frac{\partial \psi}{\partial \phi}\boldsymbol{e}_\phi \tag{B.55}
$$

Hence the $\boldsymbol{\nabla}$ operator in $\boldsymbol{\nabla}\psi$ in r,θ,ϕ-frame is given by

$$\boldsymbol{\nabla} = \boldsymbol{e}_r \frac{\partial}{\partial r} + \boldsymbol{e}_\theta \frac{1}{r}\frac{\partial}{\partial \theta} + \boldsymbol{e}_\phi \frac{1}{r\sin\theta}\frac{\partial}{\partial \phi} \tag{B.56}$$

(c) *Divergence of a tensor of rank one* $\boldsymbol{\nabla}\cdot\boldsymbol{v}$ *in* r,θ,ϕ*-frame*

Using (B.56) for $\boldsymbol{\nabla}$ in r,θ,ϕ-frame and $\boldsymbol{v} = v_r\boldsymbol{e}_r + v_\theta\boldsymbol{e}_\theta + v_\phi\boldsymbol{e}_\phi$ we could verify that in r,θ,ϕ-frame

$$\boldsymbol{\nabla}\cdot\boldsymbol{v} = \frac{1}{r^2}\frac{\partial}{\partial r}(r^2\, v_r) + \frac{1}{r\,\sin\theta}\frac{\partial}{\partial\theta}(v_\theta\sin\theta) + \frac{1}{r\,\sin\theta}\frac{\partial v_\phi}{\partial\phi} \tag{B.57}$$

(d) *Gradient of a tensor of rank one or dyadic product* $\boldsymbol{\nabla v}$ *in* r,θ,ϕ*-frame*

Components $\boldsymbol{\nabla v}$ with associated dyads can be obtained in a similar fashion as shown for r,θ,z-frame using (B.56) for $\boldsymbol{\nabla}$ in r,θ,ϕ-frame and $\boldsymbol{v} = v_r\boldsymbol{e}_r + v_\theta\boldsymbol{e}_\theta + v_\phi\boldsymbol{e}_\phi$ in which each term has dyads associated with it. Also $\text{tr}(\boldsymbol{\nabla v}) = \boldsymbol{\nabla}\cdot\boldsymbol{v}$.

B.4 Some examples: r,θ,z-frame

If we consider the r,θ,z-frame, then the velocity vector and the differential operator are given by

$$\begin{aligned}\boldsymbol{v} &= v_r\boldsymbol{e}_r + v_\theta\boldsymbol{e}_\theta + v_z\boldsymbol{e}_z \\ \boldsymbol{\nabla} &= \boldsymbol{e}_r\frac{\partial}{\partial r} + \boldsymbol{e}_\theta\frac{1}{r}\frac{\partial}{\partial\theta} + \boldsymbol{e}_z\frac{\partial}{\partial z}\end{aligned} \tag{B.58}$$

Hence, following Equation (B.51) we can write

$$\begin{aligned}\boldsymbol{\nabla v} = {} & \boldsymbol{e}_r\boldsymbol{e}_r\frac{\partial v_r}{\partial r} + \boldsymbol{e}_r\boldsymbol{e}_\theta\frac{\partial v_\theta}{\partial r} + \boldsymbol{e}_r\boldsymbol{e}_z\frac{\partial v_z}{\partial r} \\ & + \boldsymbol{e}_\theta\boldsymbol{e}_r\left(\frac{1}{r}\frac{\partial v_r}{\partial\theta} - \frac{v_\theta}{r}\right) + \boldsymbol{e}_\theta\boldsymbol{e}_\theta\left(\frac{1}{r}\frac{\partial v_\theta}{\partial\theta} + \frac{v_r}{r}\right) + \boldsymbol{e}_\theta\boldsymbol{e}_z\frac{1}{r}\frac{\partial v_z}{\partial\theta} \\ & + \boldsymbol{e}_z\boldsymbol{e}_r\frac{\partial v_r}{\partial z} + \boldsymbol{e}_z\boldsymbol{e}_\theta\frac{\partial v_\theta}{\partial z} + \boldsymbol{e}_z\boldsymbol{e}_z\frac{\partial v_z}{\partial z}\end{aligned} \tag{B.59}$$

and

$$\begin{aligned}(\boldsymbol{\nabla v})^T = {} & \boldsymbol{e}_r\boldsymbol{e}_r\frac{\partial v_r}{\partial r} + \boldsymbol{e}_r\boldsymbol{e}_\theta\left(\frac{1}{r}\frac{\partial v_r}{\partial\theta} - \frac{v_\theta}{r}\right) + \boldsymbol{e}_r\boldsymbol{e}_z\frac{\partial v_r}{\partial z} \\ & + \boldsymbol{e}_\theta\boldsymbol{e}_r\frac{\partial v_\theta}{\partial r} + \boldsymbol{e}_\theta\boldsymbol{e}_\theta\left(\frac{1}{r}\frac{\partial v_\theta}{\partial\theta} + \frac{v_r}{r}\right) + \boldsymbol{e}_\theta\boldsymbol{e}_z\frac{\partial v_\theta}{\partial z} \\ & + \boldsymbol{e}_z\boldsymbol{e}_r\frac{\partial v_z}{\partial r} + \boldsymbol{e}_z\boldsymbol{e}_\theta\frac{1}{r}\frac{\partial v_z}{\partial\theta} + \boldsymbol{e}_z\boldsymbol{e}_z\frac{\partial v_z}{\partial z}\end{aligned} \tag{B.60}$$

B.4.1 Symmetric part of the velocity gradient tensor $\boldsymbol{D}$

If $\boldsymbol{D}$ is the symmetric part of the velocity gradient tensor defined by

$$\boldsymbol{D} = \frac{1}{2}(\boldsymbol{\nabla v} + (\boldsymbol{\nabla v})^T) \tag{B.61}$$

then the symmetric part of the velocity gradient tensor

$$\begin{aligned}\boldsymbol{D} = &\boldsymbol{e}_r\boldsymbol{e}_r D_{rr} + \boldsymbol{e}_r\boldsymbol{e}_\theta D_{r\theta} + \boldsymbol{e}_r\boldsymbol{e}_z D_{rz}\\ &+ \boldsymbol{e}_\theta\boldsymbol{e}_r D_{\theta r} + \boldsymbol{e}_\theta\boldsymbol{e}_\theta D_{\theta\theta} + \boldsymbol{e}_\theta\boldsymbol{e}_z D_{\theta z}\\ &+ \boldsymbol{e}_z\boldsymbol{e}_r D_{zr} + \boldsymbol{e}_z\boldsymbol{e}_\theta D_{z\theta} + \boldsymbol{e}_z\boldsymbol{e}_z D_{zz}\end{aligned} \tag{B.62}$$

is given by

$$\begin{aligned}\boldsymbol{D} = &\boldsymbol{e}_r\boldsymbol{e}_r\frac{\partial v_r}{\partial r} + \boldsymbol{e}_r\boldsymbol{e}_\theta\frac{1}{2}\left(\frac{\partial v_\theta}{\partial r} + \frac{1}{r}\frac{\partial v_r}{\partial \theta} - \frac{v_\theta}{r}\right) + \boldsymbol{e}_r\boldsymbol{e}_z\frac{1}{2}\left(\frac{\partial v_r}{\partial v_z} + \frac{\partial v_z}{\partial r}\right)\\ &+ \boldsymbol{e}_\theta\boldsymbol{e}_r\frac{1}{2}\left(\frac{1}{r}\frac{\partial v_r}{\partial \theta} - \frac{v_\theta}{r} + \frac{\partial v_\theta}{\partial r}\right) + \boldsymbol{e}_\theta\boldsymbol{e}_\theta\left(\frac{1}{2}\frac{\partial v_\theta}{\partial \theta} + \frac{v_r}{r}\right) + \boldsymbol{e}_\theta\boldsymbol{e}_z\frac{1}{2}\left(\frac{1}{r}\frac{\partial v_z}{\partial \theta} + \frac{\partial v_\theta}{\partial z}\right)\\ &+ \boldsymbol{e}_z\boldsymbol{e}_r\frac{1}{2}\left(\frac{\partial v_r}{\partial v_z} + \frac{\partial v_z}{\partial r}\right) + \boldsymbol{e}_z\boldsymbol{e}_\theta\frac{1}{2}\left(\frac{1}{r}\frac{\partial v_z}{\partial \theta} + \frac{\partial v_\theta}{\partial z}\right) + \boldsymbol{e}_z\boldsymbol{e}_z\frac{\partial v_z}{\partial z}\end{aligned} \tag{B.63}$$

We note that component D_{rr} has dyad $\boldsymbol{e}_r\boldsymbol{e}_r$, component $D_{r\theta}$ has dyads $\boldsymbol{e}_r\boldsymbol{e}_\theta$ etc. Clearly $\boldsymbol{D}$ is symmetric as expected. In fluid flow, components of $\boldsymbol{D}$ represent strain rates.

B.4.2 Skew-symmetric part of the velocity gradient tensor $\boldsymbol{W}$

If $\boldsymbol{W}$ is the skew-symmetric part of the velocity gradient tensor defined by

$$\boldsymbol{W} = \frac{1}{2}(\boldsymbol{\nabla v} - (\boldsymbol{\nabla v})^T) \tag{B.64}$$

then the skew-symmetric part of the velocity gradient tensor

$$\begin{aligned}\boldsymbol{W} = &\boldsymbol{e}_r\boldsymbol{e}_r W_{rr} + \boldsymbol{e}_r\boldsymbol{e}_\theta W_{r\theta} + \boldsymbol{e}_r\boldsymbol{e}_z W_{rz}\\ &+ \boldsymbol{e}_\theta\boldsymbol{e}_r W_{\theta r} + \boldsymbol{e}_\theta\boldsymbol{e}_\theta W_{\theta\theta} + \boldsymbol{e}_\theta\boldsymbol{e}_z W_{\theta z}\\ &+ \boldsymbol{e}_z\boldsymbol{e}_r W_{zr} + \boldsymbol{e}_z\boldsymbol{e}_\theta W_{z\theta} + \boldsymbol{e}_z\boldsymbol{e}_z W_{zz}\end{aligned} \tag{B.65}$$

is given by

$$\begin{aligned}\boldsymbol{W} = &\boldsymbol{e}_r\boldsymbol{e}_r(0) + \boldsymbol{e}_r\boldsymbol{e}_\theta\frac{1}{2}\left(\frac{\partial v_\theta}{\partial r} - \frac{1}{r}\frac{\partial v_r}{\partial \theta} + \frac{v_\theta}{r}\right) + \boldsymbol{e}_r\boldsymbol{e}_z\frac{1}{2}\left(\frac{\partial v_z}{\partial r} - \frac{\partial v_r}{\partial v_z}\right)\\ &+ \boldsymbol{e}_\theta\boldsymbol{e}_r\frac{1}{2}\left(\frac{1}{r}\frac{\partial v_r}{\partial \theta} - \frac{v_\theta}{r} - \frac{\partial v_\theta}{\partial r}\right) + \boldsymbol{e}_\theta\boldsymbol{e}_\theta(0) + \boldsymbol{e}_\theta\boldsymbol{e}_z\frac{1}{2}\left(\frac{1}{r}\frac{\partial v_z}{\partial \theta} - \frac{\partial v_\theta}{\partial z}\right)\\ &+ \boldsymbol{e}_z\boldsymbol{e}_r\frac{1}{2}\left(\frac{\partial v_r}{\partial v_z} - \frac{\partial v_z}{\partial r}\right) + \boldsymbol{e}_z\boldsymbol{e}_\theta\frac{1}{2}\left(\frac{\partial v_\theta}{\partial z} - \frac{1}{r}\frac{\partial v_z}{\partial \theta}\right) + \boldsymbol{e}_z\boldsymbol{e}_z(0)\end{aligned} \tag{B.66}$$

We note that component W_{rr} has dyad $\boldsymbol{e}_r\boldsymbol{e}_r$, component $W_{r\theta}$ has dyads $\boldsymbol{e}_r\boldsymbol{e}_\theta$ etc. Clearly $W_{\theta r} = -W_{r\theta}$, $W_{zr} = -W_{rz}$ and $W_{z\theta} = -W_{\theta z}$. In fluid flow, $\boldsymbol{W}$ is referred to as the spin tensor.

B.5 Summary

The material presented here is useful in transforming mathematical models from x_1, x_2, x_3-frame to r, θ, z- or r, θ, ϕ-frames. This may be necessary in specific applications.

BIBLIOGRAPHY

[1] Adhikari, S. *Damping Models for Structural Vibration.* Ph.D. Dissertation, The University of Cambridge, 2000.

[2] Alijani, Farbod and Amabili, Marco and Balasubramanian, Prabakaran and Carra, Silvia and Ferrari, Giovanni and Garziera, Rinaldo. Damping for large-amplitude vibratiosn of plates and curved panels, part 1: modeling and experiments. *International Journal of Non-linear Mechanics*, 85:23–40, 2016.

[3] Amabili, Marco. *Nonlinear vibrations and stability of shells and plates, 2nd edition.* Cambridge University Press, 2014.

[4] Amabili, Marco and Alijani, Farbod and Delannoy, Joachim. Damping for large-amplitude vibrations of plates and curved panels, part 2: identification and comparisions. *International Journal of Non-linear Mechanics*, 85:226–240, 2016.

[5] Amabili, Marco. Nonlinear damping in large-amplitude vibrations: modeling and experiments. *Nonlinear dynamics*, 93, 2016.

[6] Amabili, Marco and Balasubramanian, Prabakaran and Ferrari, Giovanni. Traveling wave and non-stationary response in nonlinear vibrations of water-filled circular cylindrical shells: experiments and simulations. *Journal of Sound and Vibration*, 381:220–245, 2016.

[7] Amabili, Marco. Nonlinear damping in large-amplitude vibrations: modeling and experiments. A tool to identify damping during large amplitude vibrations of viscoelastic structures. *ASME International Mechanical Congress and Exposition, Volume 4B: Dynamics, Vibration and Control*, 2017.

[8] Amabili, Marco. Nonlinear damping in nonlinear vibrations of rectangular plates: derivation from viscoelasticity and experimental validation. *Journal of the Mechanics and Physics of Solids*, 118:275–292, 2018.

[9] Amabili, Marco. *Nonlinear mechanics of shells and plates in composite, soft and biological materials.* Cambridge University Press, 2018.

[10] Amabili, Marco. Derivation of nonlinear damping from viscoelasticity in case of nonlinear vibrations. *Nonlinear Dynamics*, 97, 2019.

[11] Anand, L. and Govindjee, S. *Continuum Mechanics of Solids.* Oxford University Press, 2020.

[12] Anthony, R.L. and Caston, R.H. and Guth, Eugene Equations of state for natural and synthetic rubber-like materials. I. Unaccelerated natural soft rubber. *Journal of Physical Chemistry*, 46,8:826–840, 1942.

[13] Balasubramanian, Prabakaran and Ferrari, Giovanni and Amabili, Marco. A tool to identify damping during large amplitude vibrations of viscoelastic structures. *ASME International Mechanical Congress and Exposition, Volume 4B: Dynamics, Vibration and Control*, 2017.

[14] Bazant, Z. P. *Stability of Structures: Elastic, Inelastic, Fracture and Damage Theories.* Oxford University Press, 1991.

[15] Becker, M. *The Principles and Applications of Variational Methods.* MIT Press, 1964.

[16] Bell, B. and Surana, K. S. *p*-Version Least Squares Finite Element Formulation for Two-Dimensional, Incompressible, Non-Newtonian Isothermal and Non-Isothermal Fluid Flow. *International Journal for Numerical Methods in Fluids*, 18:127–162, 1994.

[17] Bell, B. and Surana, K. S. *p*-Version Space-Time Coupled Least Squares Finite Element Formulation for Two-Dimensional Unsteady Incompressible, Newtonian Fluid Flow. *ASME Winter Annual Meeting*, 1993.

[18] Belytschko, T., Liu, W. K. and Moran, B. *Non-Linear Finite Elements for Continua and Structures.* John Wiley and Sons, 2000.

[19] Boehler, J. P. On Irreducible Representations for Isotropic Scalar Functions. *Journal of Applied Mathematics and Mechanics/Zeitschrift für Angewandte Mathematik und Mechanik*, 57:323–327, 1977.

[20] Bird, R. B., Armstrong, R. C. and Hassager, O. *Dynamics of Polymeric Liquids, Volume 1, Fluid Mechanics, Second Edition.* John Wiley and Sons, 1987.

[21] Bird, R. B., Armstrong, R. C. and Hassager, O. *Dynamics of Polymeric Liquids, Volume 2, Kinetic Theory, Second Edition.* John Wiley and Sons, 1987.

[22] Bridges, C. *Implicit Rate-Type Models for Elastic Bodies: Development, Integration, Linearization & Application.* PhD thesis, Texas A&M University, 2011.

[23] Brown, R. A Brief Account of Microscopical Observations Made in the Months of June, July, and August, 1827, on the Particles Contained in the Pollen of Plants; and on the General Existence of Active Molecules in Organic and Inorganic Bodies. *Edinburgh New Philosophical Journal*, 5:358–371, 1828.

[24] Brown, R. Additional Remarks on Active Molecules. *Edinburgh Journal of Science*, 1:314–319, 1829.

[25] Burshtein, A. I. *Introduction to Thermodynamics and Kinetic Theory of Matter.* Wiley-Interscience, 2008.

[26] Christensen, R.M. *Theory of Viscoelasticity, An Introduction.* Academic Press, 1971.

[27] Cotter, B. A. and Rivlin, R. S. Tensors Associated with Time-Dependent Stress. *Quarterly of Applied Mathematics*, 13(2):177–182, 1955.

[28] Dunn, J. E. and Rajagopal, K. R. Fluids of Differential Type: Critical Reveiw and Thermodynamic Analysis. *International Journal of Engineering Science*, 33(5):689–729, 1995.

[29] Einstein, A. Die Grundlage der Allgemeinen Relativitätstheorie. *Annalen der Physik*, 49:769–822, 1916.

[30] Eringen, A. C. *Mechanics of Continua.* John Wiley and Sons, 1967.

[31] Eringen, A. C. *Nonlinear Theory of Continuous Media.* McGraw-Hill, 1962.

[32] Fischer, F. D., Oberaigner, E. R., Tanaka, K. and Nishimura, F. Transformation Induced Plasticity Revised an Updated Formulation. *International Journal of Solids and Structures*, 35(18):2209–2227, 1998.

[33] Gelfand, I.M. and Fomin, S.V. *Calculus of Variations.* Dover Publications, 2000.

[34] Giesekus, H. Das reibungsgesetz strukturviskosen flüssigkeit. *Kolloid Z.*, 147:29–45, 1956.

[35] Giesekus, H. Die Rheologische Zustandsgleichung Elasto-Viskoser Flüssigkeiten - Insbesondere Von Weissenberg-Flüssigkeiten - für Allgemeine und Stationäre Fließvorgänge. *Journal of Applied Mathematics and Mechanics / Zeitschrift für Angewandte Mathematik und Mechanik*, 42:32–61, 1962.

[36] Giesekus, H. A Simple Constitutive Equation for Polymer Fluids Based on the Concept of Deformation-Dependent Tensorial Mobility. *Journal of Non-Newtonian Fluid Mechanics*, 11:69–110, 1982.

[37] Gray, W. G. and Hassanizadeh, S. M. Macroscale Continuum Mechanics for Multiphase Porous-Media Flow Including Phases, Interfaces, Common Lines and Common Points. *Advances in Water Resources*, 21(4):261–281, 1998.

[38] Grmela, M. and Ottinger, H. C. Dynamic and Thermodynamics of Complex Fluids I - Development of a General Formalism. *Physical Review E*, 56(6):6620–6632, 1997.

[39] Gurtin, M. E. and Sternberg, E. On the Linear Theory of Viscoelasticity. *Archive for Rational Mechanics and Analysis*, 11:291–356, 1962.

[40] Gurtin, M. E. *An Introduction to Continuum Mechanics.* Academic Press, 1981.

[41] Gurtin, M. The Continuum Mechanics of Coherent Two-Phase Elastic Solids with Mass Transport. *Proceedings of the Royal Society A: Mathematical, Physical, and Engineering Sciences*, 440(1909):323–343, 1993.

[42] Gurtin, M.E. and Fried, E. and Anand, L. *The Mechanics and Thermodynamics of Continua.* Cambridge University Press, 2010.

[43] Hassanizadeh, S. M. Mechanics and Thermodynamics of Multiphase Flow in Porous Media Including Interphase Boundaries. *Advances in Water Resources*, 13(4):169–186, 1990.

[44] Haupt, P. *Continuum Mechanics and Theory of Material, Second Edition.* Springer-Verlag, 2002.

[45] Hill, R. On Constitutive Inequalities for Simple Materials - I. *Journal of the Mechanics and Physics of Solids*, pages 229–242, 1968a.

[46] Hill, R. On Constitutive Inequalities for Simple Materials - II. *Journal of the Mechanics and Physics of Solids*, pages 315–322, 1968b.

[47] Hill, R. Constitutive Inequalities for Isotropic Elastic Solids under Finite Strain. *Proceedings of the Royal Society of London. Series A*, 314:457–472, 1970.

[48] Hill, R. Aspects of invariance in solid mechanics. *Advances in Applied Mechanics*, pages 1–72, 1978.

[49] Houlsby, G. T. and Puzrin, A. M. A Thermomechanical Framework for Constitutive Models for Rate-Independent Dissipative Materials. *International Journal of Plasticity*, 16(9):1017–1047, 2000.

[50] Houlsby, G. T. and Puzrin, A. M. Rate-Dependent Plasticity Models Derived from Potential Functions. *Journal of Rheology*, 46(1), 2002.

[51] Humari, Mithlesh and Daas, Narsingh. An equation of state applied to plastics, rubbers, glasses, and polymers. *Journal of Applied Physics*, 70:1863–1865, 1991.

[52] Jaumann, G. *Grundlagen der Bewegungslehre.* Leipzig, 1905.

[53] Jeffreys, H. *The Earth.* Cambridge University Press, 1929.

[54] Landel, R. F. and Peng, S. T. J. Equations of State and Constitutive Equations. *Journal of Rheology*, 30(4):741–765, 1986.

[55] Leitman, M. J. and Fisher, G. M. C. The Linear Theory of Viscoelasticity. In S. Flügge and C. Truesdell, editors, *Encyclopedia of Physics*, volume VIa/3, Mechanics of Solids III, pages 1–124. Springer-Verlag, 1973.

[56] Leonov, A. I. Analysis of Simple Constitutive Equations for Viscoelastic Liquids. *Journal of Non-Newtonian Fluid Mechanics*, 42:323–350, 1992.

[57] Lubarda, V. A. On Thermodynamic Potentials in Linear Thermoelasticity. *International Journal of Solids and Structures*, 41(26):7377–7398, 2004.

[58] Lustig, S. R., Shay Jr., R. M. and Caruthers, J. M. Thermodynamic Constitutive Equations for Materials with Memory on a Material Time Scale. *Journal or Rheology*, 40(1), 1996.

[59] Malven, L. E. *Introduction to the Mechanics of a Continuous Medium.* Prentice-Hall, 1969.

[60] Maire, J. F. and Chaboche, J. L. A New Formulation of Continuum Damage Mechanics (CDM) for Composite Materials. *Aerospace Science and Technology*, 1(4):247–257, 1997.

[61] Maxwell, J. C. On the Dynamical Theory of Gases. *Philosophical Transactions of the Royal Society of London*, A157:49–88, 1867.

[62] mikhlin, G. *Variational Methods in Mathematical Physics.* Pergamon Press, 1964.

[63] Morro, A. A thermodynamic Approach to Rate Equations in Continuum Physics. *Journal of Physical Science and Application*, 7(6):15-23, 2017.

[64] Murnaghan, F. D. *Finite Deformation of an Elastic Solid.* John Wiley and Sons, 1951.

[65] O'Reilly, O. M. On Constitutive Relations for Elastic Rods. *International Journal of Solids and Structures*, 35(11):1009–1024, 1998.

[66] Oldroyd, J. G. On the Formulation of Rheological Equations of State. *Proceedings of the Royal Society of London*, A200:523–541, 1950.

[67] Oldroyd, J. G. Non-Newtonian Effects in Steady Motion of Some Idealized Elastico-Viscous Liquids. *Proceedings of the Royal Society of London*, A245:278–297, 1958.

[68] Panton, R. L. *Incompressible Flow, Third Edition.* John Wiley and Sons, 2005.

[69] Pennisi, S. and Trovato, M. On the Irreducibility of Professor G. F. Smith's Representations for Isotropic Functions. *International Journal of Engineering Science*, 25(8):1059–1065, 1987.

[70] Phan-Thien, N. and Tanner, R. I. A New Constitutive Equation Derived from Network Theory. *Journal of Non-Newtonian Fluid Mechanics*, 2:353–365, 1977.

[71] Phan-Thien, N. and Tanner, R. I. A Nonlinear Network Viscoelastic Model. *Journal of Rheology*, 22:259–283, 1978.

[72] Poling, B. E., Prausnitz, J. M. and O'Connell, J. P. *The Properties of Gases and Liquids.* McGraw-Hill, 2001.

[73] Prager, W. Strain Hardening under Combined Stresses. *Journal of Applied Physics*, 16:837–840, 1945.

[74] Prud'homme, R. K. and Bird, R. B. The Dilatational Properties of Suspensions of Gas Bubbles in Incompressible Newtonian and Non-Newtonian Fluids. *Journal of Non-Newtonian Fluid Mechanics*, 3:261–279, 1978.

[75] Quinzani, L. M., Armstrong, R. C. and Brown, R. A. Use of Coupled Birefringence and LDV Studies of Flow Through a Planar Contraction to Test Constitutive Equations for Concentrated Polymer Solutions. *Journal of Rheology*, 39:1201–1227, 1995.

[76] Rajagopalan, D., Armstrong, R. C. and Brown, R. A. Finite Element Methods for Calculation of Steady Viscoelastic Flow Using Constitutive Equations with a Newtonian Viscosity. *Journal of Non-Newtonian Fluid Mechanics*, 36:159–192, 1990.

[77] Rajagopalan, D., Phillips, R. J., Armstrong, R. C., Brown, R. A. and Bose, A. The Influence of Viscoelasticity on the Existence of Steady Solutions in Two-Dimensional Rimming Flow. *Journal of Fluid Mechanics Digital Archive*, 235:611–642, 1992.

[78] Rajagopal, K.R. and Tao, L. Mechanics of Mixtures. World Scientific, 1995.

[79] Rajagopal, K. R. and Srinivasa, A. R. A Thermodynamic Framework for Rate Type Fluid Models. *Journal of Non-Newtonian Fluid Mechanics*, 88:207–227, 2000.

[80] Rajagopal, K. R. and Srinivasa, A. R. On the Development of Fluid Models of the Differential Type within a New Thermodynamic Framework. *Mechanics Research Communications*, 35:483–489, 2008.

[81] Rajagopal, K. R. and Srinivasa, A. R. A Gibbs-Potential-Based Formulation for Obtaining the Response Functions for a Class of Viscoelastic Materials. *Proceedings of the Royal Society A: Mathematical, Physical and Engineering Sciences*, 467(2125):39–58, 2010.

[82] Rajagopal, K. R. and Srinivasa, A. R. A Gibbs Potential Based Formulation for Obtaining the Response Functions for a Class of Viscoelastic Materials. *Proceedings of the Royal Society A*, 467:39–58, 2011.

[83] Reddy, J. N. *An Introduction to Continuum Mechanics: with Application.* Cambridge University Press, 2008.

[84] Reiner, M. A Mathematical Theory of Dilatancy. *American Journal of Mathematics*, 67:350–362, 1945.

[85] Rektorys, K. *Variational Methods in Mathematics, Science and Engineering.* Reidel, 1977.

[86] Ricci, G. *Atti Deta Reale Academia Nazconale dei Lincer*, 5, 1889.

[87] Rivlin, R. S. Further Remarks on the Stress-Deformation Relations for Isotropic Materials. *Journal of Rational Mechanics and Analysis*, 4:681–702, 1955.

[88] Rivlin, R. S. and Ericksen, J. L. Stress-Deformation Relations for Isotropic Materials. *Journal of Rational Mechanics and Analysis*, 4:323–425, 1955.

[89] Schapery, R. A. Application of Thermodynamics to Thermomechanical, Fracture, and Birefringent Phenomena in Viscoelastic Media. *Journal of Applied Physics*, 35(5):1451–1465, 1964.

[90] Rivlin, R. S. and Ericksen, J. L. Stress-Deformation Relations for Isotropic Materials. *Journal of Rational Mechanics and Analysis*, 4:323–425, 1955.

[91] Schecter, R.S. *The Variational Methods in Engineering.* McGraw Hill, 1967.

[92] Shapiro, N. Z. and Shapley, L. S. Mass Action Laws and the Gibbs Free Energy Function. *Journal of the Society for Industrial and Applied Mathematics*, 13(2):353–375, 1965.

[93] Smith, G. F. On a Fundamental Error in two Papers of C.C. Wang, "On Representations for Isotropic Functions, Part I and Part II". *Archive for Rational Mechanics and Analysis*, 36:161–165, 1970.

[94] Smith, G. F. On Isotropic Functions of Symmetric Tensors, Skew-Symmetric Tensors and Vectors. *International Journal of Engineering Science*, 9:899–916, 1971.

[95] Sokolnikoff, I. S. *Tensor Analysis, Second Edition.* John Wiley and Sons, 1964.

[96] Spencer, A. J. M. and Rivlin, R. S. The Theory of Matrix Polynomials and its Application to the Mechanics of Isotropic Continua. *Archive for Rational Mechanics and Analysis*, 2:309–336, 1959.

[97] Spencer, A. J. M. and Rivlin, R. S. Further Results in the Theory of Matrix Polynomials. *Archive for Rational Mechanics and Analysis*, 4:214–230, 1960.

[98] Spencer, A. J. M. *Theory of Invariants.* Chapter 3 "Treatise on Continuum Physics, I" edited by A. C. Eringen. Academic Press, 1971.

[99] Spencer, A. J. M. *Continuum Mechanics.* Longman, 1980.

[100] Stevens, R. N. and Guiu, F. Energy Balance Concepts in the Physics of Fracture. *Proceedings of the Royal Society A: Mathematical, Physical, and Engineering Sciences*, 435(1893):169–184, 1991.

[101] Surana, K. S., Ahmadi, A. R. and Reddy, J. N. The k-Version of Finite Element Method for Self-Adjoint Operators in BVPs. *International Journal of Computational Engineering Science*, 3(2):155–218, 2002.

[102] Surana, K. S., Ahmadi, A. R. and Reddy, J. N. The k-Version of Finite Element Method for Non-Self-Adjoint Operators in BVPs. *International Journal of Computational Engineering Sciences*, 4(4):737–812, 2003.

[103] Surana, K. S., Ahmadi, A. R. and Reddy, J. N. The k-Version of Finite Element Method for Non-Linear Operators in BVPs. *International Journal of Computational Engineering Science*, 5(1):133–207, 2004.

[104] Surana, K. S., Allu, S., Tenpas, P. W. and Reddy, J. N. k-Version of Finite Element Method in Gas Dynamics: Higher Order Global Differentiability Numerical Solutions. *International Journal Numerical Methods in Engineering*, 69:1109–1157, 2006.

[105] Surana, K. S., Reddy, J. N. and Allu, S. The k-Version of Finite Element Method for IVPs: Mathematical and Computational Framework. *International Journal for Computational Methods in Engineering Science and Mechanics*, 8(3):123–136, 2007.

[106] Surana, K. S., Allu, S., Reddy, J. N. and Tenpas, P. W. Least Squares Finite Element Processes in h, p, k Mathematical and Computational Framework for a Non-Linear Conservation Law. *International Journal of Numerical Methods in Fluids*, 57(10):1545–1568, 2008.

[107] Surana, K. S., Nunez D., Reddy, J. N. and Romkes, A. Rate Constitutive Theory for Ordered Thermoelastic Solids. *Annals of Solid and Structural Mechanics*, 3:27–54, 2012.

[108] Surana, K. S., Nunez, D., Reddy, J. N. and Romkes, A. Rate Constitutive Theory for Ordered Thermofluids. *Journal of Continuum Mechanics and Thermodynamics*, 25:625–662, 2013.

[109] Surana, K. S., Nunez, D., Reddy, J. N. and Romkes, A. Rate Constitutive Theory for Ordered Thermoviscoelastic Fluids—Polymers. *Journal of Continuum Mechanics and Thermodynamics*, 26:143–181, 2014.

[110] Surana, K. S., Ma, Y., Reddy, J. N. and Romkes, A. The Rate Constitutive Equations and Their Validity for Progressively Increasing Deformation. *Mechanics of Advanced Materials and Structures*, 17:509–533, 2010.

[111] Surana, K. S., Ma, Y. T., Romkes, A. and Reddy, J. N. Fluid-Solid Interaction of Incompressible Media in h, p, k Mathematical and Computational Framework. *International Journal of Computational Methods in Engineering Science and Mechanics*, 13:357–379, 2012.

[112] Surana, K.S. and Mendoza, Y. and Reddy, J.N. Constitutive theories for thermoelastic solids in Lagrangian description using Gibbs potential. *Acta Mechanica*, 224:1019–1044, 2013.

[113] Surana, K. S., Nunez, D and Reddy J. N. Giesekus Constitutive Model for Ordered Thermoviscoelastic Fluids Based on Ordered Rate Constitutive Theories. *Research Updates in Polymer Science*, 2:232–260, 2013.

[114] Surana, K. S., Reddy J. N. and Nunez, D. Ordered Rate Constitutive Theories for Thermoviscoelastic Solids without Memory in Lagrangian Description Using Gibbs Potential. *Journal of Continuum Mechanics and Thermodynamics*, 27:409–431, 2015.

[115] Surana, K. S., Moody, T. C. and Reddy, J. N. Ordered Rate Constitutive Theories in Lagrangian Description for Thermoviscoelastic Solids without Memory. *Acta Mechanica*, 24:2785–2816, 2013.

[116] Surana, K. S. and Powell, M. J. and Reddy, J. N. A More Complete Thermodynamic Framework for Solid Continua. *Journal of Thermal Engineering*, 1:1–13, 2015.

[117] Surana, K. S. and Powell, M. J. and Reddy, J. N. A More Complete Thermodynamic Framework for Fluent Continua. *Journal of Thermal Engineering*, 1:14–30, 2015.

[118] Surana, K. S., Reddy, J. N. and Nunez, D. Ordered Rate Constitutive Theories for Thermoviscoelastic Solids with Memory in Lagrangian Description Using Gibbs Potential. *Journal of Continuum Mechanics and Thermodynamics*, 27:1019–1038, 2015.

[119] Surana, K. S. Moody, T. C. and Reddy, J. N. Ordered Rate Constitutive Theories in Lagrangian Description for Thermoviscoelastic Solids with Memory. *Acta Mechanica*, 226:157–178, 2015.

[120] Surana, K. S., Moody, T. C. and Reddy, J. N. Rate Constitutive Theories of Order Zero in Lagrangian Description for Thermoelastic Solids. *Mechanics of Advanced Materials and Structures*, 6:440–450, 2015.

[121] Surana, K. S. and Reddy, J. N. *Mathematics of Computations and the Finite Element Method for Boundary Value Problems.* CRC Press, 2015.

[122] Surana, K. S. and Reddy, J. N. *Mathematics of Computations and the Finite Element Method for Initial Value Problems.* CRC Press, 2017.

[123] Surana, K. S. and Shanbhag, R. R. and Reddy, J. N. Necessity of Balance of Moment of Moments Balance law in Non-Classical Continuum Theories for Solid Continua. *Meccanica*, 53:2939–2972, 2017.

[124] Surana, K. S. and Long, S. W. and Reddy, J. N. Necessity of Law of Balance/Equilibrium of Moment of Moments in non-classical Continuum Theories for Fluent Continua. *Acta Mechanica*, 22:2801–2833, 2018.

[125] Surana, K.S. and Mathi, S.S.C. A thermodynamically consistent non-linear mathematical model for thermoviscoelastic plates/shells with finite deformation and finite strain based on classical continuum mechanics. *International Journal of Non-linear Mechanics*, 126:103565, 2020.

[126] Todd, J. A. Ternary Quadratic Types. *Philosophical Transactions of the Royal Society of London. Series A: Mathematical and Physical Sciences*, 241:399–456, 1948.

[127] Tran, L. and Udaykumar, H. S. A Particle-Level Set-Based Sharp Interface Cartesian Grid Method for Impact, Penetration, and Void Collapse. *Journal of Computational Physics*, 193:469–510, 2004.

[128] Truesdell, C. A. and Toupin, R. A. *The Classical Field Theories of Mechanics.* Handbuch der Physik Vol 3/1, 1963.

[129] Truesdell, C. A. and Noll, W. *The Non-Linear Field Theories of Mechanics.* Handbuch der Physik Vol 3/3, 1965.

[130] Udaykumar, H. S., Tran, L., Belk, D. M. and Vanden, K. J. An Eulerian Method for Computation of Multimaterial Impact with ENO Shock-Capturing and Sharp Interfaces. *Journal of Computational Physics*, 186:136–177, 2003.

[131] Veblen, O. Invariant of Quadratic Differential Forms. *Cambridge, Tract*, 24, 1927.

[132] Voyiadjis, G.Z. and Yaghoobi, M. Size Effects in Plasticity. *Elsevier Inc.*, 2019.

[133] Voyiadjis, G.Z. and Song, Y. Gradient-enhanced Continuum Plasticity: Theories, Experiments and Numerical Methods. *Elsevier Inc.*, 2020.

[134] Wang, C. C. On Representations for Isotropic Functions, Part I. *Archive for Rational Mechanics and Analysis*, 33:249–267, 1969.

[135] Wang, C. C. On Representations for Isotropic Functions, Part II. *Archive for Rational Mechanics and Analysis*, 33:268–287, 1969.

[136] Wang, C. C. A New Representation Theorem for Isotropic Functions, Part I and Part II. *Archive for Rational Mechanics and Analysis*, 36:166–223, 1970.

[137] Wang, C. C. Corrigendum to "Representations for Isotropic Functions". *Archive for Rational Mechanics and Analysis*, 43:392–395, 1971.

[138] Weyl, H. *Mathematische Zeitschrift*, 23:271–301, 1925.

[139] White, F. M. *Fluid Mechanics, Seventh Edition.* McGraw-Hill, 2010.

[140] Winterscheidt, D. and Surana, K. S. p-Version Least Squares Finite Element Formulation for Two-Dimensional, Incompressible Fluid Flow. *International Journal of Numerical Methods in Fluids*, 18:43–69, 1994.

[141] Zaremba, S. Sur une forme perfectionnée de la théorie de la relaxation. *Bulletin International de l'académie des Sciences de Cracovie*, 8:594–614, 1903.

[142] Zaremba, S. Sur une conception nouvelle des forces intérieures dans un fluide en mouvement. *Mémorial des Sciences Mathématiques*, 82:1–85, 1937.

[143] Zhao, J., Sheng, D. and Collins, I. F. Thermomechanical Formulation of Strain Gradient Plasticity for Geomaterials. *Journal of Mechanics of Materials and Structures*, 1(5), 2006.

[144] Zheng, Q. S. On the Representations for Isotropic Vector-Valued, Symmetric Tensor-Valued and Skew-Symmetric Tensor-Valued Functions. *International Journal of Engineering Science*, 31:1013–1024, 1993.

[145] Zheng, Q. S. On Transversely Isotropic, Orthotropic and Relatively Isotropic Functions of Symmetric Tensors, Skew-Symmetric Tensors, and Vectors. *International Journal of Engineering Science*, 31:1399–1453, 1993.

INDEX

D

F

G

H

For Product Safety Concerns and Information please contact our EU representative GPSR@taylorandfrancis.com
Taylor & Francis Verlag GmbH, Kaufingerstraße 24, 80331 München, Germany

www.ingramcontent.com/pod-product-compliance
Lightning Source LLC
Chambersburg PA
CBHW060422220326
41598CB00021BA/2253
9780367615215